Handbook of Research on the Conservation and Restoration of Tropical Dry Forests

Rahul Bhadouria
University of Delhi, India

Sachchidanand Tripathi
University of Delhi, India

Pratap Srivastava
University of Allahabad, India

Pardeep Singh
University of Delhi, India

A volume in the Practice, Progress, and Proficiency in Sustainability (PPPS) Book Series

IGI Global
DISSEMINATOR OF KNOWLEDGE

Published in the United States of America by
IGI Global
Engineering Science Reference (an imprint of IGI Global)
701 E. Chocolate Avenue
Hershey PA, USA 17033
Tel: 717-533-8845
Fax: 717-533-8661
E-mail: cust@igi-global.com
Web site: http://www.igi-global.com

Library of Congress Cataloging-in-Publication Data

Names: Bhadouria, Rahul, 1982- editor.
Title: Handbook of Research on the Conservation and Restoration of Tropical Dry Forests / Rahul Bhadouria; Sachchidanand Tripathi; Pratap Srivastava; Pardeep Singh, editors.
Description: Hershey, PA : Engineering Science Reference, [2020] | Includes bibliographical references.
Identifiers: LCCN 2019015487| ISBN 9781799800149 (hardcover) | ISBN 9781799800163 (ebook)
Subjects: LCSH: Tropical dry forests. | Forests and forestry. | Forest conservation. | Forest restoration.
Classification: LCC QH541.5.T66 C66 2020 | DDC 577.3--dc23 LC record available at https://lccn.loc.gov/2019015487

This book is published in the IGI Global book series Practice, Progress, and Proficiency in Sustainability (PPPS) (ISSN: 2330-3271; eISSN: 2330-328X)

British Cataloguing in Publication Data
A Cataloguing in Publication record for this book is available from the British Library.

Practice, Progress, and Proficiency in Sustainability (PPPS) Book Series

Ayman Batisha
International Sustainability Institute, Egypt

ISSN:2330-3271
EISSN:2330-328X

Mission

In a world where traditional business practices are reconsidered and economic activity is performed in a global context, new areas of economic developments are recognized as the key enablers of wealth and income production. This knowledge of information technologies provides infrastructures, systems, and services towards sustainable development.

The **Practices, Progress, and Proficiency in Sustainability (PPPS) Book Series** focuses on the local and global challenges, business opportunities, and societal needs surrounding international collaboration and sustainable development of technology. This series brings together academics, researchers, entrepreneurs, policy makers and government officers aiming to contribute to the progress and proficiency in sustainability.

Coverage

- Global Content and Knowledge Repositories
- Outsourcing
- Eco-Innovation
- Innovation Networks
- ICT and knowledge for development
- Socio-Economic
- Global Business
- Sustainable Development
- Strategic Management of IT
- E-Development

IGI Global is currently accepting manuscripts for publication within this series. To submit a proposal for a volume in this series, please contact our Acquisition Editors at Acquisitions@igi-global.com or visit: http://www.igi-global.com/publish/.

The Practice, Progress, and Proficiency in Sustainability (PPPS) Book Series (ISSN 2330-3271) is published by IGI Global, 701 E. Chocolate Avenue, Hershey, PA 17033-1240, USA, www.igi-global.com. This series is composed of titles available for purchase individually; each title is edited to be contextually exclusive from any other title within the series. For pricing and ordering information please visit http://www.igi-global.com/book-series/practice-progress-proficiency-sustainability/73810. Postmaster: Send all address changes to above address.

Titles in this Series

For a list of additional titles in this series, please visit:
https://www.igi-global.com/book-series/practice-progress-proficiency-sustainability/73810

Intellectual, Scientific, and Educational Influences on Sustainability Research
Rosario Adapon Turvey (Lakehead University, Canada) and Sreekumari Kurissery (Lakehead University, Canada)
Engineering Science Reference • copyright 2019 • 358pp • H/C (ISBN: 9781522573029) • US $195.00 (our price)

Driving the Development, Management, and Sustainability of Cognitive Cities
Kiran Ahuja (DAV Institute of Engineering and Technology, India) and Arun Khosla (Dr. B.R. Ambedkar National Institute of Technology, India)
Engineering Science Reference • copyright 2019 • 337pp • H/C (ISBN: 9781522580850) • US $225.00 (our price)

Technology-Driven Innovation in Gulf Cooperation Council (GCC) Countries Emerging Research and Opportunities
Anna Visvizi (Deree – The American College of Greece, Greece & Effat University, Saudi Arabia) Saad Bakry (King Saud University, Saudi Arabia) Miltiadis D. Lytras (Deree – The American College of Greece, Greece & Effat University, Saudi Arabia) and Wadee Alhalabi (Effat University, Saudi Arabia)
Business Science Reference • copyright 2019 • 180pp • H/C (ISBN: 9781522590125) • US $195.00 (our price)

Novel Technologies and Systems for Food Preservation
Pedro Dinis Gaspar (University of Beira Interior, Portugal) and Pedro Dinho da Silva (University of Beira Interior, Portugal)
Engineering Science Reference • copyright 2019 • 416pp • H/C (ISBN: 9781522578949) • US $345.00 (our price)

Bioeconomical Solutions and Investments in Sustainable City Development
José G. Vargas-Hernández (University of Guadalajara, Mexico) and Justyna Anna Zdunek-Wielgołaska (Warsaw University of Technology, Poland)
Engineering Science Reference • copyright 2019 • 306pp • H/C (ISBN: 9781522579588) • US $195.00 (our price)

Optimizing the Use of Farm Waste and Non-Farm Waste to Increase Productivity and Food Security Emerging Research and Opportunities
Leighton Naraine (Clarence Fitzroy Bryant College, St. Kitts-Nevis)
Engineering Science Reference • copyright 2019 • 223pp • H/C (ISBN: 9781522579342) • US $155.00 (our price)

Handbook of Research on International Collaboration, Economic Development, and Sustainability in the Arctic
Vasilii Erokhin (Harbin Engineering University, China) Tianming Gao (Harbin Engineering University, China) and Xiuhua Zhang (Harbin Engineering University, China)
Business Science Reference • copyright 2019 • 703pp • H/C (ISBN: 9781522569541) • US $235.00 (our price)

IGI Global
DISSEMINATOR OF KNOWLEDGE

701 East Chocolate Avenue, Hershey, PA 17033, USA
Tel: 717-533-8845 x100 • Fax: 717-533-8661
E-Mail: cust@igi-global.com • www.igi-global.com

Editorial Advisory Board

List of Contributors

Table of Contents

Section 3
Forest Diversity, Degradation, and Management

Detailed Table of Contents

Forests cover almost one-third of the Earth's land surface. Tropical dry forests are the second-most-important forest type in the world covering approximately 42% of tropical and sub-tropical forest area. The main features of these forests are their deciduousness, a prolonged dry period extending 3-9 months, and little annual precipitation of 250-2,000 mm. Tropical dry forests are found in five of the eight realms in the world. More than half of the forests are distributed in the Americas, with other portions in Africa, Eurasia, Australia, and Southeast Asia. The forests are unique in nature, and provide shelter to a huge number of endemics and endangered species. Among woody plant species, about 40% are not found anywhere in the world. These forests are now the most threatened among all forest types. The conservation status of these forests is endangered. Deforestation, rapid civilization, land conversion, fire, and climate change are the major threats. Proper management with time-oriented policy could be helpful to restore these forests and protect the existing remnant areas.

Around 1.6 billion people in the world are directly dependent on forests for food, fodder, fuel, shelter, and livelihood, out of which 60 million are entirely dependent on forests. Forests silently provide us with ecosystem services such as climate regulation, carbon sequestration, harbouring biodiversity, synchronizing nutrient cycling, and many more. Tropical Dry Forests (TDF's) occupy around 42% of total forest area of the tropics and subtropics and facilitate sustenance of world's marginalized populations. Change in vegetation composition and distribution, deflected succession, carbon sequestration potential, nutrient cycling and symbiotic associations would affect TDF at ecosystem level. At species level, climate change will impact photosynthesis, phenology, physiognomy, seed germination, and temperature-sensitive physiological processes. In order to mitigate the effects of climate change, specific mitigation and adaptation strategies are required for TDF that need to be designed with concerted efforts from scientists, policy makers and local stakeholders.

Tropical dry forests is one of the most unique forest types. It differs from other tropical forests with its climatic behavior like a prominent dry period, little annual rainfall, and high evapotranspiration. Out of six global bioclimatic zones, the forests are distributed in four. Climate change is now the most challenging issue regarding the fate of tropical dry forests. A severe climatic change is estimated to occur between 2040 and 2069 that could drastically change the precipitation pattern, temperature, aridity, and distribution of biodiversity. It could alter the forest type permanently. With a large number of heat-tolerant species, tropical dry forests have a great potentiality to conservationists with the prediction of a large area that could attain the climatic condition favorable for extension of tropical dry forests. But many of the species of tropical dry forests could be extinct due to changing climate at the same time. Proper adaptation and mitigation techniques could minimize the severity of climate change effects.

Plants have certain characteristics which allow them to respond to various environmental conditions, like changes in climate, water scarcity in the soil, lack of minerals; among others. In some of these traits, the responses to climatic phenomena such as drought can be evidenced through morphological adaptations (spines, succulent tissues, trichomes) or physiological adaptations (regulation of water potential at the cellular level, the concentration of nutrients, etc.). A systematic literature review was performed to study plant functional traits (PFTs) in tropical dry forests (TDFs). The chapter suggests the role of functional traits in community dynamics and processes. The authors will also highlight the limitations of PFTs in TDFs and how they can be improved.

Tropical dry forests (TDFs) are characterized by pronounced seasonality in precipitation, with several months of prolonged drought, 80% of annual precipitation occurring during a four- to six-month rainy season, and high interannual rainfall variability. Surprisingly, there are relatively few studies addressing patterns of functional trait in tropical dry forest (TDF) ecosystems. Functional trait analysis across plant species and the environment is a rapidly developing research field with many possible applications for forest restoration practice. Trait-based ecological research within TDFs will advance our understanding of how these ecosystems interact with and differ from other tropical ecosystems.

Section 3
Forest Diversity, Degradation, and Management

Tropical dry forests occur as patches in Tamil Nadu distributed along the East Coast, Eastern Ghats, and plains of the Indian Peninsula. The floristic studies of these regions are of great national relevance as plant resources in a tropical climate contribute to national wealth. Dry forests of the plains in Tamil Nadu have been neglected and the area under study has remained practically unexplored. This chapter studies distribution of tropical dry forests, especially in Pudukkottai district in Tamil Nadu. In total, 187 sacred groves were surveyed for their distribution and floristic composition. The GPS position of each grove was noted and their distribution maps were prepared. The groves were classified based on conservation status, namely well conserved, moderately conserved and degraded. Extensive botanical explorations were carried out periodically during 2012–2016 in these groves and 812 species belonging to 480 genera under 124 families were recorded. The endemic, threatened species of these groves were also documented.

The woody flora of the National Park of Manda in the Sudanian area of Chad has been characterized between October and December 2016 to know its floristic diversity, and to quantify its aerial woody biomass. The transect and quadra method (1m x 1m) were simultaneously adopted for this study. The pan-tropical equation of Chave et al. made it possible to evaluate the carbon stocks in different sites. The study of the flora species identified 45 species divided into 37 genus and 21 families for an average population density of 355 individuals/ha. Three classes of the diameter dominate the settlement: class $\leq$ 10 cm; class of 10-20 cm and class of 20-30 cm. The height classes belong to the class of plants $\leq$ 4 m; and at last having a height $\leq$ 7 m. The basal area was 5.86 m2 / ha. It appears that the woody components store 23.82 $\pm$ 0.01 tC / ha, the undergrowth 0.14 $\pm$ 0.01 tC / ha and the litter 0.56 $\pm$ 0.01 tC / ha. This research is a contribution to the REDD+ process (Reducing Emissions from Deforestation and Forest Degradation).

DISTANCE protocols and MIKE Survey Standards were used in the field to determine critically-endangered Sumatran elephant (Elephas maximus sumatranus) occupancy and abundance in Gunung Leuser National Park (GLNP). Forest and habitat type, age, character, and integrity were the most significant factors affecting elephant occupancy. Principal forage types relative to elephant activity were palms and lianas, which dominated significantly in undisturbed primary forest. DISTANCE model density D=0.167 elephants/km-2 (95% CI=0.106–0.262), best-fitting occupancy Ψ=0.6321 (SE±0.0010) and detection probability p=0.6225 (SE±0.0001) estimates combined yielded N=407 elephants (95% CI: 258–638) in GLNP. The most parsimonious occupancy model estimated N=392.82 elephants (SE:±30.65; 95% CI: 332.78-452.95) in GLNP. Based on the study, forest restoration, ecosystem protections, and conservation plans for Asian elephants, biodiversity, and forests are suggested in this study.

Forest degradation is attributed to the excessive use of forest resources and extraction, whether for subsistence or commercial purposes. With an increase in human population pressure on forests, forest degradation is becoming a concern for the conservation of biodiversity. The high human dependence on tropical dry forests underscores the need for a complete understanding of the interaction of humans and these forests to ensure their persistence and the wellbeing of the people who depend on these resources. This chapter examines forest resource use and degradation to provide a nuanced understanding of forest degradation and the impact of forest resource use.

Plant litter production and decomposition is a crucial ecosystem process that defines and governs the plant-soil relationships by regulating the nutrient turnover and the build-up of soil organic matter. Litter is the principal source of organic matter for soils in the forest ecosystem. The litter, upon decomposition, makes available essential nutrients for the growth and development of a forest stand. Different tree components contain different amounts of nutrients; and build up of soil organic matter. The amount of nutrients added through litter decomposition varies with forest types, species, stand attributes, and variation in seasonal environmental conditions. Nutrient return from organic matter is estimated by the physico-chemical properties of the litter. Moreover, the rate of decomposition and the nutrient releases are highly influenced by magnitude of litter produced, litter quality and nutrients release, as well as, by climatic conditions and existing microbial communities in the soil system. Ecological impact of carbon and nutrient dynamics in the litter layer is considerable in a forest ecosystem.

This chapter encompasses a literature survey and strategic analysis to understand the elaboration and implementation of Participatory Forest Management (PFM) in Benin, with a focus on the case of the "Forêt des Trois Rivières". By analyzing the historical background of forest management systems in Benin, we highlighted two major turning points. The first relates to the creation and autocratic management of protected forests, which took place from 1940 to 1990. The second change took place after the Rio conference in 1992, and this emphasized the importance of local communities in natural resources management. Moreover, the results of our strategic analysis of stakeholders involved in the specific case of Participatory Forest Management Plan (PFMP) of the "Forêt des Trois Rivières" showed that it is important to emphasize on active community participation while designing a participatory management plan and for decision making at the implementation stage. We also observed that alliances between foresters and timber loggers are likely to hinder the achievement of the PFM objectives.

The sandstone quarrying and mining in Vidhyan region of Uttar Pradesh has severely devastated the floral biodiversity of the adjacent forest area. It is necessary to conserve these areas for the protection and sustainable use of forest resources. In most of the conservation program, microbial deficiency creates problems in establishment of the vegetation and delays the natural succession. Therefore, probiotic interventions were applied to conserve these areas promptly. The probiotic beneficial microbes' interventions, due to their multifarious beneficial characters, may facilitate the upcoming of flora through enrichment of the soil, better nutrient absorption, providing resistance against different stresses, etc. Probiotic interventions may positively impact the conservation of floral diversity and restoration of stone mining-affected forest area.

Data and information are required to manage a renewable natural resource (such as forest) sustainably. This information is largely obtained through forest inventories. Forest inventory is very important in forest management since it provides the data for planning, monitoring, evaluation, research, growth and yield, biodiversity, and timber sale. Inadequate and/or irregular forest inventory has resulted to paucity of data on which forest management decisions can be based. This study aimed at enlightening conservationists, ecologists, environmentalists, and forest managers on the need to carryout diversity assessment and inventory of flora and fauna resources within their forest ecosystems. This will help to ensure sustainable yield and prevent the extinction of economic trees and endangered animal species in the tropical dry forests. The study therefore recommended that inventory should be carryout among the tropical dry forests at regular intervals.

In fire-influenced ecosystems, some plant species have the ability to recover, germinate, and to establish after a fire; however, their proportion and dominance varies between sites. The objective of this work was to evaluate natural regeneration following a fire in a tropical dry forest located in the Upper Magdalena River Valley in Colombia. In that way, all seedlings and saplings of woody species were recorded, 1.5 years after a fire, in 75 2x2-m plots installed in burned and unburned forest sites, as well as in forest gaps. Results showed that although abundance was higher in the burned sites, the species richness was lower than in unburned areas. Based on the regeneration response of the species, we identified three groups of plants: 1) fire-stimulated, 2) fire-tolerant, and 3) fire-sensitive species, which means that this tropical dry forest has species with the ability to recover, germinate, and establish after a fire. These three groups of plant species should be considered in restoration programs in light of future and more frequent forest fires due to climate change.

Recovery of African dry forests and woodlands is through either sexual or vegetative means. A sufficient number of regeneration pool (root suckers, sprouts, and seedlings) tend to occur in the herbaceous layer of the African dry forests. However, the prevailing environmental conditions such as fires and competition for nutrients and light influence the number of surviving individuals to enter the next size class. Developing restoration strategy for the African savannas requires the knowledge of regeneration mechanisms and dynamics of these woodlands which will be helpful in their management.

Foreword

Given the vast global extent (about 42%) of tropical dry forests, their progressive rate of deforestation, the dependency on their natural resources and the ecosystem services they render, attest the significance of their preservation, under the currently changing and challenging environmental scenario. There has been a quantum jump in research on forest biodiversity inventory and resource assessment and sustainable use with conservation focus in the recent decades and they have advanced our knowledge substantially. Despite these, there is a need for better understanding of threatened ecosystem such as the tropical dry forests, from ecological, resource conservation and sustainability point of view. Towards this end, in this book titled "Handbook of Research on the Conservation and Restoration of Tropical dry forests", the editors have sought chapter contributions from wide geographical locations across the world covering various aspects of forest science. The chapters include individual case studies, regional syntheses and review articles also. The nineteen chapters of this book cover a range of topics which include global overview of tropical dry forest biodiversity, floristics, inventorying and monitoring, bio-geography, ecology, recourse values, Impact of climate on forest functional ecology, forest dynamics, climate change adaptation and mitigation techniques, plant functional traits, restoration of degraded ecosystems, effect of fire, forest regeneration, patterns of resource utilization and finally the role of international governance in protecting natural environment. Many titles are on contemporary areas and fall in the priority agenda for global scientific community. The contributions in this book are expected to promote regional and global level discussion and would potentially intensify research in frontier areas of forest biodiversity, functional ecology, dynamics and sustainable use of tropical dry forests in years to come and thus help in nature protection. I sincerely hope that the topics covered in this book would be immensely useful to readers from various branches of academia and to conservation practitioners and thus help in maintaining global dry forests in a stage supportive to all including humans.

N. Parthasarathy
Pondicherry University, India

Preface

Globally, tropical forest constitutes 52% of the total forests in which about 42% are dry forests. Worldwide, tropical dry forests (TDFs) are located in countries such as Africa, Australia, Central and South America, India and Southeast Asia. These forests experience varying environmental conditions throughout the annual cycle. For example, they are characterized by a relatively warm to hot climate with a limited precipitation period of 3–4 months in a year. It consists of 2–6 months of drought each year, during which the potential evaporation to rainfall ratio (e/r) is greater than one and the mean annual temperature is more than 17 °C. TDFs are considered as the most exploited and endangered ecosystems of the world. Deforestation in these TDFs, particularly over the last 50 years has increased at such an alarming rate that the long-term sustainability of these indispensable natural resources is questionable. Further, according to some reports, tropical dry and subtropical forests and woodlands once covered more than half of the world's tropics. However, a marked decrease has been reported during the past few decades. Worldwide, degradation and deforestation of these forests has been associated with severe anthropogenic (mostly uncontrolled and illegal harvesting for food and fuel, agriculture and industrial purposes and poor forestry practices) and biotic (such as overgrazing) pressures apart from natural forces (such as climate change, fire, flood and drought). As tropical dry forests are not only the storehouses of biodiversity but also provide essential and valuable services to mankind, its extensive clearing is being recognized as the biggest contributing factor in biodiversity crises and declining ecosystem services, we are witnessing today. Further, the clearance of these forests escalates soil erosion and exotic plant invasion, which potentially leads to the habitat degradation. Under such conditions, successful restoration of the degraded tropical dry forests could be possible by the plantation of native tree species. Tropical Dry forests are likely to recover more quickly to a mature state than the wet forests. This is due to its relative simplicity and small structure of mature forests. Moreover, these forests have a considerable impact on the global carbon cycle due to their higher net primary productivity.

In order to combat and mitigate these emerging challenges, significant efforts from various levels of communities (i.e. community, domestic, regional and international) among various stakeholders (i.e. community leading group, university, research institute, government agency and international organizations) have been proceeded from last few decades. In particular, together with international research communities, research groups in forest ecology have established relatively large scale of plot-based integrated research to investigate long-term responses of forest ecosystem to natural and anthropogenic perturbations and environmental changes over broad spatial and temporal scales. The outcome of such research has contributed in better understanding of forest structure and species composition as well as ethno-botanical aspects, and species habitat requirements, apart from providing quantitative data for testing theories and hypothesis in population and community ecology.

Sustainable forest management has been proposed as a conservation strategy for tropical forests, with the idea that forests that are not managed sustainably will eventually lose their economic and ecological value and are likely to be converted to non-forest land uses. Further, a sound ecological knowledge is a prerequisite to establish the best management practices for tropical forests, and therefore the ecologists have to play a proactive role while making management decisions, which must essentially be ecologically viable and environmentally sustainable. Worldwide, ecosystems are being transformed by changes at local and global levels. Therefore, the emerging challenge is to understand such transformation with accurate prediction that how they would affect the local communities to develop effective mitigation or adaptation strategies. .. It is now agreed upon that solutions to these emerging challenges are needed to be framed within an interdisciplinary, social-ecological, biocultural context. Multiple disciplines can synergistically produce scientific knowledge that can be used by or for communities to improve livelihoods, while ensuring ecological resilience, economic viability, and social equity. In this book, we are discussing various issues, challenges therein and opportunities pertaining to conservation, restoration and management of tropical dry forests across the world along with existing knowledge gaps, which holds crucial importance in designing adaptation and mitigation strategies in the current scenario of changing climate. The book will be helpful to address the queries of environmental scientists, researchers, planners, policymakers and general people alike. The chapters included in the book cover a myriad of issues pertaining to TDFs at global level, a brief introduction is as below:

Chapter 1, 'Global Overview of Tropical Dry Forests', by G. N. Tanjina Hasnat and Mohammed Kamal Hossain from Bangladesh gives a global overview on distribution of tropical dry forests, its biodiversity, importance, climatic conditions along with soil characteristics. It also gives a brief account on threats to these forests along with the management practices.

Chapter 2, 'Effect of Climate Change on Tropical Dry Forests', authored by Pooja Gokhale Sinha from India discusses upon how structure, composition, phenology, community dynamics are changing in tropical dry forests under changing climate citing strong evidences. The chapter also gives an overview on importance of TDFs in mitigating the changing climatic scenario.

Chapter 3, 'Climate Change Effects, Adaptation, and Mitigation Techniques in Tropical Dry Forests', by G. N. Tanjina Hasnat from Bangladesh focuses on the geographic and climatic distribution of tropical dry forests. This chapter emphasized upon the impact of climate change on TDFs along with adaptations developed by plant and animal species living in these forests. Further, it discusses the mitigation measures of GHGs such as carbon sequestration, curbing deforestation, reduced forest fragmentation, restoration and management of fire and invasion.

Chapter 4, 'Plant Functional Traits in Tropical Dry Forests: A Review', by Shipra Singh and Abhishek K. Verma from India highlighted the importance of plant functional traits in processes and community dynamics of TDFs. As plants have certain characteristics which allow them to respond to various environmental conditions, like changes in climate, water loss in the soil, lack of minerals; among others, thereby characterizing a vegetation form. Though, leaf related traits are most extensively studied, other traits like seed and roots have been given less attention, which have been highlighted in this chapter. Authors also emphasized on knowledge gap in the studies of plant functional traits and future recommendation that may be helpful in the development of management strategies for TDFs.

Chapter 5, 'An Overview of the Role of Plant Functional Traits in Tropical Dry Forests', authored by Abhinav Yadav, Pramit Verma and Akhilesh Singh Raghubanshi from India provided scientific description of PFTs and their importance in community assembly, ecosystems processes, plant soil interaction and trades-off within tropical dry forests. Role of PFTs in adaptations against the variable environmental conditions have also been discussed here. They primarily emphasized upon the leaf traits and need for systems modeling using mathematical approaches in order to further the role of PFTs in forest ecology.

Chapter 6, 'Diversity and Distribution of Tropical Dry Forests: A Case Study From Pudukkottai District of Tamil Nadu, India', by Raja P, Balaguru Balakrishnan and Soosairaj S from India, explores the diversity of Pudukkottai district of Tamil Nadu in Eastern Ghats and plains of Indian Peninsula, which are practically unexplored so far. Also, the endemic, threatened species of these groves were discussed in this chapter.

Chapter 7, 'Floristic Diversity and Carbon Stock in the Dry Forests of Chad: The Case of Manda National Park – Diversity and Carbon Sequestration of the Manda National Park Flora', by Ngaba Waye Taroum Caleb and others from Chad and Cameroon, estimated the woody flora diversity of the National Park of Manda in the Sudanian area of Chad. This study was the part of REDD+ program (Reducing Emissions from Deforestation and Forest Degradation).

Chapter 8, 'Biogeographic and Anthropogenic Effects on Asian Elephants in Tropical Forests of Sumatra, Indonesia', by David M Augeri from US discusses DISTANCE protocols and MIKE Survey Standards, which were used in the field to determine critically endangered Sumatran elephant (*Elephas maximus* sumatranus) occupancy and abundance in Gunung Leuser National Park (GLNP). It also suggested plans for forest restoration, ecosystem protections and conservation for Asian elephants, biodiversity, and forests.

Chapter 9, 'Examining Patterns and Impacts of Forest Resource Extraction and Forest Degradation in Tropical Dry Forests', by Pooja Choksi from US, provided scientific evidence of resource exploitation by human interference in the form of fuel wood, fruits and edible flowers, leaves and medicinal plants tropical dry forests across the world. It highlighted that high human dependence on tropical dry forests underscores the need for a complete understanding of the interaction of humans and these forests to ensure their persistence. Therefore, it provided a holistic understanding of not just the ecological but also the socio-economic and cultural aspects and impacts of forest resource use and extraction.

Chapter 10, 'Litter Production and Decomposition in Tropical Forest', authored by Sumit Chakravarty and others from India, explores the importance of plant litter production and decomposition process along with its ecological impact of carbon and nutrient dynamics in the litter layer in tropical dry forest ecosystem.

Chapter 11, 'Participatory Management of Tropical Dry Forests in Benin', authored by Ramoudane Orou Sannou and others from Benin and Germany provides the scientific evidence of participation of local communities for the conservation of forest in Benin. They explores the political and economic liberalization policies engaged in 1990 and, the Rio de Janeiro Summit held in 1992 that recognized the importance of local communities' participation in natural resource management.

Chapter 12, 'Issues and Challenges for Stone Mining Affected Forest Area Restoration Through Probiotic Interventions', by Kumud Dubey and K. P. Dubey from India explores the need of conservation and restoration in mining affected forest area of Vindhyan region of Uttar Pradesh. This chapter emphasized the application of probiotic interventions to conserve and restore the stone mining affected forest area.

Chapter 13, 'Scenarios of the Tropical Dry Forest of Purulia District, West Bengal: A CA-MARKOV Model Approach', by Masuma Begum and others from India. In this study, use of LANDSAT satellite images after pre-processing (i.e. geometric correction and atmospheric correction using FLAASH, LULC) for the assessment of change dynamics have been done to estimate the changes in total forest cover in Purulia district through an unsupervised K-means classification scheme. Also, forest health and degradation status has been estimated via change in vegetation indices (namely NDVI, SAVI, and CVI). Moreover, future predictions have been explored by implementing a CA-MARKOV model.

Chapter 14, 'Geospatial Solutions for Forest Management: A Case Study From Nepal', authored by Him Lal Shrestha from Nepal, highlights the use of geospatial approach through GIS, Remote Sensing and GPS technologies to assess the geographical distribution of forests. This chapter focused mainly on the spatial analysis of the resources and their associating factors using geospatial approach.

Chapter 15, 'Trends in Management of Tropical Forests: Application of Remote Sensing/Geographic Information System', by Jacinta U. Ezenwenyi and Onyekachi Chukwu from Nigeria, explores various approaches for the conservation, restoration and sustainable management of tropical forest using remote sensed data and GIS. The use of these space technologies were assessed for study in biodiversity conservation, forest eco-physiology, forest trees' disease and insect interactions, forest mensuration, forest resources monitoring and evaluation, forest fires, land use/cover dynamics and vegetation cover.

Chapter 16, 'Forest Inventory: Challenges, Trend, and Relevance on Conservation and Restoration of Tropical Forests', by Onyekachi Chukwu and Japheth H. Dau from Nigeria, explores the importance of forest inventory in management, planning, monitoring, evaluation, research, growth, yield, biodiversity and timber sale of the forests. This study aimed at enlightening the conservationists, ecologists, environmentalists and forest managers the need to carryout assessment and inventory of flora and fauna resources and their diversity within the ecosystems. This will help to ensure sustainable yield and prevent the extinction of economic trees and endangered animal species in the tropical dry forests.

Chapter 17, 'Post-Fire Regeneration of Woody Species in a Colombian Tropical Dry Forest', authored by Anghy Gutiérrez-Rincón and Angela Parrado-Rosselli from Columbia' described the differential ability of plant species to recover, germinate and establish after a fire across the sites via field experimentation. Three groups (1) fire-stimulated, (2) fire-tolerant and (3) fire sensitive species) were tested for their regeneration response and recommended for the restoration programs in Columbian dry tropical forests were done based on the result output, in light of more frequent forest fires due to climate change.

Chapter 18, 'Ecology of Natural Regeneration of Tropical Dry Forests of Africa and Its Implications for Their Sustainable Management', by Stephen Syampungani, Annie Namuuya Sikanwe and Paxie W C Chirwa from Zambia explores the natural regeneration strategies of tree seedlings under prevailing environmental conditions (such as, fires, competition for nutrients and light) which influences the number of surviving individuals to enter the next size class. This chapter emphasized on developing restoration strategy for the African savannas and tropical dry forests.

Chapter 19, 'Role of Institutional Governance in Protecting the Natural Environment: A Case of National Capital Region, India', by K. K. Yadav and Kumud Dhanwantri from India provides an insight into the challenges in forests and wildlife in the Aravalli Ranges falling in different sub-regions of the National Capital Region, Delhi. It discusses requirement of some, immediate and potential government initiatives as well as how local communities can tie-up with governments in the protection of forests and wildlife.

Section 1

Global Overview and Climate Change

Chapter 1
Global Overview of Tropical Dry Forests

G. N. Tanjina Hasnat
University of Chittagong, Bangladesh

Mohammed Kamal Hossain
University of Chittagong, Bangladesh

ABSTRACT

Forests cover almost one-third of the Earth's land surface. Tropical dry forests are the second-most-important forest type in the world covering approximately 42% of tropical and sub-tropical forest area. The main features of these forests are their deciduousness, a prolonged dry period extending 3-9 months, and little annual precipitation of 250-2,000 mm. Tropical dry forests are found in five of the eight realms in the world. More than half of the forests are distributed in the Americas, with other portions in Africa, Eurasia, Australia, and Southeast Asia. The forests are unique in nature, and provide shelter to a huge number of endemics and endangered species. Among woody plant species, about 40% are not found anywhere in the world. These forests are now the most threatened among all forest types. The conservation status of these forests is endangered. Deforestation, rapid civilization, land conversion, fire, and climate change are the major threats. Proper management with time-oriented policy could be helpful to restore these forests and protect the existing remnant areas.

INTRODUCTION

Land biomes are the major habitats of the world that can be identified by different plant and animal species living there and distinguished by the existing climatic condition (Campbell, 1996; Cain, Bowman, & Hacker, 2014; Bailey, 2018). Forest is one of the most important categories of the terrestrial land biome. Forests cover about one-third of the earth's land area (Lee & Jarvis, 1996). The major forest biomes in the world are the tropical forest, temperate forest, and boreal forest (Keith, Mackey, & Lindenmayer, 2009). The tropical forest is classified typically as wet tropical forest (rain forest), moist tropical forest, and dry tropical forest (Murphy & Lugo, 1986). Tropical forests cover 40% of the land mass of tropical

DOI: 10.4018/978-1-7998-0014-9.ch001

and sub-tropical zone of which tropical dry forest covers 42%, moist forest 33% and rain forest 25% area (Holdridge, 1967; Brown & Lugo, 1982; Murphy & Lugo, 1986). Considering only tropical dry forest area, 54.2% is situated in South America; 42% in North America, Central America, Africa, and Eurasia; and only 3.8% in Australia and Southeast Asia (Miles et al., 2006).

Tropical dry forests are also called seasonally dry tropical forests (Mooney, Bullock, & Medina, 1995), tropical dry forest (Miles et al., 2006), tropical seasonal forest (Holzman, 2008), tropical broadleaf woodland, (Whittaker, 1975), and tropical seasonal forest (Whittaker, 1975). The forests consist mainly of deciduous trees and unlike the evergreen trees, deciduous species shed their foliage in the dry season. In the same dry forests, the sites with moisture content expose evergreen nature even in the dry periods. Three tropical dry broadleaf forest eco-regions namely the East Deccan dry evergreen forests, the Sri Lanka dry-zone dry evergreen forests, and the Southeastern Indochina dry evergreen forests, are characterized by evergreen trees (Ramanujam & Kadamban, 2001).

Among all types of forests, tropical dry forests are more threatened and ecosystems of these forests are most fragile and least protected (Sanchez-Azofeifa et al., 2003). About 97% of the present tropical dry forest land is at risk (Miles et al., 2006). According to Janzen (1988), the dry forest cover in the Western Hemisphere has been shrinking from 550,000 square kilometers to 480 square kilometers and exist only 0.09% of the original forested area. This scenario is also similar to Australia, Southeast Asia, Africa, and major parts of South America (Janzen, 1988).

Tropical dry forests have been exploited heavily for conversion to agricultural land because of its favorable climatic condition (Ewel, 1999) and minimum attack by weeds and pests (Murphy & Lugo, 1986). According to Piperno and Pearsall (2000), the early settlement of humans started from these forested areas. From their statement, the origin of crop plants and animal husbandry also initiated here. Tropical dry forests are unique in nature and species diversity. The conservation status of the forests is endangered. In comparison to tropical rain forests, though tropical dry forests have 30-90% of tree species diversity and 50-70% of primary productivity, but researches and other studies on tropical dry forests are 4-5 times less and only 3% literature found about tropical dry forest restoration than tropical rain forest restoration. It also acts as a carbon sink to mitigate climatic changes. Presently, these ecologically-distinctive type forests are under threat and still remain far distant from the concern of Ecologists, Scientists, and Environmentalists.

DISTRIBUTION

Favorable climatic condition for tropical dry forest generally found between two parallels of latitude, the Tropic of Cancer (23°27’ N) and the Tropic of Capricorn (23°27’ S) where there is a little precipitation only for a few months and a prolonged dry period, generally, ranges from three to seven months (Holdridge, 1967; Janzen, 1983; Murphy & Lugo, 1986; Luttge, 1997; Piperno & Pearsall, 2000).

The tropical dry forests extend to the drier areas between the north and south belts. Forests with most diverse in type and species occur in southern Mexico and in Bolivian lowlands (Parker, Gentry, Foster, Emmons, & Remsen, 1993; Bullock, Mooney, & Medina, 1995). A more unique category of species found in the dry forests of the Pacific Coast of northwestern South America due to their isolation (Parker & Carr, 1992; WWF & IUCN, 1994; Bullock, Mooney, & Medina, 1995). A large number of diversified endemics are sheltered by the subtropical forests of Maputoland-Pondoland in southeastern Africa (Cowling & Hilton-Taylor, 1994; WWF & IUCN, 1994). The dry forests of central India and Indochina

Figure 1. Major habitat types in the world (Source: WWF)

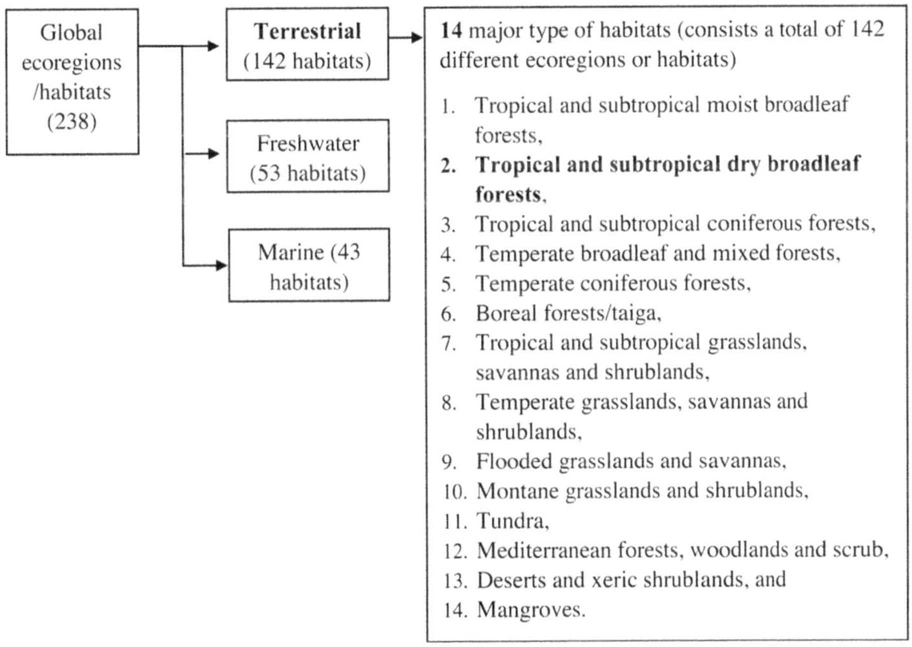

provide habitats for a large number of varied vertebrate faunas (Corbet & Hill, 1992; Stewart-Cox, 1995). A distinctive wide range of taxa and higher taxonomic levels of endemics along with a large number of relictual taxa have found their habitats and shelter in the dry forests of Madagascar and New Caledonia (IUCN/UNEP/WWF, 1987; Preston-Mafham, 1991; WWF & IUCN, 1994).

Globally, there are three major types of ecoregions/habitats – terrestrial (142), freshwater (53), and marine (43) (Olson & Dinerstein, 2002). The terrestrial eco-regions are subdivided into 14 prioritized habitats (Figure 1). These 14 habitats (Olson et al., 2001; Miles et al., 2006; WWF, 2016a) consists of 142 types of principal areas or habitats, whereas the tropical and subtropical dry broadleaf forests ecoregions comprises of 10 different areas (Olson et al., 2000) under five realms (Table 1) out of eight in the world (Udvardy, 1975; Richard, 1980; Olson et al., 2001).

Madagascar Dry Forests

The forests are situated at the Western coast of Madagascar and covered a total of 151,000 sq. km (58,000 sq. miles). The conservation status of these tropical and subtropical broad-leaved forests is endangered (WWF, 2016b; Olson et al., 2000). These forests are the habitat of hundreds of endemic plant and animal species (Goodman & Benstead, 2005, Randriamalala & Liu, 2010; Raharimahefa, 2012). The most important species are Angonoka tortoise (*Angonoka yniphora*), the world's most endangered tortoise, and the rare Aye-Aye (*Daubentonia madagascariensis*). Along with the featured species Giant jumping rat (*Hypogeomys antimena*), other important species are Lemur (*Lepilemur edwardsi*), Van Dam vanga (*Xenopirostris damii*), Appert'sgreenbul (*Phyllastrephus apperti*), and Flat-tailed tortoise (*Pyxis*

Table 1. Global extent of tropical and subtropical dry broadleaf forests

Realm	Ecoregions/Habitats (Tropical and Subtropical Dry Broadleaf Forests)	Distribution	Region
Afrotropical	(1) Madagascar Dry Forests	Madagascar	Africa
Australasia	(2) Nusu Tenggara Dry Forests	Indonesia	Southeast Asia
	(3) New Caledonia Dry Forests	New Caledonia (France)	Australasia
Indo-Malayan	(4) Indochina Dry Forests	Cambodia, Laos, Thailand, Vietnam	Southeast Asia
	(5) Chhota-Nagpur Dry Forests	India	South Asia
Neotropical	(6) Mexican Dry Forest	Guatemala, Mexico	North and Central America
	(7) Tumbesian-Andean Valleys Dry Forests	Colombia, Ecuador, Peru	South America
	(8) Chiquitano Dry Forest	Bolivia, Brazil	South America
	(9) Atlantic Dry Forests	Brazil	South America
Oceania	(10) Hawaii's Dry Forests	Hawaii (United States)	Oceania

Source: WWF (2016a)

planicauda). Some species recently have disappeared like Giant lemurs or Rodentia as documented by subfossil records (Goodman, Vasey, & Burney, 2007; Burney et al., 2008; Buckley, 2013).

Flora of Madagascar dry forests are rich in number. About 13,000-14,000 species found here (Phillipson, Schatz, Lowry, & Labat, 2006). Dominant species include *Adansonia* spp., *Dalbergia* spp. *Stereospermum* spp., *Albizia* spp., *Alchornea* spp., *Brachylaena* spp., *Dichrostachys* spp., *Euphorbia* spp., and *Grewia* spp. (Moat & Smith, 2007). Among tree species, Baobab (*Adansonia* spp.), the long-lived (about 5,000 years) tree is most important (Cowen, 1952; Swart, 1963). The tree is well-known for its fruit production capability during the dry period with highly-stored water in its vast trunk, and it is known as 'The tree of life' (Bash, 2002). Out of nine Baobab species in the world, seven are found in Madagascar dry forests. The major threats to the region are deforestation and uncontrolled burning. Deforestation is mainly for slash-and-burn agriculture, pasture, firewood, or construction materials and secondary grasslands. Considering deforestation, Madagascar dry forests are highly under threatened condition (Miles et al., 2006).

Nusu Tenggara Dry Forests

These forests are situated in Indonesia in Southeast Asia and cover a total landmass of 73,000 sq. km (28,000 sq. miles). The conservation status of these tropical and subtropical, moist broadleaf forests is endangered (Olson et al., 2000; WWF, 2016b). Two terrestrial ecoregions are existed here, one is Lesser Sundas deciduous forests, and another is Timor and Wetar deciduous forests. The number of animal species is low here but the forests are richest with mammals found in Southeast Asian islands. The Lesser Sundas deciduous forests are the habitat of 50 mammal species including five ecoregional endemics. These forests also give shelter to the world's most restricted large carnivore, the threatened Komodo dragon (*Varanus komodoensis*), the critically endangered Flores shrew (*Suncus mertensi*) and the vulnerable Komodo rat (*Komodomys rintjanus*). Others mammal species are Flores giant tree-rat

(*Papagomys armandvillei*), White-toothed shrew (*Crocidura neglecta*) and Flying fox (*Pteropus lombocensis*) (WWF, 2016b).

Among the bird species Cinnamon-banded kingfisher (*Todirhamphus australasia*), White-rumped kingfisher (*Caridonax fulgidus*), Bare-throated whistler (*Pachycephala nudigula*), Golden-rumped flowerpecker (*Dicaeum annae*), Crested white-eye (*Lophozosterops dohertyi*), Thick-billed white-eye (*Heleia crassirostris*), Scaly-crowned honeyeater (*Lichmera lombokia*), Sumba flycatcher (*Ficedula harterti*), Apricot-breasted sunbird (*Nectarinia buettikoferi*), and Yellow-spectacled white-eye (*Zosterops wallacei*) are important. Timor and Wetar deciduous forests are the harbor of near 250 different bird species, 24 of which are endemic, and five are threatened (WWF, 2016b). The threatened species are the Black cuckoo-dove, Wetar ground-dove, Timor green-pigeon, Timor imperial-pigeon, and Iris lorikeet. The Timor is the board of the rare Timor python. The two major threats to Nusu Tenggara dry forests are fire and logging (WWF, 2016b).

New Caledonia Dry Forests

The forests are located in New Caledonia, a large island northeast of Australia covering 4,400 sq. km (1,700 sq. miles) of land mass. The conservation status of these tropical and subtropical dry broadleaf forests is endangered (Olson et al., 2000; WWF, 2016b). New Caledonia can be divided into three major native vegetation types: Maquis scrub, Rain forest, and Tropical dry forest (Jaffré, Bouchet, & Veillon, 1998). Among the three, the tropical dry forests are the most endangered one (Bouchet, Jaffré, & Veillon, 1995).

New Caledonia dry forests are the most unique ecoregion on the earth. The forests are the hotspot with a diversity of 3261 native plant species, 74.3% of which are endemic (Myers, Mittermeier, Mittermeier, Da Fonseca, & Kent, 2000; Jaffré, Morat, Veillon, Rigault, & Dagostini, 2001) and not found anywhere in the world. These tropical dry forests contain the highest number of threatened species based on IUCN global conservation categories (Gillespie & Jaffré, 2003). Moreover, the forests are the harbor of 407 woody species, 243 of which are endemic to New Caledonia and 60 are endemic and restricted to tropical dry forest. Among the endemic species, yellow-flowered *Pittosporum pancheri*, the fragrant *Gardenia urvillei*, the thin-leaved *Codiaeum peltatum*, a variety of wild rice *Oryza neocalidonea*, and the rare plant *Captaincookia margaretae* are important. Intentional fire, land clearance, trampling by cattle (Bouchet, Jaffré, & Veillon, 1995), and introduction of alien species are the major threats here.

The forests appear to be the world's most endangered tropical dry forest based on the extent of forest, the number of reserves, and threatened species (Gillespie & Jaffré, 2003). The forests are deforested and fragmented, and the species are destroyed in such a way that now it is reduced to 2% of the original size (Bouchet, Jaffré, & Veillon, 1995).

Indochina Dry Forests

The forests are located in Cambodia, Laos, Thailand, and Vietnam of Eastern Indochina and covered a total of 444,000 sq. km (171,000 sq. miles) land mass. The conservation status of these tropical and subtropical dry broadleaf forests is endangered (Olson et al., 2000; WWF, 2016b). There are two terrestrial habitats found in this ecoregion, one is Southeastern Indochina dry evergreen forests, and another is Central Indochina dry forests. Indochina dry forests are globally well-known for its species richness including a large number of vertebrates and 160 mammal species. The forests provide habitat to the highly endangered wild cattle called Kouprey (*Bos sauveli*), the featured species of the forests. A large number of species such as the beautiful Red-shanked douc langur (*Pygathrix nemaeus nemaeus*), the threatened species Asian elephant (*Elephas maximus*), tiger (*Panthera tigris*), Crested gibbon (*Hylobates concolor*), and the only Javan rhinos (*Rhinoceros sondaicus*) are also living here. The forests are also shelter of 455 bird species, among which two are near-endemic and one is endemic. The only endemic endangered bird species is Orange-necked partridge (*Arborophila davidi*) (WWF, 2016b).

The forests are under diverse types of threats. The major threats are logging, land clearing for agriculture, burning and conversion to teak plantations, hunting, trapping, shifting cultivation, excessive non-timber forest product (NTFP) harvesting, wildlife poaching, and illegal logging (WWF, 2016b).

Chhota-Nagpur Dry Forests

The forests are situated in Eastern India and cover 122,000 sq. km (49,000 sq. miles). The conservation status of these tropical and subtropical dry broadleaf forests is endangered. The forests lie between the moist, deciduous forests of the Eastern Ghats and Satpura Range, and the lower reaches of the Gangetic Plains (Olson et al., 2000; WWF, 2016b). Extent of the forests is distributed through the eastern Indian states of Bihar, Madhya Pradesh, and West Bengal.

The forests are rich in numerous rare and endemic species and provide shelter to 77 mammal species. Providing habitats to some important tiger species and last populations of Asiatic elephants, the forests are unique in nature. Moreover, this area served as a refuge to a huge number of fauna at the time of the last ice age that make the present forests asylum of many rare and distinctive species. Some important species found in the forests are Tiger (*Panthera tigris*), Asian elephant (*Elephas maximus*), Sloth bear (*Ursus ursinus*), Leopard (*Panthera pardus*), Blackbuck (*Antilope cervicapra*), and the featured species of the forests Chinkara (*Gazella bennettii*). About 280 bird species are found here and the globally threatened lesser Florican (*Eupodotis indica*) is one of them. The dominated plant species of these forests are *Tectona grandis, Shorea robusta, Anogeissus latifolia, Terminalia alata, Lagerstroemia parviflora*, and *Phoenix robusta*. The endemic Cycad (*Cycas beddomei*), a critically endangered species is also grown here. Logging, clearing, overgrazing, quarrying, mining, monocultures, and hydroelectric projects are the major threats in this ecoregion (WWF, 2016b). Of the 10 tropical and subtropical dry forest ecoregions identified by Olson et al. (2000) as priorities for conservation, the Indochina, Chhota-Nagpur (India), Mexico, and Chiquitano (Bolivia) dry forest ecoregions are the most threatened ones (Miles et al., 2006).

Mexican Dry Forests

The Mexican Dry Forests is situated in Guatemala and Mexico and covered a total of 315,000 sq. km (121,000 sq. miles) land mass. The conservation status of these tropical and subtropical dry broadleaf forests is endangered (Olson et al., 2000; WWF, 2016b). This broad category dry forests include 8 terrestrial ecoregions named Jalisco dry forests, Balsas dry forests, Bajío dry forests, Chiapas Depression dry forests, Sonoran-Sinaloan transition subtropical dry forest, Southern Pacific dry forests, Sinaloan dry forests, and Sierra de la Laguna dry forests.

There are 1000 different plant species are growing in various habitats of Chiapas Depression dry forests. More than 750 plants species, 70 mammal species and 27 species of termites are found in The Jalisco dry forests. These forests are exceptional from the other seven for rarely burn nature and having trees that shed leaves for a longer period of time during the dry season. In Balsas dry forests, 24 out of

60 *Bursera* tree species are grown here. Two rare bat species found in these forests are the California myotis (*Myotis californicus*) and the Long-legged myotis (*Myotis volans*). Chapala Lake, the largest lake of Mexico is situated in the Bajío dry forests and a sanctuary of an enormous fish community, but in danger now. Sinaloan dry forests are well known for its huge number of bird species. The Sierra de la Laguna dry forests provide shelter to 224 species of plants, 50% reptiles and amphibians, and 96% mammals of the entire Cape Region. Many other species are also found in the Mexican dry forests especially the Red-knee tarantula spider (*Brachypelma smithi*), Orange-breasted bunting (*Passerina leclancherii*), White-throated magpie jay (*Calocitta formosa*), and the West Mexican chachalaca (*Ortalis poliocephala*) etc. The flagship species of these forests is Mexican Red-Knee Tarantula (*Brachypelma smithi*) (WWF, 2016b).

The major threat of the Mexican dry forests is the conversion of forest land to agricultural land. The massive land use change occurred from 1977 to 1992 and the average deforestation rate was 1.9% per year (Anonymous, 2011). At the end of 1980, 44% of the original area of tropical dry forests and by 2009, 71% tropical dry forests of Mexico have already been destroyed by human interference. Now in Mexico, it is only 0.2% of tropical dry forests are protected (Anonymous, 2011). Urbanization, increasing tourism, exploitation of wildlife, road construction, cattle ranching, annual and perennial plantations, logging, and hunting are other major threats in this ecoregion (WWF, 2016b).

Tumbesian-Andean Valleys Dry Forests

These forests generally situated in Colombia, Ecuador, and Peru and covered a total of 103,000 sq. km (40,000 sq. miles). The conservation status of these tropical and subtropical dry broadleaf forests is endangered (Olson et al., 2000; WWF, 2016). This Global ecoregion consists of 6 terrestrial ecoregions named Tumbes-Piura dry forests, Ecuadorian dry forests, Patía Valley dry forests, Magdalena Valley dry forests, Cauca Valley dry forests, and Marañón dry forests.

The forests are rich with different types of flora and faunal diversity such as flowering plants like ébano (*Zizyphus thyrisflora*), Charanblanco (*Pithecelobium* spp.), and Cedrela (*Cedrella* spp.). Among bird species most important ones are - the featured species of this forests Pacific parrotlet (*Forpus coelestis*), the endemics Watkins' antpitta (*Grallaria watkinsi*) and Velvet-fronted euphonia (*Euphonia concinna*). More interestingly, after many years of being considered as extinct, the Yellow-eared parrot, presently a rare and highly endangered species was discovered in Magdalena Valley. Logging, agricultural expansion, burning, overgrazing, hunting, invasion, increasing human settlement, pollution, and crop cultivation are the major threats in this ecoregion (WWF, 2016b).

Chiquitano Dry Forest

The forests are located in Bolivia and Brazil, and covered with a land mass of 230,500 sq. km (89,000 sq. miles). The conservation status of these tropical and subtropical dry broadleaf forests is endangered (Olson et al., 2000; WWF, 2016b). Most part of the forest is situated in Bolivia and smaller patches exist in Brazil. The forest is the land of the indigenous inhabits Chiquitanos (Gott, 1993) and named from that ethnic group.

Chiquitano dry forest recognized as a unique ecoregion (Olsen & Dinerstein, 1998; Ibisch, Columba, & Reichle, 2002) with the largest extant patch that contains a complex of forests communities, now, broadly known as the Neotropical seasonal dry tropical forest complex (Prado & Gibbs, 1993; Prado,

2000; Pennington et al., 2004). The deciduousness of the forest depends on the amount of precipitation and variation of geographic location. The ranges of the forest is complete deciduous in the South to semi-deciduous in the North (Killeen et al., 2006). Most tree species of the forest are slow growing and economically valuable with its high wood density. The most common tree species are *Tabebuia impetiginosa* (C. Mart. ex A. DC.) Standley, *Machaerium scleroxylon* Tul., *Astronium urundeuva* (Alemão) Engl. and *Schinopsis brasilensis* Engl. *Schinopsis brasilensis* Engl. is the largest tree in this forest, and many of them are more than 500 years old (Killeen et al., 2006).

The forests provide shelters to a large number of endangered and vulnerable species. Among these, a critically endangered reptile, the Broad-snouted caiman (*Caiman latirostris*); an endangered bird, the Black-and-tawny seedeater (*Sporophila nigrofufa*); and 3 mammals - the Giant armadillo (*Priodontes maximus*), Maned wolf (*Chrysocyon brachyurus*) and Giant otter (*Pteronura brasiliensis*) are most important. Moreover, 3 other bird species, 1 reptile, and 12 mammals are listed as vulnerable. The other species important here are the Barefaced curassow (*Crax fasciolata*), Puma (*Felis concolor*), Jaguar (*Panthera onca*), and the Lianas (*Bignoniaceae spp.*).Puma is the featured species of the forests.

Flooding and fires are common phenomena in this forest. Trees made them adaptable with flood and temporary water logging conditions. They also can cope with another catastrophe by having fire-resistant bark. Other significant threats are pollution, wildlife exploitation, agricultural expansion, and grazing. Areas adjacent to Chiquitania in Bolivia have experienced considerably greater rates of deforestation (Steininger et al., 2001a, b; Pacheco & Mertens, 2004) and conservation is critical here. Other most important threats are habitat degradation due to uncontrolled logging (Killeen et al., 2006), and habitat fragmentation due to civilization and rapid urbanization.

Atlantic Dry Forests

The forests are situated in Northeastern Brazil, South America and covered 115,000 sq. km (45,000 sq. miles). The conservation status of these tropical and subtropical dry broadleaf forests is endangered (Olson et al., 2000; WWF, 2016b). This forest is a border between the Cerrado Grasslands and Caatinga Scrub ecoregions of Eastern Brazil. The forests include both deciduous and semi-deciduous plant species (Leal & de Gusmão Câmara, 2003; Thomas & Britton, 2008). The neighboring moist forests and grasslands provide the forests a unique habitat characteristics that aboard diversified species (Mauro, Dantas, & Roso, 1982). This fairly dense forests provide shelter to a number of tree species grown generally 25-30 m high. The most common tree species are *Cavanillesia arborea, Cedrela fissilis, Schinopsis brasiliensis, Astronium urundeuva, Aspidosperma macrocarpa,* and *Tabebuia* spp. (Ratter, Askew, Montgomery, & Gifford, 1978; Andrade-Lima, 1981). Among all species, *Cavanillesia arborea* is significant for its extraordinary feature. This bottle-shaped large tree can reached its maximum diameter (about 3 meters above ground level) up to 1.5 meters or more (Ratter, Askew, Montgomery, & Gifford, 1978).

The forests are well-known for its bird species (Da Silva & Oren, 1992, 1997; Raposo, 1997). The bird Gilt-edged tanager (*Tangara cyanoventris*) is the featured species of the forests. The other bird species Great xenops (*Megaxenops parnaguae*), Hooded visorbearer (*Augastes lumachellus*), Caatinga nighthawk (*Chordeiles vielliardi*), Pale-throated serra finch (*Sporophila frontalis*), and Narrow-billed antwren (*Formicivora iheringi*) are also important endemics of the forests. According to IUCN, most of the threatened and near-threatened vertebrate endemics and semi-endemics are found in Atlantic dry forests. The major threat of the forests is land cover change due to agricultural practices (Da Silva & Oren, 1997). In comparison with its original size, only 30% of forests are remaining. Moreover, the forests ares the main source of fuels for steel and pig iron industries of Brazil (Da Silva & Oren, 1997). Thus, the deforestation and habitat destruction shrinkage the forests and make it less rich in biodiversity.

Hawaii's Dry Forests

The forests are situated in the Hawaiian Islands, Central North Pacific Ocean and covered an area of 10,000 sq. km (3,900 sq. miles). The conservation status of these tropical and subtropical dry broadleaf forests is endangered (Olson et al., 2000; WWF, 2016b). This broad global terrestrial ecoregion consists of Hawaii tropical low shrublands, Hawaii tropical dry forests, and Hawaii tropical high shrublands. The forests are the shelter of more than 200 plant species. The forests are dominated by the tree genera *Acacia*, *Chamaesyce*, *Metrosideros*, *Sapindus*, *Sophora*, *Pritchardia*, *Pandanus*, *Diospyros*, *Nestegis*, *Erythrina*, and *Santalum* (Sohmer & Gustafson, 1987). With other common species like *Erythrina sandwicensis, Diospyros sandwicensis, Reynoldsia sandwicensis* and *Nothocestrum* spp. the forests also provide habitat to many extremely rare and endemic plants (Bruegmann, 1996) such as *Gouania* spp. (Sohmer & Gustafson, 1987; Cuddihy & Stone, 1990). Lower Hawaiian dry forests are the home of several forest birds, such as Honeycreepers, Fly catchers, Flightless rails, Flightless birds, and the Hawaiian owl. The Palila (*Psittirostra bailleui*), an endangered endemic Finch-like bird, ecologically relate to Mamane trees. The birds became endangered due to the deforestation and destruction of the Mamane trees (Gon & Olson, 2019). The feature species of the forests is the Nene (or Hawaiian) Goose (*Branta sandvicensis*), the state bird of Hawaii.

Dry forests of Hawaii are under serious threat. The forest coverage exists only 10% of its original size, whereas, the native forest cover only extent to 1% (Bruegmann, 1996; Juvik & Juvik, 1998; Allen, 2000; Cabin et al., 2000), compared with 60% coverage of Hawaii's rain forests (Allen, 2000). Land development, fire, non-native ungulate grazing, and invasions by non-native plant species have been identified as the most serious threats to Hawaii's dry forests, causing severe degradation and fragmentation (Cuddihy, 1989; Cuddihy & Stone, 1990; Stone, Smith, & Tunison, 1992; Bruegmann, 1996; Blackmore & Vitousek, 2000). With civilization and the introduction of alien species, the forests lost its habitat. Human interference, clearing, burning, and conversion of forest land to agriculture (Kirch, 1982; Burney & Burney, 2003) and pasture make these forests more vulnerable. Fire is another major threat. The deforested and disturbed habitat provides spaces for the invasion of exotic fountain grass that makes the native plant species more vulnerable to frequent fire (Burney et al., 1995; Bruegmann, 1996; Burney & Burney, 2003).

CLIMATIC CONDITION

Dry tropical forests are occurred in the areas where the mean annual temperature is higher than 17°C and the mean annual rainfall is 250-2000 mm (Murphy & Lugo, 1986). There is a prolonged dry period with 3-9 months in a year and the forests sometimes receive less than 10 cm rainfall. In that places, some or all trees are deciduous in nature and lose their leaves in the period of drought (Pennington, Lewis, & Ratter, 2006). These forests generally occur in a warm to hot climate where potential evapo-transpiration and precipitation ratio (e/v) is more than unit (Barradas & Fanjul, 1985). The forests deal with a long dry season that lasts several months (Vieira & Scariot, 2006) and varies with geographic region (WWF, 2019). According to Holdridge life zone system (Holdridge, 1967) the favorable climatic area for global dry forests is 111,599,269 km^2 (Leemans, 1992). Out of the area, 94% is situated in the tropical region. Depending on annual rainfall, temperature, durability of dry period, ratio of mean annual temperature and mean annual precipitation, tropical dry forests are classified into several types – deciduous, semi-deciduous, mixed deciduous, dipterocarp savanna, monsoon forest-savanna ecotone, dry monsoon, deciduous savanna, mixed dry, dry evergreen, deciduous woodland, monsoon forest, dry forest, seasonal forest, semi-evergreen forest etc. (Murphy & Lugo, 1986).

SOIL CONDITION

The tropical dry forests contain highly weathered reddish soils with a high concentration of iron and aluminum, and low content of mineral nutrients. The biome contains much more fertile (Pennington, Prado, & Pendry, 2000) and rich but easily eroded the soil. Main soil orders found in these forests are Oxisols, Ultisols and Alfisols (Soil Survey Staff, 1999; Brady & Weil, 2002).

NATURE

Tropical dry forests occur in tropical regions characterized by pronounced seasonality in rainfall distribution, resulting in several months of drought (Richards, 1952; Mooney, Bullock, & Medina, 1995). The forests mainly occurred on oligotrophic, reddish soil such as Oxisols and Ultisols. Sometimes the forests occurred on Mesotrophic soils too. Tropical dry forests mostly deciduous, sometimes evergreen (Mooney, Bullock, & Medina, 1995); either open or closed. Deciduous species shed their leaves in a dry period (Hubbell, 1979). Many authors (Oliveira-Filho, Jarenkow, & Rodal, 2006; Pennington, Lewis, & Ratter, 2006) include thorn woodlands as part of a broadly defined "tropical dry forest" or "seasonally dry tropical forest" (SDTF). Tropical dry forest is dominated by broad-leaved drought-deciduous trees or, more rarely, small-leaved evergreen trees or broad-leaved sclerophyllous-leaved trees (Lugo, Medina, Trejo-Torres, & Helmer, 2006). The forests are medium to high forests with an average tree height of 15-25 m. There are several structural layers observed such as upper canopy, several sub-canopy levels of pole, sapling, seedlings, shrubs, and woody vines. The lower stratum is occupied by irregular low herbaceous plants. Tropical dry forests mainly include semi-evergreen, partially sclerophyllous, semi-deciduous and deciduous, mostly dense to semi-dense forest (Whittaker, 1975). Though both savannahs and dry forests can be grown in the same climatic condition but the dry forests mainly occur in more fertile soil (Mooney, Bullock, & Medina, 1995; Ratter, Richards, Argent, & Gifford, 1973).

Plant species diversity found in tropical dry forest regions bears both the characteristics of tropical regions and temperate regions (Gentry, 1982, 1988; Clinebell, Phillips, Gentry, Stark, & Zuuring, 1995; Trejo & Dirzo, 2002). In general, it seems that plant species of tropical dry forests are less diversified than tropical rain forests (Gentry, 1995), but, in some areas, species diversity found richer than that of some moist forests (Hubbell, 1979; Gentry, 1982, 1988; Janzen, 1988). Tropical dry forests are unique in nature as it retains two completely contrasted seasonal attitude, one is wet and another is dry. The forest is important for its unique species diversity. Because of its behavioral nature of wet as well as dry, the forest is more beneficial to humans. The forest is much more important from the point of carbon storage. Except tropical rain and moist forests, this forest biomes act like a greater carbon sink and play a most important role in the carbon cycle. These forests can hold a significant amount of carbon in the biomass and the soil.

There are two different types of flora and fauna composition here. One type whereas the need to survive or hibernate during the dry period, others enjoy their bountiful time then (Janzen, 1988). During the dry period, many wildlife migrates to the moist areas like a buffer zone of rain forests and dry forests. In this season most plant species shedding their leaves and terminate their vegetative activities to adapt with the dry condition. This deciduous type forest starts a leafless period during the dry season to retain moisture content. It is because trees release moisture through their leaves and shedding leaves allows them to hold water during the drought periods. But phenological activities of some woody plant species like flowering, fruiting and dispersing of seeds continued (Janzen, 1988). Some wildlife lives on these flowers, fruits, and seeds.

The forest revealed a distinct behavior in comparison to the tropical rain forest. Deciduousness is the main feature of these forests (Medina, 1983). Remarkably many trees which are evergreen in nature in the tropical rain forests behaved as deciduous in tropical dry forests. Places where drought is extreme, tree canopy becomes less dense. As the forests shaded during the dry period and the undergrowth vegetation receives greater sunlight, so the undergrowth becomes dense somewhere. Drought resistant plants like orchid, cacti, and bromeliads grow in these forests. To adapt with the crisis condition, trees have thicker or ribbed bark, deep roots without any buttress, variable leaves and sometimes thorns. In the rainy season the vegetation becomes greener. But during the dry period that continued from 3 to 9 months, most of the trees shed their leaves and sun penetrate to the forest floor, which makes the soil, watercourses drier. In the daytime the relative humidity ranges from 20-60% (Janzen, 1988). In this season the continuous green figure of this type of forests becomes a sporadic complex mosaic type.

The growth period is shorter in tropical dry forests and annual growth increment is little. Storage of P, N, and K is high in soil and these are 98%, 91%, and 96% respectively (Murphy & Lugo, 1986). The phenology of the forests is varied and not easy to understand. Tropical dry forests are vulnerable to stress than any other forests. As plant growth is slower so succession is slower here. People are much more related to tropical dry forests than other forests from the political, geographical, biological and ecological point of view (Murphy & Lugo, 1986). The environment near dry forests is suitable for human beings as well as livestock, as diseases dispersed less in dry conditions. Weed invasion is less than rain forest as the soil is dry, but sometimes fertility is high because of its less leaching of nutrients.

DIVERSITY

Tropical forests are rich in species, genetic, and ecological diversity. A lowland dry forest adjacent to a lowland rain forest sustains a fauna and flora about 50-100% species as exists in the neighboring rain forest (Janzen, 1986). Fauna is more similar in richness rather than flora. Among fauna maximum similarity observed in mammals and major insect groups like butterflies, moths, bees, wasps, ants, beetles, termites, scorpions, etc. (Roper, 2018). The similarity ranges from 5%-80% on an area basis (Janzen, 1988). In the case of plant life forms, there is more structural and physiological diversity in dry forests in comparison to wet forests.

Though species diversity is lower than tropical rain forests, but on a world scale, it poses a standard position. In Central America and India tropical forests are important as the habitat of migrating birds in their non-breeding season (TDF, 2019). Large animals like white rhino and giraffe found in the tropical dry forests of Africa. Dry forests of Indochina are the habitat of other types of massive highly endangered mammals, Asian elephant, and Kouprey. Unlike rain forests, the majority of mammals in the world's dry forests are typically small in size such as pumas, deer, monkeys, jaguars, squirrels, rodents. In dry forests of Madagascar, a special type of mammal is also found named giant jumping rat sized like a rabbit and legs behind modified in such that it could jump as like as kangaroo that helps this type of animal to save themselves at the time of the attack by local predators. The animal can jump up to three feet in height. Angonoka or Ploughshare tortoise, the world's most endangered tortoise is found in the Madagascar western dry forests. A large number of chameleon species are also found there. The rare Timor python (*Python timoriensis*) and the world's largest living lizard Komodo dragon (*Varanus komodoensis*) are found in the dry forests of the Lesser Sunda Islands in the Java Sea (Roper, 2018).

The tropical forests are very rich both in flora and fauna. Native important plant species are Agaves, Acacias, Cacti, etc. This is the habitat of different types of mammals, butterflies, moths, bees, wasps, ants, birds, reptiles, amphibians, etc. Tropical dry forests are home of various wildlife like large cats, monkeys, parrots, rodents, wild birds, sloths, black howler, monkeys, foxes, raccoons, badgers, bobcats, mountain lions, etc. From more than 1600 inventories in the tropical dry forests, it is found that about 6958 species of woody plants are grown in tropical dry forests and 40% of which are not found anywhere in the world. Some of the species are ancient of more than 10 million years. If the species loss once then it will be a great loss to the world (Kinver, 2016). In the difficult climatic condition, many of the species demonstrate various adaptation criteria. Tropical dry forests are also the harbor of different types of reptiles. The species living here are able to survive with the marginal hotter temperature. Birds in tropical dry forests are becoming endangered day by day. Species like Black-and-tawny seedeater (*Sporophila nigrofufa*) of the Chiquitano forest of South America, Orange-necked partridge (*Arborophila davidi*) of Indochina, and the lesser florican (*Eupodotis indica*) of the Indian subcontinent are rare now (Roper, 2018).

Dry forests are rich in species diversity and endemism, but the populations are low in density (Janzen, 1988). Thus loss in small amounts may cause species extinction or near to extinction. The dry forests comprise different types of small habitats such as springs, dry ridge tops, marshes, edaphic outcrops, temporary streams, and pickets (Janzen, 1988) for giving shelter to a variety of species.

THREATS

Tropical dry forests provide ecosystem services to millions of people. But unfortunately, these forests are gradually deforested and converted to other land uses. Tropical dry forests are the most vulnerable forest type (Bhadouria, Singh, Srivastava, & Raghubanshi, 2016). Threats in tropical dry forests are numerous in one hand and multifarious on the other side (Janzen, 1988). Though relative threats are difficult to identify precisely, but a number of different threats that tropical dry forests faces are identified by Miles et al. (2006). All the threats broadly classified as, climate change, habitat fragmentation, fire, human population density and conversion of cropland. Threats differ at region to region. Climate change found as the major threat to tropical dry forests in the Americas, while habitat fragmentation in African dry forests. Agricultural conversion and human population density are significant problems in Eurasia. Fire is a common phenomenon in every dry forest. All the threats directly or indirectly resulted from human activities (Miles et al., 2006).

Figure 2. The extent of tropical dry forest in the Americas showing areas under different threats (brown portion) and areas without any pressure (green portion)
(Source: Miles et al., 2006)

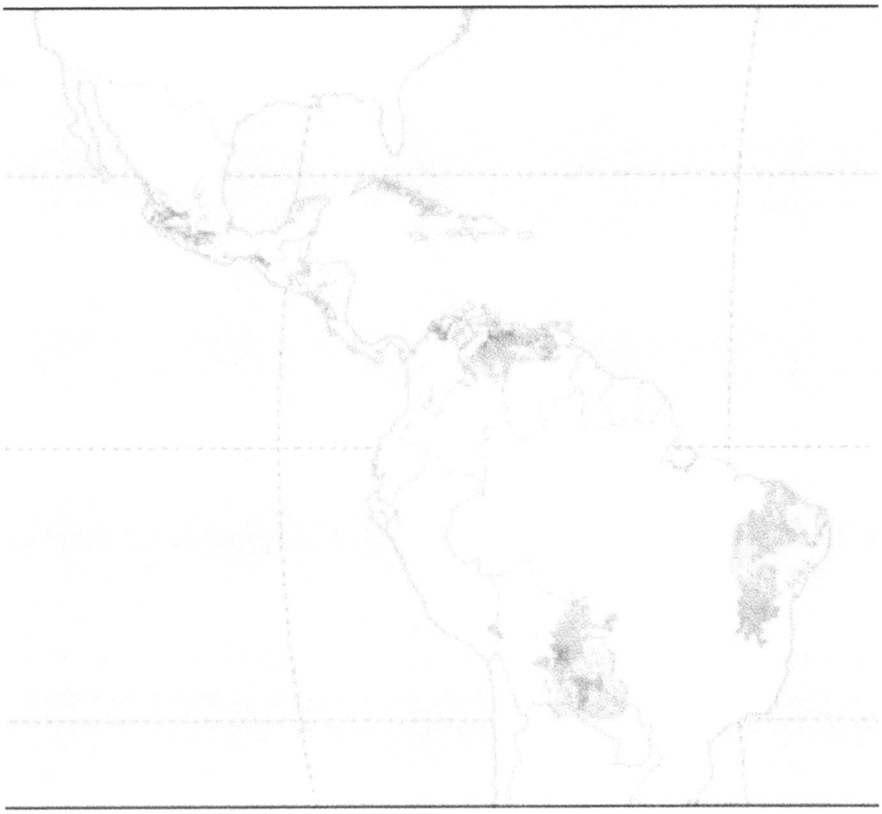

An analysis was carried out in tropical dry forests by Miles et al. (2006) separately for five major regions - North and Central America, South America, Eurasia, Africa, and Australasia/Southeast Asia (Figures 2 & 3). A high risk of climate change was faced by North and Central America (39.8%), and in South America (37%). In all other regions, the severity was less than 20% (Miles et al., 2006). Of the total dry forest area, most high forest fragmentation was observed in Africa (88.2%), Southeast Asia and Australasia (77.9%), and in North and Central America (49.9%). Considering the high risk of agricultural conversion, 60% forest area were recorded for South America, Eurasia and Africa; 41.4% for North and Central America; and 33.9% for Southeast Asia and Australasia. Considering the global tropical dry forests area, an average of 96.7% forests are at high risk (Miles et al., 2006).

Globally an average of 48.5% tropical dry forest land already has converted to other land uses (Hoekstra, Boucher, Ricketts, & Roberts, 2005). A study by Portillo-Quintero and Sánchez-Azofeifa (2010) revealed data about tropical dry forest losses in North, Central and South American, and Caribbean islands. In North and Central America, 72% of dry forest land has destroyed, whereas in South America it was 60%. Forests lost in Caribbean islands were 66%. Some countries faced a high degree of forest loss due to deforestation and conversion of forest lands into other purposes. The percentage of forest loss in Guatemala was 86%, in Peru 95% (Portillo-Quintero & Sánchez-Azofeifa, 2010) and on

Figure 3. The extent of tropical dry forest in Eurasia, Africa and Australasia/Southeast Asia showing areas under different threats (brown portion)
(Source: Miles et al., 2006)

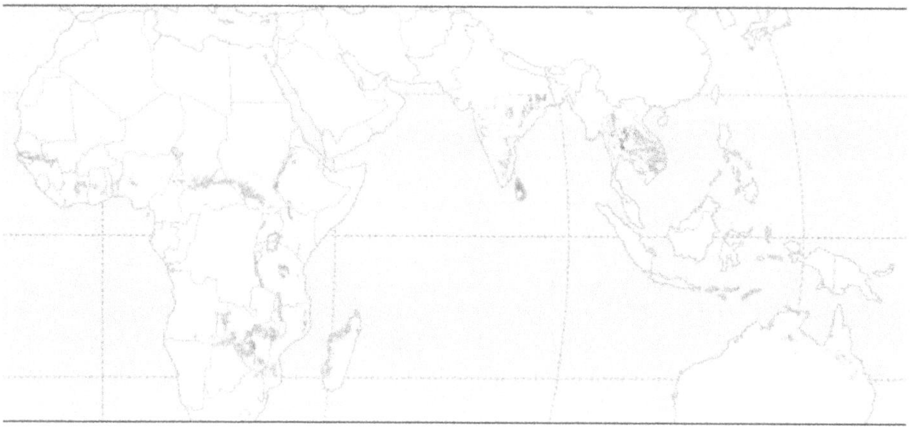

the Pacific side of Central America, it was 99.9% due to excessive deforestation (Gillespie, Grijalva, & Farris, 2000; Janzen, 1988).

In the case of fire, 17.4% (Eurasia) to 26.9% (Africa) were affected. Fire is common in the dry season and a significant threat to all tropical dry forests. Soil condition, environmental situation, plant species grow here are more prone to the fire attack than moist or wet forests, so the forests are easily susceptible to fire. In tropical dry forests, its 3-9 months a continuous rain-free dry season. In the dry season leaves shaded and as the sunlight penetrates to the forest floor the leaf litters become too dry and make the environment perfect for fire attack. Moreover, if it attacked by fire once then due to its environmental amiability to burn it is quite difficult to control fire in comparison to the tropical moist and wet forests. Fire does not stay only that area it occurred rather than many areas are cleared unintentionally within a very short period of time (Janzen, 1988; Bond, Woodward, & Midgley, 2005). The main reason of fire is unknown. But, it is estimated that fire generally occurred by unconsciousness of human being and lightning. The fire occurred by lightning can be mitigated by rain, but if it is made by unconsciousness of a human being then it is disastrous. Invasion of exotic species (WWF, 2016b) is another major threat that incites fires in tropical dry forests.

Besides these, hunting, cattle ranching, and other free-ranging perturbations are also important threats. The forests are also prominently vulnerable to pests, weeds, invasion by grasses, etc. With the initiation of civilization these forests started to destroy than any other forests as its suitability for human habitat, agriculture and livestock (Murphy & Lugo, 1986; Maass et al., 2005, Miles et al., 2006). People are much more correlated with and dependent on tropical dry forests by various means, that's why the forests are exploited drastically. Intensive cleaning up of land for agricultural production and human settlement creates habitat destruction and loss of forest area. The forests are the main source of fuelwood and charcoal.

One of the main destroying impacts created by a human is pollution. Dumping in some tropical dry forests (Jaime & Linares, 2004) causes many species to be unable to pollinate. Animals that live on consuming fruits of these species need to find other sources for living. Moreover, the species that could not produce fruits become unable to reproduce and much plant population goes through under threatened conditions. Game hunting by a human is another reason for destroying the forest population. Hunting of main predators makes the prey overpopulated and thus damages the ecosystem. Moreover, illegal poaching in the tropical dry forests reduces the forest animals at an alarming rate. The wood of deciduous trees is ideal for building construction that attracts human and deforestation of a large patch of trees are done on a regular basis. Human disturbed the tropical dry forests and fragmented it by changing of land use cover.

CONCLUSION

Two different types of behavior in nature make the forests more beneficial to humans. In the rainy season these forests provoke humans with their high productivity, when in dry period it also provokes humans by providing open spaces and makes a favorable environment for humans and domestic stock. People clear the lands easily and use for their house making, pasture land or agricultural land. Due to its geographical and environmental condition the forests are much more attractive for civilization and human settlement that make the lands most threatened. Disturbances caused by natural and human lead to changes in forest land cover. Forests that already converted to other uses are difficult to acquire for restoration and expansion the forests of its original size. Moreover, some people abuse the converted land so severely that is very much challenging to restore. Through improving existing forest policies and by formulating time oriented new policies could be helpful for tropical dry forest land conservation. Preparation of short term and long term forest management plans could be supportive strategies for tropical dry forest management.

REFERENCES

Allen, W. (2000). Restoring Hawaii's Dry Forests: Research on Kona slope shows promise for native ecosystem recovery. *Bioscience*, *50*(12), 1037–1041. doi:10.1641/0006-3568(2000)050[1037:RHSDF]2.0.CO;2

Andrade-Lima, D. (1981). The caatingas dominium. *Revista Brasileira de Botanica. Brazilian Journal of Botany*, *4*, 149–163.

Anonymous. (2011). Current Human Impacts. Tropical Dry Forests of Mexico. Retrieved from https://tropicaldryforestsofmexico.blogspot.com/

Bailey, R. (2018). Land Biomes: The World's Major Habitats. Retrieved from https://www.thoughtco.com/land-biomes-373501

Barradas, V., & Fanjul, L. (1985). Equilibrio hídrico y evapotranspiración en una selva baja caducifolia de la costa de Jalisco, México. *Biótica*, *10*, 199–218.

Bash, B. (2002). Tree of life: The world of the African Baobab. Layton, UT: Gibbs Smith.

Bhadouria, R., Singh, R., Srivastava, P., & Raghubanshi, A. S. (2016). Understanding the ecology of tree-seedling growth in dry tropical environment: A management perspective. *Energy. Ecology & Environment, 1*(5), 296–309.

Blackmore, M., & Vitousek, P. M. (2000). Cattle grazing, forest loss, and fuel loading in a dry forest ecosystem at Pu'u Wa'aWa'a Ranch, Hawai'i 1. *Biotropica, 32*(44a), 625–632. doi:10.1646/0006-3606(2000)032[0625:CGFLAF]2.0.CO;2

Bond, W. J., Woodward, F. I., & Midgley, G. F. (2005). The global distribution of ecosystems in a world without fire. *The New Phytologist, 165*(2), 525–538. doi:10.1111/j.1469-8137.2004.01252.x PMID:15720663

Bouchet, P. H., Jaffré, T., & Veillon, J. M. (1995). Plant extinction in New Caledonia: Protection of sclerophyll forests urgently needed. *Biodiversity and Conservation, 4*(4), 415–428. doi:10.1007/BF00058425

Brady, N. C., & Weil, R. R. (2002). *The nature and properties of soils* (13th ed.). Upper Saddle River, NJ: Prentice Hall.

Brown, S., & Lugo, A. E. (1982). The storage and production of organic matter in tropical forests and their role in the global carbon cycle. *Biotropica, 14*(3), 161–187. doi:10.2307/2388024

Bruegmann, M. M. (1996). Hawaii's dry forests. *Endangered Species Bulletin, 11*, 26–27.

Buckley, M. (2013). A molecular phylogeny of Plesiorycteropus reassigns the extinct mammalian order 'Bibymalagasia'. *PLoS One, 8*(3). doi:10.1371/journal.pone.0059614 PMID:23555726

Bullock, S. H., Mooney, H. A., & Medina, E. (Eds.). (1995). *Seasonally dry tropical forests*. Cambridge, UK: Cambridge University Press. doi:10.1017/CBO9780511753398

Burney, D. A., DeCandido, R. V., Burney, L. P., Kostel-Hughes, F. N., Stafford, T. W. Jr, & James, H. F. (1995). A Holocene record of climate change, fire ecology and human activity from montane Flat Top Bog, Maui. *Journal of Paleolimnology, 13*(3), 209–217. doi:10.1007/BF00682765

Burney, D. A., Vasey, N., & Godfrey, L. R., Jungers, W. L., Ramarolahy, M., & Raharivony, L. (2008). New findings at Andrahomana Cave, southeastern Madagascar. *Journal of Caves and Karst Studies, 70*(1), 13–24.

Burney, L. P., & Burney, D. A. (2003). Charcoal stratigraphies for Kaua'i and the timing of human arrival. *Pacific Science, 57*(2), 211–226. doi:10.1353/psc.2003.0011

Cabin, R. J., Weller, S. G., Lorence, D. H., Flynn, T. W., Sakai, A. K., Sandquist, D., & Hadway, L. J. (2000). Effects of long-term ungulate exclusion and recent alien species control on the preservation and restoration of a Hawaiian tropical dry forest. *Conservation Biology, 14*(2), 439–453. doi:10.1046/j.1523-1739.2000.99006.x

Cain, M., Bowman, W., & Hacker, S. (2014). The biosphere. Ecology (3rd ed.). Sunderland, MA: Sinauer, 51.

Campbell, N. A. (1996). *Biology* (4th ed.). Menlo Park, CA: The Benjamin/Cummings Publishing Company

Clinebell, R. R. II, Phillips, O. L., Gentry, A. H., Stark, N., & Zuuring, H. (1995). Prediction of neotropical tree and liana species richness from soil and climatic data. *Biodiversity and Conservation*, *4*(1), 56–90. doi:10.1007/BF00115314

Corbet, G. B., & Hill, J. E. (1992). *The mammals of the Indomalayan region: a systematic review*. Oxford, UK: Oxford University Press.

Cowen, D. V. (1952). *Flowering Trees and Shrubs in India*. Bombay, India: Thacker and Co., Ltd.

Cowling, R. M., & Hilton-Taylor, C. (1994). Plant diversity and endemism in southern Africa: an overview. In B. J. Huntley (Ed.), Botanical Diversity in Southern Africa (pp. 31-52). National Botanical Institute, Kirst-enbosch.

Cuddihy, L. W. (1989). Vegetation zones of the Hawaiian Islands. In C. P. Stone & D. B. Stone (Eds.), Conservation biology in Hawaii (pp. 27-37). Cooperative National Park Resources Studies Unit, University of Hawaii, Honolulu.

Cuddihy, L. W., & Stone, C. P. (1990). *Alteration of native Hawaiian vegetation. Cooperative National Park Resources Studies Unit*. Honolulu, HI: University of Hawaii.

Da Silva, J. M. C., & Oren, D. C. (1992). Notes on *Knipolegus franciscanus* Snethlage 1928 (Aves: Tyrannidae), an endemic of central Brazilian dry forests. *Goeldiana Zoologia*, *16*, 1–9.

Da Silva, J. M. C., & Oren, D. C. (1997). Geographic variation and conservation of the Moustached Woodcreeper *Xiphocolaptes falcirostris*, an endemic and threatened species of north-eastern Brazil. *Bird Conservation International*, *7*(3), 263–274. doi:10.1017/S0959270900001568

De Urioste, J., & Linares, M. (2004). Canary Islands dry woodlands and forests. In N. Burgess, J. D'Amico Hales, E. Underwood, E. Dinerstein, D. Olson, I. Itoua, … K. Newman (Eds.), Terrestrial ecoregions of Africa and Madagascar: a conservation assessment (pp. xxiii + 501). World Wildlife Fund, United States. Washington, DC: Island Press.

Ewel, J. J. (1999). Natural systems as models for the design of sustainable systems of land use. *Agroforestry Systems*, *45*(1-3), 1–21. doi:10.1023/A:1006219721151

Gentry, A. H. (1982). Patterns of Neotropical plant species diversity. In M. K. Hecht, B. Wallace, & G. T. Prance (Eds.), *Evolutionary Biology* (pp. 1–84). Boston, MA: Springer. doi:10.1007/978-1-4615-6968-8_1

Gentry, A. H. (1988). Changes in plant community diversity and floristic composition on environmental and geographical gradients. *Annals of the Missouri Botanical Garden*, *75*(1), 1–34. doi:10.2307/2399464

Gentry, A. H. (1995). Diversity and floristic composition of Neotropical dry forests. In S. H. Bullock, H. A. Mooney, & E. Medina (Eds.), *Seasonally Dry Tropical Forests* (pp. 146–194). Cambridge, UK: Cambridge University Press. doi:10.1017/CBO9780511753398.007

Gillespie, T. W., Grijalva, A., & Farris, C. N. (2000). Diversity, composition, and structure of tropical dry forests in Central America. *Plant Ecology*, *147*(1), 37–47. doi:10.1023/A:1009848525399

Gillespie, T. W., & Jaffré, T. (2003). Tropical dry forests in New Caledonia. *Biodiversity and Conservation*, *12*(8), 1687–1697. doi:10.1023/A:1023649831355

Gon, S., & Olson, D. (2019). Hawaii tropical dry forests. Tropical and subtropical dry broadleaf forests. World Wildlife Fund (WWF). Washington. Retrieved from https://www.worldwildlife.org/ecoregions/oc0202

Goodman, S. M., & Benstead, J. P. (2005). Updated estimates of biotic diversity and endemism for Madagascar. *Oryx, 39*(1), 73–77. doi:10.1017/S0030605305000128

Goodman, S. M., Vasey, N., & Burney, D. A. (2007). Description of a new species of subfossil shrew tenrec (Afrosoricida: Tenrecidae: Microgale) from cave deposits in southeastern Madagascar. *Proceedings of the Biological Society of Washington, 120*(4), 367–377. doi:10.2988/0006-324X(2007)120[367:DOANSO]2.0.CO;2

Gott, R. (1993). Land without evil: utopian journeys across the South American watershed. Brooklyn, NY: Verso.

Hoekstra, J. M., Boucher, T. M., Ricketts, T. H., & Roberts, C. (2005). Confronting a biome crisis: Global disparities of habitat loss and protection. *Ecology Letters, 8*(1), 23–29. doi:10.1111/j.1461-0248.2004.00686.x

Holdridge, L. R. (1967). *Life Zone Ecology*. San Jose, Costa Rica: Tropical Science Center.

Holzman, B. A. (2008). Tropical forest & woodland biomes. In S. L. Woodward (General Ed.), Greenwood guide to biomes of the world. Westport, CT: Greenwood Press.

Hubbell, S. P. (1979). Tree dispersion, abundance, and diversity in a tropical dry forest. *Science, 203*(4387), 1299–1309. doi:10.1126cience.203.4387.1299 PMID:17780463

Ibisch, P. L., Columba, K., & Reichle, S. (2002). *Plan de Conservación y Desarrollo Sostenible para el Bosque seco Chiquitabno, cedrrado y Pantanal Boliviano. Fundación para la Conservacióndel Bosque Chiquitano, Editoral F.A.N.* Santa Cruz , Bolivia.

IUCN/UNEP/WWF. (1987). *Madagascar, an environmental profile*. Gland, Switzerland: IUCN.

Jaffre, T., Bouchet, P., & Veillon, J. M. (1998). Threatened plants of New Caledonia: Is the system of protected areas adequate? *Biodiversity and Conservation, 7*(1), 109–135. doi:10.1023/A:1008815930865

Jaffré, T., Morat, P., Veillon, J. M., Rigault, F., & Dagostini, G. (2001). Composition et caractérisation de la flore indigène de Nouvelle-Calédonie= Composition and characterisation of the native flora of New Caledonia, Centre IRD, 121. (Documents Scientifiques et Techniques - IRD: II; 4). Retrieved from http://horizon.documentation.ird.fr/exl-doc/pleins_textes/pleins_textes_7/divers3/010027139.pdf

Janzen, D. H. (1983). *Costa Rican Natural History*. Chicago, IL: University of Chicago Press. doi:10.7208/chicago/9780226161204.001.0001

Janzen, D. H. (1986). *Guanacaste National Park: tropical ecological and cultural restoration (No. AS 50907). Fundación Tinkes, New York (EUA) FPN.* San José, CA: UNED.

Janzen, D. H. (1988). Tropical dry forest: the most endangered major tropical ecosystems. In E. O. Wilson (Ed.), *Biodiversity* (pp. 130–137). Washington, DC: National Academy Press.

Juvik, S. P., & Juvik, J. O. (1998). *Atlas of Hawaii* (3rd ed.). Honolulu, HI: University of Hawaii Press.

Keith, H., Mackey, B. G., & Lindenmayer, D. B. (2009). Re-evaluation of forest biomass carbon stocks and lessons from the world's most carbon-dense forests. *Proceedings of the National Academy of Sciences of the United States of America, 106*(28), 11635–11640. doi:10.1073/pnas.0901970106 PMID:19553199

Killeen, T. J., Chavez, E., Peña-Claros, M., Toledo, M., Arroyo, L., Caballero, J., ... Soria, L. (2006). The Chiquitano dry forest, the transition between humid and dry forest in eastern lowland Bolivia. In R. T. Pennington & J. A. Ratter (Eds.), *Neotropical savannas and seasonally dry forests: Plant Diversity, Biogeography, and Conservation* (pp. 225–245). Boca Raton, FL: CRC Press. doi:10.1201/9781420004496-9

Kinver, M. (2016). Dry tropical forests 'overlooked and under threat'. Science and Environment, BBC News. Retrieved from https://www.bbc.com/news/science-environment-37440485

Kirch, P. V. (1982). The impact of prehistoric Polynesian on the Hawaiian ecosystem. *Pacific Science, 36*(1), 1–14.

Leal, C. G., & de Gusmão Câmara, I. (Eds.). (2003). *The Atlantic Forest of South America: biodiversity status, threats, and outlook* (Vol. 1). Washington, DC: Island Press.

Lee, H. S., & Jarvis, P. G. (1996). Effects of tree maturity on some responses to elevated CO_2 in Sitka Spruce (*Picea sitchensis* Bong. Carr.). In G. W. Koch & H. A. Mooney (Eds.), *Carbon dioxide and terrestrial ecosystems* (pp. 53–70). San Diego, CA: Academic Press. doi:10.1016/B978-012505295-5/50005-7

Leemans, R. (1992). Global holdridge life zone classifications. Digital raster data on a 0.5 degree cartesian orthonormal geodetic (lat/long) 360x720 grid. In Global ecosystems database version 2.0. Boulder, CO: NOAA National Geophysical Data Center. Two independent single-attribute spatial layers. 537,430 bytes in 8 files. [first published in 1989]

Lugo, A. E., Medina, E., Trejo-Torres, J. C., & Helmer, E. (2006). Botanical and ecological basis for the resilience of Antillean dry forests. In R. T. Pennington, G. P. Lewis, & J. A. Ratter (Eds.), *Neotropical savannas and seasonally dry forests: Plant diversity, biogeography and conservation* (pp. 359–381). Boca Raton, FL: CRC Press.

Luttge, U. (1997). Physiological ecology of tropical plants (2nd ed.). Berlin, Germany: Springer-Verlag. doi:10.1007/978-3-662-03340-1

Maass, J. M., Balvanera, P., Castillo, A., Daily, G. C., Mooney, H. A., Ehrlich, P., ... Martínez-Yrizar, A. (2005). Ecosystem services of tropical dry forests: insights from long term ecological and social research on the Pacific Coast of Mexico. *Ecology and society: a journal of integrative science for resilience and sustainability, 10*(1), 1-23.

Mauro, C. A., Dantas, M., & Roso, F. A. (1982). Geomorfologia. In S. D. Folha (Ed.), *Brasil, MME/ SG/ Projeto RADAMBRASIL, Brasília: geologia, geomorfologia, pedologia, vegetação e usopotencial da terra* (pp. 205–296). Rio de Janeiro, Brazil: Ministério das Minas e Energia.

Medina, E. (1983). Adaptations of tropical trees to moisture stress. In F. B. Golley (Ed.), *Tropical rain forest ecosystems/ A structure and function* (pp. 225–237). Amsterdam, The Netherlands: Elsevier Scientific Company Publishing.

Miles, L., Newton, A. C., DeFries, R. S., Ravilious, C., May, I., Blyth, S., ... Gordon, J. E. (2006). A global overview of the conservation status of tropical dry forests. *Journal of Biogeography, 33*(3), 491–505. doi:10.1111/j.1365-2699.2005.01424.x

Moat, J., & Smith, P. P. (2007). *Atlas of the vegetation of Madagascar*. Kew, UK: Royal Botanic Gardens.

Mooney, H. A., Bullock, S. H., & Medina, E. (1995). Introduction. In S. H. Bullock, H. A. Mooney, & E. Medina (Eds.), *Seasonally dry tropical forests* (pp. 1–8). Cambridge, UK: Cambridge University Press. doi:10.1017/CBO9780511753398.001

Murphy, P. G., & Lugo, A. E. (1986). Ecology of tropical dry forest. *Annual Review of Ecology and Systematics, 17*(1), 67–88. doi:10.1146/annurev.es.17.110186.000435

Myers, N., Mittermeier, R. A., Mittermeier, C. G., Da Fonseca, G. A., & Kent, J. (2000). Biodiversity hotspots for conservation priorities. *Nature, 403*(6772), 853–858. doi:10.1038/35002501 PMID:10706275

Oliveira-Filho, A. T., Jarenkow, J. A., & Rodal, M. J. N. (2006). Floristic relationships of seasonally dry forests of eastern South America based on tree species distribution patterns. In R. T. Pennington, G. P. Lewis, & J. A. Ratter (Eds.), *Neotropical savannas and seasonally dry forests: Plant diversity, biogeography and conservation* (pp. 159–192). Boca Raton, FL: CRC Press. doi:10.1201/9781420004496-7

Olson, D. M., & Dinerstein, E. (1998). The Global 200: A representation approach to conserving the Earth's most biologically valuable ecoregions. *Conservation Biology, 12*(3), 502–515. doi:10.1046/j.1523-1739.1998.012003502.x

Olson, D. M., & Dinerstein, E. (2002). The Global 200: Priority ecoregions for global conservation. *Annals of the Missouri Botanical Garden, 89*(2), 199–224. doi:10.2307/3298564

Olson, D. M., Dinerstein, E., Abell, R., Allnutt, T., Carpenter, C., McClenachan, L., ... Thieme, M. (2000). *The global 200: a representation approach to conserving the Earth's distinctive ecoregions*. Washington, DC: Conservation Science Program, World Wildlife Fund-US.

Olson, D. M., Dinerstein, E., Wikramanayake, E. D., Burgess, N. D., Powell, G. V., Underwood, E. C., ... Loucks, C. J. (2001). Terrestrial ecoregions of the world: A new map of life on earth. *Bioscience, 51*(11), 933–938. doi:10.1641/0006-3568(2001)051[0933:TEOTWA]2.0.CO;2

Pacheco, P., & Mertens, B. (2004). Land use change and agricultural development in Santa Cruz, Bolivia. *CEP, 66095*, 780.

Parker, T. A., III, & Carr, J. L. (Eds.). (1992). Status of forest remnants in the Cordillera de la Costa and adjacent areas of southwestern Ecuador. RAP Working Papers 2 (p. 172). Conservation International, Washington, DC.

Parker, T. A., III, Gentry, A. H., Foster, R. B., Emmons, L. H., & Remsen, J. V., Jr. (1993). The lowland dry forests of Santa Cruz, Bolivia: A global conservation priority. RAP Working Papers 4 (p. 104). Conservation International, Washington, DC.

Pennington, R. T., Lavin, M., Prado, D. E., Pendry, C. A., Pell, S. K., & Butterworth, C. A. (2004). Historical climate change and speciation: Neotropical seasonally dry forest plants show patterns of both tertiary and quaternary diversification. *Philosophical Transactions of the Royal Society of London. Series B, Biological Sciences, 359*(1443), 515–538. doi:10.1098/rstb.2003.1435 PMID:15212100

Pennington, R. T., Lewis, G. P., & Ratter, J. A. (2006). An overview of the plant diversity, biogeography and conservation of neotropical savannas and seasonally dry forests. In R. T. Pennington, G. P. Lewis, & J. A. Ratter (Eds.), *Neotropical savannas and seasonally dry forests: Plant diversity, biogeography and conservation* (pp. 1–29). Boca Raton, FL: CRC Press. doi:10.1201/9781420004496-1

Pennington, T., Prado, D. E., & Pendry, C. A. (2000). Neotropical seasonally dry forests and quaternary vegetation changes. *Journal of Biogeography, 27*(2), 261–273. doi:10.1046/j.1365-2699.2000.00397.x

Phillipson, P. B., Schatz, G. E., Lowry, P. P., II, & Labat, J. N. (2006). A catalogue of the vascular plants of Madagascar. In S. A. Ghazanfar & H. J. Beentje (Eds.), Taxonomy and ecology of African plants: their conservation and sustainable use (pp. 613–627). *Proceedings xVIIth AETFAT Congress*. Kew, UK: Royal Botanic Gardens.

Piperno, D. R., & Pearsall, D. M. (2000). *Origins of Agriculture in the Neotropics*. Academic Press.

Portillo-Quintero, C. A., & Sánchez-Azofeifa, G. A. (2010). Extent and conservation of tropical dry forests in the Americas. *Biological Conservation, 143*(1), 144–155. doi:10.1016/j.biocon.2009.09.020

Prado, D. E. (2000). Seasonally dry forests of tropical South America: From forgotten ecosystems to a new phytogeographic unit. *Edinburgh Journal of Botany, 57*(3), 437–461. doi:10.1017/S096042860000041X

Prado, D. E., & Gibbs, P. E. (1993). Patterns of species distributions in the dry seasonal forests of South America. *Annals of the Missouri Botanical Garden, 80*(4), 902–927. doi:10.2307/2399937

Preston-Mafham, K. (1991). *Madagascar: A natural history*. Oxford, UK: Facts on File.

Raharimahefa, T. (2012). Geoconservation and geodiversity for sustainable development in Madagascar. *Madagascar Conservation and Development, 7*(3), 126–134.

Ramanujam, M. P., & Kadamban, D. (2001). Plant biodiversity of two tropical dry evergreen forests in the Pondicherry region of South India and the role of belief systems in their conservation. *Biodiversity and Conservation, 10*(7), 1203–1217. doi:10.1023/A:1016637627623

Randriamalala, H., & Liu, Z. (2010). Rosewood of Madagascar: Between democracy and conservation. *Madagascar Conservation and Development, 5*(1), 11–22. doi:10.4314/mcd.v5i1.57336

Raposo, M. A. (1997). A new species of Arremon (Passeriformes: Emberizidae) from Brazil. *Ararajuba, 5*(1), 3–9.

Ratter, J. A., Askew, G. P., Montgomery, R. F., & Gifford, D. R. (1978). Observations on forests of some mesotrophic soils in central Brazil. *Revista Brasileira de Botânica1, 1*(1), 47-58.

Ratter, J. A., Richards, P. W., Argent, G., & Gifford, D. R. (1973). Observations on the vegetation of northeastern Mato Grosso: I. The woody vegetation types of the Xavantina-Cachimbo Expedition area. *Philosophical Transactions of the Royal Society of London. Series B, Biological Sciences*, *266*(880), 449–492. doi:10.1098/rstb.1973.0053

Richard, P. (1980). PIELOU, E. C. (1979): *Biogeography*. John Wiley & Sons, ix + 351 p., ill., index, 26, $90. *Géographie physique et Quaternaire*, *34*(2), 263–263. doi:10.7202/1000411ar

Richards, P. W. (1952). *The Tropical Rain Forest: an Ecological Study*. Cambridge, UK: Cambridge University Press.

Roper, J. E. (2018). What animals live in the tropical dry forest? USA Today. Retrieved from https://traveltips.usatoday.com/animals-live-tropical-dry-forest-100072.html

Sanchez-Azofeifa, A., Kalacska, M. E., Quesada, M., Stoner, K. E., Lobo, J. A., & Arroyo-Mora, P. (2003). Tropical dry climates. In M. D. Schwartz (Ed.), *Phenology: an integrative environmental science* (pp. 121–137). Dordrecht, The Netherlands: Springer. doi:10.1007/978-94-007-0632-3_9

Sohmer, S. H., & Gustafson, R. (1987). Plants and flowers of Hawaii. Honolulu, HI: University of Hawaii Press.

Soil Survey Staff. (1999). *Soil taxonomy: A basic system of soil classification for making and interpreting soil surveys* (2nd ed.). Washington, DC: USDA Natural Resources Conservation Service.

Steininger, M. K., Tucker, C. J., Ersts, P., Killeen, T. J., Villegas, Z., & Hecht, S. B. (2001a). Clearance and fragmentation of tropical deciduous forest in the Tierras Bajas, Santa Cruz, Bolivia. *Conservation Biology*, *15*(4), 856–866. doi:10.1046/j.1523-1739.2001.015004856.x

Steininger, M. K., Tucker, C. J., Townshend, J. R., Killeen, T. J., Desch, A., Bell, V., & Ersts, P. (2001b). Tropical deforestation in the Bolivian Amazon. *Environmental Conservation*, *28*(2), 127–134. doi:10.1017/S0376892901000133

Stewart-Cox, B. (1995). *Wild Thailand*. Cambridge, MA: MIT Press.

Stone, C. P., Smith, C. W., & Tunison, J. T. (Eds.). (1992). *Alien plant invasions in native ecosystems of Hawaii: management and research. Cooperative National Park Resources Studies Unit*. Honolulu, HI: University of Hawaii.

Swart, E. R. (1963). Age of the Baobab tree. *Nature*, *198*(4881), 708–709. doi:10.1038/198708b0

TDF. (2019). Tropical Dry Forest. Slater Museum of Natural History, University of Puget Sound, Tacoma. Retrieved from https://www.pugetsound.edu/academics/academic-resources/slater-museum/biodiversity-resources/world-biomes/characteristics-of-bioclimatic/tropical-dry-forest/

Thomas, W. W., & Britton, E. G. (2008). Atlantic coastal forest of Northeastern Brazil. Bronx, NY: The New York Botanical Garden Press.

Trejo, I., & Dirzo, R. (2002). Floristic diversity of Mexican seasonally dry tropical forests. *Biodiversity and Conservation*, *11*(11), 2063–2084. doi:10.1023/A:1020876316013

Udvardy, M. D. F. (1975). A classification of the biogeographical provinces of the world. IUCN Occasional Paper 18. International Union of Conservation of Nature and Natural Resources, Morges, Switzerland.

Vieira, D. L., & Scariot, A. (2006). Principles of natural regeneration of tropical dry forests for restoration. *Restoration Ecology*, *14*(1), 11–20. doi:10.1111/j.1526-100X.2006.00100.x

Whittaker, R. H. (1975). *Communities and ecosystems* (2nd ed.). New York, NY: Macmillan Publishing Co.

WWF, & IUCN. (1994). Centres of plant diversity: A guide and strategy for their conservation. 3 Volumes. Cambridge, UK: IUCN Publications Unit.

WWF. (2016a). Major habitat types. Retrieved from https://web.archive.org/web/20161017075416/http://wwf.panda.org/about_our_earth/ecoregions/about/habitat_types/

WWF. (2016b). Tropical and subtropical dry broadleaf forest ecoregions. Retrieved from https://web.archive.org/web/20161023161710/http://wwf.panda.org:80/about_our_earth/ecoregions/about/habitat_types/selecting_terrestrial_ecoregions/habitat02.cfm

WWF. (2019). Tropical and subtropical dry broadleaf forests, World Wildlife Fund, Washington, DC. Retrieved from https://www.worldwildlife.org/biomes/tropical-and-subtropical-dry-broadleaf-forests

Chapter 2
Effect of Climate Change on Tropical Dry Forests

Pooja Gokhale Sinha
University of Delhi, India

ABSTRACT

Around 1.6 billion people in the world are directly dependent on forests for food, fodder, fuel, shelter, and livelihood, out of which 60 million are entirely dependent on forests. Forests silently provide us with ecosystem services such as climate regulation, carbon sequestration, harbouring biodiversity, synchronizing nutrient cycling, and many more. Tropical Dry Forests (TDF's) occupy around 42% of total forest area of the tropics and subtropics and facilitate sustenance of world's marginalized populations. Change in vegetation composition and distribution, deflected succession, carbon sequestration potential, nutrient cycling and symbiotic associations would affect TDF at ecosystem level. At species level, climate change will impact photosynthesis, phenology, physiognomy, seed germination, and temperature-sensitive physiological processes. In order to mitigate the effects of climate change, specific mitigation and adaptation strategies are required for TDF that need to be designed with concerted efforts from scientists, policy makers and local stakeholders.

INTRODUCTION

Forests are complex ecosystems that have a delicate balance of biotic and abiotic components that interact, influence, modify and adapt to each other. The term forest is a very widely used but an ill-defined term and globally there are around 800 ways in which forests have been defined (Lund, 2012). The Food and Agricultural Organization (FAO) defines forest as a 'land spanning more than 0.5 ha with trees higher than 5m and a canopy cover of more than 10%, or trees able to reach these thresholds *in situ*' (FAO, 2010). According to the Intergovernmental Panel on Climate Change (IPCC) forest is defined as vegetation type that is dominated by trees, and is defined in different parts of the world according to the variation in biogeophysical conditions, social structure and economics of the region (IPCC, 2014). Of the many parameters used to define forests such as type of vegetation, physiognomy, species composition, canopy cover is considered to be an important parameter. The way a country defines its forests largely

DOI: 10.4018/978-1-7998-0014-9.ch002

depends upon the vegetation type, physiognomy, forest structure, ownership of land, and economic use. In Australia, the canopy cover of typical forest should be greater than 20% of the total area, whereas in South Africa the value should be greater than 60% (Reddy et al., 2013). In India, the Forest Survey of India (FSI) defines forest as 'all lands more than one hectare in area, with a tree canopy density of >10%, irrespective of ownership and legal status' (FSI, 2011). This means that regardless of the nature of the tree cover, whether it is natural or planted, or the trees species are alien or native, any area having a minimum tree cover density of 10% is a forest. Based upon canopy density, forests in India have further been classified into five classes namely; Very Dense Forests, moderately dense forests, Open forests, Scrub and Non-forest (ISFR, 2017). Classification of forests has received much attention in the past and there have been several classification systems based upon climate, vegetation type, physiognomy and floristics. The system of life zones was the first forest classification system based on climate (Holdridge, 1947). Mueller-Dombois & Ellenberg (1974) proposed a classification system that laid emphasis on physiognomy of vegetation. Westhoff & van der Maarel (1978) integrated floristics and physiognomy to classify forests. Champion & Seth (1968) recognized four major climatic zones in India that harbours 16 major forest types comprising 221 minor types. On the basis of species associations and bioclimate, Gadgil Meher-Homji (1986) defined 42 forest types in India. Recently, Reddy et al. (2015) used IRS resourcesat-2 advanced wide field sensor to classify forest and scrub types. Seventeen percent of the global standing tropical forests are represented by tropical dry forests (UNEP, 2011). Irrespective of their type, forests across the world are under threat due to global climate change (Deb et al., 2018).

CLIMATE CHANGE

Climate change is one of the most serious environmental threats faced by the world today. It was recognized as a significant global environmental challenge a couple of decades back. International efforts to address this issue began with the establishment of the Intergovernmental Panel on Climate Change (IPCC) by the World Meteorological Organisation (WMO) and United Nations Environment Programme (UNEP) in 1988. IPCC defines climate change as a 'change in climate over time, whether due to natural variation or due to human induced activity' (IPCC, 2001). This definition differs from that of the United Nations Framework Convention on Climate Change (UNFCCC), which refers to climate change as "change in climate attributed only to anthropogenic activities which is in addition to the natural climatic variability observed over comparable time periods" (UNFCCC, 1992). UNFCCC was adopted in 1992, with an objective of stabilizing the concentration of greenhouse gases (GHG) in the atmosphere.

Factors that cause or contribute to climate change are known as 'climate forcings' that can be either natural or anthropogenic. Natural factors include volcanic eruptions, alteration in sun's intensity, and very slow changes in the oceanic circulation or land surface. Anthropogenic activities are fossil fuel combustion, industrial activities, emissions from agricultural systems and waste decomposition. In the fifth assessment report the IPCC has stated that anthropogenic climate change is very much real and is bound to have widespread impacts on natural systems. The report also says that each of the last three decades have been successively warmer than any preceding decade since 1850, and the period from 1983 to 2012 has been warmest 30-year period in the last 1400 years in the northern hemisphere. Rise in the global mean surface temperature may reach up to 4.8°Cby the end of the 21st century, whereas maximum increase is predicted in the Polar Regions (IPCC, 2014). In addition to rise in temperature and disturbed precipitation patterns, global climate change will also increase the frequency of extreme events such as droughts, cyclones, storms and hurricanes. Correspondingly rise in the severity of extreme rainfall and heat waves is also predicted. As a result, natural ecosystems including forests will be adversely influenced and may become more vulnerable.

Forests and climate are closely coupled and any alteration in one may have serious implications on the other. They influence each other and share a symbiotic relationship in which climate influences the type of vegetation and distribution of forests and forests in turn act as natural climate regulators (IUCN, 2017). The IPCC has projected that forests across the globe will be severely impacted due to climate and non-climate stressors that may lead to forest die-back, loss of biodiversity and compromised ecological benefits (IPCC, 2014). In fact deforestation is one of the key drivers of climate change as it reduces carbon uptake potential of an ecosystem. Forests provide a plethora of goods and services to mankind. Ecosystem services provided by forests include carbon sequestration, regulation of nutrient and water cycles, prevention of soil erosion and harbouring biodiversity. However, change in land use pattern due to urbanization compounded by overexploitation has adversely affected forests around the world.

EFFECT OF CLIMATE CHANGE ON FORESTS

Forests across the world have been a victim of human greed and overexploitation. According to estimates, forests covered around 45% of the earth around 10,000 years ago, and due to ruthless anthropogenic interferences they have been reduced to 31% in 2010 (FAO, 2010). FAO's Global Forest Resources Assessment (FRA) reports that we are continuously losing our forests and in the period between 1990 and 2015 world's forested area has declined from 31.6% to 30.6% (FAO, 2018).The rate of deforestation is skewed in the world and countries that are heavily dependent on forest show higher rate of deforestation. Developing countries of Latin America, sub-Saharan Africa and Southeast Asia have reported maximum loss of forest cover primarily due to conversion into agricultural farms. The threats to our forests will be augmented by climate change and all forests will not be able to adapt to the changing climate. It is projected that global environmental change may result in shifting of ecosystems and extinction of some species, particularly those who will not be able to adapt. The most vulnerable would be polar ecosystems and coral reefs.

There is an increasing consensus that increased incidents of floods, drought, storms and heatwaves have a link to climate change (Thomas & Lopez, 2015).It is predicted that the incidents as well as the intensity of extreme climate events and natural disasters will intensify in future climatic scenarios (IPCC, 2014). It is important to point out that these issues also have economic ramifications as in many parts of the world people are heavily dependent on forests. This would have adverse impacts on environment, ecosystems and also the socioeconomic life of the communities that are dependent on forests for their sustenance and livelihood. Between 1996 and 2015 more than 800 million hectares of forested land has been affected due to climate related disasters (FAO, 2015a). Natural disasters have serious economic implications and in the decade spanning 2003-2013, 26 such events lead to loss of 737 million USD (FAO, 2015b).

EFFECT OF CLIMATE CHANGE ON TROPICAL DRY FORESTS

The tropical dry forests are one of the important biomes of the world and comprise slightly less than half of the global tropical and subtropical forests (Murphy & Lugo, 1986). They provide ecosystem services such watershed protection, flood control, maintaining soil fertility and regulate climate (Maass et al., 2005). According to the Food and Agricultural Organization tropical dry forests are defined as those experiencing a tropical climate, with summer rains, a dry period of 5 to 8 months and annual rainfall ranging from 500 to 1500 mm. These are primarily found in Africa, South America and Asia and are particularly concentrated in parts of Southeast Asia, northern Australia, Pacific region, Central America and the Caribbean (FAO 2001, 18). They exist in regions that experience mean annual temperature of 25°C and annual precipitation between 700 to 2,000 mm. Due to such climatic conditions 50% of the plant species are seasonally deciduous (Sanchez-Azofeifa et al., 2005). According to Pennington et al. (2006), the ecological definition of TDFs includes savannas, riparian forests, coastlines and mangroves among the associated vegetation types. Vegetation in these forests varies from tall, closed canopy trees to short scrub vegetation that does not form a closed canopy, particularly in areas that receive less rainfall (Pennington et al., 2000). The predominant vegetation include deciduous trees that reach up to a height of 30 m. Nine major life forms are represented in TDFs which include: evergreen woody plants, deciduous woody plants (40% of the woody species), sclerophyllous woody plants, succulent woody plants (including cacti), herbs, rosettes, lianas, CAM epiphytes and hemiparasites (Medina, 1995). Dense and entangled undergrowth is also commonly observed in these forests as there is optimal light penetration. Species such as cacti, bromeliads are abundant as they are resistant to drought. TDFs may occur in a variety of structure that may range from a tall, closed canopy forest to short scrub vegetation. These forest experiences seasonal drought stress and that makes it different from moist forest (Murphy & Lugo, 1986). Lack of forest fires and seasonal drought stress further make TDFs a unique biome markedly different from savannas or moist forests.

The TDFs share a close association with human culture and economic development particularly in the neotropical region (Trejo & Dirzo, 2000; Fajardo et al., 2005). They provide several advantages for agriculture by influencing pedogenesis, improving soil fertility and supporting short-cycled crop plants. In a review on tropical dry forests of the neotropics it has been reported that globally these forests may facilitate sequestration of up to 22 Pg of carbon as above ground biomass. Additionally, they also help in conservation of freshwater sources and support biodiversity by harbouring endemic species (Portillo-Quintero et al., 2015). It is also known that the dry climate limits the propagation of pathogens transmitted by mosquitoes (Ewel, 1999; Murphy & Lugo, 1986; Fajardo et al., 2005).In addition to providing several ecosystem services, tropical dry deciduous forests also provide source of livelihood. It supports some of the world's poorest people (Cunningham et al., 2008; Waeber et al., 2012).Their trees are important source of firewood, medicinally important compounds, and also harbour fauna. The goods and services provided by these forests are different from other forest types and hence their conservation and management strategies need to be different (Gumbo & Chidumayo, 2010). Despite their high economic and social significance, dry forests haven't received much attention and are among the most threatened and least understood forest ecosystems (Miles et al., 2006; Portillo-Quintero & Sanchez-Azofeifa, 2010). Lack of concern and scientific studies have put these forests at high risks than their humid counterparts (Aide et al., 2012).

There is lot of regional imbalance on the studies of TDFs. While there are several reports of analyses of effect of climate change on dry forests in neotropics, there are few reports for the same from Asia and Africa (Walker & Desanker, 2004; Glenday, 2008; Williams & Aquino, 2010). In Asia, there have been reports on dry forests of Thailand. However, the focus has been on structure and composition of the forests without any emphasis on climate change (Bunyavejchewin, 1983; Setalaphruk & Price, 2007). Asian dry forests that include dry forests of Indochina as well as those of Central India are not particularly well studied, despite being regionally extensive (Blackie et al. 2014). Up to 30% of forests in mainland Southeast Asia and 60% of Indian forests are classified as dry forest (Waeber et al., 2012). In Asia the major dry forest type found are deciduous Dipterocarp forests, mixed deciduous forests and dry evergreen forests (Wohlfart et al., 2014). Although the FAO launched an Asian Dry Forests Initiative, as compared to those in the neotropics, the dry forests of Asia have received lesser attention. The trend is worrisome because Asian countries are more prone to the deforestation and have less adapting capacity and increasing population pressure (Bawa & Dayanandan, 1997). In India, excessive deforestation has resulted in patched distribution of TDF wherein they appear as mosaics of open canopied, and closed canopied in some undisturbed regions (Raghubanshi & Tripathi, 2009). TDF are also found in the Pacific region and according to Gillespie et al. (2012) the expanse of dry forests in some Pacific islands has been reduced to less than 10% of their original area due to anthropogenic activities and predominance of invasive species.

There are three major features of climate change that directly influence growth and behaviour of TDF: Higher global atmospheric CO_2 concentration, rise in global mean temperature and erratic rainfall pattern. At species level, increasing atmospheric CO_2 and temperature influence key physiological processes such as photosynthesis, respiration, seed germination and other temperature sensitive physiological processes. Indirectly these process influence carbon sequestration potential, soil organic carbon, nutrient cycling and symbiotic associations. At community and ecosystem levels, change in climate parameters alters phenology, succession and plant-animal interactions (Figure 1).

Shift in Vegetation

Global warming and alteration in precipitation patterns may lead to shifting of major ecosystems and biomes across the world. Increase in aridity has resulted in retreating of some tropical and temperate forests, while the savannas have expanded. The IPCC predicts that climate change would lead to boreal tipping point that would lead to thawing of permafrost resulting in threatening arctic ecosystems. Also it would result in increased shrub vegetation in the tundra and rise in incidents of forest fires in the boreal forests (IPCC, 2014).Studies have also indicated that terrestrial ecosystems in the alpine regions may shift towards the poles (Haywood et al., 2016). The shift is not restricted to latitude alone and Settele et al. (2014) have reported that vegetation and ecosystems show significant altitudinal shifts. Scientists have been using various models to predict the behaviour of vegetation to changing climate parameters. Analyses have been done to predict and project alteration in ecosystems by using models that use multiple parameters (Warszawski et al., 2013, 2014; Frieler et al., 2017). Analysing shifting patterns in the northern hemisphere, Thuiller (2007) concluded that terrestrial plants have shifted northwards on an average of 6.1 km per decade. Also, they have shown an advancement of phenoevents by 2.3 to 5.1 days in the last 50 years. In India, Telwala et al. (2013) have observed a climate change induced shift in the plants of TDFs in the Sikkim. Bates et al. (2008), have predicted a 5–15% reduction in soil moisture availability and runoff by the late twenty-first century across tropical Latin America, and studies have also reported the alteration in the length of the dry and growing season during the past decades (Sanchez- Azofeifa et al., 2013).

Figure 1. Diagrammatic representation of effect of climate change on tropical dry forests

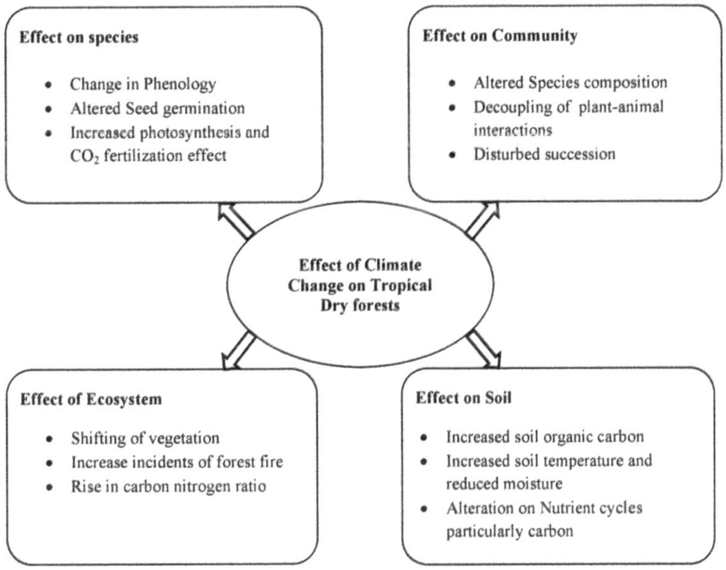

Distribution of Forests

Miles et al. (2006) compared global distribution maps of TDF and concluded that climate change is a significant factor that affects the TDF of Americas in particular. In Africa TDF are under threat due to habitat fragmentation and fire. Conversion of forest land into agricultural fields and increasing population pressure is a matter of concern for Eurasian TDF. Using dynamic global vegetation model, Meir & Pennington (2011) predicted the effect of climate change on neotropical dry forests, and concluded that warming rates for TDF regions are consistent with, or slightly above, the global predicted means of approximately 2–4°C warming by 2100. In a review of impact of climate change on seasonally dry tropical forests Allen et al. (2017) concluded that rise in frequency and severity of droughts will significantly affect distribution of species and other ecosystem processes. They concluded that temperature variation has serious implications for both above ground as well below ground processes of the forests. Mehta et al. (2008) reported that poor soil conditions may lead to a conversion of tropical dry forests into scrub vegetation. After analysing the TDF of southern India they concluded that distribution of trees, their phenology, physiology and growth of dry tropical forests is affected due to change in rainfall patterns.

Phenology

Phenology, the science of recurring events in nature, is a sensitive indicator of global climate change. It usually refers to the annual course of developmental events. Plant phenology is one of the most responsive and easily observable traits in nature that changes in response to any change in climate. Alteration in plant developmental events brought about by the current anthropogenic global climate change may have major impacts on plant productivity, competition among plant species and their interaction with other organism. Phenology also plays a crucial role in maintaining the carbon balance of terrestrial ecosystems

(Keeling et al., 1996) and in determining shifts in agricultural zones (Fischer et al., 2002). Any changes in the timing of flowering may alter processes at species, community and ecosystem levels (Springer & Ward, 2007). At the community level, differential effects of elevated CO_2 on flowering time may lead to change in community composition by altering interaction among species. Alteration in flowering time may have different implications for different plant groups. Advancement or delay in flowering time in annual plants would lead to either shortening or lengthening of the plant cycle, which in turn may disrupt the sowing of the subsequent crop. In perennials, altered flowering time may influence the resource availability for seed production during specific reproductive events (Stearns, 1992). TDFs are composed of species that show different degree of adaptation to seasonal droughts. Events such as leaf fall, leaf emergence, and extend of deciduousness is well synchronized with periodicity of rainfall and availability of soil moisture. Any alteration in the precipitation and heat periods would significantly affect process such as duration of deciduousness, bud break and flowering time of tree species in Indian forests (Singh & Kushwaha, 2005). In a review of TDF of the Caribbean region by Nelson et al. (2018) observed that change in precipitation and temperature may lead to change in species composition, growth and reproductive capacity of trees. Khayani et al. (2016) have projected that in Puerto Rico climate change may lead to shift in the forest type from subtropical dry forest to dry forest by 2099. It is important to mention that any change in phenology of dominant tree species of forests may have rippling effects on several other plant and animal species that are dependent on them. Also phenological change in many species simultaneously may change the course of evolution of the forest as an entity. It will alter succession and in turn will have serious consequences for ecosystem functioning. What is worrisome is that there are no extensive studies on understanding climate change impacts of phenology of TDF species.

Carbon Dynamics

Forests are a crucial and critical link in the global carbon cycle. Irrespective of the forest types all forests are involved in cycling of carbon and its conversion from inorganic form i.e. atmospheric CO_2 to organic form via the process of photosynthesis. They also act as carbon sinks and make carbon a part of biological cycle. Forests are integral part of the carbon cycle that maintains delicate balance between inorganic and organic forms of carbon in atmosphere, biosphere, hydrosphere and lithosphere. Needless to say, that depletion of forests leads to disruption of an important link in the carbon cycle. Alteration in Carbon sequestration potential of forests may lead to the loss in carbon trapped in biosphere that may in turn accelerate climate change. Forests are the largest storehouses of carbon after oceans and can absorb and store carbon in their biomass, soils and products, equivalent to about one tenth of carbon emissions projected for the first half of this century. At the same time, deforestation and land-use changes account for 17 per cent of human-generated carbon dioxide emissions. Global climate change may have direct effects on terrestrial carbon reserves, and may indirectly lead to increased incidents of forest fires and pest and pathogen attacks. The 21st century may witness increased trees mortality and forest dieback (IPCC, 2014). Such events may have adverse impact on carbon storage, biodiversity and wood production. According to the International Union of Forest Research organization in predicted climate change scenario, the forests of the world may end up losing all their carbon regulating services and terrestrial ecosystems may start to behave as source of carbon rather than sinks (Seppala et al., 2009). Any change in land use leads to altered carbon: nitrogen ratio in different soil fractions. In tropics, change in land use results in releasing of stored carbon thereby increasing green house gases emissions (Don et al., 2011). In Mexican dry forests, it has been observed that slower turnover of carbon and other nutrient contribute to the negative relationship between soil carbon sequestration and mean annual precipitation across (Campo & Merino, 2016). In an analysis of relationship between soil organic carbon and aridity Delgado-Baquerizo et al. (2013), suggested that extremely dry conditions may adversely affect ecosystem carbon storage as the carbon inputs from plant productivity, erosion and fire decrease. Boisvenue & Running (2006) analyzed the effects of climate change on forest productivity and concluded that forest productivity will increase in the absence of water stress. On the contrary Aber et al. (2001) projected that in the US alone about 20% of forests may experience loss of carbon in case temperature and precipitation were to increase. Holm et al. (2017) observed the effect of hurricanes on Net Primary Productivity (NPP) in the Carribean and concluded that increased incidents of hurricanes will increase the NPP of dry forests in the Carribean. Using HadRM3 model Chaturvedi et al. (2011) projected that in India, the NPP will increase by 68.8% and the soil organic carbon may show a rise by 37.5%.The rise is NPP will be higher in northwest India due to higher temperature and wetter climate. The study also reported that NPP may marginally increase or may even decrease by up to 12% in central and western India which is dominated by dry forests. Therefore it may be concluded that how forests respond to changing climate will vary locally within country. Therefore, further long-term studies are warranted to generate a comprehensive understanding towards the impacts of global climate change on TDFs.

Symbiotic Associations

Among the below ground processes, plant microbe interaction is very crucial for plant and soil health. The below ground dynamics facilitate mineral nutrition for plants by conversion of bio-unavailable form of minerals to bio-available forms by microbes. Even though not much attention has been paid to below ground processes such as rhizosphere dynamics, symbiotic associations and nutrient cycling, all these processes may change due to rise in temperature in future climate scenarios. Free living and symbiotic soil microorganisms help in inter-conversion of nitrogen, phosphorous and sulphur. The symbiotic association between N fixing legumes and roots of higher plants is sensitive to abiotic stresses such as high temperature, drought and salinity. Scarcity of water has limiting effect on initiation, growth and function of roots nodules (Serraj et al., 1999) as well as on growth and survival of rhizobia (Hungria & Vargas, 2000). In TDFs the event of nodulation fluctuates with season (Gonzalez-ruis et al., 2008, Gei & Powers, 2015). Any alteration in the season may affect the phenology of nodulation or may even induce premature senescence. These changes are particularly undesirable for seedlings. Therefore, it may be concluded that scarcity of water and increased temperature due to climate change may directly influence nitrogen fixation by symbiotic N fixers. However, for better understanding of carbon nitrogen dynamics under future climate more focussed and targeted studies are required that essentially should be region specific.

Community Dynamics

Analysis of data from experimental analysis as well as modelling studies suggest that under severe drought or increased rainfall events the structure and function of seasonally dry forests would change (Chadwick et al., 2015; Feng et al., 2013; Greve et al., 2014). Structural and functional changes in forests further influence biodiversity and results in shifts in species range (Enquist, 2002). Disturbed forests would also have diminished capacities to provide ecosystem services. In fact, it would take several years to conclusively say how dry forests would behave in changing climatic scenario. TDF take longer time to regenerate from intense or prolonged droughts (Angeler & Allen, 2016) as compared to wetter tropical forests. Meta-analysis of effects of drought on temperate zone revealed that dry ecosystems took the longest to re-establish normal growth after being exposed to extreme drought (Anderegg et al., 2015). Study on long term effects on community dynamics of TDF of Costa Rica by Enquist & Enquist (2011) reported that extended drought periods lead to reduction in number of small sized tree. On the contrary Suresh et al. (2010) reported that trees with large girth were more resistant to the effects of annual climatic fluctuations. Abiotic factors such as light, water and nutrient composition of soil has significant influence on growth and survival of seedlings in TDFs (Tripathi et al., 2016). Bhadouria et al., 2016 analyzed the interactive effects of various abiotic factors on growth, survival and regeneration of tree species of TDFs. However more elaborate and long-term simulation studies are required to conclusively say how climate change affects TDF. There are many gaps in existing knowledge that needs to be filled up by conducting well-replicated drought simulations or experiments, distributed across the range of dry forest climatic variation and biogeography.

Incidents of drought and heat under projected climate change will results in increased mortality of tree species that in turn alters species diversity and composition (Allen et al. 2010). Moist forests are well equipped to withstand scarcity of water however TDFs have less resilience towards water stress. Therefore, species mortality through water deficiency was found to occur more in the tropical dry forests, and it severely affected trees of small diameter (Suresh et al., 2010). It has been observed that due to lower infiltration and evapotranspiration there is a shift in the species composition of TDF (Krishnaswamy et al., 2013). TDFs play a significant role in regulating the hydrological cycle. In a review of studies conducted on tropical dry forests in the neotropics Portillo- Quintero et al. (2015) have pointed out that climate change and change in land use may have serious implications on the natural and socioeconomic structures in the region. Water scarcity and droughts would have adverse impacts on the structure; composition and functioning of TDFs. Change in temperature and rainfall significantly influence seed health, vigour, and seedling germination. There have been studies analysing the effect of variation in rainfall on seed rain dynamics of dry forests (Martinez-Garza et al., 2013; Meave et al., 2012). Increased duration of dry periods may even lead to delayed or unsuccessful seed germination.

INDIAN SCENARIO

India is a mega biodiverse country and geographically it lies at connexion of three diverse biogeographic realms, the Indo-malayan, Eurasian and Afro-tropical (Reddy et al., 2015). We are the seventh largest country of the world and occupy an area of 32,87,263 km^2 and are bestowed upon a rich flora and fauna due to diverse climatic, topographic and geographic conditions. According to the India State forest report (2017) 21.54% of the total geographical area that is around 7.08 lakh square km is covered by forests. Highest forest cover is found in the state of Madhya Pradesh followed by Arunachal Pradesh. As a country we have added 6,778 km^2. of forests since 2015 (ISFR, 2017). Based on multi-season IRS Resourcesat-2 Advanced Wide field Sensor, Reddy et al. (2015) developed a classification system for Indian Forests and concluded that tropical dry deciduous and tropical moist deciduous forests are the predominant and occupy 34.8% and 7.72% of the total forest area. Tropical Dry deciduous forests are found in Western Himalayas, Gangetic plains and semiarid regions of Deccan peninsula. Dominant tree species are *Anogeissus latifolia, A. penduala, Albizia procera, Anthocephalus chinensis, Diospyros melanoxylon* (Reddy et al., 2015). According to Ravindranath & Sukumar (1998), rise in temperature and reduced precipitation in central India may result in reduced soil moisture that would eventually lead to increased seedling mortality. They also predicted that the moist type of forests may change to dry type in central India. Regional studies have shown that the dry deciduous forests of Western Ghats may decline in projected climate change scenario.

TDFs of peninsular India are divided into three types, namely: teak, Sal and miscellaneous types. The area is dominated by *Tectona grandis* (teak) followed by *Shorea robusta* (sal). In a study of dynamics of dry forest of South India, Schmerbeck (2011) concluded that forest fires are dominant anthropogenic factor in dry deciduous forests which are considered the most vulnerable ecosystems. In fact, the study also reported that lack of research and proper management has even converted some of the tropical dry forests into savanna and grasslands.

General Circulation Models were used initially to analyse the effect of climate change on Indian Forests (Ravindranath & Sukumar, 1998; Ravidranath et al., 1997). Using Regional Climate model of the Hadley centre (HadRM3) Chaturvedi et al. (2011) projected that on an average 39% of forest grids analysed would change under climate change scenario. The change in forest will be up to 73%, 67% and 62% in the states of Chhattishgarh, Karnataka, Andhra Pradesh, respectively. Calculation of vulnerability index for India indicated that upper Himalayas, and parts of western ghats are most vulnerable to the impacts of climate change, whereas the forests of north east are most resilient. Tropical dry deciduous forests make up to 40% of Indian forest grids and under climate change scenario they are expected to change (Chaturvedi et al., 2011). Regional studies on effect of potential climate change in Himachal Pradesh and western ghats indicated towards a shift in the type of vegetation (Ravindranath et al., 1997; Deshingkar, 1997). Ravindranath et al. (2006) used BIOME4 vegetation model and predicted that by 2085 under climate change scenario vegetation would shift towards wetter type in the north-eastern region and towards drier type in the north-western region. Dry deciduous forests of central India in the states of Madhya Pradesh and Maharashtra are projected to change into moister type due to increase in temperature and precipitation (Ravindranath & Sukumar, 1998). Agarwala et al. (2016) analysed the effect of anthropogenic changes on TDFs of central India and concluded that human interference have serious implications on abundance of tree species.

CONCLUSION

Tropical dry forests are one of the most fragile ecosystems that are facing multiple complex threats. They provide ecosystem services and also are source of livelihood to marginalized people in many parts of the world. Overexploitation, deforestation and fragmentation are the key issues that need to be addressed. These issues are being compounded by global climate change that has serious impact on vegetation composition and distribution, phenology and carbon dynamics in TDF. In order to design conservation and management strategies it is imperative to conduct region-wise research and analysis to study the impact of climate change on TDF. Conservation and restoration measures for TDF should take into consideration the local people, their aspirations and expectations. Mitigation and adaptation strategies should be designed that are country-driven, gender-responsive and address the concerns of local inhabitants. Also, they should be an amalgamation of scientific research as well as traditional knowledge.

REFERENCES

Aber, J. D., Neilson, R. P., McNulty, S., Lenihan, J. M., Bachelet, D., & Drapek, R. J. (2001). Forest processes and global environmental change: Predicting the effects of individual and multiple stressors. *Bioscience*, *51*(9), 735–751. doi:10.1641/0006-3568(2001)051[0735:FPAGEC]2.0.CO;2

Agarwala, M., Fries, R. S., Qureshi, Q., & Jhala, Y. V. (2016). Changes in the dry tropical forests in central India with human use. *Regional Environmental Change*, *16*(1), 5–15. doi:10.100710113-015-0903-1

Aide, T. M., Clark, M. L., Grau, H. R., Lopez-Carr, D., Levy, M. A., Redo, D., ... Muniz, M. (2012). Defores.tation and reforestation of Latin America and the Caribbean (2001–2010). *Biotropica*, *45*(2), 262–271. doi:10.1111/j.1744-7429.2012.00908.x

Allen, C. D., Macalady, A. K., Chenchouni, H., Bachelet, D., McDowell, N., Vennetier, M., ... Cobb, N. (2010). A global overview of drought and heat induced tree mortality reveals emerging climate change risks for forests. *Forest Ecology and Management*, *259*(4), 660–684. doi:10.1016/j.foreco.2009.09.001

Allen, K., Dupuy, J. M., Gei, M., Hulshof, C., Medvigy, D., Pizano, C., ... Powers, J. S. (2017). Will seasonally dry tropical forests be sensitive or resistant to future changes in rainfall regimes? *Environmental Research Letters*, *12*(2). doi:10.1088/1748-9326/aa5968

Anderegg, W. R. L., Schwalm, C., Biondi, F., Camarero, J., Koch, G., Litvak, M., ... Pacala, S. (2015). Pervasive drought legacies in forest ecosystems and their implications for carbon cycle models. *Science*, *349*(6247), 528–532. doi:10.1126cience.aab1833 PMID:26228147

Angeler, D., & Allen, C. (2016). Quantifying resilience. *Journal of Applied Ecology*, *53*(3), 617–624. doi:10.1111/1365-2664.12649

Bates, B., Kundzewicz, Z. W., Wu, S., & Palutikof, J. (2008). *Climate change and Water: Technical paper VI. Intergovernmental Panel on Climate Change*. IPCC.

Bawa, K. S., & Dayanandan, S. (1997). Socioeconomic factors and tropical deforestation. *Nature*, *386*(6625), 562–563. doi:10.1038/386562a0

Bhadouria, R., Singh, R., Srivastava, P., & Raghubanshi, A. S. (2016). Understanding the ecology of tree-seeding growth in dry tropical environment: A management perspective. *Energy. Ecology & Environment*, *1*, 296–309.

Blackie, R., Baldauf, C., Gautier, D., Gumbo, D., Kassa, H., Parthasarathy, N., . . . Sunderland, T. (2014). Tropical dry forests: The state of global knowledge and recommendations for future research. (Discussion paper). Bogor, Indonesia: CIFOR.</unknown>

Boisvenue, C., & Running, S. W. (2006). Impacts of climate change on natural forest productivity—Evidence since the middle of the 20th century. *Global Change Biology*, *12*(5), 862–882. doi:10.1111/j.1365-2486.2006.01134.x

Bunyavejchewin, S. (1983). Analysis of the tropical dry deciduous forest of Thailand. I. Characteristics of the dominance-types. *Natural History Bulletin of the Siam Society*, *31*, 109–122.

Campo, J., & Merino, A. (2016). Variations in soil carbon sequestration and their determinants along a precipitation gradient in seasonally dry tropical forests. *Global Change Biology*, *22*(5), 1942–1956. doi:10.1111/gcb.13244 PMID:26913708

Chadwick, R., Good, P., Martin, G., & Rowell, D. P. (2015). Large rainfall changes consistently projected over substantial areas of tropical land. *Nature Climate Change*, *6*(2), 177–181. doi:10.1038/nclimate2805

Champion, H. G., & Seth, S. K. (1968). *A revised survey of forest types of India*. New Delhi, IN: Govt. of India Press.

Chaturvedi, R. K., Gopalakrishnan, R., Jayaraman, M., Bala, G., Joshi, N. V., Sukumar, R., & Ravindranath, N. H. (2011). Impact of climate change on Indian forests: A dynamic vegetation modelling approach. *Mitigation and Adaptation Strategies for Global Change*, *16*(2), 119–142. doi:10.100711027-010-9257-7

Cunningham, A., German, L., Paumgarten, F., Chikakula, M., Barr, C., Obidzinski, K., . . . Puntodewo, A. (2008). *Sustainable trade and management of forest products and services in the COMESA region: an issue paper*. Centre for International Forestry Research, Borgor, Indonesia. Retrieved from http://www.cifor.org/publications/pdf_files/Books/BCunningham0801.pdf

Deb, J. C., Phinn, S., Butt, N., & Alpine, C. A. (2018). Climate change impacts on tropical forests: Identifying risks for tropical Asia. *Journal of Tropical Forest Science*, *30*(2), 182–194. doi:10.26525/jtfs2018.30.2.182194

Delgado-Baquerizo, M., Maestre, F. T., Gallardo, A., Bowker, M. A., Wallenstein, M. D., Quero, J. L., ... Zaady, E. (2013). Decoupling of soil nutrient cycles as a function of aridity in global drylands. *Nature*, *502*(7473), 672–676. doi:10.1038/nature12670 PMID:24172979

Deshingkar, P. (1997). *Adapting to Climate Change in a forest based land use system- A case study of Himachal Pradesh, India. Atmospheric Environmental issues in Developing Countries*. Stockholm, Sweden: Stockholm Environment Institute.

Don, A., Schumacher, J., & Freibauer, A. (2011). Impact of tropical land-use change on soil organic carbon stocks – a meta-analysis. *Global Change Biology*, *17*(4), 1658–1670. doi:10.1111/j.1365-2486.2010.02336.x

<edb>Pachauri, R. K., Allen, M. R., Barros, V. R., Broome, J., Cramer, W., Christ, R., ... Dubash, N. K. (2014). Climate change 2014: synthesis report. Contribution of Working Groups I. II and III to the fifth assessment report of the Intergovernmental Panel on Climate Change, 151.

Enquist, B. J., & Enquist, C. A. F. (2011). Long-term change within a Neotropical forest: Assessing differential functional and floristic responses to disturbance and drought. *Global Change Biology*, *17*(3), 1408–1424. doi:10.1111/j.1365-2486.2010.02326.x

Enquist, C. A. F. (2002). Predicted regional impacts of climate change on the geographical distribution and diversity of tropical forests in Costa Rica. *Journal of Biogeography*, *29*(4), 519–534. doi:10.1046/j.1365-2699.2002.00695.x

<eref>Food and Agriculture Organization of the United Nations. (2000). FRA 2000: Global ecological zoning for the global forest resources assessment 2000: Final report. Rome, Italy: FAO. Retrieved from http://www.fao.org/docrep/006/ad652e/ad652e00.HTM

Ewel, J. (1999). Natural systems as models for the design of sustainable systems of land use. *Agroforestry Systems*, *45*(1/3), 1–21. doi:10.1023/A:1006219721151

Fajardo, L., Gonzalez, V., Nassar, J., Lacabana, P., Portillo, C., Carrasquel, F., & Rodríguez, J. P. (2005). Tropical dry forests of Venezuela: Characterization and current conservation status. *Biotropica*, *37*(4), 531–546. doi:10.1111/j.1744-7429.2005.00071.x

Fajardo, L., Rodríguez, J. P., González, V., & Briceño-Linares, J. P. (2013). Restoration of a degraded tropical dry forest in Macanao, Venezuela. *Journal of Arid Environments*, *88*, 236–243. doi:10.1016/j.jaridenv.2012.08.009

Feng, X., Porporato, A., & Rodriguez-Iturbe, I. (2013). Changes in rainfall seasonality in the tropics. *Nature Climate Change*, *3*(9), 811–815. doi:10.1038/nclimate1907

Fischer, G., Shah, M., & Velthuezin, V. (2002). *Climate change and Agricultural Vulnerability. IISA report for the World Summit on Sustainable Development, Johannesburg*. Vienna, Austria: IISA Publications Department.

Food and Agriculture Organization of the United Nations. (2010). Global Forest Resources Assessment 2010: Main Report. Rome, Italy: FAO. Retrieved from http://www.fao.org/docrep/013/i1757e/i1757e.pdf</eref>

Food and Agriculture Organization of the United Nations. (2015). *Global Forest Resources Assessment 2015: How are the world's forests changing?* Second Edition. Rome, Italy. Available at http:// www.fao.org/3/a-i4793e.pdf)

Food and Agriculture Organization of the United Nations. (2015). Forests, Trees and Disasters, Unasylva, No. 243/244. Vol. 66 2015/1–2.

Food and Agriculture Organization of the United Nations. (2018). *Potential implications of corporate zero-net deforestation commitments for the forest industry.* Discussion paper prepared for the 58th session of the FAO Advisory Committee on Sustainable Forest-based Industries. Retrieved from http://www.fao.org/forestry/46928-0203e234 d855d4dc97a7e7aabfbd2f282.pdf

Forest Survey of India. (2011). *State of Forest Report.* Dehra Dun, India.

Forest Survey of India. (2017). The State of Forest Report of India, India.</edb>

Frieler, K., Lange, S., Piontek, F., Reyer, C., Schewe, J., Warszawski, L., ... Yamagata, Y. (2017). Assessing the impacts of 1.5°C global warming - simulation protocol of the Inter-Sectoral Impact Model Inter comparison Project (ISIMIP2b). *Geoscientific Model Development*, *10*(12), 4321–4345. doi:10.5194/gmd-10-4321-2017

Gadgil, M., & Meher-Homji, V. M. (1986). Localities of great significance to conservation of India's biological diversity. *Proceedings of the Indiana Academy of Sciences*, (Suppl.), 165–180.

Gei, M. G., & Powers, J. S. (2015). The influence of seasonality and species effects on surface fine roots and nodulation in tropical legume tree plantations. *Plant and Soil, 388*(1-2), 187–196. doi:10.100711104-014-2324-1

Gillespie, T., Lipkin, B., Sullivan, L., Benowitz, D., Pau, S., & Keppel, G. (2012). The rarest and least protected forests in biodiversity hotspots. *Biodiversity and Conservation, 21*(14), 3597–3611. doi:10.100710531-012-0384-1

Glenday, J. (2008). Carbon storage and emissions offset potential in an African dry forest, the Arabuko-Sokoke Forest, Kenya. *Environmental Monitoring and Assessment, 142*(1-3), 85–95. doi:10.100710661-007-9910-0 PMID:17882526

Gonzalez-ruis, T., Jaramillo, V. J., Cabriales, J., & Flores, A. (2008). Nodulation Dynamics and Nodule Activity in Leguminous Tree Species of a Mexican Tropical Dry Forest. *Journal of Tropical Ecology, 24*(1), 107–110. doi:10.1017/S0266467407004634

Greve, P., Orlowsky, B., Mueller, B., Sheffield, J., Reichstein, M., & Seneviratne, S. I. (2014). Global assessment of trends in wetting and drying over land. *Nature Geoscience, 7*(10), 716–721. doi:10.1038/ngeo2247

Gumbo, D. J., & Chidumayo, E. N. (Eds.). (2010). The Dry forests and woodlands of Africa: managing for products and services. London, UK: Earthscan Ltd., Dunstan House.

Haywood, A. M., Dowsett, H., & Dolan, A. M. (2016). Integrating geological archives and climate models for the mid-Pliocene warm period. *Nature Communications, 7*(1). doi:10.1038/ncomms10646 PMID:26879640

Holdridge, L. R. (1947). Determination of world plant formations from simple climatic data. *Science, 342*(2727), 367–368. doi:10.1126cience.105.2727.367 PMID:17800882

Holm, J. A., Bloem, S. J. V., Larocque, G. R., & Shugart, H. H. (2017). Shifts in biomass and productivity for a subtropical dry forest in response to simulated elevated hurricane disturbances. *Environmental Research Letters, 12*(2). doi:10.1088/1748-9326/aa583c

Intergovernmental Panel on Climate Change. (2001). *Climate change 2001: The Scientific Basis: Contribution of working group I to the third assessment report of the intergovernmental panel on climate change (IPCC).* Cambridge, UK: Cambridge University Press.

International Union for Conservation of Nature. (2017). Forests and Climate Change. Retrieved from https://www.iucn.org/sites/dev/files/forests_and_climate_change_issues_brief.pdf

Keeling, C. D., Chin, J. F. S., & Whorf, T. P. (1996). Increased activity of northern vegetation inferred from atmospheric CO_2 measurements. *Nature, 382*(6587), 146–149. doi:10.1038/382146a0

Khayani, A. H., Gould, W. A., Harmsen, E., Terando, A., Quinones, M., & Collazo, J. A. (2016). Climate change implications for Tropical Islands: Interpolating and interpreting statistically downscaled GCM projections for management and planning. *Journal of Applied Meteorology, 55*(2), 265–282. doi:10.1175/JAMC-D-15-0182.1

Krishnaswamy, J., Bonell, M., Venkatesh, B., Purandara, B. K., Rakesh, K. N., Lele, S., ... Badiger, S. (2013). The groundwater recharge response and hydrologic services of tropical humid forest ecosystems to use and reforestation: Support for the 'infiltration–evapotranspiration trade-off hypothesis. *Journal of Hydrology (Amsterdam)*, *498*, 191–209. doi:10.1016/j.jhydrol.2013.06.034

Lund, H. G. (2012). National definitions of forest/forestland listed by country. Forest information service. Retrieved from http://home.comcast.net/gyde/lundpub.htm

Maass, J. M., Balvanera, P., Castillo, A., Daily, G. C., Mooney, H. A., Ehrlich, P., ... Sarukhan, J. (2005). Ecosystem services of tropical dry forests: Insights from long-term ecological and social research on the pacific coast of Mexico. *Ecology and Society*, *10*(1), 17. doi:10.5751/ES-01219-100117

Martinez-Garza, C., Tobon, W., Campo, J., & Howe, H. (2013). Drought mortality of tree seedlings in an eroded tropical pasture L. *Degraded Development*, *24*(3), 287–295. doi:10.1002/ldr.1127

Meave, J., Flores-Rodriguez, C., Perez-Garcia, E., & Romero-Romero, M. (2012). Edaphic and seasonal heterogeneity of seed banks in agricultural fields of a tropical dry forest region in southern Mexico. *Botanical Sciences*, *90*(3), 313–329. doi:10.17129/botsci.393

Medina, E. (1995). Diversity of life forms plants in neotropical dry forest. In H. A. Mooney, S. H. Bullock, & E. Medina (Eds.), *Seasonally dry tropical forests*. Cambridge, UK: University of Cambridge Press. doi:10.1017/CBO9780511753398.009

Mehta, V. K., Sullivan, P. J., Walter, M. T., Krishnaswamy, J., & De Gloria, S. D. (2008). Ecosystem impacts of disturbance in a dry tropical forests in southern India. *Ecohydrology*, *1*(2), 149–160. doi:10.1002/eco.14

Meir, P., & Pennington, R. T. (2011). Climate change and seasonal dry tropical forests. In R. Dirzo, H. Young, H. Mooney, & G. Ceballos (Eds.), *Seasonally dry tropical forests: ecology and conservation* (pp. 294–315). London, UK: Island Press. doi:10.5822/978-1-61091-021-7_16

Miles, L., Newton, A., De Fries, R., Ravilious, C., May, I., Blyth, S., ... Gordon, J. (2006). A global overview of the conservation status of tropical dry forests. *Journal of Biogeography*, *33*(3), 491–505. doi:10.1111/j.1365-2699.2005.01424.x

Mueller-Dombois, D., & Ellenberg, H. (1947). *Aims and methods of vegetation ecology*. New York, NY: John Wiley and Sons.

Murphy, P. G., & Lugo, A. E. (1986). Ecology of tropical dry forest. *Annual Review of Ecology and Systematics*, *17*(1), 67–88. doi:10.1146/annurev.es.17.110186.000435

Nelson, H. P., Devenish-Nelson, E. S., Rusk, B. L., Geary, M., & Lawrence, A. J. (2018). A call to action for climate change research on Caribbean dry forests. *Regional Environmental Change*, *18*(5), 1337–1342. doi:10.100710113-018-1334-6

Pennington, R. T., Prado, D. A., & Pendry, C. (2000). Neotropical seasonally dry forests and quaternary vegetation changes. *Journal of Biogeography*, *27*(2), 261–273. doi:10.1046/j.1365-2699.2000.00397.x

Pennington, T., Lewis, G., & Ratter, J. (2006). *Neotropical Savannas and Seasonally Dry Forests: Plant Diversity, Biogeography and Conservation*. Boca Raton, FL: CRC Press. doi:10.1201/9781420004496

Portillo-Quintero, C., & Sanchez-Azofeifa, G. (2010). Extent and conservation of tropical dry forests in the Americas. *Biological Conservation*, *143*(1), 144–155. doi:10.1016/j.biocon.2009.09.020

Portillo-Quintero, C. P., Sanchez-Azofeifa, G., Calvo-Alvarado, J. C., Quesada, M., & do Espirito Santo, M. M. (2015). The role of tropical dry forests for biodiversity, carbon and water conservation in the neotropics: Lessons learned for its sustainable management. *Regional Environmental Change*, *15*(6), 1039–1049. doi:10.100710113-014-0689-6

Raghubanshi, A. S., & Tripathi, A. (2009). Effect of disturbance, habitat fragmentation and alien invasive plants on floral diversity in dry tropical forests of Vindhyan highland: A review. *Tropical Ecology*, *50*, 57–69.

Ravindranath, N. H., Joshi, N. V., Sukumar, R., & Saxena, A. (2006). Impact of climate change on forest in India. *Current Science*, *90*(3), 354–361.

Ravindranath, N. H., & Sukumar, R. (1998). Climate Change and Tropical Forests of India. *Climatic Change*, *39*(2/3), 563–581. doi:10.1023/A:1005394505216

Ravindranath, N. H., Sukumar, R., & Deshingkar, P. (1997). *Climate change and forests: Impacts and adaptations. Regional assessment for the western ghats India. Atmospheric Environmental Issues in Developing Countries*. Stockholm, Sweden: Stockholm Environment Institute.

Reddy, C. S., Dutta, K., & Jha, C. S. (2013). Analysing the gross and net deforestation rates in India. *Current Science*, *105*(11), 1492–1500.

Reddy, C. S., Jha, C. S., Diwakar, P. G., & Dadhwal, V. K. (2015). Nationwide classification of forest types of India using remote sensing and GIS. *Environmental Monitoring and Assessment*, *187*(12), 777. doi:10.100710661-015-4990-8 PMID:26615560

Sanchez-Azofeifa, A., Portillo-Quintero, C., Wilson-Fernandes, G., Stoner, K., & Shimizu, T. (2013). *The policy process for land use/cover change and forest degradation in the semi-arid Latin American/ Caribbean region: perspectives and opportunities*. A literature review prepared for the Inter-American Development Bank.

Sanchez-Azofeifa, A., Quesada, M., Rodriguez, J. P., Nassar, J. M., Stoner, K. E., Castillo, A., ... John, A. (2005). Research priorities for neotropical dry forests. *Biotropica, 37*(4), 477–8-5.

Schmerbeck, J. (2011). Linking dynamics and locally important ecosystem services of South Indian dry forests: An approach. *Resources, Energy, and Development*, *8*(2), 149–172.

Seppälä, R., Buck, A., & Katila, P. (Eds.). (2009). *Adaptation of forests and people to climate change: A Global Assessment Report* (Vol. 22). Helsinki, Finland: IUFRO World Series.

Serraj, R., Siclair, T. R., & Purcell, L. C. (1999). Symbiotic N_2 fixation response to drought. *Journal of Experimental Botany*, *50*, 143–155.

Setalaphruk, C., & Prince, L. (2007). Children's Traditional Ecological Knowledge of Wild Food Resources: A Case Study in a Rural Village, Northeast of Thailand. *Journal of Ethnobiology and Ethnomedicine*, *3*(1), 33. doi:10.1186/1746-4269-3-33 PMID:17937791

Settele, J., Scholes, R., Betts, R., Bunn, S., Leadley, P., & Nepstad, D. (2014). Terrestrial and inland water systems In: C. B. Field, V. R. Barros, D. J. Dokken, K. J. Mach, M. D. Mastrandrea, T. E. Bilir, et al. (Eds.), Climate Change 2014: Impacts, Adaptation, and Vulnerability Part A: Global and Sectoral Aspects. Contribution of Working Group II to the Fifth Assessment Report of the Intergovernmental Panel on Climate Change. Cambridge, UK: Cambridge University Press. 271-359.

Singh, K. P., & Kushwaha, C. P. (2005). Emerging paradigms of tree phenology in dry tropics. *Current Science*, *89*(6), 964–975.

Springer, C. J., & Ward, J. K. (2007). Flowering time and elevated CO_2. *The New Phytologist*, *176*(2), 243–255. doi:10.1111/j.1469-8137.2007.02196.x PMID:17822407

Stearns, S. C. (1992). *The Evolution of Life Histories*. Oxford, UK: Oxford University Press.

Suresh, H., Dattaraja, H., & Sukumar, R. (2010). Relationship between annual rainfall and tree mortality in a tropical dry forest: Results of a 19-year study at Mudumalai southern India. *Forest Ecology and Management*, *259*(4), 762–776. doi:10.1016/j.foreco.2009.09.025

Telwala, Y., Brook, B. W., Manish, K., & Pandit, M. K. (2013). Climate-Induced Elevational Range Shifts and Increase in Plant Species Richness in a Himalayan Biodiversity Epicentre. *PLoS One*, *8*(2). doi:10.1371/journal.pone.0057103 PMID:23437322

Thomas, V., & López, R. (2015). *Global increase in climate-related disasters*. ADB Economics Working Paper Series, No. 466. Asian Development Bank, Metro Manila, Philippines.

Thuiller, W. (2007). Climate change and the ecologist. *Nature*, *448*(7153), 550–552. doi:10.1038/448550a PMID:17671497

Trejo, I., & Dirzo, R. (2000). Deforestation of seasonally dry tropical forest: A national and local analysis in Mexico. *Biological Conservation*, *94*(2), 133–142. doi:10.1016/S0006-3207(99)00188-3

Tripathi, S., Bhadauria, R., Srivastava, P., Singh, R., & Raghubanshi, A. S. (2016). Abiotic determinants of tree seedling growth in tropical dry forests. *Advances in plant physiology (Bethesda, MD)*, *17*, 119–131.

UNEP-WCMC Forest Programme. (2011). *Global statistics*. Retrieved from www.unep-wcmc.org

UNFCCC. (1992). United Nations framework convention on climate change. Climate change secretariat, Bonn, Germany. Retrieved from http://unfccc.int/

<unknown>Thomas, V., & López, R. (2015). Global increase in climate-related disasters. Asian Development Bank Economics Working Paper Series, (466).

Waeber, P., Ramesh, B., Parthasarathy, N., Pulla, S., & Garcia, C. (2012). Seasonally dry tropical forests in South Asia: A research agenda. In A research agenda to contribute to the discussions on "Key Issues for the Global Dry Forests" workshop organized by CIFOR, Zurich, Switzerland.

Walker, S., & Desanker, P. (2004). The impact of land use on soil carbon in Miombo Woodlands of Malawi. *Forest Ecology and Management*, *203*(1-3), 345–360. doi:10.1016/j.foreco.2004.08.004

Warszawski, L., Frieler, K., Huber, V., Piontek, F., Serdeczny, O., & Schewe, J. (2014). The Inter-Sectoral Impact Model Intercomparison Project (ISI-MIP): Project framework. *Proceedings of the National Academy of Sciences of the United States of America, 111*(9), 3228–3232. doi:10.1073/pnas.1312330110 PMID:24344316

Warszawski, L., Friend, A., Ostberg, S., Frieler, K., Lucht, W., Schaphoff, S., & Schellnhuber, H. J. (2013). A multi-model analysis of risk of ecosystem shifts under climate change. *Environmental Research Letters, 8*(4). doi:10.1088/1748-9326/8/4/044018

Westhoff, V., & van der Maarel, E. (1978). The Braun Blanquet approach. In R. H. Whittaker (Ed.), *Classification of plant communities*. The Hague, The Netherlands: Junk. doi:10.1007/978-94-009-9183-5_9

Williams, G. L., & Aquino, C. A. (2010). Tropical Dry Forest Landscape Restoration in Central Veracruz, Mexico. *Ecological Restoration, 28*(3), 259–261. doi:10.3368/er.28.3.259

Wohlfart, C., Wegmann, M., & Limgruber, P. (2014). Research Article Mapping threatened dry deciduous Dipterocarp forest in South-east Asia for conservation management. *Tropical Conservation Science, 7*(4), 597–613. doi:10.1177/194008291400700402

Chapter 3
Climate Change Effects, Adaptation, and Mitigation Techniques in Tropical Dry Forests

G. N. Tanjina Hasnat
University of Chittagong, Bangladesh

ABSTRACT

Tropical dry forests is one of the most unique forest types. It differs from other tropical forests with its climatic behavior like a prominent dry period, little annual rainfall, and high evapotranspiration. Out of six global bioclimatic zones, the forests are distributed in four. Climate change is now the most challenging issue regarding the fate of tropical dry forests. A severe climatic change is estimated to occur between 2040 and 2069 that could drastically change the precipitation pattern, temperature, aridity, and distribution of biodiversity. It could alter the forest type permanently. With a large number of heat-tolerant species, tropical dry forests have a great potentiality to conservationists with the prediction of a large area that could attain the climatic condition favorable for extension of tropical dry forests. But many of the species of tropical dry forests could be extinct due to changing climate at the same time. Proper adaptation and mitigation techniques could minimize the severity of climate change effects.

INTRODUCTION

Tropical dry forests cover almost half (i.e. 42%) of total tropical and sub-tropical forests of the world (Brown & Lugo, 1982; Murphy & Lugo, 1986). Tropical dry forests are defined on the basis of climatic limits (Gerhardt & Hytteborn, 1992; Portillo-Quintero, 2010). Forests that occurred in tropical regions with a climatic limits of pronounced drought period of 3-9 months, an average annual temperature of more than 17°C, precipitation ranges from 200 to 2000 mm per year, and annual ratio of potential evapotranspiration (PET) to precipitation (P) exceeds one are termed as tropical dry forests (Murphy & Lugo, 1986; Mooney, Bullock, & Medina, 1995; Blasco, Whitmore, & Gers, 2000; Sánchez Azofeifa

DOI: 10.4018/978-1-7998-0014-9.ch003

et al., 2005; Pennington, Lewis, & Ratter, 2006). Though the characteristics of the forests are difficult to be defined precisely, seasonally dry tropical forests are tree-dominated ecosystems unlike savannas. Trees of tropical dry forests are deciduous in nature, and generally grow on fertile soils (Ratter et al., 1973; Furley, 1992; Vargas, Allen, & Allen, 2008). Customarily having a closed canopy with woody flora these forests are dominated by the Leguminosae and Bignoniaceae (Pennington, Prado, & Pendry, 2000). Generally, tropical dry forests are found in the region where most of the months are dry and wet period is short. A long period without any precipitation characterizes the species of the forests, forcing them to adapt with and survive in the warm dry condition.

Tropical dry forests are the most threatened ecosystems in the world (Janzen, 1988; Hoekstra et al., 2005; Miles et al., 2006). These forests are the habitat of a large number of flora and fauna, but with the shrinkage of these forests to 10% from its original range (Janzen, 1988; Bruegmann, 1996; Juvik & Juvik, 1998; Allen, 2000; Cabin et al., 2000), it is now a highly vulnerable ecosystem. Though, the soil of this biome is much more fertile (Pennington, Prado, & Pendry, 2000) and these forests might help to mitigate climate change by storing carbon (Skutsch & Ba, 2010; Daniel et al., 2014), it remained overlooked by conservationists and researchers. More than 1600 inventories in the tropical dry forests revealed that these forests are the habitat of remarkably 6958 woody plant species (Kinver, 2016). Many of these species are endemic and among all woody species, about 40% are not grown elsewhere in the world, except in tropical dry forests (Kinver, 2016). Most interestingly, it has been found through molecular research that some species that grow in these forests are ancient (about >10 million years old) (Kinver, 2016). These forests are actually a museum of diversified species, particularly of the endemics (Trejo & Dirzo, 2000; Sánchez-Azofeifa et al., 2005). If any one species becomes extinct, it will be a great loss to the world. Moreover, tropical dry forests are affected by climate change since a long time (Parmesan, 2006; Phillips et al., 2010; Liu et al., 2013). Global climate change badly impacts on the species distribution (Peterson & Kluza, 2005). Owing to its biological and genetic importance, and its great potential contribution in the mitigation of climate change, tropical dry forests necessitate more concentration of the scientists and policymakers (FAO, 2001).

Tropical dry forests are rich in species, genetic and biological diversity, sometimes as rich as adjacent tropical rain forests (Janzen, 1986). Tropical forests are habitats of migratory birds, different endangered mammals like Asian elephant, the world's most endangered tortoise Angonoka, the rare Timor Python, the world's largest living lizard Komodo Dragon, etc. Moreover, these forests provide a shelter to the native important woody plant species, Agaves, Acacias, Cacti, mammals, butterflies, moths, bees, wasps, ants, birds, reptiles, amphibians, etc. Considering species diversity, these forests holds an important position on a world scale. However, only a few fragmentary studies have done about their biodiversity (Janzen, 1988; Miles et al., 2006). Recent observations suggest that tropical dry forests are more vulnerable than any other forests (Janzen, 1988; Bhadouria et al., 2016). The conservation status of the forests is highly threatened now. Tropical dry forests are under different kinds of threats such as climate change, habitat fragmentation, human interference and conversion of forest land to agricultural land are the most important threats. Among all these threats, climate change causes great negative impacts on these forests and it is more significantly observed in forests of the Americas (Miles et al., 2006). Generally, plants and animals living in tropical dry forests try to adapt themselves with the long dry season (Janzen, 1988). However, abrupt climate change could create the extinction of locally endemic species of these forests (Mason-Romo et al., 2018). Recent observation suggests that different types of adaptation and mitigation techniques could minimize the negative impacts of climate change (IPCC, 2007).

In recent years, a great emphasis is being given to the tropical dry forests (FAO, 2001). For example, International Union for Conservation of Nature (IUCN) and World Wildlife Fund-US (WWF) works on tropical dry forests. IUCN assesses the vulnerability of the individual species for extinction (Mace & Lande, 1991; Reynolds & Mace, 1999; Akcakaya et al., 2000; IUCN, 2001; 2003) and evaluating their conservation status. WWF assesses (Dinerstein et al., 1995; Ricketts et al., 1999; Wikramanayake et al., 2002) the global distribution of the habitats and their conservation status (Olson & Dinerstein, 1998; Olson et al., 2000). However, very limited researches have been done on the effects of climate change on tropical dry forests (Borchert, 1999; Miles et al., 2006; Salazar, Nobre, & Oyama, 2007; Bhadouria et al., 2016; Bhadouria et al., 2017a,b; Singh et al., 2017).

GEOGRAPHIC AND CLIMATIC DISTRIBUTION

Tropical dry forests are situated broadly in five realms out of eight of the world (Udvardy, 1975; Richard, 1980; Olson et al., 2001), namely Afro-tropical, Australasia, Indo-Malayan, Neotropical, and Oceania (WWF, 2018). The tropical dry forests in the world cover a total of 1,048,700 km^2 (Miles et al., 2006) The maximum of these forests are found in Brazil, Bolivia, Paraguay, and Argentina of South America, collectively sharing about 54.2% of total tropical dry forest land in the world (Miles et al., 2006; Portillo-Quintero &Sánchez-Azofeifa, 2010; Rodrigues et al., 2015). Of the remaining area, 16.4% area is situated in combined Europe and Asia that is called Eurasia, 13.1% in Africa, 12.5% in North and Central America, and 3.8% in Australasia and Southeast Asia (Miles et al., 2006).

Global climatic distribution of tropical dry forests is between two parallel lines: the tropic of Cancer and the tropic of Capricorn (Sánchez-Azofeifa et al., 2005). Tropical dry forests are distributed globally in four major climatic zones out of six (Table 1) and thus have a broad climatic tolerance range (Blasco et al., 2000) with rich species diversity. The annual temperature of these deciduous forests is more than 17°C, whereas the total annual rainfall ranges from 200 to 2000 mm (Murphy & Lugo, 1986; Gentry, 1995). Dry period exists here at least three months of the year (Blasco et al., 2000; Sánchez-Azofeifa et al., 2005).

CLIMATE CHANGE AND TROPICAL DRY FORESTS

At present, global climate change is the most challenging issue (Gashaw et al., 2014) and it directly relates with the change of global temperature and resultant change of precipitation (Maass et al., 2005), wind speed (Kirilenko & Sedjo, 2007; WWF, 2009; Brodie, Post, & Doak, 2012a;b), evaporation, water availability, and other meteorological phenomenon (Vose & Maass, 1999) due to human activities (Harris, 2019). It is predicted that severe climatic change could happened by 2055 when an increase of temperature (at least 2.5°C) or a decrease of precipitation (at least 50 mm per year) may occur (Miles et al., 2006). Tropical dry forests throughout the world are severely affected by climatic changes. The severity of climate change in different dry forests was predicted by Miles et al. (2006). Among all tropical dry forest areas, forests of America where 66.7% of total tropical dry forests located (Miles et al., 2006) are in serious threat due to climatic change (Sánchez-Azofeifa et al., 2005). Approximately 39.8% risks due to climate change found in North and Central America, while 37% in South America. This is because of a decrease in precipitation at a large scale. It is estimated that between 2040 and 2069, a severe climatic

Table 1. Distribution of tropical dry forests under different worldwide climatic zones

Sl. no.	Bioclimatic types	Continents	Elevation (m)	Rainfall (mm)/year	Dry months	Temperature (°C)	Forest types (dry forests)
1	Warm perhumid and very humid	Asia, Southeast Asia, Oceania, South America, Africa	≤ 1800 to 3000	1500 to > 2000	0-3	10 to > 15	-
2	Warm humid lowland	Asia, **Southeast Asia, Oceania**, South America, Africa	1000 to 1800	1000 to > 2000	3-6	15 to > 20	- Dry evergreen forest and semi-deciduous forests
3	Warm subhumid lowland	**Asia, Southeast Asia, Oceania, South America, Africa**	≤ 900	1000 to < 2000	5-6	> 20	- Dry deciduous forest - Tropical dry broadleaf forest
4	Warm dry lowland	**Asia**, Southeast Asia, Oceania, **South America**, Africa	≤ 800	500 to 1500	5-9	15 to > 20	- Tropical dry deciduous forest - Tropical dry broadleaf forest
5	Warm very dry lowland	South America, **Africa**	≤ 800	200 to 500	8-9	> 15	- Dry deciduous forest
6	Warm arid and subdesertic lowland	Asia, South America, Africa	≤ 800	300-500	9-11	> 15	-

Source: Blasco, Whitmore, & Gers, 2000

changes will happen, precipitation pattern will be changed, temperature will arise, aridity will increase, and distribution of biodiversity and ecosystems will be changed (Miles et al., 2006; Rodrigues et al., 2015). It would be harmful to some endemic species in tropical dry forests and could start the extinction of some locally endemic species (like *Peromyscus perfulvus* and *Liomys pictus*). On the contrary, changing climate may favor some unwanted harmful invasive species like *Oryzomys melanotis* at the same time (Mason-Romo et al., 2018).

Change in temperature with the change of era is a common phenomenon. The Earth faces warm age followed by ice age repeatedly as a natural event. However, during the last few decades, increase in temperature occurred abruptly within a short period of time and the rate of change is unprecedented (Gashaw et al., 2014). From the scenario, it is predicted that the global climate is changing toward a warm condition at a relatively faster rate. The changing climate is significantly altering the tropical dry forest ecosystems as well. The evidences of impacts of climate change on tropical dry forests are alteration in species distribution and composition, phenology, and forest structure.

Global warming assessment by IPCC observed that average surface temperature of the earth has increased 0.6 ± 0.2°C in the 20th century and, which is projected to increase by 1.4 to 5.8°C in 2100 (Parry, Hammill, & Drexhage, 2005). One of the significant effect of climate change is that it makes the dry region even drier and the wet region more wetter. Change in any part of the forests can change the entirely natural process of the earth. On the other side, changes in precipitation in one part of the world due to climate change could change precipitation or any other climatic factors in other parts of the world. A prediction stated that, unlike the present time, seasonal forests of South America could face less annual rainfall with a lengthy dry period, while, seasonal forests in Asia, Africa, and India could face more rainfall with a reduced dry period (Hulme & Viner, 1995).

According to IPCC (2007), aridity will be increased throughout the world. Global warming could be at a high rate that species hardly cope with this changing situation (Alley et al., 2003). Some species though are stout to survive (Oliveira et al., 2005; Lloyd & Farquhar, 2008; Krause et al., 2010) but many species have low thermal tolerance (Tewksbury, Huey, & Deutsch, 2008; Laurance et al., 2011). Mainly, the species of lowland tropical dry forests areas might be incapable to persist with increasing temperature (Colwell et al., 2008; Wright, Muller-Landau, & Schipper, 2009). This situation could create a high species mortality and extinction (Phillips et al., 2010). Global climate changes have significant effects on biological diversity, ecosystem balance (Buytaert, Cuesta-Camacho, & Tobón, 2011), species composition and distribution (Santos et al., 2014). Considering the change in climate at relatively higher pace than in previous 100 years (Araújo & Rahbek, 2006), some species are found under great threats (Rodrigues et al., 2015). The vegetation types and ecosystem of the tropical dry forests can tolerate a longer dry period. However, a minimum rainfall is crucial for the species survival and existence (Mooney, Bullock, & Medina, 1995; Sánchez-Azofeifa et al., 2005). A longer dry period with a shorter rainy season is vital for species reproduction and ecosystem balance. Changing climate could change the duration of the dry period as well as the effect on the amount of rainfall in coming years. Species usually live within the specific defined climatic conditions. Thus, change in temperature and precipitation in any amount could collapse the ecosystem balance by any means (Marengo et al., 2009). Due to climate change, any species could disperse, adapt, extinct (Peterson et al., 2001), or migrate (Werneck et al., 2011) with the severity or rate of change. Furthermore, species extinction is a chain of extinction process, as one species depends on another species for their existence. Rodrigues et al. (2015) observed that a number of species of tropical dry forests could become permanently extinct in near future (by 2080) due to the extended dry period that will convert the lands into arid and desert areas.

Tropical dry forests are the early indicator of potential danger of future climatic effects (Pascoe, 2018). Human activities such as slashing, burning, conversion of forest land to agricultural and urban uses, fragmentation of forests are the main causes of climate change and ecosystem degradation (Miles et al., 2006). Climate change brings shrinkage (Rodrigues et al., 2015) and alteration in the location of tropical dry forests. On the other side, the changing climate could also favor the extension of tropical dry forests (Leemans, 1999; Miles et al., 2006; Rodrigues et al., 2015). Some researchers opined that the rainforests could be replaced with dry deciduous forests, and savannas in near future (Moutinho, 2006; Pennington, Lewis, & Ratter, 2006; Rodrigues et al., 2015; Vico, 2017). Thus, only tropical dry forests have much more feasibility than any other ones according to some reports.

Deforestation and land cover change (Brodie, 2012b) reduces rainfall (Halley, 1694; Grove, 1994). It is estimated by Wilkie and Trexler (1993) that tropical deforestation reduces evapo-transpiration from tropical forests that could make up 20-80% of the incident rainfall. In order to halt the abrupt climate change, forest restoration has taken places in some parts of the world, but exotic species are given preference rather than native species (Lugo & Helmer, 2004). Thus, the destructed and fragmented habitat continued, that added to the climatic changes as well as global warming. Moreover, besides deforestation and land use change, human-induced fires in tropical dry forests also exacerbate the severity and extent of dry season months. Many tree species of tropical dry forests even have resilience power to adapt with increased temperature (Sharkey & Schrader, 2006) and moderate water stress (Nepstad et al., 2004; Oliveira et al., 2005), however, only a few have the ability to survive in the fire (Barlow & Peres, 2004). Human beings are the dominating agents for global climate change and the resultant impact on the ecosystems as well as behavior of individual species (Schmitz, 2003; Suttle, 2007; Post, 2009).

The warmer and drier condition and reduced annual precipitation will increase the frequency, extent, and duration of uncontrollable wild-fires in the future (Nepstad et al., 2008). Forest loss due to anthropogenic pressure also reduces evapotranspiration, and thus moisture content in the atmosphere. This situation makes forest areas more dry and favorable for more fires (Nepstad et al., 2008). Among all tropical forests, lowland forests are most susceptible to deforestation, land cover change and climate change. Many conservation strategies can be applied to ameliorate the causes of climate change such as fire regulation, controlling human-induced fires, expand protected areas, community involvement and education, payments for ecosystem services, etc. (Laurance, 2008).

The forest loss and climate change are interlinked bi-directionally. Loss of forest coverage due to human interference, population growth, civilization, deforestation, logging, wood used as fossil fuel, forest land conversion to agricultural land provide spaces for producing green-house effects on one side, and sequester less carbon in comparison of high volume of emitted carbon on other side. Moreover, the mismanagement of the degraded parts of forests further produces ingredients of rapid climatic change. An abrupt change in temperature makes the habitat more vulnerable to the species. Species found in tropical dry forests are well-known for their endemism and diversification. Moreover, these forests provide shelter to a large number of endangered species. Increase in temperature or change in precipitation pattern for a little bit may force the endemic species to become extinct. Extinction of any species could completely break down an ecosystem. Sometimes, this situation force to other species to become extinct.

FACTORS OF CLIMATE CHANGE AND CONSEQUENT IMPACTS

Climate change is the burning issue now in dry tropical forests (Bhadouria et al., 2017a). Climate change mainly occurred by two basic factors, natural processes and human activities (Akpodiogaga-a & Odjugo, 2010). Human-induced climate change is most destructive that causes unequivocal climate change (IPCC, 2007). People are more responsible for producing CO_2 (Singh et al., 2017) and other greenhouse gases through deforestation, conversion of forest land to other land use, burning fossil fuels, industrialization, gas flaring, urbanization, and agriculture (Akpodiogaga-a & Odjugo, 2010; Change, 2007). Increasing CO_2 and greenhouse gases are primarily liable for the changing climatic conditions (Crowley, 2000; Hasnat, Kabir, & Hossain, 2018). There are a number of evidences of notable climate change impacts around the world. Surface air temperature, sea surface air temperature, humidity, ocean heat, sea surface level, Earth's troposphere temperature are increasing due to climate change. At the same time, Arctic sea ice, glaciers, snow on hilltop, etc. are decreasing (Change, 2007; Ding et al., 2007). The changing climate and increasing heat made changes in the precipitation patterns (Ding et al., 2007).

For the growing of a forest and restoration of forest ecosystems, natural regeneration is important. Seedlings are the regeneration elements and this is the most sensitive stage of plant lifecycle (Bhadouria et al., 2017a,b; Bhadouria et al., 2016). The growth of seedlings mainly regulated by climatic factors such as temperature, light, and precipitation (Bhadouria et al., 2017b; Bhadouria et al., 2016). With the abrupt change in climate, it is difficult to cope with by seedlings that may cause the failure of regeneration. This condition may be more aggravating by slowing down the renewal process of forests (Bhadouria et al., 2017a).

ADAPTIVE NATURE OF TROPICAL DRY FORESTS SPECIES

Tropical dry forests have more adaptation capability to the heat and drought in contrast with other tropical forests. These forests are special and unique in nature ("Tropical Dry Forests," 2019). Two different types of seasonal variability are predominant here –wet and dry. The forests harbors two different types of species. One is active in the rainy season and another in the drier period. Many resident plant and animal species show a dormant behavior in the extreme weather conditions. However, many plant species continue to produce flowers, fruits, and leaves in the severe dry period (Janzen, 1967; 1982a,b) and provide supports to some faunal species for their living.

With the changing climatic condition, it will be difficult for rain forests or temperate forests to survive, leading to the extinction of a number of species. However, tropical dry forests also have plant and animal species that could cope with both conditions, and can genetically adapt themselves with dry nature. For example, deciduous plant species of the forests are adaptable to the changing climate. It suggests that there is a great potentiality of species to survive in an adverse fluctuating climatic situation as well.

Plant Adaptation

Plants of tropical dry forests face a wide variability in water availability each year, including 3-9 months of dry period (Blasco, Whitmore, & Gers, 2000; Mooney, Bullock, & Medina, 1995; Janzen, 1988). Species growing here are highly adaptive to drought (Amissah et al., 2018; Pulla et al., 2015). Plants need sufficient water to survive during the dry spells that they get during the rainy season and store ("What Are Some Tropical Dry Forest Plants?", 2019). The forests have two different types of forest covers: plants showing deciduous nature and plants showing evergreen nature (Mooney, Bullock, & Medina, 1995). Basically, the forests are deciduous in nature. During the dry season, most of the species when loses their leaves, allowing the sunlight to directly penetrate to the forest floor (Janzen, 1988). Unlike the rain/wet forests, the dry forests receive much more heat creating a situation very difficult to survive for plants and animals. To cope with the elongated dry period, dry tropical species promote various adaptation mechanisms to survive (Sunderland et al., 2015). The most common and important adaptation process in tropical dry forests is deciduousness of the plant species ("Dry Forest Ecology", 2019; Medina, 1983). Most tree species in dry tropical forests are deciduous in nature (Pennington, Lewis, & Ratter, 2006) which shed their leaves during the dry period to halt photosynthesis process and other vegetative activities (Janzen, 1988). Some plant species modify their plant parts into spines (Durán, 2004), photosynthetic barks (Wagner, 2018), storage leaves, waxy leaves or any other suitable forms to collect and store water during the rainy season for adverse dry time ("Tropical Dry Forests", 2019). For example, species like *Ceiba trichastandra, Bursera simarouba* continue their photosynthesis in the dry period with their green bark having chlorophyll even though the species shed their leaves (Septer, 2019; Wagner, 2018).

Acacia species adopt several numbers of the adaptation processes. They produce many spikes and thorns between the leaves to protect them from herbivores (Septer, 2019; Wagner, 2018). Tiny leaves hold themselves vertically to reduce water loss. The tree produces a complex root system that contains a deep taproot to penetrate the lower water table and spread out horizontally to find the ground, surface, or sub-surface water around (Wachman, 2019). Plants of dry forests store water during the short rainy season and survive with the stocked water during the prolonged hot and dry season ("What Are Some Tropical Dry Forest Plants?", 2019). The Baobabs are the most common tree species that adorn themselves with a specialized trunk. This strange featured long-lived tree holds a large amount of water in their vast sized stem (Discover, 2019; Julian, 2018). Moreover, some other plant species live on the tree during the dry season. Animals like elephant, eland, etc. chew the stem bark for water to cope with the excessive heat ("Baobab tree facts", 2015).

Many species develop water storage tissues such as swollen roots to reserve water for the hot dry periods. Some other species grow a waxy layer outside their leaves to reduce the evaporation rate. Similarly, some species clasping their leaves during the night so that the exposed amount of leaf surface area be reduced. Many plants embellish themselves with thicker barks as an anti-fire adaptation process ("Tropical Dry Forest", 2019). Some species such as lotus, ironwood, and acacias produce smaller and thicker leaves as anti-desiccation adaptation (Medina, 1983; Wagner, 2018). Some furnished themselves with spikes or thorns as anti-herbivore adaptation. Some other species produce longer roots to tap water from the deeper water table, while some produce spread root systems to find the nearest surface or sub-surface watercourses ("Tropical Dry Forest", 2019). Species acclimate with one or more of these adaptation systems to survive during the dry period. Some plants are epiphytic in nature such as orchids, bromeliads, and cacti. They grow on other plant species rather than soil. This system provides them water, nutrition, and food from the host plants during dry periods. The host plants stock water for the dry season in their roots and shoot ("What Are Some Tropical Dry Forest Plants?", 2019). Columnar cacti, Agave produce spines to survive during the dry period. Moreover, they store water and nutrient during the rainy period to cope with excessive heat in the dry period (Septer, 2019).

Animal Adaptation

Animals of tropical dry forests are adaptive to the water scarcity period ("Tropical Dry Forest", 2019). Most species undergo hibernation and remain in such condition until the rains start (Janzen, 1988). Similarly, amphibians and insects burrow themselves deep into muds and rest there until the rainy season begin. With the onset of rainy season, they come out and engage in the breeding process. However, some species move to the buffer zone of rain forests or river bank for their food, water, and shelter ("Dry Forest Ecology", 2019). Some lives in the dry condition and survive via flowers, fruits, or other parts of some tree species. The reproductive cycle mainly depends on season and most groups reproduce during the rainy season. Though, the dry period is difficult to survive for most of the plant and animal species, it is the reproduction season for some other species (Janzen, 1988). During the dry period, some species go to the hibernation, some try to live in dry condition by storing or collecting foods and the rest species migrate to the wetter area, mainly the nearby rain forest, gallery forest, wet lowlands, or buffer zone ("Dry Forest Ecology", 2019; Janzen, 1988). Crested guan, a large chicken-like bird, and the Magpie-jay store foods at the time of the rainy season for the dry and hot months. Animals such as Merriam's kangaroo, a small nocturnal rat stays in the burrows during the dry and hot period (Wagner, 2018).

A minimum amount of rainfall per annum is essential for tropical dry forests. Passing through a long dry period, the whole forests await for the rains. With first rains after a long duration, the leaf flash starts, plants engage to photosynthesis and reproduction process. After a long hibernation and migration, animals come back to their homes and indulge in their biological activities. Bacteria and insects start to decompose the leaf litter on the forest floor. All again furnish themselves for the upcoming dry period again ("Dry Forest Ecology", 2019).

ADAPTATION AND MITIGATION TECHNIQUES FOR CLIMATE CHANGE

Two different but interlinked approaches to cope with and minimize the abrupt climate change adaptation and mitigation has been advocated by IPCC (2007). Adaptation to climate change is the adjustment with the changing climate and its effects (Adger et al., 2007; IPCC, 2007), while mitigation is the management part to halt the climate change. Adaptation could be autonomous, reactive, preventive, or planned (Murthy, Tiwari, & Ravindranath, 2011). It is a part of risk management and planning for sustainable development. This is the strategy to increase the resilience of nature and scheme to cope with future climatic risks to the environment. Adaptation is necessary to cope with short term or long term effects of climate change and global warming. Mitigation is directly linked with the reduction of GHG emissions and the enhancement of carbon sequestration in forests and soils. Carbon trading is a good and crucial practice to mitigate climate change throughout the world (Locatelli & Pramova, 2011; Gashaw et al., 2014). Though, mitigation is a new concept, it received a greater consideration across the world from the researchers and policymakers rather than adaptation.

Adaptation and mitigation through different strategies could be the best way to combat the emerging climatic situations. Adaptation of species is to make them adjusted to cope with the impacts of climate changes and equip them to be less vulnerable (Shanahan et al., 2013). Plant and animal species of tropical dry forests are considerably adaptive to heat in comparison to other forest dwelling species. With rapidly increased temperature, it will be difficult to survive all of the species. Thus, besides adaptation, it needs effective mitigation techniques to minimize the climate change effects to conserve the biodiversity and ecosystems of tropical dry forests.

Carbon Sequestration

The major cause of climate change is greenhouse gas concentrations in the atmosphere. The resultant impacts create natural calamities, insect outbreaks, reduced forest growth and production, change in temperature, precipitation, and evaporation, etc. Researchers found that approximately 30% of carbon dioxide has increased during the last 150 years which is the main culprit of global warming (Stavins & Richards, 2005). For carbon sequestration, terrestrial forests are identified as the major carbon sink. Tropical forests act as carbon sink and naturally store atmospheric CO_2 in vegetation and soils. Tropical dry forests play a vital role in reducing the effects of climate change through the carbon sequestration process (Becknell, Kucek, & Powers, 2012; Dexter et al., 2015). Though, these forests store less carbon within vegetation in comparison to the rain forests (Becknell, Kucek, & Powers, 2012; Day et al., 2014), a higher amount of carbon is stored into its soils (Nepstad et al., 1994). A large scale of carbon sequestration programs have been put in action in the recent times. Reduce Emissions from Deforestation and forest Degradation, and foster conservation, sustainable management of forests, and enhancement of forest carbon stocks (REDD+), Reducing Emissions from Deforestation and Degradation (REDD) (Murdiyarso, Hergoualc'h, & Verchot, 2010), Carbon Capture and Sequestration Technologies (CC&ST), Carbon capture and sequestration (CCS) Programs are the major examples of offsetting carbon emissions (Appropedia, 2012).

Halt Deforestation

Forests also act like a major source of carbon when trees are cut down and forest land converted to other land uses. Thus deforestation and alteration of forest coverage accentuates the global climate change. Moreover, forests play a significant role in the water cycle and global precipitation. It transfers water from forest land to the atmosphere through transpiration and evapotranspiration process that enhance precipitation. Large deforestation causes a significant reduction in regional and global rainfall. In tropical dry forests, a minimum rainfall is essential for the survival of species. Deforestation severely affects the dry season rainfall that is very much important for the existence of life in the seasonal tropical dry forests (Malhi et al., 2008). Deforestation decrease cloudiness and rainfall, increase temperature, change wind speed, create an environment more smoky and dusty, produce greenhouse gases, increase atmospheric heat absorption by different means (Bala et al., 2007). In compared to tropical rain forests, tropical dry forests are highly deforested and converted to other land uses. Tropical dry forests extents only 10%, somewhere 0.1%-1.0% of its original size (Janzen, 1988; Bruegmann, 1996; Juvik & Juvik, 1998; Allen, 2000; Cabin et al., 2000), while tropical rain forests cover 60% of its original extent at some regions (Allen, 2000). Considering the interrelation between deforestation and climate change, the most important mitigation and conservation approach is to considerably reduce the deforestation.

Reduced Forest Fragmentation

Species migration is the early signals of climate change. Fragmentation of forest coverage due to civilization shrinks total forest area inviting species extinction, breakdown the species corridor promoting species migration and ecosystem shift towards the more humid region (maybe to the rain forest areas) (Malhi et al., 2008). The migration of ecosystems reduce local transpiration services, and thus induce climate change. To mitigate climate change (i.e., to reduce the greenhouse gas concentrations), a continuous forest coverage without fragmentation is essential.

Fire Management

Forest fire is another cause as well as the resultant of climate change. Intense fires in tropical dry forests reduce the storage capacity of carbon and emit a large amount of carbon as well (Silver, 1998). Reduction in a minor amount of annual precipitation can increase the risk of forest fire (Locatelli & Pramova, 2011). Species naturally growing in the tropical dry forests have adaptive features to cope with regular forest fires, and sometimes act as the mitigator. But, the introduction of exotic species by human beings creates irregular forest fires. These species have no adaptive features like tropical dry forest species and also are inflammable in nature (Mueller-Dombois, 1981), thus exhort fires to burn and increase damages. Proper management of forest fires could reduce climate change effects (Appropedia, 2012). Forest fire could be mitigated by preparing firebreaks, fire suppression, etc. Proper concern to the man-made forest fire is another way to reduce the warming of the micro-climatic area.

Management of Invasion

To acclimate with the changing climate, management of invasive species, insects and diseases is important. To minimize the bad impacts of climate change by prevention of migration and growth of invasive species is another strategy. There are different types of management strategies for invasion. A variety of control methods can be used to manage invasive species – prevention, mechanical control, physical control, cultural control, chemical control, and biological control methods ("Control Invasive Plants", 2019; Hussner et al., 2017; Myers & Bazely, 2003). Prevention control is applicable at the time of introducing any invasive species to a new area. Once the invasion has taken place, then other control measures are applicable. In mechanical control measures, an infestation of invasive species is controlled by cutting, girdling, chopping or trimming the mother tree just before seeding. If there is small invasion, then physical control is appropriate. This control measures include hand picking and digging of shallow-rooted plants. In cultural control, the competitive, rapid growth native species could be selected for plantation to dominate invasive alien species. In order to enhance the growth of native species, proper fertilization and irrigation can be applied. Various herbicides can be applicable to control invasive species. Biological control is a slow process and living organisms are used to control invasive species ("Control Invasive Plants", 2019; Control methods, 2019; Hussner et al., 2017). Besides the management of invasive species, enhancing landscape connectivity by creating corridors or buffers, conservation of biodiversity hotspots and ecosystems across environmental gradients are also needed to mitigate climate change effects.

Restoration

Restoration is the actions to recover forest structure and functioning and different ecological and biodiversity levels towards a forest succession ("Forest restoration," 2019; Elliott, Blakesley, & Hardwick, 2013). Restoration is another strategy to mitigate climate change. Restoration in tropical dry forests and rehabilitation of degraded land by planting species that uptake more carbon dioxide and store for the longest period of time can be effective. More trees uptake and store more carbon dioxide, thus more forested areas will be a storehouse of more carbon dioxide. Moreover, afforestation after the acquisition of wastage, barren and degraded lands around the natural forest areas with the heat-tolerant trees could be helpful to mitigate climatic adverse effects. The other methods include assisted natural regeneration, seed bank restoration, preventing fires, removing cattle, retaining largest and oldest trees, preserving wildlife habitat and riparian areas, protecting erodible soils, maintaining slope stability etc. ("Forest restoration," 2019; Shono, Cadaweng, & Durst, 2007; Moline, 1999).

Integrated Management

Integrated management, through maintaining natural disturbance regimes and by assisting migration (Locatelli & Pramova, 2011), could be helpful to adapt with upcoming climate change effects. Conservation of forests biodiversity and ecosystems by reducing human-induced pressures through increasing monitoring, *ex situ* and *in situ* conservation, increasing awareness and creating knowledge, reducing socioeconomic pressures on forests, building partnerships among stakeholders could be effective to adapt and mitigate climate change effects (Locatelli & Pramova, 2011). Modeling tools and new policies need to be developed to adapt and mitigate climate change effects. Adaptation strategies, planning and implementation process to the effects of climate change should be time oriented, considering economic development (Murdiyarso & Kauffman, 2011). To minimize the negative effects of climate change, the conservation of total biodiversity and ecosystem services can be used as an important part of overall adaptation and mitigation strategies (CBD, 2009).

Policymakers have a vital role to mitigate adverse impacts in tropical dry forests. Increasing the boundaries of the protected area, creating the buffer zone around the protected area and declaring the core forests as stringently reserved areas, increasing the numbers and extents of protected and reserved zones could be some strategies to cope with the changing climate. The design of the protected area needs to be redesigned and remodeled again. After declaring the area reserved and protected, directed efforts and emphasis from the respective government and policymakers would also be required. As forest land conversion is highest in tropical dry forests (Portillo-Quintero & Sánchez-Azofeifa, 2010), strict policy with strong punitive action should be developed for all tropical dry forests to reduce the land use conversion.

CONCLUSION

Tropical dry forests have a potential to nullify the climate change effects such as change of one kind of eco-region to another (i.e. other types of forests could be converted into tropical dry forests). However, such as shift could imbalance the total ecological systems. Climate change could shift a forest type and force species extinction that creates irreversible damages to the forests and the world. Moreover, deforestation and conversion of forest land to another purposes shrinks and limits the possible extent of forest expansion. Improvement of existing policies and generating time-oriented new policies, strategies and practices could mitigate the impacts of climate change and could embrace the dry tropical forests to adapt to the changing climate. Effective policies and short and long term strategies may be checking deforestation, halt forest and ecosystem fragmentation, expand protected areas, promote natural regeneration, improve silvicultural practices, restoration with native species, stop restoration with exotic species, promote *in situ* and *ex situ* conservation with rare native species. Formulation of time-oriented long-term forest policy and effective forest management plan could address the abrupt change of climate by integrating climate change mitigation and adaptation goals into forest policy, balancing forest management objectives with climate change objectives, revising relevant forest legislation, enforcing forest laws, making changes to the relevant organizations, and adjusting people's participation.

REFERENCES

Adger, W. N., Agrawala, S., Mirza, M. M. Q., Conde, C., Brien, K. O., Pulhin, J., ... Takahashi, K. (2007). Assessment of adaptation practices, options, constraints and capacity [Impacts, Adaptation and Vulnerability. Contribution of Working Group II to the Fourth Assessment Report of the Intergovernmental Panel on Climate Change.]. *Climatic Change*.

Akçakaya, H. R., Ferson, S., Burgman, M. A., Keith, D. A., Mace, G. M., & Todd, C. R. (2000). Making consistent IUCN classifications under uncertainty. *Conservation Biology*, *14*(4), 1001–1013. doi:10.1046/j.1523-1739.2000.99125.x

Akpodiogaga-a, P., & Odjugo, O. (2010). General overview of climate change impacts in Nigeria. *Journal of Human Ecology (Delhi, India)*, *29*(1), 47–55. doi:10.1080/09709274.2010.11906248

Allen, W. (2000). Restoring Hawaii's Dry Forests: Research on Kona slope shows promise for native ecosystem recovery. *Bioscience*, *50*(12), 1037–1041. doi:10.1641/0006-3568(2000)050[1037:RHSDF]2.0.CO;2

Alley, R. B., Marotzke, J., Nordhaus, W. D., Overpeck, J. T., Peteet, D. M., Pielke, R. A., ... Wallace, J. M. (2003). Abrupt climate change. *Science*, *299*(5615), 2005–2010. doi:10.1126cience.1081056 PMID:12663908

Amissah, L., Mohren, G. M., Kyereh, B., Agyeman, V. K., & Poorter, L. (2018). Rainfall seasonality and drought performance shape the distribution of tropical tree species in Ghana. *Ecology and Evolution*, *8*(16), 8582–8597. doi:10.1002/ece3.4384 PMID:30250725

Appropedia. (2012). Amazon deforestation, climate change, and possible mitigation. Retrieved from http://www.appropedia.org/Amazon_deforestation,_climate_change,_and_possible_mitigation

Araújo, M. B., & Rahbek, C. (2006). How does climate change affect biodiversity? *Science*, *313*(5792), 1396–1397. doi:10.1126cience.1131758 PMID:16959994

Bala, G., Caldeira, K., Wickett, M., Phillips, T. J., Lobell, D. B., Delire, C., & Mirin, A. (2007). Combined climate and carbon-cycle effects of large-scale deforestation. *Proceedings of the National Academy of Sciences of the United States of America*, *104*(16), 6550–6555. doi:10.1073/pnas.0608998104 PMID:17420463

Baobab Tree Facts. (2015). 9 Fascinating baobab tree facts, Africa Geographic Editorial, Research, Travel & Lifestyle, Wildlife. *The Decoding Science*. Retrieved from https://africageographic.com/blog/9-fascinating-baobab-tree-facts/

Barlow, J., & Peres, C. A. (2004). Ecological responses to El Niño–induced surface fires in central Brazilian Amazonia: Management implications for flammable tropical forests. *Philosophical Transactions of the Royal Society of London. Series B, Biological Sciences*, *359*(1443), 367–380. doi:10.1098/rstb.2003.1423 PMID:15212091

Becknell, J. M., Kucek, L. K., & Powers, J. S. (2012). Aboveground biomass in mature and secondary seasonally dry tropical forests: A literature review and global synthesis. *Forest Ecology and Management*, *276*, 88–95. doi:10.1016/j.foreco.2012.03.033

Bhadouria, R., Singh, R., Srivastava, P., & Raghubanshi, A. S. (2016). Understanding the ecology of tree-seedling growth in dry tropical environment: A management perspective. *Energy, Ecology & Environment*, *1*(5), 296–309.

Bhadouria, R., Singh, R., Srivastava, P., Tripathi, S., & Raghubanshi, A. S. (2017b). Interactive effect of water and nutrient on survival and growth of tree seedlings of four dry tropical tree species under grass competition. *Tropical Ecology*, *58*(3), 611–621.

Bhadouria, R., Srivastava, P., Singh, R., Tripathi, S., Singh, H., & Raghubanshi, A. S. (2017a). Tree seedling establishment in dry tropics: An urgent need of interaction studies. *Environment Systems & Decisions*, *37*(1), 88–100. doi:10.100710669-017-9625-x

Blasco, F., Whitmore, T. C., & Gers, C. (2000). A framework for the worldwide comparison of tropical woody vegetation types. *Biological Conservation*, *95*(2), 175–189. doi:10.1016/S0006-3207(00)00032-X

Borchert, R. (1999). Climatic periodicity, phenology, and cambium activity in tropical dry forest trees. *IAWA Journal*, *20*(3), 239–247. doi:10.1163/22941932-90000687

Brodie, J., Post, E., & Doak, D. (Eds.). (2012). *(a). Climate change and wildlife conservation*. Chicago, IL: University Chicago Press. doi:10.7208/chicago/9780226074641.001.0001

Brodie, J., Post, E., & Laurance, W. F. (2012). (b). Climate change and tropical biodiversity: A new focus. *Trends in Ecology & Evolution*, *27*(3), 145–150. doi:10.1016/j.tree.2011.09.008 PMID:21975172

Brown, S., & Lugo, A. E. (1982). The storage and production of organic matter in tropical forests and their role in the global carbon cycle. *Biotropica*, *14*(3), 161–187. doi:10.2307/2388024

Bruegmann, M. M. (1996). Hawaii's dry forests. *Endangered Species Bulletin*, *11*, 26–27.

Buytaert, W., Cuesta-Camacho, F., & Tobón, C. (2011). Potential impacts of climate change on the environmental services of humid tropical alpine regions. *Global Ecology and Biogeography*, *20*(1), 19–33. doi:10.1111/j.1466-8238.2010.00585.x

Cabin, R. J., Weller, S. G., Lorence, D. H., Flynn, T. W., Sakai, A. K., Sandquist, D., & Hadway, L. J. (2000). Effects of long-term ungulate exclusion and recent alien species control on the preservation and restoration of a Hawaiian tropical dry forest. *Conservation Biology*, *14*(2), 439–453. doi:10.1046/j.1523-1739.2000.99006.x

CBD. (2009). Connecting Biodiversity and Climate Change Mitigation and Adaptation: Report of the Second Ad Hoc Technical Expert Group on Biodiversity and Climate Change. Technical Series No. 41, Secretariat of the Convention on Biological Diversity, Montreal, Canada, 126.

Change, G. C. (2007). Impacts and Adaptation. *Nature Conservancy*, *434*, 951–0569.

Colwell, R. K., Brehm, G., Cardelús, C. L., Gilman, A. C., & Longino, J. T. (2008). Global warming, elevational range shifts, and lowland biotic attrition in the wet tropics. *Science*, *322*(5899), 258–261. doi:10.1126cience.1162547 PMID:18845754

Control Invasive Plants. (2019). Retrieved from https://www.coastalisc.com/how-to-control-invasive-plants

Control methods. (2019). Retrieved from https://dnr.wi.gov/topic/Invasives/control.html

Crowley, T. J. (2000). Causes of climate change over the past 1000 years. *Science*, *289*(5477), 270–277. doi:10.1126cience.289.5477.270 PMID:10894770

Daniel, T., Jim, O., Jeff, W., & Ollivier, G. (2014). Tropical dry forest under threat & under-researched. CIFOR and CGIAR. Retrieved from http://copa.acguanacaste.ac.cr:8080/bitstream/handle/11606/590/Tropical%20Dry%20Forest%20Under%20Threat%20&%20Under-Researched.pdf?sequence=1

Day, M., Gumbo, D., Moombe, K., Wijaya, A., & Sunderland, T. (2014). Zambia Country Profile: Monitoring, reporting and verification for REDD+.CIFOR Occasional Paper # 113.

Dexter, K. G., Smart, B., Baldauf, C., Baker, T. R., Bessike Balinga, M. P., & Brienen, R. J. W., & Pennington, R. T. (2015). Floristics and biogeography of vegetation in season-ally dry tropical regions. International Forestry Review – Special Issue. *Global Dry Forests, 17*(S2), 10–32.

Dinerstein, E., Olson, D. J., Graham, D. M., Webster, A. L., Primm, S. A., Bookbinder, M. P., & Ledec, G. (1995). *A conservation assessment of the terrestrial ecoregions of Latin America and the Caribbean*. Washington, DC: The World Bank. doi:10.1596/0-8213-3295-3

Ding, Y., Ren, G., Zhao, Z., Xu, Y., Luo, Y., Li, Q., & Zhang, J. (2007). Detection, causes and projection of climate change over China: An overview of recent progress. *Advances in Atmospheric Sciences, 24*(6), 954–971. doi:10.100700376-007-0954-4

Discover. (2019). The Baobab Tree: Africa's Iconic "Tree of Life". Retrieved from https://aduna.com/blogs/learn/the-baobab-tree

Dry Forest Ecology. (2019). Retrieved from http://staging.ceiba.org/wp-content/uploads/Dry-Forest-Ecology.pdf

Durán, E. (2004). Estructura, diversidad y mortalidad de los árboles del dosel del bosque tropical seco. (Dissertation). Universidad Nacional Autónoma de México, México, D.F.

Elliott, S. D., Blakesley, D., & Hardwick, K. (2013). *Restoring tropical forests: a practical guide* (p. 344). Kew, UK: Royal Botanic Gardens.

FAO. (2001). FAO Forestry Paper: Vol. 140. *Global forest resources assessment. Main report*. Rome, Italy: Food and Agriculture Organization of the United Nations.

Forest restoration. (2019). Retrieved from https://en.wikipedia.org/wiki/Forest_restoration#cite_note-10

Furley, P. (1992). Edaphic changes at the forest-savanna boundary with particular reference to the Neotropics. In P. Furley, J. Ratter, & J. Proctor (Eds.), *Nature and Dynamics of Forest-Savanna Boundaries* (pp. 91–117). London, UK: Chapman and Hall.

Gashaw, T., Mebrat, W., Hagos, D., & Nigussie, A. (2014). Climate change adaptation and mitigation measures in Ethiopia. *J Biol Agric Healthc*, 2225-093.

Gentry, A. H. (1995). Diversity and floristic composition of Neotropical dry forests. In S. H. Bullock, H. A. Mooney, & E. Medina (Eds.), *Seasonally Dry Tropical Forests* (pp. 146–194). Cambridge, UK: Cambridge University Press. doi:10.1017/CBO9780511753398.007

Gerhardt, K., & Hytteborn, H. (1992). Natural dynamics and regeneration methods in tropical dry forests-an introduction. *Journal of Vegetation Science, 3*(3), 361–364. doi:10.2307/3235761

Grove, R. H. (1994). A historical review of early institutional and conservationist responses to fears of artificially induced global climate change: The deforestation-desiccation discourse 1500–1860. *Chemosphere, 29*(5), 1001–1013. doi:10.1016/0045-6535(94)90165-1

Halley, E. (1694). An Account of the Watery Circulation of the Sea, and of the Cause of Springs. *Philosophical Transactions of the Royal Society, 142*(7), 468–472.

Harris, A. (2019). How Does Climate Change Affect the Temperature in a Deciduous Forest? Hearst newspaper, Hearst Seattle Media, LLC. Retrieved from https://education.seattlepi.com/climate-change-affect-temperature-deciduous-forest-6755.html

Hasnat, G. N. T., Kabir, M. A., & Hossain, M. A. (2018). Major environmental issues and problems of South Asia, Particularly Bangladesh. Handbook of Environmental Materials Management, 109-148.

Hoekstra, J. M., Boucher, T. M., Ricketts, T. H., & Roberts, C. (2005). Confronting a biome crisis: Global disparities of habitat loss and protection. *Ecology Letters*, *8*(1), 23–29. doi:10.1111/j.1461-0248.2004.00686.x

Hulme, M., & Viner, D. (1995). *A climatic change scenario for assessing the impact of climate change on tropical rain forests* (p. 34). US: Climatic Research Unit for WWF.

Hussner, A., Stiers, I., Verhofstad, M. J. J. M., Bakker, E. S., Grutters, B. M. C., Haury, J., ... Anderson, L. W. J. (2017). Management and control methods of invasive alien freshwater aquatic plants: A review. *Aquatic Botany*, *136*, 112–137. doi:10.1016/j.aquabot.2016.08.002

Intergovernmental Panel on Climate Change (IPCC) (2007). The fourth assessment report (AR4). Synthesis report for policymakers.

IPCC. (2007). Summary for policy makers. In S. Solomon et al. (Eds.), Climate Change 2007: the Physical Science Basis (pp. 1-18). Contribution of working group I to the fourth assessment report of the Intergovernmental Panel on Climate Change. Cambridge, UK: Cambridge University Press.

IUCN. (2001). IUCN Red List categories and criteria: Version 3.1. IUCN Species Survival Commission. Gland, Switzerland: IUCN.

IUCN. (2003). Guidelines for application of IUCN Red List criteria at regional levels: Version 3.0. IUCN Species Survival Commission. Gland, Switzerland: IUCN.

Janzen, D. H. (1967). Synchronization of sexual reproduction of trees within the dry season in Central America. *Evolution; International Journal of Organic Evolution*, *21*(3), 620–637. doi:10.1111/j.1558-5646.1967.tb03416.x PMID:28563685

Janzen, D. H. (1982). (a). Cenizero tree (Leguminosae: *Pithecellobium saman*) delayed fruit development in Costa Rican deciduous forests. *American Journal of Botany*, *69*(8), 1269–1276. doi:10.1002/j.1537-2197.1982.tb13372.x

Janzen, D. H. (1982). (b). Variation in average seed size and fruit seediness in a fruit crop of a Guanacaste tree (Leguminosae: *Enterolobium cyclocarpum*). *American Journal of Botany*, *69*(7), 1169–1178. doi:10.1002/j.1537-2197.1982.tb13361.x

Janzen, D. H. (1988). Tropical dry forest. The most endangered major tropical ecosystems. In E. O. Wilson (Ed.), *Biodiversity* (pp. 130–137). Washington, DC: National Academy Press.

Julian, A. (2018). The Adaptations of the Baobab Tree. Retrieved from https://sciencing.com/adaptations-baobab-tree-8344865.html

Juvik, S. P., & Juvik, J. O. (Eds.). (1998). *Atlas of Hawaii* (3rd ed.). Honolulu, HI: University of Hawaii Press.

Kinver, M. (2016). Dry tropical forests 'overlooked and under threat'. Science and Environment, BBC News. Retrieved from https://www.bbc.com/news/science-environment-37440485

Kirilenko, A. P., & Sedjo, R. A. (2007). Climate change impacts on forestry. *Proceedings of the National Academy of Sciences of the United States of America, 104*(50), 19697–19702. doi:10.1073/pnas.0701424104 PMID:18077403

Krause, G. H., Winter, K., Krause, B., Jahns, P., García, M., Aranda, J., & Virgo, A. (2010). High-temperature tolerance of a tropical tree, *Ficus insipida*: Methodological reassessment and climate change considerations. *Functional Plant Biology*, *37*(9), 890–900. doi:10.1071/FP10034

Laurance, W. F. (2008). Can carbon trading save vanishing forests? *Bioscience*, *58*(4), 286–287. doi:10.1641/B580402

Laurance, W. F., Useche, D. C., Shoo, L. P., Herzog, S. K., Kessler, M., Escobar, F., ... Hietz, P. (2011). Global warming, elevational ranges and the vulnerability of tropical biota. *Biological Conservation, 144*(1), 548–557. doi:10.1016/j.biocon.2010.10.010

Leemans, R. (1999). *Possible changes in natural vegetation patterns due to global warming*. Laxenburg, Austria: International Institute of Applied Systems Analysis.

Liu, H., Park Williams, A., Allen, C. D., Guo, D., Wu, X., Anenkhonov, O. A., ... Badmaeva, N. K. (2013). Rapid warming accelerates tree growth decline in semi-arid forests of Inner Asia. *Global Change Biology*, *19*(8), 2500–2510. doi:10.1111/gcb.12217 PMID:23564688

Lloyd, J., & Farquhar, G. D. (2008). Effects of rising temperatures and [CO_2] on the physiology of tropical forest trees. *Philosophical Transactions of the Royal Society of London. Series B, Biological Sciences, 363*(1498), 1811–1817. doi:10.1098/rstb.2007.0032 PMID:18267901

Locatelli, B., & Pramova, E. (2011). Beyond mitigation: Forest-based adaptation to climate change. FAO and UNEP Meeting on Forests and Climate Change Adaptation in Asia, Oct. 26, 2011, Bangkok, Thailand. Retrieved from http://agritrop.cirad.fr/564596/1/document_564596.pdf

Lugo, A. E., & Helmer, E. (2004). Emerging forests on abandoned land: Puerto Rico's new forests. *Forest Ecology and Management*, *190*(2-3), 145–161. doi:10.1016/j.foreco.2003.09.012

Maass, J. M., Balvanera, P., Castillo, A., Daily, G. C., Mooney, H. A., Ehrlich, P., Quesada, M., ... Martínez-Yrizar, A. (2005). Ecosystem services of tropical dry forests: insights from longterm ecological and social research on the Pacific Coast of Mexico. *Ecology and society: a journal of integrative science for resilience and sustainability, 10*(1), 1-23.

Mace, G. M., & Lande, R. (1991). Assessing extinction threats: Toward a reevaluation of IUCN threatened species categories. *Conservation Biology*, *5*(2), 148–157. doi:10.1111/j.1523-1739.1991.tb00119.x

Malhi, Y., Roberts, J. T., Betts, R. A., Killeen, T. J., Li, W., & Nobre, C. A. (2008). Climate change, deforestation, and the fate of the Amazon. *Science*, *319*(5860), 169–172. doi:10.1126cience.1146961 PMID:18048654

Marengo, J., Nobre, C. A., Betts, R. A., Cox, P. M., Sampaio, G., & Salazar, L. (2009). Global warming and climate change in Amazonia: Climate-vegetation feedback and impacts on water resources. *Amazonia and Global Change*, *186*, 273–292. doi:10.1029/2008GM000743

Mason-Romo, E. D., Ceballos, G., Lima, M., Martínez-Yrízar, A., Jaramillo, V. J., & Maass, M. (2018). Long-term population dynamics of small mammals in tropical dry forests, effects of unusual climate events, and implications for management and conservation. *Forest Ecology and Management*, *426*, 123–133. doi:10.1016/j.foreco.2018.05.058

Medina, E. (1983). Adaptations of tropical trees to moisture stress. In F. B. Golley (Ed.), *Tropical rain forest ecosystems, A structure and function* (pp. 225–237). Amsterdam, The Netherlands: Elsevier Scientific Company Publishing.

Miles, L., Newton, A. C., DeFries, R. S., Ravilious, C., May, I., Blyth, S., ... Gordon, J. E. (2006). A global overview of the conservation status of tropical dry forests. *Journal of Biogeography*, *33*(3), 491–505. doi:10.1111/j.1365-2699.2005.01424.x

Moline, A. (1999). *Tropical dry forest restoration in the Guanacaste Conservation Area*. Costa Rica.

Mooney, H. A., Bullock, S. H., & Medina, E. (1995). Introduction. In S. H. Bullock, H. A. Mooney, & E. Medina (Eds.), *Seasonally dry tropical forests* (pp. 1–8). Cambridge, UK: Cambridge University Press. doi:10.1017/CBO9780511753398.001

Mueller-Dombois, D. (1981) Fire in tropical ecosystems. In *Proceedings of the Conference: Fire Regimes and Ecosystem Properties*. Washington, DC: U.S. Department of Agriculture, General Technical Report WO-26, 137–176.

Murdiyarso, D., Hergoualc'h, K., & Verchot, L. V. (2010). Opportunities for reducing greenhouse gas emissions in tropical peatlands. *Proceedings of the National Academy of Sciences of the United States of America*, *107*(46), 19655–19660. doi:10.1073/pnas.0911966107 PMID:21081702

Murdiyarso, D., & Kauffman, J. B. (2011). *Addressing climate change adaptation and mitigation in tropical wetland ecosystems of Indonesia* (Vol. 41). Center for International Forestry Research (CIFOR). Retrieved from http://citeseerx.ist.psu.edu/viewdoc/download?doi=10.1.1.372.448&rep=rep1&type=pdf

Murphy, P. G., & Lugo, A. E. (1986). Ecology of tropical dry forest. *Annual Review of Ecology and Systematics*, *17*(1), 67–88. doi:10.1146/annurev.es.17.110186.000435

Murthy, I. K., Tiwari, R., & Ravindranath, N. H. (2011). Climate change and forests in India: Adaptation opportunities and challenges. *Mitigation and Adaptation Strategies for Global Change*, *16*(2), 161–175. doi:10.100711027-010-9261-y

Myers, J. H., & Bazely, D. (2003). *Ecology and control of introduced plants* (Vol. 313). Cambridge, UK: Cambridge University Press. doi:10.1017/CBO9780511606564

Nepstad, D., Lefebvre, P., Lopes da Silva, U., Tomasella, J., Schlesinger, P., Solórzano, L., ... Guerreira Benito, J. (2004). Amazon drought and its implications for forest flammability and tree growth: A basin-wide analysis. *Global Change Biology*, *10*(5), 704–717. doi:10.1111/j.1529-8817.2003.00772.x

Nepstad, D. C., de Carvalho, C. R., Davidson, E. A., Jipp, P. H., Lefebvre, P. A., Negreiros, G. H., ... Vieira, S. (1994). The role of deep roots in the hydrological and carbon cycles of Amazonian forests and pastures. *Nature*, *372*(6507), 666–669. doi:10.1038/372666a0

Nepstad, D. C., Stickler, C. M., Filho, B. S., & Merry, F. (2008). Interactions among Amazon land use, forests and climate: Prospects for a near-term forest tipping point. *Philosophical Transactions of the Royal Society of London. Series B, Biological Sciences*, *363*(1498), 1737–1746. doi:10.1098/rstb.2007.0036 PMID:18267897

Oliveira, R. S., Dawson, T. E., Burgess, S. S., & Nepstad, D. C. (2005). Hydraulic redistribution in three Amazonian trees. *Oecologia*, *145*(3), 354–363. doi:10.100700442-005-0108-2 PMID:16091971

Olson, D. M., & Dinerstein, E. (1998). The Global 200: A representation approach to conserving the Earth's most biologically valuable ecoregions. *Conservation Biology*, *12*(3), 502–515. doi:10.1046/j.1523-1739.1998.012003502.x

Olson, D. M., Dinerstein, E., Abell, R., Allnutt, T., Carpenter, C., McClenachan, L., ... Thieme, M. (2000). *The global 200: a representation approach to conserving the Earth's distinctive ecoregions*. Washington, DC: Conservation Science Program, World Wildlife Fund-US.

Olson, D. M., Dinerstein, E., Wikramanayake, E. D., Burgess, N. D., Powell, G. V. N., Underwood, E. C., ... Kassem, K. R. (2001). Terrestrial ecoregions of the world: A new map of life on Earth. *Bioscience*, *51*(11), 933–938. doi:10.1641/0006-3568(2001)051[0933:TEOTWA]2.0.CO;2

Parmesan, C. (2006). Ecological and evolutionary responses to recent climate change. *Annual Review of Ecology Evolution and Systematics*, *37*(1), 637–669. doi:10.1146/annurev.ecolsys.37.091305.110100

Parry, J., Hammill, A., & Drexhage, J. (2005). *Climate Change and Adaptation. International Institute for Sustainable Development*. Winnipeg, Canada: IISD.

Pascoe, J. (2018). Tropical dry forests 'canary in the coal mine' on climate change. Retrieved from https://phys.org/news/2018-03-tropical-forests-canary-coal-climate.html

Pennington, R. T., Lewis, G. P., & Ratter, J. A. (2006). An overview of the plant diversity, biogeography and conservation of neotropical savannas and seasonally dry forests. In R. T. Pennington, G. P. Lewis, & J. A. Ratter (Eds.), *Neotropical savannas and seasonally dry forests: Plant diversity, biogeography and conservation* (pp. 1–29). Boca Raton, FL: CRC Press. doi:10.1201/9781420004496-1

Pennington, T., Prado, D. E., & Pendry, C. A. (2000). Neotropical seasonally dry forests and Quaternary vegetation changes. *Journal of Biogeography*, *27*(2), 261–273. doi:10.1046/j.1365-2699.2000.00397.x

Peterson, A. T., & Kluza, D. A. (2005). Ecological niche modeling as a new paradigm for large-scale investigations of diversity and distribution of birds. In J. C. Ralph & T. D. Rich (Eds.), Proceedings of the Third International Partners in Flight Conference, 2002, Asilomar (pp. 1201-1204). Albany: U.S. Department of Agriculture Forest Service.

Peterson, A. T., Sánchez-Cordero, V., Soberón, J., Bartley, J., Buddemeier, R. W., & Navarro-Sigüenza, A. G. (2001). Effects of global climate change on geographic distributions of Mexican Cracidae. *Ecological Modelling*, *144*(1), 21–30. doi:10.1016/S0304-3800(01)00345-3

Phillips, O. L., Van Der Heijden, G., Lewis, S. L., López-González, G., Aragão, L. E., Lloyd, J., ... Amaral, I. (2010). Drought–mortality relationships for tropical forests. *The New Phytologist*, *187*(3), 631–646. doi:10.1111/j.1469-8137.2010.03359.x PMID:20659252

Portillo-Quintero, C. A., & Sánchez-Azofeifa, G. A. (2010). Extent and conservation of tropical dry forests in the Americas. *Biological Conservation*, *143*(1), 144 155. doi:10.1016/j.biocon.2009.09.020

Portillo-Quintero, C. A., & Sánchez-Azofeifa, G. A. (2010). Extent and conservation of tropical dry forests in the Americas. *Biological Conservation*, *143*(1), 144–155. doi:10.1016/j.biocon.2009.09.020

Post, E., Forchhammer, M. C., Bret-Harte, M. S., Callaghan, T. V., Christensen, T. R., Elberling, B., ... Ims, R. A. (2009). Ecological dynamics across the Arctic associated with recent climate change. *Science*, *325*(5946), 1355–1358. doi:10.1126cience.1173113 PMID:19745143

Pulla, S., Ramaswami, G., Mondal, N., Chitra-Tarak, R., Suresh, H. S., Dattaraja, H. S., ... Sukumar, R. (2015). Assessing the resilience of global seasonally dry tropical forests. *International Forestry Review*, *17*(2), 91–113. doi:10.1505/146554815815834796

Ratter, J. A., Richards, P. W., Argent, G., & Gifford, D. R. (1973). Observations on the vegetation of northeastern Mato Grosso: I. The woody vegetation types of the Xavantina-Cachimbo Expedition area. *Philosophical Transactions of the Royal Society of London. Series B, Biological Sciences*, *266*(880), 449–492. doi:10.1098/rstb.1973.0053

Reynolds, J. D., & Mace, G. M. (1999). Risk assessments of threatened species. *Trends in Ecology & Evolution*, *14*(6), 215–217. doi:10.1016/S0169-5347(99)01629-8

Richard, P. (1980). PIELOU, E. C. (1979): *Biogeography*. John Wiley & Sons, ix + 351 p., ill., index, 26,90$. *Géographie physique et Quaternaire*, *34*(2), 263–263. doi:10.7202/1000411ar

Ricketts, T. H., Dinerstein, E., Olson, D. M., Eichbaum, W., Loucks, C. J., Kavanaugh, K., ... Carney, K. (1999). *Terrestrial ecoregions of North America: a conservation assessment* (Vol. 1). Washington, DC: Island Press.

Rodrigues, P. M. S., Silva, J. O., Eisenlohr, P. V., & Schaefer, C. E. G. R. (2015). Climate change effects on the geographic distribution of specialist tree species of the Brazilian tropical dry forests. *Brazilian Journal of Biology*, *75*(3), 679–684. doi:10.1590/1519-6984.20913 PMID:26465729

Salazar, L. F., Nobre, C. A., & Oyama, M. D. (2007). Climate change consequences on the biome distribution in tropical South America. *Geophysical Research Letters*, *34*(9). doi:10.1029/2007GL029695

Sánchez-Azofeifa, G. A., Quesada, M., Rodríguez, J. P., Nassar, J. M., Stoner, K. E., Castillo, A., ... Cuevas-Reyes, P. (2005). Research priorities for Neotropical dry forests. *Biotropica: The Journal of Biology and Conservation*, *37*(4), 477–485.

Santos, M. G., Oliveira, M. T., Figueiredo, K. V., Falcão, H. M., Arruda, E. C. P., Almeida-Cortez, J., ... Antonino, A. C. D. (2014). Caatinga, the Brazilian dry tropical forest: Can it tolerate climate changes? *Theoretical and Experimental Plant Physiology*, *26*(1), 83–99. doi:10.100740626-014-0008-0

Schmitz, O. J., Post, E., Burns, C. E., & Johnston, K. M. (2003). Ecosystem responses to global climate change: Moving beyond color mapping. *Bioscience, 53*(12), 1199–1205. doi:10.1641/0006-3568(2003)053[1199:ERTGCC]2.0.CO;2

Septer, J. D. (2019). Tropical dry forest plants. Retrieved from https://www.hunker.com/13427136/tropical-dry-forest-plants

Shanahan, M., Shubert, W., Scherer, C., & Corcoran, T. (2013). Climate change in Africa: a guidebook for journalists. In F. Banda (Ed.), *UNESCO Series on Journalism Education.*

Sharkey, T. D., & Schrader, S. M. (2006). High temperature stress. In *K. V. M. Rao, A. S. Raghavendra, & K. J. Reddy (Link), Physiology and molecular biology of stress tolerance in plants* (pp. 101–129). Dordrecht, The Netherlands: Springer. doi:10.1007/1-4020-4225-6_4

Shono, K., Cadaweng, E. A., & Durst, P. B. (2007). Application of assisted natural regeneration to restore degraded tropical forestlands. *Restoration Ecology, 15*(4), 620–626. doi:10.1111/j.1526-100X.2007.00274.x

Silver, W. L. (1998). The potential effects of elevated CO_2 and climate change on tropical forest soils and biogeochemical cycling. *Climatic Change, 39*(2/3), 337–361. doi:10.1023/A:1005396714941

Singh, R., Singh, H., Singh, S., Afreen, T., Upadhyay, S., Singh, A. K., ... Raghubanshi, A. S. (2017). Riparian land uses affect the dry season soil CO_2 efflux under dry tropical ecosystems. *Ecological Engineering, 100*, 291–300. doi:10.1016/j.ecoleng.2017.01.002

Skutsch, M. M., & Ba, L. (2010). Crediting carbon in dry forests: The potential for community forest management in West Africa. *Forest Policy and Economics, 12*(4), 264–270. doi:10.1016/j.forpol.2009.12.003

Stavins, R., & Richards, K. (2005). *The cost of U.S. forest-based carbon sequestration.* USA: Prepared for the Pew Center on Global Climate Change.

Sunderland, T., Apgaua, D., Baldauf, C., Blackie, R., Colfer, C., Cunningham, A. B., ... Ickowitz, A. (2015). Global dry forests: A prologue. *International Forestry Review, 17*(2), 1–9. doi:10.1505/146554815815834813

Suttle, K. B., Thomsen, M. A., & Power, M. E. (2007). Species interactions reverse grassland responses to changing climate. *Science, 315*(5812), 640–642. doi:10.1126cience.1136401 PMID:17272720

Tewksbury, J. J., Huey, R. B., & Deutsch, C. A. (2008). Putting the heat on tropical animals. *Science, 320*(5881), 1296–1297. doi:10.1126cience.1159328 PMID:18535231

Trejo, I., & Dirzo, R. (2002). Floristic diversity of Mexican seasonally dry tropical forests. *Biodiversity and Conservation, 11*(11), 2063–2084. doi:10.1023/A:1020876316013

Tropical Dry Forest. (2019). Retrieved from https://www.pugetsound.edu/academics/academic-resources/slater-museum/biodiversity-resources/world-biomes/characteristics-of-bioclimatic/tropical-dry-forest/

Tropical Dry Forests. (2019). Retrieved from https://ceiba.org/conservation/dry-forests/

Udvardy, M. D. F. (1975). A classification of the biogeographical provinces of the world. IUCN Occasional Paper 18. International Union of Conservation of Nature and Natural Resources, Morges, Switzerland.

Vargas, R., Allen, M. F., & Allen, E. B. (2008). Biomass and carbon accumulation in a fire chronosequence of a seasonally dry tropical forest. *Global Change Biology*, *14*(1), 109–124.

Vico, G. (2017). Climate changes in the tropics may be an advantage to drought-deciduous plants, Tropical dry forests. Sveriges lantbruks universitet (SLU) news. Retrieved from https://www.slu.se/en/ew-news/2017/7/tropical-dry-forests/

Vose, J. M., & Maass, J. M. (1999). A Comparative analysis of hydrologic responses of tropical deciduou s and temperate deciduous watershed ecosystems to climatic change. In C. Aguirre-Bravo, & C. Rodríguez-Franco (Eds.), *Proceedings of the North American Science Symposium: Toward a Unified Framework for Inventorying and Monitoring Forest Ecosystem Resources. RMRSP-12*:292–298. Guadalajara, México: USDA Forest Service.

Wachman, M. (2019). Tropical Dry Forest Biome Plants. Retrieved from https://www.gardenguides.com/12451445-tropical-dry-forest-biome-plants.html

Wagner, N. (2018). Types of Plants & Animals in a Tropical Dry Forest. USA Today. Retrieved from https://traveltips.usatoday.com/types-plants-animals-tropical-dry-forest-61382.html

Werneck, F. P., Costa, G. C., Colli, G. R., Prado, D. E., & Sites, J. W. Jr. (2011). Revisiting the historical distribution of Seasonally Dry Tropical Forests: New insights based on palaeodistribution modelling and palynological evidencegeb. *Global Ecology and Biogeography*, *20*(2), 272–288. doi:10.1111/j.1466-8238.2010.00596.x

WhatA. S. T. D. F. P. (2019). Retrieved from https://www.reference.com/science/tropical-dry-forest-plants-9b9362d1a1fd6f2a

Wikramanayake, E., Dinerstein, E., Loucks, C. J., Olson, D. M., Morrison, J., Lamoreux, J., ... Hedao, P. (2002). *Terrestrial ecoregions of the Indo-Pacific. A conservation assessment. Conservation Science Program, WWF-US, Washington, DC*: Island Press.

Wilkie, D. S., & Mark, C. T. (1993). Biogeophysical Setting and Global Climate Change. In Central Africa: Global Climate Change and Development. Technical Repon. Washington, DC: Biodiversity Support Program (a consortium of World Wildlife Fund, The Nature Conservancy, and World Resources Institute).

Wright, S. J., Muller-Landau, H. C., & Schipper, J. A. N. (2009). The future of tropical species on a warmer planet. *Conservation Biology*, *23*(6), 1418–1426. doi:10.1111/j.1523-1739.2009.01337.x PMID:20078642

WWF. (2009). *Climate Change: Confronting a Global Challenge*. USA: World Wildlife Fund.

WWF. (2018). *Building a future in which humans live in harmony with nature*. USA: World Wildlife Fund.

ADDITIONAL READING

Bhadouria, R., Singh, R., Srivastava, P., & Raghubanshi, A. S. (2016). Understanding the ecology of tree-seedling growth in dry tropical environment: A management perspective. *Energy. Ecology & Environment, 1*(5), 296–309.

Bhadouria, R., Singh, R., Srivastava, P., Tripathi, S., & Raghubanshi, A. S. (2017b). Interactive effect of water and nutrient on survival and growth of tree seedlings of four dry tropical tree species under grass competition. *Tropical Ecology, 58*(3), 611–621.

Bhadouria, R., Srivastava, P., Singh, R., Tripathi, S., Singh, H., & Raghubanshi, A. S. (2017a). Tree seedling establishment in dry tropics: An urgent need of interaction studies. *Environment Systems & Decisions, 37*(1), 88–100. doi:10.100710669-017-9625-x

Blasco, F., Whitmore, T. C., & Gers, C. (2000). A framework for the worldwide comparison of tropical woody vegetation types. *Biological Conservation, 95*(2), 175–189. doi:10.1016/S0006-3207(00)00032-X

Gentry, A. H. (1995). Diversity and floristic composition of Neotropical dry forests. In S. H. Bullock, H. A. Mooney, & E. Medina (Eds.), *Seasonally Dry Tropical Forests* (pp. 146–194). Cambridge, UK: Cambridge University Press. doi:10.1017/CBO9780511753398.007

IPCC. (2007). Summary for policy makers. In S. Solomon et al., (Eds.), Climate Change 2007: the Physical Science Basis (pp. 1-18). Contribution of working group I to the fourth assessment report of the Intergovernmental Panel on Climate Change, Cambridge University Press.

Janzen, D. H. (1988). Tropical dry forest. The most endangered major tropical ecosystems. In E. O. Wilson (Ed.), *Biodiversity* (pp. 130–137). Washington, DC: National Academy Press.

Miles, L., Newton, A. C., DeFries, R. S., Ravilious, C., May, I., Blyth, S., ... Gordon, J. E. (2006). A global overview of the conservation status of tropical dry forests. *Journal of Biogeography, 33*(3), 491–505. doi:10.1111/j.1365-2699.2005.01424.x

Mooney, H. A., Bullock, S. H., & Medina, E. (1995). Introduction. In S. H. Bullock, H. A. Mooney, & E. Medina (Eds.), *Seasonally dry tropical forests* (pp. 1–8). Cambridge: Cambridge University Press. doi:10.1017/CBO9780511753398.001

Singh, R., Singh, H., Singh, S., Afreen, T., Upadhyay, S., Singh, A. K., ... Raghubanshi, A. S. (2017). Riparian land uses affect the dry season soil CO_2 efflux under dry tropical ecosystems. *Ecological Engineering, 100*, 291–300. doi:10.1016/j.ecoleng.2017.01.002

Section 2

Plant Functional Traits

Chapter 4
Plant Functional Traits in Tropical Dry Forests:
A Review

Shipra Singh
Jawaharlal Nehru University, India

Abhishek K. Verma
Jawaharlal Nehru University, India

ABSTRACT

Plants have certain characteristics which allow them to respond to various environmental conditions, like changes in climate, water scarcity in the soil, lack of minerals; among others. In some of these traits, the responses to climatic phenomena such as drought can be evidenced through morphological adaptations (spines, succulent tissues, trichomes) or physiological adaptations (regulation of water potential at the cellular level, the concentration of nutrients, etc.). A systematic literature review was performed to study plant functional traits (PFTs) in tropical dry forests (TDFs). The chapter suggests the role of functional traits in community dynamics and processes. The authors will also highlight the limitations of PFTs in TDFs and how they can be improved.

INTRODUCTION

Plants influence ecosystem functioning and processes across local to global scales and are one of the primary organisms that perform photosynthesis via solar radiation. They also play a substantial role in nutrient cycling and contribute to provisioning, regulatory, cultural, and supporting services (Grime 1998; Naeem et al. 2012). Plants have certain characteristics that allow them to respond to various environmental conditions like changes in climate, water loss in the soil, and lack of minerals, among others. In some of these traits, plant responses to climatic phenomena, such as drought, can be evidenced through morphological adaptations (e.g. regulation of water potential at the cellular level, concentration of nutrients, etc.). Plants also contribute to a more biodiverse ecosystem, which comprises three main

DOI: 10.4018/978-1-7998-0014-9.ch004

components that complement ecosystem functioning and services: (i) taxonomic diversity that accounts for species composition and abundance; (ii) phylogenetic diversity that accounts for the evolutionary history of species, and (iii) functional diversity that accounts for the ecological traits of species. In a forest, taxonomic diversity, functional diversity and phylogenetic diversity support ecosystem functions (e.g., primary productivity, litter decomposition, nutrient cycling, etc.) and promote a wide range of ecosystem services (e.g., food production, carbon sequestration, climate regulation, etc.); (Mori et al. 2017). At the community level, functional properties are being extensively used to quantify ecosystem functions and services (Kuster et al. 2010).

The study of these functional properties that drive forest structure and dynamics is a major challenge in tropical ecosystems. Difficulties are primarily due to higher species richness, smaller populations, inadequate infrastructure, and fewer researchers conducting studies compared to the temperate zone. Tropical forests are also highly threatened by human activities, such as logging and conversion to agricultural lands (Stibig et al. 2014; Both et al. 2019). In Southeast Asian forests, a few species are extensively used for their commercial value, resulting in higher rates of degradation of these species. These extensively exhausted forests reflect slow vegetation recovery due to complex interactions among species, propagules and climatic conditions (Khurana and Singh 2001). The composition and successional traits of these forests (i.e. pioneer and non-pioneer) along with species' nitrogen fixing ability, tree size, habitat preferences, etc. may respond differently to the gradient of disturbance (e.g., logging, herbivory, grass competition) and available resources (e.g. light, water and nutrients) (Chapin et al. 2000; Tilman 1987) due to nutrient poor soils (Singh et al. 1989; Bhadouria et al. 2017). Water availability (Hulshof et al. 2013) and soil fertility (Buzzard et al. 2016) are two major features of tropical dry forests that can influence the functional composition of woody plant communities (Sfair et al. 2018). In the last decade, use of advanced techniques in spatial distribution of plant species to understand the functional properties as well as community dynamics, contributed immensely in the field of plant functional ecology (Getzin et al. 2006).

Tropical Dry Forests are considered to be the most exploited forests in the world due to increased anthropogenic pressure and also increased temperature leading to drought and stressful environment. This chapter as a whole, will add information regarding roles played by different traits and their effect, as well as how they are being affected by other factors in Tropical dry forest environment. The chapter will be helpful to readers in providing a comprehensive understanding of role of plant functional traits in Tropical Dry forest ecosystems. This chapter focuses on the traits that are commonly measured in tropical dry forests (i.e., leaf traits, stem traits, and regenerative traits). Topics that will be covered in this chapter are:

- Environmental gradients within Tropical Dry Forest
- Trait-based approaches
- Plant functional traits and ecosystem processes
- Leaf Economic Spectrum
- Wood Economic Spectrum
- Root Economic spectrum
- Seed Economic Spectrum
- Future trends and conclusions
    1. Present study is an attempt to highlight the importance of plant functional traits (PFTs) in Tropical dry forests.

2. PFTs are less explored in Tropical dry forests (TDFs) despite covering the substantial area in the world.
3. The interaction between plants and environmental variables could be estimated robustly through PFTs.
4. PFTs could be treated as an indicator of a plant's response to the environmental variable (biotic and abiotic).
5. The PFTs are an indicator of climatic impact (temperature, precipitation, moisture, etc.) on plants.
6. Leaf traits are one of the most studied traits because of its accessibility and being an indicator of several responses of plant towards climatic variables.
7. Woody traits are the most crucial factor to analyze carbon stock. Additionally, it is also used to study assemblage of species.
8. Root traits are the primary index for below-ground plant health, carbon allocation, and nutrient dynamics as well as an indicator of climatic response.
9. Seed traits are important to study the survival and establishment capacity of any community.

ENVIRONMENTAL GRADIENTS WITHIN TROPICAL DRY FOREST

Forests are a major component of the Earth's vegetation cover. The Forest survey of India (FSI) defines a forest as all the types of lands comprising an area of more than 1 ha, with a tree canopy density of more than 10%, irrespective of ownership and legal status (Ravindranath et al. 2014). Tropical forests account for about 52% of the world's forest cover (Holdridge 1967) and 30% of the world's land area (FAO 2008). Tropical Dry Forests (TDF) are the most dominant forest type, occupying 42% of the world's tropical forest area (Galicia et al. 2008). Out of which, India shares about 38.2% of the total Tropical Dry Forest cover in the world (MoEF 1999). TDFs are defined as forest vegetation that experiences a minimum dry season of 5–6 months per year with strong seasonal ecological processes and functions (Pennington et al. 2006). Long annual dry seasons in TDFs causes change in soil moisture availability, which is the primary driving variable for the functioning and processes of the forest ecosystem (Balvanera et al. 2010). Soil moisture is highly linked with diversity, distribution patterns, and biomass productivity of plant communities in TDFs (Enquist and Enquist 2011). However, trait-dependent strategies for dealing with water availability may also vary with other environmental variables (Allen et al. 2017). This creates significant differences in spatial heterogeneity that cause changes in important components of nutrient uptake strategies, such as Carbon and Nitrogen cycling and resource utilization methods like leaf phenology.

The differences in deciduousness to seasonal moisture variability are well observed within TDFs due to changes in land use history, forest composition and structure, and topography (Powers et al. 2009). A similar pattern occurs among TDFs due to differences in precipitation patterns, temperature, and incoming solar radiation. Topography plays a major role, causing changes in insolation and hydrological processes (Martínez-Yrízar et al. 2000). This affects intra- and inter-annual patterns of water, carbon and energy balance in TDFs (Bohlman 2010; Chaturvedi et al. 2011a). It also creates an uneven distribution of resources across the region, which is largely available at lower elevations (Daws et al. 2002; Markesteijn et al. 2010). Changes in precipitation patterns across time are also important processes causing variation in moisture availability. This causes maximum availability of moisture in the top soil layer during the wet season and in deep layers during the dry season (Markesteijn et al. 2010). As a result of the spatio-temporal heterogeneity in soil moisture availability, functional shifts occur in water use across different soil moisture gradients (Sterck et al. 2011).

Tropical areas have significant human population levels resulting in intense global losses of biodiversity across taxonomic and functional levels (Vitousek et al. 1997). As a result, a concept was developed for global scales (Diaz et al. 2007; Duffy et al. 2007), categorising species on a functional basis rather than taxonomic classifications. (Mooney et al. 1995). This enables analyses about the role and function of species in a community and ecosystem rather than how they evolve. Changes in land use history, the rate of disturbance (anthropogenic), and topographic factors are notably responsible for variation in the functional diversity and association of plant communities in TDFs (Lebrija-Trejos et al. 2011). Understanding spatial and temporal ecosystem functions will improve predictions and management of successional forests for restoring key ecosystem services (Alvarez-Yépiz et al. 2008).

TRAIT-BASED APPROACHES

Plant functional traits are morphological, biochemical, physiological, structural, behavioral, or phenological characteristics, usually measured at the tissue-level of organisms that influence their performance or fitness (Lavorel et al. 1997; Violle et al. 2007). Choosing specific traits will help with analyzing ecosystem processes. Traits can be studied either as an effect on ecosystem processes or as a response to environmental variables. Trait values represent how and why key 'functional' traits are related and how they are affected by biotic and abiotic factors. Functional traits can be compared within a species as well as between different species. Also, functional traits can be compared among species and communities, irrespective of the presence of overlap in the composition and abundance of species across the communities (Lavorel and Garnier 2002).

Trait-based approaches are appropriate for predicting how communities and ecosystems will respond to future climate change and other abiotic factors (Webb et al. 2002). Various studies suggest that the assessment of plant biodiversity at community levels based on individual species is not a true representation of the complicated mechanisms in that community (Dubuis, 2013; García-Gutiérrez et al. 2018). Dominant species in a community, which are often the most abundant and comprise the maximum biomass in the community, exert key effects on several ecosystem functions and processes. Hence, to study the "effect-response framework", it is necessary to analyze plant functional groups rather than individual species in order to understand plant community responses to environmental factors. Therefore, species should be categorized on the basis of their functional role.

Functional traits are important in analyzing the relationship of species composition patterns and community structure with local climatic conditions (Casanoves et al. 2011). This balance of all the factors are critical for determining the success rate of a species (Naeem and Wright 2003) and the contribution of that species to ecosystem level fluxes of carbon, nutrients, and resources in an ecosystem (Petchey and Gaston 2006).

Plants express a variety of traits (Figure 1) and trait-based approaches commonly use easily measurable morphological characteristics, such as canopy height, specific leaf area, and seed mass. Plant traits have often been separated into "easy" and "hard" traits. Hard traits are more directly linked to a certain plant functions, but are often difficult to measure, e.g. Relative Growth Rate (Gibson 2015). Easy traits are less directly related to a given plant function, but are easier to measure. For example, specific leaf area

Figure 1. List of plant functional traits

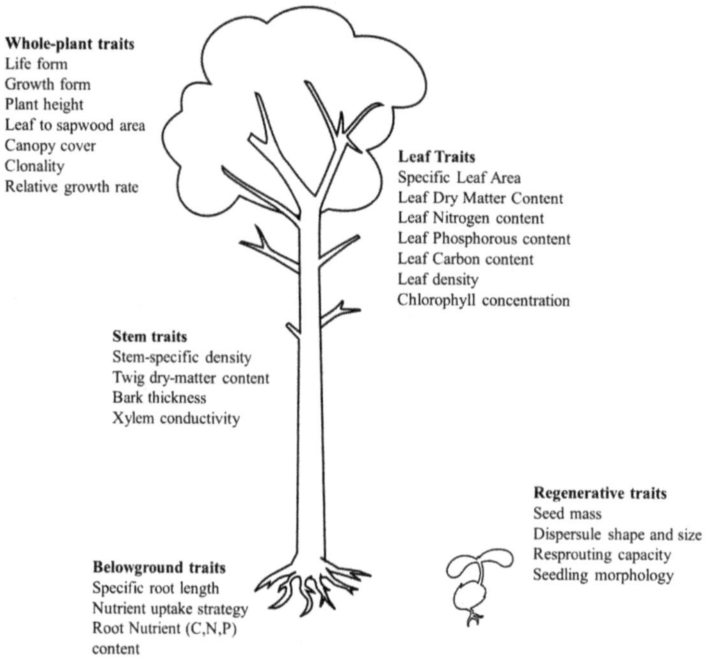

may be measured to understand a plant's strategy because it has been observed to be highly correlated with relative growth rate and leaf longevity (Weiher et al. 1999). These traits can be of two types, i.e. qualitative and quantitative. Qualitative or categorical data are associated with multi-state variables, for example, growth forms or life forms. However, quantitative data can be obtained from measurements that are expressed in continuous units, for example, plant height, leaf area, and etc. (Salgado-Negret 2015).

The number of traits in plants and plant communities can play a substantial role in altering the level of functional redundancy that a community assembly exhibits (Petchey and Gaston 2002). Increased numbers of traits, on one hand, can lead to a less redundant assemblage (i.e. change in species richness causes a change in functional diversity) while fewer traits can enhance redundant assemblages i.e. changes in species richness are insensitive to functional diversity (Petchey and Gaston 2006). Therefore, the selection of an adequate number of traits for a study is important in understanding the specific functions of interest.

Traits can be studied both at the species level as well as at the aggregated community level (Kleyer et al. 2012). At the community level traits can be averaged, which is commonly done by weighting the species by their abundance (Lavorel et al. 1997). The basic assumption underlying the abundance-weighted community trait means (CWM) is the 'mass ratio' theory, which states that the effect of a species' trait is in proportion to associated primary production (Grime 1998). Besides using the community-level response in mean trait values, within-community trait patterns can be quantified by different measures of functional diversity (FD) (Villeger et al. 2008). The functional diversity of a plant community is defined by the number and relative abundance of species in plant groups that are ecologically different (Dubuis 2013). This provides information about the mean trait variance while accounting for differences in richness (Díaz and Cabido 1997).

PLANT FUNCTIONAL TRAITS AND ECOSYSTEM PROCESSES

Disentangling ecosystem processes in heterogeneous and complex tropical forests is a major challenge. In this context, plant functional traits are becoming more useful. Functional traits are more widely used to study demographic drivers and their effects on the diversity of tropical tree species (Poorter et al. 2008; Liu et al. 2015). Various studies have examined soft traits that are easily measurable (Westoby 1998; Westoby et al. 2002) and known to represent fundamental trade-offs in life history strategies along with biotic and abiotic interactions of species and individuals (Funk et al. 2017; Amahowe et al. 2018). These measurable traits indicate correlations with ecosystem functions, such as biomass productivity, decomposition, and nutrient cycling, and enable more accurate predictions of associated ecosystem services.

Changes in species composition due to environmental drivers can alter functional trait composition, which affects ecosystem processes (McLaren and Turkington 2010). Figure 2 represents the role of plant functional traits used in decision-making such as policy making for conservation and restoration purposes. Individual or associated traits often govern several ecosystem processes which are discussed in detail in next sections, but many of these traits vary independently from one another (Bello et al. 2010). Currently, large databases of plant functional traits are being constructed and are considered a preliminary step in understanding the distribution of species with changing environmental conditions (Westoby and Wright 2006). Additionally, standard methods are being developed to study hard and soft traits in different forest ecosystems (Cornelissen et al. 2003; Perez-Harguindeguy et al. 2016).

Functional traits play an important role in influencing ecosystem processes in TDFs (Figure 3). Various traits are used to study different ecosystem properties. Whole plant traits, such as maximum life span, indicate population persistence and a plant's tolerance to climate change. Growth form is generally associated with ecophysiological adaptations, such as maximum photosynthetic production and adaptations to grazing by particular herbivores. Plant height represents reproductive size, fecundity, potential life span, and succession rate. Clonal behavior (i.e., ability of a plant species to reproduce itself vegetatively)

Figure 2. Schematic representation of the role of plant functional traits used in decision-making

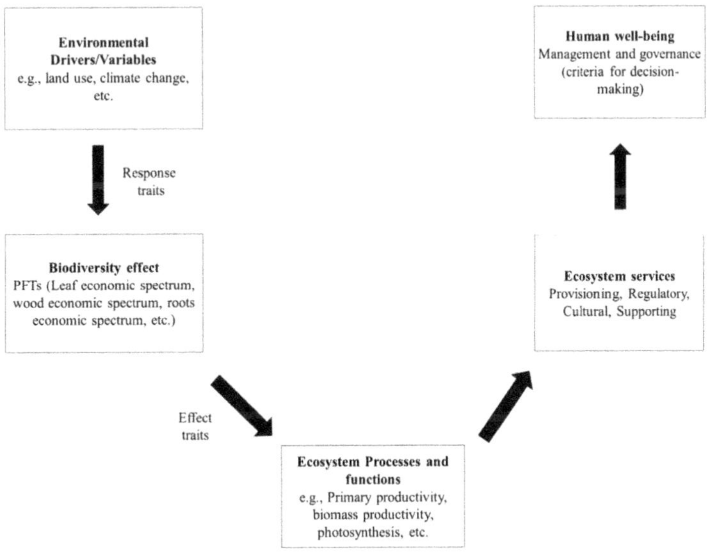

represents a plant's ability to exploit patches rich in nutrients, light, and water while it may promote persistence after environmental disturbances (Cornelissen et al. 2003). Spinescence and flammability are adaptations to tolerate disturbances (Perez-Harguindeguy et al. 2016). Leaf traits such as Specific Leaf Area (SLA) and Leaf Area (LA) are used to study relative growth rate among species. The latter is also related to photosynthetic rate, leaf life span, nutrient concentration (Nitrogen and Carbon), and assembly of species across gradients of light and water availability (Poorter 2009). Leaf Dry Matter Content (LDMC) is associated with resistance to physical hazards and litter decomposition and, thus, soil fertility. Stem traits such as Stem Specific Density (SSD) is an important trait for the stability, defense, architecture, and carbon accumulation. Bark thickness and other traits are known to establish defense mechanism against pathogens. Root traits play an important role in nutrient acquisition and resource translocation. Regenerative/reproductive traits are associated with survival and succession, maybe after disturbances like drought, fire, and insect and pathogen outbreaks, etc.

Many studies in TDFs have focussed on nitrogen and water use efficiency; photosynthetic characteristics and associated traits; variation in leaf traits due to light interception; nitrogen fixation; drought tolerance in leaf traits; herbivore grazing; and phenology-related traits (Sobrado 1991; Kitajima et al. 1997; Niinemets and Tenhunen 1997; Niinemets 1999; Brodribb et al. 2003; Brodribb and Holbrook 2003a,b; Rozendaal et al. 2006; Markesteijn et al. 2007; Niinemets et al. 2009; Posada et al. 2009; Brenes-Arguedas et al. 2009; Fallas-Cedeño et al. 2010; Freitas et al. 2010).

Relationships between plant functional traits and ecosystem functions and services under tropical dry forests are given in Table 1. Unfortunately, most of these studies were conducted on a limited number of traits and do not consider different growth forms like leguminous plants, which are dominant in TDFs and are important in N fixation (Pennington et al. 2009). These studies are restricted to moist regions and, consequently, fail to generalize the pattern of functional traits globally. Tropical forests, which are mainly characterized by high species richness, have been less studied compared to temperate forests (Wright et al. 2007; Worthy and Swenson 2019).

The traits are affected by various environmental as well as biological factors, some of which are studied and some are yet to identify. Soil moisture is one of the most important factors in TDFs, it has not been widely considered as a major determinant of plant functional traits (Worthy and Swenson 2019). Variation in moisture availability regulates intra- and inter-annual patterns of growth within seasonal forests (Baker et al. 2003). For example, in a semi-deciduous forest of Ghana, the growth rates of pioneer species associated with high rainfall and less fertile soils were substantially lower than similar species in low rainfall forests with more fertile soils (Baker et al. 2003). These results indicate that associations of species with particular environmental conditions are useful indicators of maximum growth rates.

Spatial and temporal variation in soil water availability acts as a limiting factor for overall rates and temporal patterns of growth. In the case of herbaceous species, it is considered to be the major regulatory factor along with light intensity, nutrient concentration, and canopy cover (Chaturvedi et al. 2011a, b; Sagar et al. 2012; Singh et al. 2017). This also may lead to phenotypic differences between individuals and species. Functional diversity can be an important measure to represent phenotypic differences across the spatial, temporal, and taxonomic level in tropical dry forests, but remains underexplored. Therefore, it is important to understand the mechanism of the drivers of tropical biodiversity (Swenson 2013).

Figure 3. Traits associated with ecosystem functioning in Tropical dry forest. Soil moisture and disturbances act as the primary driver regulating ecosystem processes in the region

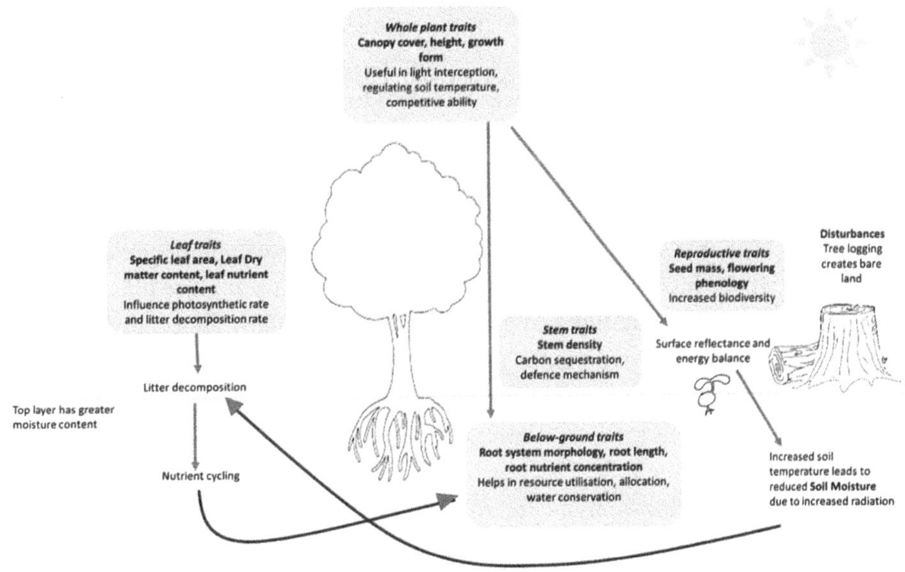

LEAF ECONOMIC SPECTRUM

Leaf traits are the most extensively studied traits in TDFs because of its accessibility. Major traits used to characterize leaf economic spectrum (LES) are leaf area (LA), specific leaf area (SLA), leaf dry matter content (LDMC), leaf nutrient content (N, P, K), and leaf lifespan. Selection of traits should be done following Grime (1977) and Westoby (1998) who suggest considering traits through three major axes of their function: resource exploitation, competition ability, and response to disturbance. LA, SLA, LDMC, and LNC (Leaf Nitrogen Content) are considered appropriate to study those major axes.

Leaf area (LA) is defined as the one-sided surface area of a leaf, which is helpful in analyzing canopy architecture and determination of leaf area index (LAI). LA is correlated with light interception, photosynthetic efficiency, carbohydrate metabolism, transpiration rate, LDMC, and Relative Growth Rate (Williams 1987; Centritto et al. 2000). The SLA of a leaf is defined as the area of one side of a fresh leaf divided by its oven-dry mass (Cornelissen et al. 2003). These traits characterize a trade-off between leaf life spans and photosynthetic rates. Leaves showing higher photosynthetic rates have shorter life spans (Reich et al. 1997; Wright et al. 2004). Trees having greater SLA and nitrogen and phosphorous content have greater photosynthesizing capacity and a shorter life spans. Thus, they have less investment in tissue (Chave et al. 2009; Swenson et al. 2012; Funk et al. 2017) and vice-versa (Donovan et al. 2011). Greater SLA, LNC, and photosynthetic rates are known to be more characteristic of tropical dry forests than tropical evergreen forest (Powers and Tiffin 2010). Species generally have larger SLA when grown in an environment with favourable conditions. Variation in SLA is often associated with changes in leaf water content, which is ultimately related to leaf thickness. Light availability plays a major role in governing the leaf economic spectrum and decreases in dense forest stand indicating conservative behavior rather than acquisitive during succession (Lohbeck et al. 2013). In contrast, light does not play a primary role in TDFs and is subordinate to temperature and moisture (Lebrija-Trejos et al. 2011). Since plants already follow a conservative behavior due to lack of resources, there is an increase in water availability during succession and plant behavior shifts from conservative to acquisitive.

Plasticity is a major phenomenon in all of the traits mentioned above, but is generally lower in TDFs (Markesteijn et al. 2007). SLA reflects a trade-off between resource capture and conservation (Poorter 2009) and shows a correlation with relative growth rate, leaf longevity, stem density, photosynthetic rate, and competitive ability (Reich et al. 1997). SLA varies with leaf phenology, with higher SLA in TDFs and lower in evergreen forests. This represents a trade-off between leaves with longer life spans and lower resource acquisition and leaves with shorter life spans and higher resource acquisition (Reich et al. 1997). Inverse of SLA (1/SLA) is often known as Leaf Mass Per Area and has relatively less variation in response to light in TDFs. Thus, again, light is not a major driver for mature plants in TDFs (Markesteijn et al. 2011). However, light availability in TDFs is considered a key factor governing the growth of seedlings (Sagar et al. 2008), which experience dense canopy cover during pre-monsoon season (Singh and Singh 1992). Thus, there is an uneven distribution of light in TDFs for seedlings (Bhadouria et al. 2018). Hard traits, such as maximum photosynthetic rate based on leaf area and stomatal conductance, have a strong positive correlation with leaf hydraulic conductivity (Santiago et al. 2004). There is also a strong correlation between leaf chlorophyll content and LNC that negatively varies with leaf thickness, which shows high stomatal density and low chlorophyll content (Loranger and Shipley 2010). Also, increased solar radiation causes a decrease in chlorophyll to nitrogen ratio (Hallik et al. 2009). Although, leaf traits are the most extensively studied traits in TDFs due to accessibility, such traits provide little evidence regarding ecosystem processes (Powers and Tiffin 2010). Therefore, wood, root, and seed economic spectrum should also be considered for more holistic studies.

Wood Economic Spectrum

Some studies suggest that woody traits are statistically independent to leaf traits (Baraloto et al. 2010; Amahowe et al. 2018). But others have shown a strong correlation between wood and leaf hydraulic traits (Santiago et al. 2004). Considering the abovementioned two contrasting sets of studies it is imperative to study woody traits for better understanding. Wood density or stem specific density (i.e., oven-dry mass of a section of the main stem of a plant divided by the volume of the same section when fresh) is considered the most essential factor for generating robust analyses of carbon stocks. The latter is the most important source for studying carbon sequestration (Chave et al. 2009) and soil water availability (Borchert 1994). Wood density has a negative correlation with mortality rates and volumetric growth (Wright et al. 2010; Swenson 2012). Bark thickness (thickness of the bark in mm) and wood density are very important for studying hazardous events (e.g., stems prone to breakage) and pathogen attack (Swenson 2012). To assess tree demography, wood density is considered to be one of the most useful traits in the Tropics as well as worldwide (Worthy and Swenson 2019).

Relative growth rate exhibits a lower correlation with wood density as compared to lumen area and vessel density (xylem traits), whereas xylem conductivity helps in determining plant water balance (Gleason et al. 2012). Higher wood density is highly correlated to smaller leaf and twig sizes (Westoby and Wright 2006). Twig Dry Matter Content (TDMC) (defined as dry mass divided by saturated mass) and bark thickness (part of the stem that is external to the wood or xylem) is associated with extreme events, such as fire-prone areas having an extremely high temperature. This trait is particularly helpful to trees and large shrubs prone to fire regimes. Studies should be focussed on soft traits that are cost-effective and able to capture below ground carbon dynamics, since above ground and below ground traits that are involved in carbon cycling are weakly linked (Deyn et al. 2008).

Xylem conductivity is defined as the efficiency of water transported from soil to leaves permitting photosynthesis, preventing the occurrence of negative water potential. Xylem vulnerability to embolism indicates the risk of loss of water transport during a drought in TDFs. It is an important aspect of drought tolerance (Perez-Harguindeguy et al. 2016) but given least priority when studying traits in TDFs. Functional traits are also used to study patterns of species richness and demography spatially and temporally. The pioneer work done by Swenson and Enquist (2007) using wood density suggests that assemblages are more diverse irrespective of their increased species richness in co-occurring tropical species. This followed a huge number of studies investigating environmental filtering versus niche partitioning using trait dispersion patterns.

ROOT ECONOMIC SPECTRUM

Dense root tissues can be either thick or thin in diameter. This freedom opens up more possibilities to develop roots with a variety of different traits combinations that may enhance fitness under different environmental conditions (Laughlin 2014; Kramer-Walter et al. 2016). Root traits play an important role in nutrient acquisition and translocating resources. Fine roots (<2mm diameter) are considered to be the primary organ for nutrient and water acquisition and are solely responsible to transfer resources between belowground and aboveground parts (McCormack et al. 2015).

The still ongoing conflict related to fine roots revolves around a single root economic spectrum like leaf economic spectrum (Reich 2014; Roumet et al. 2016). A gain in strength of the multidimensional functional approaches of root traits is assumed. For example, plants cannot develop tough, long-lived leaves that exhibit high SLA; however, plants can develop high Specific Root Length of any density.

Climate variables are suggested to be the most important regulators affecting the growth of fine roots (Freschet et al. 2017). Climate directly affects fine root growth by causing variation in plant morphology and phenology. It also provides carbon among various organs while changing vegetation composition and soil processes indirectly (Yuan and Chen 2010). Tropical species generally have roots with thicker diameter and lower specific root length as compared to temperate species (Freschet et al. 2017). An increase in soil nutrient availability causes an increase in root nutrient concentration (Liu et al. 2015). Therefore, infertile soils are mostly dominated by species with lower root nutrient concentration and higher root density (Holdaway et al. 2011).

Tropical forests exhibit high heterogeneity, which causes increased variation in root traits within communities. This might be due to the phylogenetic history of plants (Kong et al. 2014). Herbaceous root traits exhibit a high degree of plasticity as compared to woody species with rapid energetic investment and returns (Liu et al. 2015; Freschet et al. 2017). There has been an increase in a number of studies related to below-ground traits in tropical communities, but, the mechanism of root traits remains poorly understood. Since there is great plasticity between herbaceous and woody traits (Freschet et al. 2017), it is important to study the variation in different growth forms and how they respond to changing environmental conditions.

SEED ECONOMIC SPECTRUM

Seed mass forms the basis of the seed economic spectrum, which is a soft and most easily measured in tropical forest ecosystem. There are several seed traits that are important for the establishment of an individual species in TDFs. These are seed number, seed shape seed size, seed mass, seed germination time, mode of dispersion, dispersule/propagule (unit of seed as it is dispersed) size and shape, dispersule syndrome, dispersion potential, number of fruits, fruit mass, and reproductive phenology.

Seed production can result in a high number of small seeds in some plants as opposed to fewer but very large seeds in the others (Worthy and Swenson 2019). Small seeds have the capacity to reach greater distances and typically have faster growth rates in spite of having less endosperm. In contrast, large seeds have the least tendency to be dispersed (Khurana et al. 2006), often develop in nutrient-poor environments, and experience low growth and high mortality rates (Grubb 1977). However, the seedlings produced by larger seeds with greater size and mass have enhanced competitive capacities, such as enabling them to establish and survive in stressful environmental conditions. The latter can include competition, herbivory, extreme temperature, and defoliation. Larger seed sizes are related to reserve storage whereas smaller seed sizes are related to photosynthetic cotyledons. Thus, seeds inclined towards larger size are primarily responsible for the structure of TDFs (Khurana et al. 2006). Therefore, seed mass is considered the primary driver in the seed economic spectrum.

Although it is considered a very easy (soft) trait to measure (Wright et al. 2010), researchers face notable difficulties with collecting seeds from a high number of species with lower population sizes, which is a key feature of TDFs. Tropical forests experience significant disturbances. Therefore, choosing traits that typically provide more evidence of succession rates could be important characteristics in TDFs. For example, resprouting capacity after intense disturbances (Janzen 1988) can be considered as an important trait in a tropical forest facing high population loads. Sprouters generally provide a larger allocation of carbohydrates to below-ground organs. However, they have relatively slower biomass growth and lower reproductive output.

Canopy gaps are considered suitable regeneration sites in tropical forests. However, this occurs in relatively few sites due to excessive moisture reduction from intense solar radiation, which creates adverse conditions even for heliophytic species. Therefore, shaded sites are considered better for seedling survival and establishment, thereby reducing seedling desiccation (McLaren and McDonald 2003b).

FUTURE TRENDS AND CONCLUSION

Although the intrinsic values of tropical and subtropical forests are important, they also play a significant role in ecosystem dynamics. A strong knowledge base regarding the functional traits and processes linked within the tropical dry forest can improve understanding of the interdependency in the relationship between ecosystem processes and biodiversity conservation. Plant functional traits have gained momentum in the last decade. But due to scarce and fragmented information in TDFs, there can be no generalised information drawn from plant functional traits in TDFs. There is an urgent need to have plant functional trait information at local as well as global scales to implement effective restoration against the effects of climate change. Plant functional traits should be included in studies to gain better understanding of the correlation among various factors in complex ecosystem dynamics. Quantifying functional diversity

Table 1. Relationship between plant functional traits and ecosystem functions with their services in tropical dry forests

Sl. No.	Functional Traits	Ecosystem Functions	Ecosystem Services
1.	Plant Height	Access to light and its potential growth rate, competition, carbon sequestration	Carbon sequestration and Climate regulation
2.	Canopy Cover	Spatial and temporal distribution of light. It also influences local precipitation and air movements	Climate regulation, fresh air, and water supply
3.	Leaf Area	Canopy light interception, respiration, transpiration, leaf cooling, gaseous exchange and photosynthetic efficiency	Fuel production, Soil fertility (Manure), Fodder quality
4.	Specific Leaf Area	Light capture economics, net assimilation rate, relative growth rate, leaf life span, photosynthetic capacity	Fuel production, Soil fertility, Fodder quality
5.	Leaf Dry Matter Content	Biomass production, efficient resource use, resistivity, defense mechanism	Fuel production, nutrient regulation
6.	Leaf Nutrients (N,P,K) Concentration	Photosynthetic capacity, CO_2 assimilation, leaf nutrient levels, Decomposition and soil fertility	Fodder quantity, Nutrient cycling
7.	Stem Specific Density	Relationship with water and nutrient transport and support of individuals, their resistance to drought and damage by natural enemies.	Biodiversity, cultural values, soil fertility, biochemical production
8.	Specific Root Length	Transport and storage of water and nutritents and the mechanical support of individuals	Soil formation, nutrient cycling
9.	Root Nutrients (N,P,K) concentration	Nutrient acquisition strategy and soil fertility	Nutrient cycling, Fertile soil
10.	Seed Mass	Survival and establishment of young seedlings	Aesthetic and cultural value, biodiversity
11.	Seed Number	Increased regenerative capacity	Increased biodiversity

within, between and among different species of tropical dry forests will enhance our knowledge of these complicated dynamic mechanisms.

On the basis of the above selected PFTs, additional traits can be derived. They are, Leaf C:N ratio, Leaf N:P ratio, Leaf C:P ratio, LCC_a (Leaf Carbon Content on area basis), LNC_a (Leaf Nitrogen Content on area basis), LPC_a(Leaf Phosphorous Content on area basis), Root C:N ratio, Root N:P ratio, Root C:P ratio.

REFERENCES

Allen, K., Dupuy, J. M., Gei, M. G., Hulshof, C., Medvigy, D., Pizano, C., ... Waring, B. G. (2017). Will seasonally dry tropical forests be sensitive or resistant to future changes in rainfall regimes? *Environmental Research Letters*, *12*(2). doi:10.1088/1748-9326/aa5968

Álvarez-Yépiz, J. C., Martínez-Yrízar, A., Búrquez, A., & Lindquist, C. (2008). Variation in vegetation structure and soil properties related to land use history of old-growth and secondary tropical dry forests in northwestern Mexico. *Forest Ecology and Management*, *256*(3), 355–366. doi:10.1016/j.foreco.2008.04.049

Amahowe, I. O., Gaoue, O. G., Natta, A. K., Piponiot, C., Zobi, I. C., & Hérault, B. (2018). Functional traits partially mediate the effects of chronic anthropogenic disturbance on the growth of a tropical tree. *AoB Plants*, *10*(3), ply036. doi:10.1093/aobpla/ply036 PMID:29991997

Baker, T. R., Burslem, D. F., & Swaine, M. D. (2003). Associations between tree growth, soil fertility and water availability at local and regional scales in Ghanaian tropical rain forest. *Journal of Tropical Ecology*, *19*(2), 109–125. doi:10.1017/S0266467403003146

Balvanera, P., Quijas, S., & Pérez-Jiménez, A. (2011). Distribution patterns of tropical dry forest trees along a mesoscale water availability gradient. *Biotropica*, *43*(4), 414–422. doi:10.1111/j.1744-7429.2010.00712.x

Baraloto, C., Timothy Paine, C. E., Patiño, S., Bonal, D., Hérault, B., & Chave, J. (2010). Functional trait variation and sampling strategies in species-rich plant communities. *Functional Ecology*, *24*(1), 208–216. doi:10.1111/j.1365-2435.2009.01600.x

Bhadouria, R., Singh, R., Srivastava, P., Tripathi, S., & Raghubanshi, A. S. (2017). Interactive effect of water and nutrient on survival and growth of tree seedlings of four dry tropical tree species under grass competition. *Tropical Ecology*, *58*(3), 611–621.

Bhadouria, R., Srivastava, P., Singh, S., Singh, R., Raghubanshi, A. S., & Singh, J. S. (2018). Effects of light, nutrient and grass competition on growth of seedlings of four tropical tree species. *Indian Forester*, *144*(1), 54–65.

Bohlman, S. A. (2010). Landscape patterns and environmental controls of deciduousness in forests of central Panama. *Global Ecology and Biogeography*, *19*(3), 376–385. doi:10.1111/j.1466-8238.2009.00518.x

Borchert, R. (1994). Soil and stem water storage determine phenology and distribution of tropical dry forest trees. *Ecology*, *75*(5), 1437–1449. doi:10.2307/1937467

Both, S., Riutta, T., Paine, C. T., Elias, D. M., Cruz, R. S., Jain, A., ... Mielke, N. (2019). Logging and soil nutrients independently explain plant trait expression in tropical forests. *The New Phytologist*, *221*(4), 1853–1865. doi:10.1111/nph.15444 PMID:30238458

Brenes-Arguedas, T., Coley, P. D., & Kursar, T. A. (2009). Pests vs. drought as determinants of plant distribution along a tropical rainfall gradient. *Ecology*, *90*(7), 1751–1761. doi:10.1890/08-1271.1 PMID:19694125

Brodribb, T. J., & Holbrook, N. M. (2003a). Changes in leaf hydraulic conductance during leaf shedding in seasonally dry tropical forest. *The New Phytologist*, *158*(2), 295–303. doi:10.1046/j.1469-8137.2003.00736.x

Brodribb, T. J., & Holbrook, N. M. (2003b). Stomatal closure during leaf dehydration, correlation with other leaf physiological traits. *Plant Physiology*, *132*(4), 2166–2173. doi:10.1104/pp.103.023879 PMID:12913171

Brodribb, T. J., Holbrook, N. M., Edwards, E. J., & Gutierrez, M. V. (2003). Relations between stomatal closure, leaf turgor and xylem vulnerability in eight tropical dry forest trees. *Plant, Cell & Environment*, *26*(3), 443–450. doi:10.1046/j.1365-3040.2003.00975.x

Buzzard, V., Hulshof, C. M., Birt, T., Violle, C., & Enquist, B. J. (2016). Re-growing a tropical dry forest: Functional plant trait composition and community assembly during succession. *Functional Ecology*, *30*(6), 1006–1013. doi:10.1111/1365-2435.12579

Cardinale, B. J., Duffy, J. E., Gonzalez, A., Hooper, D. U., Perrings, C., Venail, P., ... Kinzig, A. P. (2012). Biodiversity loss and its impact on humanity. *Nature*, *486*(7401), 59–67. doi:10.1038/nature11148 PMID:22678280

Casanoves, F., Pla, L., Di Rienzo, J. A., & Dıaz, S. (2011). F Diversity, a software package for the integrated analysis of functional diversity. *Methods in Ecology and Evolution*, *2*(3), 233–237. doi:10.1111/j.2041-210X.2010.00082.x

Centritto, M., Loreto, F., Massacci, A., Pietrini, F., Villani, M. C., & Zacchini, M. (2000). Improved growth and water use efficiency of cherry saplings under reduced light intensity. *Ecological Research*, *15*(4), 385–392. doi:10.1046/j.1440-1703.2000.00359.x

Chapin, F. S. III, Zaveleta, E. S., Eviner, V. T., Naylor, R. L., Vitousek, P. M., Lavorel, S., ... Diaz, S. (2000). Consequences of changing biotic diversity. *Nature*, *405*(6783), 234–242. doi:10.1038/35012241 PMID:10821284

Chaturvedi, R. K., Raghubanshi, A. S., & Singh, J. S. (2011). Plant functional traits with particular reference to tropical deciduous forests: A review. *Journal of Biosciences*, *36*(5), 963–981. doi:10.100712038-011-9159-1 PMID:22116295

Chaturvedi, R. K., Raghubanshi, A. S., & Singh, J. S. (2011a). Plant functional traits with particular reference to tropical deciduous forests: A review. *Journal of Biosciences*, *36*(5), 963–981. doi:10.100712038-011-9159-1 PMID:22116295

Chaturvedi, R. K., Raghubanshi, A. S., & Singh, J. S. (2011b). Leaf attributes and treegrowth in a tropical dry forest. *Journal of Vegetation Science*, *22*(5), 917–931. doi:10.1111/j.1654-1103.2011.01299.x

Chave, J., Coomes, D., Jansen, S., Lewis, S. L., Swenson, N. G., & Zanne, A. E. (2009). Towards a worldwide wood economics spectrum. *Ecology Letters*, *12*(4), 351–366. doi:10.1111/j.1461-0248.2009.01285.x PMID:19243406

Choat, B., Jansen, S., Brodribb, T. J., Cochard, H., Delzon, S., Bhaskar, R., ... Jacobsen, A. L. (2012). Global convergence in the vulnerability of forests to drought. *Nature*, *491*(7426), 752–755. doi:10.1038/nature11688 PMID:23172141

Cornelissen, J. H. C., Lavorel, S., Garnier, E., Diaz, S., Buchmann, N., Gurvich, D. E., ... Pausas, J. G. (2003). A handbook of protocols for standardised and easy measurement of plant functional traits worldwide. *Australian Journal of Botany*, *51*(4), 335–380. doi:10.1071/BT02124

Daws, M. I., Mullins, C. E., Burslem, D. F., Paton, S. R., & Dalling, J. W. (2002). Topographic position affects the water regime in a semideciduous tropical forest in Panama. *Plant and Soil*, *238*(1), 79–89. doi:10.1023/A:1014289930621

de Bello, F., Lavorel, S., Díaz, S., Harrington, R., Cornelissen, J. H., Bardgett, R. D., ... da Silva, P. M. (2010). Towards an assessment of multiple ecosystem processes and services via functional traits. *Biodiversity and Conservation*, *19*(10), 2873–2893. doi:10.100710531-010-9850-9

De Deyn, G. B., Cornelissen, J. H., & Bardgett, R. D. (2008). Plant functional traits and soil carbon sequestration in contrasting biomes. *Ecology Letters*, *11*(5), 516–531. doi:10.1111/j.1461-0248.2008.01164.x PMID:18279352

Díaz, S., & Cabido, M. (1997). Plant functional types and ecosystem function in relation to global change. *Journal of Vegetation Science*, *8*(4), 463–474. doi:10.2307/3237198

Díaz, S., Lavorel, S., de Bello, F., Quétier, F., Grigulis, K., & Robson, T. M. (2007). Incorporating plant functional diversity effects in ecosystem service assessments. *Proceedings of the National Academy of Sciences of the United States of America*, *104*(52), 20684–20689. doi:10.1073/pnas.0704716104 PMID:18093933

Donovan, L. A., Maherali, H., Caruso, C. M., Huber, H., & de Kroon, H. (2011). The evolution of the worldwide leaf economics spectrum. *Trends in Ecology & Evolution*, *26*(2), 88–95. doi:10.1016/j.tree.2010.11.011 PMID:21196061

Dubuis, A. (2013). Predicting spatial Patterns of Plant Biodiversity: from Species to Communities. (Doctoral thesis), University of Lausanne, p. 295.

Duffy, J. E., Cardinale, B. J., France, K. E., McIntyre, P. B., Thébault, E., & Loreau, M. (2007). The functional role of biodiversity in ecosystems: Incorporating trophic complexity. *Ecology Letters*, *10*(6), 522–538. doi:10.1111/j.1461-0248.2007.01037.x PMID:17498151

Enquist, B. J., & Enquist, C. A. (2011). Long-term change within a Neotropical forest: Assessing differential functional and floristic responses to disturbance and drought. *Global Change Biology*, *17*(3), 1408–1424. doi:10.1111/j.1365-2486.2010.02326.x

Esquivel-Muelbert, A., Baker, T. R., Dexter, K. G., Lewis, S. L., Brienen, R. J., Feldpausch, T. R., ... Higuchi, N. (2019). Compositional response of Amazon forests to climate change. *Global Change Biology*, *25*(1), 39–56. doi:10.1111/gcb.14413 PMID:30406962

Fallas-Cedeño, L., Michele Holbrook, N., Rocha, O. J., Vásquez, N., & Gutiérrez-Soto, M. V. (2010). Phenology, lignotubers, and water relations of Cochlospermum vitifolium, a pioneer tropical dry forest tree in Costa Rica. *Biotropica*, *42*(1), 104–111. doi:10.1111/j.1744-7429.2009.00539.x

FAO. (2008). *FAOSTAT Online Statistical Service, Food and Agriculture Organization of the United Nations*. Rome, Italy: FAO.

Fortunel, C., Garnier, E., Joffre, R., Kazakou, E., Quested, H., Grigulis, K., ... Zarovali, M. (2009). Leaf traits capture the effects of land use changes and climate on litter decomposability of grasslands across Europe. *Ecology, 90*(3), 598–611. doi:10.1890/08-0418.1 PMID:19341132

Freitas, A. D. S., Sampaio, E. V. S. B., Santos, C. E. R. S., & Fernandes, A. R. (2010). Biological nitrogen fixation in tree legumes of the Brazilian semi-arid caatinga. *Journal of Arid Environments, 74*(3), 344–349. doi:10.1016/j.jaridenv.2009.09.018

Freschet, G. T., Valverde-Barrantes, O. J., Tucker, C. M., Craine, J. M., McCormack, M. L., Violle, C., ... Bonis, A. (2017). Climate, soil and plant functional types as drivers of global fine-root trait variation. *Journal of Ecology, 105*(5), 1182–1196. doi:10.1111/1365-2745.12769

Funk, J. L., Larson, J. E., Ames, G. M., Butterfield, B. J., Cavender-Bares, J., Firn, J., ... Wright, J. (2017). Revisiting the H oly G rail: Using plant functional traits to understand ecological processes. *Biological Reviews of the Cambridge Philosophical Society, 92*(2), 1156–1173. doi:10.1111/brv.12275 PMID:27103505

Galicia, L., Zarco-Arista, A. E., Mendoza-Robles, K. I., Palacio-Prieto, J. L., & García-Romero, A. (2008). Land use/cover, landforms and fragmentation patterns in a tropical dry forest in the southern Pacific region of Mexico. *Singapore Journal of Tropical Geography, 29*(2), 137–154. doi:10.1111/j.1467-9493.2008.00326.x

García-Gutiérrez, T., Jiménez-Alfaro, B., Fernández-Pascual, E., & Müller, J. V. (2018). Functional diversity and ecological requirements of alpine vegetation types in a bio geographical transition zone. *Phytocology, 48*, 77–89.

Getzin, S., Dean, Ch., He, F., Trofymow, J. A., Wiegand, K., & Wiegand, T. (2006). Spatial patterns and competition of tree species in a douglas-fir Vancouver Island. *Ecography, 29*, 671–682. doi:10.1111/j.2006.0906-7590.04675.x

Gleason, S. M., Butler, D. W., Ziemińska, K., Waryszak, P., & Westoby, M. (2012). Stem xylem conductivity is key to plant water balance across Australian angiosperm species. *Functional Ecology, 26*(2), 343–352. doi:10.1111/j.1365-2435.2012.01962.x

Grime, J. P. (1977). Evidence for the existence of three primary strategies in plants and its relevance to ecological and evolutionary theory. *American Naturalist, 111*(982), 1169–1194. doi:10.1086/283244

Grime, J. P. (1998). Benefits of plant diversity to ecosystems: Immediate, filter and founder effects. *Journal of Ecology, 86*(6), 902–910. doi:10.1046/j.1365-2745.1998.00306.x

Grubb, P. J. (1977). Control of forest growth and distribution on wet tropical mountains: With special reference to mineral nutrition. *Annual Review of Ecology and Systematics, 8*(1), 83–107. doi:10.1146/annurev.es.08.110177.000503

Hallik, L., Niinemets, Ü., & Wright, I. J. (2009). Are species shade and drought tolerance reflected in leaf-level structural and functional differentiation in Northern Hemisphere temperate woody flora? *The New Phytologist, 184*(1), 257–274. doi:10.1111/j.1469-8137.2009.02918.x PMID:19674334

Hérault, B., Bachelot, B., Poorter, L., Rossi, V., Bongers, F., Chave, J., ... Baraloto, C. (2011). Functional traits shape ontogenetic growth trajectories of rain forest tree species. *Journal of Ecology*, *99*(6), 1431–1440. doi:10.1111/j.1365-2745.2011.01883.x

Holdaway, R. J., Richardson, S. J., Dickie, I. A., Peltzer, D. A., & Coomes, D. A. (2011). Species-and community-level patterns in fine root traits along a 120,000-year soil chronosequence in temperate rain forest. *Journal of Ecology*, *99*(4), 954–963. doi:10.1111/j.1365-2745.2011.01821.x

Holdridge, L. R. (1967). Life zone ecology. Life zone ecology., (rev. ed.).

Hulshof, C. M., Mart'ınez-Yr'ızar, A., Burquez, A., Boyle, B., & Enquist, B. J. (2013) Functional trait variation in tropical dry forests: a review and synthesis. Tropical Dry Forests in the Americas: Ecology, Conservation, and Management. Boca Raton, FL: CRC Press. pp 129–40.

Janzen, D. H. (1988). Tropical dry forests. *Biodiversity (Nepean)*, 130–137.

Khurana, E., Sagar, R., & Singh, J. S. (2006). Seed size: A key trait determining species distribution and diversity of dry tropical forest in northern India. *Acta Oecologica*, *29*(2), 196–204. doi:10.1016/j.actao.2005.10.003

Khurana, E., & Singh, J. S. (2001). Ecology of seed and seedling growth for conservation and restoration of tropical dry forest: A review. *Environmental Conservation*, *28*(1), 39–52. doi:10.1017/S0376892901000042

Kitajima, K., Mulkey, S. S., & Wright, S. J. (1997). Decline of photosynthetic capacity with leaf age in relation to leaf longevities for five tropical canopy tree species. *American Journal of Botany*, *84*(5), 702–708. doi:10.2307/2445906 PMID:21708622

Kleyer, M., Dray, S., Bello, F., Lepš, J., Pakeman, R. J., Strauss, B., ... Lavorel, S. (2012). Assessing species and community functional responses to environmental gradients: Which multivariate methods? *Journal of Vegetation Science*, *23*(5), 805–821. doi:10.1111/j.1654-1103.2012.01402.x

Kong, D., Ma, C., Zhang, Q., Li, L., Chen, X., Zeng, H., & Guo, D. (2014). Leading dimensions in absorptive root trait variation across 96 subtropical species. *The New Phytologist*, *203*(3), 863–672. doi:10.1111/nph.12842 PMID:24824672

Kramer-Walter, K. R., Bellingham, P. J., Millar, T. R., Smissen, R. D., Richardson, S. J., & Laughlin, D. C. (2016). Root traits are multidimensional: Specific root length is independent from root tissue density and the plant economic spectrum. *Journal of Ecology*, *104*(5), 1299–1310. doi:10.1111/1365-2745.12562

Kuster, E. C., Bierman, S. M., Klotz, S., & Kuhn, I. (2010). Modelling the impact of climate and land use change on the geographical distribution of leaf anatomy in a temperate flora. *Ecography*, *12*, 2001–2012.

Laughlin, D. C. (2014). The intrinsic dimensionality of plant traits and its relevance to community assembly. *Journal of Ecology*, *102*(1), 186–193. doi:10.1111/1365-2745.12187

Lavorel, S., & Garnier, É. (2002). Predicting changes in community composition and ecosystem functioning from plant traits: Revisiting the Holy Grail. *Functional Ecology*, *16*(5), 545–556. doi:10.1046/j.1365-2435.2002.00664.x

Lavorel, S., McIntyre, S., Landsberg, J., & Forbes, T. D. A. (1997). Plant functional classifications: From general groups to specific groups based on response to disturbance. *Trends in Ecology & Evolution*, *12*(12), 474–478. doi:10.1016/S0169-5347(97)01219-6 PMID:21238163

Lebrija-Trejos, E., Pérez-García, E. A., Meave, J. A., Poorter, L., & Bongers, F. (2011). Environmental changes during secondary succession in a tropical dry forest in Mexico. *Journal of Tropical Ecology*, *27*(5), 477–489. doi:10.1017/S0266467411000253

Liu, B., Li, H., Zhu, B., Eissenstat, D. M., & Guo, D. (2015). Complementarity of absorptive fine roots and arbuscular mycorrhizal fungi in nutrient foraging strategies across 14 coexisting subtropical tree species. *The New Phytologist*, *208*(1), 125–136. doi:10.1111/nph.13434 PMID:25925733

Lohbeck, M., Poorter, L., Lebrija-Trejos, E., Martínez-Ramos, M., Meave, J. A., Paz, H., ... Bongers, F. (2013). Successional changes in functional composition contrast for dry and wet tropical forest. *Ecology*, *94*(6), 1211–1216. doi:10.1890/12-1850.1 PMID:23923479

Loranger, J., & Shipley, B. (2010). Interspecific covariation between stomatal density and other functional leaf traits in a local flora. *Botany*, *88*(1), 30–38. doi:10.1139/B09-103

Markesteijn, L., Iraipi, J., Bongers, F., & Poorter, L. (2010). Seasonal variation in soil and plant water potentials in a Bolivian tropical moist and dry forest. *Journal of Tropical Ecology*, *26*(5), 497–508. doi:10.1017/S0266467410000271

Markesteijn, L., Poorter, L., & Bongers, F. (2007). Light-dependent leaf trait variation in 43 tropical dry forest tree species. *American Journal of Botany*, *94*(4), 515–525. doi:10.3732/ajb.94.4.515 PMID:21636421

Markesteijn, L., Poorter, L., & Bongers, F. (2007). Light-dependent leaf trait variation in 43 tropical dry forest tree species. *American Journal of Botany*, *94*(4), 515–525. doi:10.3732/ajb.94.4.515 PMID:21636421

Markesteijn, L., Poorter, L., Bongers, F., Paz, H., & Sack, L. (2011). Hydraulics and life history of tropical dry forest tree species: Coordination of species' drought and shade tolerance. *The New Phytologist*, *191*(2), 480–495. doi:10.1111/j.1469-8137.2011.03708.x PMID:21477008

Markesteijn, L., Poorter, L., Bongers, F., Paz, H., & Sack, L. (2011). Hydraulics and life history of tropical dry forest tree species: Coordination of species' drought and shade tolerance. *The New Phytologist*, *191*(2), 480–495. doi:10.1111/j.1469-8137.2011.03708.x PMID:21477008

Martinez-Yrizar, A., Burquez, A., Maass, M., Robichaux, R., & Yetman, D. (2000). Structure and functioning of tropical deciduous forest in western Mexico. *The Tropical Deciduous Forest of Southern Sonora Mexico: Ecology and Conservation of a Threatened Ecosystem.*

McCormack, M. L., Dickie, I. A., Eissenstat, D. M., Fahey, T. J., Fernandez, C. W., Guo, D., ... Leppälammi-Kujansuu, J. (2015). Redefining fine roots improves understanding of below-ground contributions to terrestrial biosphere processes. *The New Phytologist*, *207*(3), 505–518. doi:10.1111/nph.13363 PMID:25756288

McGill, B. J., Enquist, B. J., Weiher, E., & Westoby, M. (2006). Rebuilding community ecology from functional traits. *Trends in Ecology & Evolution*, *21*(4), 178–185. doi:10.1016/j.tree.2006.02.002 PMID:16701083

McLaren, J. R., & Turkington, R. (2010). Ecosystem properties determined by plant functional group identity. *Journal of Ecology*, *98*(2), 459–469. doi:10.1111/j.1365-2745.2009.01630.x

McLaren, K. P., & McDonald, M. A. (2003). The effects of moisture and shade on seed germination and seedling survival in a tropical dry forest in Jamaica. *Forest Ecology and Management*, *183*(1-3), 61–75. doi:10.1016/S0378-1127(03)00100-2

Ministry of Environment and Forests. (1999) National Policy and Macrolevel Action Strategy on Biodiversity. New Delhi, India: Government of India, Ministry of Environment and Forests.

Mooney, H. A., Lubchenco, J., Dirzo, R., & Sala, O. E. (1995). Biodiversity and ecosystem functioning: basic principles. In V. H. Heywood (Ed.), *Global biodiversity assessment. United Nations Environment Programme* (pp. 275–325). Cambridge, UK: Cambridge University Press.

Mori, A. S., Lertzman, K. P., & Gustafsson, L. (2017). Biodiversity and ecosystem services in forest ecosystems: A research agenda for applied forest ecology. *Journal of Applied Ecology*, *54*(1), 12–27. doi:10.1111/1365-2664.12669 PMID:29749183

Naeem, S., Duffy, J. E., & Zavaleta, E. (2012). The functions of biological diversity in an age of extinction. *Science*, *336*(6087), 1401–1406. doi:10.1126cience.1215855 PMID:22700920

Naeem, S., & Wright, J. P. (2003). Disentangling biodiversity effects on ecosystem functioning: Deriving solutions to a seemingly insurmountable problem. *Ecology Letters*, *6*(6), 567–579. doi:10.1046/j.1461-0248.2003.00471.x

Niinemets, Ü. (1999). Research review. Components of leaf dry mass per area–thickness and density–alter leaf photosynthetic capacity in reverse directions in woody plants. *The New Phytologist*, *144*(1), 35–47. doi:10.1046/j.1469-8137.1999.00466.x

Niinemets, Ü., Díaz-Espejo, A., Flexas, J., Galmés, J., & Warren, C. R. (2009). Role of mesophyll diffusion conductance in constraining potential photosynthetic productivity in the field. *Journal of Experimental Botany*, *60*(8), 2249–2270. doi:10.1093/jxb/erp036 PMID:19395391

Niinemets, Ü., & Tenhunen, J. D. (1997). A model separating leaf structural and physiological effects on carbon gain along light gradients for the shade-tolerant species Acer saccharum. *Plant, Cell & Environment*, *20*(7), 845–866. doi:10.1046/j.1365-3040.1997.d01-133.x

Pennington, R. T., Lavin, M., & Oliveira-Filho, A. (2009). Woody plant diversity, evolution, and ecology in the tropics: Perspectives from seasonally dry tropical forests. *Annual Review of Ecology Evolution and Systematics*, *40*(1), 437–457. doi:10.1146/annurev.ecolsys.110308.120327

Perez-Harguindeguy, N., Diaz, S., Garnier, E., Lavorel, S., Poorter, H., Jaureguiberry, P., ... Urcelay, C. (2016). Corrigendum to: New handbook for standardised measurement of plant functional traits worldwide. *Australian Journal of Botany*, *64*(8), 715–716. doi:10.1071/BT12225_CO

Petchey, O. L., & Gaston, K. J. (2002). Functional diversity (FD), species richness and community composition. *Ecology Letters*, *5*(3), 402–411. doi:10.1046/j.1461-0248.2002.00339.x

Petchey, O. L., & Gaston, K. J. (2006). Functional diversity: Back to basics and looking forward. *Ecology Letters*, *9*(6), 741–758. doi:10.1111/j.1461-0248.2006.00924.x PMID:16706917

Poorter, L. (2009). Leaf traits show different relationships with shade tolerance in moist versus dry tropical forests. *The New Phytologist, 181*(4), 890–900. doi:10.1111/j.1469-8137.2008.02715.x PMID:19140935

Poorter, L., Wright, S. J., Paz, H., Ackerly, D. D., Condit, R., Ibarra-Manríquez, G., ... Muller-Landau, H. C. (2008). Are functional traits good predictors of demographic rates? Evidence from five neotropical forests. *Ecology, 89*(7), 1908–1920. doi:10.1890/07-0207.1 PMID:18705377

Posada, J. M., Lechowicz, M. J., & Kitajima, K. (2009). Optimal photosynthetic use of light by tropical tree crowns achieved by adjustment of individual leaf angles and nitrogen content. *Annals of Botany, 103*(5), 795–805. doi:10.1093/aob/mcn265 PMID:19151040

Powers, J. S., Becknell, J. M., Irving, J., & Perez-Aviles, D. (2009). Diversity and structure of regenerating tropical dry forests in Costa Rica: Geographic patterns and environmental drivers. *Forest Ecology and Management, 258*(6), 959–970. doi:10.1016/j.foreco.2008.10.036

Powers, J. S., & Tiffin, P. (2010). Plant functional type classifications in tropical dry forests in Costa Rica: Leaf habit versus taxonomic approaches. *Functional Ecology, 24*(4), 927–936. doi:10.1111/j.1365-2435.2010.01701.x

Ravindranath, N. H., Murthy, I. K., Priya, J., Upgupta, S., Mehra, S., & Nalin, S. (2014). Forest area estimation and reporting: Implications for conservation, management and REDD. *Current Science, 106*(9), 1201–1206.

Reich, P. B., Luo, Y., Bradford, J. B., Poorter, H., Perry, C. H., & Oleksyn, J. (2014). Temperature drives global patterns in forest biomass distribution in leaves, stems, and roots. *Proceedings of the National Academy of Sciences of the United States of America, 111*(38), 13721–13726. doi:10.1073/pnas.1216053111 PMID:25225412

Reich, P. B., Walters, M. B., & Ellsworth, D. S. (1997). From tropics to tundra: Global convergence in plant functioning. *Proceedings of the National Academy of Sciences of the United States of America, 94*(25), 13730–13734. doi:10.1073/pnas.94.25.13730 PMID:9391094

Roumet, C., Birouste, M., Picon-Cochard, C., Ghestem, M., Osman, N., Vrignon-Brenas, S., ... Stokes, A. (2016). Root structure–function relationships in 74 species: Evidence of a root economics spectrum related to carbon economy. *The New Phytologist, 210*(3), 815–826. doi:10.1111/nph.13828 PMID:26765311

Rozendaal, D. M. A., Hurtado, V. H., & Poorter, L. (2006). Plasticity in leaf traits of 38 tropical tree species in response to light; relationships with light demand and adult stature. *Functional Ecology, 20*(2), 207–216. doi:10.1111/j.1365-2435.2006.01105.x

Sagar, R., Archita, S., & Singh, J. S. (2008). Differential effect of woody plant canopies on species composition and diversity of ground vegetation: A case study. *Tropical Ecology, 49*, 189–197.

Sagar, R., Pandey, A., & Singh, J. S. (2012). Composition, species diversity, and biomass of the herbaceous community in dry tropical forest of northern India inrelation to soil moisture and light intensity. *The Environmentalist, 32*(4), 485–493. doi:10.100710669-012-9414-5

Salgado-Negret, B., Canessa, R., Valladares, F., Armesto, J. J., & Pérez, F. (2015). Functional traits variation explains the distribution of Aextoxicon punctatum (Aextoxicaceae) in pronounced moisture gradients within fog-dependent forest fragments. *Frontiers in Plant Science*, *6*, 511. doi:10.3389/fpls.2015.00511 PMID:26257746

Santiago, L. S., Goldstein, G., Meinzer, F. C., Fisher, J. B., Machado, K., Woodruff, D., & Jones, T. (2004). Leaf photosynthetic traits scale with hydraulic conductivity and wood density in Panamanian forest canopy trees. *Oecologia*, *140*(4), 543–550. doi:10.100700442-004-1624-1 PMID:15232729

Sfair, J. C., de Bello, F., de França, T. Q., Baldauf, C., & Tabarelli, M. (2018). Chronic human disturbance affects plant trait distribution in a seasonally dry tropical forest. *Environmental Research Letters*, *13*(2). doi:10.1088/1748-9326/aa9f5e

Singh, J. S., Raghubanshi, A. S., Singh, R. S., & Srivastava, S. C. (1989). Microbial biomass acts as a source of plant nutrients in dry tropical forests and savanna. *Nature*, *238*, 449–500.

Singh, J. S., & Singh, V. K. (1992). Phenology of seasonally dry tropical forest. *Current Science*, *63*, 684–689.

Singh, R., Sagar, R., Srivastava, P., Singh, P., & Singh, J. S. (2017). Herbaceous species diversity and soil attributes along a forest-savanna-grassland continuum in a dry tropical region. *Ecological Engineering*, *103*, 226–235. doi:10.1016/j.ecoleng.2017.04.020

Sobrado, M. A. (1991). Cost-benefit relationships in deciduous and evergreen leaves of tropical dry forest species. *Functional Ecology*, *5*(5), 608–616. doi:10.2307/2389479

Sterck, F., Markesteijn, L., Schieving, F., & Poorter, L. (2011). Functional traits determine trade-offs and niches in a tropical forest community. *Proceedings of the National Academy of Sciences of the United States of America*, *108*(51), 20627–20632. doi:10.1073/pnas.1106950108 PMID:22106283

Stibig, H. J., Achard, F., Carboni, S., Raši, R., & Miettinen, J. (2014). Change in tropical forest cover of Southeast Asia from 1990 to 2010. *Biogeosciences*, *11*(2), 247–258. doi:10.5194/bg-11-247-2014

Stibig, H. J., Achard, F., Carboni, S., Rasi, R., & Miettinen, J. (2014). Change in tropical forest cover of Southeast Asia from 1990 to 2010. *Biogeosciences*, *11*(2), 247–258. doi:10.5194/bg-11-247-2014

Swenson, N. G. (2013). The assembly of tropical tree communities–the advances and shortcomings of phylogenetic and functional trait analyses. *Ecography*, *36*(3), 264–276. doi:10.1111/j.1600-0587.2012.00121.x

Swenson, N. G., & Enquist, B. J. (2007). Ecological and evolutionary determinants of a key plant functional trait: Wood density and its community-wide variation across latitude and elevation. *American Journal of Botany*, *94*(3), 451–459. doi:10.3732/ajb.94.3.451 PMID:21636415

Swenson, N. G., Enquist, B. J., Pither, J., Kerkhoff, A. J., Boyle, B., Weiser, M. D., ... Kraft, N. J. (2012). The biogeography and filtering of woody plant functional diversity in North and South America. *Global Ecology and Biogeography*, *21*(8), 798–808. doi:10.1111/j.1466-8238.2011.00727.x

Tilman, D. (1987). Secondary succession and the patterns of plant dominance along experimental nitrogen gradients. *Ecological Monographs*, *57*(3), 189–214. doi:10.2307/2937080

Villeger, S., Mason, N. W. H., & Mouillot, D. (2008). New multidimensional functional diversity indices for a multifaceted framework in functional ecology. *Ecology*, *89*(8), 2290–2301. doi:10.1890/07-1206.1 PMID:18724739

Violle, C., Navas, M. L., Vile, D., Kazakou, E., Fortunel, C., Hummel, I., & Garnier, E. (2007). Let the concept of trait be functional! *Oikos*, *116*(5), 882–892. doi:10.1111/j.0030-1299.2007.15559.x

Vitousek, P. M., Aber, J. D., Howarth, R. W., Likens, G. E., Matson, P. A., Schindler, D. W., ... Tilman, D. G. (1997). Human alteration of the global nitrogen cycle: Sources and consequences. *Ecological Applications*, *7*(3), 737–750.

Waring, R. H., & Running, S. W. (2010). Forest ecosystems: analysis at multiple scales. Amsterdam, The Netherlands: Elsevier.

Webb, C. O., Ackerly, D. D., McPeek, M. A., & Donoghue, M. J. (2002). Phylogenies and community ecology. *Annual Review of Ecology and Systematics*, *33*(1), 475–505. doi:10.1146/annurev.ecolsys.33.010802.150448

Weiher, E., Werf, A., Thompson, K., Roderick, M., Garnier, E., & Eriksson, O. (1999). Challenging Theophrastus: A common core list of plant traits for functional ecology. *Journal of Vegetation Science*, *10*(5), 609–620. doi:10.2307/3237076

Westoby, M. (1998). A leaf-height-seed (LHS) plant ecology strategy scheme. *Plant and Soil*, *199*(2), 213–227. doi:10.1023/A:1004327224729

Westoby, M., Falster, D. S., Moles, A. T., Vesk, P. A., & Wright, I. J. (2002). Plant ecological strategies: Some leading dimensions of variation between species. *Annual Review of Ecology and Systematics*, *33*(1), 125–159. doi:10.1146/annurev.ecolsys.33.010802.150452

Westoby, M., & Wright, I. J. (2006). Land-plant ecology on the basis of functional traits. *Trends in Ecology & Evolution*, *21*(5), 261–268. doi:10.1016/j.tree.2006.02.004 PMID:16697912

Williams, L. E. (1987). Growth of 'Thompson Seedless' grapevines: I. Leaf area development and dry weight distribution. *Journal of the American Society for Horticultural Science*, *112*, 325–330.

Worthy, S. J., & Swenson, N. G. (2019). Functional perspectives on tropical tree demography and forest dynamics. *Ecological Processes*, *8*(1), 1. doi:10.118613717-018-0154-4

Wright, I. J., Ackerly, D. D., Bongers, F., Harms, K. E., Ibarra-Manriquez, G., Martinez-Ramos, M., ... Wright, S. J. (2007). Relationships among ecologically important dimensions of plant trait variation in seven neotropical forests. *Annals of Botany*, *99*(5), 1003–1015. doi:10.1093/aob/mcl066 PMID:16595553

Wright, I. J., Reich, P. B., Westoby, M., Ackerly, D. D., Baruch, Z., Bongers, F., ... Flexas, J. (2004). The worldwide leaf economics spectrum. *Nature*, *428*(6985), 821–827. doi:10.1038/nature02403 PMID:15103368

Wright, S. J., Kitajima, K., Kraft, N. J. B., Reich, P. B., Wright, I. J., Bunker, D. E., ... Zanne, A. E. (2010). Functional traits and the growth–mortality trade-off in tropical trees. *Ecology*, *91*(12), 3664–367. doi:10.1890/09-2335.1 PMID:21302837

Yuan, Z. Y., & Chen, H. Y. (2010). Fine root biomass, production, turnover rates, and nutrient contents in boreal forest ecosystems in relation to species, climate, fertility, and stand age: Literature review and meta-analyses. *Critical Reviews in Plant Sciences*, *29*(4), 204–221. doi:10.1080/07352689.2010.483579

Chapter 5
An Overview of the Role of Plant Functional Traits in Tropical Dry Forests

Abhinav Yadav
Banaras Hindu University, India

Pramit Verma
https://orcid.org/0000-0002-1913-5154
Banaras Hindu University, India

Akhilesh Singh Raghubanshi
Banaras Hindu University, India

ABSTRACT

Tropical dry forests (TDFs) are characterized by pronounced seasonality in precipitation, with several months of prolonged drought, 80% of annual precipitation occurring during a four- to six-month rainy season, and high interannual rainfall variability. Surprisingly, there are relatively few studies addressing patterns of functional trait in tropical dry forest (TDF) ecosystems. Functional trait analysis across plant species and the environment is a rapidly developing research field with many possible applications for forest restoration practice. Trait-based ecological research within TDFs will advance our understanding of how these ecosystems interact with and differ from other tropical ecosystems.

INTRODUCTION

Tropical forests share 7% of the total land's surface of the Earth and harbor a higher diversity at species and ecosystem levels (Hubbell and Foster, 1983; Wilson, 1988; Raven et al.,1993). Tropical forests are characterized by frequent variations in the temperature and precipitation. It is further categorized into moist and dry tropical forests (TDFs). TDFs generally occur under a warm to hot climate and mean annual temperature of >17 °C with >1 evaporation to rainfall ratio, due to frequent drought conditions

DOI: 10.4018/978-1-7998-0014-9.ch005

(of 2-6 months) during the year (Murphy and Lugo, 1986; Olivares and Medina, 1992). TDFs are found in the vast biogeographic region including many countries. However, due to change in climate, some of these highly biodiverse countries (such as Brazil, Egypt, etc.) are under the immense pressure of habitat loss and degradation (Raven et al., 1993). Since TDFs occurs under more arduous and less predictable environment, and are specifically sensitive to dry conditions and increased water stress, which result in increased mortality rate in these forests (Anderegg et al., 2015). It is, therefore, that plant species, with different life history traits such as tree size, successional status (i.e. pioneer and non-pioneer), leaf types (i.e., broad-leaved species and fine-leaved species), N_2 fixation ability, and habitat preferences, are predominantly controlled by the prevailing climatic conditions (Chaturvedi et al., 2011).

These highly vulnerable TDFs (Janzen,1988) are in need of development of restoration strategies for their preservation and revival. In practice, it has been observed that each forest area has its own history of disturbance and degree of resilience so that, it is very hard to determine the best strategy to recover that forest ecosystem. Since TDFs are generally characterized by different complex environmental conditions. So, slow and un-predictable vegetation recovery is often observed (Bognounou et al., 2010). Several restoration practices based on old restoration methods, especially by planting native vegetation extended in the past have recorded constrained achievement. Therefore, for a successful restoration of TDFs, the focus should be given on the responses of various species and functional groups (Dhyani and Dhyani, 2016). In previous studies, it has been established that functional ecology has the potential to turn descriptive ecology into more systematic and mechanistic science, thus getting increasing demand among the ecologist. So, for the restoration practices to be successful, it is highly important to adopt an efficient approach in the species selection for restoration planting. Plant functional traits (PFTs) may be one of the best approaches for this.

The functional trait can be defined as the "traits that can directly or indirectly affect the plant health" (see section titled "Definition of PFT"). Its analysis across the plant species and environment is a rapidly developing research field with many possible applications in forest restoration. Over the last two decades, functional trait-based ecology has been emerging as the most important advancement in ecology and evolutionary biology. In this connection, several plant functional traits (PFTs) have been identified and being promoted as indicators for the early detection of plant growth, development, and sustainability within an ecosystem. Moreover, comparison of PFTs under inter- as well as intra-species in dryland plant communities would also be a better indicator for TDFs under changing climate scenario. These traits can affect how the plant individuals interact with their surroundings or organisms in other trophic levels, which determine patterns of species interactions in a community. Several studies can be found on focusing on the plant functional traits and their capacity to predict resource use, species composition and ecosystem functioning, however, the information on the functional trait or set of such traits with ability to describe the robustness of plant species or their suitability in the changing environment with reference to TDFs, is grossly inadequate and fragmentary. Therefore, the demand for further studies on plant functional traits remains high, especially in the uncertainty caused by the effects of climate change. Plant species respond differently under diverse environmental conditions in TDFs (Chapin et al., 2003).TDFs under their early development stage are more prone to environmental stress (Murphy & Lugo, 1986). There is a need to develop a comprehensive list of ecologically significant functional traits to determine the coordination among them and the relationships between the traits and habitat conditions of the dry tropical forest ecosystem. For this, there is a need to develop a list of easily measurable important traits (Table 1). These traits can help in developing predictive models which will throw light on species distribution, productivity and plant responses to disturbances.

Functional trait-based measurements (especially, leaf traits) can be useful in providing an advanced understanding of the community structure and plant performance in dynamic environmental condition (Grime et al., 1997). Lambers & Poorter (1992) has estimated that a group of leaf traits is connected with plant growth and survival, and this leaf-related potential of PFTs has directed to a surge of relative studies on various plant trait at both communities as well as at individual level. Funk et al., (2015) found that leaf traits are associated with each other and regulate the trade-off between functionally distinct traits, resulting in dynamic ecosystem processes (fig 1). These trade-offs give the impression to allow plants to exhibit alternative combinations of functional traits with apparently equivalent fitness within a similar environment. Plant leaves are well known for contributing to climate change and regulating the various ecosystem processes. In several studies, it has been observed that the leaf related PFT shave been utilized to recognize that how and why ecosystem and their components fluctuate across environmental heterogeneity and a variety of PFTs have been proposed to understand their regeneration and performance (Garnier &Navas,2011; Laughlin et al., 2010; Wright et al., 2004). In a meta-analysis on PFTs of 278 tree species distributed in 27 sites across the world (except Antarctica), Paine et al. (2015) concluded that the PFTs (e.g., specific leaf area) may not be the best fit for estimation of the plant growth at large scale. High exposure of plants to the local environmental conditions may be the reason behind this. Therefore, PFTs based prediction and restoration strategy should be used very cautiously for the management of a forest community.

This present chapter would deal with a brief discussion about PFTs and would focus on leaf related traits with the following subsections: (1) plant functional traits definition and their importance, (2) Plant Functional Traits, (PFTs): definition,description and scope, (3) Leaf related plant functional traits, (4) Importance of leaf traits: a key indicator for plant growth and development strategy, (5) TDFs in changing climate, (6) Patterns and drivers of leaf traits in tropical dry forests, and (7) Linking leaf traits with spatial distribution in tropical dry forests. This chapter would help in advancing our knowledge about the leaf traits and the benefits and limitations of functional traits in the perspective of tropical dry forest.

PLANT FUNCTIONAL TRAITS (PFTS): DEFINITION, DESCRIPTION, AND SCOPE

Functional traits is an important topic in ecology, providing an understanding of plant life history strategies. Simultaneously, it can be exploited to study the ecological responses of plants and plant communities with a higher level of generalization. The trait-based approach is widely used in a broad scale, from individual to ecosystem level, from the stand to a global scale, to understand the plant strategies for acquisition or conservation of resources, and responses of the plant to changing environmental conditions. Functional trait analysis at the community scale provides further information on how the community changes with time. The plant has reaction and response strategies to the environmental stimuli, which varies with their resistance capacity and growth stages, disturbance as well as plant interaction with other life forms or their resource acquisitions capacity.

Definition of PFT

A plant functional trait (PFT) can be defined as "any plant characteristic which impacts fitness directly or indirectly by its consequences for development, generation, and survival" (Violle et al., 2007). The performance and resilience of a plant species, having their own fundamental significance and gaining more attention in terms of their functional roles in a natural ecosystem, can be estimated by monitoring PFTs. With this consensus, PFTs can be defined as: (1) biogeochemical or ecological parameters that can be affected by phenotypical constituents of plants, and (2) those that decide the feedback of an organism to environmental conditions (Lavorel & Garnier, 2002; Hooper et al., 2005). For example, the germination rate of the plant can be its response to the environment as it can be affected by physical factors like humidity, temperature, etc. Still, there is also a need to identify abiotic filters, which are responsible for determining community composition, and for redefining plant strategies.

Description of PFTs

Amongst the assembly of traits that can be measured for an individual plant, to be useful for comparative functional ecology, a trait should possess at least four characteristics (Lavorel et al., 2007). It should be: (1) linked to a plant function; (2) relatively easy and quick to measure (Garnier et al., 2004); (3) measurable using standardized protocols applicable to a large range of species and growth conditions; and (4) allow for the establishment of hierarchies between species, which are conserved between contrasting environments, without the absolute values of these traits necessarily remaining constant (Cornelissen et al., 2003; Garnier et al., 2001; Kazakou et al., 2014; Mokany & Ash, 2008). In fact, successional directions dependent on species structure alone are very hard to be projected (Suding et al., 2004; Zhou et al., 2014; Norden et al., 2015), yet might be progressively anticipated at the dimension of the functional groups (Connell & Slatyer, 1977; Fukami et al., 2005). In recent studies, a PFT-based practice (fig.1) has been shown to have the capacity to explore the drivers which cause changes in vegetative communities (McGill et al., 2006; Violle et al., 2007). Two statistic properties (namely, breadth development and tree mortality) for huge trees of 240 tree species from five Neotropical woodlands (Poorter et al., 2008) showed that morphological characteristics explained 41% of the variance in development rate and 54% of the variance in death rate, with wood thickness being the best indicator of statistic rates.

Previous studies of micro-climatic conditions have suggested that coordinated changes in plant functional traits are affected by the mean annual precipitation and seasonal variations over time (Poorter, 2009; Rentería& Jaramillo, 2011; Sterck et al., 2011). Various past studies using PFTs have relied on species identity and richness alone. However, it is noticeable here that these studies are based on average values of traits to estimate functional changes and do not categorizes plant species and community by their performance to dominant/prevailing environmental conditions. Based on the responses of the plant to different environmental stimuli and their ease of measurement, PFTs can be further differentiated into (1) hard and (2) soft traits. Hard traits calculate the degree of the real capacity of plants, and they are difficult to measure. It includes dispersal distance, competitive effect, and relative growth rate, etc. However, soft traits are relatively easy to quantify. The most commonly studied soft traits are leaf mass per unit area (LMA) and specific leaf area (SLA) which correlates with many other leaf attributes like relative growth rate (RGR) (Lambers & Poorter, 1992), photosynthetic rate (Wright et al., 2004), leaf nitrogen (N) content (Reich et al., 1995) and leaf lifespan (Westoby et al., 2002), and is thought to be the most useful single indicator of leaf strategy for resource utilization. As leaf related traits (i.e. leaf traits) are considered as the most appropriate and potential indicators of the plant health and developmental conditions, therefore, these traits have been widely explored by the scientific community. In this chapter, we have basically focused on various traits related to the plant leaves.

Scope of PFTs

Terrestrial plant ecology has especially benefitted from functional trait-based research (Cortois et al., 2016), as the considerable effort has been given to identifying key traits in plants. The advances in the functional trait-based ecology, especially interactions between PFTs and various natural processes, can provide essential insights into plant strategy to survive and develop. We highlight several such links (figure 1) involved in vital ecosystem services. To this end arguably, several extents of plant functional traits have been identified by several ecologists as:

1. Community assembly (Cadotte et al., 2013)
2. Ecosystem processes (Cornwell et al., 2008)
3. Global change responses (Peter et al., 2013)
4. Plant-soil interactions (Cortois et al., 2016)
5. Traits and environment (Bruelheide et al., 2018) and
6. Trade-offs (Aubin et al., 2016).

Apart from these aspects of PFTs, several challenges are faced in the understanding of the responses of plant species to drivers of change (i.e: trait and environment) and the effects of these species on whole ecosystem properties in the context of global environmental change (i.e. global change responses). Plant population and ecosystem processes are affected by Functional traits. The correlations between various Eco-physiological traits and ecosystem processes, including water availability, net productivity, litter decomposition, and nutrient cycling have the ability to predict changes in ecosystem services from changes in functional composition.

LEAF RELATED PLANT FUNCTIONAL TRAITS

This study would describe some key leaf related PFTs and their relevance in tropical dry forests along with a brief discussion of factors affecting them. In regard to plants ecosystem functioning, resource usage, and biomass production, leaf traits are often considered as one of the principal PFTs. These traits (Table 1) are easy to measure as well as reproducible for different species and throw light on the different survival strategies of plants. Leaf PFTs are generally used to describe the functional diversity of ecosystems and can be relevant in terms of biochemical and physiognomic characteristics of vegetation that control photosynthesis, nutrient, and water cycling. Various leaf traits represent trade-off between strategic allocations to construction costs (Wright et al., 2006), photosynthetic rates, leaf life spans are related to environmental or nutrient stresses and disturbances, plant-herbivore interactions, litter decomposition and nutrient cycling (Poorter & Bongers, 2006; Souza et al., 2015; Bhadouria et al., 2017). Some important variations in plant functioning and diversity can be explored by using leaf PFTs (e.g. LMA or leaf stoichiometry) due to their strong correlation with whole plant performance (Reich et al., 1995; West et al., 1999; Westoby et al., 2002). Leaf functional traits include SLA, leaf dry matter content (LDMC), leaf density, leaf thickness, photosynthesis rate, stomatal conductance, transpiration rate, water use efficiency and chlorophyll content, of which SLA and LDMC are highly related with the leaf economics spectrum (Garnier & Navas, 2011). It should be mentioned that seasonality has some impact on leaf attributes (Chaturvedi et al., 2011). Some of the key leaf related traits have been described below.

Figure 1. Illustrative representation of the role of plant functional traits as indicators for various environmental component assessments

Leaf Area (LA)

The one-sided projected surface area of the leaf can be defined as leaf area (LA) (Table 1) which is expressed in cm^2 (Cornelissen et al., 2003). LA assessment is a significant component of plant development analysis and evapotranspiration studies. LA is linked with canopy light and photosynthetic effectiveness and contributes to the metabolism of carbohydrates and the accumulation of dry matter (Centritto et al., 2000). In canopy construction analysis, LA is advantageous because it allows for the determination of the Leaf Area Index (LAI).

Leaf Area Index (LAI)

The ratio of total projected green LA to unit ground surface area is Leaf area index (Flow & Change, 2014). Leaves are regulating component of the many physical and biological processes in plant functioning along with their water balance and productivity in an ecosystem. Therefore, for accurate estimation of plant-environment interactions and Ecophysiological responses in changing the climate, LAI estimation is required. LAI can be measured by using litterfall collection or point contact (i.e. direct methods) or involving optical instruments and models (i.e. indirect methods).

$$\mathrm{LAI} = \frac{\mathrm{totalprojectedgreenLA}}{\mathrm{unitgroundsurfacearea}}$$

Specific Leaf Area (SLA)

Specific leaf area (SLA) is the ratio of the one-sided area of the fresh leaf and its oven-dry mass and is expressed in $cm^2\ g^{-1}$ (Cornelissen et al., 2003). It is linked with many key elements of plant growth and survival (Garnier et al., 2001; Shipley & Vu, 2002). It is a commonly used characteristic, as it is easily measured and often strongly correlated with numerous characteristics such as RGR, biomass increment (Bio Incr, G), photosynthesis (A), leaf longitude and palatization-related biochemical parameters (Meziane & Shipley,1999; Weiher et al., 1999). In terms of a short leaf life span high SLA provides a strong competitive ability for a high mass-based net photosynthetic rate (A_{mass}) and a low tolerance to stress (Grime, 1979; Reich et al., 1995; Wright et al., 2004).

$$\mathrm{SLA} = \frac{\mathrm{one-sidedareaofthefreshleaf}}{\mathrm{oven-drymass}}$$

Leaf Dry Matter Content (LDMC)

Leaf dry matter content (LDMC) is the ratio of leaf's oven-dry mass with its water saturated fresh mass, generally measured in percentage (%) (Cornelissen et al., 2003). It positively correlates with leaf life, while negatively correlates with potential RGR, though these associations are weaker than those involving SLA (Reich et al., 1992). It estimates the trade-off between the conversion of nutrients and biomass production. In tropical dry forests, plant resource use capacity can be estimated by measuring the LDMC. So, it can be proposed as an indicator of plant resource use. Leaves with higher physical resistance tend to show high LDMC, while in highly disturbed environments, species with lesser LDMC values are found to be linked with productive capabilities (Cornelissen et al., 2003).

$$\mathrm{LDMC} = \frac{\mathrm{ovendrymass}}{\mathrm{watersaturatedfreshmass}}$$

Leaf Mass Area (LMA)

Leaf mass per area (LMA) is the ratio of leaf density and leaf thickness (or leaf volume to the area ratio). LMA is a universally important functional trait in plants that are directly related to trade-offs among carbon assimilation, leaf life span and defence (Grime,1977; Wright et al., 2004; Reich, 2014). At a fundamental level, the allocation of resources to LMA can be modified by altering two, non-mutually exclusive (mathematically related) leaf properties: leaf thickness (Lth) and LDMC. LMA differences are determined by the nature of species and where they found, and do not necessarily exhibit a worldwide pattern (Poorter et al., 2009). This is true for both, within species, as they respond to environmental variation (Poorter et al., 2009), and between related species that differ in mean LMA values (Villar et al., 2013). LMA is associated with plant functioning especially with chemical compositions, respiratory rates, and photosynthesis.

$$LMA = \frac{\text{leafdensity}}{\text{leafthickness}}$$

Average Leaf Thickness (Lth)

Leaf thickness (Lth) is determined by the leaves' strategy to capture and utilize resources. Moreover, it is linked with the plant anatomy features like number, size and arrangement of leaf cells (Nicotra et al., 2011; Taiz & Zeiger, 2006). Therefore, it has often been used as a means to select species for "ecological performance" (Witkowski et al., 1992; Diaz et al., 2004) or productivity (White & Montes, 2005). It also plays an important role in plant functioning. In optimization theory, balancing photosynthetic benefits against transpiration and carbon costs of respiration predicts that Lth should be higher in lighter, drier and less fertile environments, as well as in long-lived leaves. These patterns are often observed, particularly in interspecific studies. It has been observed that at individual levels, outer canopy leaves are comparatively thicker than that of shaded parts of the canopy. Higher Lth is an indicator of increased resource acquisition (especially, N) and faster photosynthetic rates per unit leaf area. Thicker leaves (which are associated with the lower photosynthetic rate per unit leaf mass) often have lower % N and longer leaf life-span due to the covariance of SLA and percentage N. Further, thicker-leaved species may show lesser CO_2 diffusion due to lower mesophyll conductance, which causes longer diffusion pathways and higher optical reflectivity in combination with lower internal transmittance due to wide variations in leaf morphology (e.g. existence of specialized structures on leaf surface like protruding veins or spines and hairs).

Chlorophyll Concentration (Chl)

Chlorophyll pigments are specifically arranged in and around the leaf photosystem. The amount of light absorbed by a leaf is directly related to concentration of chlorophyll pigments in that leaf (Marenco et al., 2009; Niinemets, 2010), which further are involved in photosynthesis. Various other photochemical and non-photochemical reactions also involve chlorophyll. Leaf chlorophyll concentration is regulated by various factors, i.e. thicker leaves have high stomatal density and low Chl (Loranger & Shipley, 2010) and also by leaf nitrogen concentration (LNC) (Marino et al., 2010). Chlorophyll concentration (Chl) is estimated by measuring the absorbance (D) of the leaf extract at 645 and 663 nm wavelengths or by using chlorophyll concentration meter. Many studies have found that in low light conditions, photo-chemical part of the photosynthetic apparatus increases in comparison to the biochemical part in order to enhance light harvesting for fixing required carbon.

Photosynthetic Rate (A)

Photosynthesis is a key eco-physiological process resulting in the formation of carbohydrates by capturing atmospheric CO_2 and water in the presence of light and chlorophyll pigment and resulting in the release of oxygen as a by-product. The photosynthetic rate is generally defined in terms of oxygen production either as the amount of chlorophyll content or green plant tissues weight per unit (mass or area). It varies with plant health and growth stages as it is generally found higher in healthy leaves and decreases with leaf size and plant maturity (Santiago et al., 2004; Niinemetset al., 2009). It is further regulated by several other plant eco-physiological processes like structure, leaf N, stomatal conductance, and carboxylation capacity of the leaf (Wright et al., 2004). It is found higher for the heliophytic plant species as compared to the sciophytic species (Kitajima, 1997). The present increasing atmospheric CO_2 concentration is expected to affect several eco-physiological processes of plants (Cernusak et al., 2011). Of various eco-physiological processes, the photosynthetic rate is the major process, along with other growth (such as relative growth rate, leaf area, and total dry weight) and eco-physiological processes (such as water-use and nitrogen-use efficiencies) to be affected by enhanced atmospheric CO_2 concentration (Khurana and Singh, 2004).

Intrinsic Water Use Efficiency (WUEi)

Intrinsic Water use efficiency (WUEi) is defined as the ratio between net A and Gs (Niinemets et al., 2009), and is expressed in μmol mol^{-1}. Thus, WUEi is an indicator of plant performance in terms of water input and photosynthate production. Due to variations in the strength of correlation between net A and Gs, WUEi differs among plant species (Niinemets et al., 2009). Species with high relative growth rate exhibit low WUEi, whereas species with high WUEi have low growth rate (Angert et al., 2009). On the contrary, species which have high WUEi allocate a large proportion of leaf nitrogen to photosynthetic processes (Angert et al., 2009). Stomatal closure and site dryness increase WUEi (Ares & Fownes, 1999).

$$\text{WUEi} = \frac{\text{netphotosyntheticrate}(\text{A})}{\text{Stomatalconductance}(\text{Gs})}$$

Leaf Stomatal Conductance (Gs)

Leaf stomatal conductance (Gs) measure the rate of the flow of water vapour and CO_2 through the stomatal openings on the plant leaves. It is the measurement of diffusion of CO_2 into the mesophyll cells of leaves expressed in mmol m^{-2} s^{-1} as well as water vapour from the leaves to the atmosphere via the stomatal openings. Thus, it is regulated by several physiological mechanisms and alternatively governs various processes like transpiration rate and photosynthetic activities. Further, water stress conditions lead to a decrease in the stomatal conductance (Zhu & Cao, 2010), whereas a decrease is also reported with leaf aging and size, possibly due to a longer pathway for water transport (Niinemetset al., 2009). Interestingly, the small-sized tree species showed a prompt decline in stomatal conductance under water stress conditions (Dawson, 1996). Moreover, short duration plants showed a high Gs for maximizing the photosynthate accumulation a very short time-span (Kitajima et al.,1997). Therefore, measurement of stomatal conductance indirectly helps in identifying the plant adaptation mechanism to the forest stand composition and would further help in restoration approaches for the TDFs.

Leaf Water Potential (LWP)

Measurement of plant water status during the day is leaf water potential (LWP) and can be estimated by using the pressure chamber and the osmotic potential of the xylem sap in leaves. LWP quantification is carried out using a pressure chamber according to the procedure defined by Scholander (1966). Pure water has a water potential of 0 (Zero) MegaPascals (MPa). Water has a tendency to move in the direction of a site which has lower water potential. LWP of the leaf can be directly influenced by water potential of the atmosphere which is found to be -500 MPa at 20 percent relative humidity. Leaves of crop plants can function at -1 MPa, while some desert plants can function at leaf water potential of -10 MPa. Lesser the amount of free water in the plant, more the pressure is required to cause it to exude. Water within the plants will generally evaporate into the air from the leaf indicating that leaf has a negative potential. This water loss further helps in mineral nutrient uptake from the soil via the root system. However, excess water loss through transpiration is harmful to plants. The result is expressed in bar or MPa, always as a negative value. Due to high variability between measurements, the leaf water potential is preferably not recommended as an ideal PFT.

Leaf Litter Decompositions

Leaf litter decomposition is considered a major driver of the nutrient cycle in terrestrial ecosystems. It is also part of various fundamental ecosystem services such as nutrient availability and soil formation. Litter decomposition is affected by their inherent decomposability, changing climate conditions and community composition of soil organisms. Cornwell et al. (2008) demonstrated that the decomposition rate is highly affected by climatic variations as well as composition and intra-specific variations in leaf structure. Several leaf traits like phosphorus and/or nitrogen concentrations (Cornwell et al., 2008), which are related to leaf protective features, are found to affect the litter decomposition.

Leaf Nitrogen and Phosphorus Concentration

For photosynthetic carbon assimilation, Nitrogen (N) and phosphorus (P) is essential. These nutrients are the most common nutrients, often considered limiting for net primary productivity in terrestrial ecosystems. In tropical forests, phosphorous is generally considered as a regulating element for plant productivity because phosphorous availability is controlled by bedrock weathering. Soil P availability generally declines with bedrock weathering and soil age, while Nitrogen availability constrains plant productivity in many temperate and boreal forests by limiting leaf initiation and expansion (Vos & Biemond, 1992). It has been observed that Lowland tropical forest trees have experienced long-term decreased P status, which leads them to adapt and evolve in P limitation, though field-based evidences are lacking in this regard. A recent study in a Panamanian tropical forest showed that P limitation of plant growth is species-specific. However, it does not apply for the community-wide response, because some species are adapted to low P availability, and able to grow rapidly despite low soil P availability (Turner et al., 2018).

For a plant to survive and sustain in the changing environment, it is necessary to grow and proliferate. Simultaneously, the plant must also be resilient for future competitions. For this, plants take nutrients, water, and light energy too, for their growth and making themselves resistant from environmental stress, herbivory, and infections. In any natural ecosystem, the available nutrients for the plant are very limited. In such conditions, plants make a striking balance between their characteristics and resource utilization. Plants develop their characteristics to protect and support themselves in adverse and robust environmental conditions and competitions. This phenomenon can be better understood by considering the example given below.

In a constructed home, there is some architecture like rooftop, partitions, and corridors which are common in all the homes (analogous to leaves in plants) but their overall structure differs from other homes present in the neighboring areas. Similarly, the structure and functions of the leaves are greatly affected by their location and surroundings. If a plant has to survive in the deserted area, then it will have needle-shaped leaves to reduce evaporation water losses as well as to protect themselves from herbivores. In some plants, leaves can survive for very long durations, which mean they have used a large number of available resources to make them physically and chemically resistant. This also indicates that such leaves are robust for environmental changes. While plants with short life do not spend a large proportion of their resources into leaves rather than the root and shoot development by making carbohydrate. In such plants, leaves are weaker but more photosynthetically active. In continuation of the above example, it could be seen that the boundary of houses in some areas are stronger if there is a threat in the surroundings. On the other hand, in a harmless area, houses can be imaginable with several doors and windows to permit in light and air. Houses with the enemy (e.g. competition/herbivory in case of the plant) in the surrounding will spend more resources in constructing defence mechanism, while in the favorable condition, they will focus on internal functioning for the smooth growth and development (similar in case of plants).

The functional composition and phenology of TDFs have direct effects on the functioning of the ecosystem, particularly for seasonal variations in primary and bio-geochemical production (Jaramillo et al., 2011) as well as energy, water, and carbon cycles (Sanches et al., 2008; Maass & Burgos., 2011). Even though, very little studies in TDFs (Jaramillo et al., 2011) are available in relation to net primary productivity and biogeochemistry. Ecosystem functionality may be related to various key functional traits (Murphy & Lugo, 1986). For example, species with higher N, SLA, light saturation, photosynthetic rates and a lower lifetime of leaves have a wide-ranging ratio of phosphorus compared with deserts, evergreen tropical, constant temperate and tundra biomes (Chaturvedi et al., 2011). These characteristics affect the timing and nutrient composition of leaf fall (Jaramillo et al., 2011) and decomposition rates (Martínez-Yrízar et al., 1999; Xuluc-Tolosa et al., 2003). The N ratio is the most effective predictor of decomposition rates in leaf N, or lignin (Cornwell et al., 2008), though decomposition is significantly reduced as a result of dry conditions as compared to moist forests during the dry period (Powers et al., 2009). In addition, water availability in litter and soil in TDFs is known to regulate nutrient dynamics (Jaramillo et al., 2011).

Table 1. **Some important leaf traits, their characteristics and measurement method** *4.0 Importance of leaf traits: a key indicator for plant growth and development strategy*

Important leaf traits	Definition/ Characteristic	Measurement unit	Role	Measurement method/ Instrument	References
Area of a leaf (LA)	the one-sided or projected area of an individual leaf	mm^2	assessing growth and vigour	Leaf area meter	Poorter and Rozendaal (2008); Royer et al. (2008).
Leaf Area Index	the ratio of total projected green LA to unit ground surface area is Leaf area index (LAI) (total projected green LA/ unit ground surface area)	No unit	to predict photosynthetic primary production, evapotranspiration	(calculation)	Flow, E. S. F., & Change (2014)
Specific leaf area (SLA)	one-sided area of a fresh leaf, divided by its oven-dry mass	cm2 g-1	to estimate the reproductive strategy	(calculation)	Chen and Black (1992); Garnier et al. (2001a, 2001b)
Leaf dry-matter content (LDMC)	oven-dry mass (mg) of a leaf, divided by its water-saturated fresh mass (g)	expressed in mg g^{-1}	indicator traits of resource use strategies	(Calculation)	Garnier et al. (2001b); Vendramini et al. (2002); Vaieretti et al. (2007); Ryser et al. (2008)
Leaf Mass Area (LMA)	the ratio of leaf density and leaf thickness	gm^{-2}	plant light capture and carbon gain	(calculation)	Reich (2014)
Leaf thickness (Lth)	Thickness of leaves	mm or µm	Carbon assimilation and plant health	digital micrometre	Garnier and Laurent (1994); Wright et al. (2002); Vile et al. (2005); Poorter et al. (2009); Hodgson et al. (2011).
Chlorophyll(Chl)	The pigment in the leaf that Involved in photosynthesis	CCI unit	Part of the photosynthesis process	Chlorophyll concentration meter	Marenco et al. 2009; Niinemets (2010)
Photosynthetic rate(A)	The light-saturated photosynthetic rate (Amax) under typical field conditions	µmol Oxygen evolved or CO2 consumed per m^2 s^{-1}	Food production using light and CO_2	leaf gas-exchange system	Ellsworth and Reich (1992).
Water Use Efficiency(WUE)	the ratio between net A and Gs	µmol mol^{-1}	% of water supplied to the plant that is effectively taken up by the plant	leaf gas-exchange system or (calculation)	Niinemets et al. (2009),
Leaf Stomatal Conductance (Gs)	the rate of the flow water vapour and CO2 through the stomatal openings on the plant leaves	mmol m^{-2} s^{-1}	integral to leaf level calculations of transpiration	leaf gas-exchange system or (calculation)	
Leaf water potential as a measure of water status	The bulk leaf water potential (YL; units MPa) is a simple indicator of leaf water status; the more negative the value, the more dehydrated the leaf.	MPa	the measure of the chemical free energy of water in the leaf	pressure chamber, or Scholander bomb	Scholander (1966); Turner (1988).
Litter decomposability	Dry weight loss	percentage	In determining decomposition environment and the functional traits of the individual species	litterbags measure from 10*10cm to 20*20cm and are made of 1-mm polyester or fiberglass mesh	Taylor and Parkinson (1988); Cornelissen (1996); Freschet et al. (2012).
Leaf (N) concentration (LNC) and leaf phosphorous (P) concentration (LPC)	LNC and LPC are the total amounts of N and P, respectively, per unit of dry leaf mass	mg g^{-1}	Related with plant growth and reproduction	Macro- or micro-Kjeldahl (acidic) digestion, followed by colourimetric (flow-injection) analysis (using different reagents for N and P),	Horneck and Miller (1998); Temminghoff and Houba (2004).

In an investigation performed in a dry tropical woodland in Costa Rica (Hulshof & Swenson, 2010), leaf characteristic variability inside and among over ten coinciding tree species was analyzed, and the outcomes established that majority of variation in traits was explained by between species changes. In case of simple leaves, a trend can be observed when examining the leaf traits with higher between species variations, followed by within species variations and lastly by between leaves of different individuals. While, for the compound leaf species, this trend was followed only when individual leaflet trait value was not taken into account. For compound leaves, variations were found higher in between species and between leaflets.

The total amount of rainfall is the most limiting factor to the distribution of forests in the tropics (Holdridge,1967; Portillo-Quintero & Sánchez-Azofeifa, 2010) and water availability sets limits for the "survival and growth" of tree species in seasonally dry tropical forests. More than 80% of the annual precipitation occurs for 4 months (during the rainy season). Species native to such environments have developed different adaptations to withstand, avoid or tolerate water limitations (Portillo-Quintero & Sánchez-Azofeifa,2010). Evergreen (drought tolerant) plant species have adopted a mechanism to delay water stress by maximizing their access to water and also adopted some morphological features to reduce evapotranspiration water losses. Such species invests more biomass in roots and have an advanced defence system (such as high specific root length, small leaf area, and strong stomatal control)(Chaturvedi et al., 2011). Whereas, drought-avoiding species (i.e. deciduous species) avoid stress in the dry season by shedding and dropping-off their leaves and maximize resource capture during a limited growing season (Markesteijn & Poorter, 2009). TDFs species are successful in avoiding xylem cavitation and reduced water losses by having high SLA, high LDMC and longer tap-roots for avoiding drought conditions.

When discussing deciduous and evergreen forest, it can be noticed that deciduous species have large biomass with higher SLA, greater stem density (Markesteijn & Poorter, 2009) and large root system with a high total root length. For plant survival and growth, nutrients and light are very important, and competition for the resource can be a stress factor for the plant's growth and distributions (Khurana & Singh, 2004; Bhadouria et al., 2018). Competition for natural resources also has effects on their morphological adaptations (Craine & Dybzinski, 2013). Species in water-scarce TDFs, invest in thicker leaves to tolerate shade and drought. In these adaptations, LDMC can be found negatively correlated with SLA and Lth is positively correlated with stem density. Generally, early successional species are more light demanding than late successional species. Therefore, it is important to consider light availability when selecting species during succession.

Drought tolerance is believed to be more or less of the same pattern in early- and late-successional species, but the latter is prone to increased mortality in dry years, specifically due to lack of drought-avoiding strategies like deep rooting systems, and water storage in roots and stems (Schönbeck et al., 2015). Markesteijn & Poorter (2009) reported that TDFs species had a 62 days longer drought survival time, compared to 25 days for moist forest species. Further deciduousness explained 69% of the inter-specific variation in drought survival, and among evergreen species (drought resisters), stem density explained 20% of the drought survival. Thus, dry matter contents are positively correlated to the drought index.

TDFs IN CHANGING CLIMATE

In the changing climate, plants have to adapt against climate-related damages and increased cellular stress which will further affect their growth. Curtis et al. (2016) found that fine-scale spatial heterogeneity in physical components (i.e. water availability, light, and temperature) can provide insights of differences between habitat and leaf temperatures. An Ecological Niche Model (ENM) proposed for the tropical dry forests of Mexico has found that the TDFs will increase and occupy an area at elevated altitude away from their present niche. This kind of shifts will intensify due to climate change and result in fragmented habitats for already vulnerable species (Prieto-Torres et al., 2016). In tropical forests, especially in hot and arid environments, several leaf traits like LMA, are predicted to increase, while photosynthesis rate would decrease by increasing temperature. Thus, it can be forecasted that in tropical forests, evergreen species might increase at the cost of deciduous forest. Critical or indicator traits for different types of forests can be identified in order to monitor the health of a forest in changing climate.

"Extreme climate" and anthropogenic influences are considered as the major determinants of forest change (Álvarez-Yépiz et al., 2018). With the predicted change in climate, several physical components like temperature, precipitation, and atmospheric conditions will get affected. Extreme high-temperature events are expected to increase in both frequency and magnitude worldwide (IPCC 2014). Since plant species of tropical dry forest are already living in proximity to absolute thermal limits so they are particularly vulnerable to warming conditions which will be accompanied by drought conditions (Allen et al., 2017). Additionally, tropical plants face their limited ability to assimilate in warmer temperature because they have comparatively reduced thermal niche breadth. Thus, understanding the PFTs acquires more importance in order to understand the traits which might provide a certain resilience to plants.

PATTERNS AND DRIVERS OF LEAF TRAITS IN TROPICAL DRY FORESTS

In a restoration planting in a renounced field in Mexico, Martinez-Garza et al. (2013) assessed the connection between execution (survival and development) of 24 species and 13 hidden practical qualities that are essential for leaf show, tree design, and propagation. The observations from this examination exhibited a positive connection between development rates and survival with crown size, and negative connection with seed mass, while no relationship with leaf traits. A study in Bolivia by Poorter and Bongers (2006) found leaves had more significance in the understanding of both the light demanding and shaded plants. It was explored that species with high growth rates in low light conditions had inferior, short-lived, and physiologically dynamic leaves, while species with high survival in the understory (shade tolerant species) had comparatively superior and long-lasting leaves. A trade-off was found among development and survival, where species with enduring leaves (high survival) had low SLA, while species with quick development rates portrayed a high SLA and nitrogen focus, and were physiologically more dynamic by having high photosynthesis, breadth, and conductance rates (Poorter & Bongers, 2006). This indicated that leaves with higher growth rate rapidly died (Drenovsky et al., 2012).

The magnitude of phenological variations in the context of the trait-based approach has been less documented for the trait-level co-variations and PFTs (Wolkovich & Cleland, 2014). Evergreen species have comparatively little seasonal change with different leaf functionality (Méndez-Alonzo et al., 2012) and have a different phenological strategy. Although, it is documented that seasonality modifies the photosynthetic capacity of plants (Wilson et al., 2001), ignoring phenological variations while comparing plant groups at spatiotemporal scales may provide inconsistent as well as ambiguous results (McKown et al., 2013; Niinemets et al.,2015). It has been estimated that at least one photo-synthetically active leaf unit is always present in evergreen plant species to provide them necessary carbon supply. Whether deciduous species are different in terms of functioning and nutrient allocations during the growing season is yet to be explored (Chabot & Hicks,1982). For example, deciduous tree species produce less healthy leaves (i.e. lower LMA), when compared with evergreen species. Therefore, deciduous species are eventually more vulnerable to carbon limitations and environmental changes (Piper & Fajardo, 2014). The hypothesis that there is a link between the range of species and plant attributes, is responsible for usage traits in the community ecology (McGill et al., 2006; Weiher, 2011). To examine an ecosystem community, trait-based studies use the trait mean approach (i.e. assigning all conspecific individuals to a species' average trait value) (Kraft et al., 2008). The embedded hypothesis in many of these studies is that inter-specific trait differences are much larger than intra-specific trait differences (Garnier et al., 2011; Laughlin et al., 2014). However, there is cumulative evidence that community assembly at local scales depends critically on the extent of intra-specific trait variation (Albert et al., 2010; Paine et al., 2011).

LINKING LEAF TRAITS WITH SPATIAL DISTRIBUTION IN TROPICAL DRY FORESTS

The abiotic and biotic environment along with historical factors determine the distribution of plant species. Species interactions with their natural and physical environment have impacted the distributions of species over larger geographical ranges. The inherent variability of dry forest ecosystems, however, allows only rough generalizations of environmental character and classification of dry forest. For decades, plant ecologists are preferably examining the functional attributes of the plants to explore the coexistence and distribution (Schimper, 1898).

A major theme in plant ecology has been to quantify and define plant functional strategies. A functional trait-based approach, especially leaf traits, give an overview of functional differences across large geographical gradients between and within TDFs. The relationship between changes in PFTs, seasonal changes and changes in mean annual precipitation at local scales have been established (Poorter, 2009; Gotsch et al., 2010; Rentería and Jaramillo, 2011). However, as observed by Paine et al. (2015), PFTs based approaches may not reflect the same results at wider scales due to local scale variability in different environmental factors. Such limitations can be overcome by considering a larger sample size for species-rich tropical forest communities. However, this approach makes generating functional trait datasets more challenging under TDFs.

CONCLUSION

TDFs are gaining less attention when compared with humid forest. However, their value and vulnerability in terms of ecosystem services are globally growing. Climate change has negatively affected the ecological processes and gained much attention but the effect of more frequent and acute climatic changes in tropical regions have been less explored. Studies show that because of various factors (including local and microclimate changes), these forests are fragmented, patched or altered. In order to advance our knowledge about TDFs, an eco-physiological framework is needed to understand the responses of plant species under the changing climate. Understanding of important leaf traits i.e. LA, LDMC, SLA, leaf nitrogen and leaf phosphorous, etc can be useful to discern the constraints and relation of plant responses to changing environmental conditions. Concluding from the present discourse, it can be stated that the complex interactions between plant phenology (both leaf and whole plant phonologies) and short-term and long-term effects of changing climate need better attention of the scientific community. Nevertheless, the study of PFTs in TDFs requires a large scale field surveys which might be difficult for inaccessible regions. Also, a need for systems modeling using mathematical approaches rather than a description of traits in now required in order to further the role of PFTs in forest ecology.

ACKNOWLEDGMENT

We acknowledge the help of Mr. Rishikesh Singh in preparing the manuscript and the Head, Institute of Environment & Sustainable Development, Banaras Hindu University, Varanasi, and University Grants Commission (UGC), New Delhi, for lab facilities. I also extend my warm thanks to four anonymous reviewers for their suggestion to make this manuscript more relevant.

REFERENCES

Albert, C. H., Thuiller, W., Yoccoz, N. G., Douzet, R., Aubert, S., & Lavorel, S. (2010). A multi-trait approach reveals the structure and the relative importance of intra-vs. interspecific variability in plant traits. *Functional Ecology*, *24*(6), 1192–1201. doi:10.1111/j.1365-2435.2010.01727.x

Allen, K., Dupuy, J. M., Gei, M. G., Hulshof, C., Medvigy, D., Pizano, C., & Waring, B. G. (2017). Will seasonally dry tropical forests be sensitive or resistant to future changes in rainfall regimes? *Environmental Research Letters*, *12*(2). doi:10.1088/1748-9326/aa5968

Álvarez-Yépiz, J. C., Martínez-Yrízar, A., & Fredericksen, T. S. (2018). Resilience of tropical dry forests to extreme disturbance events.

Anderegg, W. R., Hicke, J. A., Fisher, R. A., Allen, C. D., Aukema, J., Bentz, B., ... Pan, Y. (2015). Tree mortality from drought, insects, and their interactions in a changing climate. *The New Phytologist*, *208*(3), 674–683. doi:10.1111/nph.13477 PMID:26058406

Angert, A. L., Huxman, T. E., Chesson, P., & Venable, D. L. (2009). Functional tradeoffs determine species coexistence via the storage effect. *Proceedings of the National Academy of Sciences of the United States of America*, *106*(28), 11641–11645. doi:10.1073/pnas.0904512106 PMID:19571002

Aubin, I., Munson, A. D., Cardou, F., Burton, P. J., Isabel, N., Pedlar, J. H., ... Messier, C. (2016). Traits to stay, traits to move: A review of functional traits to assess sensitivity and adaptive capacity of temperate and boreal trees to climate change. *Environmental Reviews*, *24*(2), 164–186. doi:10.1139/er-2015-0072

Bhadouria, R., Srivastava, P., Singh, R., Tripathi, S., Singh, H., & Raghubanshi, A. S. (2017). Tree seedling establishment in dry tropics: An urgent need of interaction studies. *Environment Systems & Decisions*, *37*(1), 88–100. doi:10.100710669-017-9625-x

Bhadouria, R., Srivastava, P., Singh, S., Singh, R., Raghubanshi, A. S., & Singh, J. S. (2018). Effects of light, nutrient and grass competition on growth of seedlings of four tropical tree species. *Indian Forester*, *144*(1), 54–65.

Bognounou, F., Thiombiano, A., Oden, P. C., & Guinko, S. (2010). Seed provenance and latitudinal gradient effects on seed germination capacity and seedling establishment of five indigenous species in Burkina Faso. *Tropical Ecology*, *51*(2), 207.

Bruelheide, H., Dengler, J., Purschke, O., Lenoir, J., Jiménez-Alfaro, B., Hennekens, S. M., ... & Kattge, J. (2018). Global trait–environment relationships of plant communities. *Nature Ecology & Evolution*, *2*(12), 1906.

Cadotte, M., Albert, C. H., & Walker, S. C. (2013). The ecology of differences: Assessing community assembly with trait and evolutionary distances. *Ecology Letters*, *16*(10), 1234–1244. doi:10.1111/ele.12161 PMID:23910526

Centritto, M., Loreto, F., Massacci, A., Pietrini, F., Villani, M. C., & Zacchini, M. (2000). Improved growth and water use efficiency of cherry saplings under reduced light intensity. *Ecological Research*, *15*(4), 385–392. doi:10.1046/j.1440-1703.2000.00359.x

Cernusak, L. A., Hutley, L. B., Beringer, J., Holtum, J. A., & Turner, B. L. (2011). Photosynthetic physiology of eucalypts along a sub-continental rainfall gradient in northern Australia. *Agricultural and Forest Meteorology*, *151*(11), 1462–1470. doi:10.1016/j.agrformet.2011.01.006

Chabot, B. F., & Hicks, D. J. (1982). The ecology of leaf life spans. *Annual Review of Ecology and Systematics*, *13*(1), 229–259. doi:10.1146/annurev.es.13.110182.001305

Chapin, F. S. III, Rupp, T. S., Starfield, A. M., DeWilde, L. O., Zavaleta, E. S., Fresco, N., ... McGuire, A. D. (2003). Planning for resilience: Modeling change in human–fire interactions in the Alaskan boreal forest. *Frontiers in Ecology and the Environment*, *1*(5), 255–261. doi:10.1890/1540-9295(2003)001[0255:PFRMCI]2.0.CO;2

Chaturvedi, R. K., Raghubanshi, A. S., & Singh, J. S. (2011). Leaf attributes and tree growth in a tropical dry forest. *Journal of Vegetation Science*, *22*(5), 917–931. doi:10.1111/j.1654-1103.2011.01299.x

Chen, J. M., & Black, T. A. (1992). Defining leaf area index for non-flat leaves. *Plant, Cell & Environment*, *15*(4), 421–429. doi:10.1111/j.1365-3040.1992.tb00992.x

Connell, J. H., & Slatyer, R. O. (1977). Mechanisms of succession in natural communities and their role in community stability and organization. *American Naturalist*, *111*(982), 1119–1144. doi:10.1086/283241

Cornelissen, J. H. C., Lavorel, S., Garnier, E., Diaz, S., Buchmann, N., Gurvich, D. E., & Pausas, J. G. (2003). A handbook of protocols for standardised and easy measurement of plant functional traits worldwide. *Australian Journal of Botany*, *51*(4), 335–380. doi:10.1071/BT02124

Cornwell, W. K., Cornelissen, J. H., Amatangelo, K., Dorrepaal, E., Eviner, V. T., Godoy, O., & Quested, H. M. (2008). Plant species traits are the predominant control on litter decomposition rates within biomes worldwide. *Ecology Letters*, *11*(10), 1065–1071. doi:10.1111/j.1461-0248.2008.01219.x PMID:18627410

Cortois, R., Schröder-Georgi, T., Weigelt, A., van der Putten, W. H., & De Deyn, G. B. (2016). Plant–soil feedbacks: Role of plant functional group and plant traits. *Journal of Ecology*, *104*(6), 1608–1617. doi:10.1111/1365-2745.12643

Craine, J. M., & Dybzinski, R. (2013). Mechanisms of plant competition for nutrients, water and light. *Functional Ecology*, *27*(4), 833–840. doi:10.1111/1365-2435.12081

Curtis, E. M., Gollan, J., Murray, B. R., & Leigh, A. (2016). Native microhabitats better predict tolerance to warming than latitudinal macro-climatic variables in arid-zone plants. *Journal of Biogeography*, *43*(6), 1156–1165. doi:10.1111/jbi.12713

Dawson, T. E., & Pate, J. S. (1996). Seasonal water uptake and movement in root systems of Australian phraeatophytic plants of dimorphic root morphology: A stable isotope investigation. *Oecologia*, *107*(1), 13–20. doi:10.1007/BF00582230 PMID:28307187

Dhyani, S., & Dhyani, D. (2016). Significance of provisioning ecosystem services from moist temperate forest ecosystems: Lessons from upper Kedarnath Valley, Garhwal, India. *Energy, Ecology & Environment*, *1*(2), 109–121.

Diaz, S., Hodgson, J. G., Thompson, K., Cabido, M., Cornelissen, J. H., Jalili, A., ... Band, S. R. (2004). The plant traits that drive ecosystems: Evidence from three continents. *Journal of Vegetation Science*, *15*(3), 295–304. doi:10.1111/j.1654-1103.2004.tb02266.x

Drenovsky, R. E., Grewell, B. J., D'antonio, C. M., Funk, J. L., James, J. J., Molinari, N., & Richards, C. L. (2012). A functional trait perspective on plant invasion. *Annals of Botany*, *110*(1), 141–153. doi:10.1093/aob/mcs100 PMID:22589328

Flow, E. S. F., & Change, G. (2004). Reference Module in Earth Systems and Environmental Sciences.

Freschet, G. T., Aerts, R., & Cornelissen, J. H. (2012). A plant economics spectrum of litter decomposability. *Functional Ecology*, *26*(1), 56–65. doi:10.1111/j.1365-2435.2011.01913.x

Fukami, T., Bezemer, T. M., Mortimer, S. R., & van der Putten, W. H. (2005). Species divergence and trait convergence in experimental plant community assembly. *Ecology Letters*, *8*(12), 1283–1290. doi:10.1111/j.1461-0248.2005.00829.x

Funk, J. L. (2008). Differences in plasticity between invasive and native plants from a low resource environment. *Journal of Ecology*, *96*(6), 1162–1173. doi:10.1111/j.1365-2745.2008.01435.x

Garnier, E., Cortez, J., Billès, G., Navas, M. L., Roumet, C., Debussche, M., & Neill, C. (2004). Plant functional markers capture ecosystem properties during secondary succession. *Ecology*, *85*(9), 2630–2637. doi:10.1890/03-0799

Garnier, E., & Laurent, G. (1994). Leaf anatomy, specific mass and water content in congeneric annual and perennial grass species. *The New Phytologist*, *128*(4), 725–736. doi:10.1111/j.1469-8137.1994.tb04036.x

Garnier, E., & Navas, M. L. (2011). 1 5 Assessing the Functional Role of Plant Diversity in Grasslands: a Trait-based Approach. Grassland Productivity and Ecosystem Services, 138.

Garnier, E., Shipley, B., Roumet, C., & Laurent, G. (2001). A standardized protocol for the determination of specific leaf area and leaf dry matter content. *Functional Ecology*, *15*(5), 688–695. doi:10.1046/j.0269-8463.2001.00563.x

Giraldo, J. P., & Holbrook, N. M. (2011). Physiological mechanisms underlying the seasonality of leaf senescence and renewal in seasonally dry tropical forest trees. In *Seasonally Dry Tropical Forests* (pp. 129-140). Washington, DC: Island Press.

Gotsch, S. G., Geiger, E. L., Franco, A. C., Goldstein, G., Meinzer, F. C., & Hoffmann, W. A. (2010). Allocation to leaf area and sapwood area affects water relations of co-occurring savanna and forest trees. *Oecologia*, *163*(2), 291–301. doi:10.100700442-009-1543-2 PMID:20058025

Grassi, G., & Magnani, F. (2005). Stomatal, mesophyll conductance and biochemical limitations to photosynthesis as affected by drought and leaf ontogeny in ash and oak trees. *Plant, Cell & Environment*, *28*(7), 834–849. doi:10.1111/j.1365-3040.2005.01333.x

Grime, J. P. (1977). Evidence for the existence of three primary strategies in plants and its relevance to ecological and evolutionary theory. *American Naturalist*, *111*(982), 1169–1194. doi:10.1086/283244

Grime, J. P. (1979). *Plant strategies and vegetation processes.* Plant strategies and vegetation processes.

Holdridge, L. R. (1967). Life zone ecology. Life zone ecology, (rev. ed.).

Hooper, D. U., Chapin, F. S., Ewel, J. J., Hector, A., Inchausti, P., Lavorel, S., & Schmid, B. (2005). Effects of biodiversity on ecosystem functioning: A consensus of current knowledge. *Ecological Monographs*, *75*(1), 3–35. doi:10.1890/04-0922

Horneck, D. A., & Miller, R. O. (1997). Determination of total nitrogen in plant tissue. In Handbook of reference methods for plant analysis (pp. 85–93). Boca Raton, FL: CRC Press.

Hubbell, S. P. (1983). Diversity of canopy trees in a neotropical forest and implications for conservation. *Tropical rain forest: ecology and management*, 25-41.

Hulshof, C. M., & Swenson, N. G. (2010). Variation in leaf functional trait values within and across individuals and species: An example from a Costa Rican dry forest. *Functional Ecology*, *24*(1), 217–223. doi:10.1111/j.1365-2435.2009.01614.x

India, N. B. S. A. P. (1999). *National Policy and Macro Level Action on Biodiversity*. New Delhi, India: Government of India, Ministry of Environment and Forests.

Janzen, D. H. (1988). *Tropical dry forest: the most endangered major tropical ecosystem. In E. O. Wilson (Ed.), Biodiversity.*

Jaramillo, V. J., MartÍnez-YrÍzar, A., & Sanford, R. L. (2011). Primary productivity and biogeochemistry of seasonally dry tropical forests. In Seasonally dry tropical forests (pp. 109-128). Washington, DC: Island Press.

Kazakou, E., Violle, C., Roumet, C., Navas, M. L., Vile, D., Kattge, J., & Garnier, E. (2014). Are trait-based species rankings consistent across data sets and spatial scales? *Journal of Vegetation Science*, *25*(1), 235–247. doi:10.1111/jvs.12066

Khurana, E., & Singh, J. S. (2004). Germination and seedling growth of five tree species from tropical dry forest in relation to water stress: Impact of seed size. *Journal of Tropical Ecology*, *20*(4), 385–396. doi:10.1017/S026646740400135X

Kitajima, K., Mulkey, S. S., & Wright, S. J. (1997). Decline of photosynthetic capacity with leaf age in relation to leaf longevities for five tropical canopy tree species. *American Journal of Botany*, *84*(5), 702–708. doi:10.2307/2445906 PMID:21708622

Kraft, N. J., Valencia, R., & Ackerly, D. D. (2008). Functional traits and niche-based tree community assembly in an Amazonian forest. *Science*, *322*(5901), 580–582. doi:10.1126cience.1160662 PMID:18948539

Lambers, H. A. N. S., & Poorter, H. (1992). Inherent variation in growth rate between higher plants: A search for physiological causes and ecological consequences. [Academic Press]. *Advances in Ecological Research*, *23*, 187–261. doi:10.1016/S0065-2504(08)60148-8

Laughlin, D. C. (2014). The intrinsic dimensionality of plant traits and its relevance to community assembly. *Journal of Ecology*, *102*(1), 186–193. doi:10.1111/1365-2745.12187

Laughlin, D. C., Leppert, J. J., Moore, M. M., & Sieg, C. H. (2010). A multi-trait test of the leaf-height-seed plant strategy scheme with 133 species from a pine forest flora. *Functional Ecology*, *24*(3), 493–501. doi:10.1111/j.1365-2435.2009.01672.x

Lavorel, S., Díaz, S., Cornelissen, J. H. C., Garnier, E., Harrison, S. P., McIntyre, S., & Urcelay, C. (2007). Plant functional types: are we getting any closer to the Holy Grail? In *Terrestrial ecosystems in a changing world* (pp. 149–164). Berlin, Germany: Springer. doi:10.1007/978-3-540-32730-1_13

Lavorel, S., & Garnier, É. (2002). Predicting changes in community composition and ecosystem functioning from plant traits: Revisiting the Holy Grail. *Functional Ecology*, *16*(5), 545–556. doi:10.1046/j.1365-2435.2002.00664.x

Loranger, J., & Shipley, B. (2010). Interspecific covariation between stomatal density and other functional leaf traits in a local flora. *Botany*, *88*(1), 30–38. doi:10.1139/B09-103

Maass, M., & Burgos, A. (2011). Water dynamics at the ecosystem level in seasonally dry tropical forests. In *Seasonally dry tropical forests* (pp. 141–156). Washington, DC: Island Press. doi:10.5822/978-1-61091-021-7_9

Marenco, R. A., Antezana-Vera, S. A., & Nascimento, H. C. S. (2009). Relationship between specific leaf area, leaf thickness, leaf water content and SPAD-502 readings in six Amazonian tree species. *Photosynthetica*, *47*(2), 184–190. doi:10.100711099-009-0031-6

Marino, G., Aqil, M., & Shipley, B. (2010). The leaf economics spectrum and the prediction of photosynthetic light–response curves. *Functional Ecology*, *24*(2), 263–272. doi:10.1111/j.1365-2435.2009.01630.x

Markesteijn, L., & Poorter, L. (2009). Seedling root morphology and biomass allocation of 62 tropical tree species in relation to drought-and shade-tolerance. *Journal of Ecology*, *97*(2), 311–325. doi:10.1111/j.1365-2745.2008.01466.x

Martínez-Garza, C., Bongers, F., & Poorter, L. (2013). Are functional traits good predictors of species performance in restoration plantings in tropical abandoned pastures? *Forest Ecology and Management*, *303*, 35–45. doi:10.1016/j.foreco.2013.03.046

Martínez-Yrízar, A., Núñez, S., Miranda, H., & Búrquez, A. (1999). Temporal and spatial variation of litter production in Sonoran Desert communities. *Plant Ecology*, *145*(1), 37–48. doi:10.1023/A:1009896201047

McGill, B. J., Enquist, B. J., Weiher, E., & Westoby, M. (2006). Rebuilding community ecology from functional traits. *Trends in Ecology & Evolution*, *21*(4), 178–185. doi:10.1016/j.tree.2006.02.002 PMID:16701083

McKown, A. D., Guy, R. D., Azam, M. S., Drewes, E. C., & Quamme, L. K. (2013). Seasonality and phenology alter functional leaf traits. *Oecologia*, *172*(3), 653–665. doi:10.100700442-012-2531-5 PMID:23212534

Méndez-Alonzo, R., Paz, H., Zuluaga, R. C., Rosell, J. A., & Olson, M. E. (2012). Coordinated evolution of leaf and stem economics in tropical dry forest trees. *Ecology*, *93*(11), 2397–2406. doi:10.1890/11-1213.1 PMID:23236911

Meziane, D., & Shipley, B. (1999). Interacting determinants of specific leaf area in 22 herbaceous species: Effects of irradiance and nutrient availability. *Plant, Cell & Environment*, *22*(5), 447–459. doi:10.1046/j.1365-3040.1999.00423.x

Mokany, K., Ash, J., & Roxburgh, S. (2008). Functional identity is more important than diversity in influencing ecosystem processes in a temperate native grassland. *Journal of Ecology*, *96*(5), 884–893. doi:10.1111/j.1365-2745.2008.01395.x

Murphy, P. G., & Lugo, A. E. (1986). Ecology of tropical dry forest. *Annual Review of Ecology and Systematics*, *17*(1), 67–88. doi:10.1146/annurev.es.17.110186.000435

Nicotra, A. B., Leigh, A., Boyce, C. K., Jones, C. S., Niklas, K. J., Royer, D. L., & Tsukaya, H. (2011). The evolution and functional significance of leaf shape in the angiosperms. *Functional Plant Biology*, *38*(7), 535–552. doi:10.1071/FP11057

Niinemets, U. (2010). A review of light interception in plant stands from leaf to canopy in different plant functional types and in species with varying shade tolerance. *Ecological Research*, *25*(4), 693–714. doi:10.100711284-010-0712-4

Niinemets, U., Díaz-Espejo, A., Flexas, J., Galmés, J., & Warren, C. R. (2009). Role of mesophyll diffusion conductance in constraining potential photosynthetic productivity in the field. *Journal of Experimental Botany*, *60*(8), 2249–2270. doi:10.1093/jxb/erp036 PMID:19395391

Niinemets, U., Keenan, T. F., & Hallik, L. (2015). A worldwide analysis of within-canopy variations in leaf structural, chemical and physiological traits across plant functional types. *The New Phytologist*, *205*(3), 973–993. doi:10.1111/nph.13096 PMID:25318596

Norden, N., Angarita, H. A., Bongers, F., Martínez-Ramos, M., Granzow-de la Cerda, I., Van Breugel, M., & Finegan, B. (2015). Successional dynamics in Neotropical forests are as uncertain as they are predictable. *Proceedings of the National Academy of Sciences of the United States of America*, *112*(26), 8013–8018. doi:10.1073/pnas.1500403112 PMID:26080411

Olivares, E., & Medina, E. (1992). Water and nutrient relations of woody perennials from tropical dry forests. *Journal of Vegetation Science*, *3*(3), 383–392. doi:10.2307/3235764

Paine, C. T., Amissah, L., Auge, H., Baraloto, C., Baruffol, M., Bourland, N., ... Doust, S. (2015). Globally, functional traits are weak predictors of juvenile tree growth, and we do not know why. *Journal of Ecology*, *103*(4), 978–989. doi:10.1111/1365-2745.12401

Paine, C. T., Baraloto, C., Chave, J., & Hérault, B. (2011). Functional traits of individual trees reveal ecological constraints on community assembly in tropical rain forests. *Oikos*, *120*(5), 720–727. doi:10.1111/j.1600-0706.2010.19110.x

Peters, E. B., Wythers, K. R., Zhang, S., Bradford, J. B., & Reich, P. B. (2013). Potential climate change impacts on temperate forest ecosystem processes. *Canadian Journal of Forest Research*, *43*(10), 939–950. doi:10.1139/cjfr-2013-0013

Piper, F. I., & Fajardo, A. (2014). Foliar habit, tolerance to defoliation and their link to carbon and nitrogen storage. *Journal of Ecology*, *102*(5), 1101–1111. doi:10.1111/1365-2745.12284

Poorter, H., Niinemets, Ü., Poorter, L., Wright, I. J., & Villar, R. (2009). Causes and consequences of variation in leaf mass per area (LMA): A meta-analysis. *The New Phytologist*, *182*(3), 565–588. doi:10.1111/j.1469-8137.2009.02830.x PMID:19434804

Poorter, L., & Bongers, F. (2006). Leaf traits are good predictors of plant performance across 53 rain forest species. *Ecology*, *87*(7), 1733–1743. doi:10.1890/0012-9658(2006)87[1733:LTAGPO]2.0.CO;2 PMID:16922323

Poorter, L., Wright, S. J., Paz, H., Ackerly, D. D., Condit, R., Ibarra-Manríquez, G., & Muller-Landau, H. C. (2008). Are functional traits good predictors of demographic rates? Evidence from five neotropical forests. *Ecology*, *89*(7), 1908–1920. doi:10.1890/07-0207.1 PMID:18705377

Portillo-Quintero, C. A., & Sánchez-Azofeifa, G. A. (2010). Extent and conservation of tropical dry forests in the Americas. *Biological Conservation*, *143*(1), 144–155. doi:10.1016/j.biocon.2009.09.020

Powers, J. S., Montgomery, R. A., Adair, E. C., Brearley, F. Q., DeWalt, S. J., Castanho, C. T., & González-Iturbe, J. A. (2009). Decomposition in tropical forests: A pan-tropical study of the effects of litter type, litter placement and mesofaunal exclusion across a precipitation gradient. *Journal of Ecology*, *97*(4), 801–811. doi:10.1111/j.1365-2745.2009.01515.x

Prieto-Torres, D. A., Navarro-Sigüenza, A. G., Santiago-Alarcon, D., & Rojas-Soto, O. R. (2016). Response of the endangered tropical dry forests to climate change and the role of Mexican Protected Areas for their conservation. *Global Change Biology*, *22*(1), 364–379. doi:10.1111/gcb.13090 PMID:26367278

Raven, J. A. (1993). The evolution of vascular plants in relation to quantitative functioning of dead water-conducting cells and stomata. *Biological Reviews of the Cambridge Philosophical Society*, *68*(3), 337–363. doi:10.1111/j.1469-185X.1993.tb00735.x

Reich, P. B. (2014). The world-wide 'fast–slow' plant economics spectrum: A traits manifesto. *Journal of Ecology*, *102*(2), 275–301. doi:10.1111/1365-2745.12211

Reich, P. B., Walters, M. B., & Ellsworth, D. S. (1992). Leaf life-span in relation to leaf, plant, and stand characteristics among diverse ecosystems. *Ecological Monographs*, *62*(3), 365–392. doi:10.2307/2937116

Reich, P. B., Walters, M. B., Kloeppel, B. D., & Ellsworth, D. S. (1995). Different photosynthesis-nitrogen relations in deciduous hardwood and evergreen coniferous tree species. *Oecologia, 104*(1), 24–30. doi:10.1007/BF00365558 PMID:28306909

Rentería, L. Y., & Jaramillo, V. J. (2011). Rainfall drives leaf traits and leaf nutrient resorption in a tropical dry forest in Mexico. *Oecologia, 165*(1), 201–211. doi:10.100700442-010-1704-3 PMID:20607297

Sanches, L., Valentini, C. M. A., Júnior, O. B. P., de Souza Nogueira, J., Vourlitis, G. L., Biudes, M. S., & de Almeida Lobo, F. (2008). Seasonal and interannual litter dynamics of a tropical semideciduous forest of the southern Amazon Basin, Brazil. *Journal of Geophysical Research. Biogeosciences, 113*(G4).

Santiago, L. S., Goldstein, G., Meinzer, F. C., Fisher, J. B., Machado, K., Woodruff, D., & Jones, T. (2004). Leaf photosynthetic traits scale with hydraulic conductivity and wood density in Panamanian forest canopy trees. *Oecologia, 140*(4), 543–550. doi:10.100700442-004-1624-1 PMID:15232729

Schimper, A. F. W. (1898). Pflanzen-geographie auf physiologischerGrundlage. Germany: Gustav Fischer.

Scholander, P. F., Bradstreet, E. D., Hammel, H. T., & Hemmingsen, E. A. (1966). Sap concentrations in halophytes and some other plants. *Plant Physiology, 41*(3), 529–532. doi:10.1104/pp.41.3.529 PMID:5906381

Schönbeck, L., Lohbeck, M., Bongers, F., Ramos, M., & Sterck, F. (2015). How do light and water acquisition strategies affect species selection during secondary succession in moist tropical forests? *Forests, 6*(6), 2047–2065. doi:10.3390/f6062047

Shipley, B., & Vu, T. T. (2002). Dry matter content as a measure of dry matter concentration in plants and their parts. *The New Phytologist, 153*(2), 359–364. doi:10.1046/j.0028-646X.2001.00320.x

Souza, M. C., Franco, A. C., Haridasan, M., Rossatto, D. R., Araújo, J. F. D., Morellato, L. P. C., & Habermann, G. (2015). The length of the dry season may be associated with leaf scleromorphism in cerrado plants. *Anais da Academia Brasileira de Ciências, 87*(3), 1691–1699. doi:10.1590/0001-376520150381 PMID:26221991

Sterck, F., Markesteijn, L., Schieving, F., & Poorter, L. (2011). Functional traits determine trade-offs and niches in a tropical forest community. *Proceedings of the National Academy of Sciences of the United States of America, 108*(51), 20627–20632. doi:10.1073/pnas.1106950108 PMID:22106283

Suding, K. N., Gross, K. L., & Houseman, G. R. (2004). Alternative states and positive feedbacks in restoration ecology. *Trends in Ecology & Evolution, 19*(1), 46–53. doi:10.1016/j.tree.2003.10.005 PMID:16701225

Taiz, L., & Zeiger, E. (2006). Plant Physiology. Sunderland, MA: Sinauer Associates.

Taylor, B., & Parkinson, D. (1988). A new microcosm approach to litter decomposition studies. *Canadian Journal of Botany, 66*(10), 1933–1939. doi:10.1139/b88-265

Temminghoff, E. E., & Houba, V. J. (Eds.). (2004). *Plant analysis procedures* (Vol. 179). Dordrecht, The Netherlands: Kluwer Academic Publishers. doi:10.1007/978-1-4020-2976-9

Turner, B. L., Brenes-Arguedas, T., & Condit, R. (2018). Pervasive phosphorus limitation of tree species but not communities in tropical forests. *Nature*, *555*(7696), 367–370. doi:10.1038/nature25789 PMID:29513656

Turner, N. C. (1988). Measurement of plant water status by the pressure chamber technique. *Irrigation Science*, *9*(4), 289–308. doi:10.1007/BF00296704

Vaieretti, M. V., Díaz, S., Vile, D., & Garnier, E. (2007). Two measurement methods of leaf dry matter content produce similar results in a broad range of species. *Annals of Botany*, *99*(5), 955–958. doi:10.1093/aob/mcm022 PMID:17353207

Vile, D., Garnier, E., Shipley, B., Laurent, G., Navas, M. L., Roumet, C., ... Midgley, G. F. (2005). Specific leaf area and dry matter content estimate thickness in laminar leaves. *Annals of Botany*, *96*(6), 1129–1136. doi:10.1093/aob/mci264 PMID:16159941

Villar, R., Ruiz-Robleto, J., Ubera, J. L., & Poorter, H. (2013). Exploring variation in leaf mass per area (LMA) from leaf to cell: An anatomical analysis of 26 woody species. *American Journal of Botany*, *100*(10), 1969–1980. doi:10.3732/ajb.1200562 PMID:24107583

Violle, C., Navas, M. L., Vile, D., Kazakou, E., Fortunel, C., Hummel, I., & Garnier, E. (2007). Let the concept of trait be functional! *Oikos*, *116*(5), 882–892. doi:10.1111/j.0030-1299.2007.15559.x

Vogelman, T. C., Nishio, J. N., & Smith, W. K. (1996). Leaves and light capture: Light propagation and gradients of carbon fixation within leaves. *Trends in Plant Science*, *1*(2), 65–70. doi:10.1016/S1360-1385(96)80031-8

Vos, J., & Biemond, H. (1992). Effects of nitrogen on the development and growth of the potato plant. 1. Leaf appearance, expansion growth, life spans of leaves and stem branching. *Annals of Botany*, *70*(1), 27–35. doi:10.1093/oxfordjournals.aob.a088435

Weiher, E. (2011). A primer of trait and functional diversity. *Biological diversity: frontiers in measurement and assessment*, 175-193.

Weiher, E., Van Der Werf, A., Thompson, K., Roderick, M., Garnier, E., & Eriksson, O. (1999). Challenging Theophrastus: A common core list of plant traits for functional ecology. *Journal of Vegetation Science*, *10*(5), 609–620. doi:10.2307/3237076

West, G. B., Brown, J. H., & Enquist, B. J. (1999). A general model for the structure and allometry of plant vascular systems. *Nature*, *400*(6745), 664–667. doi:10.1038/23251

Westoby, M., Falster, D. S., Moles, A. T., Vesk, P. A., & Wright, I. J. (2002). Plant ecological strategies: Some leading dimensions of variation between species. *Annual Review of Ecology and Systematics*, *33*(1), 125–159. doi:10.1146/annurev.ecolsys.33.010802.150452

White, J. W., & Montes-r, C. (2005). Variation in parameters related to leaf thickness in common bean (Phaseolus vulgaris L.). *Field Crops Research*, *91*(1), 7–21. doi:10.1016/j.fcr.2004.05.001

Wilson, E. O. (1988). The current state of biological diversity. *Biodiversity (Nepean)*, *521*(1), 3–18.

Wilson, K. B., Baldocchi, D. D., & Hanson, P. J. (2001). Leaf age affects the seasonal pattern of photosynthetic capacity and net ecosystem exchange of carbon in a deciduous forest. *Plant, Cell & Environment*, *24*(6), 571–583. doi:10.1046/j.0016-8025.2001.00706.x

Witkowski, E. T. F., Lamont, B. B., Walton, C. S., & Radford, S. (1992). Leaf demography, sclerophylly and ecophysiology of two Banksias with contrasting leaf life spans. *Australian Journal of Botany*, *40*(6), 849–862. doi:10.1071/BT9920849

Wolkovich, E. M., & Cleland, E. E. (2014). Phenological niches and the future of invaded ecosystems with climate change. *AoB Plants*, 6. PMID:24876295

Wright, I. J., Reich, P. B., Westoby, M., Ackerly, D. D., Baruch, Z., Bongers, F., & Flexas, J. (2004). The worldwide leaf economics spectrum. *Nature*, *428*(6985), 821–827. doi:10.1038/nature02403 PMID:15103368

Wright, I. J., Westoby, M., & Reich, P. B. (2002). Convergence towards higher leaf mass per area in dry and nutrient-poor habitats has different consequences for leaf life span. *Journal of Ecology*, *90*(3), 534–543. doi:10.1046/j.1365-2745.2002.00689.x

Wright, J. W., Davies, K. F., Lau, J. A., McCall, A. C., & McKay, J. K. (2006). Experimental verification of ecological niche modeling in a heterogeneous environment. *Ecology*, *87*(10), 2433–2439. doi:10.1890/0012-9658(2006)87[2433:EVOENM]2.0.CO;2 PMID:17089652

Zhou, H. C., Tam, N. F. Y., Lin, Y. M., Ding, Z. H., Chai, W. M., & Wei, S. D. (2014). Relationships between degree of polymerization and antioxidant activities: A study on proanthocyanidins from the leaves of a medicinal mangrove plant Ceriopstagal. *PLoS One*, *9*(10). doi:10.1371/journal.pone.0107606 PMID:25313980

Zhu, S. D., & Cao, K. F. (2010). Contrasting cost–benefit strategy between lianas and trees in a tropical seasonal rain forest in southwestern China. *Oecologia*, *163*(3), 591–599. doi:10.100700442-010-1579-3 PMID:20191291

Section 3

Forest Diversity, Degradation, and Management

Chapter 6
Diversity and Distribution of Tropical Dry Forests:
A Case Study From Pudukkottai District of Tamil Nadu, India – Sacred Groves

Raja Prakasam
J. J. College of Arts and Science (Autonomous), India

Balaguru Balakrishnan
Jamal Mohamed College (Autonomous), India

Soosairaj Sebastian
St. Joseph's College (Autonomous), India

ABSTRACT

Tropical dry forests occur as patches in Tamil Nadu distributed along the East Coast, Eastern Ghats, and plains of the Indian Peninsula. The floristic studies of these regions are of great national relevance as plant resources in a tropical climate contribute to national wealth. Dry forests of the plains in Tamil Nadu have been neglected and the area under study has remained practically unexplored. This chapter studies distribution of tropical dry forests, especially in Pudukkottai district in Tamil Nadu. In total, 187 sacred groves were surveyed for their distribution and floristic composition. The GPS position of each grove was noted and their distribution maps were prepared. The groves were classified based on conservation status, namely well conserved, moderately conserved and degraded. Extensive botanical explorations were carried out periodically during 2012–2016 in these groves and 812 species belonging to 480 genera under 124 families were recorded. The endemic, threatened species of these groves were also documented.

DOI: 10.4018/978-1-7998-0014-9.ch006

INTRODUCTION

Tropical forest ecosystems are of global importance and interest for several reasons. Tropical forests, with their incomparable variety of plant and animal species, offer an irreplaceable source of biological diversity. The tropical habitats have been described as the most species rich formation on earth and Mittermier & Weiner (1990) while identifying mega diversity regions of the world observed that more than 80% of the diversity is stored in the tropics. Indian forests exhibit ample diversity in their community structure owing to the existence of varied climatic and topographic conditions. Most of the tropical dry forests have the highest number of endemic species with fragmented landscapes (Myers *et al.* 2000; Mittermeier *et al.* 2011). Hence, tropical dry forests are one of the most threatened ecosystems due to their limited remaining extent and the high threat from human activities (Janzen 1988; Sanchez-Azofeifa *et al.* 2005). Detailed analyses of tropical dry forest coverage remain scarce and are considered fundamental to improving knowledge and management of tropical forests (Carrión & Southworth 2018).

Tropical dry forests of Tamil Nadu are well known for their rich biodiversity encompassing varied vegetation types which range from dry evergreen, semi evergreen, dry mixed deciduous, dry grass lands, woody savannah, thorn forests and deciduous scrub. Gamble (1915-1936) enumerated 4,516 species in the then Madras Presidency, while after the State's reorganization 2,260 species belonging to 983 genera and 173 families have been described in Tamil Nadu and Carnatic region (Matthew 1981). All these studies focused on the forests in the hills, but less work concentrated in the tropical forests in the plains which includes the coastal regions of Tamil Nadu and vegetation in sacred groves. Sacred groves are an important places in which biodiversity is preserved mostly in undisturbed condition because of certain taboos and religious beliefs (King *et al.* 1997; Khan *et al.* 2008). The traditional conservation practices of indigenous people in many parts of the world as small forest patches by dedicating them to deity, contributed to the conservation and protection of biodiversity (Yadav *et al.* 2010). Sacred groves vary in size from a few hectares to few kilometres protected by local communities. They are ancient natural sanctuaries that have supported the growth of several interesting rare and endemic flora and fauna. Sacred groves serve as valuable storehouse of biodiversity as it harbour genotypes of future importance that may be very vital for breeding programs (Henry 2008). C.P. Ramaswamy Environmental Education Centre (CPREEC) reports 100,000–150,000 sacred groves all over India that harbour and act as repositories of rare flora and fauna (Henry 2008). Indian sacred groves represent a diverse spectrum of ecosystem with their presence in barren landscape, grassland, hill slope, agricultural landscape, coastal plain and desert. Sacred groves are protected by the belief, hence kept in a relatively undisturbed state. It is expressive of important relationship of human beings with nature (Bhagwat *et al.* 2005). Sacred sites are of conservation value because of their spiritual meaning, as cultural heritage (Jackel *et al.* 2013). Optimal and efficient management of forest resources calls for reliable technologies with provision to store, update, retrieve and analyze. Towards this, tools like Geographic Information System (GIS) and remote sensing have been used for plant resources conservation and management (Udayalakshmi *et al.* 1998; Natarajan *et al.* 2004). This article is focused towards the diversity and distribution of tropical dry forests in Tamil Nadu with special emphasis on distribution status of sacred groves in Pudukkottai district as a case.

STUDY AREA

Pudukkottai district is bestowed with natural resources of land and sea, situated at the eastern part of Tamil Nadu in Peninsular India (Figure 1). The district lies between North latitudes 9°50′ and 10°40′ and Eastern longitudes 78°25′ and 79°15′ covering an area of 4663.29 sq. km, occupying 3.5% of the total area of Tamil Nadu. It has a coastline of 42 km which includes 32 coastal villages and land boundary of 4621.29 km. The district is bordered, on the south by Sivaganga and Ramanathapuram, on the east by Bay of Bengal, on the north and west by Tiruchirappalli district and in the north by Thanjavur district (Envis 2015).

Administratively, the district is composed of 3 revenue divisions and 11 taluks. They are Aavudaiyarkoil, Alangudi, Aranthangi, Gandarvakkottai, Illupur, Karambakudi, Kulathur, Manamelkudi, Ponnamaravathi, Pudukkottai and Tirumayam. Topographically the landmass of the district can be divided broadly into two natural divisions, (a) the coastal plains that run parallel to the sea along coastal villages like Kattumavadi, Manamelkudi, Kottaipattinam, Mimisal, Muthukkuda and (b) the interior southern plains that harbours open scrub jungles, aquatic and semi aquatic vegetation and sacred groves.

The wealth of minerals such as limestone, lignite, granite, clay, gypsum, feldspar and graphite are found abundantly in Tamil Nadu (Kabeer & Nair 2009). Stone quarries exist in Pudukkottai, Tirumayam and Kulathur taluks. Multi-colour stones are quarried in Narthamalai, Sittanavasal and Kudimyanmalai regions. Small amount of clay deposits occur in Gandarvakkottai and Pudukkottai taluks and small quantities of quartz and limestone are also reported in Adhanakkottai area (Envis 2015). Major portion of the interior plains possess red sterile soil besides black, red loamy and red sandy soil. Sandy coastal alluvium is found along the coastline of Pudukkottai and a stretch of coastal sandy soil is found chiefly in Kodiakkarai in Manamelkudi taluk (Envis 2015).

Generally the climate of Pudukkottai district is very hot and dry thorough the year. In summer season average temperature is 31.5°C and during the winter season it is 25.3°C (Envis 2015). The average rainfall in Pudukkottai district is 821 mm. The rain is received from both monsoons. Highest rainfall with 397 mm receives during north-east monsoon followed by south-west monsoon with 303 mm of rainfall. Summer and winter rainfalls are 80 and 41 mm, respectively. Coastal taluks of Manamelkudi and Avudaiyarkoil receives the highest rainfall in the district (Envis 2015). Agniyar, Ambuliyar and South Vellar are non-perennial rivers. Vellar is the major river of the district. The district has a rich wetland ecosystem that comprises lakes, ponds, puddles. Also, seasonally waterlogged areas are found here. Particularly Aranthangi, Aavudaiyarkoil, Manamelkudi taluks have more number of wetlands and is the main source of water for irrigation and drinking purposes (Envis 2015).

METHODOLOGY

Field Survey

Extensive botanical explorations were conducted periodically during 2012–2016 in Pudukkottai district for thorough plant collections. Every season such as post-monsoon, summer, pre-monsoon and monsoon was covered and carefully plants with flower or fruit or both were collected and field numbers were tagged.

Figure 1. Study area: Taluks in Pudukkottai district

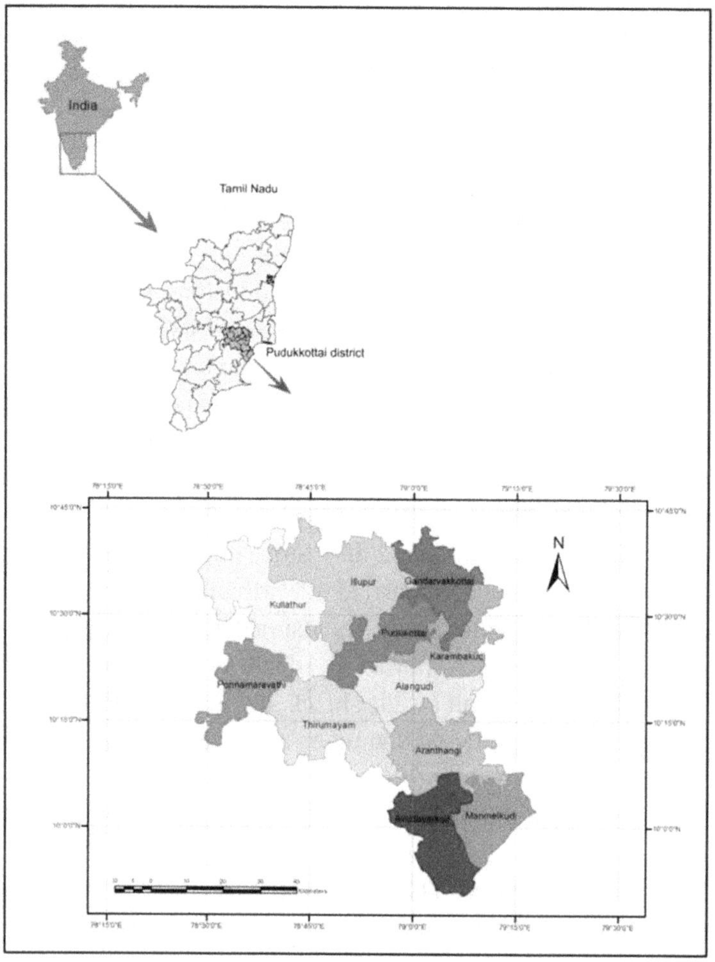

Mapping of Sacred Groves

JUNO SB Handheld GPS receivers were used to collect the Latitude and Longitude coordinates (in degree decimal) of the location of the sacred groves. Taluk map of Pudukottai were Geo-rectified, digitized with attribute as taluk name and projected as UTM44 N datum WGS 84 in the QGIS 2.8.3 version. The GPS data were entered into Microsoft XL sheet and saved in .csv format. The .csv/txt format were plotted using QGIS 2.8.3 software, generated points for x, y coordinates of the sacred groves. This map overlay on the respective taluk maps and thus generated the sacred grove distribution map. The maps were visualized with annotation, scale bar, direction and legends.

Tropical Dry Forests of Tamil Nadu

Tropical dry forests are distributed in mid elevation of Western Ghats, Eastern Ghats and Coromandal coast. The Western Ghats of Tamil Nadu, highly elevated mountain range is comprised of the Anthiyur and Sathyamangalam forest in the north to Nilgiris, Anaikatti, Megamalai, Palani hills, Sathuragiri, Agasthiyamalai, Kalakad Mudumalai and Mahendrahiri hills in the south. The Eastern Ghats of Tamil Nadu, broken and undulated elevated hills comprised of Javadi hills, Elagiri, Shevroy hills, Kalrayan hills, Chetteri hills, Kolli hills, Bodamalai, Pacchaimalai, Sirumalai and Karandamalai. The plains and the Coromandal coast is distributed in the central region of Tamil Nadu and towards the East Coast characterized by randomly distributed hillocks and reserved lands. The distribution of forest types were given below.

Dry Evergreen Forest

This forest type corresponds to Tropical Dry evergreen forest of 8A and 7C1 described by Champion & Seth (1968). In mid elevation of Western Ghats the composition includes *Nothopegia colebrookiana, Olea dioica, Phobe wightii, Vibrunum punctatum, Syzygium cumini, Celtis phillipensis, Celtis tetrandra, Prunus ceylanica, Chinanthus ramiflorus, Scolopia crenata, Lionociera ramiflora, Nothapodytes nimmoniana, Persea macrantha, Schefflera racemosa* and *Vitex altissima*. This forest type mostly distributed in hill plateau of the Eastern Ghats at an elevation above 900m msl (mean sea level) and the composition dominant trees include *Chionanthus ramflorus, Psydrax dicoccos, Memecylon edule, Ligustrum perrottetti, Olea dioica, Syzygium cumini, Olea paniculata, Vibrunum punctatum, Celtis philipensis, Aglaia roxburghiana, Canarium strictum,* and *Vitex altissima* (Balaguru *et al.* 2006). Distribution of *Gnetum ula* in this forest type need special mention as this is the only gymnosperm representative. Distribution of this forest in the plains is restricted to Corromandal coast where the rainfall is relatively high and the composition of vegetation includes *Manilkara hexandra, Pterospermum suberifolium, Syzygium cumini, Drypetes sepiaria* and *Pleiospermium alatum*. These forests consist of mostly small leathery-leaved evergreen trees with short trunks and spreading crowns (Venkateswaran & Parthasarathy 2005).

Dry Mixed Deciduous Forest

This type corresponds to Southern Tropical Dry mixed deciduous forest (5A/ C3). The composition of vegetation is dominant with deciduous tree species such as *Albizia amara, A. lebbeck, Chloroxylon swietenia, Wrightia tinctoria, Dalbergia paniculata, Gyrocarpus americanus* and *Commiphora caudata*. This forest type is mostly distributed in high slope regions between 400 to 800 m msl of the Western and the Eastern Ghats of Tamil Nadu (Natarajan *et al.* 2004). In plains this is common in undisturbed vegetation of sacred groves in Caudalore and Pudukkottai districts of Tamil Nadu.

Southern Thorn Forest

This forest is more prominent in the dry areas with low rainfall and high temperatures mostly at foothills of the Western Ghats, Eastern Ghats and the plains. Common species in this type are *Albizia amara, Morinda pubescens, Commiphora berryi* and *Atalantia racemosa*. This forest is characteristic with thorny species such as *Carissa carandas, Pterolobium hexapetalum, Acacia caesia, Capparis zeylanica, Zizyphus oenoplia, Z. mauritiana, Maytenus emerginata* and mostly very dense in nature (Soosairaj *et al.* 2004). This type is found as patches along the Corromandal coast and in the plains where rainfall is very scarce.

Dry Deciduous Scrub

This forest type is quite conspicuous throughout the Ghats and shows prominence of *Euphorbia antiquorum*, *Albizia amara, Wrightia tinctoria, Drypetes sepiaria, Diospyros montana,* and *Gyrocarpus americanus*. The trees are scarce and dormant with multiple branching. This type is very common in the rain shadow regions of the Western and the Eastern Ghats. In plains, reserve lands are dominated by this forest where the annual average rainfall is scanty. This forest is highly disturbed due to animal grazing, fuel wood extraction and other anthropogenic activities. The soil is red and mostly gravely with poor humus.

Dry Savannah Forest

This type is characterized by the tall grass like *Crysopogon* sp., *Cymbopogon* sp., *Heteropogon* sp., *Eulalia* sp., and *Eragrostis* sp., densely covering the ground. Savannah forest is distributed both in the Eastern Ghats and in the Western Ghats where the rainfall is relative poor and the terrain with steep slope. The prominent trees are *Givotia moluccana, Grewia tiliifolia, Ziziphus xylopyrus, Ficus microcarpa* and *Eriolaena hookeriana*. Such savannah forests are not found distributed in the plains and the coastal region.

Diversity and Distribution of Sacred Groves

The present study is an exploration and documentation of floristic diversity in Pudukkottai district of Tamil Nadu. The district lies between North latitudes 9°50′ and 10°40′ and Eastern longitudes 78°25′ and 79°15′ covers an area of 4663.29 sq. km including 42 km of coastline. The district is bordered, in by Sivaganga and Ramanathapuram the south, in Bay of Bengal the east Thanjavur district in the north and Tiruchirappalli in the north and west. Topographically the landmass is divided into 2 parts, namely coastal plains and interior plains. The vegetation of interior plains are scrub jungles, aquatic and semi aquatic, epiphytic, parasitic and carnivore and sacred grove vegetation.

RESULTS

Extensive botanical explorations were carried out periodically during 2012–2016. Totally 799 species of angiosperms, 12 species of pteridophytes and 1 species of gymnosperm were recorded and identified from Pudukkottai district. Eight hundred and twelve species belong to 480 genera and 124 families were collected and identified. The plants were arranged based on Angiosperm Phylogeny Group IV classification. Maximum numbers of species were in the orders of eudicots and core eudicots. Based on life form, composition the flora represents 56.88% herbaceous plants and 30.30% of woody plants such as shrubby and tree species and 12.82% of climbing plants.

Dominant families in the flora were Fabaceae (93) and Poaceae (91) species followed by Cyperaceae (49), Acanthaceae (35), Malvaceae (31), Apocynaceae (30), Euphorbiaceae and Lamiaceae (24 each), Rubiaceae (23) and Asteraceae (22). *Cyperus* and *Fimbristylis* were the top most genera with 18 and 15 species, respectively followed by *Ipomoea* (9), *Acacia*, *Crotalaria* and *Indigofera* (8 each), *Euphorbia*, *Senna*, *Utricualaria*, *Lindernia*, *Eragrostis* and *Oldenlandia* (7 species each). *Cleome*, *Ficus* and

Figure 2. Distribution of Sacred Groves in Pudukkottai district (Based on area)

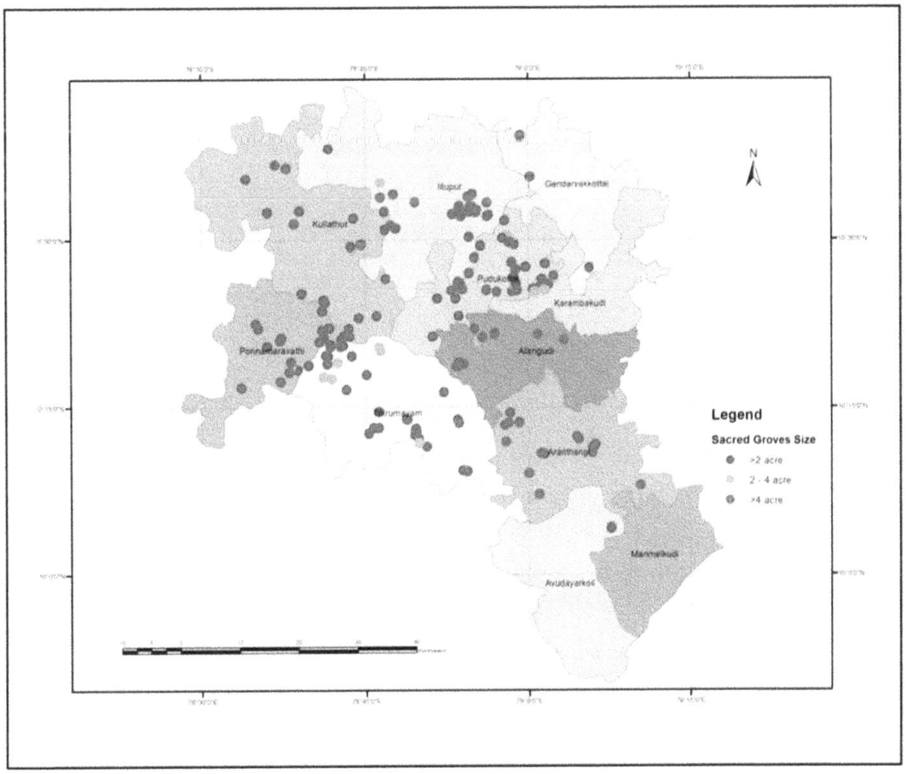

Phyllanthus (6 each), *Barleria, Justicia, Caralluma, Heliotropium, Capparis, Commelina, Murdannia, Acalypha, Rhynchosia, Brachiaria, Panicum* and *Solanum* (5 species each).

Sacred groves have a variety of plants in various life forms including rare and endemic plants. The most common plants were *Albizia amara, Chloroxylon swietenia, Strychnos nuxvomica, Memecylon umbellatum, Salvadora persica, Manilkara hexandra, Morinda coreia, Drypetes sepiaria, Putranjiva roxburghii, Mimusops elengi, Madhuca longifolia* var. *latifolia Sapindus emarginatus, Gyrocarpus americanus, Acacia leucophloea, Pterospermum suberifolium, Borassus flabellifer, Ficus benghalensis F.* microcarpa, *Mitragyna parvifolia, Limonia acidissima, Pleiospermium alatum, Mangifera indica, Corypha umbraculifera, Cordia dichotoma, Ehretia laevis, Crateva adansonii* subsp. *odora, Cassine glauca, Alangium salviifolium, Vitex altissima, Butea monosperma, Cassia roxburghii, Pongamia pinnata, Tamarindus indica, Azadirachta indica, Aegle marmelos* and *Atalantia monophylla* were very common trees. *Albizia lathamii, Terminalia bellirica* and *Dalbergia coromandeliana* were rare trees. *Tarenna asiatica, Clausena dentata, Glycosmis mauritiana, Dodonaea viscosa* subsp. *angustifolia, Flueggea leucopyrus, Barleria prionitis* and *Senna auriculata* were common shrubs in most of the sacred groves.

The present study reports a new taxa *Derris gamblei*. New records to Tamil Nadu, *Drimia wigthii* and *Scleria tesellata*, were reported and one rediscovery, *Hygrophila madurensis*, was recorded. Pudukkottai district consist of several endemic and threatened plant species. The present study revealed 35 plants as endemic, about 173 plant species fall under IUCN category and around 168 plants were identified as wild ornamental and edible plants.

Figure 3. Conservation status of Sacred Groves in Pudukkottai district

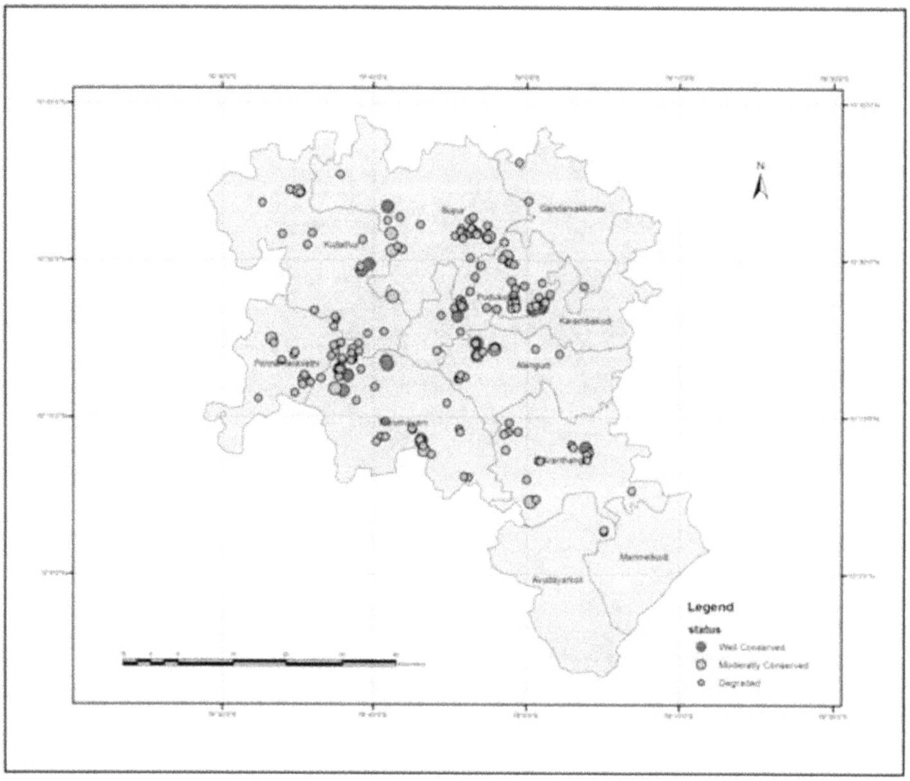

A total of 187 sacred groves were reported during the study (Figure 2). The coordinates of the groves were noted using GPS and distribution maps were generated for taluk using QGIS 2.8.3 software. The groves were classified based on conservation status namely well conserved, moderately conserved and degraded. Among the 187 sacred groves, a minimum number of well conserved and moderately conserved groves were recorded **(Figure 3)** and most of the groves were found highly degraded due to temple construction, encroachments, cattle grazing, etc. by human beings.

CONCLUSION

The present study revealed that the Pudukkottai district is bestowed with high diversity of angiosperms with endemic and threatened plants and faunal diversity. The diversity of plants and animals were under anthropogenic threats like deforestation, over exploitation of natural resources, human habitation, urbanization, industrialization and hunting activities. Developmental activities influenced the nature and composition of sacred grove to a large extent. There are number of sacred groves that have been encroached by farmers of surrounding land. Besides, the groves are frequently visited by cattle from nearby villages that lead to the introduction of invasive weeds and exhibit heavy competition to native species and exclusion of the same. On the other hand, the vegetation is being cleared for construction of temple within the core grove area thus fragmenting the vegetation which also affects the regeneration capacity of trees and shrubs. Hence, there is an immediate need for conservation activities to preserve the valuable genetic resources and sustainable utilization of plants and animal resources for future generation.

REFERENCES

Balaguru, B., Britto, S. J., Nagamurugan, N., Natarajan, D., & Soosairaj, S. (2006). Identifying conservation priority zones for effective management of tropical forests in Eastern Ghats of India. *Biodiversity and Conservation*, *15*(4), 1529–1543. doi:10.100710531-004-6678-1

Bhagwat, S. A., Kushalappa, C. G., Williams, P. H., & Brown, N. D. (2005). A landscape approach to biodiversity conservation of sacred groves in the Western Ghats of India. *Conservation Biology*, *19*(6), 1853–1862. doi:10.1111/j.1523-1739.2005.00248.x

Carrión, H. X., & Southworth, J. (2018). Understanding Land Cover Change in a Fragmented Forest Landscape in a Biodiversity Hotspot of Coastal Ecuador. *Remote Sensing*.

Champion, H. G. & Seth, S. K. (1968). *Revised survey of the forest types of India*. New Delhi, India: Manager of Publications.

Environment Information System (2015). *Pudukkottai district*. Environmental Information System Centre, State Environment and related issues, Department of Environment, Government of Tamil Nadu.

Gamble, J. S. & Fischer, C.E.C. (1915–1936). *Flora of the Presidency of Madras*. London, UK: Newman and Adlard.

Henry, J. K. (2008). Sacred groves – the need for their conservation in Eastern Ghats region. *ENVIS newsletter*, *14*(3), 3–5.

Jackel, H., Rudner, M., & Deil, U. (2013). Density, special pattern and relief features of sacred sites in northern Morocco. *Landscape Online*, *32*, 1–16. doi:10.3097/LO.201332

Janzen, D. H. (1988). Management of Habitat Fragments in a Tropical Dry Forest: Growth. *Annals of the Missouri Botanical Garden*, *75*(1), 105–116. doi:10.2307/2399468

Kabeer, K. A. A., & Nair, V. J. (2009). *Flora of Tamil Nadu–Grasses* (pp. 1–524). Kolkata, India: Botanical Survey of India.

Khan, M. L., Khumbongmayum, A. D., & Tripathi, R. S. (2008). Sacred groves and their significance in conserving biodiversity: An overview. *International Journal of Ecology and Environmental Sciences*, *34*, 277–291.

King, E. D. I. O., Viji, C., & Narasimhan, D. (1997). Sacred groves: Traditional ecological heritage. *International Journal of Ecology and Environmental Sciences*, *23*, 463–470.

Matthew, K. M. (1981). *Materials for the flora of Tamil Nadu Carnatic. The Rapinat Herbarium*. Tiruchirappalli, India: St. Joseph's College.

Mittermeier, R. A., Turner, W. R., Larsen, F. W., Brooks, T. M., & Gascon, C. (2011). Global Biodiversity Conservation: The Critical Role of Hotspots. In F. E. Zachos & J. C. Habel (Eds.), *Biodiversity Hotspots* (pp. 3–22). Berlin, Germany: Springer. doi:10.1007/978-3-642-20992-5_1

Myers, N., Mittermeier, R. A., Mittermeier, C. G., Da Fonseca, G. A., & Kent, J. (2000). Biodiversity hotspots for conservation priorities. *Nature*, *403*(6772), 853–858. doi:10.1038/35002501 PMID:10706275

Natarajan, D., Britto, S. J., Balaguru, B., Nagamurugan, N., Soosairaj, S., & Arockiasamy, D. I. (2004). Identification of conservation priority sites using remote sensing and GIS – A case study from Chitteri Hills, Eastern Ghats, Tamil Nadu. *Current Science*, *86*(9), 1316–1323.

Sanchez-Azofeifa, G. A., Quesada, M., Rodriguez, J. P., Nassar, J. M., Stoner, K. E., Castillo, A., ... Kalacska, M. E. R. (2005). Research Priorities for Neotropical Dry Forests. *Biotropica*, *37*, 477–485.

Soosairaj, S., Britto, S. J., Balaguru, B., Nagamurugan, N., & Natarajan, D. (2004). Mapping forest types of Pacchaimalai hills, Eastern Ghats, India – using remote sensing and GIS. *Eco. Env. & Cons.*, *10*(2), 131–135.

Udayalakshmi, V., Murthy, M. S. R., & Dutt, C. B. S. (1998). Efficient forest resources management through GIS and remote sensing. *Current Science*, *75*, 272–282.

Venkateswaran, R., & Parthasarathy, N. (2005). Tree population changes in a tropical dry evergreen forest of south India over a decade (1992–2002). *Biodiversity and Conservation*, *14*(6), 1335–1344. doi:10.100710531-004-9649-7

Yadav, S., Yadav, J. P., Arya, V., & Panghal, M. (2010). Sacred groves in conservation of plant biodiversity in Mahendergarh district of Haryana. *Indian Journal of Traditional Knowledge*, *9*(4), 693–700.

Chapter 7
Floristic Diversity and Carbon Stock in the Dry Forests of Chad:
The Case of Manda National Park – Diversity and Carbon Sequestration of the Manda National Park Flora

Ngaba Waye Taroum Caleb
University of Yaounde 1, Chad

Djekota Christophe Ngarmari
University of N'Djamena, Chad

Kabelong Banoho Louis-Paul Roger
University of Yaounde I, Cameroon

Zapfack Louis
University of Yaounde I, Cameroon

Mbayngone Elisée
University of N'Djamena, Chad

ABSTRACT

The woody flora of the National Park of Manda in the Sudanian area of Chad has been characterized between October and December 2016 to know its floristic diversity, and to quantify its aerial woody biomass. The transect and quadra method (1m x 1m) were simultaneously adopted for this study. The pan-tropical equation of Chave et al. made it possible to evaluate the carbon stocks in different sites. The study of the flora species identified 45 species divided into 37 genus and 21 families for an average population density of 355 individuals/ha. Three classes of the diameter dominate the settlement: class ≤ 10 cm; class of 10-20 cm and class of 20-30 cm. The height classes belong to the class of plants ≤ 4 m; and at last having a height ≤ 7 m. The basal area was 5.86 m^2 / ha. It appears that the woody components store 23.82 ± 0.01 tC / ha, the undergrowth 0.14 ± 0.01 tC / ha and the litter 0.56 ± 0.01 tC / ha. This research is a contribution to the REDD+ process (Reducing Emissions from Deforestation and Forest Degradation).

DOI: 10.4018/978-1-7998-0014-9.ch007

INTRODUCTION

Central African countries are in a process of integrating the reduction of greenhouse gas (GHG) emissions from the forest sector into their national policies, including the promotion of measures (compensation for environmental services) to encourage the conservation of forested areas (Banoho, 2014). These forests which are considered as lungs of the planet are very threatened by the effects of climate change in comparison to tropical forests around the world (Zapfack, 2005). In addition to the world's tropical forests and those of Central Africa, there are savannas and steppes that are vulnerable areas (Ngomanda et al. 2013).

Chad, where more than half of the geographical area is desert, has a flora threatened by the creation of agricultural land, the extension of cities, the search for firewood, pastoral pressure, etc (Ouya, 2010). These activities are responsible for desertification and food insecurity (Bertrand & Lagnaba, 2011)which lead into a decrease in biological diversity and promoting the appearance of invasive species (Saradoum et al. 2012). In order to conserve the large biodiversity areas around the country, Chad has opted for the development of a network of Protected Areas (PA) consisting of 30 forest entities classified and representing an area of **15,787,200 ha** or the 12. 3% of the national territory (Worgue, 2012).

Manda National Park (MNP) is part of this PA network (Kolmagne, 2000). This PA network initiative is very significant knowing that that between 1990 and 2000, the annual deforestation rate determined by FAO for Chad was 0.6% per year (Ouya, 2010).

Very few studies has been undertaken in the Manda National Park (MNP) which is very rich in terms of natural resources, faunal and floristic diversity (Saradoum et al, 2012). However, its flora, well preserved till present, is under the threat of urbanization of the third city of Chad: Sarh (Ouya, op.cit).

In tropical forest ecosystems, carbon is stored in different reservoirs, including plant biomass and soil (Asase et al. 2008). Studies to date on the assessment of carbon stocks in the Congo Basin forests did not include savannas and steppes in the Sudano-Sahelian zones. This study aims to fill this knowledge gap and to bring current knowledge on woody biodiversity and the capacity of these environments to store carbon.

The goal of this study is to determine the floristic diversity of the MNP as well as to evaluate the carbon stocks. This work is part of the process that contributes, on the one hand, to the evaluation of sequestration by the MNP flora, with a view to identifying the orientations aiming to reconcile the conservation and the rational use of natural resources, and on the other, to acquire bases that will facilitate the inscription of the MNP in the REDD+ process.

MATERIAL AND METHODS

Introduction of the Study Area

The MNP is located at 25 kilometers in the North West of the city of Sarh in the Department of Barh Kôh, Region of Moyen-Chari, about 450 km in the South East of N'Djamena, and 80 km from the Central African border. It is between latitude 9°20' and 9°35' in the North and longitude 17°45 'and 18°20' in the East; its altitude varies from 344 m to 691 m (Figure 1). With an estimated area of 1,14,000 ha, this protected area is located in one of the most populated areas of Chad, with a density of 21.8 inhabitants/km in the four peripheral cantons of the MNP (Zouglou, 2010). The population of these four cantons surrounding the MNP has increased fivefold from 17,914 to 85,592. The MNP is limited in its southern part by Barh Sara, in the West by the Sarh-N'Djamena national road, in the East by the Chari river and in the North by the crossing of the Sarh-N'Djamena national road Located in the Sudanian zone, the MNP enjoys a humid tropical climate in two seasons: a dry season that runs from November to March and a rainy season from April to October. Average monthly temperatures range from 21°C to 35°C (annual average: 28°C). The MNP vegetation is composed of mixed ecological types namely, open forest, wooded savanna, tree savanna, shrub savanna and grassy savanna. Agriculture is the main activity around the border of the MNP (Ouya, 2010). Wood and coal needs, the main sources of domestic energy in the population, are growing, and more than 0.77% of the tree savannas are deforested each year. This peripheral zone is highly populated and the land degradation increases of 0.3% each year (Ouya, op.cit).

Methods

Epigee and Hypogeum Carbon Sampling Device

The inventory made during this study is systematic with a fixed size transect as recommended by Picard (2006) when the massif has many trees of smaller sizes. Several authors had already used this method under similar conditions in Central Africa (Nkongmeneck, 1999); Hardy and Sonké, 2004; Zapfack, 2005; Ntonmen, 2013; Kabelong, 2013). Identification of vegetation types was done using visual discriminatory features that were: spacing between trees, lighting, and species along transect. The importance of each type of land use (TLU) along transects allowed for the assessment of vegetation cover changes in the North West of the MNP.

In summary 7 layons each of 1200 m were used for inventorying the flora. Thus, along the side these transects, 126 quadrats of variable depth were arranged in the North Western part of the MNP to delineate the permanent survey plot. It was: 42 quadrats of the dimension (1 x 1) m; 42 of the dimension (0.5 x 0.5) m and 42 of the dimension (20 x 50) cm.

Data collection (counting) was done along transects in the 2.4 ha (1.2 km long and 20 m wide) plots. For each tree encountered, identification was done based on several distinct parameters, and the diameter of the tree trunk was measured at breast height (dbh in cm), 1.3 m above from ground using dbh meter and at last the height of the shaft (in m). Inventoried trees are marked so that they are not counted twice. The buttresses are measured 30 cm above these foothills.

Estimation of the Biomass

The estimation of the woody biomass Epigee was done according to two methods:

- The non-destructive method used to estimate the carbon content of DHP trees DHP $\geq$ 4 cm (Ngomanda et al., 2013). It takes into account two accessible parameters namely, diameter and density by the allometric equation of Chave et al., (2005). These parameters were obtained at the end of the floristic surveys;
- The destructive method was used for the estimation of the Epigee and hypogeum carbon of the lower strata (undergrowth, litter, fine soil roots).

Figure 1. Manda National Park map (source: CNAR, 2012)

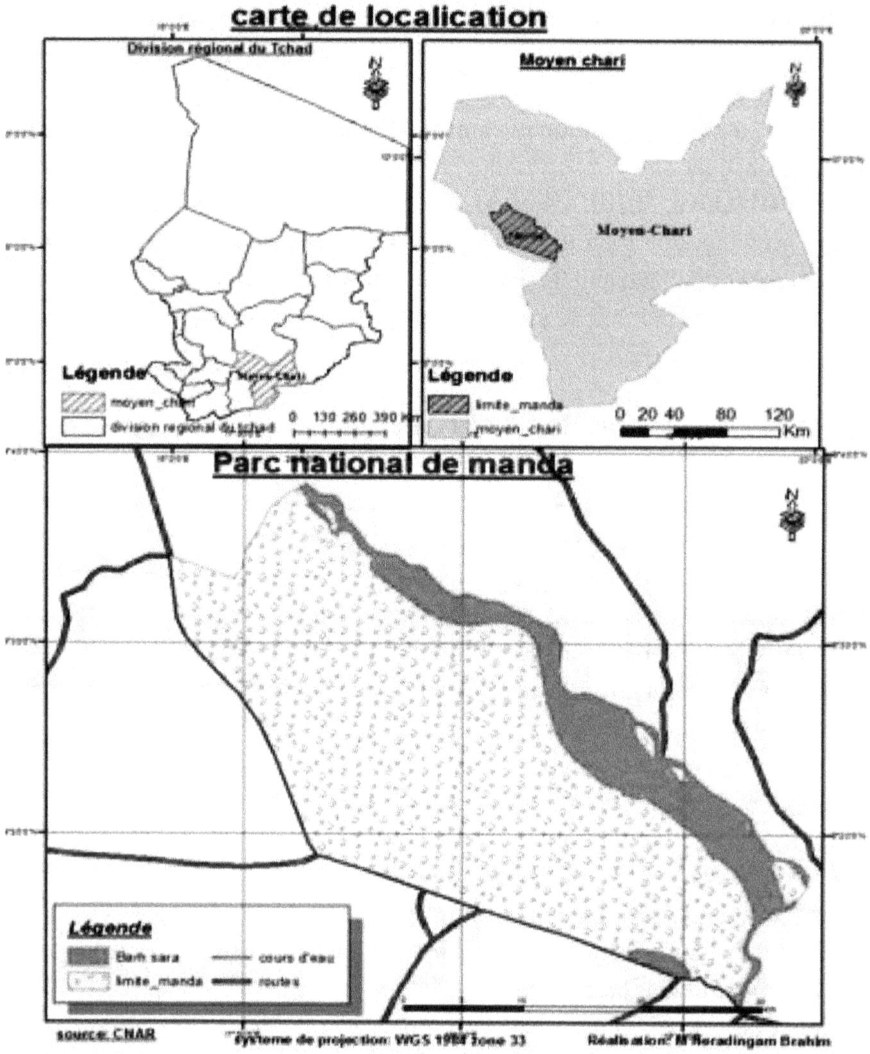

Hypogeum Biomass

The estimation of root biomass of populating trees was evaluated using the method indicated in the guidelines established by the Intergovernmental Panel on Climate Change (GIEC, 2006). In fact, according to the latter, the equivalence in root biomass of populating trees is found by multiplying the value of the aboveground biomass (AGB) by a coefficient R (stem / root ratio) whose value is estimated at 0.24.

BGB = AGB x R Equation 1

To estimate the biomass of small lignified roots and rootlets, the soil containing them is extracted in frames from 20 x 20 cm (length x breadth) to 50 cm depth. Strains as well as the main roots are avoided. Along each strip six blocks of 20 x 20 x 50 cm were taken. Soil samples are taken every 10 cm deep of the 50 cm depth (Boulmane et al., 2013). As the descent into the field took place during the dry season;

Table 1. Allometric equations for the determination of aboveground biomass

Local rainfall (mm/ yr)	Allometric equation (Kg/ tree)	Field of validity	Authors
1000 - 1500	AGB (Kg) = 0,0509 x pD^2H	5≤DHP≤156 cm	Chave *et al,* (2005)
Allometric equation (Kg / tree)			**Authors**
AGB (Kg) = ρ*exp(-1,499 +2,148lnD + 0,207 $\ln(D)^2$)-0,0281$(\ln D)^3$)*0.001			Djomo *et al.,* (2010)

ρ: Specific density of wood (g / cm^3)

roots are extracted from the soil samples using a sieve and gloves, using a sieve of sufficient size so as not to lose the vegetal material. After several sieves, the roots are completely removed and the soil was discarded (Zapfack, 2005).

The weighing series of the packaging was made using a 2000 g capacity scale, brand *Sartorius*. The fresh and chlorophyllous biomass conditioned in 107 labeled samples: herbaceous, litter, fine roots and rootlets were packaged in envelopes and dried in an oven at 75 ° C. Before drying in the oven, the weight of the empty envelopes was measured, and then the weight of the packaged fresh biomass was measured as well. Once in the oven, the samples were kept for 48 hours. Thereafter 10 samples randomly selected from the batch wereweighed in order to check the variation between fresh weight and that being dried. After this routine weighing, the samples were kept in the oven for 48 hours until a constant dry weight is obtained. The constant weight makes it possible to deduce that all the water contained in the material has completely evaporated (Ngomanda et al., 2013). The value of carbon stocks is estimated by multiplying the value of dry biomass by the coefficient 0.47 (Zapfack, 2005).

Total Biomass of Populating Woody Species

The total biomass (Total Biomass (kg)) of populating trees is estimated as follows:

TB = AGB + BGB Equation 2

Some Ecological Parameters Evaluated

Tree Density and Distribution (N ha / 1)

$N = \frac{n}{S}$ Equation 3 has been applied to obtain relative diversity:

N: density (in/trees/ha), n: number of trees present on the surface considered and S: area considered (ha).s

- Relative density = (Number of species/Total number of species in the sample) × 100;
- Relative diversity = (Number of species within a family/Total number of species). × 100;

The two indices above show the increase in the number of families or trees according to a growing area (Doucet et al., 1996).

Basal Area G (m^2 / ha)

The basal area of a tree is the cross-sectional area of this tree at 1.30 m above the ground (Rondeux, 1993). This is the area occupied by a tree ***i*** per hectare. The basal area of a population (G), therefore, represents the sum of the basal areas of individuals taken individually (g), which make up this population (Pardré and Bouchon, 1998). They are calculated using the following relationships:

$$g_i = \frac{\pi D_i^2}{4} \text{ Equation 4}$$

D_i= Diameter at chest height
gi = basal area of ligneous (m^2/ha)
Di: Diameter of woody ***i***;
gi: Woodland basal area ***i*** (m^2/ha).

$$G = \sum_{i=1}^{n} g_i x \frac{d}{n} \text{ Equation}$$

d: population density per hectare;
n: Population size in the plot.

This size depends both on the size and the number of trees and is correlated with the covert of trees, which makes it possible to quantify the level of the competition within the population, and the conditions of illumination of the soil. According to CRPF and FOGEFOR (2011):

- A dense and old population may have a high basal area of 25 to 50 m^2 / ha;
- A younger or clearer population will have a low basal area: 5 to 15 m^2 / ha.

Estimated Carbon Stock and Atmospheric CO_2 Equivalent

The total biomass estimated from the different equations was converted to the corresponding sequestered carbon stock by multiplying it by 0.47 according to the GIEC (2006). Concerning to the sequestered atmospheric CO_2 stock, it is recognized that the atomic mass of Carbon (MaC) is 12 and 16 for the Oxygen. The molecular mass of CO_2 is 44.

Thus, the combination ratio of carbon (C) to dioxygen (O_2) is 3.67. The atmospheric CO_2 equivalent stock is estimated by multiplying the carbon stock from biomass by 3.67. This method of estimating of the stock of atmospheric CO_2 equivalent was also used by Tsoumou (2016).

To calculate our carbon stock, we have chosen two equations mentioned above because they integrate three parameters: the Diameter at 1.3 m named in French Diamètre a Hauteur de Poitrine (DHP), the height and the specific gravity of each species. In addition, the allometric equation of Chave et al. (2005) has been recognized as the most accurate compared to the other existing equations namely that of the GIEC (2006), and that established by Djomo et *al*, (2010) (Mugnier et *al*, 2009).

Data Processing and Analysis

The data was analysed for the assessment of species richness: relative density and relative diversity; ArcView 3.2 software was used to representing the device in the plot.

The dendrometric parameters (diameter and length) of the species collected in the field were captured and compiled using Microsoft EXCEL 2007 for the treatment and calculation of biomass and sequestered carbon. This method of data processing and analysis was also used by Tsoumou et al., (2016). Knowing that the sequestering capacity of an ecosystem is related to its specific diversity, this coefficient has made it possible to define the reliability of the MNP vegetation in terms of carbon sequestration by establishing the following ratio:

$$CG = \frac{\text{Numberofthegender}}{\text{NumberofSpecies}} x100$$

Natural regeneration concerns trees less than 20 cm in diameter. The regeneration rate makes it possible to estimate the renewal potential of the species. It is calculated by the formula below (Anonymous, 2007; Banoho, 2013).

$$Tr = \frac{\text{Numberofindividualsofcirconferencelessthan20cm}}{\text{TotalNumberofindividuals}} x100$$

RESULTS

Characterization of the Floristic Diversity of the MNPs

The floristic inventory of 5964 DHP trees greater than or equal to 10 cm revealed the presence of 22 families, 37 genera and 45 species. The most represented families are the Rubiaceae with 6 genera and 6 species, the Combretaceae with 4 genera and 5 species, the Mimosaceae with 3 genera and 4 species and the Caesalpiniaceae with 3 genera and 3 species. In terms of diversity indices, Figure 2 shows that the Rubiaceae family is the most represented 16%, followed by those of Combretaceae (11%), Minosaceae (8%), and Caesalpiniaceae (8%).

Regeneration Rate and Population Structure

The average regeneration rate of the species is 24.58%. This rate is relatively low because of the short duration of the rainy season, which is sometimes only 4 months on average (end of May to August), but also deficiency of soil water reserves and excess insolation which causes delay in seed germination till the rainy season.

The generic coefficient of the site is 82.22%. The Shannon-Wiener index is 3.51 bits/individual, representing the specific structure of the sample. This index is high, reflecting a large number of species contributing to land use and poor population organization. The equitability of Pielou Evenness index

Table 2. Diversity indices

Index	Value
Absolute density	355 individuals/ha
Generic diversity	82.22%
Shannon's index	3.51 bits
Equitability of Pielou	0.64
Generic Coefficient	84.09
Natural Regeneration	24.58%

is 0.64. This indicates that the populating distributed evenly, that is, the distribution of individuals per species represented in the population is approximately the same. It shows the participation of species such as: *Terminalia laxiflora* (1351 stems), *Combretum collinum* (1088 stems), *Hymenocardia acida* (935 stems) and *Anogeissus leiocarpus* (588 stems) to represent the floral background of the populating.

Diversity Indices

The diversity indices in Table 2 provide an overview of the items that allowed the characterization of vegetation types in the MNP.

Absolute Density of the MNP

The absolute density varies from one transect to another. The transects surveyed towards northwest of the MNP (T1, T2, T3, T4, T5, T6 and T7) as observed in Figure 2, the highest density was in T1 transect i.e. 451 individuals/ha. This transect is located more than 2 km after the road separating the village of Nguéré from the MNP. However, the lowest density was in T5 transect i.e. 300 individuals/ha. This transect is located less than one km from the road separating the MNP from Nangnda village. There were also reported Pasture indices, poachers, and firewood sales, apart from roadside charcoal. There was also a large area decimated when the fire passed.

The average density is 355 individuals/ha. The finding is that the wood resource varies within the MNP, according to its accessibility to local populations. Further, there were intrusions of transhumant pastoralists in search of grazing and poachers who often use wood to smoke their prey, with the temporary tent setup for their clandestine stay in the MNP.

Absolute density also quantifies the competition between trees in a forest population. In present study, the basal area of the population is 5.86 m^2 / ha. As a result, the population is clear and gives easier access to light.

Treatment of circumference class data at 1.3 m from the ground (Figure 3) shows that the majority of individuals are in the 20-40 cm class. This class represents 40.59% of the individuals surveyed. For other classes the numbers are distributed as follows: 24.58% for the 0-20 cm class; 20.87% for the 40-60 cm class; 7.69% for the class of 60-80 cm. for the class of 80-100 cm 3.68%; 0.97% for the class 100-120 cm; 0.75% for the class 120-140 cm and 0.33% for the class 140-160 cm. The following classes, namely: class 160-180 cm, class 180-200 cm, class 220-240 cm have a percentage of individuals ranging

Figure 2. Absolute density per transect northwest of MNP

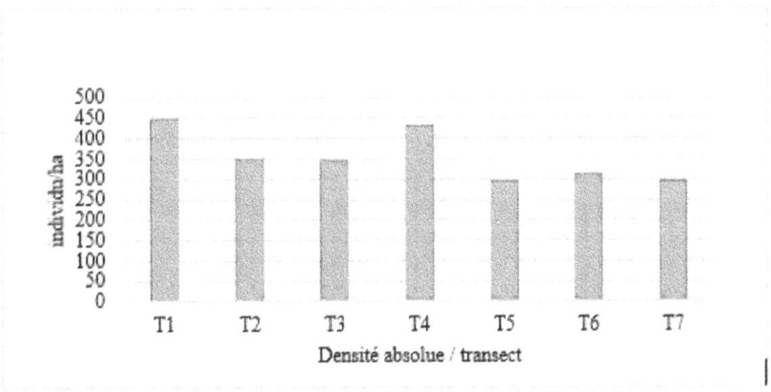

from 0.10 to 0.13%. Finally, individuals greater than 240 cm are 0.067%. The modal classification of the circumferences shows that only 6.25% of the inventoried individuals reach circumferences > 80 cm.

Horizontal Structure of the Population

The horizontal structure was obtained from the circumferences at 1.3 m above from the soil of each of the 5964 woody individuals inventoried. The distribution of the stems by diameter classes on the 16.8 ha prospected in the north-west of the MNP shows some structural divergences clearly discriminated by the number of stems. Thus, figure 3 presents the circumference class ≤ 10 cm as better represented in the population studied with 51.12% of the population; or 3049 stems. Individuals belonging to the class of circumferences [10-20 cm with 36.77% of the number; or 2193 stems followed by the class [20-30 [cm with 9.03% being poorly represented in this park. The class of large circumferences (more than 60 cm) was very small. This horizontal structure of population makes it possible to understand that the MNP has several paths to ensure its reconstitution in the future. The general distribution shows that MNP population are characteristic of the Sahelo-Sudanian bioclimatic area.

Figure 3. Modal classification of population circumferences

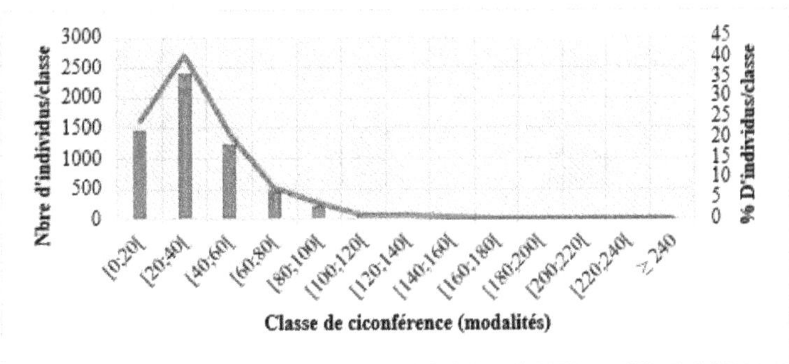

Figure 4. Population structure

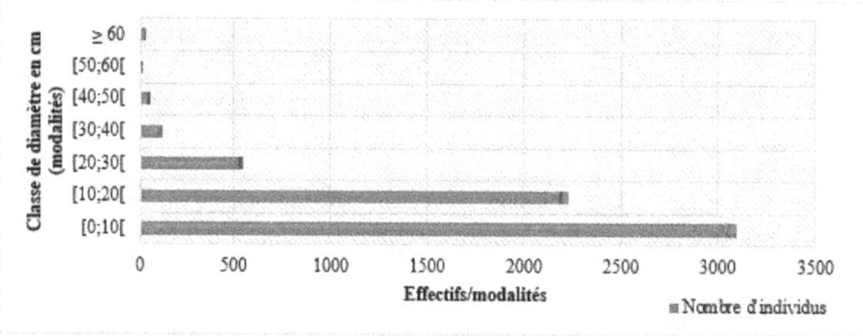

Vertical Structure of the Populating

The woody species in the study area were classified in 3 modalities of 10 m. The height class of plants ≤ 4 m represents 13.29% of the species. In this class dominants species were *Anogeissus leiocarpus, Combretum collinum, Gueira senegalensis, Terminalia laxiflora and Hymenocardia acida.* The class of 4-7 m represents 19.29% of the species, where dominant species were: *Daniella oliveri, Piliostigma reticulatum and Prosopis africana.* However, more than 67.4% of the species have a height greater than or equal to 7m (Figure 5). The dominant species by their height were: *Khaya senegalensis, Parkia biglobosa* and *Bombax costatum.* The class of heights ≤ 4m is characterized by several savanna species which, although mature, retain a relatively small size, such as: *Flugia virosa, Strychnos spinosa, Ziziphus mauritiana* and many other species (Figure 5).

MNP Carbon Stocks

By applying the pan-tropical equation of Chave et *al,* (2005) the carbon stocks obtained according to the reservoirs give: a stock of 23.82 ± 0.004 tC/ha for woody trees, a stock of 0.144 ± 0.007 tC/ha for undergrowth and grasses, and a stock of 0.560 ± 0.006 tC/ha for litter.

Figure 5. Vertical structure of MNP populating

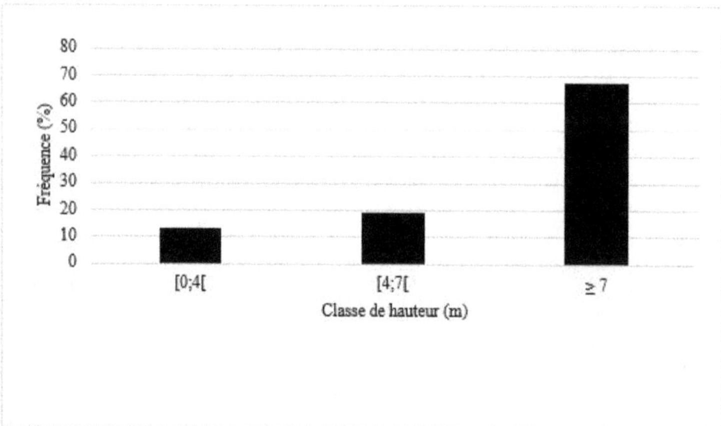

Table 3. Summary of total carbon stock

Carbon well	Carbon stock tC/ha
Undergrowth	0.144 ± 0,007
Litter	0. 560 ± 0,006
Root Woody	3.129 ± 0,007 23.82 ± 0,004
Total	27.653 ± 0,006

The hypogeum carbon stock corresponds to the values obtained by drying and weighing the fine roots of the soil and rootlets. The big roots have not been removed so as not to affect the nutrition of the trees. The value of the stock obtained is 3.129 ± 0.007 tC/ha.

The total carbon stock obtained is equivalent to the sum of Carbomass and those of litter and undergrowth.

Change in Carbon Stock According to Tree Circumference Classes

The slip of the annual average pluviometry was more than 100 mm, with suitable periods from three to five years i.e. from 1970 afterwards till 1980, and at last from 2000 to 2005, which initiated a "Sahel area" of the biotope, there in the north of the park more clearly.. This explains the area constancy of two types of the growth in the National Park Of Manda during the last thirty years, namely: the shrub savannah occupying an average of 32.3% and the grassy savanna with an average of 28.4% of the total area of the National Park Manda.

In the course of the last thirty years, the vegetable biomass of the National Park of Manda (consisting the sum of the different types of vegetation) occupies an average 78.75% of the total area of the National Park of Manda. This biomass represents a high potential of carbon sequestration.

DISCUSSION

Woody Diversity of the MNP

The floristic inventory of woody trees has revealed a considerable species diversity in the MNP. The most important species in terms of their abundance in the MNP were: *Anogeissus leiocarpus, Combretum collinum, Gueira senegalensis, Terminalia laxiflora and Hymenocardia acida.* These results are similar to those of Kabelong (2013) which obtains 49 woody species, divided into 42 genera and belonging to 25 families in the Waza National Park (WNP) located in the Sudano-Sahelian zone in Cameroon, with a generic coefficient of 83.67%. It thus appears that the MNP is generically diversified. This generic diversity is the expression of a significant potential of carbon storage epigeous as well as hypogeum This is because greater the species diversity, greater are the chances of population to resist the variability of factors that influence its distribution. For instance, the species *Anogeissus leiocarpus*, *Hymenocardia acida* and *Gueira senegalensis* are resistant to bush fires, so their presence in hedge protects the other species of the population and thus contributes to the increase of the life span of the individuals, thereby increasing the organic matter requirements that further require carbon uptake by plants. From the point of view of specific abundance, the results obtained in this study differ from those of Jiagho et al. (2016) where dominant species in WNP were obtained as: *Acacia seyal, Guiera senegalensis, Combretum soft and Balanites aegyptiaca*. They also differ from those of Boussim (2010) (cit. Lansina, 2013) in savanna zone in Burkina, which had dominant species such as: *Adansonia digitata, Faidherbia albida, Lannea microcarpa, Parkia biglobosa, Vitellaria paradoxa.*

This difference is supported by the Shannon-Weaver index of 3.51 bits for this study, this difference being related to the factors that govern the distribution of species within the MNP. These factors are climatic i.e. related to the variability of the climate, edaphic i.e. related to the nutrient content of the soil, along with the capacity of dominant species to multiply and to adapt to the variability of these factors. This indicates a great variation in environmental factors, however, a weak organization of its population, or the imbalance of itself.

Population Structure

The average density of MNP population is 355 individuals/ha. These results are different from the results generally obtained in the PAs of the Sahelo-Sudanian bioclimatic zone. Kabelong (2013) has an average density of 30 to 50 individuals/ha and Jiagho et al. (2016) obtains an average density of 40 stems / ha in the WNP. Sani (2009) in Niger shows an average density of ligneous trees, 163 individuals / ha in a reverdi site in Marriah. In contrast, Lansina (2013) obtains an average density of 305.25 individuals / ha in the strict Sahelian sector in Burkina Faso. The average density obtained in the MNP is characteristic of the Sudanian zone; the climatic and edaphic conditions are quite favorable to the growth of the species. This is also explained by the good conservation of the flora in this park due to the presence of traditional cult sites and rites but also to the action of eco-guards (Anonymous, 2011a).

The basal area of the population were 5.86 m^2 / ha. These results were contrary to those of Boulmane et al. (2013) in the dry ecosystems of Morocco, which obtains 19.3 m^2 / ha in the province of Ifrane. On the other hand, Traoré (2013) obtained in the Sahelo-Sudanese zone in Burkina a value of 3.25 m^2 / ha, a value that is similar to that of this study. Trees $\geq$7 m tall that dominate NMP population were quite distant from each other because of the predominance of grassland and shrub savanna areas, resulting in lack of competition among individuals for low basal area obtained. These results suggest that NMP population are clear and provide easier access to light.

The study of the distribution of circumferences on the 16.8 ha inventories shows that 40.59% of the individuals belong to the class 20-40 cm. Kabelong (2013) obtained a similar result in Cameroon's WNP with 34.22% of individuals in the 20-40 cm class. This result is characteristic of savanna ecosystems. As for the diametric distribution, the study shows that 51.12% of the population belongs to the diameter class 0-10 (cm). It also shows that the representation of the different diameter classes has a decreasing exponential dynamic distribution with 36.77% belonging to the class 10-20 (cm) and only 9.03% to the class 20-30 (cm). It should be noted that the number of individuals of the class of large diameters (more than 60 cm) is very small. This difference in population is explained by the fact that forest groves develop through a natural succession process that goes through several stages of maturity (Muller et al. 2002, Ngomanda et al. 2012). According to Jiagho et al. (2016) this is a sign of ecological vigor and guarantee of the sustainability of the population, the young individuals spontaneously ensure the replacement of old individuals..

Estimation of Carbon Stocks

The total carbon stock obtained averages 27.65 ± 0.01 tC / ha. It corresponds to the sum of the Carbomass (ligneous and roots), the litter, the herbaceous and the undergrowth. The equation of Chave et al. (2005) was used for ligneous and the destructive method for other pools. using the same allometric equation of Chave et al. (2005) in the Sahel-Sudanian forest in Tiendaga, Mali, the carbon stock obtained averages 37.478 tC / ha (Anonymous, 2013 d) slightly higher than this study. On the other hand, for a sampling carried out on 3700 ha in the rehabilitated savanna of the Bagombé plateau in Gabon Ngomanda et al. (2013) obtained a carbon stock of 978113 tC, it means an average of 264.35 tC / ha, allowing it to claim that the dry biomass of the undergrowth of ecosystems savannas is greater than those in the undergrowth of forest ecosystems. This assertion is due to the fact that in savanna the ecosystem being very open allows the light energy to come directly to the grassy strata, which leads to a high biomass production in savanna as compared to closed environments or forest, hence the high rate of carbon stored.

The low carbon stock obtained in this study may be explained by the reduced sample size of 16.8 ha, the season during which the assessment was made (dry season) and also because some part of stored carbon was not estimated namely: soil and large roots more than 2 mm diameter. These studies were not conducted due to lack of financial and logistical resources.

The carbon stock from Chave et al. (2005) compared to that found by Mugnier et al. 2009 at Salongo National Park (Ecuador, Democratic Republic of Congo) in a young secondary forest: 79.0 tC / ha has similarities.

Dynamics of MNP Vegetation

The vegetation cover has increased from an average total area of 88% to an average area of 78.75% of the extent of the NIP. These results were inferior to those of Anonymous (2015d) in Lagdo in the extreme north of Cameroon. For the period from 1987 to 2014, it obtained a degradation of nearly 30% of the Lagdo forest. Anonymous (2012a) justifies the density of the plant cover of the MNP by concluding that its vegetation is very well preserved till present. However, the clandestine intrusion of riparian populations into the park for various activities (fishing, hunting, cutting firewood and works) and pastoralists seeking grazing are the causes of the increase in the area of bare ground and sandbanks. Banoho (2013) in the WNP states that the proximity of vegetation cover to residential areas poses a threat to vegetation growth. The observation is the same for this study. This state of affairs constitutes a cause likely to influence the carbon sequestration capacity of the NMP vegetation and its periphery, by destocking the NMP carbon pools in the medium term.

CONCLUSION

The floristic diversity of the MNP consists of 45 woody species divided into 37 genera and 22 families. However, the flora of the MNP especially that of its periphery is threatened by the demographic pressure which prevails. The carbon stocks obtained were 27.65 ± 0.01 tC / ha. In the context of the REDD+ process to be implemented in the NIP, deforestation should be combated, the estimated quantities of carbon should be extended to the whole NIP according to the different types of vegetation and a system of rotational exploitation of the massif should be initiated from the periphery of the MNP to allow a consequent regeneration time for deforested surfaces.

REFERENCES

Anonyme. (2012a). Qualitative study on the causes of deforestation and forest degradation in the Democratic Republic of Congo Final Report REDD+ Task Force. 149p

Anonyme. (2015d). Project idea note "Conservation of the Ouro-doukoudje massive and reforestation of the western banks of Lake Lagdo" Sept. 7, 2015. 24 p

Anonymous. (2011a). *Manda National Park Development Plan 2011-2021 Provisional version November 2010.* National Parks, Wildlife Reserves and Hunting Department.

Asase, A., Wade, S. A., Ofori-Frimpong, K., Hadley, P., & Norris, K. (2008). Carbon storage and the health of cocoa agroforestry ecosystems in southeastern Ghana. *FAO Journal World Soil Resources Report 104 "Africa and the Carbon Cycle"*. pp. 131-144.

Banoho, K. (2013). *Influences of human activities on the wood resource in the periphery of Waza National Park.* (Master's thesis in Biodiversity and Forest Productions), University of Yaoundé I. 89p

Bertrand, G., & Lagnaba, K. (2011). *Potentialities and constraints of rural development in the central, eastern and southern regions of Chad. Internal review on the rural sector in Chad.* French Development Agency-World Bank.

Boulmane, M., Santa Regina, I., Khia, A., & Oubrahim, H. (2012). Estimation of the organic carbon stock in iliasis of the Moroccan Middle Atlas. 24pp.

Chave, J., Brown, S., Cairns, M. A., Chambers, J. Q., Eamus, D., Folster, H., Fromard, F., … Yamakura, T. (2005). Tree allometry and improved estimates of carbon stock and balance in tropical forest. *Oecologia, 145*, 87-99.

Decree N° 243/PR/CSPS/PNR of 23 October 1967 modifying the boundaries of the Manda National Park

Djomo, A. N., Ibrahima, A., Saborowski, J., & Gravenhorst, G. (2010). Allometric equations for biomass estimates in Cameroon and tropical moist equations including biomass data from Africa. *Forest Ecology and Management, 260*, 1873–1885. doi:10.1016/j.foreco.2010.08.034

IPCC. (2006). Guide for national greenhouse gas inventory agriculture, forestry and other land use. *Institute for Global Environmental Strategies, Japan, 4*, 46–52.

Jiagho, E., Zapfack L., Banoho, L., Tsayem-Demaze, M., Corbonnois, J. & Tchawa, P. (2016). Diversity of woody flora on the outskirts of Waza National Park (Cameroon). *VertigO - the electronic journal in environmental sciences [Online], 16*(1). Retrieved from http://vertigo.revues.org/17249

Kolmagne, M. N. (2000). *Study of the settlement of fauna and flora in Manda Park Report of the Ministry of Environment and Water.*

Lansina, T. (2013). *Influence of climate and protection on woody vegetation in western Burkina Faso.* (Doctoral thesis), Applied Biological Sciences, specialty: Botany and Phytoecology, University of Ouagadougou.

Mugnier, A., Cassagne, B., Bayos, N., & Lafon, C. (2009). Estimation of carbon stocks in Congo Basin forests for REDD: Comparative study conducted on 22 forest types, 4 countries and a management plan 4.8 million ha. Retrieved from https//www.observatoire-comifac.net/µdocs/edf2008/EN/state-of-forests

Muller, E. (2002). Reintegrate secondary forests into the landscape. ITTO.

Ngomand, A., Lebamba, J., Engone-Obiang, N., Lepengue, N., & M'Batchi, B. (1997). 2013. Characterization of Dry Biomass of forest-savannah mosaics in Gabon. *Journal of Applied Bioscience, 68*, 5417–5425.

Nkongmeneck, B. A. (1999). *The Boumba-Bek and Nki forest reserves: botany and ethnobotany WWF-Cameroon report.*

Ntonmen, Y. (2013). Biodiversity and carbon stock of plant formations from Akomnyada and Nkologok (Mbalmayo). (Master thesis), University of Yaounde I.

Ouaba. (2006). Phytoecological and socio-economic study of the classified forests of Burkina Faso: Case of the Niangoloko classified forest. (Doctoral thesis), University of Ouagadougou.

Ouya Bondoro. (2010). *Conservation and Sustainable Use of Biodiversity Around and in Protected Areas in South-East Chad: The Case of Manda Park and the Djoli-Kera Forest Uses and Risks, What Strategies for Upgrading and Local Development?* (Doctoral thesis in geography and spatial planning), University Paul-Valery of Montpelier III.

Picard, N. (2006). Forest inventory method. Participatory Rural Development Project in the Middle Atlas Central (Khenifra Project).

Sani, A. R. (2009). Biophysical Characterization of Wood Resources in a Reverdi Site and a Degraded Site in the Department of MIRRIAH. (End of Cycle Master's Degree in Water and Forest Engineering Diploma), Abdou Moumounide Niamey University.

Saradoum, G., Ngom, Fm., Diallo, A., & Guissé, A. (2012). Characterization of the herbaceous vegetation of the National Park of Manda in Chad. *International Journal of Science and Advanced Technology.*

Zapfack, L. (2005). The impact of shifting cultivation on burning on plant biodiversity and carbon sequestration. (Doctoral Thesis), University of Yaoundé I, Yaoundé.

Chapter 8
Biogeographic and Anthropogenic Effects on Asian Elephants in Tropical Forests of Sumatra, Indonesia

David M. Augeri
Biodiversity Unlimited Research and Consulting Group, USA

ABSTRACT

DISTANCE protocols and MIKE Survey Standards were used in the field to determine Critically Endangered (CR) (A2c) Sumatran elephant (Elephas maximus sumatranus) occupancy, density and abundance in Gunung Leuser National Park (GLNP). Forest and habitat type, age, character, and integrity were the most significant factors affecting elephant occupancy. Principal forage types relative to elephant activity were palms and lianas, which dominated significantly in undisturbed primary forest. DISTANCE model density D=0.167 elephants/km-2 (95% CI=0.106–0.262), best-fitting occupancy Ψ=0.6321 (SE±0.0010) and detection probability p=0.6225 (SE±0.0001) estimates combined yielded N=407 elephants (95% CI: 258–638) in GLNP. The most parsimonious occupancy model estimated N=392.82 elephants (SE: ±30.65; 95% CI: 332.78-452.95) in GLNP. Forest restoration, ecosystem protections, and conservation plans for Asian elephants, biodiversity, and forests are suggested in this study.

INTRODUCTION

Asian elephant (*Elephas maximus*) populations have declined and fragmented significantly during the second half of the 21st century (Olivier 1978; Sukumar 1989; Choudhury et al. 2008; IUCN 2019) and are most endangered of the remaining extant Proboscidea (Shoshani & Tassy 2005). *E. maximus* is listed as Endangered (EN) (A2c) with a declining population trend on the IUCN Red List (Choudhury et al. 2008; IUCN 2019). It is also protected as an Appendix I species with CITES (2019). The Sumatran subspecies, *E. m. sumatranus*, is listed as Critically Endangered (CR) (A2c) with a declining population trend across its range along with a continuing decline in area, extent, and quality of habitat (Gopala et

DOI: 10.4018/978-1-7998-0014-9.ch008

al. 2011; IUCN 2019). Mitochondrial DNA variation patterns indicate that *E. m. sumatranus* is monophyletic (Fleischer et al. 2001) and, consequently, this subspecies has been defined as "evolutionarily significant" (Choudhury et al. 2008; IUCN 2019). Given its status and the increasing rates of critical habitat loss, degradation, and fragmentation along with increased poaching and retribution killings, *E. m. sumatranus* has been given "particularly high priority" for its conservation in the wild (Choudhury et al. 2008; Gopala et al. 2011; IUCN 2019).

Tropical and subtropical dry and moist broadleaf lowland forests are critical habitats for *E. m. sumatranus* and, therefore, are of "major importance" (Gopala et al. 2011; IUCN 2019). Consequently, the biogeographically diverse Gunung Leuser National Park (GLNP) and Leuser Ecosystem (LE) in the provinces of North Sumatra and Aceh, Indonesia (Figure 1) are two of the most important refuges of these ecosystem types remaining for this subspecies. Sumatran elephants coevolved with primary lowland tropical and subtropical moist and dry forests and they are principally adapted to, and reliant on, resources from these ecosystem types. Grasslands and scrub are important for *E. maximus* in other regions (Choudhury et al. 2008), but such habitats supply lesser degrees of elephant forage in Sumatra. Tropical and subtropical dry and moist broadleaf lowland forests provide the highest forage diversity and quality for *E. m. sumatranus* (Gopala et al. 2011; IUCN 2019), including palms and lianas, which are important items in the *E. m. sumatranus* diet (Oliver 1978; Sukumar 1990, 2000; Steinheim et al. 2005).

Indonesia's tropical forest cover is the third largest in the world. Primary lowland forests are most biodiverse in Sumatra and provide critical habitat for *E. m. sumatranus*. However, 85% of elephant habitat in Sumatra lies outside of protected areas (Gopala et al. 2011) and has experienced some of the highest rates of deforestation in the world (Uryu et al. 2010; Margono et al. 2012; 2014). During 2001-2017, Indonesia lost 24.4 Mha of tree cover with 89% resulting in permanent deforestation (WRI 2019). The Leuser Ecosystem, including portions of GLNP, lost approximately 167,000 ha during this period (WRI 2019; D. M. Augeri pers. obs.) (Figure 1). During 1990–2010, primary forest declined by 40% across Sumatra while overall forest cover was reduced by 36% (Margono et al. 2012, 2014). Timber harvests on the island have been at least four times higher than sustainable levels (Brown et al. 2005), subsequently reducing these forests by 2/3rd to 4/5th of their original extent (Whitten et al. 2000; Kinnaird et al. 2003; Hedges et al. 2005). As a result, more than 69% of *E. m. sumatranus* habitat was lost between 1985-2011 (Gopala et al. 2011; IUCN 2019).

Most forest cover in Sumatra outside of protected areas is fragmented into blocks smaller than 250 km² (Figure 1) (Gopala et. al. 2011; IUCN 2019). Such fragments are too small and restricted for viable elephant populations to survive over the long-term (Gopala et. al. 2011; IUCN 2019). This type of isolation and insularization increase the most significant impacts of habitat loss and fragmentation (Lovejoy et al. 1986; Laurance & Bierregaard 1997; Bierregaard et al. 2001; Wilson 2016), particularly for large wide-ranging species like Asian elephants (Choudhury et al. 2008; Gopala et. al. 2011). These and other anthropogenic disturbances (i.e., poaching and retribution killings) continue at increasing rates in Sumatra and the elephant population is "severely fragmented" (Gopala et al. 2011; IUCN 2019).

Population estimates of *E. maximus* vary greatly. Rigorous sampling-based surveys are minimal and previous population figures are still considered educated guesses. The latter includes the widely cited estimates of a) 41,410–52,345 worldwide with as many as 2,400–3,400 elephants in Indonesia (Choudhury et al. 2008; IUCN 2019) and b) the most recent estimates of 45,697-48,534 with ca. 1,784 elephants in Indonesia (AERSM Final Report 2017). Such estimates lack sufficient data and should be considered cautiously (Blake & Hedges 2004; Hedges 2006; Choudhury et al. 2008; Gopala et al. 2011; IUCN 2019). Prior to the present study, rigorous data and population estimates did not exist for northern Sumatra, but evidence indicates elephant subpopulations in the north and across Sumatra are increasingly isolated and the total population is a fraction of its historic levels (Olivier 1978; Blake & Hedges 2004; Hedges et al. 2005; Choudhury et al. 2008; Gopala et al. 2011; IUCN 2019). Indeed, the suspected population decline in just one generation from 1985 to 2007 was as much as 50% and over the previous three generations to be at least 80% (Gopala et al. 2011; IUCN 2019).

By 2008, elephants had become locally extinct in 23 of their 43 ranges in Sumatra, indicating a "very significant decline" of at least 53% (Gopala et al. 2011; IUCN 2019). The population in Riau province declined from approximately 1,342 individuals in 1984 to just 210 individuals in 2007 and the number of fragmented subpopulations almost doubled (Uryu et al. 2008). This province had been one of the most important strongholds for the subspecies, but if forest loss rates continue the province's remaining elephants could disappear (Uryu et al. 2008; Gopala et al. 2011; IUCN 2019). The IUCN (2019) also considers *E. m. sumatranus* to be at risk of being lost from its other main stronghold in the Leuser Ecosystem in North Sumatra province (Gopala et al. 2011).

Figure 1. Map of northern Sumatra with the Leuser Ecosystem and Gunung Leuser National Park. The Leuser Ecosystem lost 145,000 ha of tree cover from 2001-2014, including ca. 63,000 ha of primary forest, accounting for ca. 3% of its primary forest cover. Another 22,000 ha of forest cover were lost between January 2015 and April 2017. (From World Resources Institute (WRI) via Global Forest Watch 2019)

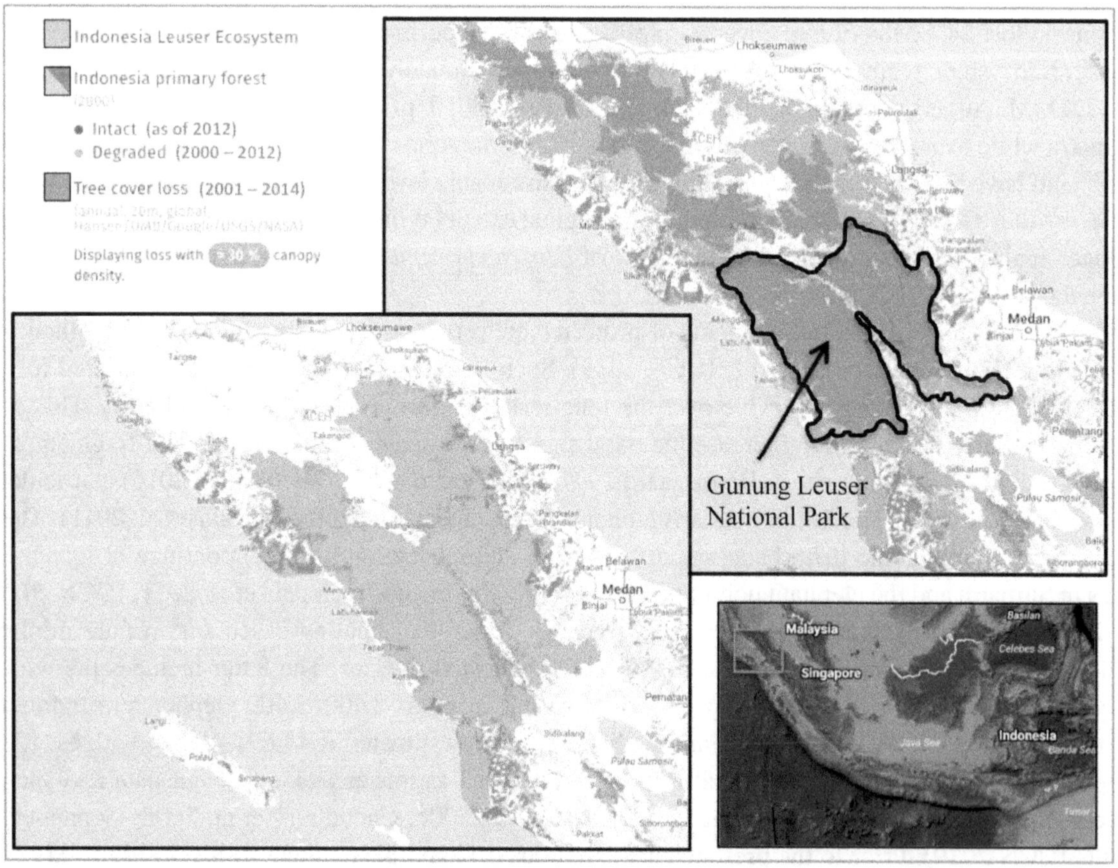

Ultimately, the persistence of *E. m. sumatranus* and *E. maximus* in general are jeopardized by increasing habitat loss and degradation, fragmentation, retribution killing, and poaching (Sukumar 1992, 2000; Sukumar & Santiapillai 1996; Leimgruber et al. 2003; Blake & Hedges 2004; Hedges et al. 2005, Hedges 2006; Choudhury et al. 2008; Gopala et al. 2011; IUCN 2019). This chapter provides results from the only empirical field study of *E. m. sumatranus* population distribution, abundance, and occupancy across Gunung Leuser National Park in the Leuser Ecosystem of North Sumatra and Aceh provinces. Robust data and analyses for conservation planning and tropical forest restoration efforts are presented.

MATERIALS AND METHODS

Study Area

This study was conducted in Gunung Leuser National Park, which is part of an IUCN World Heritage Site (Tropical Rainforest Heritage of Sumatra) in Sumatra, Indonesia (Figure 1). GLNP's 950,000 ha lie within the 2.6 million ha Leuser Ecosystem, which harbors the majority of remaining elephants in North Sumatra and Aceh provinces. Survey efforts were concentrated in 5 sample sites in a ca. 396.2 km^2 area of primary and secondary lowland tropical forests in the east-central region (97°59'00"E to 98°10'00"E and 4°0'00"N to 3°42'00"N) of GLNP. Primary lowland forests in the LE were dominated by economically valuable Dipterocarpaceae species. Secondary forest composition varied, with early successional species dominating in study sites that were cleared within a few weeks to 26-30 years earlier.

This study area was chosen primarily for two reasons. First, one location included an Elephant Patrol Unit in Aras Napal where a broad representation of habitat types in the LE exists. Consequently, a retrospective *in situ* dung decay study (Laing et al. 2003; Hedges & Lawson 2006) could be conducted there of known dung and generation times by patrol elephants consuming the same forage as wild elephants in the same diverse habitats and conditions. Second, the elephant population in this area was essentially closed. Significant population barriers existed due to: i) the steep Barisan mountain range to the west-southwest and north, which is largely impassable for elephants and provides limited elephant forage and habitat and ii) complete conversion of forests to agriculture and development in the east-southeast, also impassible for most elephants (Figure 1). There may be a few limited immigration and/or dispersal events, but the study population was effectively closed, thereby being most suitable for population models.

Elephants outside of the study area were isolated within the GLNP and LE boundaries due to extensive logging, agriculture, and development encompassing the LE (Figure 1). Extensive field surveys and interviews with local communities indicated that, due to biogeographic and resource restrictions, *E. m. sumantranus* range was primarily <1,000 m ASL in the LE, with significant predominance in lowland forests <500 m ASL (Augeri 2005; LIF unpub. data). These data appear to be consistent with other areas in Sumatra (Hedges et al. 2005; Gopala et al. 2011).

The LE is considered one of the most biogeographically diverse expanses of tropical and subtropical forests in the world (Whitten et al. 2000). This ecosystem extends from the Indian Ocean eastward to the Malacca Straits and includes topographically diverse lowland wet, moist and dry broadleaf forests, montane forests, moss and heath forests, mangroves, peat swamps, and subalpine meadows (Whitten et al. 2000). Two extensive mountain ranges >3,000 m ASL exist within the LE along with two major volcanoes, three principal lakes, and nine river systems that flow to the east and west coasts (Whitten et al. 2000). Average annual rainfall is 2,544 mm. In the eastern portion of the Ecosystem rainfall is typically more intense during the second half of the year (Whitten et al. 2000; van Schaik & Supriatna 1996). Mean daily temperature beneath primary lowland forest canopy throughout the year rarely exceeds 26°C during the day or drops below 22°C at night (Whitten et al. 2000). The driest months are February and June/July and the wettest months are March/April and September/October/November (Whitten et al. 2000).

Dung Decay Study

Following MIKE dung survey standards (Hedges & Lawson 2006) a retrospective *in situ* dung decay study was conducted in representative habitats and condition of our survey area in GLNP. One week per month during a 12-month period was spent locating and marking fresh dung piles in biogeographically divergent areas prior to the main survey. GPS locations for each dung pile were recorded, marked with flagging ribbon, and reassessed each month over the 12-month period according to the MIKE S-Classification system (Hedges & Lawson 2006). Ages of marked dung were estimated based on: i) the MIKE S-Classification system with five decay stages (Hedges & Lawson 2006) and ii) baseline study of dung with known ages from working elephants at the Elephant Patrol Unit in Aras Napal that had known diets similar to local wild elephants in the same area, habitats, and conditions.

Additional data were recorded beyond that suggested by Hedges and Lawson (2006). This included boli morphology and hardness; presence or absence of insects, mushrooms and other plants growing beneath and/or in the dung; degrees of ground moisture beneath boli; canopy cover; ground cover; degree of exposure (e.g. shaded, beneath vegetation, or in open areas, etc.); environmental influences (e.g. rainfall, floods, sun); and other characteristics that help determine boli age. Intact boli circumference was measured to estimate decay rates for different sized elephants, such as dung produced by adults versus calves.

Steady-state conditions were not assumed due to diverse habitat conditions and other variables possibly affecting dung disappearance among distinct forest and habitat types. Mean dung decay rate was generated along with associated SE, CV, etc. in GENSTAT v.15.2.0.8801 (VSNI 2012) using the Mean Decay.gen plug-in according to MIKE dung decay rate monitoring experiment procedures (Hedges & Lawson 2006). For comparison, models were run in DUNGSURV (Hilby & Lovell 1991), which also does not rely on steady-state assumptions (Hilby & Lovell 1991; Hedges et al. 2005). Other statistical procedures were conducted in SAS v.9.2 (SAS 2011), XLSTAT v.2012.6.02 (2012), and StatPlus v.5.8.3.8/Universal (2010).

Preliminary (Reconnaissance) Survey

A one month reconnaissance ("Recce") survey was conducted to i) determine the approximate distribution of elephant activity within the study area and ii) stratify the study site based on forest type, topography, habitat type, disturbance, etc. for the main survey. Recce transects were randomly located within a 2 x 2 km grid system across the 396.2 km^2 study area based on forest type, biogeographic features, and reported/potential elephant activity. Recce transects varied in length between 500–2,000 m to test optimum length for principal survey sample size and data generation. All Recce transects were conducted according to stratified random sampling, random compass bearings, MIKE dung survey standards (Hedges & Lawson 2006), DISTANCE protocols (Buckland et al. 1993, 2001), and protocols developed by Augeri (2005) specifically for this ecosystem. In addition to elephant signs (e.g. spoor, trails, feeding and rutting areas, dung, etc.), the extent of human activity and habitat disturbance (e.g. logging, forest clearings, agriculture, trails, squatting, poaching, non-timber forest product harvesting, etc.) were recorded.

Recce and dung decay data were analyzed in GENSTAT v.15.2.0.8801 (VSNI 2012). The total transect length needed to obtain a target Coefficient of Variation (CV) of 10% (Buckland et al. 2001; Hedges & Lawson 2006) was determined for our main surveys using the equation by Buckland et al. (2001):

$L= (b/\{cv_t(E)\}2)(Lo/no)$, where:

L = estimate of total line length to be surveyed to achieve target CV
b = dispersion factor (we used a factor of 3)
cv_t = target coefficient of variation
$\hat{E}$ = density estimate
L_o = total length of all pilot transects combined
n_o = total number of dung piles found on all pilot transects combined

A total of 16 Recce transects were completed in primary and secondary forests ranging in length from 535 m to 2,000 m. Of the 23.97 km of Recce transects in total, 12.53 km were located in secondary forest and 10.44 km were in primary forest. Total length of transects needed to reach a CV of 10% for the principal survey was 76.57 km as determined by mean frequency of classes S1-S4 of 3.92 dung piles/km. The optimum length of principal transects determined for this study area was 1,200 m, which enabled n=32 transects in each of two main test forest stratums (primary and secondary forests).

Principal Survey -- Elephant Distribution

The principal survey was conducted using standard line-transect (DISTANCE) methods (Buckland et al. 2001) suitable for dense tropical forests according to Hedges and Lawson (2006) and Augeri (2005). In addition, a stratified-random survey design was used with a randomly numbered 2 x 2 km grid across two main strata (i.e. macro-habitat types): primary and secondary forests (Augeri 2005). Both strata consisted of diverse biogeographic conditions and micro-habitat types (e.g. broadleaf lowland forest, swamp, scrub, grassland, sub-montane, montane, subalpine, etc.). Surveys were conducted in five study sites ranging in size between 74-184 km^2 to correspond with i) known elephant distribution in this study area and ii) home range estimates of Asian elephants via VHF radio tracking ($\bar{x}$ =115.2 ± 64 km^2; ♀: 30–160 km^2; ♂: 53–345 km^2) in Sri Lanka (Fernando et al. 2005, 2008). The latter was the nearest Asian elephant population to our study site with home range estimates in similar habitats and conditions. Final survey cells were selected using random numbers.

The starting point of each 1,200 m transect was randomly located within each cell and all transects were spaced a minimum of 1.5 km apart. Transect direction was determined by a random compass bearing directed in a straight line; in some cases slope contours were followed due to geographic barriers. Transects were surveyed without trails and replicate censuses were not done on the same transects due to possible redundant sign counts (Augeri 2005). Thus, a 10% sub-sample of parallel replicates ca. 500 m apart was conducted randomly for transects within the same habitat blocks.

A modified version of the Hedges and Lawson (2006) Dung Count Population Survey Standards was used according to protocols developed by Augeri (2005). Two teams each with three members surveyed each stratum simultaneously and conducted systematic censuses on all transects. A local researcher with hunting and elephant tracking experience on either side of the centerline followed a modified Adjusted Angle Zig-Zag Sampling Design (Thomas et al. 2006) with a sinusoidal search pattern (Anderson et al. 2001; Buckland et al. 2001; Augeri 2005). Searches were conducted at an average of <0.13 km/hour back and forth from the centerline to >25 m out (Augeri 2005). The central observer maintained the straight-line compass bearing, recorded all information, and searched 0–2.5 m on either side of the centerline (Buckland et al. 2001; Augeri 2005). Transect lengths were measured with a hip-chain to 0.1 m accuracy and the beginning, mid-point, and ends of each transect were marked with GPS coordinates. All transects were marked and subdivided into twelve 100 m survey units.

In addition to dung, all independent elephant sign events were recorded. The latter were defined as all elephant signs of different ages and types, such as sightings, dung, spoor, trails, wallows, ages and types of food species eaten, and rutting sites, etc. Metal retractable measuring tapes were used to record distances in centimeters from all signs to the transect centerline. All signs were aged and classified according to the MIKE S-Classification system (Hedges & Lawson 2006) and data from our dung decay and Recce studies. We also aged signs based on results from specific sign aging methods and tests by Augeri (2005) and colleagues (LIF unpub. data) of known dung, feeding, spoor, etc. from patrol elephants and other species at this particular site for this study. In addition, we used conventional scientific methods to age signs (see: Hedges & Lawson 2006; Augeri 2005; Murie et al. 2005; Rezendes 1999; Brown 1999; Elbrock 2003; Halfpenny 1988).

Distance functions in DISTANCE v.5.0 (Thomas et al. 2006) were fitted to i) only elephant dung and ii) all independent elephant sign events including dung. Akaike's Information Criteria (AIC) were used for objective selection of the most parsimonious model (Thomas et al. 2006). All dung and sign data were transformed using square root or log-based transformations as needed in SAS v.9.1 (SAS 2002). Differences in elephant sign and dung frequencies relative to forest and habitat type and age, forage availability, tree stand characteristics, disturbance, and so forth were examined via General Linear Model (GLM), Chi-square, and Logistic Regression. All elephant dung and sign encounter rates, detection probabilities, and densities were modeled in program DISTANCE for all forest or habitat types and ages. Differences were tested via Asymptotic Quadratic Chi-square tests in CONTRAST v.2 (Sauer & Williams 1989; Hines 2004).

Habitat Condition, Forage Availability, and Anthropogenic Disturbance

A 10 m x 10 m plot was measured at the start, end, and mid-point (600 m) along each transect. Standard vegetation and habitat characteristics were recorded for each plot, including micro- and macro-habitat types, total individuals and species of principal elephant forage items, tree composition, tree height, canopy and ground covers, and topography. A frequency of occurrence was obtained for vegetation types relative to age class for each 100 m^2 plot. In addition, at each 100 m point along the transect the main topographic and vegetation characteristics as well as the presence of principal elephant food species within a 10 m x 10 m area were recorded, along with the presence and width of important biogeographical features, such as water bodies. We also recorded all signs, ages, extents, and distances to land clearing, logging, roads, trails, agriculture, poaching, non-timber forest product harvesting (e.g. gaharu), and other forms of habitat disturbance and human activity within each plot, transect, survey grid, block, and region.

Elephant Density and Abundance

Following Hedges et al (2005), dung pile density estimates were converted to estimates of elephant density (D) and abundance (N) for each habitat stratum and forest age class in DISTANCE v.5.0 (Thomas et al. 2006). Standard multipliers for this study were the overall mean dung disappearance time of 188.1 days (df=405, SE ± 4.19, CV=7.34%) from our 12-month retrospective *in situ* dung decay study (see results below) and Tyson et al.'s (2002) dung production rate of 18.15 dung piles/day/elephant (df=11, SE ± 0.055, CV=13.94%) from elephants in southern Sumatra. (Note that the SE for this latter production rate was estimated from its associated 95% CI as reported by Tyson et al. 2002). Only 0.03% (n=25) of all elephant signs encountered were observed at distances >20 m from the transect centerline. Thus, all DISTANCE models were post-hoc truncated at 20 m. All biogeographic variables were also tested relative to elephant dung and signs. All models for the detection function were run in DISTANCE and the best-fitting model was determined based on Delta AIC values and lowest variance (Thomas et al. 2006). Statistical analyses were conducted in SAS v.9.1 (SAS 2002).

In addition to the botanical surveys noted above, we used logging records, land clearing records, a DEM model, and forest cover classification analysis generated from LANDSAT Thematic Mapper satellite images in ARC GIS v.9.1 (ESRI 2005). These data were used to determine the areal extent of i) transect and survey areas and ii) general macro-habitat (e.g. primary forests and secondary forests), micro-habitat, logged, and disturbed areas, etc. Results were used to estimate habitat, biogeographic, and disturbance, etc. effects on elephant habitat selection and abundance. All non-normal data were transformed with log, square root, or reciprocal transformations as needed in SAS v.9.1.

Elephant Area of Occupancy

Elephant area of occupancy (Ψ) and detection probability (p) rates along with abundance (N) were modeled in PRESENCE v.8.3 (Hines 2006, updated 2015) following methods by MacKenzie et al. (2002, 2006), Royle and Nichols (2003), and Augeri (2005) using i) only dung and ii) all elephant signs including dung. For landscape-level occupancy analyses, the five 74-184 km^2 study sites were modeled as survey plots and the 1,200 m transects were modeled as individual sampling occasions within their respective sites (Augeri 2005). To examine local occupancy and intensity of use along with micro-habitat variance effects, the randomly-located 20 m x 1,200 m transects were modeled as survey "plots" within a study area and each 100 m segment on a transect was modeled as a sampling occasion without time constraint. This semi-replicate structure provided good intra-plot variability measures, particularly for local intensity of use (Williams et al. 2002; Augeri 2005). Methodologically, this condition was addressed by i) the study's randomized sampling design, ii) random transect placements and compass bearings, iii) random/semi-random elephant movements, iv) complex topographic and biogeographic limitations (e.g. downed trees, ravines, impassable areas, etc.) on straight-line movements by elephants off trails in the forest, and v) censusing of fixed objects (e.g. elephant dung and signs). Consequently, potential independence conflicts could be satisfied (Williams et al. 2002; Augeri 2005; D. MacKenzie pers. comm.; J.D. Nichols pers. comm.; J.A. Royle pers. comm.).

To test the latter assumptions, Likelihood Ratio tests were used on this relationship to determine independence in the distribution of sign presence-absence (Karanth et al. 2011) between 100 m segments on all transects across all habitat types in the five study areas. Spatial dependence of elephant sign presence-absence between 100 m segments was examined via lagged correlation tests using the

Spearman (ρ) Rank Correlation Coefficient. Probabilities that sign presence–absences vary in the same or opposite directions were tested via the Kendall (τ) Rank Correlation Coefficient.

Moran's I and Geary's C tests were also conducted using the GPS locations (e.g. latitude/longitude) of elephant signs and transects to determine possible spatial autocorrelation across transects. In addition, Spatial Dependent models in PRESENCE v.8.3 (Hines 2006, updated 2015) were run to account for any possible spatial correlation among transects at both the landscape level and between individual 100 m transect segments. These latter spatial models tested i) the site parameter Theta$_0$ (Θ_0), which is the probability that the species is present, given for example that the species was not present in the adjacent transect or 100 m segment, and ii) site parameter Theta$_1$ (Θ_1), which is the probability that a species is present, given it was present in the adjacent transect or 100 m segment (Hines et al. 2010; Hines 2006). Differences among individual site parameters were tested using Asymptotic Quadratic Chi-square tests in CONTRAST v.2 (Sauer & Williams 1989; Hines 2004).

Data matrices of detection-non-detection (1, 0) histories and count histories were made in Excel pivot tables and imported into PRESENCE. For detection-non-detection, no sign or boli event was recorded as "0" and ≥1 independent event was recorded as "1". Actual independent sign event counts were also modeled for abundance-induced effects (Royle and Nichols 2003). All single season models were run in PRESENCE for single and multiple groups with constant *p* and survey-specific *p*. In addition, all other models were run, including the Abundance-Induced Heterogeneity model, Repeated Count Royle Biometrics model, Spatial Dependence model, and False-Positive Detection models. Each model was run with and without habitat, forest age, and forest type (e.g. primary vs. secondary forest), etc. covariates. Delta AIC values were used for selection of the most parsimonious model (Burnham & Anderson 2002; Hines 2006, updated 2010). Differences among occupancy and detection rates were tested using Asymptotic Quadratic Chi-square tests in CONTRAST v.2 (Sauer & Williams 1989; Hines 2004).

RESULTS

Dung Decay Rate

For the retrospective dung decay study, n=406 dung plies were encountered with a minimum of n=20 fresh piles/month. A total of n=313 dung piles were in sufficient condition to be marked and aged for this decay rate study. The number of fresh dung piles <48 hours old varied greatly between months. Disappearance rates were higher in primary forests than in secondary forests and the overall mean decay rate produced in GENSTAT according to MIKE procedures was 188.1 days (*df*=405, SE ± 4.19, CV=7.34%), while the DUNGSURV program estimated 189.6 days (CV=37.12%).

Principal Field Survey

Elephant Dung and Sign Frequencies

The total length of transects completed was 77.46 km, with 41.46 km in primary forest and 36 km in secondary forest. A total of n=237 distinct elephant dung pile events were encountered (n=225 piles in classes S1-S4 and n=12 piles in class S5) for a mean frequency of 2.9 dung piles/km in classes S1-S4. The significant majority (70.2%) of dung piles and other elephant signs encountered were in primary forests and healthy secondary forests >26 years old that retained key primary forest traits (GLM: r^2=0.41, *df*=6, F=5.85, P<0.0001) (Figure 2).

Figure 2. Total number of E. m. sumatranus dung piles by classification group encountered relative to average forest age of a transect

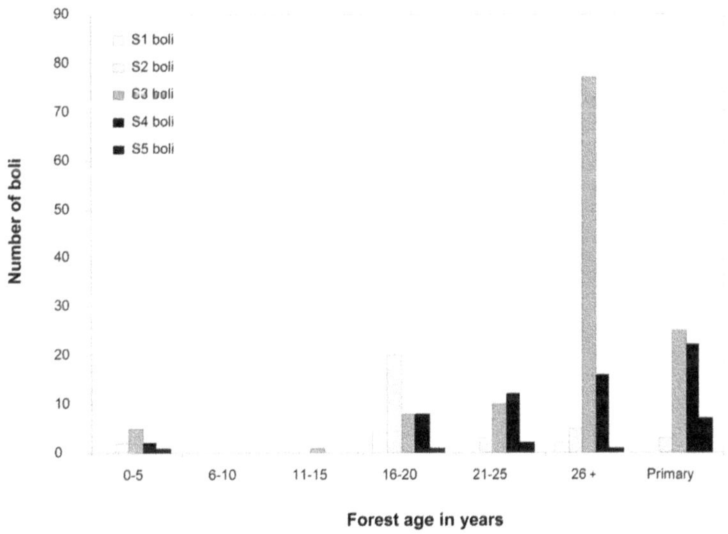

Boli ages varied greatly, with a mean of $\overline{x}$ =155.9 days old (SE ± 5.5), ranging from <24 hours old (n=4) to >1 year old (n=14). Differences in forest age appeared to have no effect on dung age and appearance (P>0.05). The most predominant elephant signs other than boli were feeding signs (n=212) and distinct track set events (n=347). There was non-significant inter-site variation in dung pile frequency and no significant difference in mean boli age relative to site location was observed (GLM: r^2=0.023, F=1.38, *df*=4, P=0.2415). However, there was a slight difference in the age of non-boli elephant signs relative to site location (GLM: r^2=0.037, F=3.3, *df*=4, P=0.0114).

Chi-square tests of sign distribution showed that elephant signs were randomly distributed in primary forests (X^2=23.685, *df*=14, P=0.332). However, elephant signs were significantly clustered in secondary forests by an average of 62% (X^2=90.84, *df*=1, P<0.0001). Sign clustering in disturbed and fragmented forests along with forests <21 years old ranged from 26% to 91% (X^2=165.77, *df*=17, P<0.0001).

Elephant Signs Relative to Habitat Type and Age

Dung detection probabilities were not significantly different among forest age classes (X^2=1.635, *df*=6, P=0.8025) (Figure 3); however, signs were more easily located in more open young forests with less debris. The majority of all elephant signs (65.3%) were in primary forests and secondary forests >20–26+ years old, where 92.4% of secondary forest signs were in healthy and minimally disturbed intact heterogeneous forests >26-30+ years old. At least 51.2% of these were in robust forests that retained characteristic primary forest traits and were adjacent to undisturbed primary forests. Disturbance in these latter sites was minimal and mainly from selective harvesting with reduced impact (Sist et al. 2002) more than 26 years earlier. The majority of older secondary forest sites corresponding with elephant activity were healthy and robust (GLM: r^2=0.41, *df*=6, F=5.85, P<0.0001). These forests had more retention of characteristic primary forest traits, including 79.6% more palms, >10% higher degrees of canopy cover,

Figure 3. E. m. sumatranus dung detection probabilities across forest age classes

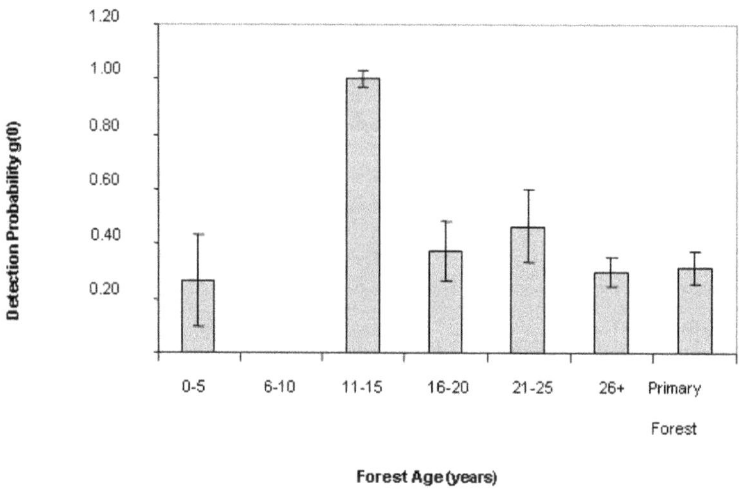

tree heights, and basal areas, and more mature and diverse fruiting tree species than younger and more degraded forests.

The only independent variable with a consistently significant effect on the variance of elephant signs was forest age (GLM: r^2=0.119, *df*=14, F=2.18, P=0.0093). There was a moderate negative correlation among forest age classes within 100 m of elephant signs (Pearson's r= -0.233), with significantly fewer signs in areas <20 years old (GLM: r^2=0.204, *df*=6, F=40.29, P<0.0001). The influence of general habitat type was non-significant relative to forest age (P=0.445). Study site location was a minor influence on sign variance due to increasing levels of human activity (GLM: r^2=0.054, *df*=4, F=3.37, P=0.0105), but forest age together with location and proximity to disturbance appeared to have a greater effect on sign variance (GLM: r^2=0.197, *df*=14, F=3.96, P=0.0001). The combination of forest age, general habitat type (primary or secondary forest), and site proximity to disturbance were the most significant interactive effects (GLM: r^2=0.202, *df*=15, F=3.79, P=0.0001).

Botanical Effects

A total of n=199 10 m x 10 m vegetation plots (n=99 in secondary and n=100 in primary forests) were surveyed and 1,019 trees were recorded from >300 species. Signs of elephants feeding were observed on >62 plant species, including among others: >20 species of Arecaceae, >20 Calameae species, >12 liana species, and >30 different tree species. The highest counts of main elephant forage species observed were in primary forests. Mean frequencies of main elephant forage groups, as well as the number of 100 m sections consisting of these groups, were notably higher in primary forests and robust intact secondary forest >26 years old, regardless of scale (Figure 4).

The only statistically significant floristic effect relative to the presence of elephants was the presence of palms (Arecaceae) (X^2=5.78, *df*=1, P=0.0161). The majority (76.5%) of palms were located in primary forest and secondary forest plots >20–26+ years old (GLM: r^2=0.195, *df*=6, F=37.85, P=0.0001). At least 57.8% of these palms were in undisturbed primary forests. No other vegetation or topographic variables were statistically significant in relation to forest age and the presence or absence of elephant

Figure 4. Mean frequencies of primary E. m. sumatranus forage groups observed at the macro-habitat scale (transect-level) (top) and micro-habitat scale every 100 m (bottom)

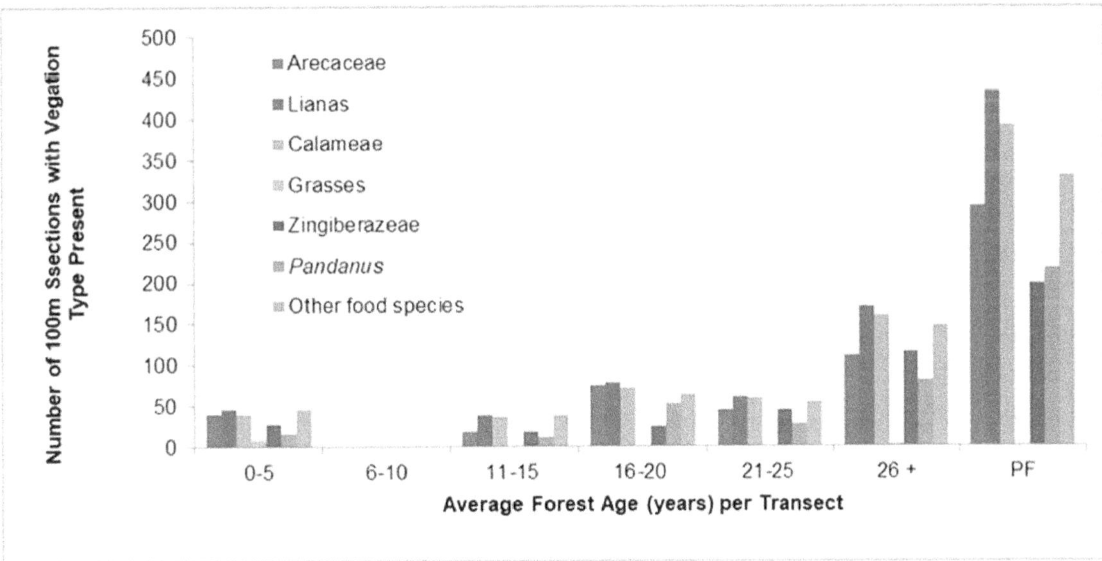

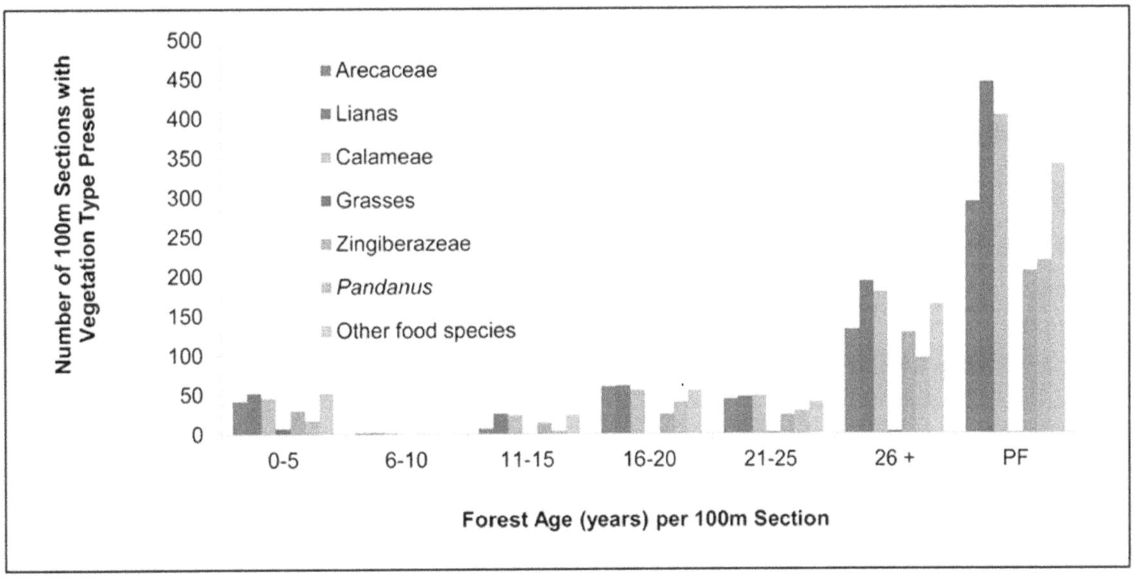

activity. But, the majority (64–81%) of all key elephant food species were in primary forest and robust secondary forest plots >26+ years old with minimal disturbance.

More than 20 species of Arecaceae were encountered. Species presence and composition varied greatly depending on forest age class. Mean frequencies/plot of palms and lianas were highest in plots aged 16-20 years ($\bar{x}$ =8.7 palms and $\bar{x}$ =30 lianas per 100 m^2). However, this difference was statistically non-significant at the P=0.05 level relative to primary forests ($\bar{x}$ =5.5 palms and $\bar{x}$ =29.3 lianas

Figure 5. Total counts of primary E. m. sumatranus forage groups relative to average forest age

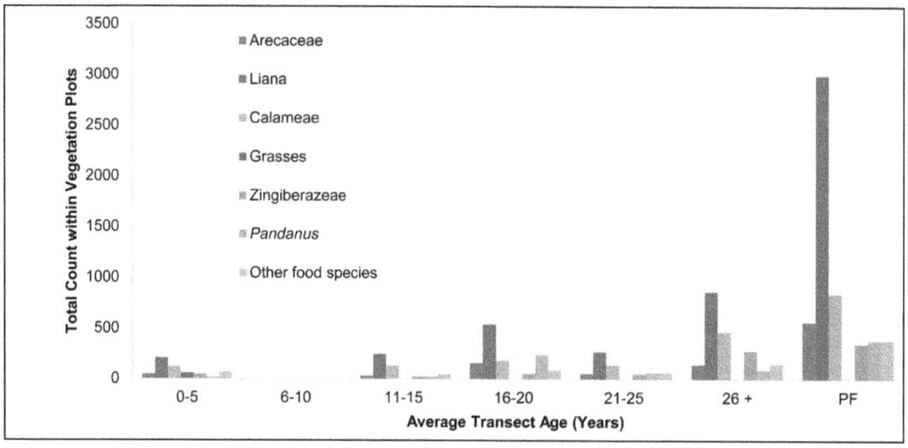

per 100 m^2), where mean palm and liana frequencies were higher than when combining all secondary forest age classes ($\bar{x}$ =4.6 palms and $\bar{x}$ =22.8 lianas).

Lianas were the most common elephant forage observed relative to the presence of elephants in over 90% of the 100 m point count plots (X^2=11.68, *df*=5, P=0.0394). Twelve liana species were identified, though many more lianas were encountered that could not be positively identified. The significant majority were observed in primary forests and robust secondary forests >26+ years old (GLM: r^2=0.043, *df*=6, F=6.27, P<0.0001). For example, total lianas in plots with average forest ages 11-15 years old was n=243 plants, compared to n=861 in the 26+ year age class, and n=2,993 plants in primary forests (Figure 5). The mean number of lianas/plot were: $\bar{x}$ =27.0 (SE ± 2.87) for forest age class 11-15 years old, $\bar{x}$ =22.1 (SE ± 0.876) for 26+ year age class, and $\bar{x}$ =29.3 (SE ± 0.902) for primary forest.

More than 20 species of Calameae (rattan) were observed. With the exception of plots <5 years old, Calameae were found in >90% of plots. The highest counts of Calameae were observed in primary forests and secondary forests >26+ years old (Figure 5). Grasses, which we could not positively identify to the species level, were the least commonly observed vegetation and were most prevalent in forested areas <5 years old (12% grass cover/plot). Nevertheless, grasses were also found in minor proportions in plots 21-25 years old (4% grass cover/plot) and in plots >26 years old (2% grass cover/plot).

Elephant Density and Abundance Based on Distance Models

The 12-month mean dung disappearance rate of 188.1 days (*df*=405, SE ± 4.19, CV=7.34%) with the lower CV was used from GENSTAT and was model-estimated to be 0.52938E^{-02} (SE=0.17085E^{-03}) in DISTANCE. The dung production rate of 18.15 (*df*=11, SE≈0.055) by Tyson et al. (2002) was estimated in our DISTANCE models to be 0.55249E^{-01} (SE=0.23198E^{-03}). The best-fitting DISTANCE model for the overall study area for all transects without stratification was the Half-Normal Cosine model (Delta AIC=7.22), resulting in an abundance estimate of N=74.8 elephants (95% CI: 45.2–120.5) (table1). This model generated an estimated density of D=0.187 elephants/km^{-2} (95% CI: 0.115–0.307) with a detection probability p=0.3468 (95% CI: 0.304–0.395) in the 396.2km^2 study area.

Two of the 26+ year old forest transects had the highest dung counts (n=18 and n=26) in the study and were randomly located ≤3 km from the Elephant Patrol Unit in the Sekundur study site. Some patrol

Table 1. Density and abundance model results from program DISTANCE. Abundance estimates for forest stratified by age groups are not practical given lack of exact area sizes. (= principal estimate, which excludes potentially confounding counts of wild and working elephant dung on two transects (anomalous boli counts n=26 and n=18) in the Sekundur site. † = alternative estimate, which includes all boli encountered.)*

Stratification	Abundance (*N*)	95% CI (*N*)	df	D/km-2	Encounter Rate (n/L)	Detection Probability (*p*)	Density Probability *f*(0)
* Study Area (396.2 km²)	* 65.6	*41.6 – 102.8	* 22.26	* 0.16706 (0.106–0.262)	* 0.006984	* 0.30591 (0.261– 0.358)	* 0.16345 (0.139–0.192)
† alt. estimate for study area	74.8	45.2 – 120.5	24.27	0.18780 (0.115-0.307)	0.008907	0.34679 (0.304– 0.395)	0.14418 (0.127– 0.164)
Forest Age Classes							
0-5 yrs old	--	--	26.79	0.0062764 (0.0020 - 0.0196)	0.0003409	0.39720	0.12588
6-10 yrs old	--	--	--	--	--	--	--
11-15 yrs old	--	--	20.05	0.0003116 (0.00006 - 0.0017)	0.0000426	--	--
16-20 yrs old	--	--	22.38	0.034746 (0.0114 - 0.106)	0.0017047	0.35874	0.13938
21-25 yrs old	--	--	24.30	0.022326 (0.00985 - 0.0509)	0.0011507	0.37686	0.13268
26+ yrs old	--	--	22.11	0.053530 (0.0248 - 0.115)	0.0030685	0.41914	0.11929
Primary Forest	--	--	26.74	0.058446 (0.0239 - 0.143)	0.0025997	0.32524	0.15373

elephant activity occured in this area, which also attracted wild elephants, and dung from patrol and wild elephants could have occurred at the same sites. Consequently, total boli counts could have been artificially raised on these transects. Thus, these two highest boli count transects may be confounding anomalies from possible mixing of wild and patrol elephant activity and they were removed from analyses. The resulting population estimate using the best-fitting model in DISTANCE (Half-Normal Cosine model [Delta AIC=0.95]) was stratified by habitat age. This model generated an estimate of N=65.6 elephants (95% CI: 41.6–102.8) at a model density of D=0.167 elephants/km^{-2} (95% CI: 0.106–0.262) and detection probability p=0.3059 (95% CI: 0.261–0.358) for the study area (Table 1).

Elephant densities/ha estimated in DISTANCE were significantly different among forest ages (X^2=33.52, *df*=5, P<0.0001). Detection probabilities (p) were not significantly different (X^2=2.6, *df*=4, P=0.6151) among the strata. Estimated elephant densities per unit area were highest in contiguous primary forests (Table 1). At least 65% of the estimated population was equally distributed between primary forests and healthy, robust secondary forests ≥26-30+ years old (X^2=0.0213, *df*=1, P=0.8840). Elephant densities in primary forests were 40.6% higher than smaller and more isolated secondary forests 16-20 years old and 61.8% higher than secondary forests 21-25 years old, but only slightly higher (8.4%) than healthy secondary forests ≥26+ years old. Corresponding encounter rates (n/L) generally fit this same pattern, i.e., sign encounter rates in primary forests were significantly higher (X^2=27.75, *df*=5, P<0.00001) than in young secondary forests 0-25 years old. Sign encounter rates in primary forests were slightly lower

(Table 1) than secondary forests ≥26+ years old, but were not significantly different (X^2=0.0868, *df*=1, P=0.7862).

Elephant Area of Occupancy

Tables 2 and 3 provide the most parsimonious occupancy models. Results were consistent with DISTANCE model estimates and were robust at both the transect and 100 m section levels with habitat and forest age covariates. The principal estimates that were most parsimonious with lowest variances, lowest Delta AIC values, and stronger congruence also excluded the anomalous boli counts of n=26 and n=18 boli from the Sekundur study site.

All statistical spatial dependence and autocorrelation tests revealed independence and negative correlations in elephant sign presence-absence among transects at the landscape scale. Similar results occurred among 100 m segments for local occupancy and intensity of use along individual transects (Table 4). This included Geary's C test, which is more sensitive to local spatial autocorrelation.

I) Landscape Level Occupancy

The most parsimonious model (Delta AIC=0.00) at the landscape level across all sample areas for all habitat types and forest ages was the Repeated Count Royle Biometrics model, which generated an occupancy rate of Ψ=0.6321 (SE ± 0.0010) with a detection probability of p=0.6225 (SE ± 0.0001) (Table 2). This model produced an overall abundance estimate for GLNP of N=392.82 elephants (SE: ± 30.65; 95% CI: 332.78-452.95) with an average abundance/sampling unit (4 km^2) of λ=6.14 (SE: ± 0.48; 95% CI: 5.20-7.08) in the study area.

Both Spatial Dependent models were not suitable at the landscape level, resulting in weak comparison scores (Delta AIC=298.24 and Delta AIC=298.29) and rendering the models improbable (Burnham

Table 2. Primary results for landscape level occupancy: (*) excludes potentially confounding counts of wild and working elephant dung on two transects (anomalous boli counts n=26 and n=18) in the Sekundar site; (†) Alternative results, which include anomalous boli counts n=26 and n=18.

Model	Delta AIC	Occupancy Rate		Detection Probability		Abundance in Study Site			Abundance in GLNP		
		Ψ	SE ±	*p*	SE ±	N	SE ±	95% CI	N	SE ±	95% CI
* All Elephant Signs and Boli from All Sites excluding anomalous boli counts n=26 and n=18 from Sekundur											
Repeated Count Royle Biometrics Model	0.00	0.6321	0.0010	0.6225	0.0001	--	--	--	392.82	30.65	332.78 - 452.95
Abundance-Induced Heterogeneity Model	21.30	0.7107	0.0551	0.2562	0.0273	78.14	11.99	54.64 – 101.64	--	--	--
† All Elephant Signs and Boli from All Sites including anomalous boli counts n=26 and n=18 from Sekundur											
3 Group Constant *p* Model	0.00	0.6995	0.0596	0.5195	0.1166						
Abundance-Induced Heterogeneity Model	50.98	0.7267	0.0516	0.2666	0.0265	85.62	12.46	61.21 – 110.04	--	--	--
1 Group Survey-Specific *p* Model	63.39	0.6823	0.0574	0.022– 0.6662	0.021– 0.7451	--	--	--	--	--	--

Figure 6. Estimated E. m. sumatranus occupancy rates by forest age class and associated mean habitat disturbance as % area disturbed (e.g. logging, road(s), cut trails, agriculture, etc.) per ha

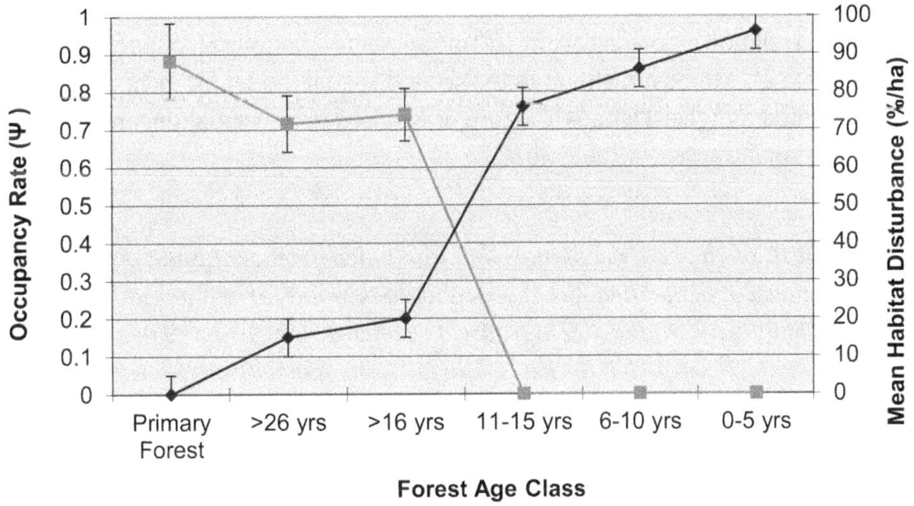

and Anderson 2002) (Table 3a). Individual site occupancy parameters Θ_0 and Θ_1 for these two Spatial Dependent models were also incongruous. Thus, results indicated no spatial dependence among transects at the landscape level (Hines 2006; Hines et al. 2010; Hines pers. comm.).

When stratified by habitat type and age, elephant area of occupancy was significantly higher in primary forests (X^2=347.03, *df*=6, P<0.0001) and corresponded with minimal to no habitat disturbance (Figure 6). The occupancy rate for primary forest across all sites was Ψ=0.8822 (SE ± 0.1069) and for primary and secondary forest >26 years old combined was Ψ=0.7162 (SE ± 0.0745). The Abundance-Induced Heterogeneity model stratified by habitat type was the most parsimonious in both cases (Delta AIC=0.00). Occupancy rates for forests <16 years old were negligible to non-existent (Ψ=0.0000).

II) Local Occupancy and Intensity of Use

The Single Season Spatial Dependent model of initial $\Theta_0 \neq$ all other Θ_0 was the most parsimonious (Delta AIC=0.00; model likelihood=1.00) (Table 3). This model measured local occupancy and intensity of use for elephant signs among 100 m sections along individual transects for all habitat types and forest ages. Its local occupancy rate and associated detection probability were not significantly different (X^2=1.6508, *df*=1, p=0.1988 and X^2=0.0535, *df*=1, P=0.8171 respectively) from the Royle Biometrics model at thc landscape scale.

Site occupancy parameters in the Single Season Spatial Dependent model were significantly different between Θ_0 and Θ_1 (X^2=166.92; df=1; P<0.00001) and between the initial 100 m segment Θ_0 and all other Θ_0 samples (individual 100 m segments) (X^2=29.49; *df*=1; P<0.00001). This is the only model tested in these analyses implying the occurrence of elephant sign(s) in a 100 m segment was possibly dependent on sign(s) occurring in adjacent 100 m segments (Hines 2006; Hines et al. 2010). However, the preponderance of results to the contrary indicates this model is inconclusive.

Examination of exact sign positions revealed that some elephant signs on or near 100 m segment borderlines were <10 m from signs in adjacent correlated segments. This situation occurred on a minority of transects (<5%) and creates a false-positive result of spatial dependence due to a minor fraction of random elephant signs' extreme proximity to random segment border locations. Furthermore, Likelihood Ratio, Spearman Rank (lagged) Correlation, and Kendal Rank Correlation tests all showed no spatial dependence or autocorrelation among adjacent 100 m segments or among elephant signs within adjacent 100 m segments (Table 4). Moreover, although the second spatial model for local occupancy (initial Θ_0 = all other Θ_0) presented a higher Delta AIC score, it resulted in no spatial dependence among 100 m segments along individual transects (Table 3b).

*Table 3. Models and tests of elephant occupancy and sign independence. * Spatial Dependent models in PRESENCE v.8.3 (Hines 2006, updated 2015) were run to account for any possible spatial correlation among transects and among individual 100 m transect segments. These spatial models test i) the site occupancy parameter Theta$_0$ (Θ_0), which is the probability that the species is present, given for example that the species was not present in the previous 100 m segment or transect, and ii) site occupancy parameter Theta$_1$ (Θ_1), which is the probability that a species is present, given it was present in the previous 100 m segment or transect. Note: potentially confounding counts of wild and working elephant dung on two transects (anomalous boli counts n=26 and n=18) in the Sekundar site were excluded from these analyses.*

A) Spatial dependent occupancy model tests for landscape level independence of elephant signs among transects.							
Test	**Parameter**	**Delta AIC**	**X²**	**DF**	**P-value**	**Spatial Independence**	
* Spatial Dependent Model Θ_0 = rest of Θ_0	Θ_0 *vs.* Θ_1	298.24	X²=0.0387	1	0.8441	Yes	
* Spatial Dependent Model $\Theta_0 \neq$ rest of Θ_0	Θ_0 *vs.* Θ_1	298.29	X²=0.0469	1	0.8285	Yes	
	Θ_0 *vs.* all Θ_0	298.29	X²=0.0257	1	0.8726	Yes	

B) Occupancy models for intensity of use and spatial dependent model tests for independence of elephant signs at the local level among 100 m segments on individual transects.

		Occupancy Rate		**Detection Probability**		**Local Intensity of Use**					
Model	**Delta AIC**	**Ψ**	**SE ±**	***p***	**SE ±**	**Θ_0**	**SE ±**	**Θ_1**	**SE ±**	**P-value**	**Spatial Independence**
* Single Season Spatial Dependent Model of Initial $\Theta_0 \neq$ all other Θ_0	0.00	0.7295	0.0758	0.6330	0.0454	0.1469	0.0433	0.8935	0.0334	0.00001	No
1 Group Survey-Specific *p* Model	24.18	0.6673	0.0758	0.3952	0.0706	--	--	--	--	--	--
* Single Season Spatial Dependent Model of initial Θ_0 = all other Θ_0	32.77	0.6732	0.0601	1.0000	0.0000	0.3003	0.0262	0.5349	0.0340	0.2562	Yes
False-Positive Detection Model	39.29	0.5569	0.0671	0.3399	0.0265	--	--	--	--	--	--

Table 4. Statistical spatial model and autocorrelation tests of elephant signs

Test	Parameter	Variable	Observed	SD	Z	P	Autocorrelation	Spatial Independence
Moran's I	Transect GPS Latitude-Longitude Coordinates	Elephant Boli	I = -0.0138	0.00659	0.198	0.8428	No	Yes
Gary's C	Transect GPS Latitude-Longitude Coordinates	Elephant Boli	C =1.0085	0.01796	0.472	0.6367	No	Yes
Moran's I	Transect GPS Latitude-Longitude Coordinates	All Elephant Signs	I = -0.0136	0.00659	0.238	0.8117	No	Yes
Gary's C	Transect GPS Latitude-Longitude Coordinates	All Elephant Signs	C = 0.9967	0.01796	-0.182	0.8555	No	Yes
Likelihood Ratio	Distribution of Elephant Sign Presence-Absence Between 100m Segments	All Elephant Signs	G^2 = 23.685 (df=14)	--	--	0.219	No	Yes
Spearman (ρ) Rank Lagged Correlation	Spatial Dependence of Sign Presence-Absence Between 100m Segments	All Elephant Signs	ρ = -0.038	--	--	0.267	No	Yes
Kendall (τ) Rank Correlation	Probability of Sign Presence–Absence Varies In Same or Opposite Direction Between 100m segments	All Elephant Signs	τ = -0.032	--	--	0.268	No	Yes

DISCUSSION

Results from this study clearly demonstrate that healthy intact primary lowland forests are critically important for Sumatran elephants. Statistical, DISTANCE, and occupancy models revealed that forest and habitat type, age, extent, integrity, and structural characteristics were the most important variables. These factors were critical for determining i) elephant distribution, abundance, and occupancy rates and ii) the degree to which elephants used or did not use various habitat types, particularly secondary forests and disturbed sites.

The significant majority of elephant abundance and activity, including feeding, was in primary forests. Healthy and intact heterogeneous secondary forests ≥26-30+ years old that were ecologically robust, minimally disturbed, and both ecologically similar and adjacent to primary forests ranked second in elephant habitat selection. At least 92% of all secondary forest elephant signs were in these latter forests. Any logging disturbance in these older secondary forest sites was from limited and minimally invasive selective harvesting ≥26+ years earlier that had significantly reduced impact practices. Consequently, these sites had a higher likelihood of retaining key primary forest character and regeneration, rather than other forms of selective or clear-cut logging (Sist et al. 2002).

These data verify that undisturbed primary forests were the most important habitat for elephants in this ecosystem. Secondary forests ≥26-30+ years old that were selected by elephants retained more critical primary forest characteristics than younger and more disturbed forests, particularly in terms of greater tree height, basal area, canopy cover, hiding cover ("refugia"), and diversity of mature fruiting tree species. However, undisturbed primary forests provided significantly more of these key traits compared to all other forest types and ages. In addition, primary forests provided significantly more elephant ranging habitat with greater ease of movement as well as higher abundances of key elephant forage, such as palms and lianas, which have been shown to be important food items for forest elephants in this region (Oliver 1978; Sukumar 1990, 2000; Steinheim et al. 2005). Elephant selection of forests younger than 16 years old was minimal or non-existent, which supports previous surveys of this and other species in this ecosystem (*see* Augeri 2005; LIF unpub. data; Augeri in prep.).

Although elephant activity and non-boli sign encounters were higher in primary forests than in forests <21 years old, we observed fewer fresh and younger age class (S1-S2) boli in primary forests. This may indicate higher dung predation in primary forests where disappearance rates were also greater than in secondary forests. Dung disappearance rates and sign visibility vary between and depend on habitat type, disturbance, fragmentation, and the presence of other species, such as dung beetles (Estrada & Coates-Estrada 2002; Hedges & Tyson 2002: Hedges et al. 2005; Augeri 2005; Nichols et al. 2007). Fragmentation can significantly increase biodiversity loss, particularly of forest dependent species like forest dung beetles. Nichols et al. (2007) revealed that there was a "strong and negative response" throughout tropical forest-dwelling dung beetle communities to increased forest fragmentation and modification. Estrada and Coates-Estrada (2002) also found a clear declining trend in dung beetle captures from primary forests through mosaic forests to fragments. Consequently, dung count studies can be affected, especially if forest age, health, structure, and character are not considered.

Availability of elephant resources was limited and compressed in young secondary and disturbed areas. Elephant activity in these sites was clumped and constrained with clustering of signs by as much as 91%. Such characteristics may indicate insularization and compression effects (Augeri 2005). In contrast, although densities and occupancy rates were higher in primary forests, elephant signs and activity were more widely and randomly dispersed.

It has been shown in these habitats on Sumatra and Borneo that patch isolation can create insularity effects as well as habitat compression (Augeri 2005). Such conditions can lead to false perceptions of animal density through higher concentrations of signs in smaller, degraded, or isolated and fragmented patches (Augeri 2005). Individuals ranging within smaller areas tend to generate more signs per unit area, which can create overestimates and false assumptions of higher animal activity, densities, and/or populations. In disturbed and fragmented patches or areas with aggregated resources, such as those found in Oliver's (1978) elephant study, a form of habitat compression can induce density- or forage-dependent spatial patterning among individuals and subsequently aggregate activity.

Augeri (2005) found in this ecosystem and on Borneo that this dynamic can create higher incidences and densities of signs in more concentrated areas, even by relatively fewer individuals. Such dynamics can still occur despite significantly lower animal occupancies and densities. Indeed, Oliver (1978) notes that elephant groups of comparable size were more mobile in primary forests whereas home ranges and movements were significantly smaller and food availability was significantly compacted in secondary forests. The latter can aggregate signs and activity patterns in fragmented secondary forest patches and create pressure, insularity, and compression effects on those populations. This can result in i) inflated density and abundance estimates and ii) individuals being obliged to range outside of smaller patches, fragmented areas, and compressed sites in search of resources, mates, and higher quality habitats (Augeri 2005). Such conditions often lead to human-wildlife conflicts and retribution killings.

This study strongly indicates that the underlying patterns and processes associated with island biogeography are significant for isolated populations of forest elephants. Insularity, compression, fragmentation, and distance-area effects are pervasive biogeographic influences, whether on true islands or in insular mainland patches (MacArthur & Wilson 1963, 1967; Lovejoy et al. 1986; Diamond 1975; Soulé & Wilcox 1980; Harris 1984; Hanski & Gilpin 1997; Wiens 1997; Wilson 1999, 2016; Bierregaard et al. 2001; Bierregard et al. 2001a; Augeri 2005). These factors can significantly impact the ecology, population dynamics, fitness, and evolutionary potential of individual species in increasingly fragmented and modified landscapes (MacArthur & Wilson 1967; MacArthur 1972; Soulé 1980; Soulé & Wilcox 1980; Frankel & Soulé 1992), including for *E. m. sumatranus*.

Some tropical species appear to have naturally irregular distributions due to habitat, competition, predation, or climate, etc. (MacArthur 1972), but many more species (e.g. forest elephants) have abnormally patchy distributions and densities caused by habitat fragmentation, insularization, and compression (Laidlaw 2000; Bierregaard et al. 2001; Meijaard et al. 2005; Augeri 2005). Such dynamics can significantly affect the distribution, habitat selection and, therefore, occupancy rates of species in Sumatra, including for *E. m. sumatranus*, as well as for myriad species in other tropical forest areas (Lovejoy et al. 1986; Laurance & Bierregaard 1997; Laidlaw 2000; Bierregaard et al. 2001; Bierregaard et al. 2001a; Meijaard et al. 2005; Augeri 2005). Consequently, this can affect density and abundance estimates if the species' consistent rates and areas of occupancy are not considered.

Large contiguous primary tropical and subtropical dry and moist broadleaf forests provide more diverse resources and habitat for forest elephants than other habitats in this region. These forests have been the principal and most important habitats for the co-evolution and long-term persistence of *E. m. sumatranus*. However, Asian elephants and many other species are susceptible to negative genetic, demographic, and population level impacts of isolation. Such influences can limit demographic exchange among populations, lower patch carrying capacity, limit the availability of otherwise suitable habitat and, therefore, shift their areas and rates of occupancy.

Furthermore, the majority of alleles in populations in small habitat "islands" or patches distant from source genes can be lost (Soulé 1980; Soulé & Wilcox 1980; Johnson et al. 2000). When disturbance is added to the system, negative effects on the equilibrium of populations and species can occur (Augeri 2005; Augeri in prep.). Autecological traits of individual species are an important component in this relationship. Villa et al. (1992) showed that when slowly developing organisms (e.g. elephants and other large mammals) are isolated they would not reach equilibrium at higher levels of disturbance. These factors will influence the persistence and survivorship of individuals and species and their capacity to consistently occupy particular areas.

Occupancy-Based Elephant Distribution and Abundance

All statistical autocorrelation and spatial tests as well as the significant majority of spatial models demonstrated spatial independence and no autocorrelation among samples for both landscape and local occupancies (Tables 3-4). The principal DISTANCE abundance estimate of N=66 elephants (95% CI: 42.4–104.8) at a model density of D=0.167 elephants/km^{-2} (95% CI: 0.106–0.262) for the 396.2 km^2 study area was the most robust and ecologically realistic (Table 1). Furthermore, this density estimate and the alternative D=0.187 elephants/km^{-2} (95% CI: 0.115–0.307) are congruous with the area-weighted mean density estimate of D=0.18 elephants/km^{-2} by Hedges et al. (2005) for Bukit Barisan Seltan National Park in southern Sumatra.

At the landscape scale, the Repeated Count Royle Biometrics model performed best with an occupancy rate of Ψ=0.6321 (SE $\pm$ 0.0010) (Table 2). The Abundance-Induced Heterogeneity models ranked lower in AIC scores, but their abundance estimates of N=78.14 elephants (95% CI=54.64–101.64) and N=85.62 elephants (95% CI: 61.21–110.04) for the 396.2 km^2 study area are consistent with our principal DISTANCE model estimate (Table 2). These estimates are also consistent with the alternative DISTANCE model of N=74.8 elephants (95% CI: 45.2–120.5) for the study area. In all cases, the 95% confidence intervals overlapped and the differences between means were less than the Least Significant Difference (LSD). In addition, the density and abundance estimates among these models were not statistically different at the P=0.05 level.

It is important to note that the best-fitting elephant density estimate modeled in DISTANCE cannot be generalized across the entire Park without accounting for several key influences. For example, if this principal density estimate of D=0.167 elephants/km^{-2} (95% CI: 0.106–0.262) is generalized across the entire 9,500 km^2 of GLNP, which includes areas where elephants clearly do not range or exist, this population could be assumed to comprise about N=1,586 elephants (95% CI: 1,007–2,489). However, this estimate relies on several tenuous assumptions. It is critical that population models account for i) variances in actual elephant habitat selection, consistent use, and distributions, including along an elevation gradient (Figure 7), ii) the extent of suitable habitat, iii) unequal site and habitat carrying capacities, and iv) consistent area of occupancy by elephants.

Even in optimal conditions very few species or taxa, if any, occupy 100% of their available range and have perfectly uniform distributions (MacArthur & Pianka 1966; MacArthur 1972; Augeri 2005; MacKenzie et al. 2006). This study demonstrates that this principle holds true for elephants in Sumatra. The reality of animal distributions is that they are patchy and most individuals (or herds) have different ranges that vary across seasons, years, habitats, and elevations, often with gaps between specific ranges. This is particularly true for large mammals, such as forest elephants. Putting aside external factors such as disturbance, it is certain that species abundance, security, mate distribution, topography, and the distribution and availability of suitable resources and habitat, among other factors, naturally determine where species like forest elephants will occur in particular densities and, importantly, at particular occupancy rates.

Thus, a generalized park-wide population estimate would be misleading if it does not include an occupancy estimate that accounts for degraded areas, unoccupied sites, influences of forest or habitat age and character, etc. on elephant distribution, and obvious areas where elephants do not occur in this ecosystem on a consistent basis, such as tropical montane and subalpine habitats >1,500 m ASL (Augeri 2005; LIF unpub. data). Elephants today are increasingly forced to transit through lower quality areas to access more suitable and higher quality habitats. However, this study demonstrates that lower quality areas, high elevation sites, etc. do not provide the resources necessary for consistent occupancy and persistence. In Leuser, the preponderance of evidence also indicates consistent elephant range in GLNP is limited to <1,000 m ASL. Our surveys showed the significant majority of elephant activity in the study area was <500m ASL and more so at <200 m ASL (r^2=0.443, *df*=288, F=1.49, P=0.0001) (Figure 7). In fact, the submontane, montane, and subalpine habitats in Leuser provide substantially less or no forage resources or suitable habitat for elephants (Augeri 2005, LIF unpub. data).

GIS analyses revealed that the extent of area for all forest and habitat types <1,000 m ASL in GLNP is ca. 3,854 km^2. Therefore, given the significant majority of elephants in GLNP range <1,000 m ASL in an essentially closed population at a density of 0.167 elephants/km^{-2} (95% CI: 0.106–0.262) and occupancy rate of Ψ=0.6321 for all lowland habitats, the most robust and ecologically realistic abundance estimate for GLNP is N=407 elephants (95% CI: 258–638). By comparison, the Repeated Count Royle Biometrics occupancy model generated an abundance estimate for GLNP of N=392.82 elephants (SE:

Figure 7. E. m. sumatranus sign frequency per 100 m along an elevation gradient in the study area

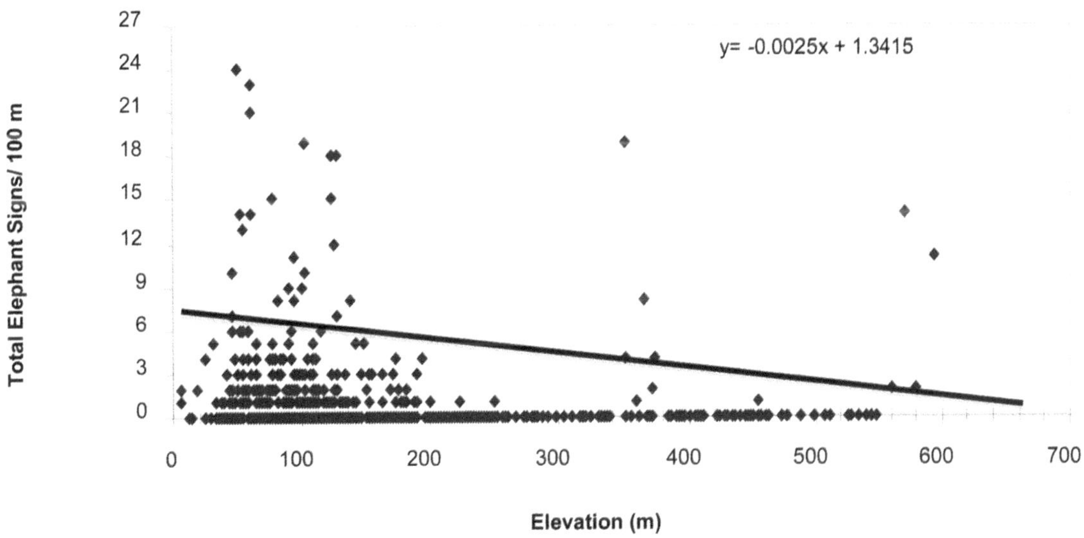

± 30.65; 95% CI: 332.78-452.95). These estimates are congruent and also appear consistent with the Hedges et al. (2005) estimate of N=498 elephants (95% CI: 373-666) with an area-weighted mean density of 0.18 elephants/km^{-2} in Bukit Barisan National Park in southern Sumatra.

Despite equal sampling effort among forest types, results clearly showed that elephants did not occur in all lowland areas of our study site. Elephant area of occupancy was significantly higher in primary forests (Ψ=0.8822; SE ± 0.1069) and corresponded with minimal to no habitat disturbance (Figure 6). GIS analyses revealed that about 66% (2,548 km^2) of the lowland forest <1,000 m ASL in GLNP consisted of primary forests and secondary forests >26-30+ years old, of which approximately 72% were occupied by elephants (Ψ=0.7162; SE ± 0.0745). A majority of the frequency (65.3%) of all elephant signs also corresponded with these forest types. Occupancy was minimal to non-existent in younger and/or degraded forests and areas >1,000 m ASL. Occupancy models estimated that about 63% (2,436 km^2) of the landscape <1,000 m ASL was occupied by elephants on a consistent level. Therefore, it is likely that the area occupied by elephants on a consistent basis in this region is below 1,000 m ASL and corresponds with the availability of primary and minimally-disturbed healthy and robust secondary forests >26-30+ years old that retained key primary forest traits.

CONCLUSION

The preponderance of data and results indicate that these analyses are highly robust. This study reveals that the Sumatran elephant population size and area of occupancy may be significantly lower than previously assumed. Data also clearly demonstrate that secondary forests need to be healthy, robust, diverse, at least >26-30+ years old, and retain critical primary forest traits for *E. m. sumatranus* to persist. The latter traits, particularly forest and habitat type, age, character, extent, integrity, diversity, and structure significantly influence *E. m. sumatranus* habitat selection and occupancy rates and are critically important in secondary forests. These biogeographic factors along with consistent area of occupancy must be included in analyses when examining the dynamics and estimating population sizes of forest-dependent species. This is particularly true for *E. m. sumatranus* as well as for global *E. maximus* populations, which also may be significantly lower than previously assumed. Such analyses are critical given forest clearing, poaching, and human-elephant conflicts have increased across *E. maximus* range.

It is necessary to caution that not all secondary forests are healthy, including some 26-30+ year old forests, and that many often remain degraded, some to substantial degrees. Therefore, forest age and ground surveys of forest condition, extent, structure, character, integrity, and diversity are important factors when analyzing the distribution of forest cover and type across the landscape, especially via remote sensing and GIS. These characteristics are imperative for forest restoration efforts and are particularly important for analyses, conservation planning, and management of biodiversity and forest-dependent species like *E. m. sumatranus*.

It is well recognized that habitat loss and its associated fragmentation form the greatest threats to biological diversity. Numerous tropical and subtropical species respond negatively to these disturbances, including forest elephants in Sumatra. Insularity is a significant and frequent cause of these responses and is likely impacting *E. m. sumatranus* and countless other species. Habitat loss, conversions, and fragmentation that do not account for cumulative effects on individual sites across wider temporal and spatial scales can affect the ecology, population dynamics, fitness, and long-term persistence of species like *E. m. sumatranus*.

The rise and fall of species like *E. maximus* and their ecological communities depend on the fitness of individuals and their ability to persist, reproduce, and recruit new reproductively successful members into their population. Their population persistence and evolution as a species will depend on the carrying capacity or ability of an area's habitat(s) to consistently support a population's individual members through multiple generations. Considering both the autecological and synecological dynamics of Asian elephants, it is clear this species is an important indicator of tropical and subtropical forest ecosystem health. This is particularly true given their broad and diverse biogeographical and foraging needs along with the specific requirements and environmental tolerances of individuals to their geographic distribution. Ultimately, the principle processes and influences of island biogeography, insularization, and distance-area effects resulting from habitat disturbance, loss, fragmentation, and/or compression are significant. Such factors help explain a substantial portion of the variance in the population size, density, occupancy, and biogeographic ecology of Asian elephants in tropical and subtropical forests in Sumatra.

It is certain that the Leuser Ecosystem and this subpopulation of *E. m. sumatranus* are exceptionally important for both the subspecies and global *E. maximus* populations– as well as for biodiversity and numerous other endangered and threatened species. This is especially significant given i) *E. m. sumatranus* is critically endangered with high priority status and ii) the Leuser Ecosystem is one of the largest and most diverse areas of intact primary forest remaining that could support a viable population of Asian elephants in the region. Therefore, it is vital that the accelerating and extensive amount of forest loss and degradation in the region is halted.

RECOMMENDATIONS

Currently, there are no specific conservation plans or projects for elephants in the Leuser Ecosystem and the situation is tenuous. Although Sumatran elephants are protected under Indonesian law, 85% of their critical habitats are outside of protected areas and will likely be converted and lost to human uses (Gopala et al. 2011; IUCN 2019). Significant expanses of elephant habitat in Sumatra are being cleared, burned, and destroyed daily for unsustainable and illegal development projects, including for oil palm and other agriculture, settlements, and roads– all of which provide greater access for, and increases in, poaching as well as higher incidences of human-elephant conflicts. GIS and ERDAS analyses of satellite data by Laumonier et al. (2010) revealed that, during 1985-2007, approximately 12 Mha of forest on Sumatra were cleared at an average of 550,000 ha/year. This amounted to a loss of 48% of Sumatra's forests and by 2007 only 30% (13 Mha) of forest cover remained on the island (Laumonier et al. 2010). During 2001-2017, the Leuser Ecosystem lost 167,000 ha (WRI 2019). Furthermore, spatial plans proposed by local authorities are absent of conservation strategies and would legalize such forest conversions inside the LE. As a result, the elephant population and biodiversity in general in the Leuser Ecosystem and across Sumatra will continue to decline (Choudhury et al. 2008; Gopala et al. 2011; IUCN 2019).

Comprehensive forest restoration efforts must occur in the Leuser Ecosystem and comprise only species native to the ecosystem, including key forest elephant forage species. Such efforts must i) connect fragmented forest patches via broad and ecologically robust corridors and ii) fully restore and reconnect the landscape to reestablish a contiguous undisturbed forest ecosystem. Biological corridors that are <5 km long should be ≥1 km wide when connecting primary forest patches that are large enough to sustain at least one elephant herd. However, longer corridors should be significantly more broad and ecologically robust to sustain elephant herd(s) while transiting seasonally and/or for at least 1-2 months. Realistically, such corridors should have a minimum length:width ratio of 5:1, preferably wider. Additional areas adjacent to corridors and protected patches should consist of buffer zones ≥3-5+ km wide each and should be in place on all sides of every corridor and patch. Such transition zones can be lower quality habitats that are undeveloped and uninhabited by humans and relatively unattractive for elephants. Importantly, these zones should be as large as possible to increase the extent of range in the landscape while providing sufficient buffer and transition zones from high quality protected habitat to lower quality areas, thereby reducing incursions and human-wildlife conflicts. This corridor mosaic will increase survivorship of forest elephants along with most other species and biodiversity in the area while minimizing human-wildlife conflicts (Augeri 2005).

The exceptionally high conservation value of the Asian elephant as an indicator, flagship, and umbrella species cannot be overestimated for biodiversity and tropical/subtropical forest restoration efforts. Comprehensive species, restoration, and landscape conservation plans should be developed based on data of *Elephas maximus* biogeographic ecology throughout its range. This strategy will not only help conserve Asian elephants, but also a broad and extensive degree of floral and faunal biodiversity across tropical and subtropical forest landscapes. Such efforts should be led by an advisory board with leading members of the scientific, NGO, government, and local communities with a focus on elephant conservation, which can serve as a critical umbrella for biodiversity conservation across Southeast and South Asia.

ACKNOWLEDGMENTS

This study was funded by the Denver Zoological Foundation and notably aided by Mike Griffiths and the Leuser International Foundation (LIF) with in-kind support. I would like to especially thank Susanna Wingard Brian for her outstanding field assistance and work as well as her co-authorship of the original report for this project. We are indebted to numerous local field assistants, colleagues, and staff at the

LIF Elephant Patrol Unit in Aras Napal and the FFI Elephant Camp in Tangkahan along with staff at the Teman Nasional Gunung Leuser (TNGL) offices in Medan and Besitang. I am also very grateful to James Zumbrunnen, Dr. Jeffrey (Andy) Royle, and James Hines for their invaluable modeling and statistical advice. Finally, I would like to thank the anonymous reviewers of this manuscript for their valuable peer review and advice along with Michael Brehm and the editorial team at IGI Global for their editorial help.

REFERENCES

AERSM Final Report (Asian Elephants Range State Meeting Final Report). (2017). Facilitated by IUCN/SSC Asian Elephant Specialist Group. https://www.asesg.org/PDFfiles/2017/AsERSM%20 2017_Final%20Report.pdf

Anderson, D. R., Burnham, K. P., Lubow, B. C., Thomas, L., Corn, P. S., Medica, P. A., & Marlow, R. W. (2001). Field trials of Line Transect Methods Applied to Estimation of Desert Tortoise Abundance. *The Journal of Wildlife Management, 65*(3), 583–597. doi:10.2307/3803111

Augeri, D. M. (2005). On the Biogeographic Ecology of the Malaysian Sun Bear. (Doctoral thesis), School of Biological Sciences, University of Cambridge, UK.

Augeri, D. M. (n.d.). In Prep. Malayan Sun Bear (*Helarctos malayanus*) Occupancy and Distribution Across the Landscape Matrix in Indonesia. *Ursus. International Association of Bear Research and Management,* Missoula, MT.

Bierregaard, R. O., Jr., Laurance, W. F., Gascon, C., Benitez-Malvido, J., Fearnside, P. M., Fonseca, C. R., Ganade, G., … Williamson, B. (2001a). Principles of forest fragmentation and conservation in the Amazon. In: R. O. Bierregaard, Jr., C. Gascon, T. E. Lovejoy, & R. Mesquita (Eds.), Lessons from the Amazon: The Ecology and Conservation of a Fragmented Forest. New Haven, CT: Yale University Press.

Bierregaard, R. O. Jr, Gascon, C., Lovejoy, T. E., & Mesquita, R. (2001). *Lessons from the Amazon: The Ecology and Conservation of a Fragmented Forest*. New Haven, CT: Yale University Press.

Blake, S., & Hedges, S. (2004). Sinking the Flagship: The Case for Forest Elephants in Asia and Africa. *Conservation Biology, 18*(5), 1191–1202. doi:10.1111/j.1523-1739.2004.01860.x

Brown, T. (1999). Tom Brown's Science and Art of Tracking. New York, NY: Berkley Trade.

Brown, T. H., Simangunsong, B. C. H., Sukadri, D., Brown, D. W., Subarudi, S., Dermawan, A., & Rufi'ie, A. (2005). Restructuring and Revitalization of Indonesia's Wood-based Industry: Synthesis of Three Major Studies. Ministry of Forestry, CIFOR, and DFID-MFP. Jakarta.

Buckland, S. T., Anderson, D. R., Burnham, K. P., & Laake, J. L. (1993). *Distance Sampling: Estimating Abundance of Biological Populations*. London, UK: Chapman and Hall.

Buckland, S. T., Anderson, D. R., Burnham, K. P., Laake, J. L., Borchers, D. L., & Thomas, L. (2001). Introduction to Distance Sampling: Estimating Abundance of Biological Populations. Oxford, UK: Oxford University Press.

Burnham, K. P., & Anderson, D. R. (2002). *Model Selection and Multimodel Inference: A Practical Information-Theoretic Approach* (2nd ed.). New York, NY: Springer-Verlag.

Choudhury, A., Lahiri Choudhury, D. K., Desai, A., Duckworth, J. W., Easa, P. S., Johnsingh, A. J. T., Fernando, P., … Wikramanayake, E. (IUCN SSC Asian Elephant Specialist Group). (2008). *Elephas maximus*. The IUCN Red List of Threatened Species.

CITES. (2019). *Conventional International Trade of Endangered Species of Wild Fauna and Flora*. Geneva, Switzerland.

Diamond, J. M. (1975). Assembly of Species Communities. In M. L. Cody & J. M. Diamond (Eds.), *Ecology and Evolution of Communities* (pp. 342–444). Cambridge, MA: Harvard University Press.

Elbrock, M. (2003). Mammal Tracks and Sign: A Guide to North American Species. Mechanicsburg, PA: Stackpole Books.

ESRI. (2005). *ARC GIS, version 9.1*. Redlands, CA: ESRI.

Estrada, A., & Coates-Estrada, R. (2002). Dung Beetles in Continuous Forest, Forest Fragments and in an Agricultural Mosaic Habitat Island at Los Tuxtlas, Mexico. *Biodiversity and Conservation*, *11*(11), 1903–1918. doi:10.1023/A:1020896928578

Fernando, P., Wikramanayake, E., Weerakoon, D., Jayasinghe, L. K. A., Gunawardena, M., & Janaka, H. K. (2005). Perceptions and Patterns of Human Elephant Conflict in Old and New Settlements in Sri Lanka: Insights for Mitigation and Management. *Biodiversity and Conservation*, *14*(10), 2465–2481. doi:10.100710531-004-0216-z

Fernando, P., Wikramanayake, E. D., Janaka, H. K., Jayasinghe, L. K. A., Gunawardena, M., Kotagama, S. W., ... Pastorini, J. (2008). Ranging Behavior in Sri Lankan Elephants. *Mammalian Biology*, *73*(1), 2–13. doi:10.1016/j.mambio.2007.07.007

Fleischer, R. C., Perry, E. A., Muralidharan, K., Stevens, E. E., & Wemmer, C. M. (2001). Phylogeography of the Asian elephant (*Elephas maximus*) Based on Mitochondrial DNA. *Evolution; International Journal of Organic Evolution*, *55*(9), 1882–1892. doi:10.1111/j.0014-3820.2001.tb00837.x PMID:11681743

Frankel, O. H., & Soulé, M. E. (1992). *Conservation and Evolution*. Cambridge, UK: Cambridge University Press.

Gopala, A., & Hadian, O., *Sunarto, S. A., Williams, A., Leimgruber, P., Chambliss, S. E., & Gunaryadi, D. (2011). Elephas maximus ssp. sumatranus*. The IUCN Red List of Threatened Species.

Halfpenny, J. C. (1988). *A Field Guide to Mammal Tracking in North America* (1st ed.). Boulder, CO: Johnson Books.

Hamid, A. F., & Darjo, L. (1993). Laporan kegiatan Monitoring Proses Pelapukan Kotoran Gajah di Stasiun Penelitian Sekundur Ecosystem Leuser. Program Pengembangan Leuser, EU, and the Indonesian government, Division of Research, Monitoring, and Information, Medan.

Hanski, I. A., & Gilpin, M. E. (1997). *Metapopulation Biology: Ecology, Genetics, and Evolution*. Academic Press.

Harris, L. D. (1984). The Fragmented Forest: Island Biogeography Theory and the Preservation of Biotic Diversity. Chicago, IL: University of Chicago Press. doi:10.7208/chicago/9780226219950.001.0001

Hedges, S. (2006). Conservation. In M. E. Fowler & S. K. Mikota (Eds.), *Biology, Medicine and Surgery of Elephants* (pp. 475–490). Oxford, UK: Blackwell Publishing. doi:10.1002/9780470344484.ch36

Hedges, S., & Lawson, D. (2006). Dung Survey Standards for the MIKE Programme. CITES MIKE Programme, Central Coordinating Unit, PO Box 68200, Nairobi, Kenya.

Hedges, S., & Tyson, M. (2002). *Some Thoughts on Counting Elephants in SE Asian forests, with Particular Reference to the CITES Monitoring of the Illegal Killing of Elephants Program.* New York, NY: Wildlife Conservation Society.

Hedges, S., Tyson, M. J., Sitompul, A. F., Kinnaird, M. F., Gunaryadi, A., & Aslan. (2005). Distribution, Status, and Conservation Needs of Asian elephants (*Elephas maximus*) in Lampung Province, Sumatra, Indonesia. *Biological Conservation, 124*(1), 35–48. doi:10.1016/j.biocon.2005.01.004

Hedges, S. (Ed.). (2005). Monitoring the Illegal Killing of Elephants; Dung Count Population Survey Standards for the MIKE Programme. Final draft (V1d) July 18, 2005. CITES MIKE Programme, Nairobi, Kenya.

Hiby, L., & Lovell, P. (1991). DUNGSURV– A Program for Estimating Elephant Density from Dung Density Without Assuming 'Steady State'. Censusing Elephants in Forests. In *Proceedings of an International Workshop (Eds.* U. Ramakrishnan, J. A. Santosh,, & R. Sukumar*)*, pp. 73–79. Asian Elephant Conservation Centre, Bangalore, India.

Hines, J. (2004). Contrast, v.2. Patuxent Wildlife Research Center, United States Geological Survey. US Department of Interior, Laurel, MD.

Hines, J. (2006). PRESENCE v.8.3 (updated 2015). Software to Estimate Patch Occupancy and Related Parameters. USGS Patuxant Wildlife Research Center, Laurel, MD.

Holmes, D. A. (2002). The Predicted Extinction of Lowland Forests in Indonesia. In E. Wickramanayake, E. Dinerstein, C. J. Loucks, D. M. Olson, J. Morrison, J. Lamoreux, & ... P. Hedao (Eds.), *Terrestrial Ecoregions of the Indo-Pacific: A Conservation Assessment* (pp. 7–13). Washington, D.C.: Island Press.

IUCN. (2019). The IUCN Red List of Threatened Species 2011: e.T199856A9129626. doi:. doi:10.2305/IUCN.UK.2011-2.RLTS.T199856A9129626.en

Johnson, K. P., Adler, F. R., & Cheery, J. L. (2000). Genetic and Phylogenetic Consequences of Island Biogeography. *Evolution; International Journal of Organic Evolution, 54*(2), 387–396. doi:10.1111/j.0014-3820.2000.tb00041.x PMID:10937215

Karanth, K. U., Gopalaswamy, A. M., Kumar, N. S., Vaidyanathan, S., Nichols, J. D., & MacKenzie, D. I. (2011). Monitoring Carnivore Populations at the Landscape Scale: Occupancy Modelling of Tigers from Sign Surveys. *Journal of Applied Ecology, 48*(4), 1048–1056. doi:10.1111/j.1365-2664.2011.02002.x

Kemf, E., & Santiapillai, C. (2000). *Asian Elephants in the Wild. A WWF Species Status Report.* Gland, Switzerland: World Wide Fund for Nature.

Kinnaird, M. F., Sanderson, E. W., O'Brien, T. G., Wibisono, H. T., & Woolmer, G. (2003). Deforestation Trends in a Tropical Landscape and Implications for Endangered Large Mammals. *Conservation Biology*, *17*(1), 245–257. doi:10.1046/j.1523-1739.2003.02040.x

Laidlaw, R. K. (2000). Effects of Habitat Disturbance and Protected Areas on Mammals of Peninsular Malaysia. *Conservation Biology*, *14*(6), 1639–1648. doi:10.1111/j.1523-1739.2000.99073.x

Laing, S. E., Buckland, S. T., Burn, R. W., Lambie, D., & Amphlett, A. (2003). Dung and Nest Surveys: Estimating Decay Rates. *Journal of Applied Ecology*, *40*(6), 1102–1111. doi:10.1111/j.1365-2664.2003.00861.x

Laumonier, Y., Uryu, Y., Stüwe, M., Budiman, A., Setiabudi, B. & Hadian, O. (2010). Eco-Floristic Sectors and Deforestation Threats in Sumatra: Identifying New Conservation Area Network Priorities for Ecosystem-Based Land Use Planning. *Biodiversity Conservation*, 19, 1153-1174.

Laurance, W. F., & Bierregaard, R. O. (Eds.). (1997). Tropical Rainforest Remnants: Ecology, Management, and Conservation of Fragmented Communities. Chicago, IL: University of Chicago Press.

Leimgruber, P., Gagnon, J. B., Wemmer, C. M., Kelly, D. S., Songer, M. A., & Selig, E. R. (2003). Fragmentation of Asia's remaining wildlands: Implications for Asian elephant conservation. *Animal Conservation*, *6*(4), 347–359. doi:10.1017/S1367943003003421

LIF (Leuser International Foundation). (n.d.). Unpublished Data.

Lovejoy, T. E., Bierregaard, B. O. Jr, Rylands, A. B., Malcolm, J. R., Quintela, C. E., Harper, L. H., ... Hays, M. B. (1986). Edge and Other Effects of Isolation on Amazon Forest Fragments. In M. E. Soulé (Ed.), *Conservation Biology; the Science of Scarcity and Diversity* (pp. 257–285). Sunderland, MA: Sinauer Associates

MacArthur, R. H. (1972). *Geographical Ecology. Patterns in the Distribution of Species*. Princeton, NJ: Princeton University Press.

MacArthur, R. H., & Pianka, E. R. (1966). On Optimal use of a Patchy Environment. *American Naturalist*, *100*(916), 604–609. doi:10.1086/282454

MacArthur, R. H., & Wilson, E. O. (1963). An Equilibrium Theory of Insular Zoogeography. *Evolution; International Journal of Organic Evolution*, *17*(4), 373–387. doi:10.1111/j.1558-5646.1963.tb03295.x

MacArthur, R. H., & Wilson, E. O. (1967). *The Theory of Island Biogeography*. Princeton, NJ: Princeton University Press.

MacKenzie, D. I., Nichols, J., & Royle, J. A., Pollock, K. H., Bailey, L., & Hines, J. (2006). Occupancy Estimation and Modeling. Inferring Patterns and Dynamics of Species Occurrence. Academic Press.

Margono, B. A., Potapov, P. V., Turubanova, S., Stolle, F., & Hansen, M. C. (2014). Primary forest cover loss in Indonesia over 2000–2012. *Nature Climate Change*, *29*(June), 1–6. doi:10.1038/NCLIMATE2277

Margono, B.A., Turubanova, S., Zhuravleva, I., Potapov, P., Tyukavina, A., Baccini, A., Goetz, S., & Hansen, M. C. (2012). Mapping and Monitoring Deforestation and Forest Degradation in Sumatra (Indonesia) Using Landsat Time Series Data Sets from 1990 to 2010. Environ. Res. Lett. doi:. doi:10.1088/1748-9326/7/3/034010

Meijaard, E., Sheil, D., Nasi, R., Augeri, D., Rosenbaum, B., Iskandar, D., ... O'Brien, T. (2005). *Life After Logging: Reconciling wildlife conservation and production forestry in Indonesian Borneo*. Bogor, Indonesia: Center for International Forestry Research.

Murie, O. J., Elbrock, M., & Peterson, R. T. (2005). Peterson Field Guide to Animal Tracks (3rd ed.). Boston, MA: Houghton Mifflin Harcourt.

Nichols, E., Larsen, T., Spector, S., Davis, A. L., Escobar, F., Favila, M., & Vulinec, K. (2007). Global Dung Beetle Response to Tropical Forest Modification and Fragmentation: A Quantitative Literature Review and Meta-Analysis. *Biological Conservation, 137*(1), 1–19. doi:10.1016/j.biocon.2007.01.023

Olivier, R. C. D. (1978). *On the ecology of the Asian elephant, Elephas maximus Linn. With particular reference to Malaya and Sri Lanka.* [Doctoral thesis], University of Cambridge, Cambridge.

Rezendes, P. (1999). Tracking and the Art of Seeing: How to Read Animal Tracks and Sign. New York, NY: Collins Reference.

Royle, J. A., & Nichols, J. D. (2003). Estimating Abundance from Repeated Presence-Absence Data or Point Counts. *Ecology, 84*(3), 777–790. doi:10.1890/0012-9658(2003)084[0777:EAFRPA]2.0.CO;2

SAS. (2011). *SAS version 9.2*. Cary, NC: SAS Institute

Sauer, J. R., & Williams, B. K. (1989). Generalized Procedures for Testing Hypotheses About Survival or Recovery Rates. *The Journal of Wildlife Management, 53*(1), 137–142. doi:10.2307/3801320

Shoshani, J., & Tassy, P. (2005). Advances in Proboscidean Taxonomy and Classification, Anatomy and Physiology, and Ecology and Behavior. *Quaternary International, 126-128*, 5–20. doi:10.1016/j.quaint.2004.04.011

Sist, P., Sheil, D., Kartawinata, K., & Priyadi, H. (2002). Reduced-impact logging in Indonesian Borneo: Some results confirming the need for new silvicultural prescriptions. In Forest, Science and Sustainability: The Bulungan Model Forest, Technical Report Phase I 1997-2001. ITTO Project PD 12/97 REV.1 (F).

Soulé, M. E. (1980). Thresholds for Survival: Maintaining Fitness and Evolutionary Potential. In M. E. Soulé & B. A. Wilcox (Eds.), *Conservation Biology: An Evolutionary-Ecological Perspective* (pp. 153–169). Sunderland, MA: Sinaur Associates.

Soulé, M. E., & Wilcox, B. A. (Eds.). (1980). *Conservation Biology: An Evolutionary-Ecological Perspective*. Sunderland, MA: Sinaur Associates

StatPlus. (2010). StatPlus version. Walnut, CA: AnalystSoft

Steinheim, G., Wegge, P., Fjellstad, J. I., Jnawali, S. R., & Weladji, R. B. (2005). Dry Season Diets and Habitat use of Sympatric Asian elephants (*Elephas maximus*) and Greater One-horned Rhinoceros (*Rhinocerus unicornis*) in Nepal. *Journal of Zoology (London, England)*, *265*(4), 377–385. doi:10.1017/S0952836905006448

Sukumar, R. (1989). Ecology of the Asian Elephant in Southern India. I. Movement and Habitat Utilization Patterns. *Journal of Tropical Ecology*, *5*(1), 1–18. doi:10.1017/S0266467400003175

Sukumar, R. (1990). Ecology of the Asian Elephant in Southern India. II. Feeding habits and Crop Raiding Patterns. *Journal of Tropical Ecology*, *6*(1), 33–53. doi:10.1017/S0266467400004004

Sukumar, R. (1992). *The Asian Elephant: Ecology and Management* (2nd ed.). Cambridge, UK: Cambridge University Press.

Sukumar, R. (2000). Asian Elephant. In R. P. Reading & B. Miller (Eds.), Endangered Animals (pp. 34–39). Westport, CT: Greenwood Press.

Sukumar, R., & Santiapillai, C. (1996). *Elephas maximus*: Status and Distribution. In J. Shoshani & P. Tassy (Eds.), *The Proboscidea: Evolution and Paleoecology of Elephants and Their Relatives* (pp. 327–331). Oxford, UK: Oxford University Press.

Thomas, L., Laake, J. L., Strindberg, S., Marques, F. F. C., Buckland, S. T., Borchers, D. L., ... Marques, T. A. (2006). *Distance 5.0. Release 2. Research Unit for Wildlife Population Assessment*. UK: University of St. Andrews.

Tyson, M. J., Hedges, S., & Sitompul, A. F. (2002). Elephant Defecation Rate Study, Way Kambas National Park 2000/2001: Final Report to WWF, WCS, and PHKA, Jan. 5, 2002. Wildlife Conservation Society — Indonesia Program, Bogor, Indonesia.

Uryu, Y., Mott, C., Foead, N., Yulianto, K., Budiman, A. S., Takakai, F., Nursamsu, S., … Stüwe, M. (2007). Deforestation, forest degradation, biodiversity loss and CO2 emissions in Riau, Sumatra, Indonesia. WWF Indonesia, Jakarta.

Uryu, Y., Purastuti, E., Laumonier, Y., Sunarto, S., Budiman, A., Yulianto, K., Sudibyo, A., … Stüwe, M. (2010). Sumatra's Forests, Their Wildlife and the Climate. Windows in Time: 1985, 1990, 2000 and 2009. A quantitative assessment of some of Sumatra's natural resources submitted as technical report by invitation to the National Forestry Council (DKN) of Indonesia. WWF Indonesia, Jakarta.

van Schaik, C., & Supriatha, J. (1996). *Leuser: a Sumatran Sanctuary*. Depok, Indonesia: Yayasan Leuser Internasional.

Villa, F., Rossi, O., & Sartore, F. (1992). Understanding the Role of Chronic Environmental Disturbance in the Context of Island Biogeographic Theory. *Environmental Management*, *16*(5), 653–666. doi:10.1007/BF02589019

VSNI. (2012). *Genstat, version 15.2.0.8801. VSN International*. Hemel Hempstead, UK: LTD.

Whitten, T., Damanik, S. J., Anwar, J., & Hisyam, N. (Eds.). (2000). *The Ecology of Sumatra. Periplus Editions*. Singapore: Berkeley Books Pte. Ltd.

Wiens, J. A. (1997). Metapopulation Dynamics and Landscape Ecology. In I. A. Hanski & M. E. Gilpin (Eds.), *Metapopulation Biology: Ecology, Genetics, and Evolution* (pp. 43–62). San Diego, CA: Academic Press. doi:10.1016/B978-012323445-2/50005-5

Williams, B. K., Nichols, J. D., & Conroy, M. J. (2002). *Analysis and Management of Animal Populations* (1st ed.). San Diego, CA: Academic Press.

Wilson, E. O. (1999). Preface. In R. H. MacArthur and E. O. Wilson (Eds.), The Theory of Island Biogeography. Princeton, NJ: Princeton University Press.

Wilson, E. O. (2016). *Half-Earth: Our Planet's Fight for Life*. New York, N.Y.: Liveright Publishing Corporation.

WRI (World Resources Institute). (2019). Global Forest Watch. Retrieved from http://globalforestwatch.org/Dashboards/country/IDN?treeLossTsc=eyJoaWdobGlnaHRlZCI6ZmFsc2V9

XLSTAT. (2012). XLSTAT version 2012.6.02. Addinsoft, New York, NY.

Chapter 9
Examining Patterns and Impacts of Forest Resource Extraction and Forest Degradation in Tropical Dry Forests

Pooja Choksi
Columbia University, USA

ABSTRACT

Forest degradation is attributed to the excessive use of forest resources and extraction, whether for subsistence or commercial purposes. With an increase in human population pressure on forests, forest degradation is becoming a concern for the conservation of biodiversity. The high human dependence on tropical dry forests underscores the need for a complete understanding of the interaction of humans and these forests to ensure their persistence and the wellbeing of the people who depend on these resources. This chapter examines forest resource use and degradation to provide a nuanced understanding of forest degradation and the impact of forest resource use.

INTRODUCTION

Tropical dry forests have historically had a high rate of resource exploitation across the world (Janzen, 1988; Miles et al., 2006). Presently, only 17% of the tropical forests that remain across the world are dry forests as the rate of clearing tropical dry forests and forest degradation continue to remain high and present unique challenges to forest conservation (Miles et al., 2006). Of this 17% that remain, 54% of the dry forests are found in the Americas (Miles et al., 2006). This rate of exploitation is primarily due to the affinity humans have for these forests. Dry forests are mainly located in relatively flat terrains, which receive seasonal rainfall(Fajardo et al., 2005). Further, a lower biomass and structural complexity in comparison to tropical humid forests makes them ideal for clearing by humans to engage in short-cycle crop agriculture, livestock farming (Fajardo et al., 2005). Humans have relied on tropical dry forests for several resources including biomass for construction and fuel, fruits and edible flowers, leaves and

DOI: 10.4018/978-1-7998-0014-9.ch009

medicinal plants and continue to do so even today (Castillo et al. 2005; Portillo-Quintero et al. 2014; Portillo-Quintero & Smith, 2018).

Due to the historic exploitation of these forests and rapid human population growth, large contiguous areas of tropical dry forests presently only remain in Brazil, Bolivia and Paraguay in South America (Gillespie et al., 2012). While tropical dry forests thrive in some parts of the world, they are among the least protected and possibly the most vulnerable to anthropogenic disturbances in the world (Gillespie et al., 2012). Deforestation for agricultural purposes has been the largest threat to tropical dry forests, however, forest degradation is also a cause for concern; however, forest degradation, which is more cryptic is also a major threat to this forest type (Gillespie et al., 2012; Puyravaud, Davidar, & Laurance, 2010; Thompson et al., 2013). Numerous reports have attempted to estimate the extent of global forest degradation across the world, with one estimate by the International Tropical Timber Organization (ITTO; ITTO, 2002) at 850 million hectares across the tropics.

The high human dependence on tropical dry forests underscores the need for a complete understanding of the interaction of humans and these forests to ensure their persistence. In the recent past, ecosystem conservation has taken the form of 'fortress conservation' (Chan et al., 2007). However, humans have relied on tropical dry forests for centuries, making forests an integral part of their lifestyle and livelihoods and a model of fortress conservation may not provide the intended impact (Kareiva et al., 2007). In order to successfully conserve and manage tropical dry forests, it is critical to understand the impacts of human resource use and what constitutes forest degradation. Further, using this knowledge of the impacts of forest resource use, natural resource managers and scientists may create a model of forest conservation and management that fulfils the needs of people who rely on forests and the environmental conservation agenda.

This chapter examines forest resource use in tropical dry forests, forest degradation and ecological restoration. Questions such as what constitutes forest degradation, how degradation is defined and quantified, how forest resource extraction affects flora and fauna, how local communities manage resource extraction and how decisions to restore degraded forests are taken are explored in this chapter. This chapter aims to provide a holistic understanding of not just the ecological but also the socio- economic and cultural aspects and impacts of forest resource use and extraction.

BACKGROUND

What Is Forest Degradation?

The term forest degradation has only recently been coined and refers to '*the reduction of the capacity of a forest to provide goods and services*' (FAO, 2002). Another definition, by the United Nations Framework Convention on Climate Change (UNFCCC), refers to forest degradation as the '*changes in forest structure, dynamics or functionality as a result of human causes*' (UNFCC, 2002). These goods and services may include any or all of the following: biogeochemical processes, non- timber forest products, recreation and cultural value, and taxonomic biodiversity to name a few (Sasaki & Putz, 2009). Similarly Intergovernmental Panel on Climate Change (IPCC) defines forest degradation as '*direct human-induced long-term loss...of at least Y% of forest carbon stocks...since time T*'(Penman et al., 2003). A further elaborate definition of forest degradation by the Convention on Biological Diversity (CBD) is '*a degraded forest is a secondary forest that has lost, through human activities, the structure, function, species composition or productivity normally associated with a natural forest type expected on that site'*(Secretariat of the Convention on Biological Diversity, 2002). Over 50 definitions of forest degradation is evidence of the not only the varying ecological features of forest but also the different perceptions, objectives and values of what constitutes a forest and its quality (Ghazoul et al., 2015).

Why Is Defining Forest Degradation Problematic?

Forest are spatially and temporally dynamic in nature (Chazdon, 2014; DeFries & Pandey, 2010; Ghazoul et al., 2015). The dynamism of forests is a result of climatic, natural and anthropogenic causes and allow a forest to remain in different stable states (Chazdon, 2014; Scheffer, Carpenter, Foley, Folke, & Walker, 2001). Additionally, the spatial and temporal resolution of a study will also determine how the health of a forest is perceived. For example, the temporal changes in a forest also make quantification of forest degradation difficult on a meaningful time scale of forest recovery.

Ecological questions often require a baseline to quantify changes to the resource or system being studied. When defining forest degradation structurally, scientists refer to an 'undisturbed', 'natural' or 'pristine' forest as a baseline or reference state to determine the impact of various anthropogenic uses (Ghazoul et al., 2015; Thompson et al., 2013). The reference to a 'pristine' or 'natural' forest promotes the idea of a former state of a forest that is undisturbed by humans. Moreover, the consideration of a reference state inserts a bias of an expected state of forest. Measuring forest degradation with a reference state results not only poses a problem in correctly assessing forest degradation, but also allows for the interpretation of forest states that maybe used by various stakeholders for their vested interests (Ghazoul et al., 2015).

Conceptualization of Forest Degradation

Forest degradation, being inherently subjective is difficult to conceptualize and subsequently measure. Scientists have put forward approaches to conceptualize and subsequently study forest degradation. The following sections present these two approaches to conceptualize on forest degradation. The first approach conceptualizes forest degradation using the ecological concept of alternative stable states and ecosystem resilience. The second approach encourages a broader view of forest use and conservation with the aim of simplifying policy making with respect to forest health. Both the approaches recognize that humans are a significant part of an ecosystem and interact with forests.

Resilience and Alternative Stable Steady-States

Ghazoul et al. (2015) provide a conceptual framework of degradation that depends on the resilience of the ecosystem. Using the analogy of basins of attraction, the authors explain the dynamism of forest states that is dependent on its resilience. A stable state is not static and allows for a forest to respond to different disturbances and return to a state of pre- disturbance through natural progression. When disturbances are large or frequent, a forest may change to a steady- state, which is unlike its previous pre- disturbance state. Resilience of system refers to its ability to remain in one stead- state when exposed to a disturbance of any kind, natural or anthropogenic. In this context, Ghazoul et al. (2015) define forest degradation as a '*human-induced loss of resilience, which prevents natural recovery to the pre-disturbance state*'.

The basins of attraction represent different stable steady states (Figure 1). In the larger basin are multiple steady states of 'degraded' forest, which comprise varying species compositional and structural possibilities. In the smaller basin is the non- forest steady state, which represent different species compositions of non- forest landscapes. The depth of the basin reflects the attractiveness of the basin or the resilience to prevent a change of stable- steady states despite disturbances.

While the forest species composition and structural changes are visible and easier to monitor to quantify forest degradation, another 'cryptic' form of degradation occurs when the resilience of a forest declines due to natural or anthropogenic factors. For example, extreme climatic events may make a forest may be more vulnerable to changes to anthropogenic disturbances in the future. In this case, the basins of attraction become shallower to depict the decreased resilience of a system, which allows it to move to an alternative stable steady- state more easily in the event of a disturbance.

Novel Ecosystems

Hobbs et al., (2009) provide an approach to conceptualise forest degradation based on biotic and abiotic changes. The authors use a gradient of biotic and abiotic changes to present the state of an ecosystem from a 'historic' state to a 'novel' one (Figure 2). The historic state or configuration is considered to be the state with the least amount of change at a given point of time. The hybrid state represents a change in either abiotic or biotic conditions or both. Instead of calling an ecosystem degraded, they term a novel ecosystem as one in which measurable traits have altered from historical ones.

Hobbs et al. (2009) concept of a novel ecosystem attempts to change the negative view of forest use, which in many cultures has taken place for time unknown. In this manner, the changing of abiotic and biotic elements results in an ecosystem, which is entirely new as opposed to degraded, thus acknowledging the dynamic nature of ecosystems.

Applying this concept to forests, a forest can be viewed as novel when considerable alterations to it have taken place. For example, changes to the abiotic elements include selective removal of certain tree or grass species, which result in a compositional change over time or after a disturbance. Further, a common form of abiotic change in forests is the invasion by alien species, which are associated with altered species composition and soil characteristics (Veldman & Putz, 2010; Wilson, Desai, Sim, Gruber,

Figure 1. The conceptualization of forest degradation as alternative stable states (adapted from Ghzoul et al., 2015)

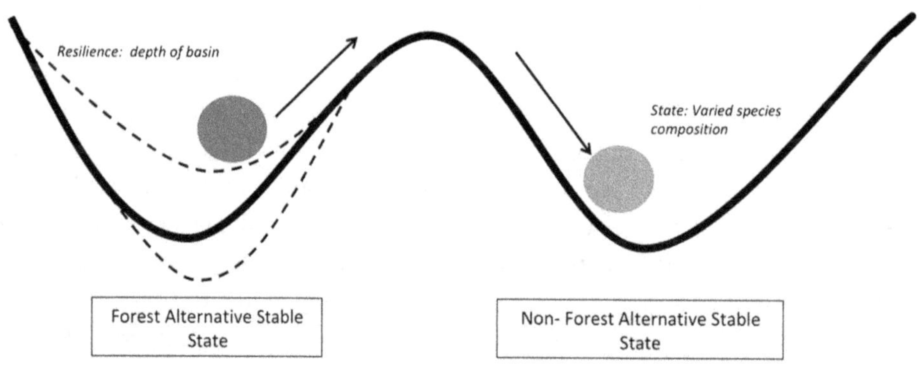

Figure 2. The conceptualization of a novel ecosystem; this figure shows a novel ecosystem as a change in flora and fauna

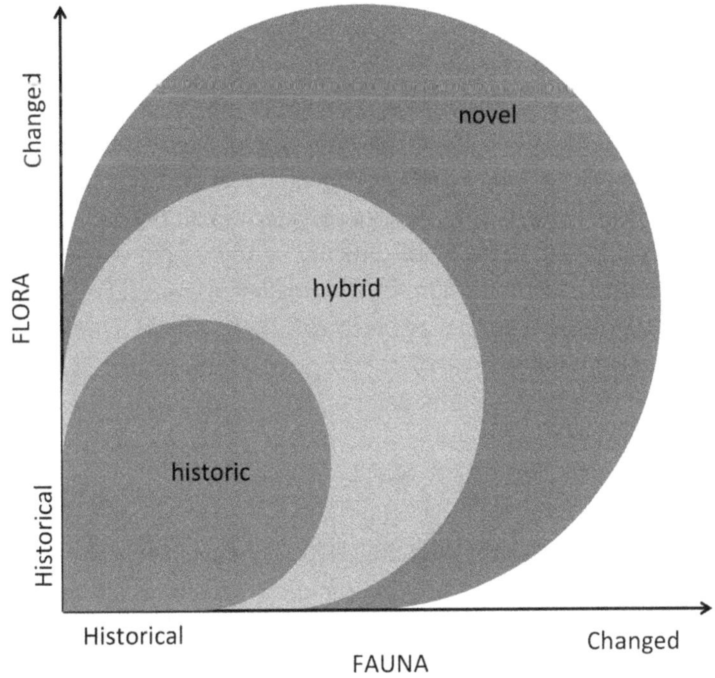

& Lester, 2014). Biotic changes to a forest may result in modified trophic interactions. For example, the removal of predators leads to an increase in prey and subsequent increase in herbivory, thus resulting in a change in structural and compositional changes to a forest.

While the concept of novel ecosystems presents a neutral view of structural and compositional changes, it notes the role of these changes in the provision of ecosystem functions. Abiotic changes in a forest, such as the proliferation of an invasive species, may result in the reduction in the capacity of the forest to provide flood attenuation function.

Both the approaches to view forest degradation – as alternative stable steady- states or as novel ecosystems – are unique ways of studying structural and compositional changes to a forest. However, there remain two grey areas with both approaches of conceptualization of forest degradation. First, both approaches require a steady- state as a reference or a historic state. Second, there remains a gap in our knowledge about the thresholds at which changes in a forest occur. Without the knowledge of what a forest looks like at the threshold at which it changes stable- states makes it difficult to determine the trajectory of change a forest has taken and subsequently, conserve or manage a forest.

Measuring Forest Degradation

Given the various definitions and approaches to conceptualize what forest degradation looks like, there has been much debate amongst scientists with respect to the criteria used to quantify forest degradation. Unlike deforestation, which has a binary response degradation takes place on a continuum, and may often have a non- linear response(Putz & Redford, 2010). This dynamism in the state of a forest renders the measurement of degradation challenging. Scientific papers have quantified forest degradation using the following criteria: production, biodiversity, protective function, carbon storage and disturbances such as invasive species (Baland et al., 2010; Heltberg et al., 2000; Marinho et al., 2016; Piana & Marsden, 2014).

Thompson et al. (2013) attempt to standardize methods of forest degradation quantification to bring more clarity to the concept of forest degradation, which continues to remain debated. The authors consider the five criteria listed and provide a list of indicators for each criterion. For example, biodiversity may be measured by the species richness and relative abundance of flora and fauna, or the amount of fragmentation of the forest. Production may be measured by the amount of growing stock it has. Similarly, protective function of a forest is quantified by the amount of soil erosion or water volume it possesses. Thompson et al. (2013) recommend a minimum of 7 indicators across all five criteria when measuring degradation. While this is a positive step towards reconciling the complexities of quantifying degradation, there is space and need for improvement. The criteria and their respective indicators only measure ecological characteristics of a forest. People depend on forests, and by virtue of their interaction with forests, it is necessary to include certain social and cultural criteria when assessing forest degradation. This especially applies to tropical dry forests, which provide livelihoods to large populations across the world.

Forest degradation has mainly been studied using two methods, remote sensing technology, such as satellite imagery and field surveys (Agarwala et al., 2016; Thompson et al., 2013). Satellite imagery is efficient and the quality of the imagery is improving rapidly with time enabling scientists. Higher resolution images allow scientists to carry out studies at finer spatial scales (Pimm et al., 2015). Detecting deforestation using satellite imagery is easier than detecting forest degradation, which is cryptic and thus requires field surveys (Davidar et al., 2010; Lambin, 1999). Given the difficulty of detecting forest degradation in satellite imagery, monitoring it remotely is often not a viable option (Peres et al., 2016; Puyravaud et al., 2010). Thus, field surveys and ground truthing are necessary in such situations (Pimm et al., 2015). However, field surveys are time consuming and labour intensive. When studying forest degradation, choosing methods to quantify degradation is dependent on the research question and the spatial scale and extent of the study.

In the following section, there are some examples of forest resource use and extraction, which are often considered activities that drive forest degradation. It is important to understand the social as well as ecological context of forest resource use, before it is labelled as forest degradation.

PATTERNS AND IMPACT OF FOREST RESOURCE USE

Forest Resource Use and Extraction

Forest use, whether subsistence or commercial in nature, is associated with forest degradation (Chazdon, 2003; Ghazoul et al., 2015; Hosonuma et al., 2012; Sasaki & Putz, 2009). Forest degradation, which lends an inherently negative connotation to any kind of forest use, is a relatively recent environmental conservation concern. However, humans have long domesticated landscapes and modified ecosystems to provide for themselves (Cincotta et al., 2000; Kareiva et al., 2007). A long history of extensive resource use and human habitation in tropical forests suggests that the forest structure and composition may be a result of this long-term interaction (Heywood & Iriondo, 2003). Thus, forest use is a significant means to a lifestyle and culture in many parts of the world, an important factor to take into consideration when conducting research or planning environment conservation (Agarwala et al., 2016; Gardner et al., 2009; Nagendra et al., 2010).

Subsistence on forests includes reliance on forested landscape for timber, energy, livestock grazing, and other non- timber forest product (NTFP) needs (Hosonuma et al., 2012). Forest use and extraction for commercial purposes include illegal sale of fuel-wood in local markets and selective logging (Hosonuma et al., 2012). Each of the forest resource uses impacts a forest structurally and functionally. However, given the long history of human interaction with tropical forests, it is difficult to accurately determine the impact of current forest resource use on structural and compositional differences as it may reflect historic forest use (Agarwala et al., 2016; Nagendra, 2012). The following sections focus on different forest resource uses and extraction in tropical dry forests across the world to highlight the impact of such activities on local flora and fauna.

Biomass Extraction

Selective Logging

Selective logging in primary and secondary forests for commercial purposes is considered one of the largest drivers of forest degradation (Asner et al., 2018). Selective logging is associated with lower tree species richness (Ahrends et al., 2010; Clark & Covey, 2012) and residual damage to the forest ecosystem (e.g. Edwards et al., 2011; Veldman & Putz, 2010). This in turn impacts faunal diversity and the functionality of the forest, including the forest's potential for water infiltration and carbon storage. For example, a study by Ahrends et al. (2010) observed 'waves' of forest degradation in the tropical dry forests around Dar es Salaam (DES) to meet increasing timber and other woody resources needs in DES through selective logging from 1991 to 2005. The first wave of selective logging removed the low and medium value timber followed by a wave of high value timber extraction. The authors examined the patterns of the wave and marked an increase in the tree species richness in forests with an increase in distance from DES.

The second form of degradation due to selective logging is the residual impact of logging operations. Logging, primarily in frontier landscapes, requires dirt roads, which once created lead to more forest access for people and the expansion of agriculture. Logging also disturbs soil, opens forest canopies and makes a forest more fire prone (Laurance, 2003; Peres et al., 2016). Further, residual impacts may also be in the form of a change in vegetation species composition or the invasion by an alien species. Veldman and Putz (2010) noted the invasion of alien grass species in a tropical dry forest in eastern Bolivia following selective logging operations there. They concluded that the invasion was result of the seed dispersal by logging vehicles and the opening of the forest canopy, which provided ideal conditions for the alien grass species, *Uruchloa maxima*. An estimated 1800 seeds of the alien species were dispersed by logging trucks, when the seeds were lodged in the tyres or the wheel wells. This led to a change in the understory of the forest following the selective logging.

In addition to an impact of selective logging on the forest itself, there are parallel impacts on people who depend on the forests for non-timber forest products and fauna. Rist et al. (2012) found that following consecutive selective logging operations in Ipixuna, Brazil, the number of fruit available to local communities declined. From 93 fruits per family per year in 1993, the number of fruits available to a family reduced to 15. These local communities also relied on game species from the forest as a source of protein. Due to the anthropogenic disturbance of selective logging operations on the fauna, the communities found that they had to go further into the forest to hunt game species.

Initiatives to reduce the detrimental impacts of logging, with respect to above ground biomass carbon, soil nutrient degradation or faunal occupancy of a logging site, were introduced a few decades ago (Putz et al., 2012). These initiatives took the form of forest timber certifications and guidelines on reduced impact logging (RIL). Putz et al. (2008) find that logging operations following the RIL guidelines indeed have reduced detrimental impact on soil health and aboveground carbon. Putz et al. (2012) advocate for sustainable selective logging through a suite of recommendations to improve logging operations by adopting voluntary third-party certifications of the timber, devolution of control over forests to indigenous or local communities and creating a tighter link between climate change mitigation programs and RIL to name a few.

Fuelwood Extraction

One of the most common forms of forest biomass extraction across the world is the collection of fuelwood for cooking and heating of households (Brooks et al., 2016; Heltberg et al., 2000; Jeuland & Pattanayak, 2012). An estimated 40% of the world's population relies on biomass for cooking purposes (Bonjour et al., 2013). Extraction of fuelwood has several impacts on the condition of a forest, which is explained through studies referenced below. It is noteworthy, that biomass extraction in tropical dry forests across the world has had different impacts depending on the nature, frequency and quantity of fuelwood extraction.

A result of forest biomass extraction is the change in the vegetative community composition in tropical dry forests. Specht et al. (2015) studied fuelwood consumption patterns in villages in the Brazilian Atlantic Forest. They noted that 88% of the surveyed households selectively harvested certain native species in remnant forest fragments or in regeneration stands after a past wave of deforestation in the region. The fuelwood extracted consisted of tree species effective for cooking and heating purposes and those species at an early successional stage. This fuelwood extraction pattern was associated with a change in the species composition in forests, with an abundance of the preferred tree species extracted for fuelwood. Similarly, Agarwala et al. (2016) studied vegetative species compositions with high intensity of resource use, including fuelwood collection in the state of Madhya Pradesh, India. The authors found a significant difference in the species richness, biomass and abundances of trees at different size classes between the treatment and control sites associated with human population densities and frequency of resource use. Further, the alteration of species transition to higher size classes may alter the long- term forest species composition with the retention of only those species that are of utility value to local communities.

Several efforts to reduce biomass extraction have been implemented across the world (International Energy Agency, 2016). From 1980 to 2010, the number of households reliant on firewood as a primary cooking source declined from 62% to 40% (Bonjour et al., 2013). Substitutes for fuelwood include liquefied petroleum gas (LPG), biogas, and electricity amongst others.

In India, an estimated 70% of population relies on biomass fuel for cooking and heating purposes (Government of India, 2011). Efforts to reduce the use of fuelwood include the provision of subsidized natural gas (DeFries & Pandey, 2010) and biogas (Agarwala et al., 2017) by governmental and non-governmental institutions. Agarwala et al. (2017) studied the impact of biogas digesters on aboveground biomass in Chikkaballapur district in the state of Karnataka, India, which is dominated by dry deciduous and scrub forests. The local forest department introduced biogas digesters in 885 villages in 2005 and the authors test the assumption that the provision of a biogas digester for cooking purposes reduces the firewood extraction. The results of this study indicated an association of the treatment (villages where biogas digesters were provided) with higher aboveground biomass in contrast to the control (villages where no biogas digesters were provided), which correlated with lower aboveground biomass. However, the study also noted a significant correlation of the above ground biomass with soil sulphur content and slope, making it challenging to quantify the exact impact of the adoption of biogas digesters in the district. The authors discovered an interesting correlation between aboveground biomass and soil sulphur, which is related to frequent fires in the forest and inversely proportional to the abundance of lantana (*Lantana camara*).

While aboveground biomass is associated with adoption of biogas as an alternative to fuelwood for cooking purposes in Chikkaballapur district, Karnataka, such an association is not consistent with other studies across the world. For example, a study in Nepal by Nepal et al., (2010) observed a different relationship between the adoption of improved cookstoves (ICS) and fuelwood extraction. Nepal et al. (2010) in their study asked if the use of an ICS is associated with lower amount of fuelwood consumption in comparison to traditional cookstoves, which require more fuelwood. Nepal introduced the ICS program in 1950s and the authors use two national living standards survey to determine whether ICS has had the intended impact of a reduction of fuelwood consumption. The study concluded that households with open and ICS collect more fuelwood, which is contrary to the expectations of the ICS program. One of the hypotheses for this trend is that ICS did not meet the cooking requirements in comparison to a traditional open fire cookstoves in Nepal. This example demonstrates the complexity, often arising due to socio-cultural factors, in promoting such programs that are intended to reduce forest dependence.

Non-Timber Forest Product Extraction

Another driver of forest degradation in dry forests is the extraction of non-timber forest products (NTFP). NTFP is defined as all biological material except timber, which commonly includes fruits, nuts, seeds, or leaves. The impacts of NTFP collection on a forest are varied as illustrated in the following examples.

Brazil nut (*Bertholletia excelsais*) is a seed collected from natural forests and traded internationally (Peres et al., 2004). Peres et al. (2004) found that long-term harvest of the Brazil nut in the Brazilian, Peruvian and Bolivian Amazon lead to 'recruitment bottlenecks', with a lower number of older trees in stands that were continuously harvested. The authors' field observations and model demonstrate that in consistently exploited forest stands, there are fewer older trees and the juveniles have not replaced them in recent times. In addition to other external pressures, such as deforestation, overexploitation threatens the long-term viability of the resource.

A prior study on Brazil nut extraction in the Madre de Dios region in the Peruvian Amazon by Escobal and Aldana (2003) showed the unintended environmental impact of Brazil nut extraction. The study noted the collateral pressures of the harvest in the form of illegal hunting of mammals due to access to forested areas. Collateral damage from NTFP collection contributes significantly to ecological changes in a forest. For example, Agarwala et al. (2016) noted that low- intensity ground fire was regularly used in the study site in Madhya Pradesh, India to augment the production of NTFP such as *tendu*(*Diospyros melanoxylon*) leaves and *mahua*(*Madhuca latifolia*) flowers. While these tree species persistent in forests subjected to controlled fires, as they are more tolerant to it, the use of fire suppressed the growth of other species with lower resistance. Long- term use of fire and NTFP resource extraction will eventually lead to a change in the forest's species composition. Approximately 60% of the population in this region of

India relies on NTFP revenue for a livelihood (Saigal, 2008). Given the high proportion of the population reliant on NTFP collection, sustainable NTFP harvest must be advocated for in such regions.

Varghese et al. (2015) studied commonly exploited species, *amla* (*Phyllanthus emblica*) and *nellikai* (*Phyllanthus indofischeri*) and *hirda* (*Terminalia chebula*). (Combretaceae) in the tropical dry forests of Nilgiri Biosphere Reserve (NBR). In this region, 40% of the local communities' income is generated from NTFP collection (Pain, 2009). The authors arrived at results inconsistent with other local studies of these species in the region. For example, their results indicated that fruit harvest, whether low or high intensity, did not predict the recruitment of *Phyllanthus* spp. However, with an increasing use of fires, the authors found a lower recruitment of *P. emblica*. The study concluded that NTFP collection has no significant reduction to the recruitment of the species of interest. Instead, the use of fire and the proliferation of invasive species such as the mistletoe (*Loranthes* spp.) have a stronger association with the reduced recruitment of the species studies.

Livestock Grazing and Fodder Extraction

A forest use that is often for subsistence purposes is the grazing of livestock in forested landscapes and fodder collection. In several cultures around the world, livestock is allowed to roam free in forests. Livestock is often left in a forest for an entire day and then herded back to their sheds at night. Excessive pressures of grazing result in soil compaction and could also lead to the proliferation of weed species, which affects the forest's nutrient cycle (Schulz et al., 2016; Silori & Mishra, 2001). Due to the preference of livestock for certain vegetation, consistent grazing in patches may affect the rate of tree establishment and thus, affect the successional stage of a forest (Agarwala et al., 2016; Ganesan, 2000). The evidence on grazing remains conflicting, as noted in the examples discussed in the following paragraphs.

For example, in the North West Biosphere Reserve (NWBR) in Peru, livestock grazing has both positive and negative impacts on biodiversity. Piana & Marsden (2014) study the impact of livestock grazing on the forest structure and raptor population distribution across three protected areas in NWBR. Livestock clearing usually clears the understory of a forest, which is beneficial for raptors, which predate on smaller ground dwelling prey. The authors found that raptor species abundances were positively associated with cattle densities of 20-50 individuals per km^2. However, the abundance of rare raptors decreased significantly beyond a cattle density of 60 individuals per km^2. Thus, cattle grazing, while impacts the understory may be beneficial for species that use the higher canopies for predation such as raptors.

Livestock grazing is associated with soil compaction and a change in soil organic carbon due to changes in vegetation. Schulz et al (2016) studied soil organic carbon (SOC) in Caatinga, Brazil to determine the impacts of different livestock grazing intensities on soil health in the region. The Caatinga region has faced heavy deforestation in the recent past, and only 46% of the tropical dry forest remains, of which most is subjected to livestock grazing. The authors found that the SOC in the upper 5 cms of the soil was significantly lower in heavily grazed areas as compared to areas with little to no grazing. However, at greater depths, the SOC did not vary significantly depending on the intensity of grazing.

Livestock grazing and fodder collection for cattle often takes place in conjunction with other resource uses and the use of low- intensity fires to promote vegetation regeneration, making it difficult to assess the impact of only grazing activities (Agarwala et al., 2016; Thompson et al., 2013).

It is noteworthy that each of the studies used as examples above have arrived at different conclusions of anthropogenic use of forests. Thus, it is imperative to understand that not all results and conclusions from these studies may be extrapolated to tropical dry forests in other parts of the region, country or world. Each location is unique and thus, scientists must not hold a priori expectations with respect to the impacts of forest resource use and extraction.

Disentangling the Impact of Forest Resource Use and Extraction

The effects of each of the forest resource use and extraction are usually confounded as these activities take place simultaneously. Decoupling the effects of each of these resource uses is a challenge. For example, Varghese et al. (2015) and Agarwala et al. (2016) highlighted the confounding effects of the use of fire and grazing making forest management difficult. The study of forest resource use is often complex due to the inability to disentangle the exact impact of each activity. However, for most efficient and targeted tropical dry forest conservation and management impact, it is necessary to understand the effect of each forest resource use.

Ticktin et al. (2012) attempted to disentangle the effect of invasive species, the mistletoe and lantana, which have proliferated due to anthropogenic disturbances, and NTFP collection in Biligiri Rangaswamy Wildlife Sanctuary in the state of Karnataka, India on the growth rate of *amla* and *nellikai*. They concluded that when mistletoe populations reached moderate levels of invasions, the population growth rates of *amla* and *nellikai* reduced when compared to pre-invasion growth rates. Furthermore, simulations of fruit harvest showed no impact on *amla* in control plots or the plots with both the invasive species. The authors were thus, able to disentangle the effects of NTFP harvest and invasion to disprove the common misconception in the region that fruit harvest drove the decline of the species populations. The invasion of alien species had a larger impact on the tree species in question. While such studies that disentangle the impacts of each anthropogenic activity in tropical dry forests are few and challenging to carry out, their contribution to our understanding of the interaction of humans and forests is of significant value.

What Drives Forest Resource Use and Degradation?

Environmental Associations

Factors such as the distance to forested areas, tourist hotspots, local markets and urban centres are also determinants of the amount of forest resource extraction that takes place in forests (Ahrends et al., 2010; Davidar et al., 2010; Ganesan, 2000; Heltberg et al., 2000; Specht et al., 2015). For example, Davidar et al. (2010) studied five protected areas in India and regions to determine the patterns and impact of forest resource use. The authors noted that the areas with highest fuelwood extraction are Bandipur and Sigur, which have high tourism and a local market, where fuelwood may be sold illegally. The demands for fuelwood had increased due to an increase in tourism in the region. In this case, it appears that forest degradation had been fueled by an increase in tourism and the creation of a local market, as opposed to simply the population growth or increase in subsistence resource use by local communities in the region. The consideration of such external factors driving forest degradation is important to arrive at sound policy with respect to forest resource use and extraction.

With respect to distance to urban centres or local markets, Ahrends et al. (2010) found a pattern of selective logging in DES whereby the forests most proximate to the city were the most exploited, while the forests farther away had a higher tree species richness and aboveground biomass. The first wave of degradation in 1991 in DES was for charcoal needs in the city, followed by low value timber needs for construction, leading to the extraction of high value timber as forests become more degraded closer to DES over the period of 1991 to 2005.

Cultural Associations

Fuelwood and fodder collection, NTFP extraction and grazing, whether for commercial sale in local markets or for subsistence, are associated with certain socio- economic and cultural factors.

The geographic coincidence of poverty, protected areas and a dependence on natural capital has led to the creation of innumerable initiatives to alleviate poverty in such regions and promote a shift away from the use of forest resources (Barrett et al. 2011; Velho et al., 2018). However, it remains inconclusive if such initiatives to create upward economic mobility create the intended impact of a shift away from a reliance on forest resources(Wright et al., 2015). This inconsistency is a result of factors beyond economics. As conservation planners and researchers continue to study forest degradation to create solutions, a consideration of socio-economic and cultural attributes of a region is imperative, as has been demonstrated by several examples presented in this chapter.

An example of the inconsistency of the theory that upward economic mobility is associated with less extraction and resource use pressures on forests is demonstrated in Velho et al. (2018). The authors use the number of assets a household possesses as a measure of their economic wealth to determine the patterns of forest resource use. They do not note a strong relationship between number of assets and forest resource use. For example, an increase in number of assets leads to a slight decrease in fuelwood and NTFP collection. Conversely, an increase in the number of assets owned is correlated with greater livestock ownership; cattle, which are often grazed in unprotected forests. Similarly, five-year predictions for forest resource use based on number of assets show that households prefer to increase NTFP collection with an increase in the number of their assets in the future. This preference for a lifestyle, which is dependent on a forest, may be driven by social or cultural beliefs rather than a household's economic situation.

For the success of environmental conservation programs and programs aimed to better the wellbeing of communities dependent on forest resources, the consideration of socio-economic and cultural factors is imperative. Further, these factors are also critical when considering forest management, explained in the sections below.

Forest Management and Degradation

Local communities often manage forest for their own use and extraction. These forests may be referred to as commons. Commons are resources collectively owned by a group of people. Local communities often manage unprotected forests as a common pool of resources available to the entire community (Agrawal, Chhatre, & Hardin, 2008; Ostrom, 2009). The tragedy of the commons refers to an outcome of a resource when resources which are owned collectively are inevitably doomed as human greed leads to excessive exploitation of the resource (Hardin, 1968). It is necessary to recognize that communities have sustainably managed their forests for generations and a question that intrigues many scientists is: *how do communities successfully and sustainably manage their common pool of resources?*

To answer this question, scientists have studied natural resource management systems around the world and developed frameworks to understand resource management. For example, Ostrom (2009) developed the concept of socio- ecological systems, within which humans and nature are two sub systems, which interact with each other. These subsystems of humans and nature consist of the following components, the natural resource, the resource unit, governance body and the users. Ostrom (2009) identified certain key factors in each of these components, which allowed for successful management of a natural resource. For example, Ostrom (2009) noted that the governing body must only consist of members from the user group. This governing body is the entity that creates the guidelines for the resource use and extraction, which is honestly followed by all users. Further, Ostrom (2009) observed that the governing body must address grievances of users in a timely and economic fashion to enable successful management. Another key aspect of successful commons management is the nature of the resource unit and the spatial extent of the natural resource. If the unit is stationary, such as trees or grasses, it is easier to manage as compared to wildlife, which is mobile.

Community forest management, while once was considered an informal system of governance, is now a legitimate form of providing protection and managing the forests. Increasingly, this form of management is being given the political and scientific importance, it is due, creating space for different forms and strategies of forest conservation. Several of the studies used as examples in the sections above are studies that have been carried out in forest commons around the world. As scientists, it is crucial to keep in mind the importance of rights to natural resources and inclusivity in in scientific research and environmental conservation.

SOLUTIONS AND RECOMMENDATIONS

Restoration of Degraded Forests

Forest degradation due to lack of sustainable management or overexploitation of resources is commonplace in several forests around the world. As forests face increased human population pressures, degradation of forests is a major concern. However, in order to make timely and impactful decisions with regards to the ecological restoration of tropical dry forests, it is first important to interpret the health of forest correctly.

Misinterpretation of Degraded Lands

As humans in most parts dominate tropical dry forests, it is imperative to consider socio-economic and cultural factors associated with forest resource use and the perceptions of the health of a forest. Studies and conservation efforts on forest degradation are prone to misinterpretations of the condition of a forest (Hobbs, 2016). Hobbs (2016) highlights the preconceived notions of scientists and policy makers with regards to forest degradation, which shape the agenda of forest conservation and management. Stating that the perception of forest degradation is 'in the eye of the beholder', he notes that different people view degradation different. A wildlife biologist views the lack of large mammals in a forest as a sign of forest degradation, where as a forester does not. Further, what a local community perceives as a healthy forest, where they collect fuelwood, NTFP and fodder, may appear to be a degraded forest to an ecologist due to the reduction of biomass and change in species composition. However, do these changes necessarily translate into the categorization of a forest as 'degraded'? This question is problematic as it depends on knowledge of the history of a landscape, information that we seldom have and one's perception. This issue of subjectivity in the definition of forest degradation, the measurement of degradation and the interpretation of the impact of forest resource use and extraction is a challenge yet to be resolved.

A solution to the subjectivity in the approach to studying forest resource use and degradation is the creation of interdisciplinary teams. Many scientists have advocated for interdisciplinary approaches to studying interactions between forests and humans to incorporate the socio- economic, cultural and ecological aspects of a landscape (Chazdon, 2008; Hobbs, 2016). Using an interdisciplinary approach is important as such misinterpretations of forest conditions entail consequences, largely for local people who depend on the forest.

The Restoration Staircase

Decisions to restore degraded forest depend on several factors, including the objective of ecological restoration, the economic limitations, the availability of tree species seedlings, support of local stakeholders and the political will to ensure the successful restoration. Chazdon (2008) provided a framework to assist conservation planners and scientists making restoration decisions for degraded forests in order to restore ecosystem services. Using a ‘restoration staircase’, which while does not accommodate the cultural significance of forest resource use, continues to remain useful a decade later as people make decisions to restore very degraded lands. The staircase consists of three criteria, biodiversity and ecosystem services, and time and cost of restoration.

At the bottom of the staircase is reclamation and rehabilitation of degraded lands, which are the costliest steps in the process of restoration and provide no/few ecosystem services. At the top of the staircase is the natural regeneration of land, which involves the least cost and time and provides the highest biodiversity and ecosystem services to humans. Such staircases are useful when planning restoration of tropical dry forests, which are now being restored around the world due to their high historic exploitation.

Ecological restoration is a costly and lengthy process. When undertaking such ecological processes, it is vital to consider the objectives of restoration for biodiversity and people dependent on such landscapes. This chapter provides an overview of forest degradation, subsistence and commercial forest resource uses and extraction, which are crucial considerations when undertaking forest management and restoration.

FUTURE RESEARCH DIRECTIONS

This chapter provides on overview of significant research carried out in the field of forest degradation and forest resource use. As an emerging concern, forest degradation is a complex problem that faces social and natural scientists. While the last two decades have seen an increase in the number of publications with respect to forest degradation and subsistence forest resource use and extraction, there continue to remain many unanswered questions. For example, what are the thresholds at which fuelwood collection becomes detrimental to fauna? Does economic upliftment lead to a change in forest resource use and extraction patterns? Is the impact of fuelwood collection and NTFP collection comparable? How can forest degradation measurement be standardized across tropical dry forests around the world?

The wellbeing and lifestyles of people around the world is closely linked to forest health and continued forest access to collect and use resources. Thus, any research on forest degradation or forest resource use must take into consideration the socio- economic and cultural aspects of forests.

CONCLUSION

Forest degradation is a major cause for concern as human population pressures on forests increase around the world. However, the measurement of degradation remains problematic. Different cultures use forests differently, and this makes standardized measurement and monitoring of degradation complex. Forest degradation is attributed to subsistence and commercial use of timber and non- timber forest products. However, our understanding of the exact impacts of these uses, especially subsistence uses, is incomplete. As a result, policy making with regards to forest resource use and extraction and biodiversity conservation difficult under such circumstances. As further research is carried out in this emerging field in forest ecology, crucial understanding of forest dynamics will enable more informed decision on forest resource use and extraction and ecological restoration.

REFERENCES

Agarwala, M., DeFries, R. S., Qureshi, Q., & Jhala, Y. V. (2016). Changes in the dry tropical forests in Central India with human use. *Regional Environmental Change*, *16*(1), 5–15. doi:10.100710113-015-0903-1

Agarwala, M., DeFries, R. S., Qureshi, Q., & Jhala, Y. V. (2016). Factors associated with long-term species composition in dry tropical forests of Central India. *Environmental Research Letters*, *11*(10). doi:10.1088/1748-9326/11/10/105008

Agarwala, M., Ghoshal, S., Verchot, L., Martius, C., Ahuja, R., & DeFries, R. (2017). Impact of biogas interventions on forest biomass and regeneration in southern India. *Global Ecology and Conservation*, *11*, 213–223. doi:10.1016/j.gecco.2017.06.005

Agrawal, A., Chhatre, A., & Hardin, R. (2008). *Changing Governance of the World's Forests*, *320*(5882), 1460–1462.

Ahrends, A., Burgess, N. D., Milledge, S. A. H., Bulling, M. T., Fisher, B., Smart, J. C. R., & Lewis, S. L. (2010). Predictable waves of sequential forest degradation and biodiversity loss spreading from an African city. *Proceedings of the National Academy of Sciences*, *107*(33), 14556–14561. doi:10.1073/pnas.0914471107

Asner, G. P., Rudel, T. K., Aide, T. M., Defries, R., & Emerson, R. (2018). *A Contemporary Assessment of Change in Humid Tropical Forests. 23*(6), 1386–1395.

Baland, J. M., Bardhan, P., Das, S., & Mookherjee, D. (2010). Forests to the people: Decentralization and forest degradation in the Indian Himalayas. *World Development*, *38*(11), 1642–1656. doi:10.1016/j.worlddev.2010.03.007

Barrett, C. B., Travis, A. J., & Dasgupta, P. (2011). On biodiversity conservation and poverty traps. *Proceedings of the National Academy of Sciences of the United States of America*, *108*(34), 13907–13912. doi:10.1073/pnas.1011521108 PMID:21873176

Bonjour, S., Adair-Rohani, H., Wolf, J., Bruce, N. G., Mehta, S., Prüss-Ustün, A., ... Smith, K. R. (2013). Solid Fuel Use for Household Cooking: Country and Regional Estimates for 1980–2010. *Environmental Health Perspectives*, *121*(7), 784–790. doi:10.1289/ehp.1205987 PMID:23674502

Brooks, N., Bhojvaid, V., Jeuland, M. A., Lewis, J. J., Patange, O., & Pattanayak, S. K. (2016). How much do alternative cookstoves reduce biomass fuel use? Evidence from North India. *Resource and Energy Economics*, *43*, 153–171. doi:10.1016/j.reseneeco.2015.12.001

Castillo, A., Magana, A., Pujadas, A., Martinez, L., & Godı, C. (2005). *Understanding the Interaction of Rural People with Ecosystems: A Case Study in a Tropical Dry Forest of Mexico*. *8*, 630–643. doi:10.100710021-005-0127-1

Chan, K. M. A., Pringle, R. M., Ranganathan, J. A. I., Boggs, C. L., Chan, Y. L., Ehrlich, P. R., ... Macmynowski, D. P. (2007). When Agendas Collide: Human Welfare and Biological Conservation. *Conservation Biology*, *21*(1), 59–68. doi:10.1111/j.1523-1739.2006.00570.x PMID:17298511

Chazdon, R. (2014). Second Growth: The Promise of Tropical Forest Regeneration in an Age of Deforestation. Chicago, IL: University of Chicago Press. doi:10.7208/chicago/9780226118109.001.0001

Chazdon, R. L. (2003). Tropical forest recovery: Legacies of human impact and natural disturbances. *Perspectives in Plant Ecology, Evolution and Systematics*, *6*(1–2), 51–71. doi:10.1078/1433-8319-00042

Chazdon, R. L. (2008). Beyond Deforestation : Restoring Forest and Ecosystem Services on Degraded Lands. *Science*, *320*(June), 1458–1460. doi:10.1126cience.1155365 PMID:18556551

Chazdon, R. L., Harvey, C. A., Komar, O., Griffith, D. M., Bruce, G., Martínez-Ramos, M., … Philpo, S. M. (2009). Beyond Reserves : A Research Agenda for Conserving Biodiversity in Human-Modified Tropical Landscapes. Association for Tropical Biology and Conservation Stable Retrieved from http://www.jstor.org/stable/20492544 *Biotropica, 41*(2), 142–153.

Cincotta, R. P., Wisnewski, J., & Engelman, R. (2000). Human population in the biodiversity hotspots. *Nature*, *404*(April), 1305–1307. doi:10.1038/35010105 PMID:10801126

Clark, J. A., & Covey, K. R. (2012). Forest Ecology and Management Tree species richness and the logging of natural forests : A meta-analysis. *Forest Ecology and Management*, *276*, 146–153. doi:10.1016/j.foreco.2012.04.001

Davidar, P., Sahoo, S., Mammen, P. C., Acharya, P., Puyravaud, J. P., Arjunan, M., ... Roessingh, K. (2010). Assessing the extent and causes of forest degradation in India: Where do we stand? *Biological Conservation*, *143*(12), 2937–2944. doi:10.1016/j.biocon.2010.04.032

DeFries, R., & Pandey, D. (2010). Urbanization, the energy ladder and forest transitions in India's emerging economy. *Land Use Policy*, *27*(2), 130–138. doi:10.1016/j.landusepol.2009.07.003

Edwards, D., Larsen, T. H., Docherty, T. D. S., Hamer, K. C., Ansell, F. A., Hsu, W. W., … Wilcove, D. S. (2011). *Degraded lands worth protecting : the biological importance of Southeast Asia's repeatedly logged forests*, (278), 82–90. doi:10.1098/rspb.2010.1062

Fajardo, L., Gonzalez, V., Nassar, J. M., Lacabana, P., Portillo, Q. C. A., Carrasquel, F., & Rodriguez, J. P. (2005). Tropical Dry Forests of Venezuela. *Characterization and Current Conservation Status, 37*(4), 531–546.

FAO. (2002). *Proceedings: second expert meeting on harmonizing forest-related definitions for use by various stakeholders*. *Rome*.

Ganesan, B. (2000). *Extraction of Non-Timber Forest Products, including Fodder and Fuelwood, in Mudumalai, India, 47*(3), 268–274.

Gardner, T. A., Barlow, J., Chazdon, R., Ewers, R. M., Harvey, C. A., Peres, C. A., & Sodhi, N. S. (2009). Prospects for tropical forest biodiversity in a human-modified world. *Ecology Letters, 12*(6), 561–582. doi:10.1111/j.1461-0248.2009.01294.x PMID:19504750

Ghazoul, J., Burivalova, Z., Garcia-Ulloa, J., & King, L. A. (2015). Conceptualizing Forest Degradation. *Trends in Ecology & Evolution, 30*(10), 622–632. doi:10.1016/j.tree.2015.08.001 PMID:26411619

Gillespie, T. W., Lipkin, B., Sullivan, L., Benowitz, D. R., Pau, S., & Keppel, G. (2012). The rarest and least protected forests in biodiversity hotspots. *Biodiversity and Conservation, 21*(14), 3597–3611. doi:10.100710531-012-0384-1

Government of India. (2011). Census of India, 2001: Tables on houses, household amenities, and assets.

Hardin, G. (1968). The Tragedy of the Commons. *Science, 162*(June), 1243–1248. doi:10.1126cience.162.3859.1243 PMID:5699198

Heltberg, R., Arndt, T. C., & Sekhar, N. U. (2000). Fuelwood Consumption and Forest Degradation: A Household Model for Domestic Energy Substitution in Rural India. *Land Economics, 76*(2), 213. doi:10.2307/3147225

Heywood, V. H., & Iriondo, J. M. (2003). Plant conservation: Old problems, new perspectives. *Biological Conservation, 113*(3), 321–335. doi:10.1016/S0006-3207(03)00121-6

Hobbs, R. J. (2016). Degraded or just different? Perceptions and value judgements in restoration decisions. *Restoration Ecology, 24*(2), 153–158. doi:10.1111/rec.12336

Hobbs, R. J., Higgs, E., & Harris, J. A. (2009). Novel ecosystems: Implications for conservation and restoration. *Trends in Ecology & Evolution, 24*(11), 599–605. doi:10.1016/j.tree.2009.05.012 PMID:19683830

Hosonuma, N., Herold, M., De Sy, V., De Fries, R. S., Brockhaus, M., Verchot, L., ... Romijn, E. (2012). An assessment of deforestation and forest degradation drivers in developing countries. *Environmental Research Letters, 7*(4). doi:10.1088/1748-9326/7/4/044009

International Energy Agency. (2016). *World Energy Outlook*. Paris, France.

ITTO. (2002). ITTO guidelines for the restoration, management and rehabilitation of degraded and secondary tropical forests. In ITTO guidelines for the restoration, management and rehabilitation of degraded and secondary tropical forests.

Janzen, D. (1988). *Tropical dry forests. The most endangered.*

Jeuland, M. A., & Pattanayak, S. K. (2012). Benefits and Costs of Improved Cookstoves: Assessing the Implications of Variability in Health. *Forest and Climate Impacts.*, *7*(2). doi:10.1371/journal.pone.0030338 PMID:22348005

Kareiva, P., Watts, S., Mcdonald, R., & Boucher, T. (2007). Domesticated Nature : Shaping for Human Welfare. *Science*, *316*(June), 2005–2008. doi:10.1126cience.1140170 PMID:17600209

Lambin, E. F. (1999). Monitoring Forest Degradation in Tropical Regions by Remote Sensing: Some Methodological Issues. *Global Ecology and Biogeography*, *8*(3), 191–198. doi:10.1046/j.1365-2699.1999.00123.x

Laurance, W. F. (2003). *Slow burn : the insidious effects of surface fires on tropical forests. 18*(5), 209–212. doi:10.1016/S0169-5347(03)00073-9

Miles, L., Newton, A., Defries, R. S., Ravilious, C., May, I., Blyth, S., ... Gordon, J. (2006). A Global Overview of the Conservation Status of Tropical Dry Forests. *Journal of Biogeography*, *16*(3), 201–202.

Nagendra, H. (2012). Assessing relatedness and redundancy of forest monitoring and change indicators. *Journal of Environmental Management*, *95*(1), 108–113. doi:10.1016/j.jenvman.2011.10.002 PMID:22115515

Nagendra, H., Rocchini, D., & Ghate, R. (2010). Beyond parks as monoliths: Spatially differentiating park-people relationships in the Tadoba Andhari Tiger Reserve in India. *Biological Conservation*, *143*(12), 2900–2908. doi:10.1016/j.biocon.2010.04.050

Nepal, M., Nepal, A., & Grimsrud, K. (2010). *Unbelievable but improved cookstoves are not helpful in reducing firewood demand in Nepal, 16*, 1–23. doi:10.1017/S1355770X10000409

Ostrom, E. (2009). *A General Framework for Analyzing Sustainability of Social-Ecological Systems, 325*(5939), 419–422.

Pain, A. (2009). What is driving change in the Nilgiri Biosphere Reserve and what effects might such change have on the role of NTFP in the livelihoods of indigenous people? In R. Dutt, J. Seeley, & P. Roy (Orgs.), *Proceedings of the Biodiversity and Livelihoods Conference, Coonoor, India.*

Penman, J., Gytarsky, M., Hiraishi, T., Krug, T., Kruger, D., & Pipatti, R. … Wagner, F. (2003). Good practice guidance for land use, land-use change and forestry. Kanagawa Prefecture, Japan: Institute for Global Environmental Strategies.

Pereira Marinho, F., Guilherme Gerhardt, G., Pellegrini Manhaes, A., Weisser, W. W., & Ganade, G. (2016). *Effects of past and present land use on vegetation cover and regeneration in a tropical dryland forest, 132*, 26–33. doi:10.1016/j.jaridenv.2016.04.006

Peres, C. A., Baider, C., Zuidema, P. A., Wadt, H. O., Kainer, K. A., Gomes-Silva, D. A. P., … Freckleton, R. P. (2004). *Demographic Threats to the Sustainability of Brazil Nut Exploitation, 302*(December 2003).

Peres, C. A., Barlow, J., & Laurance, W. F. (2016). *Detecting anthropogenic disturbance in tropical forests. 21*(March), 3–6. doi:10.1016/j.tree.2006.03.007

Piana, R. P., & Marsden, S. J. (2014). *Impacts of cattle grazing on forest structure and raptor distribution within a neotropical protected area*, 559–572. doi:10.100710531-013-0616-z

Pimm, S. L., Alibhai, S., Bergl, R., Dehgan, A., Giri, C., Jewell, Z., ... Loarie, S. (2015). Emerging Technologies to Conserve Biodiversity. *Trends in Ecology & Evolution*, *30*(11), 685–696. doi:10.1016/j.tree.2015.08.008 PMID:26437636

Portillo-Quintero, C., Sanchez-Azofeifa, A., Calvo-Alvarado, J., Quesada, M., & do Espirito Santo, M. M. (2014). The role of tropical dry forests for biodiversity, carbon and water conservation in the neotropics: Lessons learned and opportunities for its sustainable management. *Regional Environmental Change*, *15*(6), 1039–1049. doi:10.100710113-014-0689-6

Portillo-Quintero, C., & Smith, V. (2018). Emerging trends of tropical dry forests loss in North & Central America during 2001–2013: The role of contextual and underlying drivers. *Applied Geography*, *94*(October 2017), 58–70. doi:10.1016/j.apgeog.2018.03.011

Putz, F. E., & Redford, K. H. (2010). The Importance of Defining 'Forest': Tropical Forest Degradation, Deforestation, Long-Term Phase Shifts, and Further Transitions. Association for Tropical Biology and Conservation Stable. Retrieved from http://www.jstor.org/stable/27742857 *The Imp*. *42*(1), 10–20.

Putz, F. E., Zuidema, P. A., Pinard, M. A., Boot, R. G. A., Sayer, J. A., Sheil, D., ... Vanclay, J. K. (2008). *Improved Tropical Forest Management for Carbon Retention*, *6*(7), 1368–1369. doi:10.1371/journal.pbio.0060166

Putz, F. E., Zuidema, P. A., Synnott, T., Pe, M., Pinard, M. A., Sheil, D., … Zagt, R. (2012). *Sustaining conservation values in selectively logged tropical forests: the attained and the attainable*, *5*, 296–303. doi:10.1111/j.1755-263X.2012.00242.x

Puyravaud, J. P., Davidar, P., & Laurance, W. F. (2010). Cryptic destruction of India's native forests. *Conservation Letters*, *3*(6), 390–394. doi:10.1111/j.1755-263X.2010.00141.x

Rist, L., Shanley, P., Sunderland, T., Sheil, D., Ndoye, O., Liswanti, N., & Tieguhong, J. (2012). The impacts of selective logging on non-timber forest products of livelihood importance. *Forest Ecology and Management*, *268*, 57–69. doi:10.1016/j.foreco.2011.04.037

Saigal, S. (2008). *Non-Timber Forest Products and Forest Governance Forest Governance Learning Group – India*.

Sasaki, N., & Putz, F. E. (2009). Critical need for new definitions of "forest" and "forest degradation" in global climate change agreements. *Conservation Letters*, *2*(5), 226–232. doi:10.1111/j.1755-263X.2009.00067.x

Scheffer, M., Carpenter, S., Foley, J. A., Folke, C., & Walker, B. (2001). Catastrophic shifts in ecosystems. *Nature*, *413*(6856), 591–596. doi:10.1038/35098000 PMID:11595939

Schulz, K., Voigt, K., Beusch, C., Almeida-Cortez, J. S., Kowarik, I., Walz, A., & Cierjacks, A. (2016). Grazing deteriorates the soil carbon stocks of Caatinga forest ecosystems in Brazil. *Forest Ecology and Management*, *367*, 62–70. doi:10.1016/j.foreco.2016.02.011

Secretariat of the Convention on Biological Diversity. (2002). *Review of the status and trends of, and major threats to, the forest biological diversity.*

Silori, C. S., & Mishra, B. K. (2001). Assessment of livestock grazing pressure in and around the elephant corridors in Mudumalai Wildlife Sanctuary, south India. *Biodiversity & Conservation, 10*(12), 2181–2195.

Specht, M. J., Pinto, S. R. R., Albuqueque, U. P., Tabarelli, M., & Melo, F. P. L. (2015). Burning biodiversity: Fuelwood harvesting causes forest degradation in human-dominated tropical landscapes. *Global Ecology and Conservation, 3*, 200–209. doi:10.1016/j.gecco.2014.12.002

Thompson, I. D., Guariguata, M. R., Okabe, K., Bahamondez, C., Nasi, R., Heymell, V., & Sabogal, C. (2013). An Operational Framework for Defining and Monitoring Forest Degradation. *Ecology and Society, 18*(2), 307–329. doi:10.5751/ES-05443-180220

Ticktin, T., Ganesan, R., Parmesha, M., & Shetty, S. (2012). Disentangling the effects of multiple anthropogenic drivers on the decline of two tropical dry forest trees. *Journal of Applied Ecology, 49*(4), 774–784. doi:10.1111/j.1365-2664.2012.02156.x

UNFCC. (2002). *Report of the Conference of the Parties on its Seventh Session, held at Marrakesh from Oct. 29 to Nov. 10, 2001.*

Varghese, A., Ticktin, T., Mandle, L., & Nath, S. (2015). *Assessing the Effects of Multiple Stressors on the Recruitment of Fruit Harvested Trees in a Tropical Dry Forest, Western Ghats, India, 10*(3), 1–17. doi:10.1371/journal.pone.0119634

Veldman, J. W., & Putz, F. E. (2010). *Long-distance Dispersal of Invasive Grasses by Logging Vehicles in a Tropical Dry Forest, 42*(6), 697–703.

Velho, N., Defries, R. S., Tolonen, A., Srinivasan, U., & Patil, A. (2018). Aligning conservation efforts with resource use around protected areas. *Ambio*. doi:10.100713280-018-1064-5 PMID:29949080

Wilson, G., Desai, A. A., Sim, D. A., Gruber, M. A. M., & Lester, P. J. (2014). *The association between invasive Lantana camara and seedlings/saplings of a plant community in Mudumalai Tiger Reserve, India*, 551–563. doi:10.1017/S026646741400039X

Wright, J. H., Hill, N. A. O., Roe, D., Rowcliffe, J. M., K, F., Day, M., & Booker, F. (2015). Reframing the concept of alternative livelihoods. *Conservation biology, 30*(1), 7–13. doi:10.1111/cobi.12607

ADDITIONAL READING

Ahrends, A., Burgess, N. D., Milledge, S. A. H., Bulling, M. T., Fisher, B., Smart, J. C. R., ... Lewis, S. L. (2010). Predictable Waves of Sequential Forest Degradation and Biodiversity Loss Spreading from an African City. *Proceedings of the National Academy of Sciences of the United States of America, 107*(33), 14556–14561. doi:10.1073/pnas.0914471107 PMID:20679200

Chazdon, R. L. (2003). Tropical Forest Recovery: Legacies of Human Impact and Natural Disturbances. *Perspectives in Plant Ecology, Evolution and Systematics, 6*(1–2), 51–71. doi:10.1078/1433-8319-00042

Ghazoul, J., Burivalova, Z., Garcia-Ulloa, J., & King, L. A. (2015). Conceptualizing Forest Degradation. *Trends in Ecology & Evolution*, *30*(10), 622–632. doi:10.1016/j.tree.2015.08.001 PMID:26411619

Hobbs, R. J., Higgs, E., & Harris, J. A. (2009). Novel Ecosystems: Implications for Conservation and Restoration. *Trends in Ecology & Evolution*, *24*(11), 599–605. doi:10.1016/j.tree.2009.05.012 PMID:19683830

Kareiva, P., Watts, S., Mcdonald, R., & Boucher, T. (2007). Domesticated Nature : Shaping for Human Welfare. *Science*, *316*(June), 2005–2008. PMID:17600209

Miles, L., & … . (2006). A Global Overview of the Conservation Status of Tropical Dry Forests. *Journal of Biogeography*, *16*(3), 201–202.

Ostrom, E. 2009. "A General Framework for Analyzing Sustainability of Social Ecological Systems." 325(5939): 419–22.

Portillo-Quintero, C., & Smith, V. (2017, October). 2018. "Emerging Trends of Tropical Dry Forests Loss in North & Central America during 2001–2013: The Role of Contextual and Underlying Drivers. *Applied Geography (Sevenoaks, England)*, *94*, 58–70. doi:10.1016/j.apgeog.2018.03.011

Specht, M. J., Pinto, S. R. R., Albuquerque, U. P., Tabarelli, M., & Melo, F. P. L. (2015). Burning Biodiversity: Fuelwood Harvesting Causes Forest Degradation in Human-Dominated Tropical Landscapes. *Global Ecology and Conservation*, *3*, 200–209. doi:10.1016/j.gecco.2014.12.002

Thompson, Ian D et al. 2013. "An Operational Framework for Defining and Monitoring Forest Degradation." *Ecology & Society* 18(2): 307–29. http://search.ebscohost.com/login.aspx?direct=true&db=8gh&AN=91274450&site=ehost-live

Varghese, A., Ticktin, T., Mandle, L., & Nath, S. (2015).. . *Assessing the Effects of Multiple Stressors on the Recruitment of Fruit Harvested Trees in a.*, *10*(3), 1–17.

KEY TERMS AND DEFINITION

Alternative Stable States: In the context of forests, alternative stable states refer to the different states a forest can be in, forested, with a dynamic species composition or non- forested.

Commons: A common pool of resources, which are jointly owned and managed by a group of people or a community for their own use a recreation.

Ecological Restoration: Ecological restoration refers to the restoration of structural and functional aspects of a once degraded land.

Forest Degradation: A structural or functional change in a forest due to anthropogenic disturbance.

Non-Timber Forest Products (NTFP): NTFP represent all components of a forest beyond timber. People use NTFPs, which include fruits, soils, fodder, leaves, and flowers for various purposes, such as personal and livestock nutrition, decoration or commercial sale.

Selective Logging: Unlike complete deforestation, selective logging is the removal of only certain tree species from a forest.

Spatial: Spatial refers to anything with respect to location or space. Spatial scale refers to the spatial resolution (pixel size) of an image and the extent of the image or study region.

Temporal: Temporal refers to anything with respect to time. Temporal scale refers to the time scale in question.

Chapter 10
Litter Production and Decomposition in Tropical Forest

Sumit Chakravarty
Uttar Banga Krishi Viswavidyalaya, India

Prakash Rai
Uttar Banga Krishi Viswavidyalaya, India

Vineeta
Uttar Banga Krishi Viswavidyalaya, India

Nazir A. Pala
Uttar Banga Krishi Viswavidyalaya, India

Gopal Shukla
Uttar Banga Krishi Viswavidyalaya, India

ABSTRACT

Plant litter production and decomposition is a crucial ecosystem process that defines and governs the plant-soil relationships by regulating the nutrient turnover and the build-up of soil organic matter. Litter is the principal source of organic matter for soils in the forest ecosystem. The litter, upon decomposition, makes available essential nutrients for the growth and development of a forest stand. Different tree components contain different amounts of nutrients; and build up of soil organic matter. The amount of nutrients added through litter decomposition varies with forest types, species, stand attributes, and variation in seasonal environmental conditions. Nutrient return from organic matter is estimated by the physico-chemical properties of the litter. Moreover, the rate of decomposition and the nutrient releases are highly influenced by magnitude of litter produced, litter quality and nutrients release, as well as, by climatic conditions and existing microbial communities in the soil system. Ecological impact of carbon and nutrient dynamics in the litter layer is considerable in a forest ecosystem.

DOI: 10.4018/978-1-7998-0014-9.ch010

INTRODUCTION

Tropical forests occupy about 13% of the land surface and their total world biomass is around 90 x 10^{10} tonnes (on a dry weight basis) with an annual average production of 4 x 10^{10} t/yr. The primary production of tropical rain forest is 2,000 gm^{-2} yr^{-1} (Lieth, 1975). The net primary production of tropical rain forest, rain green forest, summer green forest, warm temperate mixed forest, boreal forest, tundra scrub, desert scrub, tropical grassland, temperate grassland, agricultural fields, lake & streams, coral reefs and open oceans are 2000, 1500, 1000, 1000, 500, 140, 70, 700, 500, 650, 500, 2000 and 125g/m^2/yr, respectively. The global soil organic carbon stock in the top one meter depth amounts to 30 tons ha^{-1} in arid climate, which is the least, and 800 tons ha^{-1} in organic soils of cold regions, the highest, but generally ranges from 50 to 150 tons ha^{-1} (Lal, 2004). More than half of the net primary production is returned to the soil and 60 Pg C $year^{-1}$goes to the atmosphere through litter decomposition (Wardle *et al.*, 2004; Houghton, 2007). Soil respiration (Rsoil) releases annually 98 PgC to the atmosphere (Bond-Lamberty and Thomson, 2010) which is more than seven times the C-equivalent of fossil fuel emissions (Sotta *et al.*, 2004). It is reported that global soil respiration (Rsoil) is annually increasing by 0.1 PgC in the atmosphere due to rise in air temperature (Bond-Lamberty and Thomson, 2010), changing soil CO_2 flux dynamics and litter decomposition (IPCC, 2014). Plant litter contributes 18-48% to the total global forests Rsoil (Han *et al.*, 2015; Zhao *et al.*, 2016).

The potential of trees to sequester carbon in forests are recognized through the quantity and quality of produced litter and its subsequent decomposition and release of nutrients (Rai *et al.*, 2016). Litter production is defined as the shedding of vegetative and reproductive plant parts caused by senescence, stress, mechanical factors (e.g. wind), a combination of these factors, or by death and weathering of the whole plant in a given time period (Krishna and Mohan, 2017). Litter fall which phenologically indicates the effect of climate change on the forest (Hansen *et al.*, 2009) and also sustains nutrient cycling is a vital source of nutrients and organic matter for regulating forest productivity and energy flow (Li, 2014). The proportion of leaves in the litter is always higher than the other form of litters like twig, roots or others and in tropical may totally decompose within a year (Meentemeyer, 1984). Litter decomposes gradually after they drop. Sustained nutrient cycling with faster turnover of litter through its decomposition regulates forest productivity (Vendrami *et al.*, 2012). Decomposition of litter regulates soil fertility through soil organic matter production and nutrient cycling (Guendehou *et al.*, 2014). Decomposition of litter is influenced by soil organisms, soil and climatic conditions and the nature and composition of litter (Bargali *et al.*, 2015).

LITTER AND CLIMATE CHANGE

Litter decomposition and Rsoil has global potential in terms of climate change mediated shifts in relative abundance of vegetation and change in water availability and soil temperature which will have significant influence on global carbon budget due to litter quality and its decompositions (Pinto Jr. *et al.*, 2018). The influence of climate change on litter decomposition and Rsoil spatially and temporally is still uncertain (Rohr *et al.*, 2013).

Climate change mediated temperature, precipitation and soil moisture is expected to change forest productivity, phenology, mortality and species composition (Dalmolin *et al.*, 2015). These changes in forest also change litter decomposition and Rsoil through alteration of soil and forest floor C quantity and quality (Fang *et al.*, 2015). Moreover, fresh litter exerts a priming effect on Rsoil rate through litter quantity and thickness along with an increment of available labile C microorganisms promoting microbial degradation of soil organic matter (Fang *et al.*, 2015; Brechet *et al.*, 2018). Soil moisture and temperature varies in varying litter quantity and thickness (Prescott *et al.*, 2017). Changes in litter quantity have direct effect on Rsoil across forest types (Han *et al.*, 2015; Brechet *et al.*, 2018).

Litter Production

Litter is comprised of dead plant materials that are separated from the living plant (Krishna and Mohan, 2017). Leaf-litter and non-leaf litter constitutes total litter fall (Bisht *et al.*, 2014). The pattern of litter fall correlates with the canopy development and growth pattern which is different for different species (Carrera *et al.*, 2008). Different forests produce different amount and composition of litter as structure and plant diversity of forests vary (Indriyanto, 2009). The topsoil accumulation varies from site to site depending plant species, climate, landuse, decomposer population and their activities (Fernandes *et al.* 1997). The quantity and nature of litterfall influences the physical, chemical and biological features of soil affecting its formation and fertility and thus growth of the trees as well (Panda *et al.* 2007). Litter represents an important link between forest community composition, productivity and biomass, and ecosystem processes (Purahong *et al.*, 2014; Faboya *et al.*, 2015; Uriarte *et al.*, 2015). The material fixed in the above ground biomass (AGB) is sooner or later returned to the soil through litter fall and precipitation. In forests, the spatial distribution of vegetation and species-specific differences in litter production and quality are likely to cause spatial heterogeneity in nutrient returns to the forest floor and, therefore, in the redistribution of soil nutrients (Uriarte *et al.*, 2015).

Litterfall and throughfall are the major pathways for elements to return from forest canopy for maintaining the soil nutrient pool and continuity of the nutrient cycling processes (Rawat *et al.*, $2010_{a,\,b}$). The contribution of each of the two pathways is mainly dependent on the nature of element, foliage properties and rainwater acidity (Hagen-Thorn *et al.*, 2006). Moreover, litter production and the percentage contribution seem to vary with forest type, species, stand structural attributes and seasonal variation in climatic condition. The decomposition rates of leaf litter in tropical and subtropical broad-leaved forests and pine trees are ranged from 1.0 to 1.7 (Cuevas *et al.*, 1991).

Litter production, rate of decomposition, release and its seasonal distribution have been reported from different forests and plantations across the world (Patricio *et al.*, 2012; Bisht *et al.*, 2014; Faboya *et al.*, 2015). Higher litter inputs occur in the most humid forests of Atlantic and Amazon (Tonin *et al.*, 2017). Annual litter fall estimated for about 500 stands of Eurasian forests was 6.53 Pg dm a^{-1} (1 Pg = 10^{15} g; dm, dry matter), of which more than 50% was contributed by tropical and subtropical forests, and rest by boreal forests. About 45% of the total annual litter fall (2.94 Pg C per year) gets decomposed in these Eurasian forests (Chun-jiang *et al.*, 2003). The contribution in annual leaf litter and total aboveground litter production by different forest types in Eurasian forests adopted from Chun-jiang *et al.* (2003) is given in table 1.

The litterfall in tropical rain forests, sub-tropical rain forests and temperate forests are about 1,600, 1,200 and 300-500 $gm^{-2}\ yr^{-1}$, respectively (Rodin and Basilevic, 1968). Litter production was reported to be more in Conifer forests than deciduous forests in the Northern Hemisphere (Millar, 1974). However, at the local level broadleaf forests shed more litter than coniferous forests (Kavvadias *et al.*, 2001). The available information related to Indian forests indicate that annual production of dead organic matter in evergreen forests is of the order of 20 tonnes ha^{-1} and estimated average litter production in tropical forest

Table 1. Annual leaf litter and total aboveground litter production in Eurasian forests

Forest groups	Leaf litterfall*	Total litterfall*
Tropical		
Rain forest	0.92	1.44
Moist	0.17	0.26
Dry	0.25	0.34
Mountain	0.14	0.27
Subtotal	**1.48**	**2.31**
Subtropical		
Humid	0.32	0.42
Dry	0.02	0.02
Mountain	0.24	0.34
Dry	0.09	0.12
Subtotal	**0.67**	**0.94**
Temperate		
Continental	0.84	1.01
Mountain	0.11	0.14
Oceanic	0.08	0.10
Subtotal	**1.03**	**1.25**
Boreal	**1.49**	**2.07**
Total Eurasian	**4.66**	**6.53**

$*10^{15}$ g dm a^{-1}; adopted from Chun-jiang *et al.*, 2003

is 9.3 tonnes ha^{-1} (Lal, 1992). The most important factor globally that influence litterfall is precipitation and temperature but locally with similar climatic conditions litterfall is more influenced by differences in ecophysiology of littering species, slope, aspect and site productivity (Barnes *et al.*, 1998). Generally, high latitude forests produce less litter which gradually increases to equator because of increasing duration of growth period.

In two tropical dry evergreen forest sites on the Coromandel coast of south India, total litter production was 4.11-4.86 t ha^{-1}, the maximum of which was produced during summer and about three fourth was contributed by leaves (Pragasan and Parthasarathy, 2005). Evergreen, semi -deciduous and deciduous species respectively, contributed 36.0-42.3, 29.6-30.4 and 27.3-34.4% of the total leaf production at both the sites. This study documented higher litter production than compared to litter production by various tropical forests of the world. Litter production in the dry tropical deciduous forest of vindhyan highland, India was 4.76 ±1.21 Mg ha^{-1} yr^{-1}, majority of which was leaves and varied significantly among species as dry mix >*S. robusta*>*T. grandis*>*Butea monosperma*>*Hardwickia binnata* (Rai *et al.*, 2016). In tropical deciduous forests of northern India annual litterfall was 5.0-6.7 Mg ha^{-1} (Shukla *et al.*, 2017).

Litter Composition

Litter is comprised of many organic compounds which differ relatively among the plant parts and species (Krishna and Mohan, 2017). Sugars, phenols, hydrocarbons and glycerides are soluble organic compounds. Mono- and oligosaccharides are soluble sugars present in litter. Nutrients, lignin, cellulose and hemicellulose content determine the quality of litter (Swift *et al.*, 1979). Lignin content varies with species in the range of 15–40% of the total litter mass but sometimes it may be 4-50%. In deciduous species syringyl and guaiacyl while in conifers guaiacyl lignin is found (Esperschutz *et al.*, 2013). Cellulose and hemicelluloses occurring in ratio of 0.7: 1.2 are the common form of carbohydrates in litter, which varies among plant species, e.g., deciduous litter has higher and coniferous litter has lower ratio (Akpor *et al.*, 2005).

Litter Degradation and Decomposition

The stability of an ecosystem depends upon the orderly transfer of nutrients between the living and non-living components and one of the major links for this transfer is litter (Xu and Hirata, 2005). Nutrients stocked in the forest biomass return to the forest soil through litterfall and subsequent decomposition and leached by percolating water (Polyakova and Billor, 2007). The release of elements during decomposition completes the cycle of elements in plant litter within the soil-plant system (Rawat *et al.*, 2010_a). Therefore, decomposition of organic matter and mineralization processes are responsible for virtually all of the nutrients taken up by vegetation for plant growth which affect hydrology and ecosystem structure and function (Muoghalu and Odiwe, 2011).

Decomposition Dynamics

Decomposition of litter is a sequential biogeochemical process where complex organic compounds are broken down to simpler compounds (Table 2) and nutrients which vary with climate, season, substrate quality, litter chemistry, and type of biota and is a fundamental source of soil organic matter, energy and nutrients in forested ecosystems (Rawat *et al.*, 2010_b; Krishna and Mohan, 2017).

Different tree components contain different amount of nutrients, a fraction of which is return through leaf litter, twigs, etc., and is responsible for the cycling of nutrients in the forest ecosystem. The amount

*Table 2. Decomposition pattern of complex organic compounds in litter**

Substrate	Microorganism	Breakdown
Cellulose	Wood-decay fungus	Monomers or oligomers of cellobiose
Hemicellulose	*Bacillus, Achromobacter, Cytophaga, Pseudomonas, Sporocytophaga, Lactobacillus, Vibrio, Streptomyces*	-
Pectin	*E. chrysanthemi, E. cartovora, Pseudomonas, Bacillus, Clostridium, Lachospina, Butyrivibrio, Bacteroides*	Demethylated to pectic acid
Lignin	Wood-decay fungus	CO_2, water, carbonyl and carboxyl group; wood softening by breaking middle lamella of cell wall

*Adopted from Krishna and Mohan (2017)

of litter production, as well as the, amount of nutrients and organic matter released during the process of decomposition play an important role in nutrient cycling (Steink *et al.*, 1993), vegetation productivity and related food chain (Ashton *et al.*, 1999).

Above and below-ground litter in forests decomposes differently (Powers *et al.*, 2009). Leaf litter decomposes faster in wet forests, while roots decompose faster in dry forests (Cornwell *et al.*, 2008). A pan-tropical study in 23 tropical forests of 14 countries reported annual turnover rate of litter decomposition as one because more than 95% mass litter got decomposed within a year in most of these sites (Powers *et al.*, 2009). The study did not find any relationship between decomposition and litter quality, placement and mesofaunal exclusion, but these factors were not to be independent of each other in affecting litter decomposition with site specific magnitude of decomposition. The decomposition of organic residues is an important source of nutrients for primary production, particularly in nutrient-poor soils as those of the humid tropic (Herrera *et al.*, 1978). It is widely believed that plant litter decomposes rapidly and completely in this region because the conditions of humidity and temperature favour the microbial activity (Singh and Gupta, 1977).

Generally, moisture, temperature regimes, precipitation and pH of soil along with abundance of macro-detritivore, composition of microbial and plant species decides the rate of litter decomposition (Rawat *et al.*, $2010_{a,\ b}$). Litter decomposition globally is strongly mediated by climatic regimes by influencing microbial activity (Bradford *et al.*, 2016), i.e., doubling of microbial activity with every 10°C increase in temperature (Friedlingstein *et al.*, 2006). Mean annual temperatures of < 10°C and moisture contents of < 30% or > 80% is reported as inhibiting thresholds for litter decay (Prescott, 2010). However, litter decomposition locally is more controlled by site factors like that of soil and other local disturbances (Cornwell *et al.*, 2008). This is because locally stable long-term climatic conditions along with anthropogenic disturbances (like land use) determines the composition of plant species and ecosystem structure influencing litter composition and quality and in turn the decomposition (Djukic *et al.*, 2018). Further, forests with open canopies have lower decomposition than forest with closed canopy (Ritter, 2005). The decomposition rate of non-leaf litter is slower than leaf litter but it provides habitat and energy for fungi, bacteria, arthropods, and invertebrates (Devi and Yadava, 2007).

In general the dynamics of litter breakdown vary with species and geographical location (Ashton *et al.*, 1999) and affected by chemical composition of litter, amount of lignin, nitrogen, hemicelluloses, and secondary compounds (Sangha *et al.*, 2006). Decomposition rate and nutrient release by plant litters are mainly controlled by environmental conditions, composition and activities of soil organisms, and chemical quality of the substrate (Palacios *et al.*, 2016; Bradford *et al.*, 2016). Many bacteria and fungi are reported to be active in litter decomposition (table 3).

Of late, it is believed that forest system management practices can also significantly influence leaf litter decomposition rates and nutrient dynamics (Purahong *et al.*, 2014). The litter decomposition process is ultimately driven by specific controlling factors related to the abundance and requirement of the decomposer community and whose availability is partly determined by tree species and soil properties (Rottmann *et al.* 2010). Different species have differing decay rates, often dependent on initial nutrient content and carbon (Mascha *et al.*, 2010). Tree species can affect the decomposition process through the quality of their leaf fall and through the species-specific conditions that they generate in their environment (Aponte *et al.*, 2012). Tree species can also alter decomposition rates indirectly through effects on environmental conditions. Tree species can induce changes in soil fertility, microclimate and faunal and microbial communities in the forest floor (Aponte *et al.*, 2011), all of which influence the decomposition process (Austin and Vivanco, 2006). Tree species generate significantly different biotic and abiotic

*Table 3. Microorganisms for decomposition of various litter component**

Component	Microorganisms
Cellulose	*Alternaria, Chaetomium, Fusarium, Myrothecium, Penicillium, etc.*
Hemicellulose	*Alternaria, Aspergillus, Fusarium, Rhizopus, Trichothecium, etc.*
Lignin	*Clavaria, Clitocybe, Collybia, Flammula, Hypholoma, Lepiota, etc.*
Starch	*Aspergillus, Fomes, Fusarium, Rhizopus, Achromobacter, Bacillus, etc.*
Pectin	*Fusarium, Verticillium, Bacillus, Clostridium, etc.*
Inulin	*Penicillium, Aspergillus, Fusarium, Pseudomonas, Flavobacterium, etc.*
Chitin	*Fusarium, Mucor, Mortierella, Trichoderma, Aspergillus, etc.*
Proteins, nucleic acids	*Bacillus, Pseudomonas, Clostridium, Serratia, etc.*
Cutin	*Penicillium, Rhodotorula, Mortierella, etc.*
Tannin	*Aspergillus and Penicillium*
Humic acid	*Penicillium and Polystitus*
Fulvic acid	*Poria*

*Adopted from Krishna and Mohan (2017)

environments beneath their canopy though their distinct leaf litter nutrients return (Aponte *et al.*, 2011). The effects of individual species may differ in mixed forest conditions as a result of positive, negative or neutral interactions between litter types (Hättenschwiler and Gasser, 2005).

Differences between tree species and litter decomposition have commonly been related to distinct substrate quality with litter C:N and N:P ratios, lignin content, Ca and Mn concentrations emerging as the main rate-controlling factors (Berg *et al.*, 2010). The simultaneous effects of trees on decomposition both through their litter quality and by modifying the environmental conditions might cause positive litter-environment interactions and further increase decomposition. This interaction, termed home-field advantage, implies litter decomposes faster beneath the tree species from which it is derived than beneath other plant covers and could be explained as an adaptation of the local soil communities to the litter produced by the plant species above them (Ayres *et al.*, 2009). Despite the implications for ecosystem functioning and carbon cycling, the environmental effect of tree species on litter decomposition has barely been explored and the relative importance of the litter versus the tree species environment effect on the decomposition process still remain unclear (Vivanco and Austin 2008).

Litter Nutrient Dynamics and Release

Litter N, Ca and Mn, as well as, soil N, P and soil moisture were reported as the best predictors for decomposition rates (Patricio *et al.*, 2012). The quantity of N, P, K, Ca, Mg and C in litter decreases gradually as decomposition advances. In general, distinct patterns in the dynamics of particular chemical elements in various forest ecosystems reflect the different availabilities of nutrients to decomposers. Thus, those elements with concentrations below the limiting threshold for decomposers would be immobilized in litter (Swift *et al.*, 1979).

Stages of Decomposition

There are two marked stages of litter decomposition (Berg and McClaugherty, 2008). The litter mass loss up to 40% is the early or initial stage dominated by decomposition of solubles, non-lignified cellulose and hemicellulose along with leaching of soluble compounds (Heim and Frey, 2004). In the next stage lignified tissue degrades. The early phase is highly dynamic and is crucial to understand the whole decay process and the conditions influencing it (Djukic *et al.*, 2018). The quality of initial litter and type mainly influence the initial phase of decomposition (Bradford *et al.*, 2016; Didion *et al.*, 2016) which is due to presence of high water-soluble compounds (Didion *et al.*, 2016) leading to more leaching losses than due to microbial mineralization of soil organic C (Pouyat *et al.*, 2017; Djukic *et al.*, 2018). Initially during decomposition many species microorganisms are involved in mineralization of labile C compounds but at the later stages decomposers become specific due to substrate complexity. This explains the differences in litter mass loss between the land-use types (McGuire and Treseder, 2010). At later and more advanced stages of decomposition, Home-field advantage theory by Gholz *et al.* (2000) holds good to explain a fraction of the remaining variability.

Litter Decomposition Across Biogeographic Regions

Litter production, its decomposition and its seasonal distribution have been studied in different forests and plantations throughout the world. In general, deciduous forests are believed to have faster nutrient turnover than coniferous forest (Pastor, 1987). A coniferous stand generally accumulate larger quantity of organic matter and has longer mean nutrient resident times than deciduous stands (Vogt *et al.*, 1986), consistent with slower nutrient turnover. The positive effect of deciduous trees on nutrient cycling is usually attributed to their high quality litter causing faster litter decomposition and faster nutrient cycling (Pastor, 1987; Scott and Binkley, 1997). The deciduous litter showed a faster initial decomposition but left a larger fraction of slow decomposable biomass compared with the perennial litter; in contrast the deciduous environment impeded early decomposition while promoting further carbon loss in the latter decay stages (Aponte *et al.*, 2012).

The leaf litter of *Shorea robusta* plantation at Doon valley was 83.87% to the total litter fall. The higher rate of decomposition was recorded during rainy season (Pande, 2001). Nutrient concentration in litter of this plantation was generally higher in leaf litter than in twig. Ca and N contributed higher proportion in total nutrient return, release and accumulation. The higher nutrient was associated with the higher litter fall and litter nutrient concentrations. Further, the order of importance in term of magnitude of nutrient return was Ca > N > Mg > K > P. The higher rate of return and release of nutrients were due to higher litter input and its subsequent decomposition. Significant variations in all the nutrients were also noticed (Pande, 2001).

The total litter produced in young *Tectona grandis* plantation at Coimbatore was 4,574 kg ha^{-1} (Rajagopal *et al.*, 2005). The nutrient return at these plantations was in terms of N, P, K, Ca and Mg was 46, 3, 39, 86 and 21 kg ha^{-1} respectively. The magnitude of annual litter fall on forest floor in some tropical dry deciduous *T. grandis* forests of Satpura Plateau in Chhindwara Forest Division Madhya Pradesh was in the range of 3,276 to 4,536 kg ha^{-1} (Pande *et al.*, 2002). The nutrient return to these forest floors was in the order of C>N>K>Mg>P. The litter production in a reserve forest of the Terai zone, West Bengal was 5.37 Mgha^{-1} with turnover rate of litter decomposition of only 12 months (Shukla, 2010). The NPK content in fresh litter of this West Bengal terai forest was recorded as 1.58, 0.42 and 1.12%, respectively, while return of available NPK from litter through decomposition was 0.084, 0.023 and 0.060 $Mgha^{-1}$, respectively.

The nutrient concentrations in tropical dry miscellaneous deciduous forest of northern India in the N-P-K content was 4.46%, 0.588% and 0.773%, respectively (Shukla *et al.*, 2017). N and P release in the forest for species was *Shorea robusta*>*Diospyros melanoxylon*>*Schleichera oleosa*>*Madhuca indica* while for K was *D. melanoxylon*>*S. oleosa*>*S. robusta*>*Madhuca indica*. In a tropical dry deciduous forest of Raipur, Chhattisgarh, India, decomposition was faster in *Shorea robusta* as compared to other species (Bargali *et al.*, 2015). The nutrient content decreased gradually as decomposition advanced for all the species in this forest. The turnover rate of litter in a dry deciduous forest of Chhattisgarh, India was one as about 88-96% of litter decomposed within a year with rate of decomposition more intense during rainy and summer season (Bargali *et al.*, 2015). The rate of decomposition ranged from 0.21-0.28% day^{-1} in the forest. N concentration also increased in the decomposing litter of this forest exhibiting inverse linear correlation with the remaining litter mass. The P content in the forest litter was low following similar pattern of N immobilization and uptake by micro-organisms. K was found extensively leached in this forest also from the decomposing litter indicating its high mobility and availability to the decomposers.

Litter mass loss and mineralization differed significantly among the tree species in the dry tropical deciduous forest of Vindhyan high land, India due to different initial chemical composition of their litter (Rai *et al.*, 2016). N gradually increased with advancing litter decomposition but finally decreased for all the species in the forest, while C decreased throughout. The species having with higher initial N has lower initial lignin, lignin: N and C: N ratio in their litter, decomposed faster than the others as dry mix >*Hardwickia binnata* >*Butea monosperma* >*Tectona grandis* >*S. robusta*. Dry mix trees exhibited faster rate of decomposing litter, released more nutrients and were considered important for regulating organic matter and nutrient cycling in the forest i.e. they had higher N content in their litter and is thus best predictor of N dynamics. Greater N in litter of these tree species indicates lower lignin with faster decomposition rates of their litter increasing carbon, nitrogen and microbial activities in the soil on which it is growing. The C: N (i.e. < 25) and lignin: N ratio in the litter of dry mixed trees, *H. binnata* and *B. monosperma* was high thus aiding faster mineral N release.

Forest Soil Nutrient Dynamics and Availability

The selective absorption of nutrient elements by different tree species and their capacity to return them to the soil brings about changes in soil properties (Singh *et al.*, 1986). Soil, litter and plant organ analysis is not an efficient tool for monitoring nutritional status of vegetation. Presence of mineral elements gives good information towards the knowledge of nutrient cycling and biochemical cycle in the soil-plant ecosystem. The presence of nutrients in the soil does not mean that the tree or vegetation is satisfactorily nourished. The availability of nutrients to the tree is conditioned by the content of water in the soil, soil aeration, soil temperature, soil microorganisms and the efficiency of the root system to absorb nutrients (Raij, 1981). Proper soil management and forest nutrition is the basis for the availability of the nutrients and maintaining the productivity of planted or natural forests (Alvarado *et al.*, 2014).

Forests in general have a greater influence on soil conditions than most other plant ecosystem types, due to a well-developed 'O' horizon, moderating temperature, and humidity at the soil surface, input of litter with high lignin content, high total net primary production, and high water and nutrient demand (Binkley and Giardina, 1998). The C:N ratio indicates the availability of carbon and nitrogen in the soil or its floor material through assimilation, mineralization, denitrification and organic matter decomposition, rate of decay and the quality of the organic matter under the canopy which can also be linked to the soil microbial biomass (Hogberg, 2004). The ratio is a key variable through which all ecosystem perturbations act (Johnson and Ball, 1996). The vegetation influences the C:N ratio and the C:N ratio in turn determines the stand composition (Fisher and Binkley, 2000). Both organic C and N were significantly affected by factors such as stand type, stand age, and the interaction between age, species composition, and soil properties (Côté *et al.*, 2000).Generally the soil of a forest stand attains a steady state when the C:N ratio reaches 10 and at that level the release of nutrients is rapid due to mineralization (Kawahara and Tsutsuni, 1972). Temperate broad-leaved and conifer tree species growing in a single species dominant stands alter soil C and N dynamics (Binkley, 1995).

A relationship exists between growth and nutritional status of the vegetation (Ferreira *et al.*, 1993). Density of forest stands may influence nutrient demands from the soil and subsequent below ground productivity through differential aboveground biomass allocation patterns and tissue nitrogen concentrations (Dicus and Dean, 1998). Organic carbon, N and C: N ratio values were lowest in barren land, intermediate in cultivated well managed soil and highest in forest and cultivated unmanaged land (Gupta *et al.*, 2001). The pH of forest soil is generally acidic (Gairola *et al.*, 2012). This may be due to high organic matter content and the undisturbed nature of the forest soils (Adams and Sidle, 1987). The reduction in pH can be attributed to accumulation and subsequent slow decomposition of organic matter, which releases acids (de Hann, 1977). It has been reported that forest soils should be slightly acidic for nutrient supply to be balanced (Leskiw, 1998).

Moreover, different tree species can differ significantly in their influence on soil properties as well as soil fertility (Augusto *et al.*, 2002). Mineral soil in conifer stands (black pine and scats pine) contains lower total nitrogen stock than the broadleaved beech stands (Sariyildiz *et al.*, 2015). Differences in soil nitrogen stock between tree species were attributed to differences in litter quality and quantity of tree species. Differences in the amount of litter produced and in its biochemical properties between coniferous and deciduous species affects litter decomposition rates, and eventually influence soil carbon stocks (Sariyildiz and Anderson, 2003). Beech leaf litter contains more components that are difficult to decompose than conifers (Sariyildiz *et al.*, 2005) due to higher lignin concentrations resulting into its accumulation in the forest floor and formation of acidic compounds (Sariyildiz and Kucuk, 2008).

The age of trees can also affect the soil parameters (Sariyildiz *et al.*, 2015). Old stands have a high nitrogen stock because of its accumulation in relatively undisturbed forests, and some years are needed to change the soil chemistry (Oostra *et al.*, 2006). Land use changes such as forest clearing, cultivation and pasture/grassland conversion are known to result in changes in nitrogen content (Wu *et al.*, 2009). The soil nitrogen stock of grasslands was higher than that of black pine stands, but lower than that of beech stands and no significant differences between Scots pine and fir stands (Sariyildiz *et al.*, 2015). Differences in nitrogen storage between pasture, grassland and forest sites are attributed to variations in vegetation type, tree stand age and physical properties of soils (Osher *et al.*, 2003). The variability in soil C/N ratios and N retention can be closely linked with variation in tree species composition (Lovett *et al.*, 2002). Organic carbon, N and C: N ratio values were lowest in barren land, intermediate in cultivated well managed soil and highest in forest and cultivated unmanaged land (Gupta *et al.*, 2001).

Available nitrogen was maximum in natural forest of *Shorea robusta* (121.22 kg / ha) in terai zone of West Bengal and least in fallow land (86.10 kg / ha) that decreased with increase in soil depth (Koul, 2004). The total available N-P-K in a reserve forest of the Terai zone of West Bengal, India, up to 30 cm soil depth was 261.89, 108.73 & 152.22 Kg ha^{-1} and 561.77, 237.67 & 334.87 Mg, respectively (Shukla, 2010). In dry deciduous forests of West Bengal,, P use efficiency was higher than N and K in *Shorea robusta* stands while K use efficiency was higher than P and N in *Buchanania latifolia*, *Lagerstroemia parviflora* and *Madhuca longifolia* stands (Biswas *et al.*, 2012).The total nutrient pool of N-P-K in Yamuna Forest Division of Garhwal Himalayas was in the order K>N>P and available nutrient (ANP) pool in order of N>K>P (Jha and Dimri, 1991). Values of N, P, K, C, soil organic matter, C:N ratio and pH of moist temperate forest of Mandal-Chopta area in the Garhwal region of Uttarakhand ranged from 0.17 to 0.45%, 2.73 to 20.17 ppm, 40.67 to 261.17 ppm, 2.29 to 4.31%, 3.95 to 7.43%, 8.12 to 14.49 and 5.47 to 6.67, respectively (Gairola *et al.*, 2012).

CONCLUSION

Litter decomposition is important for proper functioning of ecosystems, sustaining the material cycle, and maintaining productivity. Decomposition of litter is a complex phenomenon of physical, chemical and biological processes differing with geographic regions. As yet, there is a paucity of knowledge on the rates of decomposition and role of various factors in different ecosystems. In today's context of climate change, it has become all the more important to understand litter production and decomposition due to anthropogenic disturbances on energy transfer and material cycles. Preparing for the global environmental change will require micro-level understanding and integration of information locally and globally. This requires extensive field studies to formulate, understand and utilize the local and global biogeochemical models in effective ways. Extensive variability exists and thus more research is required for a better mechanistic understanding of decomposition processes at various scales and for a more accurate estimation of present and future global carbon budgets.

REFERENCES

Adams, P. W., & Sidle, R. C. (1987). Soil conditions in three recent landslides in southeast Alaska. *Forest Ecology and Management*, *18*(2), 93–102. doi:10.1016/0378-1127(87)90136-8

Akpor, B. O., Okoh, A. I., & Babalola, G. O. (2005). Culturable microbial population during decomposition of *Cola nitida* leaf litters in a tropical soil setting. *Akdeniz Üniversitesi Ziraat Fakültesi Dergisi*, *18*, 313–319.

Alvarado, A., Fernández-Moya, J., Segura, J. M., Vaides, E. E., Camacho, M., Avellán, M. J., & Ávila, C. E. (2014). Nutrient Management Improvements in Forestry Species. *Better Crops with Plant Food*, *98*, 16–18.

Aponte, C., Garcia, L. V., & Maranon, T. (2012). *Tree Species Effect on Litter Decomposition and Nutrient Release in Mediterranean Oak Forests Changes Over Time*. Ecosystem. doi:10.100710021-012-9577-4

Aponte, C., Garcıa, L. V., Pe'rez-Ramos, I. M., Gutierrez, E., & Maranon, T. (2011). Oak trees and soil interactions in Mediterranean forests: A positive feedback model. *Journal of Vegetable Science*, *22*(5), 856–867. doi:10.1111/j.1654-1103.2011.01298.x

Ashton, E. C., Hogarth, P. J., & Ormond, R. (1999). Breakdown of mangrove leaf litter in a managed mangrove forest in Peninsular Malaysia. In *Diversity and function in mangrove ecosystems* (pp. 77–88). Dordrecht, The Netherlands: Springer. doi:10.1007/978-94-011-4078-2_8

Augusto, L., Ranger, J., Binkley, D., & Rothe, A. (2002). Impact of several common tree species of European temperate forests on soil fertility. *Annals of Forest Science*, *59*(3), 233–253. doi:10.1051/forest:2002020

Austin, A. T., & Vivanco, L. (2006). Plant litter decomposition in a semi-arid ecosystem controlled by photo degradation. *Nature*, *442*(7102), 555–558. doi:10.1038/nature05038 PMID:16885982

Ayres, E., Steltzer, H., Simmons, B. L., Simpson, R. T., Steinweg, J. M., Wallenstein, M. D., ... Wall, D. H. (2009). Home-field advantage accelerates leaf litter decomposition in forests. *Soil Biology & Biochemistry*, *41*(3), 606–610. doi:10.1016/j.soilbio.2008.12.022

Bargali, S. S., Shukla, K., Singh, L., Ghosh, L., & Lakhera, M. L. (2015). Leaf litter decomposition and nutrient dynamics in four tree species of dry deciduous forest. *Tropical Ecology*, *56*, 191–200.

Barnes, B. V., Zak, D. R., Denton, S. R., & Spurr, S. H. (1998). *Forest Ecology* (4th ed.). New. York, NY: John Wiley and Sons.

Berg, B., Davey, M., De Marco, A., Emmett, B., Faituri, M., Hobbie, S., ... De Virzo Santo, A. (2010). Factors influencing limit values for pine needle litter decomposition: A synthesis for boreal and temperate pine forest systems. *Biogeochemistry*, *100*(1-3), 57–73. doi:10.100710533-009-9404-y

Berg, B., & McClaugherty, C. (2008). Plant Litter: Decomposition, Humus Formation, Carbon Sequestration. Berlin, Germany: Springer Science & Business Media. doi:10.1007/978-3-540-74923-3

Binkley, D. 1995. The influence of tree species on forest soils: processes and patterns. In D. J. Mead & I. S. Cornforth (Orgs.), *Proceedings of the trees and soil workshop*, Lincoln University, Agronomy society of New Zealand special publications number 10. Lincoln, New Zealand: Lincoln University Press. pp. 1-33.

Binkley, D., & Giardina, C. (1998). Why do tree species affect soils? The warp and woof of tree-soil interactions. *Biogeochemistry*, *42*(1/2), 89–196. doi:10.1023/A:1005948126251

Bisht, V. K.; Nautiyal, B. P.; Kuniyal, C. P.; Prasad, P., & Sundriyal, R. C. (2014). Litter Production, Decomposition, and Nutrient Release in Subalpine Forest Communities of the Northwest Himalaya. *Journal of Ecosystems.* doi:. doi:10.1155/2014/294867

Biswas, S., Biswas, A., Sinha, S., & Khan, D. K. (2012). Major nutrient cycling of two different tropical dry deciduous forest of West Bengal, India. *Journal of Ecology and Environmental Sciences*, *3*, 77–81.

Bond-Lamberty, B., & Thomson, A. M. (2010). Temperature-associated increases in the global soil respiration record. *Nature*, *464*(7288), 579–582. doi:10.1038/nature08930 PMID:20336143

Bradford, M. A., Berg, B., Maynard, D. S., Wieder, W. R., & Wood, S. A. (2016). Understanding the dominant controls on litter decomposition. *Journal of Ecology*, *104*(1), 229–238. doi:10.1111/1365-2745.12507

Brechet, L. M., Lopez-Sangil, L., George, C., Birkett, A. J., Baxendale, C., Trujillo, B. C., & Sayer, E. J. (2018). Distinct responses of soil respiration to experimental litter manipulation in temperate woodland and tropical forest. *Ecology and Evolution*, *8*(7), 3787–3796. doi:10.1002/ece3.3945 PMID:29686858

Carrera, A. L., Bertilier, M. B., & Larreguy, C. (2008). Leaf litter fall, fine root production, and decomposition in shrub lands with different canopy structure induced by grazing in the Pantagonian Monte, Argentina. *Plant and Soil*, *311*(1-2), 39–50. doi:10.100711104-008-9655-8

Chun-jiang, L., Ilvesniemi, H., Berg, B., Kutsch, W., Yu-sheng, Y., Xiang-qing, M., & Westman, C. J. (2003). Aboveground litterfall in Eurasian forests. *Journal of Forestry Research*, *14*(1), 27–34. doi:10.1007/BF02856758

Cornwell, W. K., Cornelissen, J. H., Amatangelo, K., Dorrepaal, E., Eviner, V. T., Godoy, O., ... Westoby, M. (2008). Plant species traits are the predominant control on litter decomposition rates within biomes worldwide. *Ecology Letters*, *11*(10), 1065–1071. doi:10.1111/j.1461-0248.2008.01219.x PMID:18627410

Côté, L., Brown, S., Paré, D., Fyles, J., & Bauhus, J. (2000). Dynamics of carbon and nitrogen mineralization in relation to stand type, stand age and soil texture in the boreal mixed wood. *Soil Biology & Biochemistry*, *32*(8-9), 1079–1090. doi:10.1016/S0038-0717(00)00017-1

Cuevas, E., Brown, S., & Lugo, A. E. (1991). Above and below ground organic matter storage and production in a tropical pine plantation and a paired broadleaf secondary forest. *Plant and Soil*, *135*(2), 257–268. doi:10.1007/BF00010914

Dalmolin, Â. C., Lobo, F. A., Vourlitis, G., Silva, P. R., Dalmagro, P. R., Antunes, M. Z. Jr, & Ortíz, C. E. R. (2015). Is the dry season an important driver of phenology and growth for two Brazilian savannah tree species with contrasting leaf habits? *Plant Ecology*, *216*(3), 407–417. doi:10.100711258-014-0445-5

de Hann, S. (1977). Humus, its formation, its relation with the mineral part of the soil and its significance for soil productivity. In *Organic matter studies* (Vol. 1, pp. 21–30). Vienna, Austria: International Atomic Energy Agency.

Dicus, C., & Dean, T. J. (1998). Stand density effects on biomass allocation patterns and subsequent soil nitrogen demand. In *Proceedings of the Ninth Biennial Southern Silivicultural Research Conference*.

Didion, M., Repo, A., Liski, J., Forsius, M., Bierbaumer, M., & Djukic, I. (2016). Towards harmonizing leaf litter decomposition studies using standard tea bags- a field study and model application. *Forests*, *7*(12), 167. doi:10.3390/f7080167

Djukic, I. K., Kepfer-Rojas, S., Schmidt, I. K., Larsen, K. S., Beier, C., Berg, B., Verheyend, K., & & Tóth, Z. (2018). Early stage litter decomposition across biomes. *The Science of the Total Environment*, *626*. doi:10.1016/j.scitotenv.2018.01.012 PMID:30045558

Edenhofer, O., Pichs-Madruga, R., Sokona, Y., Farahani, E., Kadner, S., Seyboth, K., Adler, A., ... von Stechow, C. (2014). *Climate Change 2014: Mitigation of Climate Change. Contribution of Working Group III to the Fifth Assessment Report of the Intergovernmental Panel on Climate Change.* (Zwickel, T., & Minx, J. C., Eds.). Cambridge, UK: Cambridge University Press.

Esperschutz, J., Zimmermann, C., Dumig, A., Welzl, G., Buegger, F., Elmer, M., ... Schloter, M. (2013). Dynamics of microbial communities during decomposition of litter from pioneering plants in initial soil ecosystems. *Biogeosciences*, *10*(7), 5115–5124. doi:10.5194/bg-10-5115-2013

Faboya, I. O., Adebola, S. I., & Awotoye, O. O. (2015). Assessment of decomposition rate and soil nutrient status under different woody species combination in a tree plantation. *Agriculture, Forestry and Fisheries*, *4*, 46–54. doi:10.11648/j.aff.20150402.14

Fang, X., Zhao, L., Zhou, G., Huang, W., & Liu, J. (2015). Increased litter input increases litter decomposition and soil respiration but has minor effects on soil organic carbon in subtropical forests. *Plant and Soil*, *392*(1-2), 139–153. doi:10.100711104-015-2450-4

Fernandes, E. C. M., Biot, Y., Castilla, C., Canto, A., & Matos, J. C. S. (1997). The impact of selective logging and forest conversion for subsistence agriculture and pastures on terrestrial nutrient dynamics in the Amazon. *CieciaCultura*, *49*, 34–47.

Ferreira, C. A., Bellote, A. F. J., & da Silva, H. D. (1993). Concentracao de nutrients minerais no lenho de Eucalyptus salignaesuarelacao com aaplicacao de fertilizantes. In Congress FlorestalBrasilicro 7 curitiba. Anais. Swo Paulo: SBS/SBEF 1: 227-230.

Fisher, R. F., & Binkley, D. D. (2000). *Ecology and management of forest soils*. New York, NY: Wiley.

Friedlingstein, P., Cox, P., Betts, R., Bopp, L., Von Bloh, W., Brovkin, V., & Bala, G. (2006). Climate-carbon cycle feedback analysis: Results from the C4MIP model intercomparison. *Journal of Climate*, *19*(14), 3337–3353. doi:10.1175/JCLI3800.1

Gairola, S., Sharma, C. M., Ghildiyal, S. K., & Suyal, S. (2012). Chemical properties of soils in relation to forest composition in moist temperate valley slopes of Garhwal Himalaya, India. *The Environmentalist*, *32*(4), 512–523. doi:10.100710669-012-9420-7

García Palacios, P., Shaw, E. A., Wall, D. H., & Hättenschwiler, S. (2016). Temporal dynamics of biotic and abiotic drivers of litter decomposition. *Ecology Letters*, *19*(5), 554–563. doi:10.1111/ele.12590 PMID:26947573

Gholz, H. L., Wedin, D. A., Smitherman, S. M., Harmon, M. E., & Parton, W. J. (2000). Long-term dynamics of pine and hardwood litter in contrasting environments: Toward a global model of decomposition. *Global Change Biology*, *6*(7), 751–765. doi:10.1046/j.1365-2486.2000.00349.x

Guendehou, G. H. S., Liski, J., Tuomi, M., Moudachirou, M., Sinsin, B., & Mäkipää, R. (2014). Decomposition and changes in chemical composition of leaf litter of five dominant tree species in a West African tropical forest. *Tropical Ecology*, *55*, 207–220.

Gupta, J. P., Sharma, M. P., & Gupta, G. D. (2001). Characterization of Kandi Belt Soils of Jammu Region as Affected by Different Land Use Patterns. *Journal of the Indian Society of Soil Science*, *49*, 770–773.

Hagen-Thorn, A., Varnagiryte, I., Nihlgård, B., & Armolaitis, K. (2006). Autumn nutrient resorption and losses in four deciduous forest tree species. *Forest Ecology and Management*, *228*(1-3), 33–39. doi:10.1016/j.foreco.2006.02.021

Han, T.; Huang, W.; Liu, J.; Zhou, G., & Xiao, Y. (2015). Different soil respiration responses to litter manipulation in three subtropical successional forests. *Scientific Reports-Nature* 5: 18166.

Hansen, K., Vesterdal, L., Schmidt, I. K., Gundersen, P., Sevel, L., Bastrup-Birk, A., ... Bille-Hansen, J. (2009). Litterfall and nutrient return in five tree species in a common garden experiment. *Forest Ecology and Management, 257*(10), 2133–2144. doi:10.1016/j.foreco.2009.02.021

Hättenschwiler, S., & Gasser, P. (2005). Soil animals alter plant litter diversity effects on decomposition. *Proceedings of the National Academy of Sciences of the United States of America, 102*(5), 1519–1524. doi:10.1073/pnas.0404977102 PMID:15671172

Heim, A., & Frey, B. (2004). Early stage litter decomposition rates for Swiss forests. *Biogeochemistry, 70*(3), 299–313. doi:10.100710533-003-0844-5

Herrera, R., Jordan, C. F., Klinge, H., & Medina, E. (1978). Amazon ecosystems. Their structure and functioning with particular emphasis on nutrients. *Interciencia, 3*, 223–231.

Hogberg, N. M. (2004). *Carbon and nitrogen relations among soils, microbes and plants in boreal forests.* (Unpublished doctoral thesis), Swedish University of Agricultural Sciences.

Houghton, R. A. (2007). Balancing the global carbon budget. *Annual Review of Earth and Planetary Sciences, 35*(1), 313–347. doi:10.1146/annurev.earth.35.031306.140057

Indriyanto. (2009). Produksiseresahpadakomunitashutan yang dikelolapetanidalam Register 19 Provinsi Lampung. In *ProsidingPenelitian-penelitianAgroforestri di Indonesia.* INAFE Punlisher, Tahun Lampung. pp 75-83.

Jha, M. N., & Dimri, B. M. (1991). Soil nutrients and forest productivity in four natural stands. *Journal of the Indian Society of Soil Science, 39*, 735–738.

Johnson, D. W., & Ball, J. T. (1996). Interactions between CO_2 and nitrogen in forests: can we extrapolate from the seedling to the stand level? In G. W. Koch & H. A. Mooney (Eds.), *Carbon dioxide and terrestrial ecosystems.* San Diego, CA: Academic Press. doi:10.1016/B978-012505295-5/50017-3

Kavvadias, V. A., Alifragis, D., Tsiontsis, A., Brosfas, G., & Stamatelos, G. (2001). Litterfall, litter accumulation and litter decomposition rates in four forest ecosystems in northern Greece. *Forest Ecology and Management, 144*(1-3), 113–127. doi:10.1016/S0378-1127(00)00365-0

Kawahara, T., & Tsutsuni, T. (1972). Studies on the circulation of Carbon and Nitrogen in forest ecosystem. *Bulletin of Kyoto University of Forestry, 44*, 141–158.

Koul, D. N. (2004). *Carbon sequestration estimates of various land uses in Terai Zone of West Bengal.* (Unpublished Master of Science Thesis), Uttar Banga Krishi Viswavidyalaya, Pundibari, West Bengal.

Krishna, M. P., & Mohan, H. (2017). Litter decomposition in forest ecosystems: A review. *Energy. Ecology & Environment, 2*, 236–249.

Lal, J. B. (1992). *Forest Ecology.* Dehradun, India: Natraj Publishers.

Lal, R. (2004). Soil carbon sequestration impacts on global climate change and food security. *Science, 304*(5677), 1623–1627. doi:10.1126cience.1097396 PMID:15192216

Leskiw, L. A. (1998). *Land capability classification for forest ecosystem in the oil stands region*. Edmonton, Canada: Alberta Environmental Protection.

Li, Y., & Ye, Y. (2014). Dynamics of decomposition and nutrient release of leaf litter in *Kandelia obovata* mangrove forests with different ages in Jiulongjiang Estuary, China. *Ecological Engineering, 73*, 454–460. doi:10.1016/j.ecoleng.2014.09.102

Lieth, H. (1975). Primary production of the major vegetation units of the world. In Primary productivity of the biosphere (pp. 203-215). Berlin, Germany: Springer.

Lovett, G. M., Christenson, L. M., Groffman, P. M., Jones, C. G., Hart, J. E., & Mitchell, M. J. (2002). Insect Defoliation and Nitrogen Cycling in Forests: Laboratory, plot, and watershed studies indicate that most of the nitrogen released from forest foliage as a result of defoliation by insects is redistributed within the ecosystem, whereas only a small fraction of nitrogen is lost by leaching. *Bioscience, 52*(4), 335–341. doi:10.1641/0006-3568(2002)052[0335:IDANCI]2.0.CO;2

Mascha, J. K. V.; Andrea Polle, F. M., & Thomas. (2010). Leaf litter decomposition in temperate deciduous forest stands with a decreasing fraction of beech (*Fagus sylvatica*). *Oecologia* 164: 1083-1094.

McGuire, K. L., & Treseder, K. K. (2010). Microbial communities and their relevance for ecosystem models: Decomposition as a case study. *Soil Biology & Biochemistry, 42*(4), 529–535. doi:10.1016/j.soilbio.2009.11.016

Meentemeyer, V. (1984). The geography of organic decomposition rates. *Annals of the Association of American Geographers, 74*(4), 551–560. doi:10.1111/j.1467-8306.1984.tb01473.x

Millar, C. S. (1974). Decomposition of coniferous leaf litter. In C. H. Dickinson & G. J. F. Pugh (Eds.), *Biology of plant litter decomposition* (pp. 105–128). London, UK: Academic Press. doi:10.1016/B978-0-12-215001-2.50010-6

Muoghalu, J. I., & Odiwe, A. I. (2011). Litter production and decomposition in cacao (*Theobroma cacao*) and kolanut (*Cola nitida*) plantations. *Ecotropica (Bonn), 17*, 79–90.

Oostra, S.; Majdi, H., & Olsson, M. (2006). Impact of tree species on soil carbon stocks and soil acidity in southern Sweden. *Scandinavian Journal of Forest Research,* 21, 364-371. doi: 27580600950172. doi:10.1080/028

Osher, L. J., Matson, P. A., & Amundson, R. (2003). Effect of land use change on soil carbon in Hawaii. *Biogeochemistry, 65*(2), 213–232. doi:10.1023/A:1026048612540

Panda, T., Panda, B., & Mishra, N. (2007). Comparative study of *Penicillia* from soil, leaf, litter and air in coastal sandy belt of Orissa. *Journal of Phytological Research, 20*, 335–336.

Pande, P. K. (2001). Litter nutrient dynamics of *Shorea robusta* Gaertn. plantation at Doon Valley (Uttaranchal), India. *Indian Forester, 127*, 980–994.

Pande, P. K., Meshram, P. B., & Banarjee, S. K. (2002). Litter production and nutrient return in tropical dry deciduous teak forests of Satpura plateau in central India. *Tropical Ecology, 43*, 337–344.

Pastor, J., Gardner, R. H., Dale, V. H., & Post, W. M. (1987). Successional changes in nitrogen availability as a potential factor contributing to spruce declines in boreal North America. *Canadian Journal of Forest Research*, *17*(11), 1394–1400. doi:10.1139/x87-216

Patricio, M. S., Nunes, L. F., & Pereira, E. L. (2012). Litterfall and litter decomposition in chestnut high forest stands in northern Portugal. *Forest Systems*, *21*(2), 259–271. doi:10.5424/fs/2012212-02711

Pinto, O. B. Jr, Vourlitis, G. L., & Carneiro, E. M. (2018). Interactions between vegetation, hydrology, and litter inputs on decomposition and soil CO_2 efflux of tropical forests in the Brazilian Pantanal. *Forests*, *9*(5), 281–297. doi:10.3390/f9050281

Polyakova, O., & Billor, N. (2007). Impact of deciduous tree species on litterfall quality, decomposition rates and nutrient circulation in pine stands. *Forest Ecology and Management*, *253*(1-3), 11–18. doi:10.1016/j.foreco.2007.06.049

Pouyat, R. V., Setälä, H., Szlavecz, K., Yesilonis, I. D., Cilliers, S., Hornung, E., ... Whitlow, T. H. (2017). Introducing GLUSEEN: A new open access and experimental network in urban soil ecology. *Journal of Urban Economics*, *3*(1). doi:10.1093/jue/jux002

Powers, J. S., Montgomery, R. A., Adair, E. C., Brearley, F. Q., DeWalt, S. J., Castanho, C. T., ... Ramanamanjato, J. (2009). Decomposition in tropical forests: A pan-tropical study of the effects of litter type, litter placement and mesofaunal exclusion across a precipitation gradient. *Journal of Ecology*, *97*(4), 801–811. doi:10.1111/j.1365-2745.2009.01515.x

Pragasan, L. A., & Parthasarathy, N. (2005). Litter production in tropical dry evergreen forests of south India in relation to season, plant life-forms and physiognomic groups. *Current Science*, *88*, 1255–1263.

Prescott, C. E. (2010). Litter decomposition: What controls it and how can we alter it to sequester more carbon in forest soils? *Biogeochemistry*, *101*(1-3), 133–149. doi:10.100710533-010-9439-0

Prescott, C. E., Reid, A., Wu, S. Y., & Nilsson, M. C. (2017). Decomposition rates of surface and buried forest-floor material. *Canadian Journal of Forest Research*, *47*(8), 1140–1144. doi:10.1139/cjfr-2016-0533

Purahong, W., Kapturska, D., Pecyna, M. J., Schulz, E., Schloter, M., Buscot, F., ... Krüger, D. (2014). Influence of different forest system management practices on leaf litter decomposition rates, nutrient dynamics and the activity of ligninolytic enzymes: A case study from Central European forests. *PLoS One*, *9*(4). doi:10.1371/journal.pone.0093700 PMID:24699676

Rai, A., Singha, A. K., Ghosal, N., & Singh, N. (2016). Understanding the effectiveness of litter from tropical dry forests for the restoration of degraded lands. *Ecological Engineering*, *93*, 76–81. doi:10.1016/j.ecoleng.2016.05.014

Raij, B. (1981). *Avaliacao da fertilidade do solo* (p. 142). Piracicaba, Brazil: Instituto da Potassa and Fosfato.

Rajagopal, K., Buvaneswaran, C., Subramanian, V., & George, M. (2005). Nutrient cycling in young teak plantation I-Restitution of nutrients through litter and rain wash. *Indian Forester*, *131*, 221–228.

Rawat, N., Nautiyal, B. P., & Nautiyal, M. C. (2010). Annual nutrients budget for the grazed and un-grazed sites of an alpine expanse in North-West Himalaya, India. *The Environmentalist, 30*(1), 54–66. doi:10.100710669-009-9243-3

Rawat, N., Nautiyal, B. P., & Nautiyal, M. C. (2010). Litter decomposition rate and nutrient release from different litter forms in a Himalayan alpine ecosystem. *The Environmentalist, 30*(3), 279–288. doi:10.100710669-010-9275-8

Ritter, E. (2005). Litter decomposition and nitrogen mineralization in newly formed gaps in a Danish beech (*Fagus sylvatica*) forest. *Soil Biology & Biochemistry, 37*(7), 1237–1247. doi:10.1016/j.soilbio.2004.11.020

Rodin, L. E., & Basilevic, N. I. (1968). World distribution of plant biomass. Functioning of terrestrial ecosystems at the primary production level. UNESCO, Paris, 45-52.

Rohr, T., Manzoni, S., Feng, X., Menezes, R. S. C., & Porporato, A. (2013). Effect of rainfall seasonality on carbon storage in tropical dry ecosystems. *Journal of Geophysical Research. Biogeosciences, 118*(3), 1156–1167. doi:10.1002/jgrg.20091

Rottmann, N., Dyckmans, J., & Joergensen, R. G. (2010). Microbial use and decomposition of maize leaf straw incubated in packed soil columns at different depths. *European Journal of Soil Biology, 46*(1), 27–33. doi:10.1016/j.ejsobi.2009.09.004

Sangha, K. K., Jalota, R. K., & Midmore, D. J. (2006). Litter production, decomposition and nutrient release in cleared and uncleared pasture systems of central Queensland, Australia. *Journal of Tropical Ecology, 22*(2), 177–189. doi:10.1017/S0266467405003020

Sariyildiz, T., & Anderson, J. M. (2003). Interactions between litter quality, decomposition and soil fertility: A laboratory study. *Soil Biology & Biochemistry, 35*(3), 391–399. doi:10.1016/S0038-0717(02)00290-0

Sariyildiz, T., & Anderson, J. M. (2005). Variation in the chemical composition of green leaves and leaf litters from three deciduous tree species growing on different soil types. *Forest Ecology and Management,* 210, 303-319. doi:eco.2005.02.043. doi:10.1016/j.for

Sariyildiz, T.; Anderson, J. M., & Kucuk, M. (2005). Effects of tree species and topography on soil chemistry, litter quality and decomposition in Northeast Turkey. *Soil Biology and Biochemistry,* 37, 1695-1706. doi:.02.004. doi:10.1016/j.soilbio.2005

Sariyildiz, T., & Küçük, M. (2008). Litter mass loss rates in deciduous and coniferous trees in Artvin, northeast Turkey: Relationships with litter quality, microclimate, and soil characteristics. *Turkish Journal of Agriculture and Forestry, 32*, 547–559.

Sariyildiz, T., Savaci, G., & Kravkaz, I. S. (2015). Effects of tree species, stand age and land-use change on soil carbon and nitrogen stock rates in northwestern Turkey. *IForest (Viterbo), 9*(1), 165–170. doi:10.3832/ifor1567-008

Sarjubala Devi, A., & Yadava, P. S. (2007). Wood and leaf litter decomposition of *Dipterocarpus tuberculatus* Roxb. in a tropical deciduous forest of Manipur, Northeast India. *Current Science, 93*, 243–246.

Scott, N. A., & Binkley, D. (1997). Foliage litter quality and annual net N mineralization: Comparison across North America forest site. *Oecologia, 111*(2), 151–159. doi:10.1007004420050219 PMID:28307988

Shukla, A. K., Srivastava, P. K., Singh, B., Behera, S. K., & Thomas, T. (2017). Litterfall patterns and soil nutrient chemistry in varied tropical deciduous forests. *International Journal of Chemical Studies, 5*, 1203–1210.

Shukla, G. (2010). *Biomass production and vegetation analysis of Chilapatta reserve forest ecosystem of West Bengal*. [Published doctoral thesis], Uttar Banga Krishi Viswavidyalaya, Pundibari.

Singh, A. K., Parsad, A., & Singh, B. (1986). Availability of phosphorus and potassium and its relationship with physicochemical properties of some forest soils of Pali-range (Shahdol, MP). *Indian Forester, 112*, 1094–1104.

Singh, J. S., & Gupta, S. R. (1977). Plant decomposition and soil respiration in terrestrial ecosystems. *Botanical Review, 43*(4), 449–528. doi:10.1007/BF02860844

Sotta, E. D., Meir, P., Malhi, Y., Nobre, A. D., & Grace, J. (2004). Soil CO_2 efflux in a tropical forest in the central Amazon. *Global Change Biology, 10*(5), 601–617. doi:10.1111/j.1529-8817.2003.00761.x

Steinke, T. D., Holland, A. J., & Singh, Y. (1993). Leaching losses during decomposition of mangrove leaf litter. *South African Journal of Botany, 59*(1), 21–25. doi:10.1016/S0254-6299(16)30770-0

Swift, M. J., Heal, O. W., & Anderson, J. M. (1979). *Decomposition in terrestrial ecosystems*. Oxford, UK: Blackwell Scientific Publications.

Tonin, A. M., Gonçalves Jr.; J. F., Bambi, P., Couceiro, S. R. M., Feitoza, L. A. M., Fontana, L. E., Hamada, N., … Boyero, L. (2017). Plant litter dynamics in the forest stream interface: Precipitation is a major control across tropical biomes. *Scientific Reports, 7*. doi:10.103841598-017-10576-8 PMID:28883445

Uriarte, M., Turner, B. L., Thompson, J., & Zimmerman, J. K. (2015). Linking spatial patterns of leaf litterfall and soil nutrients in a tropical forest: A neighborhood approach. *Ecological Applications, 25*(7), 2022–2034. doi:10.1890/15-0112.1 PMID:26591466

Vendrami, J. L., Jurinitz, C. F., Castanho, C. T., Lorenzo, L., & Oliveira, A. A. (2012). Litterfall and leaf decomposition in forest fragments under different successional phases on the Atlantic Plateau of the state of Sao Paulo, Brazil. *Biota Neotropica, 12*. Retrieved from http://www.biotaneotropica.org.br/v12n3/en/abstract?article+bn03312032012.

Vivanco, L., & Austin, A. T. (2008). Tree species identity alters forest litter decomposition through long-term plant and soil interactions in Patagonia, Argentina. *Journal of Ecology, 96*(4), 727–736. doi:10.1111/j.1365-2745.2008.01393.x

Vogt, K. A., Grier, C. C., & Vogt, D. J. (1986). Production, turnover and nutrient dynamics of above and belowground detritus of world forests. *Advances in Ecological Research, 15*, 303–377. doi:10.1016/S0065-2504(08)60122-1

Wardle, D. A., Bardgett, R. D., Klironomos, J. N., Setälä, H., Van Der Putten, W. H., & Wall, D. H. (2004). Ecological linkages between aboveground and belowground biota. *Science, 304*(5677), 1629–1633. doi:10.1126cience.1094875 PMID:15192218

Wu, J.-S., Jiang, P.-K., Chang, S. X., Xu, Q.-F., & Lin, Y. (2009). Dissolved soil organic carbon and nitrogen were affected by conversion of native forests to plantations in subtropical China. *Canadian Journal of Soil Science*, *90*(1), 27–36. doi:10.4141/CJSS09030

Xu, X., & Hirata, E. (2005). Decomposition patterns of leaf litter of seven common canopy species in a subtropical forest: N and P dynamics. *Plant and Soil*, *273*(1-2), 279–289. doi:10.100711104-004-8069-5

Zhao, X., Li, F., Zhang, W., & Ai, Z. (2016). Contribution of aboveground litter to soil respiration in *Populus davdiana* Dode plantations at different stand ages. *Journal of Mountain Science*, *13*(6), 1000–1012. doi:10.100711629-015-3777-2

Chapter 11
Participatory Management of Tropical Dry Forests in Benin:
Case Study From the "Trois Rivières" Forest, Borgou Region

Ramoudane Orou Sannou
University of Parakou, Benin

Idrissou Bako
University of Parakou, Benin

Ismaïl Moumouni
University of Parakou, Benin

Mohamed Nasser Baco
University of Parakou, Benin

Adewole Olagoke
Technische Universität Dresden, Germany

ABSTRACT

This chapter encompasses a literature survey and strategic analysis to understand the elaboration and implementation of Participatory Forest Management (PFM) in Benin, with a focus on the case of the "Forêt des Trois Rivières". By analyzing the historical background of forest management systems in Benin, we highlighted two major turning points. The first relates to the creation and autocratic management of protected forests, which took place from 1940 to 1990. The second change took place after the Rio conference in 1992, and this emphasized the importance of local communities in natural resources management. Moreover, the results of our strategic analysis of stakeholders involved in the specific case of Participatory Forest Management Plan (PFMP) of the "Forêt des Trois Rivières" showed that it is important to emphasize on active community participation while designing a participatory management plan and for decision making at the implementation stage. We also observed that alliances between foresters and timber loggers are likely to hinder the achievement of the PFM objectives.

DOI: 10.4018/978-1-7998-0014-9.ch011

INTRODUCTION

Benin's natural resources are increasingly threatened by human activities due to its growing demography. As a consequence, the creation, development and management of protected areas is one of the effective strategies for biodiversity conservation (Arouna & Djogbenou, 2006). According to the International Union for Conservation of Nature (IUCN) and its World Commission on Protected Areas (WCPA), a protected area is "a clearly defined, recognized, dedicated and managed geographical area, by all effective means, legal or otherwise, to ensure the long-term conservation of nature and associated ecosystem services and cultural values". In this respect, since the 1940s and 1950s, the colonial governments of Benin have listed and classified as protected areas all the forested **massifs** throughout the country. Today, Benin has protected forests covering National Parks (869,867 ha in total), Hunting Zones (443,679 ha), classified forests (1,292,543 ha), and Reforestation Perimeters. Outside these protected areas, Benin has a range of natural resources relevant to the agro-pastoral sector. These include rivers, grazing areas, wetlands, etc. However, drastic reduction of resources and climate change in the present time is crucial to amplifying competition between actors around natural resources, and this poses with acuity the need for governance systems that are well suited to the socio-economic context of the riparian populations who strongly depend on ecosystem services provided by natural resources.

The continuous degradation of natural resources in general and agro-forestry resources in particular are now at the centre of Benin's major concerns. In order to counter this continuing degradation, the national forest policy has undergone profound regulatory changes since 1989 (MEPN, 2010). Earlier, order No. 4524 of 6 September 1949 established the ''Trois Rivières'' forest as a State Protected Forest. In 1994, a new Forest policy was adopted, which has the involvement of local riparian communities in natural resource management as a major innovation. These local riparian communities almost always include agro-pastoral communities whose livelihoods essentially depend on the exploitation of natural resources. The approach promoted by the new policy is to put all actors, including agro-pastoralists, in an organized and functional structure able to manage sustainably the available natural resources, called as participatory forest management. Despite this effort, natural resources continued to dwindle due to accelerated degradation and, paradoxically, there are disagreements between agro-pastoralists and foresters with consequences on the living conditions of the latter. Thus, we wondered about the reasons for the failure of participatory management plans to ensure better cohesion between stakeholders and effective sustainable resource management.

The initial answer to this question can be possibly found in the ideas of Crozier and Friedberg (1977), who postulated that actors in an organizational situation engage in games to defend their strategic interests. Accordingly, agro-pastoralists would perhaps find themselves in a game of actors from which they come out as a loser. Crozier and Friedberg define organized action as "the process through which the strategic interactions between a set of actors placed in a given policy area and mutually dependent for the solution of a number of common "problems" are stabilized and structured" (1995). According to Olaya and Ruess (2004), the previous approach is structured around four main axes: (1) a strategic actor (can be an individual, a group or any other collective entity) with its own interests and interactions with other actors who also act strategically; (2) a concrete system of action formed by interacting actors; (3) the game as an integration mechanism between the actor and the system where each actor has his own interests, but also the interest of keeping alive a concrete system of action; and (4) power as the capacity for action and consisting in an unbalanced exchange of possibilities for action (Crozier and Friedberg 1979, 1995). The approach proposed by Crozier and Friedberg which served as analytical framework of this chapter has been used in many studies in relation to organized situations and project management as it allows a quick reading of actors' logics.

After a comprehensive literature survey on Participatory Forest Management in Benin, Crozier and Friedberg's approach of organized actions was used in this study to analyze the shortcomings of the current forest management systems in the Trois Rivières forest. The chapter is divided into four sections. First section introduces the background of study with associated problems in participatory management of natural resources in Benin. The second section describes the methodology used to collect, process and analyze the data. The third section deals with our study results and analyses. The last section presents the conclusions and suggestions to transform the participatory forest management in Benin into an instrument that betters agro-pastoralists' livelihoods.

BACKGROUND

The world is now facing unprecedented challenges of accelerated depletion of natural resources leading primarily into poverty accentuation, food insecurity and tremendous negative social changes in some regions. Natural resources deterioration is also tightly related to climate change as deforestation leads to enhanced greenhouse gas emission. There is limited understanding of the processes leading to improvement or deterioration of natural resources, which means that scientific knowledge is needed to describe and explain complex socio-ecological systems (SESs). SESs are composed of resource system (e.g., a forest), resource units (pastures), users (pastoralists), and governance systems (organizations and rules that govern the exploitation of forest's resources) Ostrom (2009). Although climate change is rife, pastoralism remains one of the most important activities in sub-Saharan Africa. The uncertainty over forage resources due to the effects of these changes forces pastoralists to use special breeding techniques that preserve their productive capital: livestock and ecosystems (Dia & Duponnois, 2013). Two trends exist regarding the literature about relations between pastoralism and natural resources protection. On the one hand, there are studies that present pastoral mobility as highly detrimental to natural resources preservation. Pastoralism is considered as being responsible for environmental destruction by causing soil deterioration, followed by groundwater pollution and deterioration of pastoral pathways (Carriere & Toutain, 1995; Carriere, 1996; FAO, 2009). Further, pastoralism is considered as a source of greenhouse gas emissions leading to resources depletion and pollution with an impact on biodiversity (Benhammou et al., 2005; LEAD/FAO, 2006; Dumortier et al., 2013). However, the second trend goes beyond the negative image of pastoralism and shows that it can contribute to the ecological sustainability of the natural resources it uses in arid and semi-arid zones (Bonfiglioli, 1990; Marty, Bonnet, Guibert, & Swift, 2006; Blanfort, 2011). Pastoralism allows the flexible use of natural resources to avoid the risk of overgrazing and environmental degradation. It also influences economic and environmental efficiency, enabling better vegetation regeneration (Hiya Maidawa, Andres, Yamba, & Lebailly, n.d.). At least we can stand on a median position to say that pastoralists can adopt environmentally friendly practices through mobility, to sustainably manage resources and avoid environmental degradation.

In Benin, 21% of the total land area is covered by natural forests and Fulani pastoralists are often found in each of the riparian areas. Fulani, originally a nomadic tribe, are now settling down and practicing agriculture as a complement to cattle breeding. They are, therefore, involved in some ways to land use changes in those riparian areas. Besides, as stated by FAO, (2016), Benin is one of the 18 African countries with net gains in agricultural area and net losses in forest area from 2000 to 2010. To reverse this trend, the government is implementing several policies and one of them consists of drawing up Participatory Forest Management Plans (PFMP); involving both the government and local populations in protecting forests. In most cases, participatory plans require that each forest resource user pay a monetary contribution to the received ecosystem services. This policy, obviously, cannot accommodate the predominantly poor populations and lead to disagreements between State representatives (foresters) and local populations (in most cases agro-pastoralists). Worse still, this could affect the pastoralist's livelihoods by undermining the participatory process aimed by the government and leading to increased deforestation.

In the view of the above discussion, this chapter will seek to highlight major flaws in the participatory forest management approach and discuss the possible remedies to the actual governance systems that could be beneficial for pastoral communities. The chapter gives answers to the main question "why do agro-pastoral communities oppose the implementation of participatory management plans in Benin?"

As such, the following sub-questions will be addressed:

- What is the history of Participatory Forest Management in Benin?
- What are the conflicting issues for forest management in the Trois Rivières forest?
- Who are the conflicting actors for forest management in the Trois Rivières forest?
- What is the current involvement of pastoral communities in forest management as one of the stakeholders in decision-making?
- How can some of the current issues of the PFMP of the Trois Rivières forest be redressed?

CONCEPTS

Participation vs Participatory Management Plan

In agricultural extension, the concept of participation has its origin in the failure of the top-down approach through which rural communities received everything that comes from extension providers. Participatory management of natural resources refers to processes and mechanisms that enable those people who have a direct stake in resources to be part of the decision making about their management at different levels, from managing resources to formulating and implementing institutional frameworks (Schreckenberg, Luttrell, & Moss, 2006; Idrissou, 2012). Several authors laid the foundations for the participatory approach by defining a few principles, including the decision-making power that belongs to the grassroots communities, the need to have interventions that facilitate local development and the perpetual search for agreement between all the parties involved in an intervention. Accordingly, participatory management of natural resources aims at promoting sustainable exploitation of resources in order to contribute to the improvement of local populations' living conditions and consequently to reduce the pressure exerted on the natural resources.

Pastoralism and Pastoral Mobility

Pastoralism cannot be defined without stating at first what a livestock system is, as Pastoralism is a particular type of livestock system. Most scientific definitions of the term "livestock system" reflect a holistic view. For example: "A livestock system is a set of elements in dynamic interaction organized by man in order to develop resources through domestic animals to obtain varied productions (such as milk, meat, hides and skins, manure, etc.) or to meet other objectives" (Landais, 1994). Pastoralism is, therefore, understood as a system of animal husbandry, which is mobile in space. Herd mobility is at the heart of the definition of pastoralism. The latter can indeed be described as a way of breeding animals by moving them in such a way as to make the best use of natural resources (Marty et al., 2006). Pastoralism and rangeland management refer to the extensive production of livestock using pastures and rangelands located mainly in the arid and semi-arid areas. Pastoralism, as practiced in Benin, is a traditional form of extensive livestock farming, based on the movement of herds between the rich pastures of the northern pastoral areas (in rainy season) and those of the southern regions (in dry season) according to the seasonal availability of water and pasture/fodder (including residual vegetation of cultivated land) (FAO, n.d.). According to FAO, the following types can be distinguished:

1. **Nomadism***:* Nomads are livestock producers who do not farm and only depend on the sale or exchange of their animals and products for food (e.g., Tuaregs and Fulani). Their movements are opportunistic. They monitor pastures and water resources according to a model that varies from year to year, depending on the availability of these resources.
2. **Transhumance:** Is the regular movement of herds between fixed points in order to exploit the seasonal availability of pastures. A characteristic of transhumance is herd splitting, with herders taking most animals for pasture, but leaving the resident community with a core of lactating cows and/or camels (e.g., Masai and Fulani). In West Africa, governments have tried to delimit transhumance corridors and legislate for cross-border mobility.
3. **Agro-Pastoralism:** Describes settled pastoralists, who live in the villages and cultivate enough land to feed their families. They keep their livestock as a valuable asset (herds are usually smaller). The combination of crops and livestock is primarily used to minimize risks. For example, poor harvests provide forage for the animals. Agro-pastoralism is the most present system in our study area.

METHODOLOGY

Research Approach and Selection of Study Area

This study mainly adopted qualitative research approach including both literature survey and interviews. First, our work consisted of consulting literature on the participatory forest management in Benin with a particular focus on its history and different changes that could explain the current conflict situation. Secondly, we inquired a specific Participatory Forest Management Plan (PFMP) to understand its realization process. In addition, based on the focus groups conducted, we were able to analyze the current state of participation of pastoral communities involved in participatory management, taking the case of the Trois Rivières forest, which is a tropical dry forest and has it management plan funded by the Forest and Riparians Territories Management Program (Programme de Gestion des Forêts et Terroirs Riverains – PGFTR, in french).The PGFTR has a national scope. Nevertheless, drawing on the work of Djogbenou et al. (2011), three main criteria motivated the selection of the classified forest of the Trois Rivières for our study: (1) it is an officially classified forest by law, (2) it has a participatory management plan and (3) it has a recent co-management experience. The advantage of being a classified forest criterion lies in the existence of public authorities (State) and local populations having established very ancient and daily socio-economic and cultural links with the classified forest to justify a participatory management. Then, the availability of participatory management plan and the experience of co-management of the forest allow us to analyze why pastoral communities do not cooperate in the implementation of the management plan.

The predominant criterion for the choice of the study villages is the presence and involvement of agro-pastoral communities. Another important criterion is the intensity of disagreements between forest rangers and pastoral communities involved in the area. On this, we had discussions with the forest authority which made it possible to retain the villages of Péonga, Korogui, Kidaroukperou in the commune of Kalalé for our interviews and focus groups. As a case study, we concentrate on the northern part of Benin, precisely in the department of Borgou (Kalalé). We used theoretical sampling which consists of a selection process of respondents resulting from grounded theory and intended to allow conceptualization (*Table 1*). Our respondents were chosen according to the core theory that guided our study. Based on the answers we got and the analyses done, we discover new categories of respondents that helped us to establish the relationships among them.

Analytical Framework

Our study made use of the strategic analysis approach of organizations' sociology to describe the relationships that existed between the actors involved in the Participatory Forest Management Plan and determined the reasons why agro-pastoralists oppose the plan. Strategic analysis is a theory of organizational sociology developed by Crozier and Friedberg. It uses a reverse approach to traditional organizational theories of studying individuals in order to understand the general structure of an organization. According to its authors, strategic analysis has as its central object the power relations within organizations, and privileges the strategic choices of social actors. The actors here do not exist outside the system that define his own freedom and rationality that one can use for his actions. But the system exists only through the actor who alone can carry it, give it life, and who alone can change it (Crozier and Friedberg, 1977). The organization is a social construct composed of actors who develop singular strategies, as a concrete system of action where the different strategies of the actors are deployed. Opposed to determinism, this theory insists on the relative freedom that every actor has within the organizational framework and on the sources of power (control of areas of uncertainty) that he can mobilize to optimize his strategy (Desreumaux, 2014).

Table 1. Study sample

	Stakeholder	*Number of respondents*	*Place*
PGFTR Actors	Forest Rangers	03	Parakou, Kalalé, Dunkassa
	Agro-pastoralist's representatives	02	Kalalé
	Local Administration	01	Péonga
	Agro-pastoralists	07	Korogui
	Timber loggers	07	Kidaroukperou
	Farmers	05	Kalalé, Péonga
Total	--	25	--

Source: By authors using data from field work

Figure 1. Map of the study area

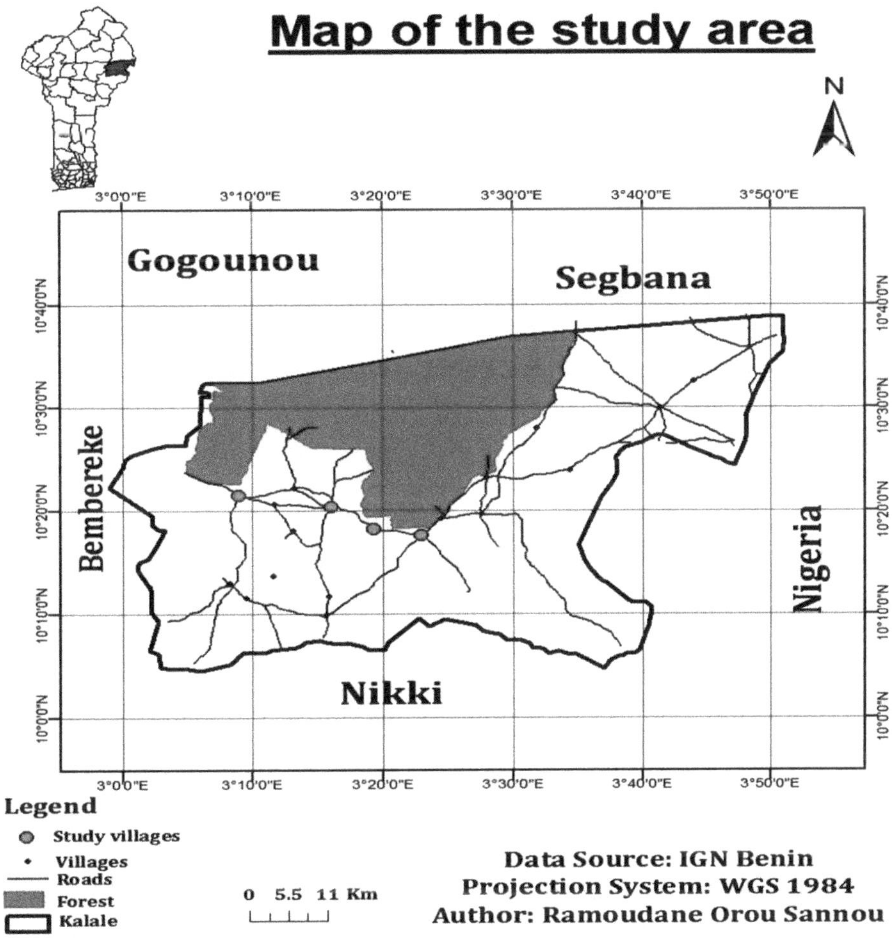

Moreover, with reference to the strategic approach, a theoretical framework for environmental management analysis emerged in the 1980s: the Strategic Analysis of Environmental Management that elaborate on the basis of very diverse case studies, a theory of environmental management practice (Mermet, 1991, 1992). This theory adopts a systemic view of management situations that finds its source in Crozier and Friedberg's (1977) systemic analysis on the one hand, and in Mintzberg's (1989) on the structure and dynamics of organizations on the other. Our use of strategic analysis will consist in placing the PFMP, we are studying in a context of action systems. It will consist in elucidating the organization constitutes by all the involved actors, clarifying relationships that emerge as well from a social point of view (i.e. actors, rules, stakes) and ecological aspects (i.e. animals, plants, environments, etc.) (Mermet et al, 2005).

RESULTS

History and Impacts of Participatory Forest Management in Benin

In Africa, community forestry in different forms has been introduced during the 1960s and 70s. At that time, the trend was to establish 'community woodlots' around Sahelian villages so that pressure on remaining scarce tree cover for firewood supplies could be reduced. These projects turned out to be less than successful because of ill-defined management plan. Nonetheless, long-term thinking had begun in recognition of the fact that local forests were disappearing as population needs for pastureland, arable land and fuel wood were growing. It became apparent that government-reserved forests could not satisfy these needs for all rural people without some kind of management plan. One of the first documented experiences of developing a management plan under the heading of 'community-based management of natural forests' in Africa came from Niger in 1983 (Polansky, 2003).

In Benin, the power of control over forest resource management changed sides from one actor to the other before an inclusive model was adopted. Indeed, during the pre-colonial period, forest management was placed under the authority of land chiefs (MEHU, 2012). Most of the country's protected areas were created during the colonial period between 1940 and 1960. About 59 protected areas were created covering 2,179,418 ha, which represents about 20% of the country's total area. The colonial administration established them by removing rural lands and putting them under governmental control without the consent of the local communities whose farm lands were expropriated. From their creation until early 1990s, these protected areas were solely managed by government officials. Two major events triggered participatory management of protected areas in Benin. First, in 1990 Benin engaged political and economic liberalization policies that led to the withdrawal of the State from the management of several economic sectors. Then, the Rio de Janeiro Summit held in 1992 recognized the importance of environmental degradation and local communities' involvement in natural resources management. Afterwards, Benin's government issued the new forest law No 93-009 on 2 July 1993 and the adoption of a new forestry policy in November 1994 (later revised in 2011), which opened the management of the protected areas to local communities (Idrissou, 2012). The participation of local populations in forest management then became a necessity to protect forests and achieve socio-economic development.

Accordingly, several projects and programs for participatory management of protected forests have been initiated and implemented. These are the Bassila Forest Resources Restoration Project (Projet de Restauration des Ressources Forestières de Bassila - PRRF) from 1988 to 2003, the Natural Resources Management Project (Programme de Gestion des Ressources Naturelles - PGRN) from 1992 to 1998, the Participatory Management of Natural Forests and Village Reforestation for Carbon Reduction Project (Projet Aménagement Participatif de forêts naturelles et reboisement villageois pour réduction de carbone - BEN-93-G13 Project) from 1993 to 1998, the Agoua, Monts Kouffé and Wari-Maro Forest Massif Management Project (Projet d'Aménagement des Massifs Forestiers d'Agoua, des Monts Kouffé et de Wari-Maro - PAMF) from 2002 to 2007 and the Forest and Riparians Territories Management Program (Programme de Gestion des Forêts et Terroirs Riverains - PGFTR) which started in 2003. The start of participatory management of the classified forests of Tchaourou - Toui - Kilibo in November 1996 marked the commencement of participatory forest management in Benin.

The evolution of participatory forest management informs us of two notable changes (*Figure 2*).

First, the expropriation of land from local communities which began with colonial governments and lasted for half a century (1940 to 1990) has had undocumented consequences on the current relations between forest administration and local populations. In fact, many conflicts emerged between both stakeholders because the protected areas were established without the consent of land chiefs and the riparian populations. The only use rights granted to riparian populations were limited to collecting dead wood and gathering fruit. A strict protection approach, supported by repressive methods was promoted to protect the forests. However, the very limited human and material resources of the Forest Administration have not really ensured the protection of classified forests and rather subjected them to various degradation factors. The second major change occurred after the 1990s and was related to the introduction of community participation in the protection of natural resources. Here again, the management approaches promoted have not met the set expectations. Studies conducted in that sense revealed that the lack of success of the forestry reforms of the early 1990s was mainly due to the failure to implement an adapted participatory approach to manage the protected areas (MDR and PGFTR 1999; Siebert and Elwert 2004). Like other forests in Benin where the participatory management approach has been implemented, the result after eight (08) years of implementation in the Trois Rivières forest is that timber resources are still illegally logged for the timber market and charcoal production, farmers continue to expand their farms deeper into the protected areas, and the pastoralists are still using the forests as grazing areas, with little respect for the regulations. Assuredly, these activities violate the agreements set during the project implementation phase. Worse still, people involved in the management of the project are equally and heavily involved in indiscriminate logging activities.

Participatory Forest Management and Agro-Pastoral Communities: The Case of the Tropical Dry Forest of Trois Rivières in Benin

Benin is a West African country located in the so called ''Dahomey Gap'' which is a dry corridor. The Dahomean Gap has particular environmental conditions, making countries located in that position drier than the other coastal countries of West Africa and consequently, the rainforest belt from Guinea to Cameroon skips Benin and Togo (Siebert and Elwert, 2004). In Benin, protected areas have been created by the colonial administration to avoid the complete depletion of ecosystems and protect very limited vegetation (Idrissou, 2012). The Trois Rivières classified forest is one of the largest protected forest in Benin with an area of about 259,600 ha. It is located in the northeast of Benin between 10°20' and 10°50' N latitude and between 2°45' and 3°40' E longitude. Given it geographical location and the type of vegetation it includes, the Trois Rivières classified forest is considered as a tropical dry forest. In essence, it is located in the northern region of Benin where a dry tropical climate prevails with a dry season from November to May, and a rainy season from June to September. It takes its name from the ''Bouli'', ''Tassiné'' and ''Sota'' rivers, and these rivers make the forest a preferred site for herd transhumance during the dry season (Toko Mouhamadou & Ozer, 2007). The forest is straddling the departments of Borgou and Alibori and has as riparian districts *i)* in the North-West, Gogounou, *ii)* in the North-East, Ségbana, *iii)* in the South-West, Bembereke and *iv)* in the South-East, Kalalé. Its cover is made up of 1.32% of dense forest, 41.55% of tree and shrub savannah, 35.40% of open forest and wooded savannah, and 14.65% of mosaic of crops and fallow land. The gallery forest covers 5.63% of the total area (MEPN, 2010).

In 2010, the World Bank-funded Program for Forests and Riparian Lands Management (*Programme de Gestion des Forêts et Terroirs Riverains-PGFTR*) established a Participatory Forest Management Plan (PFMP) for the Trois Rivières forest. The main objective of this 10-year plan was to involve local communities in sustainable forest management while promoting their own socio-economic develop-

Figure 2. Historical profile of forest management in Benin (Source: Authors with data from literature survey)

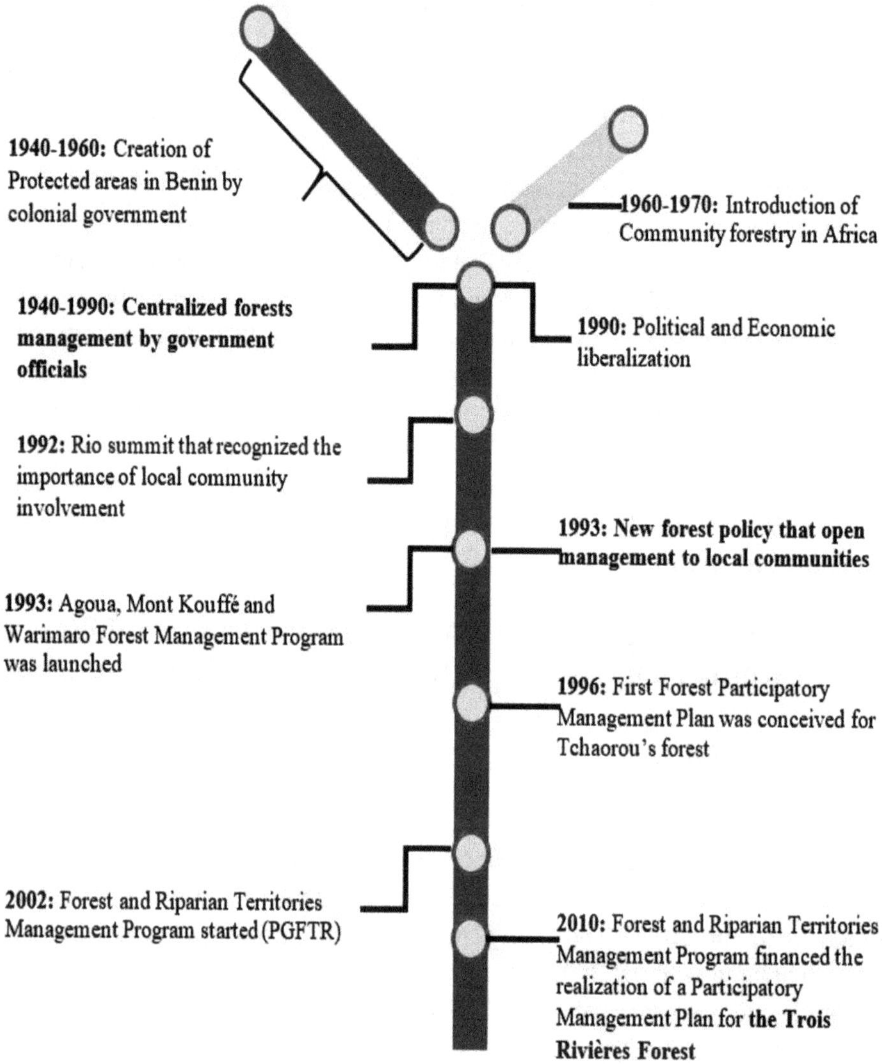

ment. According to that plan, the Trois Rivières protected forest is composed of three different zones. First, there is a buffer zone, which is a band of the protected forest estate surrounding the protected area. There are also peripheral areas, which are the areas outside the buffer zone, including the territories of all the bordering villages and hamlets. Finally, there is the rangeland, which is the forest area open to herd pastures for grazing or for watering. As the global trend is to encourage Payment for Ecosystem Services (PES) as suitable economic instruments to maintain ecosystem functions and services by rewarding benefits through payments and markets (BfN, 2012), the participatory management system in place in the Trois Rivières forest requests both sustainable exploitation and monetary contribution from local communities. After having shown the reason behind such kind of management system through its historical evolution, we wonder how it impacts local communities' livelihood and why pastoral communities refuse to participate in such a management process. As an entry point into the discourse, we will look at the relations among stakeholders involved in the PFMP.

Stakeholders and Their Role in Managing the Forest

The Participatory Forest Management Plan (PFMP) of the Trois Rivières classified forest defines eight types of actors, namely: Timber loggers; Hunters and Fishermen; Pastoralists, Non-Ligneous Forest Product (NTFP) Operators and Tourists, Private plantation owners and Farmers. For the purposes of this study, we consider these actors as stakeholders instead of being simple actors. "Stakeholder" refers to the individuals or groups of individuals who not only exercise strategic actions but also have a stake in the project (Figure 3).

Most stakeholders identified here have a stake (whether it be positive or negative) that may interact with that of Agro-pastoralists and are carrying out strategic actions to protect their interests. Below, we provide concise descriptions of some of these important stakeholders.

1. **Forest Rangers:** foresters are the State's representatives in the community. They are responsible for coordinating with grassroots stakeholders, implementing the PFMP and mainly ensuring forest protection. They encounter enormous difficulties on the ground because of the non-respect of the PFMP rules by certain agro-pastoralists. The latter sometimes settle beyond the established limits (in the buffer zone) or prune trees in the forest during the dry season to feed their animals. This kind of exploitation is in contradiction with the spirit of participatory management, especially when the agro-pastoralists do not first pay for a forest pasture exploitation permit.
2. **Agro-Pastoralists:** Agro-pastoralists use the forest pastures to feed their livestock. Most of them settled in riparian villages or camps around the forest and farming is their secondary activity. The reason given by the agro-pastoralists to justify the pressure exerted on the forest is the multiplication of extreme climatic events, particularly long dry seasons. Because of the lack of fodder in times of drought, they are forced to bring their animals into the classified forest. In fact, the PFMP allows them to bring their animals into the classified forest, but this requires a permit beforehand. Only a few agro-pastoralists pay for the permit because they feel they do not need to pay to use nature's products. Most agro-pastoralists are against the participatory management plan, some use aggression to defend themselves from forest rangers. Others, however, try to corrupt them in vain.
3. **Agro-Pastoralists' Representatives:** These are agro-pastoralists who are members of the communal participatory forest management committee. They are mostly successful agro-pastoralists who have left the villages to settle in the central district and are often contacted to represent agro-pastoralists in decision-making. They are torn between defending their agro-pastoral comrades in conflict against forest rangers and preserving their commercial activities, being close friends of forest rangers.
4. **Farmers:** They occupy riparian cropping areas in the forest and mostly grow crops and cotton. Some farmers also used to occupy cultivation areas within the classified forest. However, with the intervention of foresters, most have released these areas, complaining of declining soil fertility.
5. **Timber Loggers:** Their activity consists of cutting trees to produce timber. Timber harvesting in the Trois Rivières forest should be carried out under conditions clearly stated in the PFMP. Some of the operators have declared themselves to the forestry administration. They are referred to as ''Registered Loggers''. Others, on the other hand, carry out their activity clandestinely: these are

the "Undeclared Loggers". Registered Loggers work with foresters to harvest timber, and consequently formed an alliance against Undeclared Loggers.

6. **Local Administration:** These are the first administrative officials in the area, the District Chiefs. They act as intermediaries between foresters and their constituents. Local elected officials are aware of the antagonism between foresters and Agro-pastoralists and mediate between the two groups. They are hesitant between the desire to protect the forest and that of defending the interests of Agro-pastoralists who constitute their electorate.

Following an analysis of relevant stakeholders, we were able to identify those who have interests in upholding the PFMP objectives, and others who are less concerned. Depending on whether they have a stake for or against the project, each stakeholder takes action to defend their interests. The stake here is what the actor has to gain or lose with the implementation of the PFMP. The case of the stakeholders involved in the Trois Rivières forest is summarized in figure 3. Indeed, foresters are strongly involved in the implementation of the project, not only as civil servants, but also for the perceived ancillary benefits, especially through logging, which is a very profitable business in the country. As far as declared timber loggers are concerned, they are torn between the desire to continue their activities either clandestinely - in order to cut as much wood as they want - or to grasp the opportunity offered to them by the foresters to exploit the wood but in limited numbers. Undeclared loggers, on the other hand, still operate illegally and are therefore opposed to the objectives of the PFMP. Farmers have mostly moved away from the buffer zone and now have their fields in the crop zone. In all the cases mentioned above, the stakeholder groups concerned continue their activities without being stopped by the PFMP. On the contrary, agro-pastoralists mentioned that they have no alternative but to continue to exploit the forest to feed their animals and therefore are opposed to participatory management. According to them, the forest is a gift of nature that belongs to all. The local elected representatives, especially the district chiefs, are torn between defending the interests of the riparian populations (including agro-pastoralists) and at the same time concerned with the preservation of the forest. They mediate between foresters and agro-pastoralists, trying to convince farmers to pay for exploiting forest resources. Following the analysis of agro-pastoralists' discourses, two main causes - linked to power relations between actors - emerge to explain their refusal to participate in forest management. At first, they see themselves as powerless against foresters, since foresters represent the National authority. These excerpts testify to it:

"We have no strength. Foresters represent the government and we can do nothing against them."

"Our weakness in the project is that we didn't go to school, so we often have to accept everything the foresters tell us."

"During meetings we have nothing to impose, all we have to do is listen to the rules of the foresters. We have no rules other than those established by the foresters."

The second cause of non-participation of the agro-pastoralists in the implementation of PFMP is rather related to the frustration they feel due to the increasing development of logging activities. For them, timber loggers have sufficient means to pay their contributions (fees) and continue to operate freely while farmers and agro-pastoralists appear to be victims of oppression. This frustration can be noted in the following statement:

Figure 3. Stakeholders' map (Source: field data)

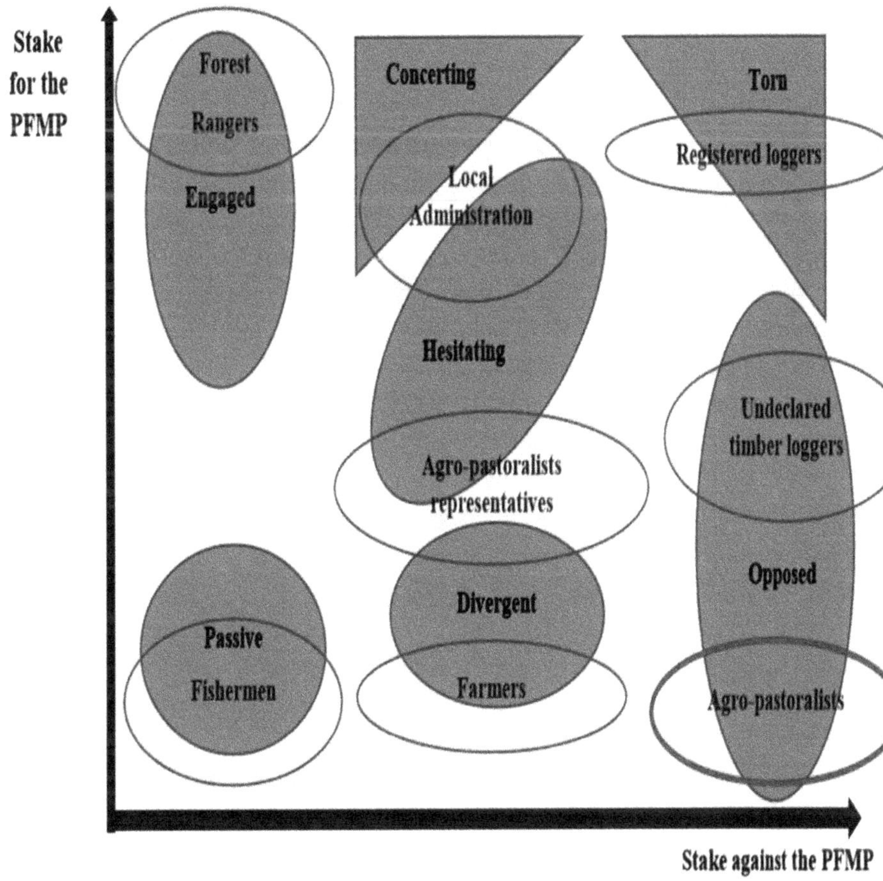

"Loggers are not our friends [...], unfortunately we can do nothing against them because logging is done in agreement with the forest administration and they are also close friends of district chiefs"

Thus, it was noted that agro-pastoralists have little motivation to participate in participatory management because for them, the PFMP has "made access to pasture difficult". They oppose to the plan because they feel aggrieved while some actors still benefit from forest resources. This is compounded by the feeling of inability to act against the public authority represented by forest rangers.

Methodological Flaws Associated With the Management of the Trois Rivières Forest

The elaboration process of the PFMP was led by the Forest Authority and involved passive participation of local communities. Passive participation is defined as participation where the local population is told what is going to happen or what has already happened. Sometimes, participation is done through information, which consists of involving the population by asking them questions without them being able to influence the process. Thus, in the case of the PFMP of theTrois Rivières classified forest, the participation of local populations - including agro-pastoral communities - in the elaboration of the plan was done through awareness raising campaigns whose sole objective was to make them adhere to the ideals of the plan. The overall approach to developing the PFMP - as stated in the document itself - has been to consider the forest as *"a business to be managed sustainably"* while respecting the existing ecosystem. This conception of the forest already suggests a capitalist approach to its management as entrepreneurial management is often profit-oriented. This philosophy can be considered as the basis of the introduced regulation that obliges all farmers to pay a contribution in order to have access to forest resources, notwithstanding the capacities of some actors such as agro-pastoralists as well as their sociological perception of the forest.

Furthermore, given the current implementation of the plan, we have noticed that the Village Forest Management Committees (VFMCs) that have been created to involve communities in forest management are inactive and therefore do not influence decision-making. Consequently, decisions are taken unilaterally by the forest authority. In some cases, the leaders of the agro-pastoral community residing in the central district are invited to serve as representatives for decision-making. Also, according to the foresters currently working at the forest office, organizing awareness campaigns for agro-pastoral communities is a method that allows them to get involved in the management process. On the other hand, the agro-pastoralists consider these awareness sessions as moments of "call to order" and maintain that their voices are never taken into account.

Also, they think that sending representatives from their community to make decisions is not enough to have their opinions taken into account. Since these representatives live in the city, they rarely report to the large mass of agro-pastoralists living in villages. These are some of their statements:

"We participate in awareness sessions, but these meetings are made to ask us to leave the forest surroundings and not to bring our cattle there."

"Sometimes we send our delegates to the meetings and they are supposed to often come and report to us, but as they live in the city, and most of them are traders, they don't have time to report to us, they don't care about our problems."

In the context of the classified forest of the three rivers, the foresters and the declared timber loggers, have reached a fate of alliance allowing the timber operators to continue their activity and the foresters to benefit equally. This causes frustration among agro-pastoralists and accentuates their antagonism towards the PFMP. Indeed, agro-pastoralists claim that due to extreme climatic events, especially long droughts, it is difficult to find enough fodder for their animals in the rangeland. Thus, they are forced to bring their herds to graze in the classified forest. However, they do not understand why they must first pay for a permit to exploit the natural heritage of their living environment. In short, agro-pastoralists refuse to pay monetary contribution for a participatory plan that benefits other stakeholders and not them. This can be deduced from the following statements:

"The foresters' project has no advantage over us. They're asking us to pay for access to the forest. This forest belongs to us all."

It also appeared that although agro-pastoralists are aware of the regulations brought by the PFMP, they are unaware of the boundaries of the classified forest. In fact, there is real confusion in determining the boundaries between the buffer zone and the cultivation area. This can be noted through the following excerpts:

"We helped reforest the forest. We thought the foresters were going to leave us this part but now they say that it is in the classified forest and that we must no longer cut these trees. Then we're asked to pay 200 FCFA (less than 50 cents USD) per head of oxen to bring our animals into the forest but we don't know what the money is for."

"A few years ago, they showed us where we could do our farms but today they still say that we are in the classified forest and we have to leave the place"

"They arrest people amongst us every day, even some of those who have paid their contribution in past years."

In summary, agro-pastoralists outrightly decided not to participate in participatory management or even to go against the objectives of the PFMP for three reasons. First, they feel alienated from the participatory process and acknowledge that they only received information about the rules to follow. This cause finds its source in the inactivity of the VFMCs and the methodological defect in the design of the PAP, which involved the population only through information campaigns. The second reason is related to the logging activities. Indeed, the loggers are considered by the agro-pastoralists as the owners of the forest, next to the foresters: ***"they continue to harvest timbers while we are denied access to the forest"***. According to agro-pastoralists, this point contributes to the third cause, which is rather linked to the scarcity of pasture due to long droughts. They ventilate that the species of trees cut down by loggers are fodder for their animals. It can therefore be assumed that logging activities in the classified forest of the three rivers considerably reduces fodder for animals in periods of drought.

Participation of Agro-Pastoralists in the PFMP of the Trois Rivières Forest

Following our analyses, the reasons that explain the refusal of agro-pastoralists to participate in the implementation of the PFMP can be categorized into three groups (*Table 2*). These are the historical reasons, those linked to the institutional set-up of the plan and those linked to the current conflicting relations with the other stakeholders. As demonstrated above, the various changes observed in forest management history in Benin have not only accentuated the impoverishment of local populations, but have also encouraged them to further encroach into the protected forest areas, especially after the 1990s. Likewise, the current setting of management that is being implemented is responsible for many disagreements, making it difficult for pastoralists to sustainably benefit from the forest. In fact, it is important to mention that the institutional framework of forest management provides for the existence of a Participatory Forest Management Committee in charge of monitoring the strategic orientations of the Participatory Forest Management Plan (PFMP). At village level, there are Village Forest Management Committees (VFMC) seconded by village sub-committees responsible for monitoring exploitation of each forest resource and serving as link between the forest administration and villagers. However, our discussion with the pastoralists showed that these committees are essentially inactive and almost non-existent. This lack of a viable communication channel potentially explains the persistence of some misunderstandings. As matter of fact, it is known that pastoralists are often present in the area because of its potentialities

Table 2. Rationales for the non-participation of agro-pastoralists in participatory management

Key issues	Consequences
Historical background	
Autocratic forest management from the colonial period to the 1990s	Impoverishment of riparian populations including agro-pastoralists
Introduction of the participatory management approach in the 1990s without adequate means for its implementation	Rush of populations towards the forest leading to its accelerated destruction due to lack of necessary means to effectively operate participatory management
Methodological shortcoming of the PFMP	
Non-active involvement of local communities in the development process of the PFMP	Inactivity of Village Forest Management Committees (VFMCs)
Introduction of a management system unsuited to the socio-economic context of local communities	Refusal of payment of participatory management contributions by agro-pastoralists leading to conflicts
Stakeholders relations	
Alliance between foresters and certain timber loggers leading to an increase of logging activity	Reduction of forage resources and frustration of agro-pastoralists who refuse to pay their monetary contribution in order to access forest resources
Hegemony of the forest administration on all other actors, due to the inactivity of participatory management committees	Potential driver of the rebellion of agro-pastoralists leading to violent conflicts

in terms of pastures and waterholes. They have settled down, practicing agriculture (agro-pastoralism) even when they have no legal right over the land because of their past mobility tradition. In addition, many farmers in the riparian villages are facing problems of declining soil fertility and climate change, leading them to make new farmlands in the forest without permission. Consequently, there is a strong pressure on the forest's resource system, leading to changes in land use patterns. Similarly, timber logging is a dominant activity around the forest and involves actors at various levels. This combined with the inactivity of local monitoring subcommittees, as stated above, to accentuate the disagreements between the different actors involved in the management of the forest. Although the Participatory Forest Management Plan (PFMP) reserves exclusive access for natural resources exploitation only to local population, they are still required to obtain access permits and pay fees for receiving ecosystem services provided by the forest. For instance, herders are required to pay 300 FCFA (around 0.50 USD) per head of cattle brought to forage in the forest. Procedures for obtaining permits and paying operating fees should normally involve participatory management committees at various levels. Here again, the inactivity of these committees leads to domination of Forest Rangers over other actors. As a result, agro-pastoralists no longer contribute in the participatory management process because they fail to understand why they have to pay for access to the forest's natural resources while timber logging activity is concurrently increasing. They believe that the pressure exerted by timber harvesters mixed with extreme climatic events (long dry seasons), greatly reduces forage for their livestock and as a consequence affects their livelihoods.

How to Address the Current Situation?

Based on our results, three major sources of problems were identified vis-à-vis the conflicting situation among pastoralists and foresters in the three rivers context. As far as historical causes are concerned, it would no longer be possible to go back in history to address them. However, solutions can be found to the methodological flaws of the plan and the current conflict situation. As a first step, it would be important to conduct a quantitative study to assess the socio-economic effects this plan has inflicted on the agro-pastoralists' livelihood and then identify appropriate corrective actions. Taking into account the identified problems pertaining to participatory forest management, we propose two possible remedies. The first relates to the governance aspects while the second is linked to the involvement of agro-pastoralists.

1. Re-Adapt the Management Plan to Local Socio-Economic Context

Considering the current management context, foresters are the decision-makers on all aspects of management, making it necessary to have active community management committees that balance power relations between local communities and foresters. Also, public participation must be preferred over appointing agro-pastoralist representatives for the decision-making process. Sanctions should be designed according to locally relevant procedures and the use of imprisonment should be avoided as much as possible. A sanction mechanisms must also be enforced for foresters involved in illegal logging. Furthermore, since it has become evident after discussions with the agro-pastoralists that the boundaries of the buffer zone are not clearly defined, we suggest a redefinition of those boundaries. Riparian communities and mainly agro-pastoralists should be involved in planting local tree species to allow biological materialization of buffer zone boundaries. Prospective tree species that could be planted in the buffer include *Tectona grandis, Vitellaria paradoxa, Khaya Senegalensis* or better, use species whose foliage are appreciated by cattle, namely *Acacia albida, Daniellia oliveri, Lophira lanceolala.*

2. Encourage Agro-Pastoralists to Preserve the Forest

Several actions can be taken to encourage agro-pastoralists to participate effectively in the participatory management. Indeed, the PFMP includes a component that should in principle support closest communities in the development of income-generating activities. Unfortunately, the implementation of this component is not yet effective. To reduce pressure on the resource, it will be beneficial to orient and train agro-pastoralists on income-generating activities that they can combine with pastoralism. Also, it would be important to train the latter on soil fertility management techniques to avoid opening of new farmlands in the forest as the decline in soil fertility has become a major problem in the region. Finally, by considering the multiplication of long dry seasons leading to lack of fodder in times of drought as one of the reasons given by the agro-pastoralists to justify the pressure exerted on the forest, it would be helpful to create community pastures for grazing or for watering. Those community pastures might be managed as common pool resources in order to facilitate grazing during the dry seasons and avoid further forest degradation.

CONCLUSION

The participatory management of forests in Benin essentially fails to ensure active involvement of riparian communities' right from the onset. This has to some extent contributed to the conflicting situations observed between the government agents in charge of forest protection and agro-pastoralists. Consequently, forest resource degradation is increasing and the participatory management plan in place tends to create inequality in the access to resources. This study tried to understand why agro-pastoralists communities refuse to take part in the management of the Trois Rivières classified forest. Through the use of strategic analysis approach, we were able to determine the key elements of the established PFMP that block agro-pastoralists' participation in forest management. Indeed, it appears that the active or passive participation of agro-pastoral communities from the design of programs often determines what happens next. In the studied case, the village committees in charge of leading the participatory process are inactive due to the absence of clear bases from their constitution. Similarly, sending representatives during the design phase of the management plans has proven to be ineffective as a way of involving or engaging these communities. Moreover, the creation of alliances between stakeholders with relevant assets not only has negative consequences on the participation of agro-pastoralists but also on the resource. In addition, we found that setting clear management rules and limits are critical to ensuring agro-pastoralists' participation in the PFMP. Although these theoretical results need to be tested in future quantitative research, we conclude that the non-active involvement of agro-pastoralists in the design of natural resource management policies and the development of unexpected relationships in their implementation are the main issues of concerns in the institutional design of PFMP. We propose that these issues be taken into account in order to ensure a better participatory forest management and avoid its perverse effects on indigenous communities, whose livelihoods depend heavily on natural resources.

REFERENCES

Arouna, O., & Djogbenou, C. P. (2006). Evaluation du plan d'aménagement participatif des forets classées de goun-goun, de la Sota et de la rôneraie de Goroubi au Benin : critères et indicateurs pertinents de réussite. In the *Colloque International sur la Gestion Concertée des Ressources Naturelles et de l'Environnement*, Cotonou, Benin.

Benhammou, F., Baillon, J., Senotier, J.-L., Chalbert, J.-P., Lécrivain, E., Meuret, M., & Reynes, A. (2005). Pastoralisme et biodiversité. *La Voie Du Loup*, *22*, 10–23.

BfN. (2012). Payment for Ecosystem Services Towards an Implementation Strategy. Florian Carius (Ed).

Blanfort, V. (2011). Impacts et services environnementaux de l'élevage en régions chaudes. In Elevage en régions chaudes. Inra prod.

Bonfiglioli, A. M. (1990). Pastoralisme, agro-pastoralisme et retour: Itinéraires sahéliens. *Cahier Des Sciences Humaines*, *26*(1–2), 255–266.

Carriere, M. (1996). *Impact des systemes d'elevage pastoraux sur l'environnement en afrique et en asie tropicale et sub-tropical aride et sub aride*. CIRAD.

Carriere, M., & Toutain, B. (1995). Utilisation des terres de parcours par l'élevage et interactions avec l'environnement. Montpellier, France: CIRAD.

Crozier, M., & Friedberg, E. (1977). L'acteur et le système (Editions du seuil). Paris, France.

Desreumaux, A. (2014, May). Présentation générale des théories des organisations. iae LILLE Ecole Universitaire de Management. Retrieved from bricks.univ-lille1.fr

Dia, A., & Duponnois, R. (Eds.). (2013). Le pastoralisme en Afrique subsaharienne. In La Grande Muraille Verte : Capitalisation des recherches et valorisation des savoirs locaux (pp. 12–31). Retrieved from http://books.openedition.org/irdeditions/3336

Djogbenou, C. P., Glèlè Kakaï, R., Arouna, O., & Sinsin, B. (2011). Analyse des perceptions locales des aménagements forestiers participatifs au Bénin. VertigO - La Revue Électronique En Sciences de l'environnement, 11(1).

Dumortier, P., Rabier, F., Backers, Y., Vanlierde, A., Jérôme, E., & Mathor, M. (2013). Elevage et gaz à effet de serre : le bilan des émissions de l ' animal à la filière. CRA-W et GxABT-Carrefour Productions Animales, pp. 41–52.

FAO. (2009). La situation mondiale de l'alimentation et de l'agriculture, le point sur l'élevage.

FAO. (2016). *State of the world's forest.* Rome, Italy: FAO.

FAO. (n.d.). Pastoralisme et gestion des parcours.

Hiya Maidawa, M., Andres, L., Yamba, B., & Lebailly, P. (n.d.). Mobilité pastorale au Sahel et en Afrique de l'Ouest: essai de synthèse.

Idrissou, L. (2012). Dynamics of Conflict in Participatory Forest Management in Benin. A Framing Perspective [Doctoral thesis]. Wageningen University, The Netherlands.

Landais, É. (1994). Système d'Elevage D'une intuition holiste à une méthode de recherche, le cheminement d'un concept. In *Blanc-Pamard et Boutais (1994): A la croisée des parcours: Pasteurs, éleveurs, cultivateurs (OSTROM). In Dynamique des Systèmes agraires (OSTROM).* Paris, France: ORSTROM.

LEAD/FAO. (2006). L'ombre Portée De l' élevage, impacts environnementaux et options pour leur atténuation.

Marty, A., Bonnet, B., Guibert, B., & Swift, J. (2006, July). La mobilité pastorale et sa viabilité, entre atouts et défis (No. 3). IRAM.

MEHU. (2012, October). Cadre fonctionnel du Programme de Gestion des Forêts et Terroirs Riverains (PGFTR).

MEPN. (2010). Plan d'Aménagement Participatif de la Forêt classée des trois rivières 2010-2019.

Mermet, L. (1991). *Dans quel sens pouvons-nous gérer l'environnement?* (pp. 68–81). Annales Des Mines/Gérer & Comprendre.

Mermet, L. (1992). *Stratégies pour lagestion de l'environnement: lanature comme jeu de société?* Paris, France: L'Harmattan.

Mermet, L., Billé, R., Leroy, M., Narcy, J.-B., & Poux, X. (2005). L'analyse stratégique de la gestion environnementale : Un cadre théorique pour penser l'efficacité en matière d'environnement. *Nature Sciences Sociétés, 13*(2), 127–137. doi:10.1051/nss:2005018

Ostrom, E. (2009). A General Framework for Analyzing Sustainability of Social-Ecological Systems. *Science, 325*(5939), 419–422. doi:10.1126cience.1172133 PMID:19628857

Polansky, C. (2003). Participatory Forest Management in Africa: Lessons not learned. *International Journal of Sustainable Development and World Ecology*, *10*(2), 109–118. doi:10.1080/13504500309469790

Schreckenberg, K., Luttrell, C., & Moss, C. (2006). Participatory forest management: An overview. Forest Policy and Environment Programme: Grey Literature.

Toko Mouhamadou, I., & Ozer, A. (2007). Évolution de l'occupation du sol dans les zones périphériques de la forêt classée des Trois Rivières (Bénin) entre 1949 et 1986. In Quelles aires protégées pour l'Afrique de l'Ouest? Conservation de la biodiversité et développement (IRD).

Chapter 12
Issues and Challenges for Stone Mining Affected Forest Area Restoration Through Probiotic Interventions

Kumud Dubey
https://orcid.org/0000-0001-6367-4617
. Forestry Research Centre for Eco-Rehabilitation, India

K. P. Dubey
Uttar Pradesh Forest Department, India

ABSTRACT

The sandstone quarrying and mining in Vidhyan region of Uttar Pradesh has severely devastated the floral biodiversity of the adjacent forest area. It is necessary to conserve these areas for the protection and sustainable use of forest resources. In most of the conservation program, microbial deficiency creates problems in establishment of the vegetation and delays the natural succession. Therefore, probiotic interventions were applied to conserve these areas promptly. The probiotic beneficial microbes' interventions, due to their multifarious beneficial characters, may facilitate the upcoming of flora through enrichment of the soil, better nutrient absorption, providing resistance against different stresses, etc. Probiotic interventions may positively impact the conservation of floral diversity and restoration of stone mining-affected forest area.

1. INTRODUCTION

Mineral deposits are one of the major natural resource which can be used beneficially for the humankind. They are regarded as non-renewable resources and are mainly used by man for (a) materials for industry (b) sustenance of life and (c) energy. The mining is an age-old economic activity, though its nature and form has been changed over times. The dependence of primitive societies upon mined products is illus-

DOI: 10.4018/978-1-7998-0014-9.ch012

trated by the nomenclature of those epochs like Stone Age, Bronze Age and Iron Age, a sequence which also depicts the increasing complexity of society's relationship with mining. In a sense, the history of mining is the history of human civilization (Khoshoo, 1991).

It is a vital sector of economy for all countries. In India, an area of about 683,671 ha is under mining leases in twenty one (21) states, maximum number of mines being in the states of Rajasthan, Jharkhand, Bihar and Orissa. Major share of the mineral production comes from the open cast mining all over the world and about 75 percent of the world mineral output comes from open cast operation (Dubey, 2010, 2018). Mining affects the environment in adverse ways depending on the type of material excavated, method of mining, surface overburden and mine spoils etc. The resultant environment is not suitable for any productive use, be it agriculture, forestry, pasture or recreation. (Saxena and Chatterji, 1988 a, 1988 b). Major environmental impacts of Mining Activities are sinking of land, removal of vegetation or deforestation, pollution hazards, change of land capability class, changes in hydrological pattern, overturning of soil-substratum sequence, degradation of land etc. In addition to these impacts, the mining operations lead to various sociological changes and adverse visual impacts (Khoshoo, 1991).

STONE MINING

There are an estimated 1,000 million tons of stone in India. India accounts for around 27% of the total natural stone production of the World. The Sandstone quarrying and mining in Vindhyan region, consisting mainly of tropical dry deciduous forest is mainly opencast type and carried out predominantly by private parties as artisanal mines and marketed as Slab stone, Millstone and Building stone and for cement production. The environmental impacts of these quarry operations can be large and far-reaching. Sandstone mining scars the landscape, disrupts ecosystems and destroys microbial communities (Srivastava, 1999; Sharma *et. al.,* 2000; Liu et. al., 2009; Sarma, 2013; Verma et. al., 2014; Chaturvedi and Singh, 2017; Dubey, 2010; 2017; 2018; 2019, Upadhyay et. al., 2016; Jitendra et. al. 2019). Most of the mining areas are associated with forest, this exploration and exploitation of mineral wealth brings in complete destruction of forest and vegetation in the area and formation of new wastelands in the form of barren dumps of mine overburdens with consequential ecological disturbances and environmental hazards.

Major Impacts of Stone Mining

It is a well-established fact that mining, underground or opencast, is an environmentally unfriendly activity affecting five environmental components, namely, society, ecology, land, water and atmosphere (MINENVIS, 2006). The impacts of mining on the environment and people are manifold. It destructs the whole bionetwork. The present intense mining activity at the expense of the environment has resulted in a global awareness towards undertaking studies for providing the means to counter the environmental damage. The indiscriminate and unscientific mining, absence of post mining treatment and management of mined areas are making the already fragile ecosystems more vulnerable to environmental degradation and leading to large scale land cover and land use changes (Bhardwaj *et al.,* 2000). There is a range of negative environmental impacts associated with extractive industries. This is especially true for opencast mining of the sort, used in extracting sandstone in India (Basu, 1986). The greatest impact is the destruction of the habitat where the quarrying occurs (Walker and del Moral, 2003; Walker, 2002 a, b, c). As mining strips the substrate down to the rock layer no biological bequest remains on the site. Major impacts of mining on the environment (Anon, 2007) are identified as:

- Land degradation.
- Degradation of forest and loss of biodiversity.
- Soil contamination
- Air pollution
- Surface and ground water pollution
- Noise and vibrations
- Deterioration of natural drainage system

These effects can have consequences that reach far from the mine site (Liu et al, 2009). Hydrological impacts can include lowering of the water table and pollution of catchments (Ibarra et. al. 2005; Bhadra *et al.* 2007). Soil erosion around and on the site can cause loss of growing substrates reducing the ability of the surrounding area to support plant life and hindering plant colonization and organization on removed substrates (Whisenant, 2005). It also has severe impacts on the economy and food production capacity of agricultural areas (Pimentel *et al.* 1995).

Impact on Nearby Tropical Forest

The main forest types in Uttar Pradesh are Tropical Semi Evergreen (0.21%), Tropical Moist Deciduous (19.68%), Tropical Dry Deciduous (50.66%), Tropical Thorn (4.61%) and Littoral and Swamp forests (2.35%) (India State of Forest Report 2017). Most of stone mines in Uttar Pradesh are located in Vindhyan region which harbor rich forest biodiversity, mainly covering Allahabad, Mirzapur and Sonabhadra districts. Mineral extraction, therefore, has excessively affected forest ecosystems and the surrounding environment and consequentially, the nearby forests. In Vindhyan region most of the forest is tropical dry deciduous type in which floral diversity is severely affected by mining activities. Thus, the impacts of stone mining and mines upon natural ecosystems and biodiversity have become a key environmental concern. In the Vidhyan region, mining is denuding hills, leading to depletion of invaluable forests. The impact of stone mining on floral diversity of nearby forest area was studied by Dubey (2010, 2017) and it was observed that the stone mining affected about 30% of the biodiversity in Allahabad, Mirzapur and Sonbhadra districts (Dubey, 2017). Similar finding was also reported in case of iron ore mining in Bellary (Karnataka Govt, 2015). The impact of damage caused by mining is not only restricted to the areas leased for mining but the areas immediately surrounding the mining lease including adjacent forest area also bear the brunt of mining works. The total extent of the Sandur forests is about 32,000 hectares out of which about 8,000 hectares have been affected by iron ore mining, although the actual area broken up and cleared for mining is much less (about 2,000 hectares) (Karnataka Govt., 2015). Sharma and Chaudhry (2018) also reported that mining affected the plant diversity in tropical thorn forest of Asola-Bhatti Wildlife Sanctuary located in northern India.

Since it is not possible to avoid destroying habitat during the mining operation, options for mitigation lie in restoration of these mining areas Mine spoils left to nature may take decades to centuries to develop any vegetation cover. However, carefully planned artificial interventions that mimic natural processes can reduce this time span (Dobson et al. 1997). (Singh *et. al.*, 2002; Singh *et. al.* 2004 a; Chatterjee and Bose, 2003; Maiti and Das, 2004; Maiti and Ghose, 2005; Maiti, 2006; Zedler, 2005; Grant, 2006, Dubey,

2010, 2018). Singh *et al.* (2004 a) studied the impact of three native tree plantations of two leguminous (*Albizia procera* and *Albizia lebbeck*) and one non leguminous *Tectona grandis* tree species on the soil development process during the early phase of mine restoration on mine spoil in a dry tropical environment. There was a general improvement in soil properties due to establishment of plantations. Plantation adopted as common practice for the restoration of overburden dump of coal mine in Madhya Pradesh and its rehabilitation benefited from the plantation because it allowed jump-starting succession. The catalytic effects of plantation were due to changes in under storey, micro-climatic conditions (increased soil moisture, reduced temperature, proper aeration etc.) and development of litter and humus layers (Singh *et al.,* 2002). According to Chaturvedi and Singh (2017), surface mining extensively alters the physical, chemical and biological characteristics of the soil, and therefore, the re-vegetation of mine spoil through natural process takes a very long time and poses a major problem because nutritionally it is a recalcitrant medium for plant growth. They reviewed important studies on the restoration of mine spoils in the tropical dry region of India.

From the literature it follows that serious and sincere restoration measures based on scientific research findings have not been taken in most of the mined areas. Moreover these restoration operations of the mining areas are not achieving desired success because of high rate of mortality and slow growth of planted species. For sustainable restoration of degraded forest area due to mining, there are following issues which should be properly addressed, before implementing the project.

ISSUES IN RESTORATION

The main issues in mined land restoration are local vegetation, site characteristics and suitability of planting species which has to be targeted during restoration In addition to this, the socio-economic status of the local people also plays an important role in sustainable reclamation of a site and without their active participation; sustainability of a reclamation programme is not possible. The nutrient deficiency, potentiality of erosion, low moisture retention and lack of biological activities are the major hurdles in such restoration. In most of the restoration programmes, microbial deficiency creates problems in the establishment of plants and their natural growth. Soil microbes not only fix nitrogen symbiotically but also microbes like the Phosphate Solubilizing Microbes (PSMs) render unavailable form of phosphorus to available form of phosphorus and thus reduce the dependence on phosphorus from external sources. Successful microbial colonization of substrates poor in microorganisms, such as mining area surface soils, may strongly stimulate vegetation development, as microbial activity releases important nutrients. For sustainable restoration, these issues have to be considered properly.

Mining of minerals causes enormous damage to the flora, fauna, hydrological relations and soil biological systems. It creates dumps of haphazardly mixed infertile, consolidated and unconsolidated materials, also known as overburdens. These overburdens when dumped in un-mined areas in the vicinity of the mines create mine spoils. This nutrient deficient sandy mine spoils are generally hostile to plant growth and the restoration of these mine spoils are major constraints. Due to its physical, chemical and biological properties, it puts several constraints and consequent intricacies for the establishment of vegetation (Dubey, 2010; 2018). These Challenges are briefly described below:

Physico-Chemical Constraints

The main physical problems with mining affected sites are thin soil cover (or often lack of it), large cavities in the very coarse-grained substrate, very high stone content, extremely coarse texture, compaction, and the limited availability of moisture (Pandey et. al., 2005).The physical challenges account for unfavourable porosity, aeration, water infiltration and percolation properties along with high bulk density and absence of proper structural aggregates in the dumps. These mine spoils represent extremely rigid substrata for plant growth and development. Colonization, establishment and maintenance of vegetation on these spoils are very difficult. The physical factors that limit plant establishment and survival include high temperature, moisture stress (Richardson, 1975), soil particle size (Down, 1974) and compaction (Holl, 2002; Richardson, 1975). In the rooting zone (0-25 cm) the physical properties are generally less favourable in mine soils. Mine soils had a higher bulk density and coarse fragments and lower porosity, clay content and water holding capacity than reference soils (Masoodi and Soni, 1999). The colonization of plant species on mine spoils is influenced by the particle size of the soil. The soil with high clay content become water logged whereas with high silt content, the soils become compact forming crust which often restricts seedling growth and entry of water and air into the soil system. The pH is a major determinant in controlling plant growth on impoverished lands such as mine spoils.

In certain areas, the main factor in preventing vegetation from becoming established is acidity. It has been learnt if specific problems hindering ecosystem re-development has been identified, a cure can be designed using natural processes or stimulating the natural nutrient cycle (Maiti, 2003; Singh *et. al.*, 2002, Dubey, 2010, 2018).

The main chemical problems are the lack of nitrogen and phosphorus due to the lack of organic-matter content, low cation exchange capacity, and base saturation (Pandey et. al. 2005). The chemical challenges involve low concentration of essential plant nutrients, poor nitrogen, phosphorus and potassium status, unfavourable pH range, deficiency of calcium, magnesium etc. and very poor organic matter. Lower organic matter levels in mine spoils than that in natural soils have been reported by several researchers (Indorante *et. al,*. 1981; Schafer *et. al.,* 1980; Toy and Shay, 1987). Lower organic matter is deficient in plant nutrients and without some amelioration, any plants grown on these spoils were susceptible to water, temperature and nutrient stresses. Presence of phytotoxic substrates, elevated levels of some elements like Cu, Ni, B etc, deficiencies of minor nutrients are responsible for the paucity of natural colonization (Piha *et. al.,* 1998) and this can only be partially compensated by the addition of fertilizer.

Biological Constraints

Such challenges refer to the absence of beneficial microorganisms which play an important role in restoring physico-chemical and biological properties of soil. The microorganisms contribute to the development of soil structure, synthesis of plant nutrients through the nitrogen fixation and amelioration of adverse chemical and physical limitations. Soil microbial ecosystems are important for the effective restoration of mined lands (Birch *et. al.*, 1991). The presence of soil microbes is essential for the germination, growth and long term survival of most plant species and for the re-establishment of natural ecosystems. Microbial community is adversely affected by mining (Srivastava, 1999). Mining also affects the various biological reactions occurring in soil. For example, the rates of nitrogen mineralisation in mines are significantly reduced in comparison to the native forest (Todd *et. al,* 2000). Mining disrupts the Phosphorous (P) cycle and affects the proportional distribution of P between soil fractions (Short *et.*

al., 2000). Mining severely affects the flow of nitrogen through a stable soil plant microbial ecosystem (Reeder and Sabey, 1987).

RESTORATION OF STONE MINING AFFECTED AREA

The present chapter is based on the study of restoration of the stone mining affected area through probiotic interventions by both ways i.e. *in situ* and *ex situ* (Dubey, 2018; 2019). Innovative strategies opted and recommended for restoration of the stone mining affected area has been described in proceeding section.

Strategies for Restoration

Mining is adversely affecting the biodiversity and productivity of the tropical forest area. Therefore, it is imperative to restore this area (Soni, 2003; Pandey et. al., 2005; Dubey, 2018; 2019). In Mirzapur District, where a restoration study was executed, has total geographical area of 4405 sq.km (km^2). Out of which, 805 km^2 area is mainly tropical dry deciduous forest and accounts for 18.27% of total geographical area (Forest Survey of India report, 2017). Further, an increment of a total 278 km^2 of forest area was reported for Uttar Pradesh which is mainly attributed to plantation activities and conservation efforts. However, in case of Mirzapur district, it was reduced by 33 sq. km which is mainly due to mining (FSI report, 2017).

A holistic strategy for restoration of mining affected forest area has been suggested to create a multifunctional and self-sustainable ecosystem. It is based on the findings of restoration of stone mining affected forest area (Dubey, 2018; 2019). It involves both *in situ* and *ex situ* following innovative restoration strategies:

For ***In situ*** **Restoration** following steps have been recommended:

- Survey of Site and Assessment of Damage
- Involving local stake holders in *in situ* restoration
- Probiotic Interventions
- Establishment of surface covering vegetation
- Broadcasting of indigenous seeds
- Training of Forest Officials.

Ex situ **Restoration:** It incorporates to establish plantation of seedlings inoculated with microbes (probiotics) that may develop into a functional ecosystem in due course of time. It should be recommended preferably within the natural affected areas and the indigenous species were planted at the affected site.

In situ Restoration

Following steps have to be executed for restoration of forest area affected due to stonemining:

Survey of Site and Assessment of Damage

For Restoration, as prelude step we have to survey the area for assessing the damage. Stone Mining scenario was also studied. Vindhyan sandstone mining is spread over covering Mirzapur, Sonbhadra,

Agra, Lalitpur, Chitrakoot, Allahabad and Varanasi districts. Mining is mainly executed by private parties and sold as Slabstone, Millstone and Building stone materials Mining is small-scale cluster type. In almost all cases, it is open cast which involves excavation. It creates an undulating landscape and leads to large-scale devastation of soils, mostly in form of sand and stone crush derived from Dolomite and Quartzite rocks. Mining leaves an unstable land, causing landslides, erosion, siltation, and polluted water. It disturbs ecosystems, scars the landscape, and destroys microbial communities resulting in degradation of forests. Quarries/pit created and artificial hills formed during mining operation such as deposition of overburden are a major problem in the area. We selected a site which was situated in Padari of Mirzapur district, harshly devastated due to stone mining. Soil was poor in moisture content, organic matter, nitrogen and other major and minor nutrients. The soil moisture content (SMC) in mining area was not found to be conducive for agricultural activity. Water holding capacity (WHC) and SMC of mining area was less in comparison to the unaffected forest area. It was mainly due to the coarse sand and sandy part in mining area sample which contributes to major portion of the soil (Dubey, 2018, 2019). Texture analysis and physico-chemical parameters of soil sample of area disturbed due to stone mining and undisturbed area were depicted through Figure 1 and Figure 2.Existing forest of the studied mining affected site was tropical dry deciduous type. The floristic composition of the affected forest area was disturbed due to mining. Plant species diversity at tree, shrub, climber and herb levels was more in unaffected forest of the area as compared to the affected area Number of plant species were reduced drastically at site disturbed due to mining (Figure 3). On undisturbed site, tree species belonging to eight families were reported, whereas on disturbed site, tree species belonging to six families were present. Plants belonging to leguminosae were more in number at undisturbed forest area (Figure 4 and Figure 5). Similar observations were also reported in case of shrubs and herbs (Figure 6, Figure 7, Figure 8 and Figure 9 respectively). At the unaffected site species like; *Boswellia serrata, Madhuca indica, Acacia catechu, Haplophragma adenophyllum, Butea monosperma, Terminalia bellirica, Diospyros melonoxylon, Pongamia pinnata, Acacia nilotica, Azadirachta indica, Lagerstroemia parviflora, Dalbergia sissoo, Albizia procera* (tree species), *Nyctanthes arbor-tristis, Ziziphus nummularia, Vernonia cinerea Solanum nigrum, Sida cordifolia, Indigofera tinctoria, Abrus precatorius, Acacia concinna, Cocculus hirsutus Dandrocalimus strictus* (shrubs), *Xanthium indicum, Tridax procumbens, Semecarpus anacardium, Phyllanthus niruri, Oxalis corniculata, Evolvulus alsinoides, Eclipta prostrate, Cannabis sativa, Boerhavia diffusa, Cynodon dactylon, Digitaria adsedense, Solanum xanthocarpum, Tribulus terrestr* (Herbs) were found. Whereas *Butea monosperma, Diospyros melonoxylon, Acacia catechu* and *Lagerstroemia parviflora* were the major tree species at disturbed site site. *Vernonia cinerea, Solanum nigrum, Sida cordifolia, Abrus precatorius Cocculus hirsutus and Ziziphus nummularia* were major shrubs and *Cynodon dactylon, Boerhavia diffusa* and *Eclipta prostrata* were major herbs at affected site. Frequency, density and abundance were also low for above mentioned species at mining affected forest site.

Variations were observed in the composition of plants in the disturbed and undisturbed sites. The plant species showed a drastic reduction in their numbers in disturbed sites, affected due to mining when compared with that of the undisturbed sites, unaffected due to mining. Among herbaceous species, the predominance of Poaceae was observed and species reported were mainly specific to poor soil. *Cynodon dactylon* is the species of tropical and warm areas and its higher importance value suggests that it can multiply rapidly in the disturbed environments. This perennial grass by virtue of its stolon and rooting at each node can bind the soil particles making the soil more stable. A good number of leguminous species was observed on undisturbed site in comparison to the disturbed site, which is an indication of enriched fertility status of the soil at undisturbed site.

Figure 1. Soil texture analysis of disturbed (affected) and undisturbed site (forest area)

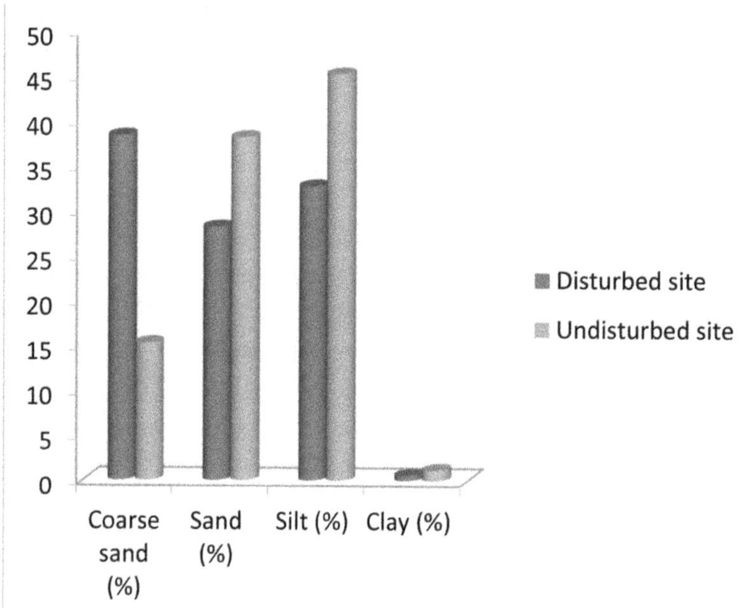

Involving Local Stake Holders In in situ Restoration

Involvement of local stake holder's i.e. local inhabitants in *in situ* restoration is essential for sustainable restoration. To involve local people in restoration process, socio-economic survey has to be conducted in the nearby human inhabitations (villages) of mining area. This is to study the existing resources of the area, social structure of the community, employment pattern, income generation, dependence on forest and species preferred by the local people along with information on other related socioeconomic

Figure 2. Soil physico-chemical properties of disturbed (mining affected) and undisturbed site (forest area)

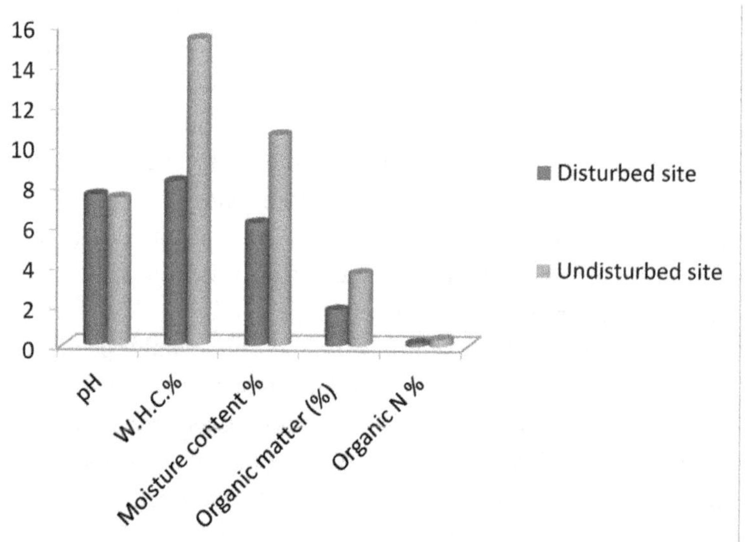

Figure 3. Plant species reported at disturbed site (mining affected) and undisturbed site (forested area)

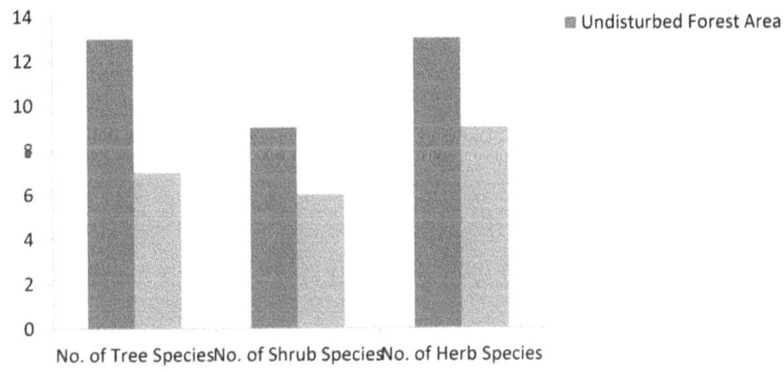

Figure 4. Tree species reported at adjacent undisturbed site (forested area) to disturbed site (mining affected)

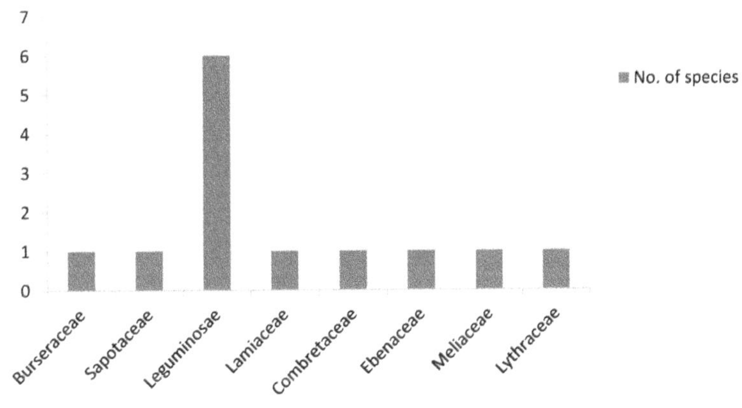

aspects. The understanding of these parameters is essential for the sustainable management of mining

Figure 5. Tree species reported at disturbed site (mining affected)

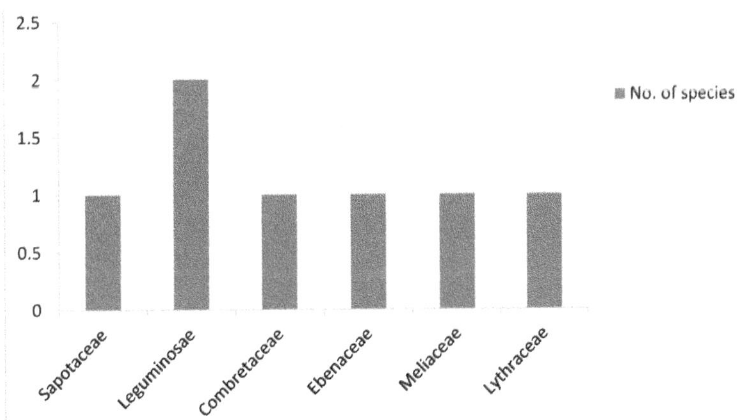

Figure 6. Shrub species reported at adjacent undisturbed (forested) site to disturbed (mining affected) site

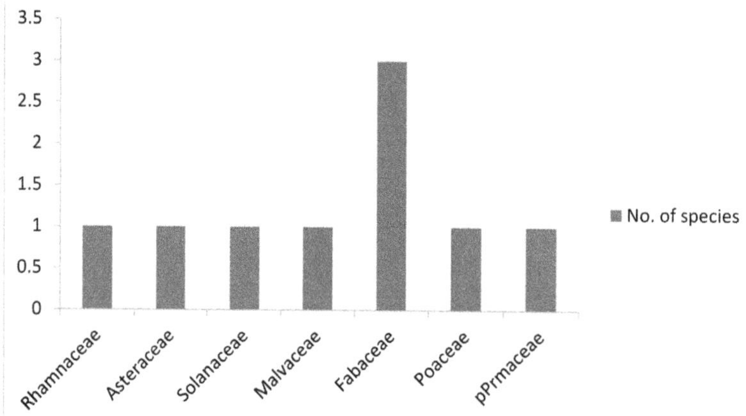

affected area. In present case, the study was performed at mining affected area in Padari of Mirzapur district by using Participatory Rural Appraisal (PRA) tools and by Questionnaire based pre-structured Interviews. Participatory Rural Appraisal (PRA) covered the village areas whereas Questionnaire based pre-structured interviews were done in town area of Padari, Mirzapur. Survey was mainly carried out to know the preference of local people for the land use of particular mining area under study and preference of plant species for the *ex situ* restoration. The survey was also aimed to motivate the villagers towards tree planting and to suggest improvement for marketability of tree and forest products. The respondents were selected so as to comprise all strata of people; economic, social and educational including all age classes and adequate representation of ladies in order to avoid age and gender biases. They preferred fodder grass and indigenous multipurpose plant species for restoration. They had been involved in restoration process. They have also been involved in protection and plantation.

Figure 7. Shrub species reported at disturbed (mining affected) site

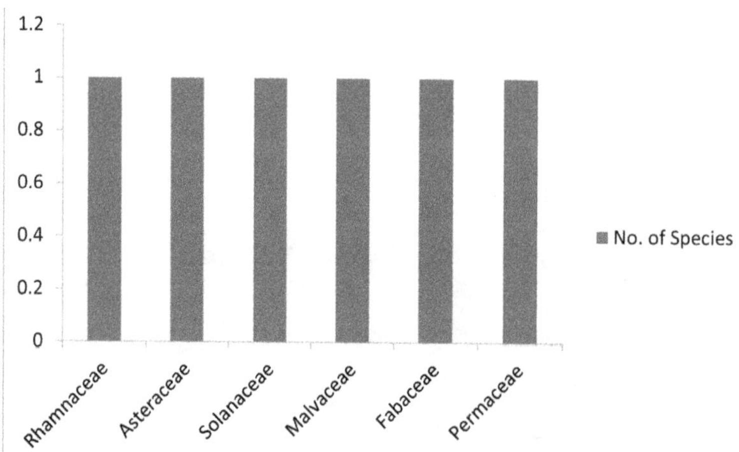

Figure 8. Herbaceous species reported at adjacent undisturbed (forested) site to disturbed (mining affected) site

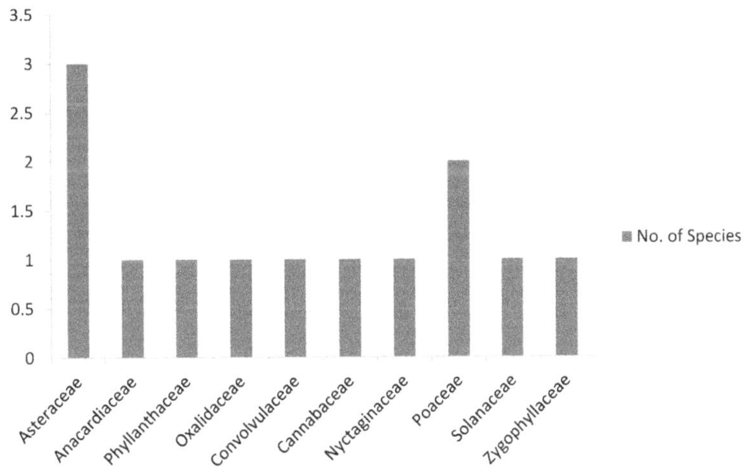

Probiotic Interventions

This step is the most important strategic step for sustainable restoration. Treatments with probiotics have been utilized for restoring the productivity of the mining affected area which has almost lost its productivity. In soil ecosystems, the probiotics are known as the microbes that are involved in primary production, decomposition, nutrient recycling and other associated processes in soil. These are microbial inoculants consisting of living cells of microorganisms like bacteria, algae, fungi, alone or a combination which may help in increasing soil productivity. Biological activities are markedly enhanced by microbial interactions in the rhizosphere of plants (Vishal and Abhishek, 2014). They also act synergistically (Dubey, 2010, 2017, 2019). Plant probiotics are classified into different types depending on the type or group of microorganisms they contain. Table 1 shows the classification of plant probiotics on the bases

Figure 9. Herbaceous species reported at disturbed (mining affected) site

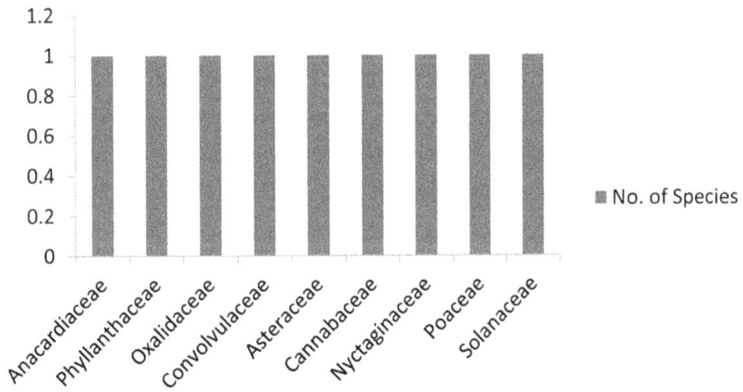

Table 1. Different microorganisms used as plant probiotics

S. No.	Group	Examples
N_2 fixing Probiotics		
i)	Symbiotic	*Rhizobium* in leguminous plants and *Frankia*in non-leguminous trees
ii).	Non-symbiotic (free-living, associative or endophytic)	*Azotobacter,Azospirillum, Cyanobacteria (Blue Green Algae),* Azolla
Phosphate Solubilizing Probiotics		
i)	Bacteria	*Bacillus subtilis, Bacillus megaterium, Bacillus circulans, Pseudomonas striata*
ii)	Fungi	*Penicilliumsp, Aspergillusawamori*
Phosphate Mobilizing Probiotics		
i)	Mycorrhiza	*Glomussp.,Gigasporasp.,Acaulospora sp., Scutellospora sp. & Sclerocystis sp.*
ii)	Ectomycorrhiza	*Laccaria sp., Pisolithus sp., Boletus sp., Amanita sp.*
Other Plant Growth Promoting Rhizobacterial Probiotics		
i)	Silicate and Zinc solubilizers	*Bacillus sp.*
ii)	Other Growth Promoting and Micro-nutrient Solubilizer	*Pseudomonas sp.*

of the different types of microorganisms used. Important probiotics, frequently used for the purpose are reported by various studies (Dubey, 2010; Ritika and Uptal, 2014; Dubey, 2017a,b, Malik *et. al.*, 2018):

Since, the microbial activity declines considerably due to poor microbial community in mining affected forest areas, which may directly affect productivity of tropical forests, microbe assisted management holds promise for *in situ* restoration of such forest area. These beneficial microbes, also termed biofertilizers, are involved in nitrogen fixation, carbon fixation, and in improving the nutrient availability to the plants in the soil. They help in improving soil fertility and enhance nutrient uptake by plants in deficient soils, thereby aiding in better establishment and growth of plants. They also secrete growth substances and antifungal chemicals, as well as improved seed germination and root growth (Dubey, 2017a).

Previous attempts to restore the area through replanting of indigenous tree species were mostly not successful due to poor SMC, poor soil health and high soil temperatures. Therefore, a probiotic approach for re-establishment of native forest was applied. There are also various supportive evidences that disturbed sites may be revegetated more promptly when the seedling of planted species are pre-inoculated with beneficial microbial inoculants or probiotics (REFERENCE). But the uses of beneficial microorganisms like Vesicular-Arbuscular Mycorrhiza (VAM), Rhizobium, Azotobacter, Phosphate Solubilizing Microorganism (PSM), etc. in the reclamation programme have received less attention and are poorly explored. Use of probiotics may also be helpful in the development of a low cost technology to restore the mining areas and increase their productivity (Dubey, 2010; 2018).

Stone mining affected site may be prepared for executing the probiotic interventions. Mining affected sites are generally undulating and having some deep pits. These pits may be used for harvesting the rainwater for irrigation purposes. Site treatment may be done by microbial seasoning of the soil of the area in rainy season through probiotics interventions. For selecting suitable microbial combinations, experiments with different combination treatments have to be conducted and best combinations may be selected for seasoning. In present case, it was conditioned with different soil treatments comprising of probiotics treatments with farm yard manure (FYM) amendments. Probiotics used were *Azotobacter sp.*, *Azospirillum sp.* and phosphate solubilising microbes (PSB). Soil parameters need to be studied regularly after the probiotic treatment. Probiotic treatments acted as a soil conditioning agent by adding organic carbon and other nutrients to the soil (Dubey, 2018). Increased organic carbon may assist in the binding of the soil sandy particles together preventing soil eradication, desertification, and erosion and also increased water retention capacity of the soil (Swathi, 2010). Improved soil quality augments the probiotic biota of soil through promoting the already existing ones (Swathi, 2010).

Flora of all the treated sites was observed regularly. It was observed that frequency and density of different species enhanced after the probiotic treatments and broadcasting of fodder grass seeds. It had a stimulatory effect on succession and restoration. Soil seasoning exhibited the improvement of site conditions and its biodiversity (Dubey, 2018, 2019).

Establishment of Surface Covering Vegetation Through Grass Fodder Species

Seeds of fodder grass were broadcasted to establish the initial surface covering in rainy season for soil binding. In case of present study, fodder seeds (Dinanath grass, Stylo and Gunie grass) were procured from IGFRI, Jhansi. Seeds of odder species were mixed with soil and broadcasted equally.

Probiotic treatments and broadcasting of fodder seeds/ herbaceous seeds may be repeated in rainy seasons for two consequent years. Site conditions may also be observed regularly. Due to microbial seasoning, soil conditions were found to be improved and pedogenisis may take place. Soil formation may be observed. Soil should be analyzed regularly. If the site conditions are improved adequately, plantation of microbes inoculated seedlings in third year may be done. . Increase in floral species and amelioration of the soil characteristic may represent the improvement of the soil characteristics and may be taken as indicator of restoration (Dubey, 2018, 2019).

Broadcasting of Indigenous Seeds

Direct seeding / broadcasting of indigenous species has been found to be useful and cost-effective restoration methodology (Singh et al. 2004b, Pandey et. al. 2005, Dubey, 2018, 2019). Generally it incorporates the seeds, locally available in plenty, easy to grow and leguminous. Seeds of *Prosopis juliflora* and *Cassia siamea* may be sown directly in rainy season in stone mining affected area. Both species are leguminous and come up easily. Being leguminous, they may promote soil microbial biomass and mineralization and simultaneously promote growth of other saplings growing in their vicinity or planted species. It accelerates natural succession also.

Training Courses for Forest Department Officials

Training courses for forest department staff to inform them about *in situ* conservation methodologies and the role played by local communities in such conservation has to be organized

Ex-situ Restoration Through Planting of Probiotics Inoculated Indigenous Tree Species

For *ex situ* restoration, plantation of probiotic inoculated seedlings has been proposed. Such plantations enhance the survival rate of the plantation. Species selection for the plantation is a crucial step. To involve local people, selection of species for restoration may be done through socio-economic survey, already discussed in previous section and on the basis of ecological requirements. The first aspect of species choice is the purpose for which the trees are intended for viz. timber, fuel wood and softwood etc. The second aspect of species choice is identification of species which will grow well on the site in question. The selection of species should, therefore, be based on the site conditions and the users' needs (Melkania, 1987, 1991). For local needs, a tree species with several attributes or a mixture of tree species can be planted to obtain multiple benefits e.g., ability to enrich soil fertility, wood suitable for fuel and poles and nutritious leaves for fodder etc. Therefore following species were selected keeping above factors in mind and in proper consultation with local people and forest department officials:

1. **Leucaena Leucocephala:** Belongs to Fabaceae sub family of Leguminosae family. It is used for a variety of purposes, such as firewood, fiber, and livestock fodder. Being leguminous, it enriches the soil.
2. **Holoptelea Integrifolia:** It is large deciduous tree belonging to Ulmaceae family and well suited to dry regions. It is used for medicinal, firewood and livestock fodder purposes .
3. **Gliricidia Sepium:** Is a medium size leguminous tree belonging to the family Fabaceae. it is considered to be the most important multi-purpose legume tree. Well suited to tropical dry conditions and used for many purposes including live fencing, fodder, firewood and green manure.
4. **Heplophragma Adenophyllum:** Commonly known as Katsagon belongs to Bignoniaceae family. The tree is extensively used in traditional medicine. As an ingredient in message oils, it is supposed to ease muscular tension.
5. **Pithecellobium Dulce:** The tree is drought resistant and can survive in dry conditions. It is leguminous tree belonging to the family Fabaceae. It has medicinal and food value.
6. **Melia Dubia:** It is a tree in the family Meliaceae. It is multipurpose tree species having various use as timber, for making plywood, as bio-fuel, as fodder and medicinal value.
7. **Jatropha Curcas:** It is a semi-evergreen shrub or small tree. It belongs to family Euphorbiaceae. It is well known biofuel plant and can be used for reclamation of dry areas. It has medicinal value also.
8. **Cassia Siamea:** It is a medium-size, evergreen tree belonging to the family Fabaceae. This plant has medicinal value, fodder and fuel wood value.
9. **Ficus Religiosa:** *Ficus religiosa* is a semi-evergreen tree belonging to family Moraceae. It has medicinal and religious value.
10. **Madhuca Indica:** This tropical tree is also known as *Mahua longifolia* belonging to family Sapotaceae. Its flower is edible and is a food item for tribals. They are used to make syrup for medicinal purposes. It has food, timber value and medicinal value.

Since generally the mining affected sites are low in soil nutrients, the self-propelling nitrogen fixing tree species are the most ideal candidates for restoration of such lands. Therefore out of selected ten species, four species were belonging to the family Fabaceae. Moreover, the selected species should establish themselves in poor soil and moisture stress conditions of the site. Utility value was also kept in mind while selection.

Probiotic inoculations may be made at the stage of nursery raising. Different probiotic combinations have to be standardized and recommended for different species (Dubey, 2018, 2019). Performances of seedlings of the selected species, inoculated with probiotic combinations has to be evaluated in the field conditions. The inoculated seedlings may perform well. They perform well as far as survival and growth is concerned.

CONCLUSION

The sandstone quarrying and mining in Vidhyan region is opencast type. Extensive quarrying and open cast mining in the area have resulted into long barren, unproductive deeply irregular lands. The restoration of this mined area becomes imperative to counter environmental hazards and restore the ecological balance. The present article is focused on the exploitation of probiotics for the restoration of this stone mining affected forest area. Though it is a long and complicated process to restore such degraded forest area, the application of probiotics may be used for re-vegetating the mine overburdens, derelict mine areas and mine spoils in a cost effective and eco-friendly manner as evident by the present study. *Riccia* sp. and mosses were also reported which may also be taken as indicator of the restoration. After the microbial inoculation, soil organic carbon was improved, and thereby the available nitrogen was also enhanced (Dubey, 2010, 2018). This was mainly due to the increase of the herbaceous species and microbial populations in the area. Soil microbes' account for significant carbon sequestration in soil. Soil seasoning play important role in enhancing the soil organic carbon. It is an indicator soil formation also. Soil organic carbon (SOC) plays an essential role in the early stages of pedogenisis and ecological restoration in reclaimed mine soils (Qu et al., 2017, Hu et al., 2015). Though probiotic treatments vary from species to species and site to site, it may certainly speed up the natural restoration of the degraded forest and other sites. However, its affectivity depends on various factors like species selections, site conditions, participation of local populace, standardization of the probiotic combinations, quality of probiotics etc. Selected species should be of local economy value. Species with multiple uses, preferred by the local people supporting their livelihoods, may be selected. Species should be suitable for the area with fast growth and dense spreading canopies. Collection and storage of seeds should be easy. Some species should be leguminous, so that they can ameliorate the soil through nitrogen fixation. Some keystone species which are also socio-culturally important like *Ficus religiosa*, may also be selected. Quality of probiotics mainly decides its performance and therefore its affectivity in restoration process. Sometimes its application does not lead to the expected positive results and this could be because of exposure to high temperature or hostile conditions of the site or of storage conditions. Other constraints limiting the use of biofertilizer technology may be environmental, human resource, unawareness, unavailability of suitable strains, and unavailability of suitable carrier and so on (Ritika and Uptal, 2014). Short shelf life, lack of suitable carrier material, susceptibility to high temperature, problem in transportation, and storage are biofertilizers bottlenecks that still need to be solved in order to obtain effective inoculation (Chen, 2006)., Therefore, through managing all these factors, stone mining affected forest area may be restored sustainably.It may also be noted that although the present study has focused on stone mining affected area, experiences learnt from the above study may hold promise for other regions and countries facing similar problems.

REFERENCES

Anonymous. (2007). RPCB Report 2007. Retrieved from www.rpcb.nic.in

Basu, A. K. (1986). Geology of parts of the Bundelkhand Granite massif, Central India. *Records of the Geological Survey of India*, *117*(II), 61–124.

Bhadra, B. K., Gupta, A. K., Sharma, J. R., & Choudhary, B. R. (2007). Mining activity and its impact on the environment: Study from Makrana marble and Jodhpur sandstone mining areas of Rajasthan. *Journal of the Geological Society of India*, *70*, 557–570.

Bhardwaj, S. D., & Pankaj, P. (2000). Impact of mining and rehabilitation of mined areas. *Advances in Forestry Research in India*, *22*, 151–172.

Birch, P., Bentham, H., & Harris, J. A. (1991). Soil microbial ecosystems: importance for the effective restoration of mined lands. In W. Oaks & J. Bosden (Orgs.), *Proc. 1991 Natl. Meet. ASSMR,* May 14-17, Durango, CO. pp. 631-640.

Chatterjee, M., & Bose, B. (2003). Marred by mines: Afforestation in nutrient deficient inhospitable lands. *Wastelands News*, *18*(3), 44–49.

Chaturvedi, R. K., & Singh, J. S. (2017). Restoration of mine spoil in a dry tropical region: a review. In Proceedings of Indian National Science Academy, 83(4), 789-844. doi:10.16943/ptinsa/2017/49123

Chen, J. (2006) The combined use of chemical and organic fertilizer and or bio-fertilizer for crop growth and soil fertility. International Workshop on "*Sustained management of the Soil-rhizosphere system for efficient crop production and fertilizer use*". 37-43.

Dobson, A. P., Bradshaw, A. D., & Baker, A. J. M. (1997). Hopes for the Future: Restoration Ecology and Conservation Biology. *Science*, *277*(5325), 515–522. doi:10.1126cience.277.5325.515

Down, C. G. (1974). The relationship between colliery waste particle sizes and plant growth. *Environmental Conservation*, *1*(4), 29–40. doi:10.1017/S0376892900004902

Dubey, K. (2017). Socio Economic Impact Study of Mining and Mining Polices on the Livelihoods of Local Population in the Vindhyan Region of Uttar Pradesh. Report. Delhi, India: NITI Aayog.

Dubey, K. (2019). Restoration of Sand Stone Mining Area through Microbial Technology. Germany: Lambert Academic Publishing.

Dubey, K, Dubey, K. P., Pandey, A., & Tripathi, P. (2017b). Microbial Biofertilizer Interventions in Augmenting Agroforestry. In Probiotics. (pp. 421-442). Berlin, Germany: Springer.

Dubey, K., Pandey, A., Tripathi, P., & Dubey, K. P. (2017a). Sustainable Management of Waterlogged Areas through Biodrainage & Microbial Agro-ecosystem. In Probiotics. (pp. 387-406). Berlin, Germany: Springer.

FSI. (2011). India State of Forest Report 2011. Retrieved from http://fsi.nic.in/cover_2011/uttarapradesh.pdf

Govt., K. (2015). Chapter 4: Land Degradation-Mining and Quarrying. Retrieved from https://www.karnataka.gov.in/empri/Documents/CHAPTER-4.%20Land%20degradation-%20Mining%20and%20Quarrying-DS.pdf

Grant, C. D. (2006). State-and-Transition Successional Model for Bauxite Mining Rehabilitation in the Jarrah Forest of Western Australia. *Restoration Ecology*, *14*(1), 28–37. doi:10.1111/j.1526-100X.2006.00102.x

Holl, K. D. (2002). Long-term vegetation recovery on reclaimed coal surface mines in the eastern USA. *Journal of Applied Ecology*, *39*(6), 960–970. doi:10.1046/j.1365-2664.2002.00767.x

Hu, Z., Fu, Y., Xiao, W., Zhao, Y., & Wei, T. (2015). Ecological restoration plan for abandoned underground coal mine site in Eastern China. *International Journal of Mining, Reclamation and Environment*, *29*(4), 316–330. doi:10.1080/17480930.2014.1000645

Ibarra, J. M. N., & de las Heras, M. M. (2005). Open-Cast Mining Reclamation. In *Forest Restoration in Landscapes*. Berlin, Germany: Springer. doi:10.1007/0-387-29112-1_53

Indorante, A. J., Jansen, I. J., & Boast, C. W. (1981). Surface mining and reclamation: Initial changes in soil character. *Journal of Soil and Water Conservation*, *36*(6), 347–350.

Jitendra, S. K. (2019). *Aravallis broken beyond repair.* Retrieved from Error! Hyperlink reference not valid.

(2004). <jrn>Singh, A. N., Raghubanshi, A. S., & Singh, J. S. (2004b). Survival and growth pattern of three tropical plantations raised on coal mine spols of central India. [<jrn>]. *Indian Forester*, (April): 376–383.

Khoshoo, T. N. (1991). Environmental concern and strategies (3rd ed.). Punjabi Bagh, Delhi, India: Ashish Publishing House.

Kumud, D. (2010). Bioreclamation of Silica Mining Area of Shankargarh, Allahabad. (Unpublished doctoral thesis). Forest Research Institute, Dehradun, India.

Liu, Z. H., Tang, Z. Z., Huang, J. P., Huang, Z. Y., Huang, H., Li, Y. S., & … . (2009). Survey on environmental lead pollution in rural areas around lead/zinc mining area in Guangxi, 2008. *Journal of Environmental Health*, *26*(8), 708–710.

Maiti, S. K. (2003). Eco restoration problems of coalmine overburden dumps. *International Conference on Ecorestoration*, Dehradun, India.

Maiti, S. K., & Das, D. (2004). Ecorestoration of coalmine overburden dumps. *Environment and Ecology*, *22*(2), 329–333.

Maiti, S. K. (2006). Ecorestoration of coalmine overburden dumps in Korba, Gevra and Kusmunda area of SECL, India. *International Symposium on Environmental issues of Mineral industries,* VNIT, India.

Maiti, S. K, & Ghose, M. K. (2005). Ecological restoration of acidic coalmine overburden dumps-An Indian case study. *Land Contamination & Reclamation, 13*(4), 1-10.

Malik, A., Shazia, L., Lone, F. A., Malik, M. A., Dar, Z. M. et al. (2018). Liquid Bio-Fertilizers and Sustainable Forestry: A Mini Review. *Open Access J. ARTOAJ 15*(3). doi:. 2018. 15. 555958 doi:10.19080/ARTOAJ

Masoodi, T. H., & Soni, P. (1999). Physical properties of lime stone mine soil restored for forest land use. *Indian Journal of Forestry*, *22*(1-2), 30–36.

Melkania, N. P. (1987). Allelopathy and its significance on the production of agroforestry plant associations. In P. K. Khosla & D. K. Khurana (Eds.), *Agroforestry for Rural Needs* (pp. 221–224). Solan, India: ISTS.

Melkania, N. P. (1991). Woody fodder species in Indian mountain ecosystems. *Range Management and Agroforestry*, *12*(2), 179–194.

MINENVIS. (2006). *Newsletter of the ENVIS Centre on Environmental Problems of Mining Areas. Number 49. June 2006*. Published by Professor Gurdeep Singh Head and Co-ordinator. ENVIS Nodal Centre of Mining Environment Indian School of Mines Dhanbad.

Pandey, D. N., Chaubey, A. C., Gupta, A. K., & Vardhan, H. (2005). Mine Spoil Restoration: A strategy combining rainwater harvesting and adaptation to random recurrence of droughts in Rajasthan. *International Network on Ethnoforestry*, p. 25.

Piha, M., Mpepereki, S. M. (Ed.), & Makonese, F. T. (1998). Use of legumes in the revegetation of mine wastes. Harnessing biological nitrogen fixation in African agriculture: challenges and opportunities. *Sixth International Conference of the African Association for Biological Nitrogen Fixation, Harare, Zimbabwe*, Sept. 12-17, 1994. Selected papers. 251-254.

Pimentel, D., Harvey, C., Resosudarmo, P., Sinclair, K., Kurz, D., McNair, M., ... Blair, R. (1995). Environmental and economic costs of soil erosion and conservation benefits. *Science*, *267*(5201), 1117–1123. doi:10.1126cience.267.5201.1117 PMID:17789193

Qu, J. F., Hou, Y. L., Meng, Y. G., Wang, K., Liu, S., Zhang, S.-L., ... Chen, F. (2017). Carbon Dynamics of Reclaimed Coal Mine Soil under Agricultural Use: A Chronosequence Study in the Dongtan Mining Area, Shandong Province, China. *Sustainability*, *9*(4), 629. doi:10.3390u9040629

Reeder, J. D., & Sabey, B. (1987). Nitrogen. pp. 155-184. In R. D. Williams & G. E. Schuman (Eds.), Reclaiming mine soils and overburden in the western United States: Analytical Parameters and procedures. Soil and Water Conservation Society, Ankeney, IA.

<edb>Dubey, K. (2018). Restoration of Sand Stone Mining Area of Vindhyan Region through Microbial Technology. In K. Dubey (Ed.), Project Report. Prayagraj, India: FRCER.</edb>

Richardson, J. A. (1975). Physical problems of growing plants o colliery wastes. In M. J. Chadwick & G. T. Goodman (Eds.), *Ecology of resource degradation and renewal* (pp. 275–285). Oxford, UK: Blackwell Scientific Publication.

Ritika, B., & Uptal, D. (2014). Bio-fertilizer a way towards organic agriculture: A Review. *Academic Journals.*, *8*(24), 2332–2342.

Sarma, K., & Yadav, P. K. (2013). Relentless Mining in Meghalaya, India. *Conservation Science*, *1*(1), 5–12. doi:10.3126/cs.v1i1.8578

Saxena, S. K., & Chatterji, P. C. (1988a). Ecological imbalances caused by mining in Rajasthan and their restoration. Mining and Environment in India. Joshi and Bhattacharya (Eds.). HRG Publication Series, Nainital, India. pp. 223-241.

Saxena, S. K., & Chatterji, P. C. (1988b). In K. A. Shankarnarayan (Ed.), *Mining activity and creation of wasteland in western Rajasthan. Wasteland development and their utilization.* (pp. 41–61). Jodhpur, India: Scientific Publisher.

Schafer, W. M., Nielsenand, G. A., & Nettleton, W. D. (1980). Mine spoil genesis and morphology I a spoil chronosequence in Montana. *Soil Science Society of America Journal*, *44*(4), 802–807. doi:10.2136ssaj1980.03615995004400040029x

Sharma, K. D., Kumar, S., & Gough, L. P. (2000). Rehabilitation of lands mined for limestone in the Indian desert. *Land Degradation & Development*, *11*(6), 563–574. doi:10.1002/1099-145X(200011/12)11:6<563::AID-LDR414>3.0.CO;2-I

Sharma, V., & Chaudhry, S. (2018). Vegetation composition and plant diversity in mining disturbed tropical thorn forest of Asola-Bhatti Wildlife Sanctuary, Northern India. *Taiwania*, *63*(3), 267–280.

Short, T. A., Menzies, N. W., Mulligan, D. R., & Loch, R. (Ed.), & Jasper, D. (2000). Mining disturbance alters phosphorus fractions in northern Australian soils. Special Issue: Soil research for mine reclamation. Australian Journal of Soil Research, 38(2): 411-421.

Singh, A. N., Raghubanshi, A. S., & Singh, J. S. (2002). Plantations as a tool for mine spoil restoration. *Current Science*, *82*(12), 1436–1441.

Singh, A. N., Raghubanshi, A. S., & Singh, J. S. (2004a). Impact of native tree plantations on mine spoil in Adry Tropical Environment. *Forest Ecology and Management*, *187*(1), 49–60. doi:10.1016/S0378-1127(03)00309-8

Soni, P. (2003). Climate change and restoration of tropical forests. *Indian Forester*, *129*, 865–873.

Srivastava, S. C. (1999). Effect of coal mining on microbial biomass and nutrient availability in dry tropical forest of Vindhyan Hill region. *Journal of Tropical Forestry*, *15*(1), 15–23.

Swathi, V. (2010). *The use and benefits of bio-fertilizer and biochar on agricultural soils.* (Unpublished thesis). Department of Chemical and Biological Engineering, Chalmers University of Technology, Sweden. 20-24.

Todd, M. C. L., Adams, M. A., Grierson, P. F., Loch, R. (Ed.), & Jasper, D. (2000). Mineralisation of nitrogen in a chronosequence of rehabilitated bauxite mines. Special Issue: Soil research for mine reclamation. Australian Journal of Soil Research, 38(2), 435-451.

Toy, J. W., & Shay, D. (1987). Comparision of some soil properties on natural and reclaimed hill slopes. *Soil Science*, *143*(4), 264–277. doi:10.1097/00010694-198704000-00004

Upadhyay, N., & Verma, SPratap Singh, ADevi, SVishwakarma, KKumar, NPandey, A… Sharma, S. (2016). Soil ecophysiological and microbiological indices of soil health: A study of coal mining site in Sonbhadra, Uttar Pradesh. *Journal of Soil Science and Plant Nutrition*, *16*(3), 778–800.

Verma, S., Singh, A. P., Devi, S., Mewaram, R. R., Sharma, S., & Dubey, K. (2014). Assessment of Microbial Community and Soil Enzyme Activity of Coal Mine Dumps of Sonbhadra Uttar Pradesh, India. In *Proc. of the Intl. Conf. on Advances In Bio-Informatics, Bio-Technology and Environmental Engineering*, pp. 12-15.

Vishal, K. D., & Abhishek, C. (2014). Isolation and characterization of Rhizobium leguminosarum from root nodules of *Pisums sativum L. Journal of Academic and Industrial Research*, *2*(8), 464–467.

Walker, L. R., & del Moral, R. (2003). *Primary Succession and Ecosystem Rehabilitation*. New York, NY: Cambridge University Press. doi:10.1017/CBO9780511615078

Walker, R. F. (2002a). Organic amendment, fertilizer, and lime effects on bare root Jeffrey pine out planted on a Sierra Nevada surface mine. *Journal of Sustainable Forestry*, *15*(3), 29–55. doi:10.1300/J091v15n03_02

Walker, R. F. (2002b). Reestablishment of Jeffrey pine on an acidic Sierra Nevada surface mine: Influence of fertilizer and lime amendments on early growth and nutrition. *Journal of Sustainable Forestry*, *15*(3), 1–27. doi:10.1300/J091v15n03_01

Walker, R. F. (2002c). Responses of Jeffrey pine on a surface mine site to fertilizer and lime. *Restoration Ecology*, *10*(2), 204–212. doi:10.1046/j.1526-100X.2002.00070.x

Whisenant, S. (2005). First steps in erosion control. In S. Mansourian, D. Vallauri, & N. Dudley (Eds.), *Forest restoration in landscapes: beyond planting trees*. New York, NY: Springer. doi:10.1007/0-387-29112-1_50

Zedler. J. B. (2005). Ecological restoration: guidance from theory. *San Francisco Estuary and Watershed Science 3*(2) [September 2005], Art. 4.

Section 4

Remote Sensing in Dry Tropical Forests

Chapter 13
Scenarios of the Tropical Dry Forest of Purulia District West Bengal: A CA-MARKOV Model Approach

Masuma Begum
Directorate of Forests, India

Niloy Pramanick
Jadavpur University, India

Anirban Mukhopadhyay
Jadavpur University, India

Sayani Datta Majumdar
Jadavpur University, India

ABSTRACT

In this chapter, satellite images of the years 1995, 2005, and 2015 of LANDSAT have been used. After pre-processing (geometric correction and atmospheric correction using FLAASH, LULC change dynamics have been assessed to estimate the changes in total forest cover in Purulia district through an unsupervised K-means classification scheme. To evaluate the health status, vegetation indices, namely NDVI, SAVI, and CVI, have been used. The increase in NDVI, SAVI, and CVI values was inferred as no significant degradation of Purulia forest cover. Moreover, future scenarios have been predicted by implementing a CA-MARKOV model. Using the land cover map of 1995 as the base map, and from 1995 to 2005 as training data, a land cover map of 2015 has been generated which in turn validated by the actual land cover of 2015. After validation, prediction of land cover was possible for the years 2035 and 2050. The prediction suggested that forest area will increase by approximately 4% from 2015 to 2035 and by 3% from 2035 to 2050.

DOI: 10.4018/978-1-7998-0014-9.ch013

INTRODUCTION

Tropical dry forests consist of deciduous broad-leaved trees located in tropical and subtropical areas. These forests thrive in regions characterized by a prolonged period of drought, rainy and humid conditions with annual precipitation of 400-1700 mm. The drought occurs during the time of limited evapotranspiration, which permits the growth of deciduous and evergreen vegetation (Hayden et al. 2010). The tropical forests are referred to as one of the species-rich terrestrial ecosystems and are known to generate a wide variety of natural resources for sustainable livelihood of local communities as well as provide habitat to some of the endangered species of the world out of an estimated 1 million sq. Km of tropical dry forests of the world, about half of it is found in South America. India and Southeast Asia, Australia, the Caribbean, Central America, and two parallel belts in Africa are some other regions where these forests flourish (Hayden et al. 2010). The trees which alter their leaves every year are considered as deciduous. The existence of Sal trees is widespread in a dry deciduous forest. Sal grows in a wide range of soil type but prefers slightly acidic to neutral (pH 5.1 – 6.8) sandy loam textured soil. The growth of Sal also depends on alluvial and lateritic soil. The maximum temperature recorded in this type of forest is 45°C. The height of trees can reach 15 to 16 m and have a girth of 1.5 m in its' favorable localities (Volume 2, Second Working Plan, Puruliya District).

India constitutes 38.2% of tropical dry forest cover characterized by some dominant species like *Shorea robusta, Anogeissus latifolia, Lagerstroemia parviflora, Terminalia tomentosa, Hardwickia binata, Boswellia serrata, Buchanania lanzan, Acacia catechu*, etc. (Champion and Seth, 1968). Once covering about 52% of the total forests of the world, tropical dry forests are now considered as one of the most vulnerable forest areas of the world (Singh and Singh 1991). Apart from the eminent consequence of climate change, other anthropogenic factors also act as threats for these types of forest. Degradation of forests in the form of felling, encroachment, grazing, and forest fire is usually accompanied by loss of primary productivity, reduction in biodiversity and extinction of species (Kumar et al. 2000, 2001, 2002; Yadav & Yadav 2008). In the future, to secure sustainable flow of the ecosystem services and for proper management of this critical ecosystem, present as well as the future scenario needs to be studied well. At present, the Joint Forest Management Committee plays a vital role in the Forest Department to protect the forest. In the year 1988, Arabari of Midnapore district was envisaged as the procreator of Joint Forest Management. There are 776 JFMC under the forest of Puruliya. It protects 737.93 sq km of forest area (Annual Administrative Report 2017-2018, Department of Forest, West Bengal).

Various empirical studies have recognized the positive relationship that often exists between habitat area and species richness (Brooks et al. 1997, Pimm and Raven 2000). Using multiple sources of satellite data, DeFries et al. (2005) examined the range of existing forest habitat and loss throughout the moist and dry tropical forest of the world over the last two decades. There are a wide variety of modelling approaches to predict landscape patterns and changes (Cheong et al.2012; Liu et al.2009; Pontius et al.2005; Brown et al.2004). Markov models are the earliest of the fitted models, which are simple to create with minimal data requirements (Tattoni et al.2011; Cabral et al.2009; Brown et al.2004]. To evaluate the impact of the National Forest Policy of 1988 on the forest cover, Adhikary and Southwort (2012) selected Bannerghatta National Park (BNP) and its buffer as their study area. Their main objectives were (1) to create e a landscape of no policy intervention by predicting the forest cover of 1999 and 2007 based on a CA Markov model; (2) to compare the observed and predicted separate land covers and changing courses. By integrating remote sensing GIS techniques, counterfactual approach, and Cellular Automata-Markov modeling approach, Mondal and Southworth (2010), estimated the significance of certain conservation intervention in a human-dominated tropical forests of Central India. Their study area witnessed several management strategies since its declaration as a protected area in the mid-1970s.

Remote sensing offers an efficient means of collecting information to map vegetation type and acreage. The spectral reflectance of vegetation always varies concerning phenology, stage type, and health, and these could be measured with the help of multi or hyper spectral sensors (Dutta, 2006). Vegetation indices (VIs) are semi-analytical measures of vegetation activity and widely vary not only with the seasonal erraticism of green foliage but also across space, thus appropriate for detecting spatial variability within the field of agriculture (Vina et al. 2011). Defined as spectral transformations of two or more bands, VIs are computed directly without any bias regarding land cover class, soil type, or climatic conditions (Huete et al. 2002).

The objectives of present study were to a) assess the future land cover scenarios of the Puruliya district of West Bengal using CA-MARKOV model approach and b) to study the changing trends of forest health over the years by using 3 vegetation indices namely- NDVI (Normalized Differential Vegetation Index), SAVI (Soil Adjusted Vegetation Index) and CVI (Chlorophyll Vegetation Index).

STUDY AREA

Tropical dry forests account for 29% of total forest cover of West Bengal. This forest cover comprises of four sub-categories namely: Dry Peninsular Sal Forest, Northern dry mixed deciduous forest, Dry deciduous scrubs, and Butea Forest. All the above-mentioned sub-categories of tropical dry forests are found in Puruliya district of West Bengal. Puruliya is the westernmost district of West Bengal with a total geographical area of 6259 sq. Km (Figure1). It comprises 1198.03 sq. Km area of dry deciduous forest with the abundance of *Shorea robusta, Pterocarpus marsupium, Madhuca indica, Terminalia tomentosa, Diospyros melanoxylon, Schleichera oleosa, Terminalia bellrica, Terminali arjuna* etc. After splitting from Manbhum district of Bihar, the district was formed in the year 1956. It encompasses, Hazaribag and Dhanbad towards the northern side,

Singbhum in the Southern and Ranchi towards the western side. Bounded by three districts namely Bankura, Burdwan and Midnapore in the eastern side, Puruliya is entwined with Damodar, Kangsabati Kumari, Darakeswar and Subarnarekha River. Forest is the most important natural resource of the district. With a total forest cover of 1857.26 sq km, the natural resource in this district is distributed in three division of Forest Department–Puruliya Division, Kangsabati North Division, and Kangsabati South Division. Kangsabati south division. (Second Working Plan, Puruliya District, 2014-2015 to 2023-2024, Volume 1). Forest –the emerald of Puruliya strew across 19 Ranges Ajodhya, Arsha, Baghmundi, Balarampur, Jhalda, Kotshila, Joypur, Matha, Hura, Kahipur, Puncha, Puruliya Para, Raghunathpur, Bandwan 1, Bandwan 2, Barabazar, Jamuna, Manbazar 1 and Manbazar 2. This forest is characterized by the *presence of Shorea robusta, Acacia auriculiformis, Eucalyptus, Phyllanthus emblica, Terminalia bellerica, Diospyros melanoxylon, Madhuca indica, Termanalia chebula, Cassia siamea, Azadirachta indica, Buchanania lanzan etc* (Figure2). The presence of common shrubs and climbers (Table 1) are also easily noticeable in the forest. (Second Working Plan, Puruliya district, 2014-2015 to 2023-2024, Volume1).

Growing stock is the volume of all living trees per hectare which is measured in a cubic meter (Second Working Plan, Puruliya District, 2014-2015 to 2023-2024, Volume 1)

Figure 1. Location map of Puruliya District

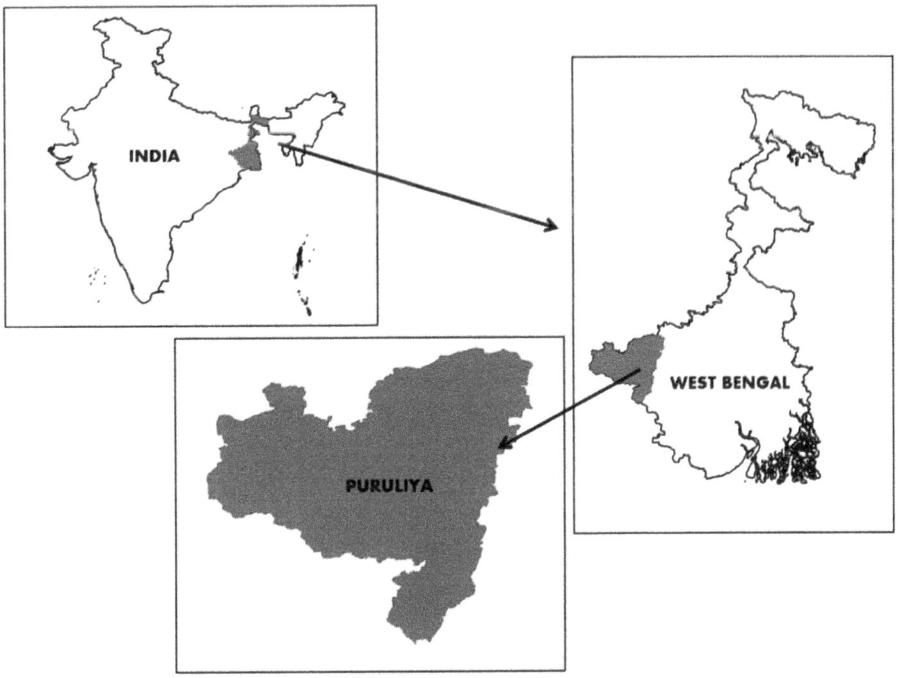

METHODOLOGY

To understand the present scenario of the dry forest areas of Puruliya district multi-temporal satellite images of the year 1995, 2005, and 2015 of LANDSAT and SENTINEL have been used. After atmospheric correction of the images using Fast Line-of-sight Atmospheric Analysis of Hypercubes (FLAASH), the images have been classified to estimate the changes of the forest areas and the surrounding land cover. Changing trend of the forest has been assessed. Different indices based studies have been applied like Normalized Difference Vegetation Index (NDVI), Soil Adjusted Vegetation Index (SAVI), Chlorophyll Vegetation Index (CVI) to assess the changes of the health of the forest (Figure 5). Normalized Difference Vegetation Index (NDVI) introduced by Rouse et al. (1973), is by far the most extensively used vegetation index with some limitations related to brightness of the soil background, in which separate NDVI relationships with biophysical properties of canopy are found over different soil and moisture conditions (Bausch, 1993; Elvidge, & Lyon, 1985; Huete et al., 1985). The value ranges from -1 to 1 nearer the value to one are the healthier vegetation (Genc et al. 2008). Normalized Difference Vegetation Index (NDVI) has some restrictions related to the brightness of the soil background (Elvidge and Chen 1995; Elvidge and Lyon, 1985; Huete et al., 1985).

$$\text{NDVI} = (\text{NIR} - \text{RED}) / (\text{NIR} + \text{RED}) \text{ Genc et al 2008}$$

To overcome this problem, Huete (1988) suggested the usage of a soil-adjustment factor, L, to account for non-linear, differential NIR and red radiative transfer through a canopy, and developed a soil-adjusted

Table 1. The tropical dry forest species of Puruliya

Trees	Scientific Names	Shrubs and climbers	Scientific Names
Akashmoni, Sonajhuri	*Acacia auriculiformis*	Alkushi	*Nucuna pruriens*
Akand	*Calotropis gigantean*	Anantamul	*Hemidesmus indicus*
Am	*Mangifera Indica*	Asamlata	*Eupatorium odoratum*
Amla	*Emblica officinalis*	Atari	*Combretum decandrum*
Amlaki	*Phyllanthus emblica*	Bagnaki	*Martynia diandra*
Amra	*Spondius pinnata*	Bainchi	*Flacourtia indica*
Ankar	*Alangium salvifolium*	Bankal or Dholkalmi	*Choisy Austin*
Arjun	*Terminali arjuna*	Bantulsi	*Ocium americanum*
Asan	*Terminalia tomentosa*	Berala	*Sida cord folia*
Aswatha	*Ficus religiosa*	Bhabri	*Lantana camara*
Bahera	*Terminaliabellerica*	Bhurru	*Gardenia gummifera*
Bandar Lathi, Sonalu	*Cassia fistula*	Chakunda	*Cassia tora*
Bat	*Ficus benghalensis*	Kalmegh	*Andrographis paniculata*
Bel	*Aegle Marmelos*	Koromcha / Bankoromcha	*Carissa spinarum*
Cashew, Kaju	*Anacardium occidentale*	Kulekhara	*Hygrophila schulii*
Challa	*Holoptela integrifoia*	Kurchi	*Holarrhena antidysenterica*
SajneSiris	*Albiziaodoratissima*	Maiynakanta	*Catunaregam spinosa*
Chhatim	*Alston cholaris*	Nakdana	*Opuntica stricta*
Dhaw	*Anogeissus latifolia*	Nilkanta	*Curcuma cassia*
Dumur	*Ficus rica*	Nisinda	*Vitex negunda*
Gabdi, Palas	*Butea monosperma*	Bankhejur	*Phoenix acaulis*
Gamar	*Gmelina arborea*	Putri	*Croton roxburghii*
Gokul	*Ailanthus excelsa*	Satamuli	*Asparagus racemosa*
JagnyaDumur/ Gular	*Ficus racemosa*	Shialkanta	*Mimosa rubicaulis*
Haldu or Karam	*Haldina cordifolia*	Talmuli	*Curculigo orchioides*
Harjora	*Cissus quadrangularis*	Dudhilata	*Ichnicarpus frutescens*
Haritaki	*Termanalia chebula*	Gulancha	*Tinospora cordifolia*
Kanthal	*Artocarpus heterophyllus*	Kantaalu	*Dioscoreaspinpentaphylla*
Jarul	*Lagerstroemia flos reginae*	Kunch	*Abris precator*
Kadam	*Anthocephalus cadamba*	Latapalash	*Butea superba*
Kaj	*Uridelia retusa*	Maljam	*Bauhinia vahlii*
Kaju	*Anacardium occidentale*	Maula	*Butea parviflora*
KantaBhuel	*Xantolis tomentosa*	Nayantara	*Catharanthus roseus*

vegetation index **(SAVI)** (Jiang et al., 2008). Soil Adjusted Vegetation Index (SAVI) takes into account the impact of soil surrounding the vegetation.

Figure 2. Contribution of the growing stock of Sal, Eucalyptus, Akashmoni, and other trees in the forest part of Puruliya. (Second Working Plan, Puruliya District, 2014-2015 to 2023-2024, Volume 1)

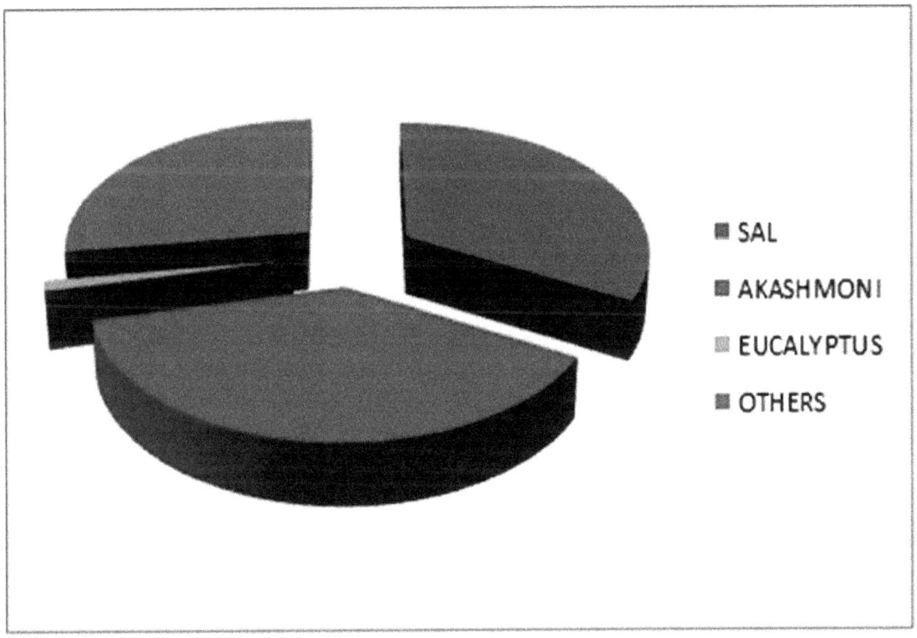

$$\text{SAVI} = (\text{NIR} - \text{RED})*(1+\text{L})/(\text{NIR} + \text{RED} + \text{L}) \text{ Alhammadi and Glenn 2008}$$

Chlorophyll Vegetation Index (CVI) gives the leaf chlorophyll concentration in canopy level. To obtain a broad-band VI Chlorophyll vegetation index (CVI) was developed by Vincini et al. 2007 by integrating the spectral information of the green band with heightened sensitivity to leaf chlorophyll content and less sensitivity to LAI variation (Vincini et al. 2008).

$$\text{CVI} = \text{NIR}\frac{\text{RED}}{\text{GREEN}^2} \text{ Vincini et al. 2008}$$

The red/green ratio (Gamon and Surfus 1999) has been used to evaluate foliage development in canopies.

To assess the future scenario of the tropical dry forest of Puruliya, a CA-MARKOV model have been used. For calculating the transition probabilities of the classes, a hybrid methodology has been adopted using a statistical model (Markov chain analysis) assemblages which have been incorporated into a cellular automata (CA) nonlinear geospatial model for projecting the future conditions (Mukhopadhyay et al., 2015). Cellular Automata is a collection of arbitrarily shaped cells arranged in grid-like structures. The cells holding diverse values, binary being the simplest ones, alter their states simultaneously, according to some rules which are applied to the system at a regular distinct time interval (Ghosh et al., 2017). Thus each cell is led to a new state which may or may not be the same as its previous state. Being a random process, Markov chain, on the other hand, is characterized by the transition of any sorts of

Table 2. Temporal change of area (in Sq.km) of different land use and land cover of Puruliya District

Class	1995	2005	2015	2035	2050
Forest	1151.527	1137.135	1174.3	1216.63	1248.378
Agricultural land	3104.703	3189.736	3279.55	3327.179	3360.901
Settlement	558.313	612.664	663.264	695.464	717.614
Fallow land	686.6791	583.0819	459.629	396.723	351.5435
Water	751.1443	729.7495	675.623	616.37	573.9303

events from one state to another state such that it depends only on the present state and not on the past that the process went through (Ghosh et al., 2017). For the base land cover class 1995 classified image have been used, and from 1995 and 2005 the futuristic scenarios of the year 2015 have been projected. From the actual 2015 land cover, the result has been validated, and future predictions of land cover change for the years 2035 and 2050 have been done.

RESULTS AND DISCUSSION

Land cover and Land use classes could be delineated from a combined unsupervised K- means the classification of a Landsat OLI image acquired in 2015. These are namely: Agricultural land, forest, water, fallow land, and settlement. It was evident from the maps that the blocks: Bundwan, Bagmundi, Jhalda 1 and 2 and Neturia have prominent forest cover. Agriculture with a total area of 3279.55 sq.km, presumes to be the significant primary occupation in this district (Figure3). The forest covers account for a total area of 1174.3 sq. Km (Table 2). The area under forest, as well as agricultural land, showed a steady increase from 1995 to 2015.

Moreover, it has been predicted that forest area will increase by approximately 4% from 2015 to 2035 and by 3% from 2035 to 2050, whereas area under agriculture is assumed to increase by only 1.5% during the same period. The rate of areal increase of settlement in Puruliya has been unprecedented since 1995. The area under urban built up is further predicted to increase by 5% from 2015 to 2035 and by 4% from 2035 to 2050 (Figure 4). It has been observed that while agricultural land increased from 1995 to 2015 fallow land also showed a steady decrease in the area over the same period. The incident of increase in agricultural land at the expense of decreasing fallow land may be attributed to the same cause as cited by Verma and Raghubanshi (2019) in their rural development and land use land cover change study of Varanasi district. At the district level, a change of green spaces into agricultural land was observed between 1995–2015, making green spaces and open/barren land the highest contributor towards agricultural expansion. Thus, open/barren spaces were reclaimed, and along with green spaces, these land uses transformed to fuel agricultural growth and build settlements. There has been an increase in an urban built-up area in the study area from 1993 to 2015. The data further indicated an increase in agricultural land (3%) and built-up area (9% increases) at the expense of green vegetation and open or barren spaces in the two decades under consideration for Puruliya district.

To determine the health of the forest cover, various vegetation indices were carried out using remote sensing techniques. The NDVI, SAVI, and CVI values varied in different years (Figure 6). The mean NDVI value dropped from 1995 to 2005 and further increased to 0.163 in the year 2015.

Figure 3. Change in the area (in Sq. km) under different Land use and Land covers of Puruliya District

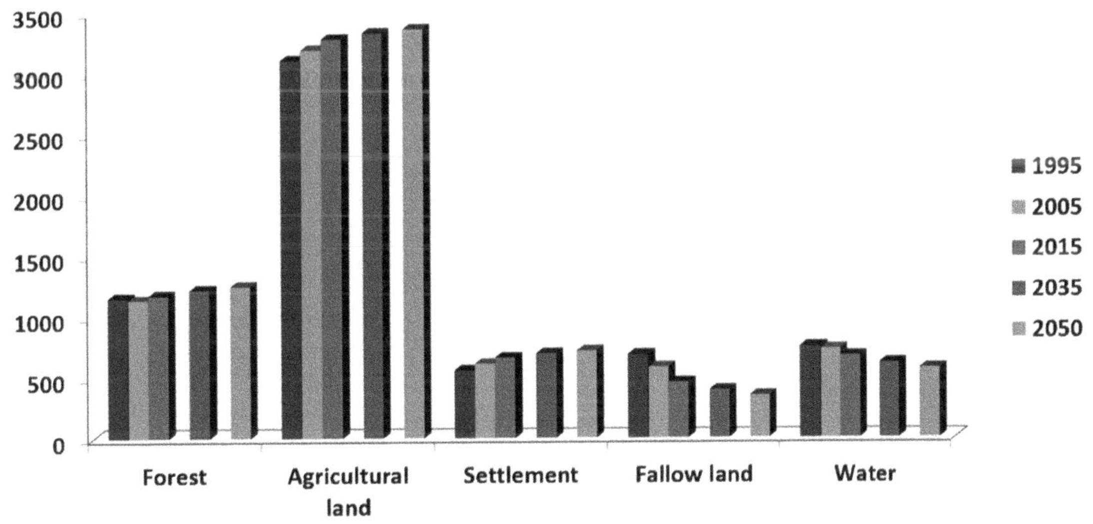

Similarly, the mean SAVI value increased from -0.185 in 1995 to -0.245 in 2015. Finally, the CVI value was found to be 0.814 in 2015. The increase in NDVI, SAVI, and CVI values through the years depict that the health of the Puruliya forest has not degraded till date. Further with an increase in the area of forest cover chlorophyll concentration in canopy cover also increased which pertain to healthy vegetation cover of the district.

Ferriera and Huete (2004) associated the use of NDVI and SAVI to study the major vegetation types encountered in Brazilian Cerrado. They concluded that while SAVI responded to near-infrared variations, NDVI showed a strong reliance on red reflectance. The NDVI values ranging from 0.50 to 0.65 was noticed in Pandey et al. (2019)'s study of the spatial distribution of mangrove species and biomass of Bhitarkanika forests, which reflected the distribution of dense, healthy mangrove forest in homogenous condition. Burapapol and Nagasawa (2016) in their study to assess wildfire fuel load distribution in Sri Lanna National Park, northern Thailand investigated the relationships between the calculated standard leaf biomass in sample plots and each potential VI, i.e., NDVI, SAVI, EVI2, RVI, different vegetation index (DVI), normalized difference water index (NDWI), and chlorophyll vegetation index (CVI), to compare the strength of correlation of each index with the calculated standard leaf biomass. The equations relating the NDVI to the calculated standard leaf biomass showed the best performance for deciduous forests, with the highest R^2 values, followed by SAVI whereas CVI results in this study presented the least significant correlation with leaf biomass. Since radiation in the visible red light part of the spectrum is strongly absorbed (or poorly reflected) by the chlorophyll in green plants, radiation in the NIR part of the spectrum is strongly reflected by the mesophyll cells of leaves. Therefore, the NDVI can serve for detecting green stands or biomass. In this study, a similar relationship between the three indices was evaluated. As observed from the data, high mean values of NDVI in the years 2005 and 2015 indicated healthy green canopies, with a high value of leaf biomass. There is a strong relationship positive relationship between near-infrared/red reflectance ratio of vegetation canopy and

Figure 4. Land use and Land cover map of the years 2015, 2035 and 2050 of Puruliya District

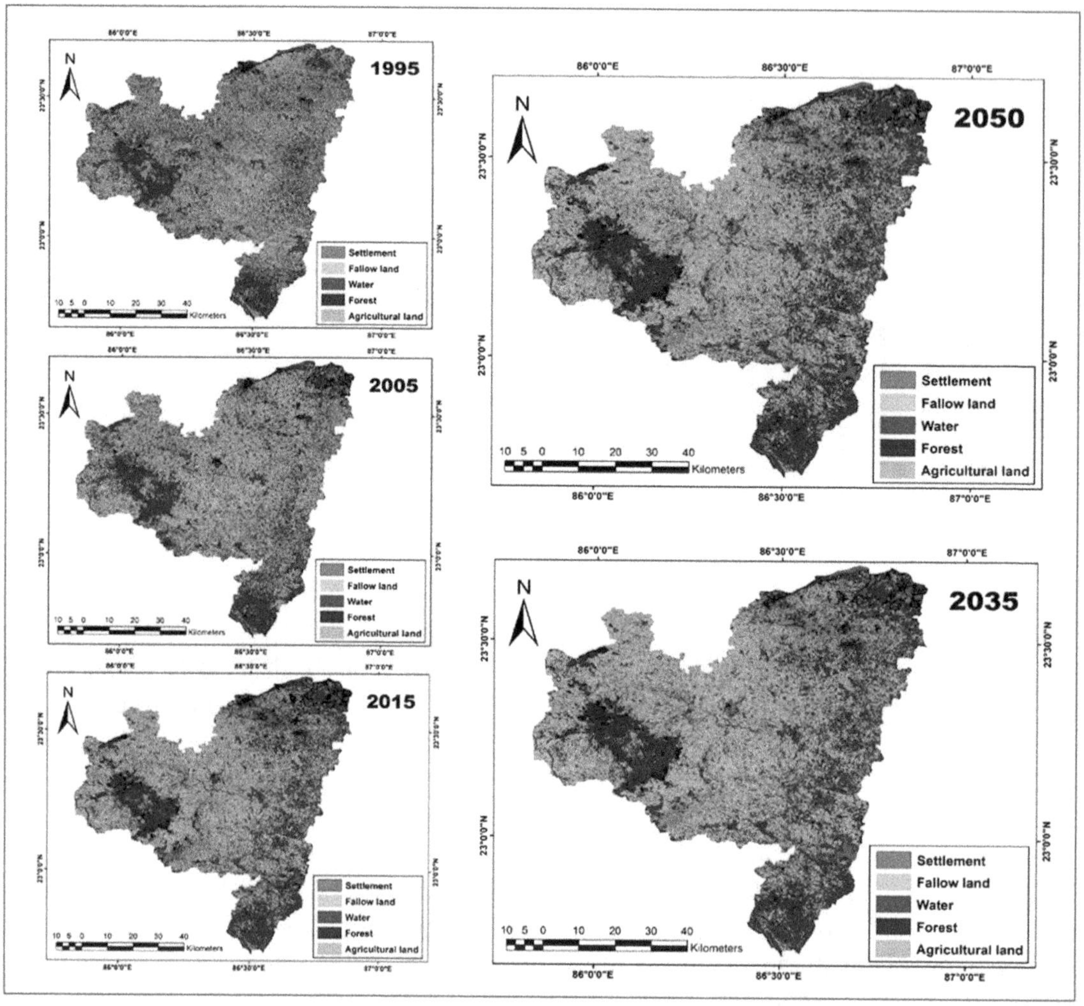

leaf biomass per unit ground area (Kanemasu et al.1974). The reflectance spectrum of a plant canopy is a blend of the spectral reflectance of plant and soil components. The soil contribution progressively decreases with growth in vegetation but may remain significant, depending on plant density, canopy geometry, wind effects Rondeaux et al. (1996). The low values of SAVI evaluated in this study may be attributed to above the cause. The high mean values of CVI in 2005 and 2015compared to other years indicated greater pigment concentration in the leaves. Since the difference between near infrared and red reflectance regulates the area of red edge peak in the reflectance derivative curve, a robust relationship between the area of the peak and total biomass is anticipated. A similar correlation could be established between leaf biomass and chlorophyll content of the forest canopy of Puruliya. Moreover, according to Collins (1978); Curran et al. (1990), the chlorophyll content in leaves is also a great determinant of nutritional stress, photosynthetic capacity, and developmental stages. Since chlorophyll content is well correlated to nitrogen availability and water stress (Filella and Peñuelas 1994) and the measurement of the red edge peak in this study could afford the information of these two parameters for better analysis.

Figure 5. Map showing a change in the health of vegetation of Puruliya district using three indices

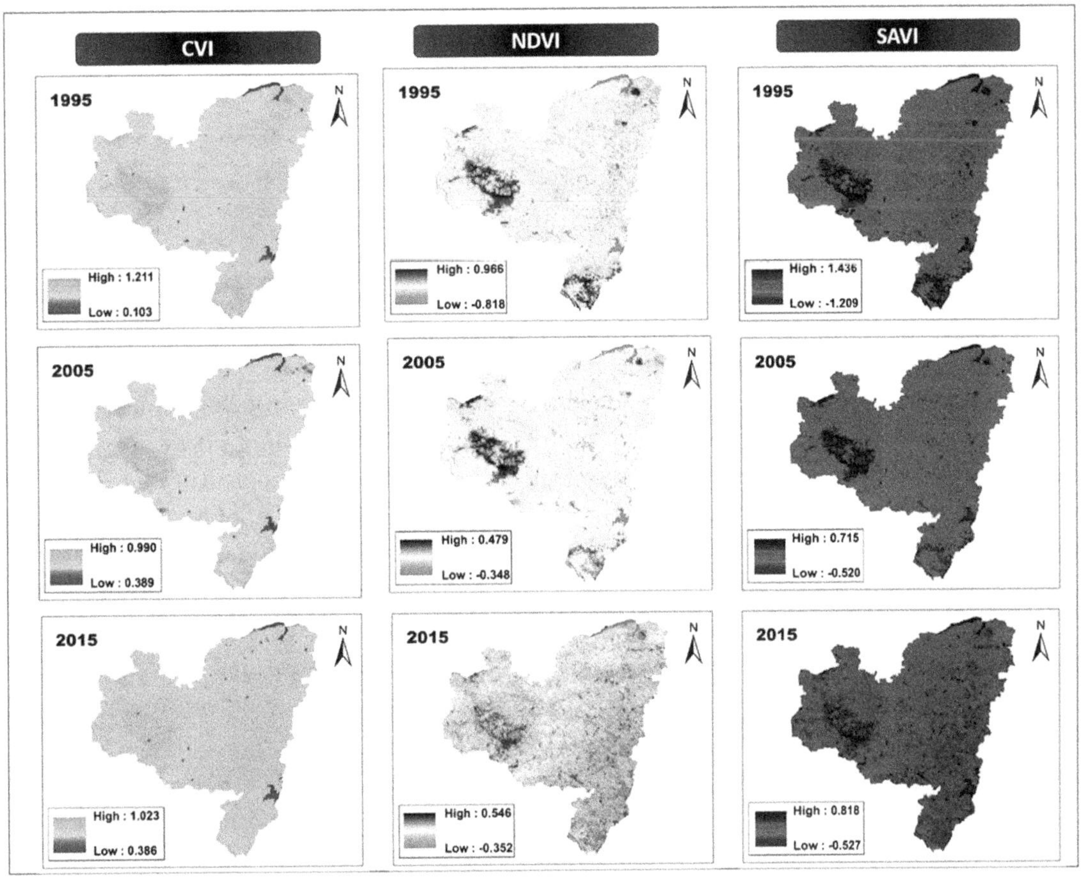

It can be surmised that although forest cover of Puruliya is not being threatened till date, the recent

Figure 6. The index values plotted against years showing the temporal change in vegetation health of Puruliya District

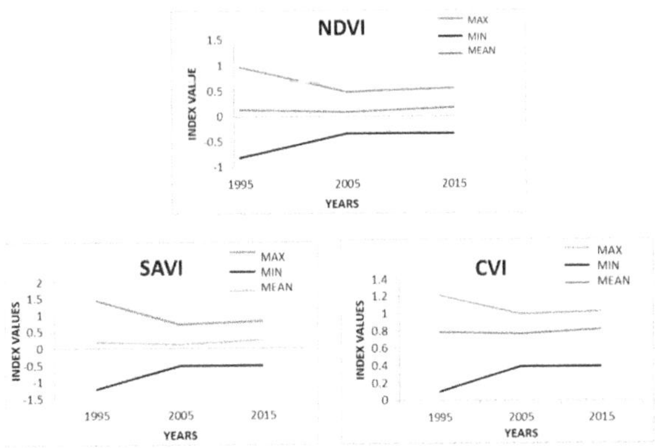

urban developmental activities may pose a threat shortly, may be in the form of forest clearing and urban sprawl or land conversion. The paces of the growth rate of the area under agriculture in the coming years are slower than forest or settlement, but if the rate of growth of cultivation since 1995 till date is considered, then it can be inferred that years before 2015 witnessed approximately 2 -3% growth rate in the agricultural area. This has been possible either by felling down of trees due to the need for timber or conversion of forest land to agriculture and settlement.

CONCLUSION

Now a day's forests are in the grip of crisis due to felling, encroachment, grazing and forest fire. Forest area has diminished because of random felling by local people. Grazing causes damage to the saplings of the plantation. In the year 2018, 4.54 sq km plantation area of Puruliya Forest Department was affected due to a forest fire. (Annual Administrative Report 2017-2018, Department of Forest, West Bengal).

The present study indicates that the temporal change in the health of the tropical dry forest of Puruliya could be analyzed using the three vegetation indices. The NDVI and CVI index values plotted against the years show a positive trend from 1995 to 2015. This further indicates the increasing trend of vegetation leaf biomass along with a positive trend of chlorophyll content in their leaves. From the Markov chain analysis, it can be concluded that there can be a significant threat to the forest cover from rapid urban built up as well as from growing cultivation, especially in the western and the northern sector of the district. Encroachment of forests due to the growth of agriculture and settlements might be a notable trend of change in the landcover maps of Puruliya district shortly.

Moreover, a steady decrease in area under fallow land indicated the transformation of barren spaces into agricultural land. This may be further attributed to recent developments in agricultural practices. Increase in the built-up area also reflected growing population pressure in the district.

The indices indicated a significant rise in vegetation health over the years and The Markov Chain model approach projected a steady rise in biomass and chlorophyll concentration in leaves but the growing land conversion practices in the form of agricultural expansion and settlement due to the increasing pressure of population might pose a threat to the forest cover of Puruliya. Sustainable agricultural practices and utilization of forest resources might lead to successful coexistence of both the land cover and land use type.

Joint Forest Management Committee (JFMC) plays a vital role in the Forest Department to protect the forest. In the year 1988, Arabari of Midnapore district was envisaged as the procreator of Joint Forest Management. West Bengal became the predecessor state of India by crowning JFMC in 618 families which are an integral part of the forest ecosystem from 11 villages. There are 776 JFMC under the forest of Puruliya district which protects 737.93 sq km of forested area.

ACKNOWLEDGMENT

The authors are thankful to the Government of West Bengal and the State forest department for all kind of help during the study period. Particularly we are indebted to the Chief Conservator of Forests, South West Circle, Conservator of Forest, Working Plan and GIS Circle, Divisional Forest Officer, Kangsabati North Division, Kangsabati South Division, and Purulia Division. We are also grateful to Jadavpur

University for extending all kind of support during the study. Authors are thankful to NASA, USGS for the satellite data. Comments from anonymous reviewers helped us improve the manuscript.

REFERENCES

Adhikari, S., & Southworth, J. (2012). Simulating forest cover changes of Bannerghatta National Park based on a CA-Markov model: A remote sensing approach. *Remote Sensing*, *4*(10), 3215–3243. doi:10.3390/rs4103215

Annual Administrative Report. (2017-2018). Department of Forest, West Bengal.

Brooks, T. M., Pimm, S. L., & Collar, N. J. (1997). Deforestation predicts the number of threatened birds in insular Southeast Asia. *Conservation Biology*, *11*(2), 382–394. doi:10.1046/j.1523-1739.1997.95493.x

Burapapol, K., & Nagasawa, R. (2016). Mapping wildfire fuel load distribution using Landsat 8 Operational Land Imager (OLI) data in Sri Lanna National Park, northern Thailand. *Journal of the Japanese Agricultural Systems Society*, *32*(4), 133–145.

Cabral, P., & Zamyatin, A. (2009). Markov processes in modeling land use and land cover changes in Sintra-Cascais, Portugal. *Dyna (Bilbao)*, *158*, 191–198.

Champion, S. H., & Seth, S. K. (1968). A revised survey of the forest types of India. *A revised survey of the forest types of India.*

Cheong, S.-M., Brown, D. G., Kok, K., & Lopez-Carr, D. (2012). Mixed Methods in Land Change Research: Towards Integration. *Transactions of the Institute of British Geographers*, *37*(1), 8–12. doi:10.1111/j.1475-5661.2011.00482.x

DeFries, R., Hansen, A., Newton, A. C., & Hansen, M. C. (2005). Increasing isolation of protected areas in tropical forests over the past twenty years. *Ecological Applications*, *15*(1), 19–26. doi:10.1890/03-5258

Dutta, R. (2006). Assessment of tea bush health and yield using geospatial techniques. (Published Master's Thesis), ITC, Enschede, The Netherlands. GIS Development: *The Geospatial Resource Portal.*

Elvidge, C. D., & Chen, Z. (1995). Comparison of broad-band and narrow-band red and near-infrared vegetation indices. *Remote Sensing of Environment*, *54*(1), 38–48. doi:10.1016/0034-4257(95)00132-K

Elvidge, C. D., & Lyon, R. J. (1985). Influence of rock-soil spectral variation on the assessment of green biomass. *Remote Sensing of Environment*, *17*(3), 265–279. doi:10.1016/0034-4257(85)90099-9

Ferreira, L. G., & Huete, A. R. (2004). Assessing the seasonal dynamics of the Brazilian Cerrado vegetation through the use of spectral vegetation indices. *International Journal of Remote Sensing*, *25*(10), 1837–1860. doi:10.1080/0143116031000101530

Filella, I., & Penuelas, J. (1994). The red edge position and shape as indicators of plant chlorophyll content, biomass and hydric status. *International Journal of Remote Sensing*, *15*(7), 1459–1470. doi:10.1080/01431169408954177

Gamon, J. A., & Surfus, J. S. (1999). Assessing leaf pigment content and activity with a reflectometer. *The New Phytologist*, *143*(1), 105–117. doi:10.1046/j.1469-8137.1999.00424.x

Genc, H., Genc, L., Turhan, H., Smith, S. E., & Nation, J. L. (2008). Vegetation indices as indicators of damage by the sunn pest (Hemiptera: Scutelleridae) to field grown wheat. *African Journal of Biotechnology*, *7*, 2.

Ghosh, P., Mukhopadhyay, A., Chanda, A., Mondal, P., Akhand, A., Mukherjee, S., ... Hazra, S. (2017). Application of Cellular automata and Markov-chain model in geospatial environmental modeling-A review. *Remote Sensing Applications: Society and Environment*, *5*, 64–77. doi:10.1016/j.rsase.2017.01.005

Hayden, B., Greene, D. F., & Quesada, M. (2010). A field experiment to determine the effect of dry-season precipitation on annual ring formation and leaf phenology in a seasonally dry tropical forest. *Journal of Tropical Ecology*, *26*(2), 237–242. doi:10.1017/S0266467409990563

Huete, A., Didan, K., Miura, T., Rodriguez, E. P., Gao, X., & Ferreira, L. G. (2002). Overview of the radiometric and biophysical performance of the MODIS vegetation indices. *Remote Sensing of Environment*, *83*(1-2), 195–213. doi:10.1016/S0034-4257(02)00096-2

Huete, A. R. (1988). A soil-adjusted vegetation index (SAVI). *Remote Sensing of Environment*, *25*(3), 295–309. doi:10.1016/0034-4257(88)90106-X

Huete, A. R., Jackson, R. D., & Post, D. F. (1985). Spectral response of a plant canopy with different soil backgrounds. *Remote Sensing of Environment*, *17*(1), 37–53. doi:10.1016/0034-4257(85)90111-7

Jiang, Z., Huete, A. R., Didan, K., & Miura, T. (2008). Development of a two-band enhanced vegetation index without a blue band. *Remote Sensing of Environment*, *112*(10), 3833–3845. doi:10.1016/j.rse.2008.06.006

Kanemasu, E. T. (1974). Seasonal canopy reflectance patterns of wheat, sorghum, and soybean. *Remote Sensing of Environment*, *3*(1), 43–47. doi:10.1016/0034-4257(74)90037-6

Liu, M., Hu, Y., Chang, Y., He, X., & Zhang, W. (2009). Land use and land cover change analysis and prediction in the upper reaches of the Minjiang River, China. *Environmental Management*, *43*(5), 899–907. doi:10.100700267-008-9263-7 PMID:19290569

Mondal, P., & Southworth, J. (2010). Evaluation of conservation interventions using a cellular automata-Markov model. *Forest Ecology and Management*, *260*(10), 1716–1725. doi:10.1016/j.foreco.2010.08.017

Mukhopadhyay, A., Mondal, P., Barik, J., Chowdhury, S. M., Ghosh, T., & Hazra, S. (2015). Changes in mangrove species assemblages and future prediction of the Bangladesh Sundarbans using Markov chain model and cellular automata. *Environmental Science. Processes & Impacts*, *17*(6), 1111–1117. doi:10.1039/C4EM00611A PMID:25719448

Nirmal Kumar, J. I., Kumar, R. N., Hiren Kumar, B. S., & Amit, N. G. (2000). Preliminary investigations of plant diversity of Khatana and Waghai forests of North western Ghats, South Gujarat. *International Journal of Ecology, Environmental Conservation*, *6*, 87–92.

Nirmal Kumar, J. I., Kumar, R. N., & Joseph, S. (2001). Tree species diversity of Khatana forest of south Gujarat. *Journal of Nature Conservation*, *13*, 149–166.

Nirmal Kumar, J. I., Kumar, R. N., & Joseph, S. (2002). Tree species diversity of Waghai forest of the Northern part of Western Ghats. *International Journal of Ecology, Environmental Conservation, 8*, 235–248.

Pandey, P. C., Anand, A., & Srivastava, P. K. (2019). Spatial distribution of mangrove forest species and biomass assessment using field inventory and earth observation hyperspectral data. *Biodiversity and Conservation*, 1–20.

Pimm, S. L., & Raven, P. (2000). Biodiversity: Extinction by numbers. *Nature, 403*(6772), 843–845. doi:10.1038/35002708 PMID:10706267

Pontius, R. G. J., & Malanson, J. (2005). Comparison of the structure and accuracy of two land change models. *International Journal of Geographical Information Science, 19*(2), 243–265. doi:10.1080/13658810410001713434

Raghuvanshi, D. S., Verma, N., & Gupta, A. (2019). I2/K2S2O8: An Unprecedented Deoxygenating System for N-Oxides and Sulfoxides. *ChemistrySelect, 4*(7), 2075–2078. doi:10.1002lct.201801838

Rondeaux, G., Steven, M., & Baret, F. (1996). Optimization of soil-adjusted vegetation indices. *Remote Sensing of Environment, 55*(2), 95–107. doi:10.1016/0034-4257(95)00186-7

Second Working Plan, Puruliya District. (2014-2015) to (2023-2024), Volume 1.

Second Working Plan, Puruliya District. (2014-2015) to (2023-2024), Volume 1.

Second Working Plan, Puruliya District. (n.d.). Volume 2.

Singh, L., & Singh, J. S. (1991). Species structure, dry matter dynamics and carbon flux of a dry tropical forest in India. *Annals of Botany, 68*(3), 263–273. doi:10.1093/oxfordjournals.aob.a088252

Tattoni, C., Ciolli, M., & Ferretti, F. (2011). The fate of priority areas for conservation in protected areas: A fine-scale Markov chain approach. *Environmental Management, 47*(2), 263–278. doi:10.100700267-010-9601-4 PMID:21190021

Viña, A., Gitelson, A. A., Nguy-Robertson, A. L., & Peng, Y. (2011). Comparison of different vegetation indices for the remote assessment of green leaf area index of crops. *Remote Sensing of Environment, 115*(12), 3468–3478. doi:10.1016/j.rse.2011.08.010

Vincini, M., Frazzi, E., & D'alessio, P. (2007). Comparison of narrow-band and broad-band vegetation indexes for canopy chlorophyll density estimation in sugar beet. In *Precision Agriculture '07: Proceedings of the 6th European Conference on Precision Agriculture* (189-196).

Vincini, M., Frazzi, E. R. M. E. S., & D'Alessio, P. A. O. L. O. (2008). A broad-band leaf chlorophyll vegetation index at the canopy scale. *Precision Agriculture, 9*(5), 303–319. doi:10.100711119-008-9075-z

Yadav, R. K., & Yadav, A. S. (2008). Phenology of selected woody species in a tropical dry deciduous forest in Rajasthan, India. *Tropical Ecology, 49*(1), 25.

Chapter 14
Geospatial Solutions for Forest Management:
A Case Study from Nepal

Him Lal Shrestha
https://orcid.org/0000-0003-3452-7428
Kathmandu Forestry College, Nepal

ABSTRACT

Nepal is experiencing forest degradation and deforestation due to diverse drivers which are more geographically distributed and having impact at different levels. Those causes and their spatial distribution can be assessed using geospatial approach through GIS, Remote Sensing, and GPS technologies. This approach involves mainly spatial analysis of the resources and their causal factors. Nepal has also implemented several successful forest conservation programs such as Community Forest Development Program, perceived as the most successful program at government initiation. The forest degradation and restoration both can be assessed using GIS technologies. Nepal has already experienced adoption of such technologies to develop the dataset and monitor the forest resource in terms of land use land cover and, forest cover change over the period as a part of National Forest Inventory.

INTRODUCTION

Nepal is considered as a country with rugged terrain and complex landscape with higher altitudinal variation due to which it has varieties of micro-climatic areas with specific landscape functions and ecosystem services. Despite of all these variations, complexities and diverse actors to determine the natural resources, Nepal has 44% forest coverage (FRA, 2014) which is higher than the forest area coverage of 39.6% in 1994 (DFRS, 1994). These positive changes in the forest cover were because of the community forestry program hugely implemented since 1990. We can also find such type of evaluation of the community forestry intervention in Nepal using functionalities of vegetation indices to use remote sensing images to monitor the forest cover changes over the period. The spatial variation in the natural resource levels, spatial diversities of the forest cover changes and geographic nature of the data

DOI: 10.4018/978-1-7998-0014-9.ch014

were not fully explored. Thus, the geospatial approach offers to define the spatial variations in natural resource, its changes over time, spatial thinking while resource management, prediction and modeling for the future scenario.

Government of Nepal, International Centre for Integrated Mountain Development (ICIMOD) and other development organizations are working in forestry, environment and development sector since long in history, however there are evident actions done by those organizations to develop the core forestry and allied data in Nepal such as to work together with Survey Department, Remote Sensing Center, National Land use Project (NLUP), Water Energy Commission Secretariat (WECS), Environment Division to work on the development of the forest coverage over the period like Forest Resource Assessment projects in 1994 and 2014, National Geographic Information Infrastructure Project (NGIIP) through Finnish Mapping Company (FINNMAP FM-International) in 2004, Land use mapping and zoning since 2010 and to date. Those attempts provide up-to-date information on forest resources in Nepal. Non-state actors such as NGOs, bilateral organizations and academic institutions are providing further science based and research-based exploration on those data such as decadal land cover data, which ultimately indicates the extent of forest cover changes. By taking input from this land cover data generated during 1990 to 2010, carbon flux from the forest was estimated considering other studies i.e. demand and supply of the timber and fuelwood, forest types, population growth etc. The study was done using the IPCC Guidelines (2006) to estimate the carbon flux from the forest from 1990 to 2010 by "Gain and Loss" method.

ICIMOD has also been conducting the study on "Low Cost Forest Monitoring" aiming to support the science and national level stakeholders by estimating the forest stock in terms of biomass and forest carbon using multi-source forest inventory techniques. This multi-source inventory technique uses the inputs from the Remote Sensing data, spatial analysis, Global Positioning System (GPS) tracking and the field measurement. The high-resolution satellite images were used to generate the templates for the quick estimation of the forest stock which can be used with Google Earth High Resolution images. Formerly Department of Forests (DOF) and now Department of Forests and Soil Conservation (DOFSC) has initiated 'Satellite based Forest Fire Monitoring and Alert System' in collaboration with ICIMOD, Nepalese Forester's Association (NFA), and Federation of Community Forest User Group Nepal (FECOFUN) which ultimately provides the forest fire information to the subscribed users on the fire location since 2012. The alert system sends the short message system (SMS) and email alerts when fire happens and satellite passes which can be used as the basis for the forest fire management. ICIMOD is also working on the analysis of forest vulnerability, fragmentation of the forest areas, species distribution mapping, and consolidating all the effort for the forest management as a knowledge base in the form of Web enabled mapping and information system i.e. Forest fire alert system, and Reducing Emission from Deforestation and forest Degradation (REDD) mechanism demands REDD Information System. These efforts can contribute nationally on the development of Monitoring/measuring, Reporting and Verification (MRV) system, Reference Emission Level (REL), National Forest Management Database and Data sharing mechanism as a part of National Forest Monitoring System (NFMS) and globally forestry sciences taking the support from the spatial domain.

The historical data on the forest cover, land use changes and the conversion rates are important while reporting carbon stock and the emission level to IPCC and UNFCCC (Herold et al., 2011). The base line information on the current rates of the forest growth and carbon emission can be quantified using multi-source forest inventory methods to achieve Tier 3 with higher precision (GOFC-GOLD, 2017) which includes remote sensing, GPS, GIS and field measurements (Gibbs et al., 2007). The geospatial approach may also have limitation on the spatial resolution and location shift in real ground.

FOREST FIRE DETECTION AND ALERT SYSTEM AND BURNT AREA ASSESSMENT

Forest fire is the detrimental factor of forest degradation and causes loss on human, livestock, and ecosystem interactions. Fire affects to the juvenile regeneration plants and over storey of the forest. The forest fire is hazardous to human and natural resources. The distribution of the forest fire occurrence is affected due to several factors such as human practices of cultivation, slash and burn, shifting cultivation and other anthropogenic activities. The outbreak of fire to the forest from agricultural areas, picnic spots and roadside burning can be prevented if the forest manager detect the ignition of fire. The remote sensing technologies coupled with Information Technology (IT) can provide solutions to detect the forest fire in time to allow forest managers to control it before becoming out of control. ICIMOD has initiated an automated forest fire detection and alert system to inform its subscriber to know the location of forest fire after the pass of Moderate Resolution Imaging Spectroradiometer (MODIS) satellite within half an hour. The system works in automated fashion acquiring MODIS images at ICIMOD and applying forest fire detection algorithm under SERVIR Himalaya project. It produces fire location information and disseminate through web maps, portable data file (PDF) maps, email alerts and SMS alerts to subscribed emails and mobile numbers. Through this project ICIMOD has included forest officials, community forest committee members and other interested persons who wants to use the information either to apply the forest fire management activities and conduct research on forest fires in Nepal (Matin et al., 2017).

The system reports the forest fire detected using MODIS and Visible Infrared Imaging Radiometer Suite (VIIRS) satellite imageries in different rendering. Fig.1 shows the active forest fires detected in Nepal from 1st January to 6th March 2019. The system has capability to locate and map for entire country or particular districts. This kind of information available to forest managers and researchers may support to go further prepare the forest fire management plans for upcoming years.

Figure 1. Forest Fire Detection and alert System
(Source: ICIMOD/RDS, http://geoapps.icimod.org/NepalForestFire/#)

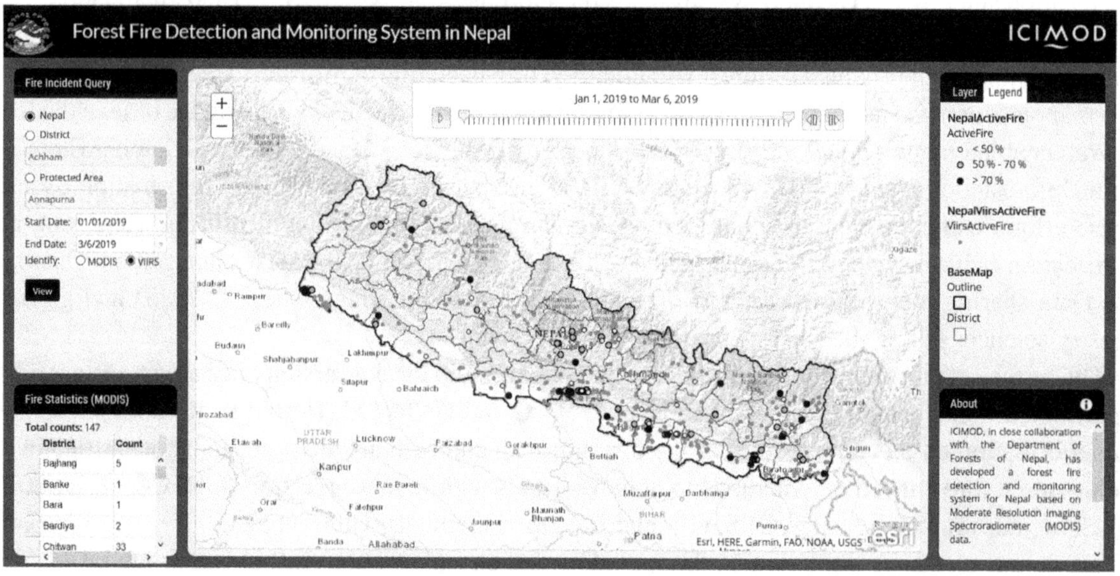

The other product available from MODIS and VIIRS satellite imageries is the burnt area product at global level. The global level burnt areas product can be further analyzed to the country or specific landscape to understand the forest fire burning area in the areas which also provides the opportunity to plan activities and human resources to manage the forest fire in those areas.

Shrestha and Dhonju (2013) and Khanal (2015) attempted to map and document the annual burnt areas inside Nepal using MODIS burnt area products. The future scope of research and quantification of the forest burning in the region and their contribution on the carbon emission from the forest due to forest fire using remote sensing data and geospatial approach are more pertinent as those technologies allow to assess the extents of forest fire and its impact on ecosystem and community. The outputs and information generated from those analysis and quantification of course increase the capability of Nepal to report forest fire burning emission to Intergovernmental Panel for Climate Change (IPCC) and United Nations Framework Convention on Climate Change (UNFCCC).

BIOMASS ESTIMATION AND CARBON QUANTIFICATION

Nepal is moving ahead with the implementation of REDD+ framework in the country targeting to 2020. On top of the forest degradation and deforestation analyses, carbon quantification is another important aspect for the REDD+ implementation. MRV is the first step to determine the carbon quantification of the baseline and repeated measurement in other word monitoring of the carbon enhancement. The MRV part is crucial in REDD+ implementation which provides baseline for the carbon budgeting and provides the scope on monitoring the carbon enhancement and additionality and permanence at the defined project sites. Nepal has prepared different instruments to prepare the REDD implementation such as Readiness Project Identification Note (RPIN), Readiness Plan Proposal (RPP), Emission Reduction Project Document (ERPD). Nepal has got financial support from UNREDD and World Bank's Forest Carbon Partnership (FCPF) initiative to do preparation for the REDD implementation.

The carbon quantification basically depends on the Tier we are targeting either Tier 1, Tier 2 or Tier 3. Tier 1 is more regional and generic information on the carbon and emission values on different forest types, forest conversion and other activities. Tier 2 speaks on the national level average of the biomass expansion factor and emission factors for the national level activities. Whereas Tier 3 is related to the more precise measurement of the carbon contents and emission factors related to the forest cover changes, forest conversion and other activities. This includes some time wall to wall estimation of the forestry parameters, precise measurements using high resolution satellite imageries supplemented by intensive field measurement such as high resolution satellite imageries and Light Detection and Ranging (LiDAR) data.

Thus, quantification biomass and carbon contents in different forests in the country using high resolution satellite imageries were practiced through ITC Netherlands, ICIMOD and NORAD which basically used high resolution satellite images of WorldView and GeoEye with intensive field measurement and LiDAR data for the quantification of forest carbon and biomass. The project piloted the techniques of use of different types of satellite images and field measured information such as Crown project area (CPA) method and 'inverse watershed' function for individual tree count (ITC) technique and generation of allometric equation to relate DBH and Height of the tree with image indices such as CPA. Some of the researcher examined to establish regression equation with the sample plot measurements of forest carbon with Normalized Difference Vegetation Index (NDVI) values which shows medium level of correlation (Shrestha et al., 2015). Karna et al. (2015) has also reported that high resolution satellite imageries can support highly precise measurement of forest carbon with wall to wall information throughout the forest areas under study.

Figure 2 shows the forest carbon quantification using WorldView PAN image and RapidEye image using the field measurements. The methods adopted for the quantification of forest carbon with WorldView PAN data is mentioned in Figure 3.

MONITORING DEFORESTATION, LAND USE CHANGE, FOREST COVER CHANGE AND CARBON FLUX

The results published by Government of Nepal after rigorous National Forest Inventory (NFI) from 2010 to 2014, show that the forest cover has been increased as compare to the NFI reporting of 39.6% of forest coverage, there are several study have been conducted over the period that there is strong evidence of forest degradation and deforestation in Nepal (Shrestha and Shrestha, 2017). There are some gray areas that the inputs, applied methodologies and the skill of the interpreter may cause those minor changes on the forest cover. The statement says that the increase of the forest cover in Nepal is attributed due to the implementation of Community Forestry Program (CFP) since 1990. There are 9 drivers identified which contributed to the deforestation and forest degradation in Nepal (Dangi, 2012). ICIMOD has initiated mapping land use land cover for the years 1990, 2000 and 2010 using temporal image data of Landsat series i.e. Landsat TM, Landsat 8 OLI. The land cover classification adopted the harmonized Land cover classification scheme (LCCS) was used to classify Landsat images with 30 m spatial resolution to represent the land cover classes in Nepal such as Broadleaved forest, conifer forest, mixed forest, grassland, shrub land, bare areas, water body, snow and glacier, urban/settlement areas etc.

The land cover classes from decadal interval further analyzed to develop the decadal changes in land use land cover in Nepal applying change detection methods. The land cover product further used to study the carbon flux in Nepal at different physiographic zones i.e. High Mountain, Middle Mountain, Hill, Siwalik and Terai region. The decadal land cover product assisted to analyze the carbon flux level using stock change method. The growth rates of the forest trees are attributed by the different tree species of

Figure 2. Biomass estimation using WorldView PAN and RapidEye MSS data

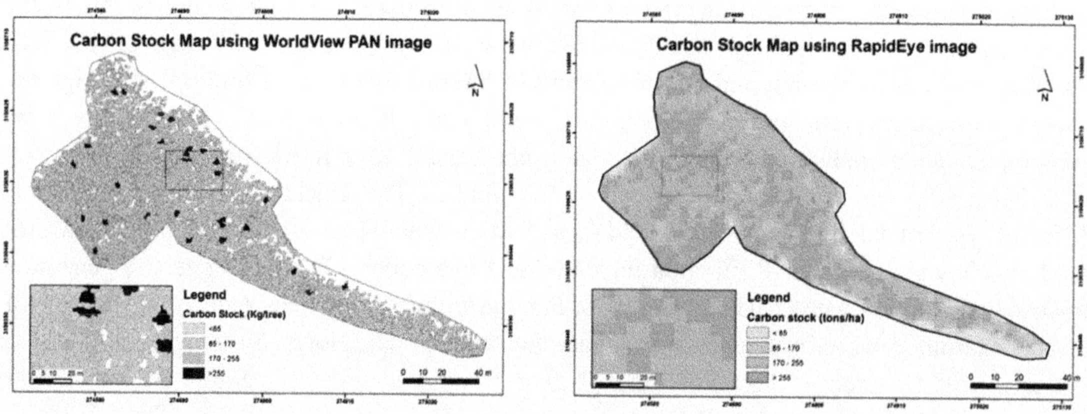

Figure 3. Methodological flow adopted for the biomass estimation using WorldView data

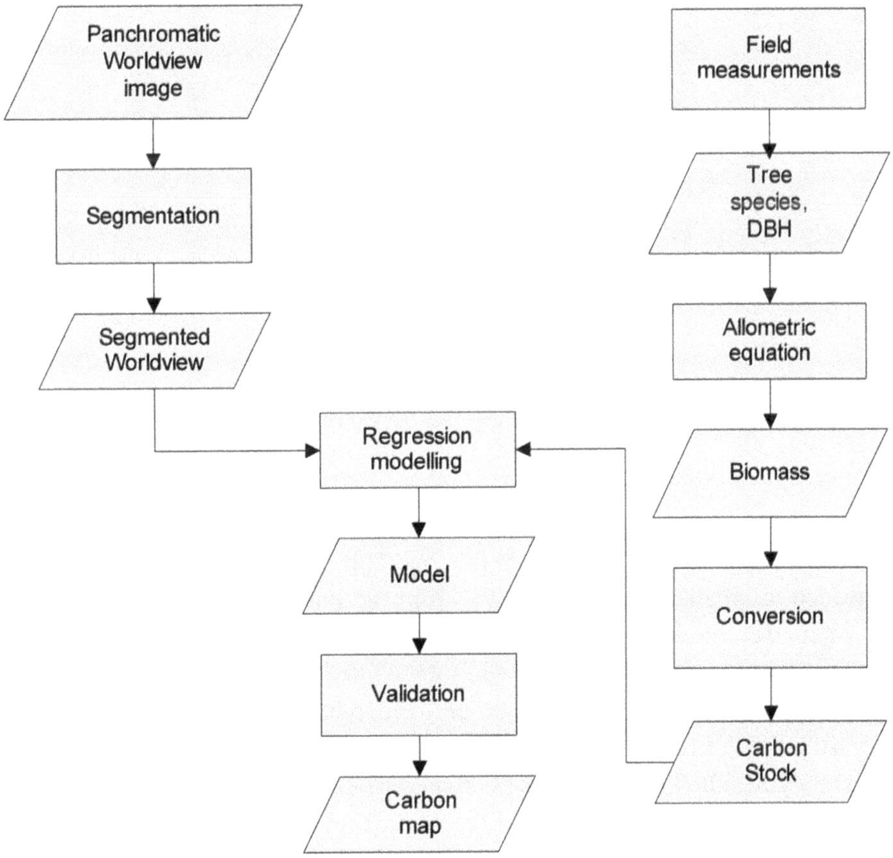

the particular areas. The change in forest coverage over the period is considered as activity data and net flux is examined by multiplying the activity data by carbon emission during the land cover conversion.

DOFSC has also initiated monitoring of deforestation caused due to the encroachment of forest areas as agricultural land and settlement areas using historical remote sensing data, aerial photographs, cadastral information and available other recent satellite imageries (Rimal et al., 2017).

TECHNOLOGICAL ADOPTION AND SPATIAL DATABASE OF FOREST MANAGEMENT

Nepal has adopted various types of forest management modalities namely Community Forest, Leasehold Forest, Protected Forest, Conservation Area, Religious Forest, Bufferzone Forest and Government managed Forest. Community based forest managements are basically handed over to the user groups which are located at different geographic domain even within district and local municipal level. Basic forest information related to the user groups and forest areas are different each other. Individual forest user groups have separate constitution and operational plans for the handed over period. The conventional way of data keeping is with traditional means of filing and storing. The recent day requirements

Figure 4. Land cover change analysis from 1990 to 2010 in Nepal: a) land cover 1990; b) land cover 2010; c) forest cover change 1990–2010; forest (Data source: ICIMOD/RDS, Uddin et al., 2015)

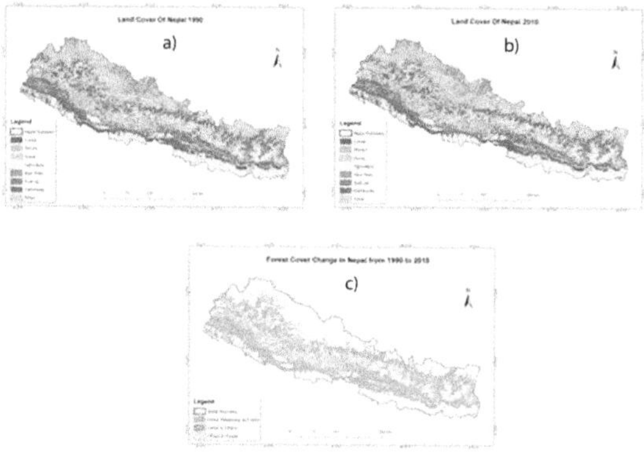

are not fulfilled by those conventional methods, these days we need digital form of data with the query functioned database and compatible database to interoperable among the systems i.e. geodatabase. The database also needed to be shared to access on it from the multiple locations and from the multiple responsibilities where web-based and internet technologies play its important role. Even the work from distributed locations is also possible if we provide such platform to the distributed analyst and experts.

For this purpose, one example work has been performed by eG-Tech Pvt. Ltd in support of Multi-Stakeholder Forestry Project (MSFP) in 2014 to prepare web enabled interactive spatial database of forest management regimes in 3 districts of western part of Terai, Nepal. The workflow applied in this work is presented in the following flow chart of the work (Fig. 6).

CLIMATE CHANGE VULNERABILITY AND ADAPTATION STRATEGY

Vulnerability and Adaptation Strategy

The geospatial approach can be adopted to document the spatial variability of the different indicators of exposure, sensitivity, adaptive capacity and ultimately vulnerability of the ecosystem and community. The geospatial approach thus documents the climate change and their impacts, temporal and spatial changes on the resources. During the formulation of National Adaptation Programme of Action (NAPA) in 2010, the Government of Nepal mapped the vulnerability of communities, taking into consideration extreme climatic events, disasters, climate change, migration of the population and institutional set up (MOE, 2010). Siddiqui et al. (2012) has also analyzed the vulnerability of communities considering water ecosystems on the basis of NAPA details for the Mid-Hill micro-watersheds. ICIMOD has implemented a geospatial approach for climate change vulnerability assessment with an interactive web-application considering forest ecosystems and their sensitivity and exposure (Chitale et al., 2014).

The vulnerability level of the Gandaki River Basin (GRB) has been described in the NAPA which considers climatic risk in terms of the following hazards: exposure, Human and ecological sensitivity and socio-economic, technology and infrastructure adaptive capacity at district level (MOE, 2010).

Figure 5. Methodological flow adopted to develop spatial database of forest management

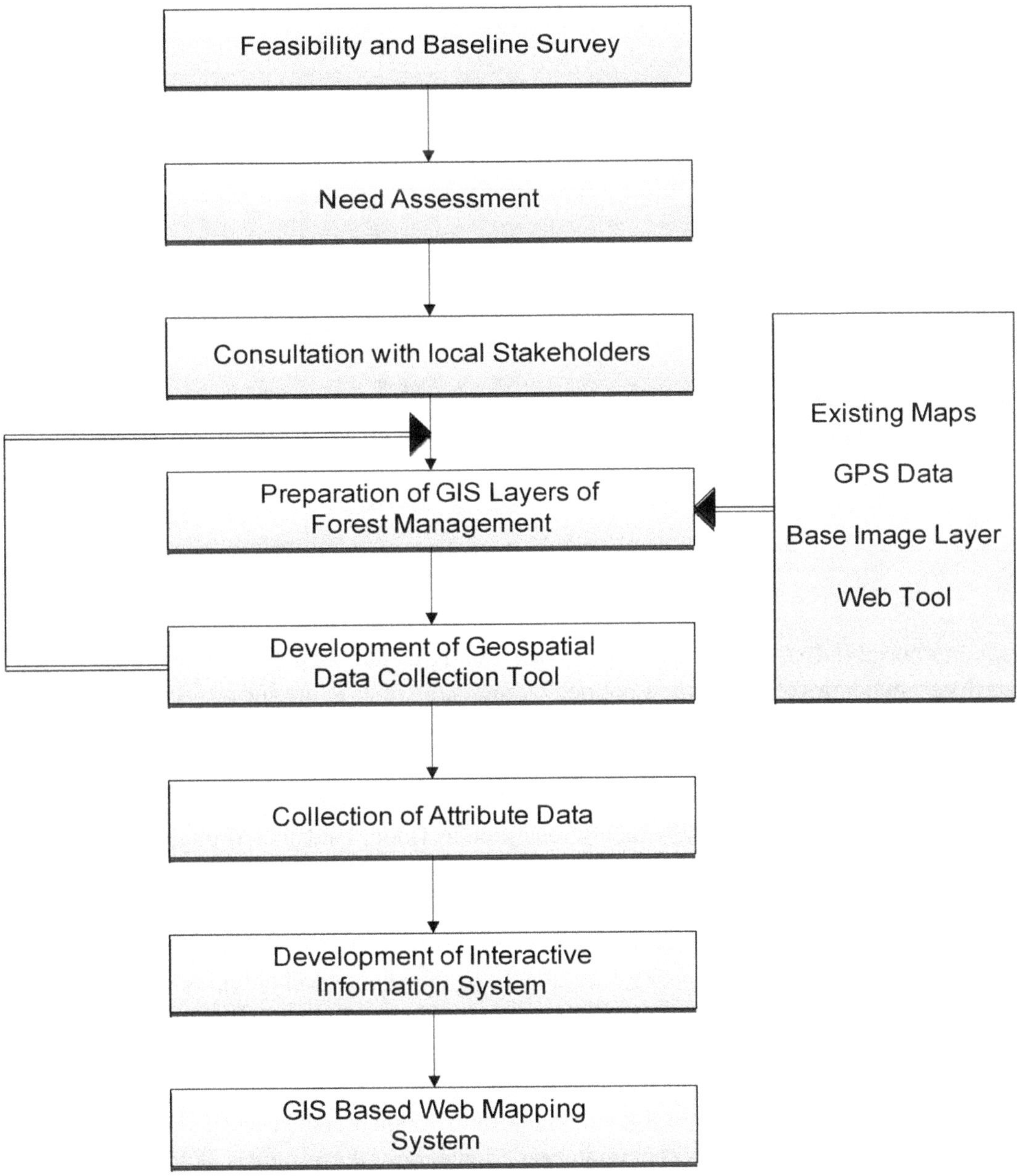

Vulnerability mapping of the GRB was carried out using inputs from the vulnerability layer generated by ICIMOD (Chitale et al., 2014). The layer used multiple inputs from the remote sensing-based indicators, spatial variables, topographic variables, climatic variables and field based bio-physical variables. Vulnerability was calculated considering the functions of exposure, sensitivity and adaptive capacity as suggested by IPCC (2006) using the 5 categories of the vulnerability scale i.e. Very Low, Low, Moderate, High and Very High. The details of the vulnerability layer generation can be viewed in the ICIMOD's

Figure 6. User Interface of the spatial database of forest management

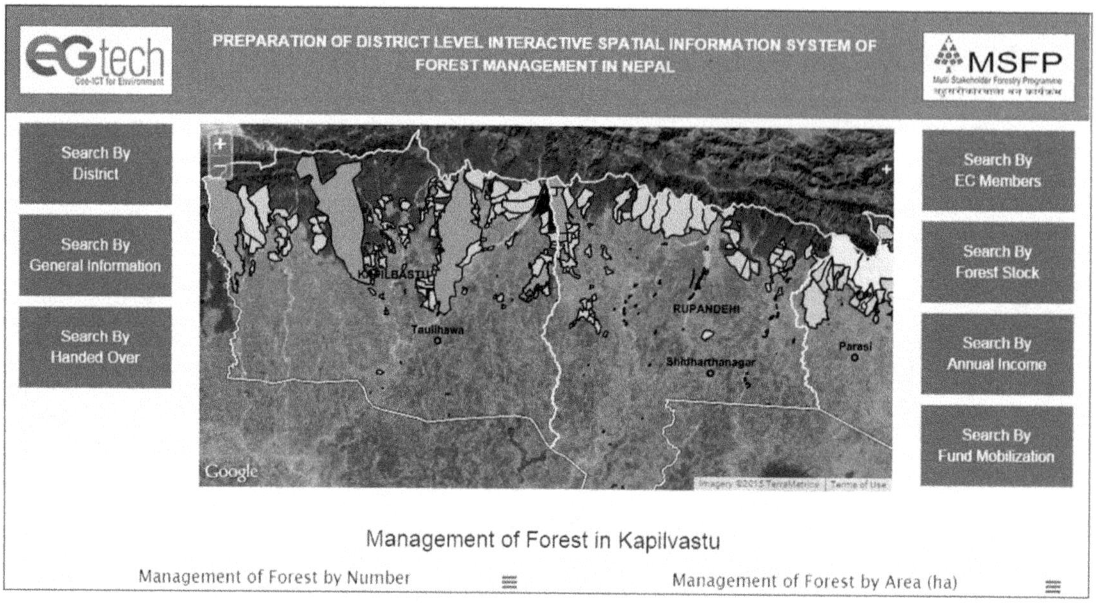

Vulnerability portal (http://geoapps.icimod.org/chalccv). ICIMOD prepared vulnerability classes for 5 and local level boundaries of Nepal's new political/administrative structure and project determined cluster level watersheds were overlaid and the vulnerability level for each cluster was generalized.

Climate change vulnerabilities are other threats to the community and to the ecosystem. Natural resource dependent communities are facing climate change vulnerabilities due to lack of resources to sustain their livelihood, climax weather conditions such as flood, landslide, drought, land degradation and depletion of soil productivity due to temperature rise and erratic and irregular rainfall pattern. Communities are exposed to those climatic factors and its resultant events. They are also sensitive due to the resource depletion, and they also have adaptive capacity to cope climate change impacts and consequences in terms of institutional capacity, traditional and indigenous technologies and advancement in the technologies to cope with the adverse climatic events. Above example of the village and municipal level of vulnerability categorization is based on the IPCC framework of climate change vulnerability of sensitivity, exposure and adaptive capacity of the community.

In other sense, ecosystems itself are also vulnerable due to climate change and anthropogenic activities ultimately trickle down to impact to the community as a depletion of the resources. ICIMOD has developed a web-based climate change vulnerability and adaptation strategy with the input of geospatial approach taking inputs from Remote Sensing data such as MODIS products and ground based adaptive capacity.

Vulnerability assessment at local unit to the complex landscape and ecosystem level is the initial step before planning for the adaptation strategy and options considering indicators, minimum units and the complexities of the ecosystem functionalities. NAPA analyzed the vulnerability using IPCC framework of vulnerability assessment considering districts as a unit considering following vulnerability indexes

International water management institute (IWMI) have accessed the watersheds including GRB in Nepal in terms of their climate change vulnerability and adaptation capacity and ranked them according to different categories at micro-watershed level (IWMI, 2012).

Figure 7. Community based vulnerability maps

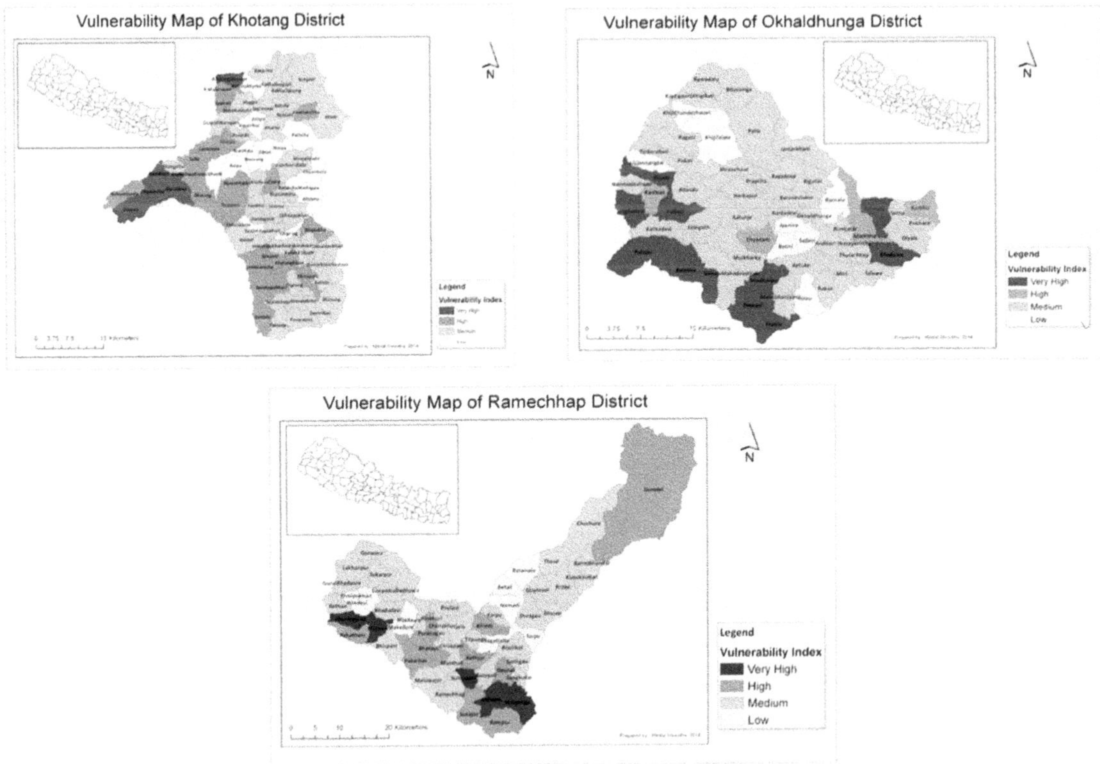

Mitigation and REDD+ Mechanism

Participation in the international REDD+ mechanism has potential for Nepal, to generate carbon revenues as well as non-carbon benefits for the country and its people. Preliminary estimates show that REDD+ may bring between \$20-86 million per year to Nepal (UN-REDD, 2014). Nepal further envisions that REDD+ implementation will assist in advancing sustainable forest management, the integral design of various sectoral policies that optimize cross-sectoral synergies, and will lead ultimately to an improvement of forest law enforcement and governance at large, with necessary amendment of act and regulations accommodating the concerns of all the stakeholders. A sound REDD+ architecture will also contribute to global low carbon emission development pathways and the global sustainable development agenda. REDD+ has emerged as a prominent climate change mitigation approach due to the potential

Table 1. Criteria for the vulnerability assessment of community

Sensitivity	Human, ecological
Exposure	Temperature, rainfall, ecological risk, flood and landslide risk, drought and GLOF
Adaptive capacity	Socio-economic condition, technological capability, infrastructure capacity

Figure 8. Ecosystem Based Vulnerability Map

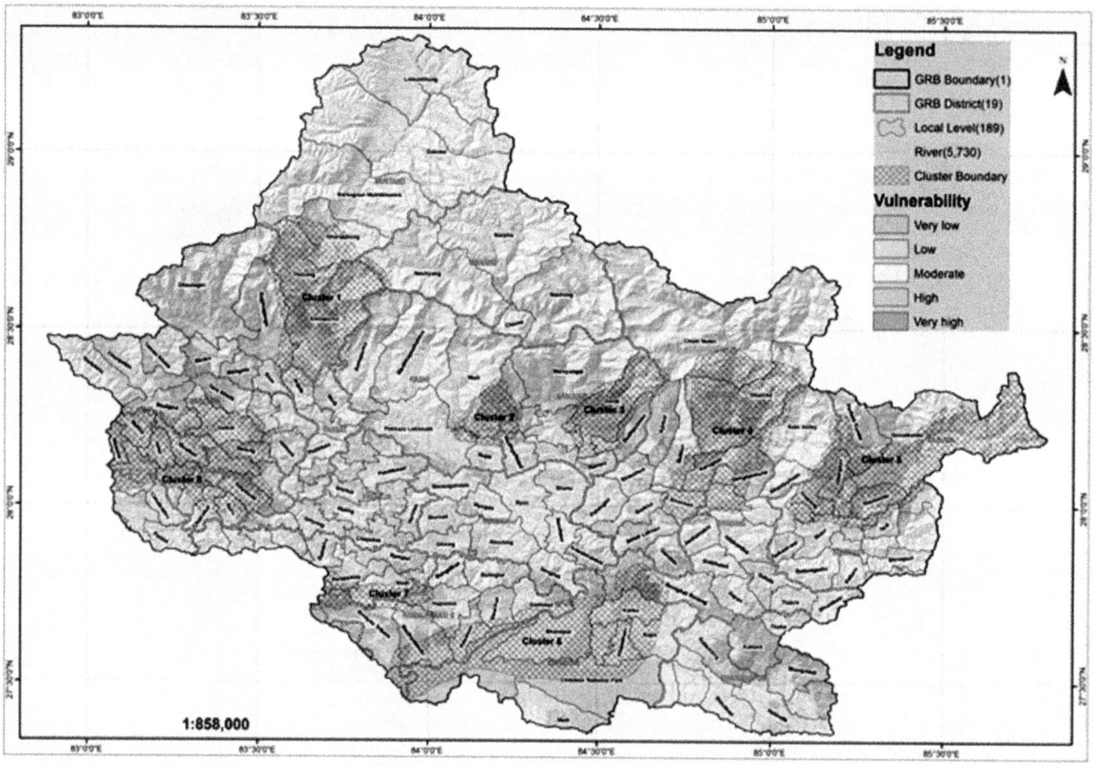

to provide multiple benefits to local and global communities. However, enhancement of carbon stock and reduction of carbon emissions for successful REDD+ implementation will pose challenges for the generation of non-carbon forest ecosystem services (ES) and conservation of forest biodiversity, thus trade-offs may take place between carbon and other non-carbon ecosystem services and forest biodiversity. Such challenges will be more prominent in community-based forest managed systems, where poor and forest-dependent people rely on forest resources for their subsistence.

Nepal has been a pioneer in implementing the REDD+ scheme. The country's engagement with REDD+ began when Nepal formally became a member of the forest carbon partnership facility (FCPF) scheme of the World Bank in 2008. Nepal initiated a REDD+ readiness process with the submission of a Readiness Project Idea Note (R-PIN) to the World Bank in March 2008, and was nominated for developing a Readiness Preparation Proposal by the FCPF/World Bank in mid-2008. With the approval of an Emission Reduction Project Idea Note (ER-PIN) from the World Bank, Nepal is now undertaking to initiate performance-based carbon emissions activities for 2015-2020 in Terai landscape areas (GoN, 2015).

The Government of Nepal has formed a separate unit – the REDD forestry and climate change cell under the Ministry of Forests and Soil Conservation (MFSC) for facilitating REDD+ related initiatives (Acharya et al., 2009; Dangi, 2012). A multi-stakeholder forum includes relevant civil society and non-government organizations that foster participation and coordination in formulating the national REDD+ strategy (Paudel and Karki, 2013).

The gross historical emissions from deforestation and forest degradation totaled 132,742,895 tCO_2e between 1990 to 2000, increasing to 293,231,645 tCO_2e in the period between 2000 and 2010. The overall trend shows an increase in emissions within the period 1990-2010. Going forward to 2020, net emissions are projected to continue in line with the reference period average at 18,842,458 tCO_2e / year, with deforestation accounting for 2,077,533 tCO_2e / year while forest degradation will account for 24,579,889 tCO_2e / year. During the same period, enhancement of forest carbon stocks is predicted to result in the removal of 7,814,964 tCO_2e / year based on a continuation of the average removals achieved during the period 2000 - 2010. Multi-stakeholder approach to benefit sharing has been stressed by REDD+ strategy. Desired outcomes for each of the objective have been mentioned. However, REDD+ implementation involves too much recentralization, and that too much revenue from collaborative forest management programs will be passed back to central government.

For ecosystem services (especially forest carbon and eco-tourism), biodiversity rich Nepal has high potential for benefitting from forest carbon sequestration and expansion of eco-tourism. While Nepal's community forests have significant potential to implement REDD+ projects geared towards cashing carbon, significant areas of barren lands or grasslands with scattered trees that cover about 1.6 million ha area are ideal for afforestation and reforestation activities. Our estimate shows that Nepal would be able to sell about 8.25 million tCO_2e per annum under conservative scenario and 19.5 million tCO_2e under optimistic scenario. For ecotourism, there is a good potential for expanding existing ecotourism products and developing new ones. The number of eco-tourists could be increased by at least 1.5 times under conservative scenario while 3 times in optimistic scenario (Subedi et al., 2015).

A study conducted by WWF Nepal, ANSAB and ICIMOD in 2015, in the 12 districts of GRB (Baglung, Dhading, Gulmi, Gorkha, Kaski Parbat, Myagdi, Manang, Mustang, Syangja, and Tanahun) shows that the land use change detection shows irregular pattern. These districts experienced a loss of 227.07 km^2 forest area from 1978 to 1990 and about another 114.28 km^2 forest area from 1990 to 2000. Whereas the geospatial analysis showed that there has been gain of 57.42 km^2 of forest area from 2000 to 2010, there was a net loss in forest area by -283.93 km^2 from 1978 to 2010. The baseline forest carbon stock was 540.1 million tCO_2e (147.17 million tC) with an average of 725.9 $tCO_2e\ ha^{-1}$. With an annual average of 12.97 $tCO_2e\ ha^{-1}$, the total annual CO_2e sequestration potential in 12 districts of the GRB was 9.66 million tCO_2e (Subedi et al, 2015).

The diverse social, cultural and ecological settings and highly dependence on forest resources have invited different anthropogenic disturbances such as firewood collection, uncontrolled grazing and lopping, illegal timber extraction and soil erosion and landslides of forest deforestation and forest degradation. Thus, the REDD+ project has enormous potential to contribute in maintenance and enhancement of forest carbon benefits and co-benefits of biodiversity conservation, livelihood generation and climate change adaptation and mitigation in the studied districts of the GRB through conversion of shrub land as well as grassland into forests promoting natural regeneration and plantation; improving value chains of non-timber forest products, sustainable management of the forest, introducing agroforestry practices, pastoralism and improved grazing; and bringing national forests under the community control or managed forests.

Assess the current carbon balance in the GRB, including current rates of GHG emissions/storage from agriculture and LULUCF, an assessment of potential storage t/ha for key habitats found in the river basin and a set of verified calculations on likely mitigation potential that will result from the project's planned interventions. Local communities are adopting against climate change and extreme climatic events using their own indigenous and traditional knowledge which seems insufficient as the impacts of climate change in present are more prevalent and severe impacts are visible on the community and ecosystem. The planned adaptation strategy seems to be needed. Considering these facts government has implemented different programs at community and local level to adapt towards climate change i.e. Community adaptation plan of action (CAPA) and local adaptation plan of action (LAPA).

CAPA basically focuses on the community level plan of action to cope with the adverse impacts of climate change looking into the future scenario of the climate changes and extreme climatic events i.e. landslide, flood, drought, GLOF etc. Whereas LAPA focuses on the local level adaptation planning confined within the local body of the administrative structure. The new local structure of local level administration demands to clump the previous attempt of local level planning to the municipal level as Government of Nepal has introduced new structure of local level from 4000VDC/Municipality to the 753 local level units.

On the other hand, considering the ecosystem vulnerability and ecological functionality of the complex landscape and watershed/basin, Ecosystem based Adaptation (EbA) has been introduced to maintain the ecosystem and serve to the community to build and enhance the resilience capacity of both the community and the ecosystem against climate change impacts. The concept of EbA includes the build the resilience of the community using ecosystem based natural resources and their services such as regulatory, provisioning and functional services of the natural systems consists of flora, fauna, land, water and their interactions. Panchase and Lamjung projects have implemented restoration of the forest through reforestation activities, enrichment plantation, species conservation, soil conservation, eco-tourism, alternative livelihood option such as bee-keeping, Cardamom plantation etc. to maintain the forest ecosystem, agro-ecosystems and to enhance the resilient capacity of the community.

FUTURE CONSIDERATIONS

Multiple stakeholders will have major role to develop the geospatial approaches to implement in forestry and natural resource management. Stakeholders for this kind of information range from decision-makers at the regional level addressing transboundary issues, to national governments, scientists, students, and the general public. There are future scopes to develop i) Geo-enabled national forest monitoring system, ii) Wall to wall carbon quantification and verification mechanism, iii) Spatial decision-making tools for climate change vulnerability assessment and adaptation strategy, iv) Spatial tools for forest management and v) Spatial tools for environmental decisions.

Acknowledgment

I would like to acknowledge institution: ICIMOD, IUCN, eG-Tech and others with whom I worked and allowed me to use the outputs in this publication. I also like to acknowledge Dr. Nabin K. Joshi, Mr. Bhaskar Shrestha, Mr. Hari Krishna Dhonju and Ms Sushmita Timilsina for their contributions during the analysis. I would also like to acknowledge Dr. Bharat M. Shrestha, Ms. Puspa Dhakal and Ms. Bhawana KC for their inputs. I would also like to acknowledge Dr. MSR Murthy for his guidance during the conceptualization and analysis.

REFERENCES

Bajracharya, S. B., Mool, P. K., & Shrestha, B. R. (2007). *Impact of Climate Change on Himalaya Glaciers and Glacial Lakes: Case Studies on GLOF and Associated Hazards in Nepal and Bhutan.* Kathmandu, Nepal: ICIMOD.

Baral, M. (2009). Water Induced Disasters, Flood Hazard Mapping & Koshi Flood Disaster of Nepal. In *Report prepared for East.* Manila, Philippines: Southeast Asia Regional Seminar on Flood Hazard Mapping.

CBS. (2004). *Handbook of Environment Statistics 2003.* Kathmandu, Nepal: Central Bureau of Statistics.

Collins, D. N., Davenport, J. L., & Stoffel, M. (2013). Climatic variation and runoff from partially-glacierised himalayan tributary basins of the Ganges. *The Science of the Total Environment*, 468–469. PMID:24296050

Cramer, W., Yohe, G. W., Auffhammer, M., Huggel, C., Molau, U., Dias, M. A. F. S., . . . Tibig, L. (2014). Detection and attribution of observed impacts. In C. B. Field, V. R. Barros, D. J. Dokken, K. J. Mach, M. D. Mastrandrea, T. E. Bilir, . . . L. L. White. (Eds.), Climate Change (2014): Impacts, Adaptation, and Vulnerability. Part A: Global and Sectoral Aspects. Contribution of Working Group II to the Fifth Assessment Report of the Intergovernmental Panel on Climate Change (pp. 979–1037). Cambridge, UK: Cambridge University Press.

Dangi, R. (2012). REDD+: Issues and challenges from a Nepalese perspective. *Climate Change and UNFCCC Negotiation Process,* 61.

DeFries, R., Achard, F., Brown, S., Herold, M., Murdiyarso, D., Schlamadinger, B., & de Souza, C. Jr. (2007). Earth observations for estimating greenhouse gas emissions from deforestation in developing countries. *Environmental Science & Policy*, *10*(4), 385–394. doi:10.1016/j.envsci.2007.01.010

Department of Forest Research and Survey. (2015). *State of Nepal's Forests.* Government of Nepal.

Dixit, A. (2009). Koshi embankment breach in Nepal: Need for a paradigm shift in responding to floods. *Economic and Political Weekly*, 70–78.

DOHS. (2005). *Annual Report 2003/2004.* Kathmandu, Nepal: Department of Health Services.

Edward, R. (2003). Dendroclimatic signals in long tree-ring chronologies from the Himalayas of Nepal. *International Journal of Climatology*, *23*(7), 707–732. doi:10.1002/joc.911

Eggleston, S., Buendia, L., Miwa, K., Ngara, T., & Tanabe, K. (Eds.). (2006). *2006 IPCC guidelines for national greenhouse gas inventories* (Vol. 5). Hayama, Japan: Institute for Global Environmental Strategies.

Garg, S., Shukla, P. R., & Kapshe, M. (2007). From Climate Change Impacts to Adaptation: A Development Perspective for India. *Natural Resources Forum*, *31*(2), 132–141. doi:10.1111/j.1477-8947.2007.00142.x

Gautam, D. K., & Dulal, K. (2013). Determination of Threshold Runoff for Flood Warning in Nepalese Rivers. *Journal of Integrated Disaster Risk Management*, *3*(1), 125–136. doi:10.5595/idrim.2013.0061

Gibbs, H. K., Brown, S., Niles, J. O., & Foley, J. A. (2007). Monitoring and estimating tropical forest carbon stocks: Making REDD a reality. *Environmental Research Letters*, *2*(4). doi:10.1088/1748-9326/2/4/045023

GOFC-GOLD. (2017). *A sourcebook of methods and procedures for monitoring essential biodiversity variables in tropical forests with remote sensing*. The Netherlands: GOFC-GOLD Land Cover Project Office, Wageningen University.

GoN/MoFSC. (2014). *Nepal Biodiversity Strategy and Action Plan 2014-2020.* Kathmandu, Nepal: Government of Nepal, Ministry of Forests and Soil Conservation.

Herold, M., Román-Cuesta, R. M., Mollicone, D., Hirata, Y., Van Laake, P., Asner, G. P., & MacDicken, K. (2011). Options for monitoring and estimating historical carbon emissions from forest degradation in the context of REDD+. *Carbon Balance and Management*, *6*(1), 13. doi:10.1186/1750-0680-6-13 PMID:22115360

Karna, Y. K., Hussin, Y. A., Gilani, H., Bronsveld, M. C., Murthy, M. S. R., Qamer, F. M., ... Baniya, C. B. (2015). Integration of WorldView-2 and airborne LiDAR data for tree species level carbon stock mapping in Kayar Khola watershed, Nepal. *International Journal of Applied Earth Observation and Geoinformation*, *38*, 280–291. doi:10.1016/j.jag.2015.01.011

Khanal, S. (2015). Wildfire trends in Nepal based on MODIS burnt-area data. *Banko Janakari*, *25*(1), 76–79. doi:10.3126/banko.v25i1.13477

Matin, M. A., Chitale, V. S., Murthy, M. S., Uddin, K., Bajracharya, B., & Pradhan, S. (2017). Understanding forest fire patterns and risk in Nepal using remote sensing, geographic information system and historical fire data. *International Journal of Wildland Fire*, *26*(4), 276–286. doi:10.1071/WF16056

NCVST. (2017). *Vulnerability through the eyes of the vulnerable: climate change induced uncertainties and Nepal's development predicaments, Institute for Social and Environmental Transition – Nepal (ISET-N).* Kathmandu & Boulder, CO: Institute for Social and Environmental Transition.

NDRI. (2010). *United Nations World Food Programme (WFP); Food Security Monitoring Task Force of National Planning Commission (NPC). The food Security Atlas of Nepal; Nepal Development Research Institute.* Patan, Nepal: NDRI.

Parajuli, A., Chand, D. B., Rayamajhi, B., Khanal, R., Baral, S., Malla, Y., & Poudel, S. (2015). Spatial and temporal distribution of forest fires in Nepal. In *XIV World Forestry Congress*, Durban, South Africa (pp. 7-11).

Regmi, B., & Paudyal, A. (2009). *Climate Change and Agro biodiversity in Nepal: Opportunities to include agrobiodiversity maintenance to support Nepal's National Adaptation Programme of Action*. NAPA.

Rimal, R. K., Maharjan, R., Khanal, K., Koirala, S., Karki, B., Nepal, S. M., & Shrestha, H. L. (2017). Detection, assessment, and updating the maps of encroached forest areas: A case study from Bara district, Nepal. *Banko Janakari*, *27*(1), 65–71. doi:10.3126/banko.v27i1.18554

Shrestha, H. L., & Dhonju, H. K. (2013). Satellite Based Forest Fire Burned Area Assessment in Nepal from 2000 to 2013. *Journal of the Institute of Forestry,* Nepal, 98.

Shrestha, H. L., Uddin, K., Gilani, H., Pradhan, S., Shrestha, B. R., & Murthy, M. S. R. (2015). Forest Carbon Flux Assessment in Nepal Using the Gain-Loss Method. In M. S. R. Murthy, S. Wesselman, & H. Gilani (Eds.), Multi-scale forest biomass assessment and monitoring in the Hindu Kush Himalayan region: A geospatial perspective. Kathmandu, Nepal: ICIMOD

Siddiqui, S., Bharati, L., Pant, M., Gurung, P., & Rakhal, B. (2012). Nepal: building climate resilience of watersheds in mountain eco-regions climate change and vulnerability mapping in watersheds in middle and high mountains of Nepal. ADB Technical Assistance Consultant's Report for Department of Soil Conservation and Watershed Management (DSCWM), Government of Nepal. Kathmandu, Nepal: Asian Development Bank (ADB). 96p.

Subedi, B. P., Gauli, K., Joshi, N. R., Pandey, A., Charmakar, S., Poudel, A., ... Khanal, S. C. (2015). *Forest Carbon Assessment in Chitwan-Annapurna Landscape. Study Report, WWF Nepal Hariyo Ban Program*. Kathmandu, Nepal: Baluwatar.

Subedi, B. P., Ghimire, P. L., Koontz, A., Khanal, S. C., Katwal, P., Sthapit, K. R., & Mishra, S. K. (2014). *Private Sector Involvement and Investment in Nepal's Forestry: Status, Prospects and Ways Forward. Study Report, Multi Stakeholder Forestry Programme - Services Support Unit*. Kathmandu, Nepal: Babarmahal.

Uddin, K., Shrestha, H. L., Murthy, M. S. R., Bajracharya, B., Shrestha, B., Gilani, H., ... Dangol, B. (2015). Development of 2010 national land cover database for the Nepal. *Journal of Environmental Management*, *148*, 82–90. doi:10.1016/j.jenvman.2014.07.047 PMID:25181944

Chapter 15
Trends in Management of Tropical Forests:
Application of Remote Sensing and Geographic Information System

Jacinta U. Ezenwenyi
Nnamdi Azikiwe University, Nigeria

Onyekachi Chukwu
https://orcid.org/0000-0002-6376-6084
Nnamdi Azikiwe University, Nigeria

ABSTRACT

Forests are important plant communities that consist of trees and other woody vegetation that performs life-supporting functions on Earth. This chapter presents the application concepts necessary to link remote sensing data and geographic information system to conservation, restoration, and sustainable management of tropical forests. Despite the rising global concern, there is still continuous destruction of tropical forests at an alarming rate. This chapter assessed various approaches for conservation, restoration, and sustainable management of tropical forests using remote sensed data and geographic information system. This involves their applications to biodiversity conservation, forest ecophysiology, forest trees' disease and insect interactions, forest mensuration, forest resources monitoring and evaluation, forest fires, land use and land cover dynamics, and vegetation cover.

INTRODUCTION

Forests play dynamic roles in providing shelter and food for wildlife, purifying the air, regulating climate and controlling water runoff (Cunningham *et al.*, 2005). Tropical dry forests are distinctive ecosystems with extraordinary stages containing many unique flora and fauna. They are forests categorized by prominent dry season throughout some parts of the year, which aggravates a diversity of adaptations in plants and animals. Various plant species of these forests are deciduous; shedding their leaves at the beginning of

DOI: 10.4018/978-1-7998-0014-9.ch015

the dry season. There are also other plants adaptations which are designed for water conservation such as spines leaves, waxy leaves, photosynthetic bark and swelling of tissues for water storage during the raining season. Forests in Nigeria are within the tropics, spanning from the tropical rain forests in the southern part to the tropical dry forests towards the north. The dry forests in the northern part of Nigeria are characterised by prolonged dry season and provide unique ecosystems which are under threat. These forests are steadily converted to grassland due to anthropogenic activities; more prominently are cattle pasture and felling of trees for fuel wood. The conversion has caused fragmentation and loss of large scale of wildlife habitat (Ali *et al.,* 2014).

Despite the rising global environmental concern, there is still continuous destruction of tropical forest at an alarming rate. Forest like any other natural resources is of immense value to man. The constant exploitations of these forests have resulted in serious reservations for their sustainability. The demands for conservation and management methods could help these forests meet the need of future generations. Many species of trees are currently faced with extinction while some are at different phases of risk. The interference of man into forests for different purposes have resulted in loss of biodiversity, deforestation, reduction in the available forest, loss and extinction of plant and animal species resulting to predictable long term effect from climate change. The prevailing interdependent association amid man and his regular environment requires an equilibrium system on the activities that impose the nature in demand to certify the sustainability of the environment. Forest ecosystems provide variety of services that help in maintaining the status of life support systems. Conservation and proper management strategies are of vast concern if the needs of future generations are to be met in a sustainable way.

Forest resources are one of the most key renewable natural resources aimed for timber production, raw material for pulp and paper and other forest resources as well as genetic resources. It is progressively documented that the loss of forest generates some amount of serious environmental and conservation problems such as: soil erosion, flood, sedimentation, and loss of valued materials which aid to ensure the peoples' lifespan (Agbelade and Akindele, 2013). It is a matter of worry that tropical dry forests are constantly disappearing at disturbing speed due to the increasing demand over ecosystem services and natural resources. In whichever zone, forest is a true indicator of ecological system dominant.

Remote sensing (RS) is an indispensible tool for providing dependable data and precise information about natural resources such as forest. Geographic information system (GIS) on the other hand, is a system for capturing, storing, analysing and managing data and associated attributes, which are spatially referenced to the earth. They can be used for scientific investigations, resource management, and asset management to mention but few. Forest managers in Nigeria and the world at large are whirling progressively to digital forest inventories and geographic information systems to help in the management of forest resources. In meeting the numerous information desires for forest conservation, restoration and management, different data sources, such as field survey, aerial photography and satellite imagery are used.

This chapter therefore, aimed at exploring the various areas of application of remote sensing and GIS in forest management such as species composition (biodiversity), forest ecophysiology, forest trees' disease and insect interactions, forest mensuration, forest resources monitoring and evaluation, forest fires, land use and land cover dynamics and vegetation cover.

Geographic Information System (GIS)

Geographic information systems are used to collect, store, analyze, distribute and control information that can be referenced to a geographical position. GIS is a tool capable of integrating, storing, editing, analysing, sharing and displaying remotely sensed data as geographically referenced information. Histories of types of forest or environmental information such as forest fire, land use and land cover, forest degradation to mention but few can be stored in a database and recorded to indicate their places of occurrences. The relatively recent development of GIS technology seems preferably suitable to conservation exertion, because they empower conservationists to expeditiously attain, store, analyze, and display spatial data on entities and their environment (Johnston, 1998). Additionally, the recent improvement of technology particularly the development of GIS plays an important role to detect immediate actual disruption.

Remote Sensing (RS)

Remote sensing is the science and art of acquiring information (spectral, spatial, and temporal) about materials, objects, area and/or phenomenon, without coming into physical contact with the resources or phenomenon under study (Natural Resources Canada, 2008). According to Franklin (2010) remote sensing is both technology (sensors, platforms, transmission and storage devices, and so on) and methodology (radiometry, geometry, image analysis, data fusion, and so on). The confirmation based on the literatures about these technologies, showed that they have played important roles in viable decision making, proper management, sound planning (short-term and long-term), and better allocation of resources (Wang *et al.,* 2010).

Remote sensing can be used for various forestry applications (qualitative and quantitative assessment) with high accuracy of some sensor systems for different purposes which include apprising of prevailing forest inventories, cultivation, terrain analysis, forest cover type discernment, the description of fire regime and mapping for forest management. Remotely sensed data are being used as the source of this information (Ahern and Leckie, 1987). Remote sensing in the form of aerial photography has delivered evidence for many years, satellite sensors have allowed information to be provided over wider areas than possible with aerial photography. Previously, aerial photography was used for forest management purposes; information was generally obtained by means of field surveys, identifying and measuring forest types and stands. Conversely, this method was very elaborate, time consuming and expensive and currently use for areas with small sample sizes.

Sensors and Platforms Used in Remote Sensing

Remote sensors are conventionally divided into two groups: passive sensors (utilizes solar radiation to illuminate the Earth's surface and detect the reflection from the surface) and active sensors (provide their own source of energy to illuminate the objects and measure the observations). A new approach in 2013 was developed with the ability to integrate active and passive infrared imaging capability into a single chip (Zhu *et al.,* 2018). This in modern remote sensing made it difficult to categorize sensors traditionally into passive and active sensors. Hence, remote sensing sensors can be classified into imaging sensors and non-imaging sensors in this chapter.

- **Imaging Sensors:** Typically employ optical imaging systems, thermal imaging systems, and/or synthetic aperture radar (SAR). Optical imaging systems use the visible (0.4–0.7 μm), near-infrared (0.7–1.3 μm), and shortwave infrared (1.3–3 μm) spectrums and typically produce panchromatic, multispectral, and hyperspectral imagery. Thermal imaging systems employ mid-to-far infrared wavelengths.

- **Non-Imaging Sensors:** Include microwave radiometers, microwave altimeters, magnetic sensors, gravimeters, Fourier spectrometers, laser rangefinders, and laser altimeters (Japan Association of Remote Sensing, 2010; Zhu *et al.,* 2018).

Remote sensing platforms can be defined as the structures or vehicles on which remote sensing instruments (sensors) are mounted. For remote sensing applications, sensors should be mounted on suitable stable platforms. As the platform height increases, the spatial resolution and observational area increases. These platforms can be ground based, air borne and/or space borne based.

1. **Ground-Borne Platforms:** Ground borne platforms are used to record detailed information about the surface which is compared with information collected from aircraft or satellite sensors i.e. for ground observation. Ground observation platforms include – handheld platform, cherry picker, towers, portable masts and vehicles etc. Portable handheld photographic cameras and spectroradiometers are largely used in laboratory and field experiments as a reference data and ground truth verification.
2. **Air-Borne Platforms:** Airborne platforms are used to collect very detailed images and facilitate the collection of data over virtually any portion of the Earth's surface at any time. Airborne platforms were the sole non-ground-based platforms for early remote sensing work. They include: balloon, drone, aircraft and high altitude sounding rockets.
3. **Space-Borne Platforms:** In space-borne remote sensing, sensors are mounted on-board a spacecraft (space shuttle or satellite) orbiting the earth. Space-borne or satellite platform are onetime cost effected but relatively lower cost per unit area of coverage, can acquire imagery of entire earth without taking permission. They provide; large area coverage, frequent and repetitive coverage of an area of interest, quantitative measurement of ground features using radiometrically calibrated sensors, semi-automated computerized processing and analysis and Relatively lower cost per unit area of coverage. There are two types of well recognized satellite platforms: manned satellite platform and unmanned satellite platform.

Presently, more sophisticated remote sensing platforms and sensors with automated image analysis techniques for consistently and repeatedly monitoring of forests over larger areas are used. Furthermore, numerous types of remote sensing data have been used by forest agencies to detect, identify, classify, evaluate and measure various forest cover types and their changes. These include multi-spectral scanner (MSS), Radio Detection and Ranging (RaDAR), Light Detection and Ranging laser (LiDAR) and Videography data (Agbelade and Akindele, 2013, Ali *et al.,* 2014). The satellite imageries have shown effectiveness for forest classification and consequently mapping for large areas. Gradually, other types of remote sensing tools were developed with which forest object properties were registered from the space. The new technologies, integrating satellite imagery, analytical photogrammetry and geographic information systems (GIS) offer new possibilities, especially for general interpretation and mapping

Currently, numerous high-resolution satellites that can be used to detect forest disruption have been launched and available with free or low cost, quality data, adequate spatial and temporal details.

Table 1. Common sensors and platforms and their areas of application in forestry

Sensor Space (S)/ Airborne (A)	Temporal resolution	Spectral and Spatial resolution	Some area of Application in Forestry and Environmental Studies
SLICER (A), LVIS (A)	N/A (SLICER, LVIS)	V/NIR. 1–10 m	**Vertical canopy structure:** Provides 3D measurements via laser pulses; provides biomass estimates and information about vegetation structure
TM/ETM+ (S), ALI (S), HYPERION (S), ASTER (S), IKONOS (S), Quickbird (S), AVIRIS (A), CASI (A)	16 days (ETM, ALI, Hyperion); 4–16 days (ASTER); 2–5 days (IKONOS); 2–4 days (Quickbird); N/A for aircraft	V/NIR, SWIR, ASTER also has TIR. <1–30 m	**Species Composition:** Tested for their ability to measure canopy community, and perhaps species, type based upon unique spectral signatures.
MODIS (S), TM/ETM+ (S), ASTER (S), ALI (S), IKONOS (S), Quickbird (S)	1–2 days (MODIS); 16 days (TM/ETM+); 4-16 days (ASTER); 2–5 days (IKONOS); 2–4 days (Quickbird)	V/NIR, SWIR, MODIS and ASTER also have TIR. <1–1000 m	**Land Use/Land cover:** Can discriminate different land surfaces at various resolutions; land cover classification is considered a first-order analysis for species occurrence.
SeaWIFS (S), MODIS (S), ASTER (S), TM/ETM+ (S), ALI (S), Hyperion, (S), IKONOS (S), Quickbird (S), AVIRIS (A), CASI (A)	1 day (SeaWiFS); 1–2 days (MODIS); 4–16 days (ASTER); 16 days (TM/ETM+), (ALI, Hyperion); 2–5 days (IKONOS); 2–4 days (Quickbird)	V/NIR, SWIR, MODIS and ASTER also have TIR. <1–1000 m	**Vegetation (Chlorophyll and Phenology):** Measure reflectance to assess greenness measures nibbling detection of water and land surface chlorophyll useful for calculating productivity and plant health. Information on leaf turnover and flowering/fruiting cycles. Provides for identification of species tied to certain phenological events.
CERES (S), AMSR-E (S)	1–2 days (CERES, AMSR-E)	Microwave. 5.4–56 km	**Rainfall and Soil moisture:** Enable detection of precipitation and surface moisture at coarse resolutions.
SRTM (S), ATM (A), ASTER (S), IKONOS (S), SLICER (A), LVIS (A)	N/A (SRTM); 4–16 days (ASTER); 2–5 days (IKONOS); N/A (SLICER, LVIS)	Microwave SRTM; V/NIR and SWIR for others. 90 m SRTM; 30 m/15 m ASTER; 1–15 m IKONOS, SLICER, LVIS	**Topography:** DEM derived from radar signals via interferometry (SRTM); image stereo pairs (ASTER) or discrete-return (usually) LIDAR signals.

Where: ALI, Advanced Land Imager; AMSR-E, Advanced Microwave Radiometer for EOS; ASTER, Advanced Spaceborne Thermal Emission and Reflection Radiometer; ATM, Airborne Topographic Mapper; AVIRIS, Airborne Visible/ Infrared Imaging Spectrometer; CASI, Compact Airborne Spectrographic Imager; CERES, Clouds and the Earth's Radiant Energy System; ETM +, (Landsat) Enhanced Thematic Mapper Plus; LIDAR, Light Detection and Ranging; LVIS, Laser Vegetation Imaging Sensor; MODIS, Moderate-resolution Imaging Spectroradiometer; SeaWiFS, Sea-viewing Wide Field-of-view Sensor; SLICER, Scanning LiDAR Imager of Canopies by Echo Recovery; SRTM, Shuttle Radar Topography Mission; SWIR, short-wave infrared (roughly corresponds to the near and middle infrared bands); TIR, thermal infrared; TM, (Landsat) thematic Mapper; V/NIR, visible/near-infrared.

Source: (Turner *et al.*, 2003; Wani *et al.*, 2016)

ISSUES IN THE MANAGEMENT OF TROPICAL DRY FOREST

Tropical dry forests are the most threatened of the major tropical forest types (Janzen 1988). Their threats mostly are due to anthropogenic activities (Gandhi and Sundarapandian, 2014) which altered the tree species diversity and population structure, destruction of wildlife habitat, forest fragmentation, and disruption of the biogeochemical cycle of the ecosystem. There is a growing demand for the measurement and monitoring of the state of forested landscapes for the purposes of conservation, biodiversity, sustainable development and management (Layton *et al.*, 2004). As a consequence, efficient and accurate methods are required for the measurement of forest resources, regardless of whether these are made by using remote sensing approaches or field survey. Unlimited efforts had been made in current years to draft and implement forest management plans wholly and efficiently. Sustainable forest management application has been limited by inadequate availability of information to forest administration in most countries. In addition, political changes also have effects on effective sustainable management of tropical dry forests.

Universally, the main problems to a better understanding of tropical forest conservation and restoration arise from dearth of information about the management application and a meagre understanding of ecosystem dynamics, particularly techniques for promoting adequate regeneration method. Remote sensing and GIS technologies help in overcoming most of these difficulties. Up-to-date information about these subjects would be useful to reinforce the basis on which sustainable forest management system are developed. Nevertheless, it is commonly believed that the results obtained to date (from numerous studies and projects) have provided a reasonably solid basis for effective management of tropical dry forests.

In practice, researchers select one or numerous forms of remotely sensed data conferring to their information needs. The information requirements are converted to particular properties of remotely sensed data, such as spatial resolution, spectral resolution, temporal resolution, etc. These sensors offer various remotely sensed data with a distinguishing structure of image resolutions.

VARIOUS AREAS OF APPLICATION OF REMOTE SENSING/ GIS IN FOREST MANAGEMENT

Species Composition (Biodiversity)

In 1998, the Centre of Biodiversity and Conservation (CBC) established Remote Sensing (RS) and Geographic Information System (GIS) services. These tools have helped in identifying possible assessment of locations, analyzing deforestation rates in important study regions, integrated spatial and non-spatial databases and generate convincing graphical supports to develop reports and applications. Stoms and Estes (1993) proposed a framework to guide the application of remotely-sensed data in mapping and monitoring biodiversity.

According to Turner *et al.* (2003) the current improvements in remote sensing, such as the availability of remotely sensed data with high spatial and spectral resolutions, make it feasible to identify key conservational parameters, which can be applied to determine the distribution and abundance of species across landscapes through ecological models. This method generally referred to as indirect remote sensing, plays a key part in of biodiversity assessment. There has been an outstanding upsurge in world observation satellites and sensors over the past few years that are being used to measure and model biodiversity from space (Gillespie *et al.*, 2008). For instance, studies have reported the use of hyper spectral image data for differentiation of several tropical species (Martin *et al.*, 1998), classification of coniferous species (Gong *et al.*, 2001), species distribution patterns (Debinski *et al.*, 1999), ecology and biodiversity in forest (Guerschman *et al.*, 2009).

Land Use and Land Cover Dynamics

In the planning and conservation of biodiversity in the dry tropical forest, the understanding of the dynamics of the forest loss is critical. However, in realizing forest dynamics to support restoration, conservation, preservation and evaluation of the effectiveness of forest management, perceptions into

Table 2. Application of RS/GIS in the field of tropical dry forest studies

RS Data	GIS	Application	Source
Multispectral satellite data obtained from Landsat 7 for the years 2000, 2005 and 2016	Change detection analysis	Land use and land cover (LULC)	Cheruto *et al.* (2016)
Earth observation (EO) data from Landsat TM 5 imagery	Multi criterion evaluation technique (modelling approach)	Forest fire probability	Srivastava *et al.* (2018)
Landsat NDVI, Ikonos imagery, and topographic information	Identifying the distribution of forests, woodlands, and shrub land, providing a natural representation of the vegetation patterns on the island.	Vegetation types	Martinuzzi *et al.* (2008)
Multi-Season Landsat OLI Data (Landsat 8 OLI 30 m imagery)	Model for burn scar mapping	Classification of Burned areas	Axel (2018)
LiDAR (Light Detection and Ranging)	Classification	Forest type and succession	Martinuzzi *et al.* (2013)
IRS-ID LISS-III satellite imagery of path 103 and row 57 during December 2001 from National Remote Sensing Agency (NRSA),	NDVI analysis	Biomass and carbon estimation	Bijalwan *et al.* (2010)
IRS-ID LISS-III satellite imagery of path 103 and row 57 during December 2001 from National Remote Sensing Agency (NRSA),	Supervised classification using maximum likelihood algorithm.	Land-use and vegetation status	Bijalwan *et al.* (2010)

the rates, forms and trends of landscape changes are required. Ramachandra and Kumar (2004) defined change detection as a procedure that is used to enumerate the changes that are related with land use and land cover in the landscape using geo-referenced multi-temporal remote sensing information. Speedy population growth and growing commercial activities have resulted in unsustainable exploitation and rapid decline in the spatial extent of forest reserves in Nigeria and other parts of the world.

Therefore, studying land use dynamics of these forests is essential for analysing various ecological and developmental consequences over time. Land use/land cover mapping, change detection and prediction are indispensible for proper managerial and execution of appropriate policy retorts concerning to land uses. For example, Adedeji *et al.* (2015) in their study for assessing and predicting changes in land use/land cover at Gambari forest reserve, Nigeria used remote sensing and GIS techniques. The study determined the degree, frequency and dynamics of change in the spatial magnitude of the forest reserve between 1984 and 2014 using multi-temporal datasets. They reported dramatic decline in the extent of the forests as a result of logging in several parts for timber, cultivation and pressure from other land uses like settlements. They concluded that increasing loss of the forest would continue hence having implications for biodiversity conservation in the study area. Thus, there is the need for involvement of different shareholders and segments to resolve the disagreeing demands on restricted resources and ensure ecosystem reliability.

Correspondingly, Diallo *et al.* (2009) used remote sensing to highlight the significance of digital change detection in capturing the environmental condition in the forested southern part of Yunnan Province, China. Severe landcover changes was observed to have occurred in crop lands (+24.90%), forest or shrub land (-18.77%) and building (+16.72%) areas. Thus, the unused area constituted the most extensive type of land use and landcover in the period under consideration. Forest cover changes around the Santchou Wildlife Reserve of west Cameroon was assessed by Meli and Meli (2015) using remotely sensed Landsat images. The study observed that a combined regrowth of degraded lowland forest and abandoned farmlands is possible if factors that eradicated this secondary plant succession process are curtailed.

Willkomm *et al.* (2016) also examined general land use and landcover structures and changes between 2010 and 2015 around Mt. Kenya region using a remote sensing method with high resolution Rapid Eye satellite images. The study detected that it is possible to detect main landcover classes and characterize the spatial and spectral mixture of different areas through remote sensing technology. Tudun-Wada *et al.* (2014) examined Remote Sensing (Landsat Thematic Mapper (TM)) for analysis of forest cover in Nimbia Forest Reserve, Kaduna State, Nigeria between 1986 and 2010 with the objective to detect the changes that may have taken place in their study area between these periods. The 1986, 1998 TM and 2010 ETM Satellite remote sensing data were used to identify, classify, assess and interpret Nimbia Forest Reserve for the year 1986, 1998 and 2010 respectively. They reported that Landsat data proved to be effective in mapping and monitoring the dynamics of land use /land cover of the study area.

Vegetation Cover

Remotely sensed data is a proven source of information for detailed characterization of vegetation type and condition (Luther e*t al.,* 2006; Rossini *et al.,* 2006). Ali *et al.* (2014) examined three decades of freely available Landsat satellite images of the north-eastern part of Nigeria using a supervised classification-based technique to create maps of vegetation change in Yobe State. The maps were then used to examine the temporal and spatial aspects of changes which have occurred in the context of previous evidence and literature. Their study therefore, investigated Yobe state (an area that falls within those examined by Odihi (2003) and Anyamba and Tucker (2005)) and one of the country's known "hotspots" for vegetation degeneration (Odiogor, 2010), using Remote Sensing and fieldwork-based findings. They reported that the remote sensing analysis clearly showed large changes in vegetation cover across the study area from 1978-2005.

Oseia *et al.* (2005) as well assessed and analyzed the temporal patterns and rate(s) of change in land use and forest coverage area; and compare the rates of change over the period 1978-2003 in a region located in South-western Nigeria, well known for its dense forest resources and fertile soils. Remotely sensed data of different sources (Landsat MSS, 1978, SPOT XS, 1986 and NigeriaSat-1, 2003) acquired in rainy seasons and (SPOT XS, 1994) obtained in dry season were used for their study. Onojeghuo and Onojeghuo (2015) utilised multi-temporal Landsat and UK-DMC-2 satellite images over three epochs (1987, 1999 and 2011) to assess forest landscape change of Okomu forest reserve, Nigeria. Defries *et al.* (2000) applied the 1km Advanced Very High Resolution Radiometer (AVHRR) to estimate and map percentage vegetation cover and associated proportions of trees with different leaf longevity (evergreen and deciduous) and leaf type (broadleaf and needle-like leaf).

Forest Resources Monitoring and Evaluation

Remote sensing (RS) is a good and generally used technique for mapping tropical forests. Through the use of RS methods, current information can be acquired for large areas at the same time. The temporal nature of remote sensing platforms, offers a possibility to monitor tropical forests at regular intervals.

Though the early inventory of forest resources deposited in a GIS is imperative, fluctuations exist that need to be monitored and documented. For instance, assessment of forest health in terms of forest degradation and silvicultural practices for timber management include composite and explicit involvement to control stand density, species composition, stand structures, health indicators, maintaining site quality, invasive species, climatic measures (such as drought, frost, storms and floods) and harvesting strategies. Additional variations could result from speedy, distinct measures and/or instabilities, such as immense deforestation or disease that recruit different growth patterns in the affected regions.

Remote sensing can be employed for a long-term monitoring of dry tropical forest over a larger area. Hence, remote sensing technology makes significant contributions. To meet up-to-date information requirements, assessments of forest destruction need to be timely and completed at numerous spatial measures. Consequently, adding to broad monitoring programmes that offer time-series information on precise form of damage and their causes, there is a need for local and regional inventories improved to specific damage measures. In this way data can be obtained to support not only general strategic decisions but also specific regional and local mitigation. To determine the causes and potential effects of deforestation, study of deforestation must also include an integration of disparate data from both human and natural science discipline, using a Geographic Information System to provide spatial Analysis capabilities.

For example, Oseiaa *et al.* (2005) noted that proper forest monitoring and management can be achieved by using remote sensing techniques and creating spatial representations such as maps to know the exact locations and extent of deforestation. Pekkarinen *et al.* (2009) provided Pan- European forest/non-forest mapping developed by Landsat ETM+ and the Coordination of information on the environment (CORINE) database. Wanger *et al.* (2003) submitted that Landsat and other high-resolution satellite images have recently been applied for land resources management in a number of countries.

Forest Trees' Disease and Insect and Air Pollution

Gradual forest decline is a type of change that needed to be monitored for proper conservation and preservation of tropical forest. Vegetation is very complex and sensitive to factors allied with changes in temperature, humidity and as well as anthropogenic influences such as forest pest, disease outbreak, air pollution and insect interactions. Remote sensing techniques together with GIS offer the means of assessing and evaluating the magnitude of deterioration. For example, in Nigeria, Landsat Thematic Mapper (TM) and Enhanced Thematic Mapper Plus (ETM+) multi-seasonal images were used in detecting the extent of environmental degradation caused by disease outbreak and insect attack (Oseia *et al.,* 2005). The researchers found that there was a decline in the land area covered with vegetation, loss of forested areas due to pressure from diseases and insects.

Similarly, in Germany, a three-year project was initiated in 1986 to establish methods to detect, classify, and map forest decline using a combination of Landsat MSS and airborne multi-spectral imagery. This study was carried out because of Landauer (1989) and Zawila-Niedzwiecki (1989) reported that forests in Germany have been dying gradually due to industrial air pollution. The researchers found that characteristic spectral signatures could be identified for different tree species (spruce, pine, and beech) depending on the degree of decline (Landauer, 1989). Insects and diseases are amid the main natural disorders that change forested lands. Appropriate update information and certified inventory databases are contemporary adequate to support forest conservation and management.

Forest Fires

Forest fire is one of the problems of the world's tropical dry forests. Apart from human cause, frequent drought occurrences in Nigeria especially towards the northern parts; have been disposing the vegetation to annual fires. The forest fires in this part are as a result of the instantaneous existence of some unfavourable phenomena such long-standing drought, deterioration of trees, sumptuous growth of grasses and the formation of slack canopy cover which are causing huge quantities of combustible materials and other human activities. Forest fire could be useful or destructive to the forest, although tree species differ in their responses to forest fires. Prescribed fire may play a significant part in sustainable management of forest subjected to wildfires (Fatubarin and Olojugba, 2014). Fire can cause the breaking of seed dormancy because the heat from the fire may break the hard seed coat and allow water absorption by the seed. Hein et *al.* (2010) in his study stated that the probability of succession after fire is low in mixed forest.

Reifsnyder (1978) and Oguntala (1982) have reported that bush fires can be predicted using climatic information. Remote sensing techniques can be used in obtaining this climatic information and then GIS can be used in processing and analysing the data into meaningful information for conservation and management purposes. The necessity to control bush fires is becoming gradually imperative in the tropics in view of the increasing cost of damages done by fire. High resolution satellite such as SPOT, ERS-SAR and Landsat Thematic Mapper combined with low resolution satellite images such as NOAA-AVHRR have been reported by Dozier (1981) to offer new possibilities to monitor forest fires. They have numerous advantages over conventional means of observation. For instance, using satellite images, it is possible to view large expanses of land. This can be accomplished repeatedly for the same expanse and documented in diverse wavelengths, therefore, providing information on the state of forest resources.

For example, Casanova *et al.* (2015) reported that the MSG-SEVIRI sensor has shown its capacity as an earth observer sensor that provides relevant information in the search and study of large forest fires in real time as well as operating fire detection. They concluded that in relative to each of the themes studied, detection and monitoring can be reached using the receiver system. These new techniques and technologies (Remote sensing data and Geographic Information Systems) can be useful by proximately finding the most direct access to endangered areas in dry tropical forest for conservation, restoration and management. In addition, hyper spectral image have been used to acquire data for wildfire management (Keramitsoglou *et al.,* 2008).

Forest Mensuration

Forest mensuration plays a key role in the practice of forest conservation, restoration and management. Forest exploration necessitates information about the growing stocks such as the size distributions of trees, species diversity, and spatial distribution of trees, stand density, stand structures, expected tree growths and yield. In order to understand the conservation dynamics and forest productivity capacity, estimation of forest growing stocks is of great important. The continuous demand for extra and improved information from forests has stimulated the development and application of new instrumentation and measurement techniques. Instead of the conventional cost-effective and time-consuming ground based measurements, remote sensing data (satellite images) and GIS are being progressively used in estimating many forest structures including forest growing stocks. For instance, Mohammad *et al.* (2015) in their study estimated forest growing stocks at Khadimnagar national park, Sylhet, Bangladesh using regression relationship of vegetation indices (VIs) of Landsat Thematic Mapper (TM) image with field-measured growing stock.

In the study, they concluded that Remote sensing technology has proven to be powerful research and management tool for inventory and assessment of the forest resources of the world. Ngueguim *et al.* (2009) used remotely sensed data generated from Landsat TM (year 1986) and Landsat ETM + (year 2000) to examine the impact of logging activities on tropical forest in Eastern Cameroon with the objectives of demonstrating the process whereby tropical forest got transformed into forest fragmentation and canopy damage. The availability and use of time series satellite images permitted the detection and quantification of land cover dynamics in the study area. This has been useful in improving understanding of the impact of industrial timber extraction on tropical forest. In addition, laser scanner data in combination with very-high-resolution satellite images, such as IKONOS, Terra Aster platform, or aerial multispectral scanner data have been used for the assessment of individual tree heights, timber volume calculations, and the detection of even single trees of different species, especially for forest inventory tasks (Shippert, 2004; Xie *et al.,* 2008).

Forest Ecophysiology and Ecological System

Information on the ecological and physiological status of forest is essential for growth diagnostics and yield prediction. Physiological activities of forest stands are influenced by the vitality and health status of the trees. The study of carbon and nitrogen cycles as well as environmental management on tropical forest requires assessment of the spatial variability of ecophysiological variables, scaling up of which is also needed for scientific and decision –making, conservation and management purposes. Remote sensing has great potential for these applications because it enables wide-area, non-destructive, and real-time acquisition of information about forest ecophysiological conditions. With recent advances in sensor technology, a variety of electromagnetic monograms, such as hyper spectral reflectance, thermal-infrared temperature, and microwave backscattering coefficients, can be acquired for both plants and ecosystems using ground-based, airborne, and satellite platforms. Their spatial and temporal resolutions have both recently been improved (Yoshio, 2003).

For example, Kokaly and Clark (1999) developed an approach to estimate the concentrations of nitrogen, lignin, and cellulose in dried and ground leaves using ban-depth analysis of absorption features (centred at 1.73 μm, 2.10 μm, and 2.30 μm) and stepwise multiple linear regression. Hyperspectral remote sensing was used to estimate the leaf pigment of sugar maple (*Acer saccharum*) in the Algoma Region, Canada, and promising results were obtained (Zarco-Tejada *et al.,* 2001). Hyper spectral imageries have been found useful in the identification of fragmented ecosystem and succession (Lee *et al.,* 2007). Quattrochi and Luvall (2009) reported that with the advancement in the ecological thermodynamics, Thermal Infrared (TIR) remote sensing is capable of uncovering the principles of ecological patterns of structure and function.

The use of remote sensing and GIS encompasses stages such as input, data processing, manipulation and analysis, output and storage. These stages are shown in Figure 1.

Figure 1. Data flow in RS/GIS from the data collection to output

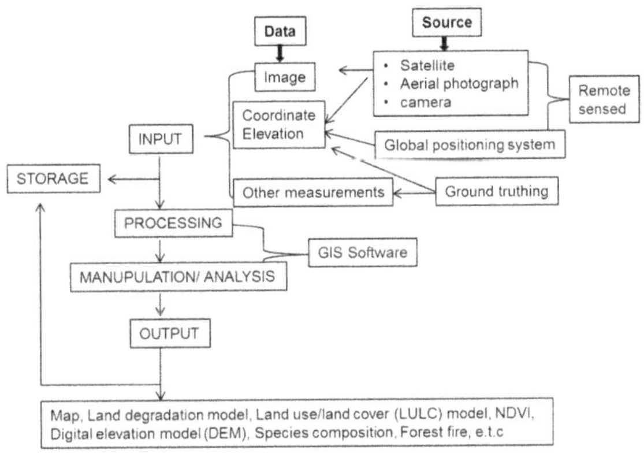

USE OF REMOTELY SENSED DATA IN DEVELOPMENT OF VEGETATION INDICES

Vegetation indices are quantifiable measurements demonstrating the vigour of vegetation (Campbell, 1987). Vegetation indices are the most frequently and commonly used indices in vegetation monitoring. These indices are designed to indicate the relative density and health of vegetation. Virtually all of these indices associate the reflectance in the red and near-infrared bands. A significant amount in the near-infrared portion of the spectrum is reflected by healthy vegetation because chlorophyll in plant leaves strongly absorbs visible light for use in photosynthesis and the cell structure of the leaves strongly reflects near-infrared light. The data from satellite can yield vegetation indices to produce "greenness" maps used to display vegetation health, effects of drought and variations in plant growth. According to Baret (1986), the importance of these indices is contingent to their practicality in the explanation of remote sensing imageries; they constitute remarkably a technique for the uncovering of terrestrial usage alterations, the appraisal of vegetative cover and density, crop discrimination and crop prediction.

An index is a figure qualifying the strength of a phenomenon which is too intricate to be disintegrated into known parameters. It is commonly difficult to make this disintegration for signs observed on remote sensing images, though it is imaginable to determine the portion of effect of numerous factors when in a renowned and controlled environment (Bannari *et al.,* 1995). The concept of vegetation index is modified for qualifying vegetation over large areas, for example, over areas covering many pixels of an image.

In an attempt to separate green vegetation from the soil background using the near-infrared (NIR)/ Red ratio was one of the earliest studies carried out by Pearson and Miller (1972). Subsequently, various vegetation indices were developed, tested, modified and used for vegetation-related studies worldwide. These indices include Normalized Difference Vegetation index (NDVI), Normalized Burn Ratio (NBR), Enhanced Vegetation Index (EVI), Ratio Vegetation Index (RVI), Vegetation Index Number (VIN), Differenced Vegetation Index (DVI), Yellow Vegetation Index (YVI), Transformed Vegetation Index (TVI), Green Vegetation Index (GVI), Misra Yellow Vegetation Index (MYVI), Misra Green Vegetation Index (MGVI), Perpendicular Vegetation Index (PVI), Adjusted Green Vegetation Index (AGVI), Normalized Difference Greenness Index (NDGI), Redness Index (RI), Normalized Difference Index (NDI), Multi-Temporal Vegetation Index (MTVI), Atmospherically Resistant Vegetation Index (ARVI), Soil Adjusted

Vegetation Index (SAVI), and so on. The importance of these indices is based on their effectiveness or efficacy in the classification or interpretation of remote sensing imageries.

Normalized Difference Vegetation Index (NDVI)

Normalized Difference Vegetation Index (NDVI) is one of the most widely used vegetation index. The NDVI is a measurement of the balance between energy received and energy emitted by objects on Earth. It is calculated from multispectral information as normalized ratio of the difference between the near-infrared band (NIR) and the red band (R) and the sum of these two bands (Rouse Jr *et al.* 1974). NDVI is mathematically expressed as:

$$NDVI = \frac{NIR - RED}{NIR + RED}$$

Where;
NIR= reflectance in the near-infrared band, and
RED = reflectance in the visible red band

The result of a given pixel of calculated NDVI ranges from a number -1 to +1. Nevertheless, values from 0.8 – 0.9 which are close to +1 indicate highest possible density of green leaves, zero (0) value means no vegetation and a value close to zero (0) indicates no green leaves. Sparse vegetation such as grasslands and shrubs may result in moderate NDVI values nearly 0.2 - 0.5. An NDVI value of ≤ 0.1 which is very low indicates areas of desolate rock, sand and/or snow. Negative NDVI value close to -1 indicates clear water and/ or bare ground. Very high NDVI values nearly 0.6 to 0.9 relate to dense vegetation (for example: temperate and tropical forests) at their highest growing phase.

There are different applications of NDVI such as the applications of NDVI for Biodiversity monitoring and conservation, Vegetation burning, Drought and drought early warning, Desertification, Land Use and Land-cover Change and Soil organic carbon and so on.

Application of NDVI in Forestry

Biodiversity Monitoring and Conservation

NDVI in combination with remote sensing tools offer opportunities for descriptions of biodiversity for large areas in an efficient or systematic, repeatable and spatially extensive way (Turner *et al.,* 2003; Duro *et al.,* 2007). According to Turner *et al.* (2003), NDVI is an important tool in the direct method or foremost directive analysis of species occurrence and in the development of land cover maps. Previous studies revealed that land cover maps offer implicit or explicit data on the composition, richness (abundance) and distribution of individual or groups of species liable to the measure, biome and environment in request (Turner *et al.,* 2003; Duro *et al.,* 2007; Pettorelli *et al.,* 2014). Oindo *et al.* (2000) used AVHRR-derived NDVI to explain the spatial inconsistency of species abundance of birds at a quarter degree spatial resolution in Kenya. The study revealed a strong positive correlation between maximum average NDVI and species abundance.

The indirect method of assessing species composition, richness (abundance) and distribution can as well be done using NDVI. Different aspects of vegetation state derived from NDVI contribute to the planning or mapping of environmental variables which offer signs of species composition, richness and distribution (Duro *et al.,* 2007; Pettorelli *et al.,* 2014). The occurrence and distribution of the disturbing *Schistocerca gregaria* (locust species) in Mauritania was described by high NDVI values resulting from NOAA/AVHRR satellite imagery which showed high resource abundance (Despland *et al.,* 2004).

Vegetation Burning

According to Mitchell and Roundtable (2010), fire is a common occurrence in many parts of the world. Wildfires in forests, savannas, mountain regions and other ecosystems are integral to the evolution of some of these systems; periodic fires are important in maintaining many grassland, shrub steppe, and savanna ecosystems. Satellite remote sensing has been used for modeling and mapping a variety of ecosystem conditions associated with fire risks, potential, and management. These include fire-fuel mapping risk estimation fire detection, post-fire severity mapping, and ecosystem recovery from fire stress (De Angelis *et al.* 2012).

Farther than the detection of active fires, remote sensing is also used to assess and quantify the spatial and temporal variations of changes of vegetation cover in areas affected by fires. NDVI can also be used to define phenological information of the area affected (Chuvieco *et al.* 2004). Lanorte *et al.* (2014) used NDVI time-series to monitor vegetation retrieval after disturbance by fire at two test sites in Spain and Greece. Leon *et al.* (2012) used Moderate-resolution Imaging Spectroradiometer (MODIS) NDVI to monitor post-fire vegetation response in New Mexico. Their study delineated the prospective of using NDVI to monitor the retrieval vegetation cover after fire disturbance.

Drought and Drought Early Warning

Drought generally refers to a substantial decline in the amount of precipitation received over a prolonged period (Mishra and Singh 2010). Droughts occur in almost all climatic regions and are documented as an unadorned threat to the atmosphere and growth (Sivakumar and Stefanski 2007). The NDVI has been used by Anyamba and Tucker (2012) to examine the persistence of drought in the Sahel in the period 1982-1993. NDVI has also been used to develop famine early warning systems (FEWS), such as the Famine Early Warning Systems Network (FEWS NET) which is an operational system for propagation of data interrelated to food production and accessibility universally. The system uses NDVI data from both NOAA AVHRR and MODIS operationally. The system was heralded by severe testing of the capability of NDVI to identify zones of impending food scarcities (Henricksen and Durkin 1986, Hutchinson 1991, Quarmby et al. 1993). It was found that NDVI in combination with pertinent environmental data have a very robust potential for projecting crop failure.

Desertification Monitoring

Desertification is defined as land degradation in arid, semi-arid, and dry sub-humid areas resulting from various factors, including climatic variations and human activities (UNCCD 1994). Herrmann *et al.* (2005) demonstrated the possibility of using NDVI as a proxy for environmental response to management. In detecting the status and trend of desertification, researchers have built on the relationship between NDVI and biomass productivity (Jensen 2007, Purkis and Klemas 2011).

Land Use and Land-Cover Change (LULCC)

In previous studies, research efforts have been devoted in the use of NDVI to assess the site and degree of land use and land cover change (Diouf and Lambin 2001, Stow *et al.,* 2004 and UNEP 2012). Measures of presentation range from universal studies of land cover classification and mapping through provincial and state measures to very local studies (DeFries and Townshend 1994, Hansen *et al.,* 2000 and Friedl *et al.,* 2002, Stow *et al.,* 2004, Lunetta *et al.,* 2006 and Sternberg *et al.,* 2011). According to Horion *et al.* (2014), long-term trends in dry-season minimum NDVI was used to assess changes in tree cover in the Sahel. The dry-season minimum NDVI was found not to be associated with dry grass filtrates from the previous growing period or with periodic fire occurrence and control over most of the Sahel, this implies that NDVI parameter can be used as a substitution for assessing variations in tree cover in such ecosystems.

Soil Organic Carbon (SOC) Assessment

The absorption of carbon by land-based ecosystems drives several means to balancing worldwide fossil fuel radiations (Piao *et al.,* 2009; Mishra and Singh, 2010; Bernoux and Chevallier, 2014). Soil organic carbon is also increasingly recognized as an excellent indicator to monitor the status and functioning of soils – hence progressively recommended in various international initiatives for monitoring soil quality (Bernoux and Chevallier 2014). Yan *et al.* (2005) used NDVI data input into the Carnegie-Ames-Stanford Approach (CASA) terrestrial ecosystem model to estimate losses of SOC resulting from wind erosion in China. A significant correlation was found between NDVI and SOC, as well as NDVI and Nitrogen (N) in the study investigating changes in soil organic C and total N in the Hexi corridor, China (Pan *et al.* 2013).

CONCLUSION AND RECOMMENDATIONS

Remote sensing and GIS application provide new management and conservation understanding. This help to resolve the present conflict between man and nature. The continued hitching of forest resources has in recent times been of distress to conservationist, forest officers and policy makers. While remote sensing provides new technologies and benefits to conservation and restoration of tropical forests with remote sensing and GIS, computational challenges can be overcome, making environmental and geospatial data more easily accessible for conservation ecologists especially in the tropical dry forests. A good understanding of forest community helps forest managers to describe their forest in detail and generate ideas on how best the forest should be managed. This study recommended the adoption of remote sensing and GIS techniques for the assessment and monitoring of tropical dry forests for ease of planning and decision-making.

REFERENCES

Adedeji, O. H., Tope-Ajayi, O. O., & Abegunde, O. L. (2015). Assessing and Predicting Changes in the Status of Gambari Forest Reserve, Nigeria Using Remote Sensing and GIS Techniques. *Journal of Geographic Information System*, *7*(03), 301–318. doi:10.4236/jgis.2015.73024

Agbelade, A. D., & Akindele, S. O. (2013). Land Use Mapping and Tree species Diversity of Federal University of Technology, Akure. *American International Journal of Contemporary Research*, *3*(2), 104–113.

Ahern, F. J., & Leckie, D. G. (1987). Digital remote sensing for forestry: Requirements and capabilities, today and tomorrow. *Geocarto International*, *3*(3), 43–52. doi:10.1080/10106048709354107

Ali, I. N., Brian, B., Richard, G. H., & Peter, C. (2014). Changing Vegetation Patterns in Yobe State Nigeria: An Analysis of the Rates of Change, Potential Causes and the Implications for Sustainable Resource Management. *International Journal of Geosciences*, *5*(01), 50–62. doi:10.4236/ijg.2014.51007

Anyamba, A., & Tucker, C. J. (2005). Analysis of Sahelian Vegetation Dynamics Using Noaa-AVHRR NDVI Data from 1981-2003. *Journal of Arid Environments*, *63*(3), 596–614. doi:10.1016/j.jaridenv.2005.03.007

Axel, A. C. (2018). Burned Area Mapping of an Escaped Fire into Tropical Dry Forest in Western Madagascar Using Multi-Season Landsat OLI Data. *Remote Sensing*, *10*(3), 371. doi:10.3390/rs10030371

Bannari, A., Morin, D., Bonn, F., & Huete, A. R. (1995). A review of vegetation indices'. *Remote Sensing Reviews*, *13*(1), 95–120. doi:10.1080/02757259509532298

Baret, F. (1986): Contribution au suivi radiometrique de cultures de cereals. These de Doctorat, Universite Paris-Sud Orsay, France, 182.

Bijalwan, A., Swamy, S. L., Shanna, C. M., Sharma, N. K., & Tiwari, A. K. (2010). Land-use, biomass and carbon estimation in dry tropical forest of Chhattisgarh region in India using satellite remote sensing and GIS. [</jrn>]. *Journal of Forestry Research*, *21*(2), 161–170.

Campbell, J. B. (1987). *Introduction to Remote Sensing* (p. 551). New York, NY: The Guilford Press.

Casanova, J. L., Calle, A., Romo, A., & Sanz, J. A. (2015): Forest fire detection and monitoring by means of an integrated MODIS-MSG system. 409–414.

Cheruto, M. C., Kauti, M. K., Kisangau, P. D., & Kariuki, P. (2016). Assessment of Land Use and Land Cover Change Using GIS and Remote Sensing Techniques: A Case Study of Makueni County, Kenya. *J. Remote Sensing and GIS*, *5*, 175. doi:10.4175/2469-4134.1000175

Chuvieco, E., Cocero, D., Riano, D., Martin, P., Martinez, J., De la Riva, J., & Perez, F. (2004). Combining NDVI and surface temperature for the estimation of live fuel moisture content in forest fire danger rating. *Remote Sensing of Environment*, *92*(3), 322–331. doi:10.1016/j.rse.2004.01.019

Cunningham, W. P., Cunningham, M. A., & Saigo, B. C. (2005). *Environmental Science, A Global Concern* (8th ed., pp. 124–125, 240). New York, NY: McGraw Hill Publishers.

De Angelis, A., Bajocco, S., & Ricotta, C. (2012). Modelling the phenological niche of large fires with remotely sensed NDVI profiles. *Ecological Modelling*, *228*, 106–111. doi:10.1016/j.ecolmodel.2012.01.003

Debinski, D. M., Jakubauskas, M. E., & Kindscher, K. (1999). A remote sensing and GIS-based model of habitats and biodiversity in the Greater Yellowstone Ecosystem. *International Journal of Remote Sensing*, *20*(17), 3281–3292. doi:10.1080/014311699211336

DeFries, R., & Townshend, J. (1994). NDVI-derived land cover classifications at a global scale. *International Journal of Remote Sensing*, *15*(17), 3567–3586. doi:10.1080/01431169408954345

Defries, R. S., Hansen, M. C., Townshend, J. R. G., Janetos, A. C., & Loveland, T. R. (2000). A new global 1-km dataset of percentage tree cover derived from remote sensing. *Global Change Biology*, *6*(2), 247–254. doi:10.1046/j.1365-2486.2000.00296.x

Despland, E., Rosenberg, J., & Simpson, S. J. (2004). Landscape structure and locust swarming: A satellite's eye view. *Ecography*, *27*(3), 381–391. doi:10.1111/j.0906-7590.2004.03779.x

Diallo, Y., Guangdao, H., & Xingping, W. (2009). Applications of Remote Sensing in LULC Change Detection in Puer and Simao Counties. *Marsland Press Journal of American Science*, *5*(4), 157–166.

Diouf, A., & Lambin, E. (2001). Monitoring land-cover changes in semi-arid regions: Remote sensing data and field observations in the Ferlo, Senegal. *Journal of Arid Environments*, *48*(2), 129–148. doi:10.1006/jare.2000.0744

Dozier, J. (1981). A method for satellite identification of surface temperature fields of subpixel resolution. *Remote Sensing of Environment*, *11*, 221–229. doi:10.1016/0034-4257(81)90021-3

Duro, D. C., Coops, N. C., Wulder, M. A., & Han, T. (2007). Development of a large area biodiversity monitoring system driven by remote sensing. *Progress in Physical Geography*, *31*(3), 235–260. doi:10.1177/0309133307079054

Fatubarin, A. R., & Olojugba, M. R. (2014). The influences of forest fire on the vegetation and some soil properties of a savanna ecosystem in Nigeria. *Journal of Soil Science and Environmental Management*, *5*(2), 28–34. doi:10.5897/JSSEM13.0413

Franklin, S. E. (2010). *Remote Sensing for Biodiversity and Wildlife Management: Synthesi and Applications*. New York, NY: McGraw-Hill Companies

Gandhi, D. S., & Sundarapandian, S. (2014). Inventory of trees in tropical dry deciduous forests of Tiruvannamalai district, Tamil Nadu, India. *Biodiversitas (Surakarta)*, *15*(2), 169–179. doi:10.13057/biodiv/d150208

Gillespie, T. W., Foody, G. M., Rocchini, D., Giorgi, A. P., & Saatchi, S. (2008). Measuring and modelling biodiversity from space. *Progress in Physical Geography*, *32*(2), 203–221. doi:10.1177/0309133308093606

Gong, P., Pu, R., & Yu, B. (2001). Conifer species recognition: Effects of data transformation. *International Journal of Remote Sensing*, *17*, 2471–2481.

Guerschman, J. P., Hill, M. J., Renzullo, L. J., Barrett, D. J., Marks, A. S., & Botha, E. J. (2009). Estimating fractional cover of photosynthetic vegetation, non-photosynthetic vegetation and bare soil in the Australian tropical savanna region upscaling the EO-1 Hyperion and MODIS sensors. *Remote Sensing of Environment*, *113*(5), 928–945. doi:10.1016/j.rse.2009.01.006

Hansen, M. C., DeFries, R. S., Townshend, J. R. G., & Sohlberg, R. A. (2000). Global land cover classification at 1 km spatial resolution using a classification tree approach. *International Journal of Remote Sensing*, *21*(6-7), 1331–1364. doi:10.1080/014311600210209

Hein, G., Justice, O., & Andrisano, A. (2010): From monospecific to mixed forest after fire? An early forecast for the montane belt of Majella, Italy. *Forest Ecology and Management.* 259, 433-439. doi:10.1016/j.foreco.2009.10.040

Henricksen, B., & Durkin, J. (1986). Growing period and drought early warning in Africa using satellite data. *International Journal of Remote Sensing*, *7*(11), 1583–1608. doi:10.1080/01431168608948955

Herrmann, S. M., Anyamba, A., & Tucker, C. J. (2005). Recent trends in vegetation dynamics in the African Sahel and their relationship to climate. *Global Environmental Change*, *15*(4), 394–404. doi:10.1016/j.gloenvcha.2005.08.004

Horion, S., Fensholt, R., Tagesson, T., & Ehammer, A. (2014). Using earth observation-based dry season NDVI trends for assessment of changes in tree cover in the Sahel. *International Journal of Remote Sensing*, *35*(7), 2493–2515. doi:10.1080/01431161.2014.883104

Hutchinson, C. (1991). Uses of satellite data for famine early warning in sub-Saharan Africa. *International Journal of Remote Sensing*, *12*(6), 1405–1421. doi:10.1080/01431169108929733

Janzen, D. H. (1988): Tropical Dry Forests The Most Endangered Major Tropical Ecosystem. In E. O. Wilson & F. M. Peter (Eds.) *Biodiversity.* Washington DC: National Academies Press. pp. 1-6. Available at https://www.ncbi.nlm.nih.gov/books/NBK219281/?report=printable

Japan Association of Remote Sensing. (2010). *Sensors.* Available at http://www.jars1974.net/pdf/03_Chapter02.pdf

Jensen, J. (2007). Remote Sensing of the Environment. Upper Saddle River, NJ: Pearson Prentice Hall.

Johnston, C. A. (1998). *Geographic information systems in ecology*. Oxford, UK: Blackwell Science.

<jrn>Bernoux, M., & Chevallier, T. (2014). Carbon in dryland soils - Multiple essential functions. Montpellier, France: CSFD/Agropolis International.

Keramitsoglou, I., Kontoes, C., Sykioti, O., Sifakis, N., & Xofis, P. (2008). Reliable, accurate and timely forest mapping for wildfire management using ASTER and Hyperion satellite imagery. *Forest Ecology and Management*, *255*(10), 3556–3562. doi:10.1016/j.foreco.2008.01.077

Kokaly, R. F., & Clark, R. N. (1999). Spectroscopic determination of leaf biochemistry using band-depth analysis of absorption features and stepwise linear regression. *Remote Sensing of Environment*, *67*(3), 267–287. doi:10.1016/S0034-4257(98)00084-4

Landauer, G. (1989): Detection, classification and mapping of forest decline by means of multispectral remote sensing. In Proceedings, Global Natural Resource Monitoring and Assessment: Preparing for the 21st Century, Vol. 1.

Layton, P. A., Guynn, S. T., & Guynn, D. C. (2004). Biodiversity metrics in Sustainable Certification Programs. *Journal of Forestry*, *102*(3), 46–52.

Lee, Z., Casey, B., Arnone, R., Weidemann, A., Parsons, R., Montes, M. J., ... Dye, J. (2007). Water and bottom properties of a coastal environment derived from Hyperion data measured from the EO-1 spacecraft platform. *Journal of Applied Remote Sensing*, *1*(1), 011502. doi:10.1117/1.2822610

Leon, J. R. R., van Leeuwen, W. J., & Casady, G. M. (2012). Using MODIS-NDVI for the modeling of post-wildfire vegetation response as a function of environmental conditions and pre-fire restoration treatments. *Remote Sensing*, *4*(3), 598–621. doi:10.3390/rs4030598

Lunetta, R. S., Knight, J. F., Ediriwickrema, J., Lyon, J. G., & Worthy, L. D. (2006). Land-cover change detection using multi-temporal MODIS NDVI data. *Remote Sensing of Environment*, *105*(2), 142–154. doi:10.1016/j.rse.2006.06.018

Luther, J. E., Fournier, R. A., Piercey, D. E., Guindon, L., & Hall, R. J. (2006). Biomass mapping using forest type and structure derived from Landsat TM imagery. *International Journal of Applied Earth Observation and Geoinformation*, *8*(3), 173–187. doi:10.1016/j.jag.2005.09.002

Martin, M. E., Newman, S. D., Aber, J. D., & Congalton, R. G. (1998). Determining forest species composition using high spectral resolution remote sensing data. *Remote Sensing of Environment*, *65*(3), 249–254. doi:10.1016/S0034-4257(98)00035-2

Martinuzzi, S., Gould, W. A., Gonzalez, O. M. R., Robles, A. M., Maldonado, P. C., Pérez-Buitrago, N., & Caban, J. J. F. (2008): Mapping tropical dry forest habitats integrating Landsat NDVI, Ikonos imagery, and topographic information in the Caribbean Island of Mona. *Rev. Biol. Trop., 56*(2), 625-639.

Martinuzzi, S., Gould, W. A., Vierling, L. A., Hudak, A. T., Nelson, R. F., & Evans, J. S. (2013). Quantifying tropical dry forest type and succession: Substantial improvement with LiDAR. *Biotropica*, *45*(2), 135–146. doi:10.1111/j.1744-7429.2012.00904.x

Meli, F., & Meli, V. (2015). Modelling Drivers of Forest Cover Change in the Santchou Wildlife Reserve, West Cameroon using Remote Sensing and Land Use Dynamic Degree Indexes. *Canadian Journal of Tropical Geography*, *2*(2), 29–42.

Mishra, A. K., & Singh, V. P. (2010). A review of drought concepts. *Journal of Hydrology (Amsterdam)*, *391*(1–2), 202–216. doi:10.1016/j.jhydrol.2010.07.012

Mitchell, T. D., & Jones, P. D. (2005). An improved method of constructing a database of monthly climate observations and associated high-resolution grids. *International Journal of Climatology*, *25*(6), 693–712. doi:10.1002/joc.1181

Mohammad, R., Romana, A., Mirajul, I., Kazi, M. M., & Mohammad, S. H. C. (2015). Estimating growing stock volume in a Bangladesh forest site using Landsat TM and field-measured data. *International Journal of Geomatics and Geosciences*, *6*(2), 1607–1619.

Natural Resources Canada. (2008): Fundamentals of Sensing. Retrieved from ww.nrcan.gc.ca

Ngueguim, J. R., Gatchui, H. C., Ayobami, S. T., & Orimoogunje, O. O. (2009). Evaluation of logging impacts on tropical rainforest in Eastern Cameroon using Remote Sensing and GIS techniques. *International Journal of Biological and Chemical Sciences*, *3*(4), 771–785. doi:10.4314/ijbcs.v3i4.47163

Odihi, J. (2003). Deforestation in Afforestation Priority Zone in Sudano-Sahelian Nigeria. *Applied Geography (Sevenoaks, England)*, *23*(4), 227–259. doi:10.1016/j.apgeog.2003.08.004

Odiogor, H. (2010). *Special Report on Desertification in Nigeria: The Sun Eats Our Land.* Vanguard Newspaper. Retrieved from http://odili.net/news/source/2010/may/4/300.html

Oguntala, A. B. (1982): *Wildland fires particularly in tropical regions.* World Meteorological Organization. CAgM Report No.10.

Oindo, B. O., de By, R. A., & Skidmore, A. K. (2000). Interannual variability of NDVI and bird species diversity in Kenya. *International Journal of Applied Earth Observation and Geoinformation*, *2*(3), 172–180. doi:10.1016/S0303-2434(00)85011-4

Onojeghuo, A. O., & Onojeghuo, A. R. (2015). Mapping forest transition trends in Okomu reserve using Landsat and UK-DMC-2 satellite data. *South African Journal of Geomatics*, *4*(4), 486–501. doi:10.4314ajg.v4i4.9

Oseia, A., Edmund C. M., & Yaw, A. T. (2005): Use of Remote Sensing Data to Detect Environmental Degradation in the Coastal Region of Southern Nigeria. 1-6.

Pan, C., Zhao, H., Zhao, X., Han, H., Wang, Y., & Li, J. (2013). Biophysical properties as determinants for soil organic carbon and total nitrogen in grassland salinization. *PLoS One*, *8*(1), e54827. doi:10.1371/journal.pone.0054827 PMID:23372776

Pearson, R. L., & Miller, L. D. (1972). Remote mapping of standing crop biomass for estimation of the productivity of the shortgrass prairie. ed. *Remote Sensing of Environment*, *VIII*, 1355.

Pekkarinen, A., Reithmaier, L., & Strobl, P. (2009). Pan-European forest/non-forest mapping with Landsat ETM+ and CORINE Land Cover 2000 data. *ISPRS Journal of Photogrammetry and Remote Sensing*, *64*(2), 171–183. doi:10.1016/j.isprsjprs.2008.09.004

Pettorelli, N., Safi, K., & Turner, W. (2014). Satellite remote sensing, biodiversity research and conservation of the future. *Philosophical Transactions of the Royal Society of London. Series B, Biological Sciences*, *369*(1643). doi:10.1098/rstb.2013.0190 PMID:24733945

Piao, S., Fang, J., Ciais, P., Peylin, P., Huang, Y., Sitch, S., & Wang, T. (2009). The carbon balance of terrestrial ecosystems in China. *Nature*, *458*(7241), 1009–1013. doi:10.1038/nature07944 PMID:19396142

Purkis, S. J., & Klemas, V. V. (2011). *Remote sensing and global environmental change*. Hoboken, NJ: John Wiley & Sons. doi:10.1002/9781118687659

Quarmby, N. A., Milnes, M., Hindle, T. L., & Silleos, G. N. (1993). The use of multi-temporal NDVI measurements from AVHRR data for crop yield estimation and prediction. *International Journal of Remote Sensing*, *14*(2), 199–210. doi:10.1080/01431169308904332

Quattrochi, D. A., & Luvall, J. C. (2009). Thermal remote sensing in Earth science research. In T. A. Warner, M. D. Nellis, & G. M. Foody (Eds.), *The SAGE handbook of remote sensing* (pp. 64–78). London, UK: SAGE Publications Ltd. doi:10.4135/9780857021052.n5

Ramachandra, T. V., & Kumar, U. (2004): Geographic Resources Decision Support System for Land Use, Land Cover Dynamics Analysis. *Proceedings of the FOSS/GRASS Users Conference*, Bangkok, Thailand. 12-14.

Reifsnyder, W. E. (1978). *Systems for evaluating and predicting the effects of weather and climate on wildland fires.* World Meteorological Organization Special Report No. 11.

Rossini, M., Panigada, C., Meroni, M., & Colombo, R. (2006). Assessment of oak forest condition based on leaf biochemical variables and chlorophyll fluroscence. *Tree Physiology*, *26*(11), 1487–1496. doi:10.1093/treephys/26.11.1487 PMID:16877333

Rouse, J. W. Jr, Haas, R. H., Schell, J. A., & Deering, D. W. (1974). Monitoring vegetation systems in the Great Plains with ERTS. *NASA Special Publication*, *351*, 309.

Shippert, P. (2004). Why use hyperspectral imagery? *Photogrammetric Engineering and Remote Sensing*, *70*(4), 377–396.

Sivakumar, M. V., & Stefanski, R. (2007): Climate and land degradation—an overview. Climate and Land Degradation. Springer, 105-135.

Srivastava, P. K., George, P. P., Manika, G., Sudhir, K. S., Tanvir, I., & Dimitra, L. (2018). Deriving forest fire probability maps from the fusion of visible/infrared satellite data and geospatial data mining. *Modeling Earth Systems and Environment*, 1–17. doi:10.100740808-018-0555-5

Sternberg, T., Tsolmon, R., Middleton, N., & Thomas, D. S. G. (2011). Tracking desertification on the Mongolian steppe through NDVI and field-survey data. *International Journal of Digital Earth*, *4*(1), 50–64. doi:10.1080/17538940903506006

Stoms, D. M., & Estes, J. E. (1993). A remote sensing research agenda for mapping and monitoring biodiversity. *International Journal of Remote Sensing*, *14*(10), 1839–1860. doi:10.1080/01431169308954007

Stow, D. A., Hope, A., McGuire, D., Verbyla, D., Gamon, J., Huemmrich, F., ... Myneni, R. (2004). Remote sensing of vegetation and land-cover change in Arctic Tundra Ecosystems. *Remote Sensing of Environment*, *89*(3), 281–308. doi:10.1016/j.rse.2003.10.018

Tudun-Wada, M. I., Tukur, Y. M., & Hussaini, Y. (2014). Analysis of forest cover changes in Nimbia Forest Reserve, Kaduna State, Nigeria using Geographic Information System and Remote Sensing techniques. *International Journal of Environmental Monitoring and Analysis*, *2*(2), 91–99. doi:10.11648/j.ijema.20140202.15

Turner, W., Spector, S., Gardiner, N., Fladeland, M., Sterling, E., & Steininger, M. (2003). Remote sensing for biodiversity science and conservation. *Trends in Ecology & Evolution*, *18*(6), 306–314. doi:10.1016/S0169-5347(03)00070-3

UNCCD. (1994): *Elaboration of an International Convention to Combat Desertification in Countries Experiencing Serious Drought and/or Desertification, Particularly in Africa.* Paris, France: United Nations Convention to Combat Desertification.

UNEP. (2012). *Sahel Atlas of Changing Landscapes: Tracing trends and variations in vegetation cover and soil condition.* Nairobi, Kenya: United Nations Environment Programme.

Wang, K., Franklin, S. E., Guo, X., & Cattet, M. (2010). Remote sensing of ecology, biodiversity and conservation: A review from the perspective of remote sensing specialists. *Sensors (Basel), 10*(11), 9647–9667. doi:10.3390101109647 PMID:22163432

Wanger, Y., Bonynge, G., Nugranad, J., Traber, M., Ngusaru, A., Lynne, H., ... Makota, V. (2003). Remote sensing of mangrove change along the Tanzania coast. *Marine Geodesy, 26*, 1–14.

Wani, R. A., Baba, U. A., Afeefa, Q. S., & Imran, K. (2016). *Conserving Biological Diversity: A Multi-scaled Approach. Published by* (pp. 133–158). New Delhi, India: Research India Publications.

Willkomm, M., Vierneisel, B., & Dannenberg, P. (2016). Land use change dynamics in the Mt. Kenya region – a remotely sensed analysis using RapidEye satellite images. *Zbl. Geol. Paläont. Teil 1, 1*, 23–40.

Xie, Y., Sha, Z., & Yu, M. (2008). Remote sensing imagery in vegetation [94] mapping: A review. *Journal of Plant Ecology, 1*(1), 9–23. doi:10.1093/jpe/rtm005

Yan, H., Wang, S., Wang, C., Zhang, G., & Patel, N. (2005). Losses of soil organic carbon under wind erosion in China. *Global Change Biology, 11*(5), 828–840. doi:10.1111/j.1365-2486.2005.00950.x

Yoshio, I. (2003). Synergy of Remote Sensing and Modeling for Estimating Ecophysiological Processes in Plant Production. *Plant Production Science, 6*, 1, 3–16. doi:10.1626/ pps.6.3

Zarco-Tejada, P. J., Miller, J. R., Noland, T. L., Mohammed, G. H., & Sampson, P. H. (2001). Scaling-up and model inversion methods with narrowband optical indices for chlorophyll content estimation in closed forest canopies with hyperspectral data. *IEEE Transactions on Geoscience and Remote Sensing, 39*(7), 1491–1507. doi:10.1109/36.934080

Zawila-Niedzqiecki, T. (1989). Satellite images for forest decline assessment. In *Proceedings, Global Natural Resource Monitoring and Assessment: Preparing for the 21st Century* (Vol. 1, pp. 473–478). Bethesda, MD: American Society for Photogrammetry and Remote Sensing.

Zhu, L., Suomalainen, J., Liu, J., Hyyppä, J., Kaartinen, H., & Haggren, H. (2018). A Review: Remote Sensing Sensors. In R. B. Rustamov, S. Hasanova, & M. H. Zeynalova (Eds.), *Multi-purposeful Application of Geospatial Data* (pp. 19–42). Haggren, IntechOpen; doi:10.5772/intechopen.71049

Chapter 16
Forest Inventory:
Challenges, Trend, and Relevance on Conservation and Restoration of Tropical Forests

Onyekachi Chukwu
https://orcid.org/0000-0002-6376-6084
Nnamdi Azikiwe University, Nigeria

Japheth H. Dau
University of Agriculture Makurdi, Nigeria

ABSTRACT

Data and information are required to manage a renewable natural resource (such as forest) sustainably. This information is largely obtained through forest inventories. Forest inventory is very important in forest management since it provides the data for planning, monitoring, evaluation, research, growth and yield, biodiversity, and timber sale. Inadequate and/or irregular forest inventory has resulted to paucity of data on which forest management decisions can be based. This study aimed at enlightening conservationists, ecologists, environmentalists, and forest managers on the need to carryout diversity assessment and inventory of flora and fauna resources within their forest ecosystems. This will help to ensure sustainable yield and prevent the extinction of economic trees and endangered animal species in the tropical dry forests. The study therefore recommended that inventory should be carryout among the tropical dry forests at regular intervals.

INTRODUCTION

Among the terrestrial ecosystems, tropical forest is one of the most species diverse ecosystems. It accommodates a variety of resources that supports the livelihood of local communities (Dash *et al.*, 2009; Gandhi and Sundarapandian, 2014). Murphy and Lugo (1986) stated that about 40% of the earth's sub-tropical land area is dominated by closed or open forests and, out of which, 42% is covered by tropical

DOI: 10.4018/978-1-7998-0014-9.ch016

dry forest (TDF), 33% is covered by moist forest, and 25% is covered by wet and rain forest. More than half of the TDF (54.2%) is located within South America and 45.8% is distributed among North and Central America, Africa, Eurasia, Australasia and Southeast Asia (Miles *et al.,* 2006). TDF was defined by Baker *et al.* (2005) as areas within the tropics where there is a drought (<30 mm of rain per month for at least four months) during the low sun period, and a rainy season of several months duration with more than 100 mm of monthly rainfall when the sun is highest in the sky. While other tropical forests tend to have at least one period of lowered water availability in a year, tropical dry forests (TDFs) experience both prolonged and severe dry season. Tropical dry forests present well-defined phenological cycles, which make them potential proxies for the biological response to climate change and specifically to the drought effects. Since TDF are ecologically, socially and economically high valued all over the world (Mooney *et al.,* 1995), therefore, comprehensive research studies are required to understand the ecosystem services provided by TDF (Maass *et al.,* 2005).

In addition to being a major source of wood, tropical forest also acts as a mean of protecting the ecosystems (via regulating stream flow, watershed protection and reducing the rate of soil erosion) for tourism, Non-Timber Forest Products (NTFPs) provision and wildlife protection among many others (Zerihun and Yemir, 2013). However, tropical dry forests are under great anthropogenic pressures and require urgent management intervention to ensure the conservation and restoration of the region for sustainability. Habitat destruction, forest fragmentation, and the adverse physical and biological consequences of edge and buffer effects are some of the effects of deforestation in the tropical dry forests.

The biological diversity of tropical forests has been poorly documented, while the available data is grossly inadequate. Most species (plants, terrestrial and aquatic invertebrate, small mammals, reptiles, amphibians, among others) in the tropical forests remain undocumented (Boboye and Jimoh, 2016). Basic data and information play key role in the conservation and restoration of tropical dry forest. This information is largely obtained through forest inventories. However, inadequate or irregular inventory of tropical dry forest has resulted to paucity of information on which forest management decisions can be based. Forest inventory provides the data on condition of tropical forests for decision-making and management (such as planning, monitoring, evaluation, research, growth and yield), information on site quality and biodiversity, and timber harvest and sale plans. This study aimed at illustrating the need to carryout regular inventories of the tropical dry forests; which will help the forest managers to make informed decisions towards the conservation and restoration of economical tree species and endangered animals in the tropical regions.

The conservation and restoration of tropical dry forest have become an issue in the past decade, out of concerns of both overexploiting the resources (Powers, 2001; Monserud, 2003), and possible climate change effects (Watson *et al.,* 1995; Schwalm and Ek, 2001). Sustainable forestry comprises a host of management regimes to maintain and enhance the long-term health and integrity of forest ecosystems and forest-dependent communities, while providing ecological, economic, social, and cultural opportunities for the benefit of present and future generations (Wilson and Wang, 1999). Therefore, the present chapter aimed at enlightening conservationists, ecologists and forest managers the need to carryout assessment or inventory of resources and flora and fauna biodiversity in the tropical forests; to ensure sustainable yield and prevent the extinction of economic tree and endangered animal species in the tropical forest ecosystems.

TREND OF TROPICAL DRY FORESTS

Tropical dry forest is a region within the tropical forests. Globally, 52% of the total forests are in tropical regions and they are known to be the most important areas in terms of biodiversity (Kacholi, 2014). Tropical forests occupy center stage in term of biological diversity. It consists of at least two-thirds of the world's biodiversity; despite covering less than 10% of the Earth's land surface (Bradshaw *et al.*, 2009). The species richness of TDF is less when compared to tropical moist forests (Gentry,1995). The negative environmental impacts of forest loss are most evident in the tropical dry forest; where majority of future deforestation is anticipated (Laurance *et al.*, 2014).The tropics host critical ecosystem services, including biodiversity and carbon storage (Adekunle *et al.*, 2013).

The rapid increase in human population near forest ecosystems has increased threats of degradation and fragmentation to these ecosystems (FAO, 2010). As forests cover in the tropical regions continue to disappear due to varying land use pressures; large numbers of economic tree species are gradually going into extinction. The diversity of tree species is decreasing due to the rate of habitat destruction and over-exploitation, which are far greater than the rate of collection and conservation of genetic diversity (Shamaki and Akindele, 2013). The tropical dry forests ecosystem is also scarcely represented in the international protection schemes, which perhaps causes increased vulnerability of this ecosystem to the tropical fingerprint of global human development.

It is clear that most tropical dry forests will be destroyed or degraded within the next few years if urgent steps are not taken; because of the size of the human population in the tropics and subtropics, extensive poverty and collective ignorance on effective ways to manage tropical ecosystems so that they will be productive on sustainable basis. Over the next few years, the confrontation between human needs and forest preservation will become higher. The protection of forest or game reserves against threat and extinction can be achieved, if there is genuine willingness to take the task by all the people that directly or indirectly benefits from the forests and forest resources.

Precise estimate of the numbers of species being extinguished in the tropical forests is unavailable, due to the unknown numbers of species originally present and dearth of information on the species present in these ecosystems. There is no doubt that extinction is increasingly far faster than it was in the 19th century. Tropical dry forests are disappearing at alarming rates worldwide, reducing annually by about 4% of their current area (Naidu and Kumar, 2016). Tropical countries were losing 127,300 km^2 of forest annually, while deforestation in Nigeria was put at about 3.5% per annum translating into a loss of 350,000 to 400,000 hectares of forest land per annum (FAO, 2008). Tropical dry forests are faced with pressures of deforestation, urbanization, agriculture and poor management (Baithalu *et al.*, 2013). The impacts of climate change will further aggravate the plights of many indigenous and exotic tree species as climatic variability may limit the ability of forest trees to quickly adapt to the changing climate. Only 9.8% of the entire tropical forest biome lies within strictly protected areas; and the long-term viability of existing reserves is strongly affected by patterns of human activity in adjacent areas (Schmitt *et al.*, 2008).

Prospects of tropical forests and the biodiversity therein are becoming increasingly bleak owing to unabated deforestation and forest alteration that stem from human activities such as logging, hunting, agricultural expansion, and human settlement (Wilcove *et al.*, 2013; D´avalos *et al.*, 2016). A threat to these plant species means a threat to the survival of man. It is clear that the future of many biodiversity in the tropical forests depends more than ever on the effective management of human factors and their impacts on landscapes (Perfecto and Vandermeer, 2008).

There is urgent need for all individuals to join hands and conserve the tropical dry forests particularly among the developing countries. The biological diversity of tropical dry forests must therefore be constantly monitored and managed in order to direct succession processes towards maintaining species and habitat diversity. For this to be achieved; the use of forest inventory as management tool is unavoidable. This chapter focus on the relevance of forest inventories, challenges and the possible ways forward for biodiversity conservation and restoration of the tropical dry forests.

CHALLENGES ON CONSERVATION AND RESTORATION OF TROPICAL DRY FORESTS

Some of the challenges of conservation and restoration of the tropical dry forest include: inaccessibility to the location, financial and technological challenges, inadequate trained personnel, and calculation/compilation challenges among others for the synthesis of adequate forest inventory. Ecologist and forest managers are faced with the problem of inadequate roads to freely access the forests. Carrying out inventory which involves field work is mostly very difficult. The funds availability and the cost of any inventory strongly influence the success of the field work, the design chosen, the type of information required, the standard of precision chosen, the total size of the area to be surveyed and the minimum size of the unit area for which estimates are required (Shuaibu and Dagba, 2013).

Many forestry services are handicapped by shortage of trained personnel both in quantity and quality. The few trained technicians are not mostly available when they are needed for inventory work (Akindele, 2001). All these have resulted in big challenges to forest inventory on conservation and restoration; therefore, led to difficulties in determination of the forest status. A change in techniques of obtaining data from forest inventory has made it difficult for most foresters to get involved in field work. Obsolete computer systems, inadequate software knowledge has made it difficult to get involved in forest inventory in developing countries within the tropical dry forests.

The use of remotely sensed data to support and enhance forest inventories for conservation purpose has become a common practice in Europe and North America (McRobertset *al.*, 2002; McRoberts and Tomppo, 2007; Tomppoet *al.*, 2008; McRoberts, 2010a; McRoberts, 2010b). Despite the importance of remote sensing in modern forest management, adequate ground truthing (inventory data) is relied upon to validate the remote sensing data.

Most countries within the tropical dry region are signatories to many international environmental-related treaties, while they have promulgated many of such laws to ensure sustainable use and conservation of natural resources within their political boundaries. But the challenge has always been the poor implementation. For example, many of the National Parks in Nigeria and neighboring countries covers large land area, yet poorly staffed such that proper monitoring of the parks is not done (WCS, 2010). There is no prospect at the moment that the scientific task will be completed before large fractions of the existing but unknown species vanish. There is a lack of professionals in the world that are competent to deal with the number of species found in the tropical dry forests. Their number might also be dropping due to decreased professional opportunities, reduced funding for research, and a relatively higher priority given to other disciplines.

The problem of tropical dry forests conservation is the lack of knowledge and the paucity of ongoing research. In order to make precise assessments and recommendations, it is necessary to know which species are present. Majority of such species has not even received a scientific name, their geographical ranges and possible vulnerability to environmental change (Wilson, 1999). Rapid human population growth rate is the major cause of plant species loss. These population growths pose a serious challenge on biodiversity due to indiscriminate exploitation by artisans and herbalists coupled with the high rate of agricultural activities; these implied that soil degradation in the area is inevitable (FAO, 2009).

FOREST INVENTORIES: AN OVERVIEW

All decision-making process needs information. In forestry, this information is acquired by means of forest inventories. Forest inventory is the systematic collection of data and forest information for assessment or analysis. It is the systems for measuring the extent, quantity and condition of forests (Penman *et al.,* 2003). Maximum benefits from forests can only be achieved if the existing forests could be properly managed. Sound forest management depends on the quantity and quality of information available on the forest. Forest inventories are basic tools for oriented sustainable management and are based on ecological principles and the conservation of the forest.

More specifically, the purpose of forest inventories is to estimate means and totals for the measures of forest characteristics over a given area. Such characteristics include the growth variables, volume of the growing stock, the area of a certain type of forest. Forest inventory does not only record the trees height, diameter at breast height (DBH) and number of trees for yield calculations; but it also records the conditions of the forest such as the geology, site conditions, tree health and other tree factors. Nowadays, the inventories also measure the important parameters concerned with forest biodiversity (Kangas *et al.,* 2006).The importance of forest inventory is to provide comprehensive information about the state and dynamics of forests for strategic and management planning. Accurate, up-to-date information about the size, distribution, composition and condition of our forests and woodlands is essential for developing and monitoring policies and guidance to support the conservation and restoration of the tropical dry forests. To gather this information and keep the forests up to date for conservation and restoration, there is need to carry out periodic inventory of forests and woodlands in the tropical dry region.

Forest inventory could be carried out based on a complete assessment i.e. measuring every tree in a given area. This type of forest inventory is usually impossible in forestry because of the large areas involved, expensiveness and involvement of high degree of errors. Thus, accurate assessment of forest estates, for the purpose sustainable forest management acquiring relevant information, is typically based on sampling (i.e. using a proportion of the population as sample to represent the whole estates or forest cover). Forest inventory may also be based on sampling and systematic taking of a proportion of a forest to represent the entire forest area. However, according to Kangas *et al.* (2006), forest inventory is more than sampling; it also involves measuring trees and sample plots.

Forest inventories cover different spatial scales from the stand level to the woodlot level, regional and country level and global level. It is evident that these varying scales also require different methodologies. The purposes of inventories may also vary. For example, stand-level inventories can be carried out in order to estimate the number of saplings after regeneration, woodlot-level inventories in order to facilitate harvesting or silvicultural decisions, and regional level inventories in order to enhance policy decisions. Global inventories may also serve certain purposes in global-level politics, such as international agreements over actions concerning biodiversity or global warming. Forest inventories may also be a way of estimating the current growing stock, but most often they are carried out at several points in time in order to analyze the temporal changes.

According to Zerihun and Yemir (2013), the first forest inventory was carried out in Europe in the 14th and 15th centuries. Carl von Carlowitz, a mining engineer who also managed the forest resources around his mine since 1712, was the first to describe the principles of sustainability. The first large area (at provincial level) inventories took place in Sweden around 1840. In the 19th century, forest inventories were seen as an established section of forest planning, which involved gathering data based on visual estimation without statistical sampling. In 1910s, national forest inventories were carried out in the Nordic countries Norway, Sweden and Finland. During this era, representative samples were used probably to reduce costs (Bellhouse, 1988; Frayer and Furnival, 1999). Statistical knowledge was gradually introduced into the forestry literature between 1900 and 1920, primarily in Scandinavia (Loetsch and Haller, 1973). Later in the 1930s, the method of cruising with continuous strips of fixed width covering a known land area became most popular because it served purposes for inventory and mapping (Kangas *et al.*, 2006).

Today, forest inventory has witnessed a significant progress as it now involves the fields of statistics (i.e. sampling and modeling), remote sensing (i.e. aerial photos and satellite images), computers and sophisticated measurement devices.

RELEVANCE OF FOREST INVENTORY IN CONSERVATION AND RESTORATION OF TROPICAL DRY FORESTS

Forest inventory serves as a very important tool in sustainable forests management. It provides the needed data for planning, monitoring, evaluation, research, growth and yield, and timber sale. It is an attempt to describe quantity, quality and stocking density (diameter distribution) of forest trees and many characteristics of land, upon which trees are growing, for the efficient and sustainable management of the forest ecosystem. Thus, forest inventory is useful in ensuring conservation and restoration of forests through the following:

Provision of Information on Stand Structure and Species Diversity

Forest inventory in a natural forest provides counts of tree species in a forest ecosystem. These count data can be further used to compute various species diversity indices. Up-to-date information on tree species is required to understand the structure of the forest and to base management decisions. According to Janzen (1998), tropical dry forests are the most used and threatened ecosystems; as such, their conservation is significant. Gandhi and Sundarapandian (2014) assessed large scale quantitative inventory on the structure and floristic composition of trees in tropical dry forests of Sathanur dam Reserve Forests in Tiruvannamalai district, Tamil Nadu, India. They provided a baseline data of tree species richness, population structure and distribution patterns, and developed some conservation measures to protect the disappearing forest resources. A wide range of variables and indices can be assessed through forest inventory (Table 1).

Forest Planning, Management, and Decision-Making

Inventory is aimed at forest management; and without forest inventory, forest management would not be possible in a sustainable manner since there would not be sufficient data for planning or decision-making and implementation. It assesses forests, forest resources, provides qualitative and quantitative informa-

Table 1.Summary of tree diversity inventory in tropical dry forests at Sathanur Reserve forests, Tamil Nadu, India

Variable	Site I (8ha)	Site II (8ha)	Site III (4ha)	Total (20ha)
Species richness	32	54	38	60
Genera	26	45	33	47
Families	18	28	23	29
Stand species richness range (no. ha^{-1})	7-21	14-25	18-28	7-28
Diversity indices				
Shannon	1.639	2.194	2.014	2.102
Simpson	0.374	0.248	0.288	0.292
Fisher's alpha	4.46	8.48	6.81	8.178
Individuals	5817	4929	1802	12548
Density (stems ha^{-1})	727+ 101	616 + 50	450.5 +22.4	627.4 +49.3
Total basal area (m^2)	150.79	162.54	108.82	422.15
Basal area (m^2 ha^{-1})	18.85+1.80	20.32+3.02	27.2+3.16	21.1+1.6
Stand tree density range (stems ha^{-1})	336-1075	400-809	406-512	336-1075
Stand basal area range (m^2ha^{-1})	9.8-25.1	8.4-30.0	19.9-34.2	8.4-34.2

(Data source: Gandhi and Sundarapandian, 2014)

tion on site quality, extent, status, use and management of the resources and enhances forest planning. It also, estimates the growth characteristics and volumes of standing timber, and comparisons of annual growth in the forest which promotes sustainable forest management.

The goal of timber management is to provide the mix of timber quality, quantity and size that will maximize owner satisfaction while meeting the imposed constraints (Shamaki and Akindele, 2013). Consequently, accurate and flexible methods are required to estimate stand, tree growth and yield for

Figure 1. Diameter distribution (structure) of trees in tropical dry forests at Sathanur Reserve forests, Tamil Nadu, India
(Data source:Gandhi and Sundarapandian, 2014)

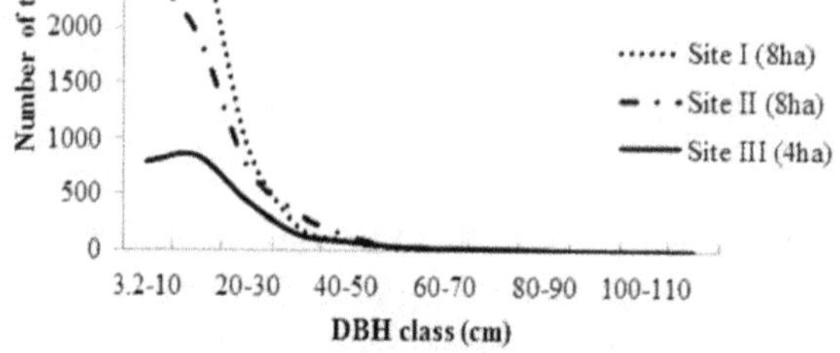

evaluating the numerous management and utilization alternatives for timber resources (Sharma *et al.*, 2000). Inventory equations of various kinds have been very useful to forest managers for sustainable forest management. The most basic equations provide an estimate of the available forest resources, their market values and how they would be sustained across generations. Figure 2 shows equation for predicting stem height from diameter at breast height for inventory and management of the *Tectona grandis* Linn. f. species in Omo Forest Reserve, Nigeria.

Assessment of Forest Growth and Yield

Forest inventory provides data on which other stand variables can be computed. One can calculate the number of trees per hectare, the basal area per hectare, the volume of trees in an area, and the volume of timbers. Clutter *et al.* (1983) and Husch *et al.* (2003) described the volume of the stem of a tree as a function of the independent variables such as diameter, height, and form. Inventory guides foresters on the estimation of volume of timber per unit area as well as in forest resource management. For sustainable forest management, there is need for reliable volume estimate of tree species. For this purpose, volume equation based on the established relationship between some measurable tree variables and tree volume is very paramount through forest inventory. Tree variables include breast height diameter (DBH), basal area and merchantable height (Ajayi and Odey, 2012). Total height and DBH are adequate for predicting tree volume for management planning.

Forests can be cruised to visually assess timber and determine the potential fire hazards. The results of this type of inventory can be used in preventative actions as well as in mass awareness (Moores *et al.*, 1996). Wildlife surveys can be undertaken in conjunction with timber inventory to determine the number and type of wildlife within a forest.

Site Quality Assessment

Site quality assessment is an essential tool required by foresters to effectively manage their forest estates. It is, therefore, important to assess the ability of a particular site to support tree growth or how well a particular species is utilizing a site. Site quality assessment is very important in forest management because a site could support one species excellently while supporting another species poorly. Age–height

Figure 2. Height-Diameter Predictive Equation for Management Purposes of Tectona grandis Species in the Tropical Forest of Omo Forest Reserve, Nigeria
(Data source: Chukwu, 2017)

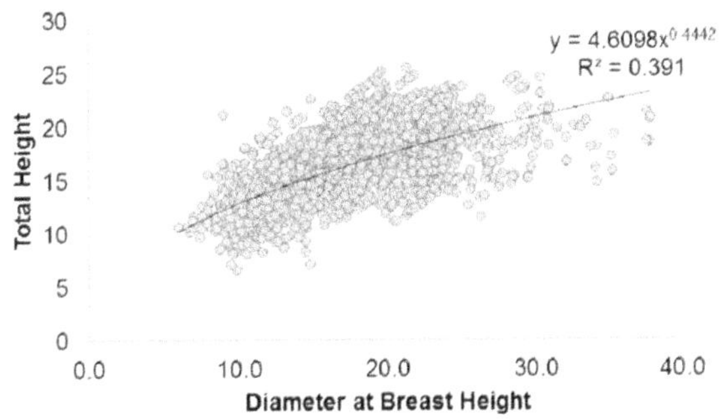

relationships of trees are commonly used to assess site potential productivity of forest stands. This relationship is known as site index curve. Information from sites index curve is used to manage the current and future indigenous and exotic species plantations (Onyekwelu, 2003; 2005). Site Index is the timber production potentials of a site for a particular species at a given index age.

The site class for *Gmelina arborea* plantations in Makurdi, Nigeria, was developed by Dau *et al.,* (2015) through forest inventory (Figure 3). Site quality assessment would be useful in identifying suitable sites for large-scale establishment of specie plantations; to quantify its growth and yield on various site conditions (Dau *et al.,* 2015). Site index curves usually reveal the best sites for future specific plantations. For example, Teshome and Petty (2000) used site index curve to assess *Cupressus lusitanica* plantation in Ethiopia, they recommended the first thinning to be carried out at the attainment of dominant height of 11 m for site classes of 27, 21 and 12 at ages;5, 7 and 13 years, respectively. While, the thinning schedules developed for Opepe plantations as recommended by Dupuy and Mille (1993) and Onyekwelu (2001) indicated that first, second and third thinning should be administered at the attainment of average dominant heights of 13, 17 and 22 m, respectively for sustainable forest management.

Predicting Growth Spacing and Control Competition Among Tree Species

To improve production, fast growth and quality of tree species in a plantation of native species, individual tree species in a stand must have unrestricted continuous free-growing space (Table 2). This requires knowledge of maximum occupancy (stocking) of sites with time and it can only be achieved through forest inventory. Inventory of stand occupancy helps in the prediction of growth space for optimum planting and sustainable management of different tree species. This is further illustrated by the possible means of predicting crown area and stocking for indigenous tree species plantation.

Figure 3.Site Index Curves for Gmelina arborea Plantations at Agan Forest, Makurdi, Nigeria
(Data Source:Dau et al., 2015)

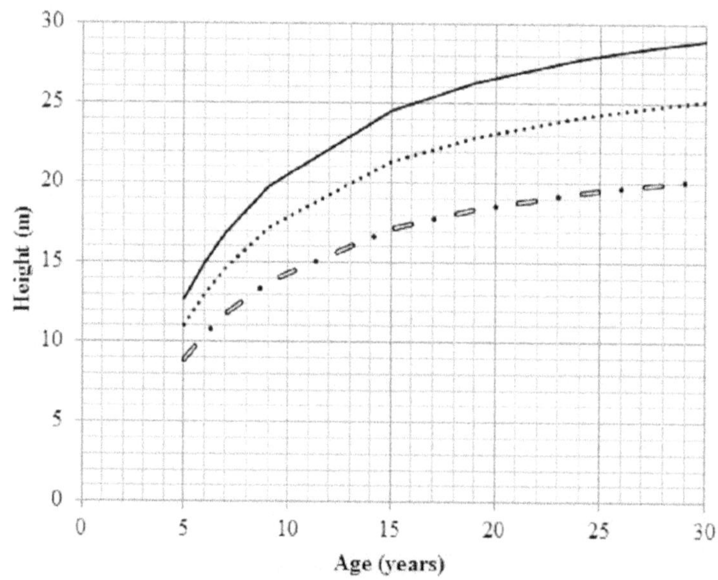

Table 2. Predicted Growth Space (S), Stocking (N) and Stand Density (D) for Parkia biglobosa and Prosopis africana in Makurdi, Nigeria

Species	DBH (cm)	Cd (m)	Cd/DBH	S (ha)	Nha^{-1} (1/S)	D (m^2ha^{-1})
	18.3	3.2	0.175	0.0015	670	2.631^{06}
Parkia biglobosa	25.4	4.7	0.185	0.0023	441	5.068^{06}
	31.1	5.9	0.19	0.003	332	7.597^{06}
	40.2	7.7	0.192	0.0044	227	1.269^{05}
	45.1	8.9	0.196	0.0053	190	1.598^{05}
	55.4	11.9	0.215	0.0073	136	2.411^{05}
	65.7	12.7	0.193	0.0097	103	3.391^{05}
	70.2	12.9	0.184	0.0109	92	3.871^{05}
	75.7	13.4	0.177	0.0124	81	4.501^{05}
	81.8	13.8	0.169	0.0142	70	5.256^{05}
	16.7	3.1	0.186	0.0013	746	2.191^{06}
Prosopis africana	20.0	3.9	0.195	0.0017	601	3.142^{06}
	25.9	5.3	0.205	0.0023	429	5.269^{06}
	35.9	7.4	0.206	0.0037	269	1.012^{05}
	40.7	7.89	0.194	0.0045	223	1.301^{05}
	45.6	8.8	0.193	0.0054	186	1.633^{05}
	52.9	9.7	0.183	0.0068	147	2.198^{05}
	60.2	11.2	0.184	0.0086	117	2.901^{05}
	66.3	12.3	0.186	0.0099	101	3.453^{05}
	70	12.5	0.179	0.0109	92	3.849^{05}

The above table was derived using *P. africana* simple linear model: *Cd*= 0.873+ 0.174 *DBH* and *P. biglobosa*:*Cd*= 2.134 + 0.126.
Where: Growing space (S) = $Cd^2\pi_{\pi}$ /40 000 and stocking (N) = 1/ S; Stand Density (D) = πD^2/ 40,000. DBH= diameter at breast height, Cd= crown diameter.
(*Source:*Dau *et al.*, 2016; Dau and Chenge, 2016).

Adequate spacing can significantly increase gross volume and growth yield while severe overcrowding of tree species could greatly restrict root and crown development, especially in the absence of thinning. The control of high density on stand through thinning and other silvicultural means is an important aspect of forest management that can enhance growth, production and yield of plantations. Stands with wider spacing in time have larger average stem and crown diameter than similar stands with closer spacing.

Competition among individual trees is a fundamental ecological process that plays a major role in population dynamics, survival, growth and species replacement. The identity of neighboring species is an important factor in the characterization of their competitive effect (Maleki *et al.*, 2015). Competition can be *conspecific* (i.e. between plants of same species), and *hetero-specific* (i.e. between plants of different species). Species-specific competition affects the ability of spatial indices, which majorly account for observed growth variability.

Assessment of Forest Health

Numerous concepts are used to describe forest health, and the definitions vary greatly, depending on the perspective (e.g., utilitarian or ecosystem), and scale (which may range from individual trees through landscapes to regions) (Kolb *et al.*, 1994). Forest health is defined as the perceived condition of a forest derived from concerns about factors such as age, structure, composition, function, vigour, presence of unusual levels of insects or diseases, and resilience to disturbance (Helms, 1998). Some factors that must be considered includes the species' growth habit, tree condition, branch attachments, resistance to decay, condition of anchoring roots, culturalor maintenance history, and previous damage. Thus, there is a clear need for accurate information on tree damage to support decision-making on the management of the trees species in the tropical dry forest.

Forest health can be assessed using different tree attributes, which serve as health indicators. These attributes include tree slenderness coefficient (TSC), tree crown-based indices etc., which are derived from forest inventory data. The vulnerability of a tree to wind-throw is highly influenced by the ratio of the tree's height to its stem diameter which is known as tree slenderness coefficient (Wang, 1998). Adeyemi and Adesoye (2016) classified TSC for tropical forests of Nigeria into three, namely; "Low" (TSC< 70), "Moderate" (TSC 70-80) and "High" (TSC > 80). Chukwu and Chenge (2018) reported the distribution of tree species found within Faculty of Agriculture and Faculty of Renewable Natural Resources, University of Ibadan, Ibadan, Nigeria according to their TSC classes (Figure 4), and recommended that trees with TSC> 80 be subjected to appropriate silvicultural treatment to improve their health.

POSSIBLE WAYS FORWARD

Knowledge on the status of tropical dry forests (both temporally and spatially) is essential for planning and managing forest ecosystems in a rational manner for short and long term goals in a sustainable manner. Protection and rational utilization of the regional forests become more important in order to

Figure 4. Distribution of tree species found within the Faculty of Renewable Natural Resources, University of Ibadan, Ibadan, Nigeria according to their TSC classes
(Data source: Chukwu and Chenge, 2018)

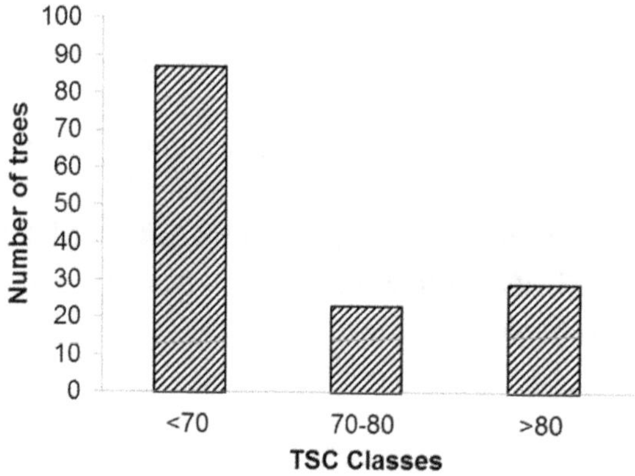

meet the increasing demand for raw materials (such as wood) and Non-Timber Forest Products (NTFPs). The conservation and restoration of tropical dry forest have become a critical issue in the contemporary world. It comprises of different management techniques to maintain and enhance the long-term health of forest ecosystems and suitable habitats for biodiversity conservation.

Positive decisions must be taken and effected to ensure conservation and restoration of endangered species in the tropical forests. The information for effective decision making in this regard can be acquired by means of forest inventories or ecological assessments; systems for measuring the extent, quantity and condition of forests and biodiversity habitats. In carrying out the appropriate management practices, forest inventory is needed to obtain the necessary information for timber evaluation (Shuaibu and Dagba, 2013); and ecosystems management. More specifically, there is need for regular inventories of the tropical dry forests. This calls for need to determine the threats and conservation status of plant species in the forests regions.

The only one answer to the problem of reduction in stands quality and quantity is to give proper management practices to the available TDFs, control deforestation and raise new trees and forests wherever possible with adequate management practices. Guards and Rangers cannot save tropical dry forests. Thus, there is need for joint management between managers of wild lands and society. The scientific community must aggressively participate in amalgamating and effectively executing the management policies for conservation and restoration of biodiversity. Without such participation and bottom-to-top decisions making, tropical dry forests would be nothing but high-degradation and gradual diminishing of forest ecological zones.

The future lies in the hands of young and agile people, who are well-educated in technologies but they have low interest in conservation and restoration of biodiversity especially in the developing countries. These youths will replace their parents in the nearest future; what will happen to the tropics biodiversity and endangered species? The collective power to turn the game around resides with policy makers. We cannot force the youths and the world to conserve tropical nature; but they can be mobilized via effective communication methods. This can be achieved by involving the young minds in decisions making and implementation of conservation policies. There is need for countries within the tropical regions to:

1. develop their national red list;
2. ensure effective protected area management and *ex-situ* conservation;
3. embark on trees domestication;
4. embark on community-based natural resources management;
5. carryout evaluation of suspected "endangered" plants; and
6. train more forest inventory personnel

The growth of the human population must be checked, if possible. It is because, if the scale of human activities continues to increase for even a few more decades, the extinction of tropical dry forests' biota cannot be avoided.

Finally, demands on placing monetary value on forest ecological zones or endangered species before they can be conserved and restored should be discouraged. Youths should be enlightened on the urgent need to conserve and restore tropical dry forests and their biodiversity. This can be done by frequent inclusion of youths in decisions making, organizing biodiversity sports competition among tropical countries and the use of non-formal education methods (such as drama, play-let, musical concerts among others) to effective communicate the message of tropical dry forests' conservation and restoration.

CONCLUSION AND RECOMMENDATIONS

Forest inventory serves as a very important tool for conservation of forests. It provides data for planning, monitoring, evaluation, research, growth and yield, timber sale among others. It is an attempt to describe quantity, quality, stocking density and distribution of forest trees and many characteristics of land, upon which the trees are growing. Thus, forest inventory is useful in ensuring conservation and restoration of forests and forest products through different ways. These include: provision of information on stand structure and species diversity, forest planning and decision-making, prediction of forest growth and yield, forest health and site quality assessment, prediction of growth spacing and control competition among tree species in the tropical dry forests. Conservation and restoration of tropical dry forests would a play vital role in addressing the environmental degradation and deforestations, while increasing the direct benefits to mankind and the environment. This study, therefore, recommended that more emphasis should be placed on assessment (inventory) of tropical dry forests and forest resources to avoid extinction of endangered flora and fauna species with greater involvement of young minds in decision making. Furthermore, inventory should be carried out in forest estates at regular intervals for ease of planning and decision-making.

REFERENCES

Adekunle, V. A. J., Olagoke, A. O., & Akindele, S. O. (2013). Tree species diversity and structure of a Nigerian strict nature reserve. *Tropical Ecology*, *54*(3), 275–289.

Adeyemi, A. A., & Adesoye, P. O. (2016). Tree slenderness coefficient and percent canopy cover in Oban Group Forest, Nigeria. *Journal of Natural Sciences Research*, *6*(4), 9–17.

Ajayi, S., & Odey, P. O. (2012). Individual tree volume equation for a plantation of *Tectona Grandis* in the Cross River University of Technology (Crutech), Obubra, Cross River State, Nigeria. *Nigerian Journal of Agriculture, Food and Environment*, *8*(2), 12–17.

Akindele, S. O. (2001). Review of urgent management requirement for Nigerian forestry. In L. Popoola, J. E. Abu, & P. I. Oni (Eds.), *Forestry and National Development: Proceedings of the 27th Annual Conference of the Forestry Association of Nigeria* (pp 75-84). Abuja, Nigeria: Forestry Association of Nigeria.

Baithalu, M. S., Anbarashan, M., & Parthasarathy, N. (2013). Two-decadal changes in forest structure and tree diversity in a tropical dry evergreen forest on the Coromandel Coast of India. *Tropical Ecology*, *54*(3), 397–403.

Baker, P. J., Bunyavejchewin, S., Oliver, C. D., & Ashton, P. S. (2005). Disturbance history and historical stand dynamics of a seasonal tropical forest in western Thailand. *Ecological Monographs*, *75*(3), 317–343. doi:10.1890/04-0488

Bellhouse, D. R. (1988). Systematic Sampling. In P. R. Krishnaiah & C. R. Rao (Eds.), Handbook of Statistics. (6:125-145). Amsterdam, The Netherlands: Elsevier Science Publishers B. V. doi:10.1016/S0169-7161(88)06008-0

Boboye, O., & Jimoh, S. O. (2016). Pattern of plant species diversity in a dry forest ecosystem of Nigeria. *Journal of Forestry Research and Management*, *13*, 31–47.

Bradshaw, C. J. A., Sodhi, N. S., & Brook, B. W. (2009). Tropical turmoil: A biodiversity tragedy in progress. *Frontiers in Ecology and the Environment*, *7*(2), 79–87. doi:10.1890/070193

Chukwu, O. (2017). *Stand variables prediction from stump diameter for Tectona grandis Linn. f. Stands in Omo Forest Reserve, Nigeria.* (Unpublished master's thesis), University of Ibadan, Nigeria.

Chukwu, O., & Chenge, I. B. (2018). Stability assessment and composition of tree species in Nigerian University. *World Scientific News: An International Scientific Journal*, *95*, 235–245.

Clutter, J. L., Fortson, J. C., Pienaar, L. V., Brister, G. H., & Bailey, R. L. (1993). *Timber Management: A Quantitative Approach.* New York, NY: John Wiley & Sons.

D'avalos, L. M., Sanchez, K. M., & Armenteras, D. (2016). Deforestation and coca cultivation rooted in twentieth-century development projects. *Bioscience*, *66*(11), 974–982. doi:10.1093/biosci/biw118

Dash, P. K., Mohapatra, P. P., & Giri Rao, Y. (2009). Diversity and distribution pattern of tree species Niyamgiri hill Ranges, Orissa, India. *Indian Forester*, *135*(7), 922–942.

Dau, J. H., & Chenge, B. I. (2016). Growth space requirements models for *Prosopis Africana* (Guill & Perr) Taub tree species in Makurdi, Nigeria. *European Journal of Biological Research*, *6*(3), 209–217.

Dau, J. H., Mati, A., & Dawaki, S. A. (2015). Role of forest inventory in sustainable forest management: A review. *International Journal of Forestry and Horticulture*, *1*(2), 33–40.

Dau, J. H., Vange, T., & Amonum, J. I. (2016). Growth space requirements models for *Daniellia Oliverii* (Rolfe) Hutch and Daviz Tree in Makurdi, Nigeria. *International Journal of Forestry and Horticulture*, *2*(3), 31–39.

Dupuy, D., & Mille, D. (1993). FAO Forestry Paper: Vol. 98. *Timber plantations in the humid tropics of Africa*. Rome, Italy.

FAO. (2005). Global forestry resource assessment. Food and Agriculture Organization (FAO) Forestry Paper 147, Rome, Italy.

FAO. (2008). REDD-Workshop: monitoring, assessment and verification. Rome, Italy: FAO. (FAO. Working paper, 2).

FAO. (2009). *State of the world's forests 2009*. Rome, Italy: Food and Agriculture Organization of the United Nations.

FAO. (2010). Global forest resources assessment 2010: main report. Rome: FAO, 2010. 340 p. *FAO Forestry Paper*, 163.

Frayer, W. E., & Furnival, G. M. (1999). Forest survey sampling designs: A history. *Journal of Forestry*, *97*(12), 4–10.

Gandhi, D. S., & Sundarapandian, S. (2014). Inventory of trees in tropical dry deciduous forests of Tiruvannamalai district, Tamil Nadu, India. *Biodiversitas (Surakarta), 15*(2), 169–179. doi:10.13057/biodiv/d150208

Gentry, A. H. (1995). Diversity and floristic composition of neotropical dry forests. In H. A. Mooney, S. H. Bullock, & E. Medina (Eds.), *Seasonally dry tropical forests* (pp. 146–194). Cambridge, UK: University of Cambridge Press. doi:10.1017/CBO9780511753398.007

Helms, J. A. (1998). *The dictionary of forestry*. Bethesda, MD: Society of American Foresters.

Husch, B., Beers, T. W., & Kershaw, J. A. Jr. (2003). *Forest mensuration* (4th ed.). Hoboken, NJ: John Wiley & Sons

Janzen, D. H. (1998). Tropical dry forests: the most endangered major tropical ecosystem. In E. O. Wilson (Ed.), *Biodiversity National*. Washington, DC: Academy Press.

Kacholi, D. S. (2014). Analysis of Structure and Diversity of the Kilengwe Forest in the Morogoro Region, Tanzania. *International Journal of Biodiversity, 2014*, 1–8. doi:10.1155/2014/516840

Kangas, A., Gove, J. H., & Scott, C. T. (2006). Introduction. In A. Kangas & M. Maltamo (Eds.), Managing Forest Ecosystems: Forest Inventory Methodology and Applications. (10: 3-11). Switzerland: Springer Nature.

Kolb, T. E., Wagner, M. R., & Covington, W. W. (1994). Concepts of forest health- Utilitarian and ecosystem perspectives. *Journal of Forestry, 92*(7), 10–15.

Laurance, W. F., Sayer, J., & Cassman, K. G. (2014). Agricultural expansion and its impacts on tropical nature. *Trends in Ecology & Evolution, 29*(2), 107–116. doi:10.1016/j.tree.2013.12.001 PMID:24388286

Loetsch, F., & Haller, K. E. (1973). *Forest inventory* (Vol. I). Mu¨nchen, Germany: BLV Verlagsgesellschaft.

Maass, J. M., Balvanera, P., Castillo, A., Daily, G. C., Mooney, H. A., Ehrlich, P., ... Sarukhán, J. (2005). Ecosystem Services of Tropical Dry Forests: Insights from Long- term Ecological and Social Research on the Pacific Coast of Mexico. *Ecology and Society, 10*(1), 1–17. doi:10.5751/ES-01219-100117

Maleki, K., Kiviste, A., & Korjus, H. (2015). Analysis of Individual Tree Competition Effect on Diameter Growth of Silver Birch in Estonia; *Forest Systems, 24*(2).

McRoberts, R. E. (2010a). Probability and model-based approaches to inference for proportion forest using satellite imagery as ancillary data. *Remote Sensing of Environment, 114*(5), 1017–1025. doi:10.1016/j.rse.2009.12.013

McRoberts, R. E. (2010b). The effects of rectification and Global Positioning System errors on satellite image-based estimates of forest area. *Remote Sensing of Environment, 114*(8), 1710–1717. doi:10.1016/j.rse.2010.03.001

McRoberts, R. E., & Tomppo, E. O. (2007). Remote sensing support for national forest inventories. *Remote Sensing of Environment, 110*(4), 412–419. doi:10.1016/j.rse.2006.09.034

McRoberts, R. E., Wendt, D. G., Nelson, M. D., & Hansen, M. H. (2002). Using a land cover classification based on satellite imagery to improve the precision of forest inventory area estimates. *Remote Sensing of Environment, 81*(1), 36–44. doi:10.1016/S0034-4257(01)00330-3

Miles, L., Newton, A. C., DeFries, R. S., Ravilious, C., May, I., Blyth, S., ... Gordon, J. (2006). A global overview of the conservation status of tropical dry forests. *Journal of Biogeography, 33*(3), 491–505. doi:10.1111/j.1365-2699.2005.01424.x

Monserud, A. R. (2003). Evaluating Forest Models in a Sustainable Forest Management Context; *Forest Biometry. Modeling and Information Sciences, 1*, 35–47.

Moores, L. J., Pittman, B., & Kitchen, G. (1996). Forest ecological classification and mapping: Their application for ecosystem management in Newfoundland. *Environmental Monitoring and Assessment, 39*(1–3), 571–577. doi:10.1007/BF00396169 PMID:24198030

Murphy, P. G., & Lugo, A. E. (1986). Ecology of dry tropical forest. *Annual Review of Ecology Evolution and Systematics, 17*(1), 67–88. doi:10.1146/annurev.es.17.110186.000435

Naidu, M. T., & Kumar, O. A. (2016). Tree diversity, stand structure, and community composition of tropical forests in Eastern Ghats of Andhra Pradesh, India. *Journal of Asia-Pacific Biodiversity, 9*(3), 328–334. doi:10.1016/j.japb.2016.03.019

Onyekwelu, J. C. (2003). Choosing appropriate index age for estimating site index of Gmelina arborea timber plantations in Oluwa forest reserve, Nigeria. [</ths>]. *Journal of Food Agriculture and Environment, 1*(3 & 4), 286–290.

Onyekwelu, J. C. (2005). Site index curves for site quality assessment of *Nauclea diderrichii* monoculture plantations in Omo Forest Reserve, Nigeria. *Journal of Tropical Forest Science, 17*(4), 532–542.

Penman, J., Gytarsky, M., Hiraishi, T., Krug, T., Kruger, D., & Pipatti, R. … Wagner, F. (Eds.) (2003). Good practice guidance for land use, land-use change and forestry. Intergovernmental Panel on Climate Change: IPCC National Greenhouse Gas Inventories Programme. Kanagawa Japan: Institute for Global Environmental Strategies (IGES) for the IPCC. Retrieved from http://www.ipcc-nggip.iges.or.jp/public/gpglulucf/gpglulucf_contents.htm

Powers, R. F. (2001). Assessing Potential Sustainable Wood Yield. In J. Evans (Ed.). The Forestry Handbook: Applying Forest Science for Sustainable Management (2: 105-128). Oxford, UK: Blackwell Science, Ltd. doi:10.1002/9780470757079.ch5

Schmitt, C. B., Belokurov, A., Besancon, C., Boisrobert, L., Burgess, N. D., Campbell, A. … Winkel, G. (2008). Global Ecological Forest Classification and Forest Protected Area Gap Analysis. Analyses and recommendations in view of the 10% target for forest protection under the Convention on Biological Diversity (CBD). In I. Perfecto & J. Vandermeer (Eds.), Biodiversity conservation in tropical agroecosystems (1134, 173–200). Annals of the New York Academy of Sciences, University of Freiburg, Germany.

Schwalm, C., & Ek, A. R. (2001). Climate change and site: Relevant mechanisms and modeling techniques. *Forest Ecology and Management, 150*(3), 241–257. doi:10.1016/S0378-1127(00)00568-5

Shamaki, S. B., & Akindele, S. O. (2013). Volume estimation models from stump diameter for teak (*Tectona grandis* Linn f.) plantation in Nimbia forest reserve, Nigeria. *Journal of Environmental Science and Water Resources*, *2*(3), 89–94.

Sharma, M., Oerwald, R. G., & Amateis, R. L. (2000). A consistent system of equations for tree and stand volume. *Northern Journal of Applied Forestry*, *54*(2), 110–125.

Shuaibu, R. B., & Dagba, B. I. (2013). Challenges of forest inventory and its effects on forest management and planning in Nigeria. In L. Popoola, O. Y. Ogunsanwo, V. A. J. Adekunle, I. O. Azeez, & N. O. Adewole (Eds.), *The green economy: Balancing: environmental sustainability and livelihoods in an emerging economy. Proceedings of the 36thAnnual Conference of Forestry Association of Nigeria.* Uyo, Nigeria: Forestry Association of Nigeria.

<ths>Onyekwelu, J. C. (2001). Growth characteristics and management scenarios for plantation-grown Gmelina arborea and Nauclea diderrichii in south-western Nigeria (Doctoral dissertation). Available from Hieronymus Publishers, Munich, Germany.

Tomppo, E., Haakana, M., Katilia, M., & Peräsaari, J. (2008). *Multi-source national forest inventory (373).* Dordrecht, The Netherlands: Springer.

Wang, Y., Titus, S. J., & Lemay, V. M. (1998). Relationship between tree slenderness coefficients and tree or stand characteristics for major species in Boreal mixed Forest. *Canadian Journal of Forest Research*, *28*(8), 1171–1183. doi:10.1139/x98-092

Watson, R. T., Zinyowera, M. C., & Moss, R. H. (1995). *Climate Change 1995: Impacts, Adaptations and Mitigation of Climate Change: Scientific-Technical Analyses; Contribution of Working Group II to the Second Assessment of the Intergovernmental Panel on Climate Change (878).* Cambridge, UK: Cambridge University Press.

WCS. (2010). Nigeria: World Conservation Society (WCS). Retrieved from www.wsc.org>wherewework>Africa.

Wilcove, D. S., Giam, X., Edwards, D. P., Fisher, B., & Koh, L. P. (2013). Navjot's nightmare revisited: Logging, agriculture, and biodiversity in Southeast Asia. *Trends in Ecology & Evolution*, *28*(9), 531–540. doi:10.1016/j.tree.2013.04.005 PMID:23764258

Wilson, B., & Wang, S. (1999). Sustainable forestry--the policy prescription in British Columbia. In A. Yoshimoto & K. Yukutake (Eds.), *Global Concerns for Forest Resource Utilization-- Sustainable Use and Management* (pp. 35–45). London, UK: Kluwer Academic Publishers. doi:10.1007/978-94-017-6397-4_4

Zerihun, A., & Yemir, T. (2013). Forest Inventory and Management in the Context of SFM & REDD. Training Manual (1-67), Hawassa University, Ethiopia: Wondo Genet College of Forestry and Natural Resources.

Section 5
Forest Regeneration and Policy

Chapter 17
Post-Fire Regeneration of Woody Species in a Colombian Tropical Dry Forest

Anghy Gutiérrez-Rincón
Universidad Distrital Francisco José de Caldas, Colombia

Angela Parrado-Rosselli
Universidad Distrital Francisco José de Caldas, Colombia

ABSTRACT

In fire-influenced ecosystems, some plant species have the ability to recover, germinate, and to establish after a fire; however, their proportion and dominance varies between sites. The objective of this work was to evaluate natural regeneration following a fire in a tropical dry forest located in the Upper Magdalena River Valley in Colombia. In that way, all seedlings and saplings of woody species were recorded, 1.5 years after a fire, in 75 2x2-m plots installed in burned and unburned forest sites, as well as in forest gaps. Results showed that although abundance was higher in the burned sites, the species richness was lower than in unburned areas. Based on the regeneration response of the species, we identified three groups of plants: 1) fire-stimulated, 2) fire-tolerant, and 3) fire-sensitive species, which means that this tropical dry forest has species with the ability to recover, germinate, and establish after a fire. These three groups of plant species should be considered in restoration programs in light of future and more frequent forest fires due to climate change.

INTRODUCTION

Fire has been considered an important disturbance agent that influences composition and structure of plant communities in several ecosystems of the world (Rodrigues, Martins, & Matthes, 2005; Alanís-Rodríguez et al., 2012; Oliveira et al., 2014; Salazar & Goldstein, 2014; Bhadouria et al., 2017; Young et al., 2018; Koontz, North, Werner, Fick, & Latimer, 2019). In fire-influenced ecosystems some plant species have the ability to resprout, germinate and establish after fire (Hoffmann, 2000; J. E. Keeley &

DOI: 10.4018/978-1-7998-0014-9.ch017

Fotheringham, 2000; Rodrigues et al., 2005; Pausas et al., 2016; Ocampo-Zuleta & Bravo, 2019). Thus, depending on the proportion of these species in comparison to others that do not have such adaptations, some ecosystems will have a greater or lesser capacity to recover from the occurrence of forest fires (Pausas & Lavorel, 2003; Pausas, Bradstock, Keith, Keeley, & Network, 2004; Hardesty, Myers, & Fulks, 2005; Myers, 2006; Pausas et al., 2016). Moreover, based on the plant's capacity to regenerate after being exposed to fire or heat, different studies have all proposed three categories of germination and regeneration: fire/heat-tolerant, stimulated and sensitive (Otterstrom, Schwartz, & Velazquez-Rocha, 2006; Griscom & Ashton, 2011; Jaureguiberry & Díaz, 2015). These classifications have also been used for seedlings and saplings found in the understory after a fire (Rocca, 2009).

Mechanisms of post-fire plant regeneration have been widely studied in Mediterranean ecosystems, temperate zones and tropical savannas (Pausas et al., 2004; Hardesty et al., 2005; Rodrigo, Retana, & Xavier, 2005; Shlisky, A. et al., 2007; Gómez-González et al., 2017). In contrast, in tropical dry forests, information is more limited (Snook, 1993; Kellman & Meave, 1997; Myers, 2006; Bhadouria et al., 2017), and most of the studies have been carried out in the Brazilian Cerrado, Mexico and Bolivia (*e.g.* Fredericksen & Mostacedo, 2000, Rodrigues et al., 2005, Alanís-Rodríguez et al., 2010, Carón, Dalmasso, Ortín, & Verheyen, 2015, Gómez-González et al., 2017). Additionally, several studies have focused on invasive species while very little is known about the native ones (Gómez-González et al., 2017, Hoffmann, 2000, Lahoreau et al., 2006).

During the last 20 years in Colombia, the majority of fire activity has been associated with the tropical dry forests (hereafter TDF) cover of the inter-Andean valleys, the Orinoco and the Caribbean region (Armenteras, González-Alonso, & Aguilera, 2009; Armenteras et al., 2011; Pizano & García, 2014; Etter, Andrade, Amaya, & Arévalo, 2015; Pausas & Ribeiro, 2017). However, little is known about the capacity of Colombian TDF species to regenerate after wildfires (Vieira & Scariot, 2006; Pizano & García, 2014; Rodríguez-Buritica, Aguilar-Garavito, & Norden, 2017). Therefore, the objective of this research was to characterize floristic composition and structure of the sexual regeneration of woody species, 1.5 years after a fire in a TDF of the upper Magdalena River basin. Firstly, floristic composition and dominance of seedlings between burned and unburned forest sites, including forest gaps were compared. Secondly, based on the abundance, frequency and size classes of seedlings, the natural regeneration index (hereafter NRI) for each species was obtained. Considering that the TDF of the study site includes many genera and species that have been described as fire-tolerant and fire-stimulated in other tropical dry forests (*e.g. Astronium graveolens, Aspidosperma polyneuron, Celtis iguanaea, Trichilia pallida and Casearia sylvestris;* Rodrigues et al., 2005; Villanueva, Melo, & Rincón, 2015; Aguilar et al., 2016; Melo, Fernandez-Méndez, & Villanueva, 2016; Pizano, González-M, Hernández-Jaramillo, & García, 2017) it is expected to find a great number of species with the ability to regenerate after a fire.

METHODS

Study Site

The study was carried out in a tropical dry forest of the Jabirú Natural Reserve of Civil Society RNSC located in the upper basin of the Magdalena River, in the municipality of Armero, Tolima, Colombia (Figure 1). The reserve has 670 ha of which 250 ha are tropical dry forest fragments (Natural National Parks, 2014). Mean annual temperature is 28 °C and mean annual precipitation is 1387 mm, with three

Figure 1. Study site, Sentinel 2A satellite imagery. The black line represents the Jabirú Natural Reserve of Civil Society. The green areas represent the unburned forest and the red areas the burned forest.

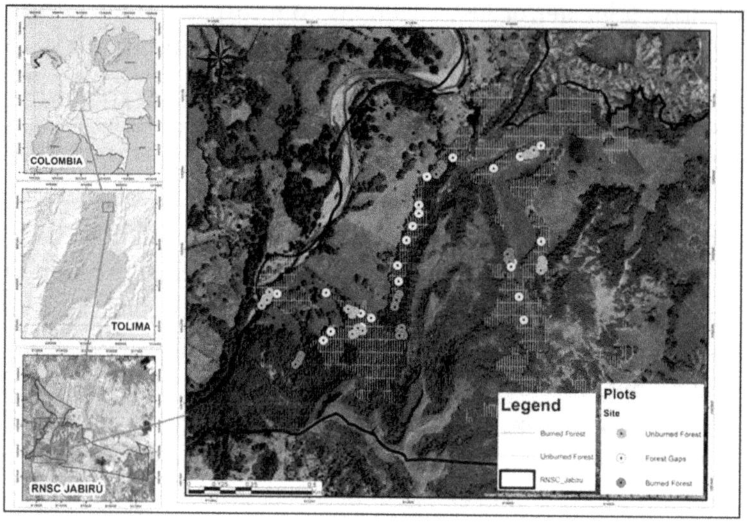

months of drought (less than 60 mm per month) between July and September. Altitude ranges between 280 - 420 m characterized by a flat to undulating topography.

On August 24, 2016, a forest fire burned approximately 90 ha of primary and secondary forests, grass and crop areas (Arenas & Noy, 2019). Through the use of a 10-m resolution Sentinel-2A satellite imagery, areas of burned forest (BF), unburned forest (UBF) and forest gaps (FG) were selected for this study (Fernández-Manso, Fernández-Manso, & Quintano, 2016; Navarro et al., 2017).

Data Collection

Eighteen months after the fire, 75 2 x 2 m plots were randomly installed (25 per site) nested in 10 x 50 m plots and spatially referenced. In each plot, all individuals of woody species regenerated from seeds, between 0.1 – 3.0 m height and diameter at the base < 2,5 cm, were recorded and identified (Mostacedo, Fredericksen, Gould, & Toledo, 1999; Kennard et al., 2002; Melo & Vargas, 2003; Norden et al., 2009; González-M., Avella, & Díaz-Triana, 2015). Height was measured from the base to the apical meristem and diameter at the base of the individual. Samples of specimens were herborized for later identification at the Herbario Forestal Colombiano (UDBC) of the Universidad Distrital Francisco Jose de Caldas. Seedlings were differentiated from resprouts through visual inspection and proximity to trunks or roots of live and dead standing trees.

Data Analysis

Sampling representativeness was estimated using various species accumulation curves (Mostacedo, et al., 1999; Alanís, et al., 2010, Chao & Jost, 2012). Diversity curves for order 0 (species richness); 1 (Shannon diversity) and 2 (the inverse of Simpson's concentration) were estimated for each site (burned forest, unburned forest, forest gaps; Chao et al., 2014; Hsieh, Ma, & Chao, 2016). Jaccard similarity coefficient was also calculated among sites (Hammer et al, 2001; Zuloaga-Aguilar, 2010). Prior to analyses, the data were tested for normality, variance and autocorrelation using Kolmogorov-Smirnov, Levenne and Fligner-Killeen, and Durbin-Watson tests, respectively (Amat, 2016). Analyses were calculated in PAST software and R-Studio (Hammer et al, 2001; R Core Team, 2018).

One-way analysis of variance and Kruskal-Wallis test were performed to evaluate whether abundance and species richness differed between sites (Amat, 2016). In order to assess vegetation structure of seedlings, a natural regeneration index -NRI- was calculated using equation (1) (Rangel-Ch & Velázquez, 1997; Cantillo-H & Rangel–CH, 2002; Lezama, 2018).

$$NRI(\%) = \frac{RA(\%) + RF(\%) + SC(\%)}{3} \tag{1}$$

This index determines the importance of a species within the community by considering its relative abundance (RA%), relative frequency (RF%) and the proportion of the number of individuals in a given size class (SC%).

Based on the differences in this index between sites, using the Duncan's Test, all species with more than twelve individuals were classified within three possible regeneration responses to the fire: fire-stimulated, fire-tolerant and fire-sensitive regeneration (Otterstrom et al., 2006; Rocca, 2009; Jaureguiberry & Díaz, 2015). Thus, if in the burned plots a particular species exhibited a NRI and abundance significantly higher than in the unburned ones, the species was classified as fire-stimulated. If the natural regeneration index was similar between burned and unburned forest plots (including forest gaps) the species was considered fire-tolerant. Finally, if the natural regeneration index was higher in the unburned plots the species was considered fire-sensitive.

RESULTS

A total of 1053 individuals belonging to 78 species, 58 genera and 33 families were recorded in both burned and unburned sites (Table 1). The families Leguminosae, Malpighiaceae and Malvaceae were the overall diverse families in terms of species richness (Appendix 1; Table 5). The highest number of individuals was found in the burned forest, while greater species richness was found in the unburned forests, followed by the forest gaps.

The Shannon Index was also greater in the UBF than in the FG and BF (Figure 2). In contrast, the inverse of Simpson's concentration index was similar between BF and FG, in terms of the probability that any two individuals drawn randomly from a sample belong to the same species (Figure 2).

Out of the 78 species, 28 were found in at least two of the three treatments (burned, unburned and forest gaps), so they were called shared species (Table 2). In that way, the unburned forest and the forest gaps, shared 23 woody species, exhibiting a higher similarity in relation to the burned forest (Figure 3). Twelve species were shared among the three sites, but abundance was significantly higher at the burned forest ($p = 0.00353$), except for *Albizia* sp. which was higher in the unburned sites (*i.e.* unburned forest and forest gaps). *Casearia corymbosa, Cordia alliodora*, *Machaerium capote* and *Astronium graveolens* were the most abundant shared species (Table 2).

Table 1. Summary of woody species richness and diversity found in the burned forest, unburned forest and forest gaps.

Indicator	Burned forest	Unburned forest	Forest Gaps	Total
No. Families	21	23	21	33
No. Genera	30	33	30	58
No. Species	34	44	40	78
No. Individuals	454	323	276	1053
No. Shared species	17	25	26	28*
No. Exclusive species	17	19	14	50

* Species that were reported in at least two of the three sites.

In terms of exclusive species (Table 3), many of them exhibited a low number of individuals; thus, it was not possible to analyze statistically for differences between the three treatments. However, for those with more than four individuals, it was found that *Bunchonsia* sp. and *Piper angustifolium* were exclusive to the unburned sites; *Aphelandra barkleyi, Celtis iguanaea* and *Handroanthus ochraceus* to the forest gaps, and *Aphelandra glabrata,* sp. to the burned forest.

The natural regeneration structure varied between sites (Figure 4). At the burned forest, 50% of the regeneration was concentrated in only three species (*Casearia corymbosa, Cordia alliodora, Machaerium capote*) that exhibited the highest values of abundance and frequency (Appendix 1; Table 6). In contrast, at the unburned forest, rare species represented 30.5% of the species recorded, while in the forest gaps 27.5%. *Casearia corymbosa* showed the highest NRI in all sites, but its abundance was significantly higher in the burned forest and forest gaps (p = 0,0455). Similarly, *Machaerium capote* showed a higher NRI in the burned forest (p = 0,0236). NRI of *Astronium graveolens* was high in both the burned and unburned forest (Figure 4).

Figure 2. Diversity indexes: richness (q=0); Shannon (q=1); Simpson (q=2). Unburned forest (squares), burned forest (circles), forest gaps (triangles)

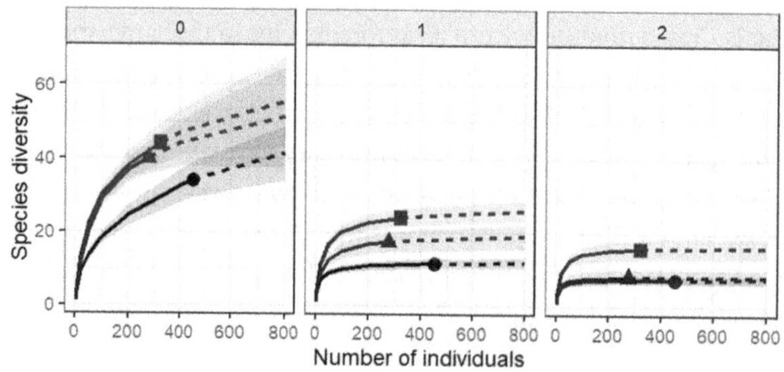

Table 2. Abundance (number of individuals/100 m²) of shared species in the burned forest, un-burned forest and forest gaps

Plant species	Burned forest	Unburned forest	Forest gaps
Casearia corymbosa	121	54	86
Machaerium capote	62	16	10
Cordia alliodora	56	8	25
Eugenia micrantha	50	5	3
Astronium graveolens	44	32	12
Eugenia procera	26	11	3
Swartzia trianae	18	14	8
Albizia sp.	10	26	21
Triplaris americana	9	5	5
Cupania latifolia	8	1	3
Bunchonsia nitida	4	3	5
Sorocea sprucei	1	4	4
Posoqueria sp.	7	18	-
Acalypha diversifolia	1	8	-
Guazuma ulmifolia	4	-	4
Piptadenia sp.	2	-	14
Ampelocera sp.	1	-	2
Triplaris melaenodendron	-	16	1
Cordia sp1.	-	9	3
Leguminoseae sp4	-	8	6
Cupania sp.	-	8	4
Cupania cinerea	-	7	1
Albizia sp2.	-	6	7
Gustavia sp1.	-	6	3
Guarea guidonia	-	3	9
Acanthaceae sp1.	-	2	4
Machaerium biovulatum	-	2	3
Cordia sp2.	-	1	1
Total	424	273	247

DISCUSSION

The results of this study show that there are important differences in the composition and structure of sexual regeneration of woody species in a TDF of the Upper Magdalena River basin between burned and unburned sites. Although seedling diversity was lower in the burned forest, the total abundance of individuals was significantly higher than in the unburned forest. This coincides with other studies of post-fire regeneration that have found greater abundances in burned sites, mainly due to certain spe-

Figure 3. Jaccard similarity coefficient between among sampling sites

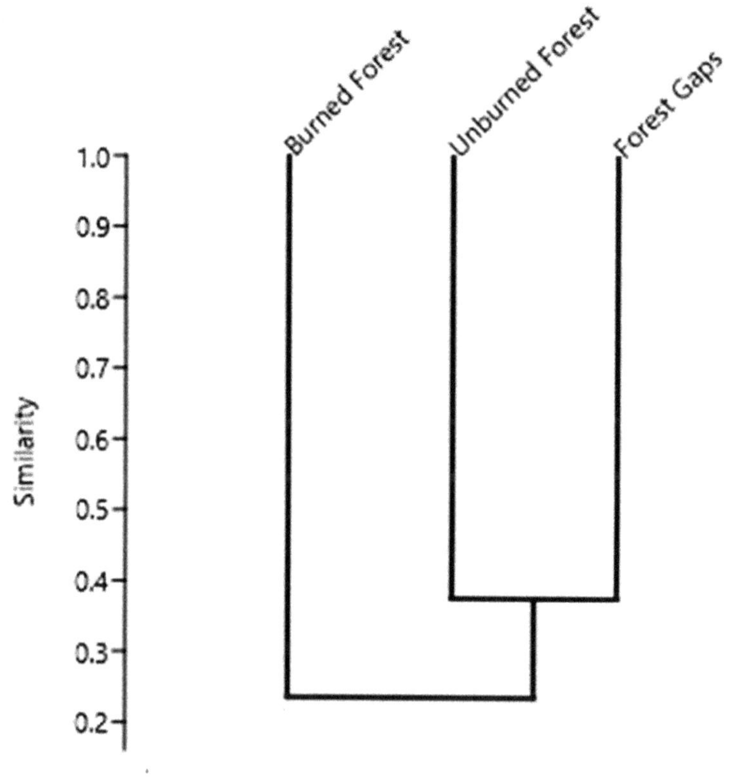

cies that are highly favored by fire (Kellman & Meave, 1997; Gould et al., 2002; Kennard et al., 2002;

Table 3. Abundance of exclusive species per site.

Species	Burned Forest	Unburned forest	Forest Gaps
Aphelandra glabrata	6	-	-
Machaerium goudotii	3	-	-
Acalypha sp.	3	-	-
Bunchosia sp.	-	11	-
Piper angustifolium	-	11	-
Nectandra sp.	-	4	-
Sapindus saponaria	-	4	-
Calliandra sp.	-	3	-
Aphelandra barkleyi	-	-	4
Celtis iguanaea	-	-	4
Handroanthus ochraceus	-	-	4
Croton schiedeanus	-	-	3
Petiveria alliacea	-	-	3

Figure 4. Natural regeneration index: NRI of shared species between sites

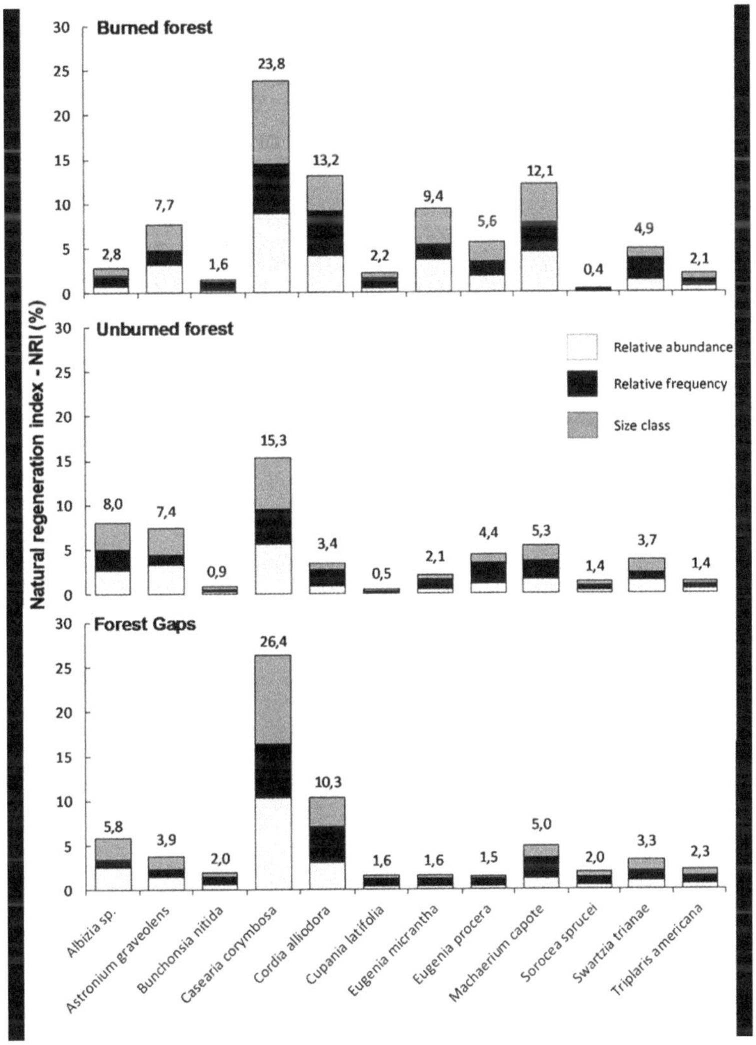

Safford & Harrison, 2004; Otterstrom et al., 2006; Alanís-Rodríguez et al., 2012; Oliveira et al., 2014; Young et al., 2018). In the studied TDF *Casearia corymbosa, Cordia alliodora, Machaerium capote* and *Astronium graveolens* were favored by the fire as their abundance was significantly higher in the burned forest plots.

Besides usual differences in the successional process (Keeley, Lubin, & Fotheringham, 2003; Rodrigues et al., 2005; Gandiwa, 2011), the presence of both shared species between burned and unburned sites, and exclusive species, suggests three distinct groups of plants regarding their regeneration response to fire: 1) fire-stimulated, 2) fire-tolerant and 3) fire-sensitive regeneration (Rocca, 2009; Griscom & Ashton, 2011; Jaureguiberry & Díaz, 2015). The first group of species was best represented in burned forests compared to unburned forests, including forest gaps (Gould et al., 2002; Kennard et al., 2002). Within this group, species such as *Cordia alliodora, Machaerium capote, Eugenia micrantha* and *Casearia corymbosa* are found (Table 4). According to Pausas & Ribeiro (2017), the advantage of this type of species is that their rapid regeneration after a fire will allow the formation of an understory capable

Table 4. Natural regeneration index of woody species per type of regeneration response in the face of fire (fire-stimulated, fire-tolerant; fire-sensitive)

Regeneration response	Species	Natural regeneration index*			P-value
		BF	UBF	FG	
	Cordia alliodora	13,18^{a}	3,45^{b}	10,32^{b}	3,15e^{-05}
Fire-Stimulated regeneration	*Machaerium capote*	12,13^{a}	5,34^{b}	4,95^{b}	0,0236
	Eugenia micrantha	9,40^{a}	2,10^{b}	1,59^{b}	0,0413
	Casearia corymbosa	23,77^{a}	15,28^{ab}	26,41^{b}	0,0445
	Eugenia procera	5,60^{a}	4,45^{a}	1,48^{a}	0,12
	Cupania latifolia	2,23^{a}	0,46^{a}	1,56^{a}	0.176
	Astronium graveolens	7,67^{a}	7,39^{a}	3,86^{a}	0,45
Fire – Tolerant regeneration	*Albizia* sp.	2,81	7,99	5,81	0,529
	Swartzia trianae	4,91^{a}	3,73^{a}	3,26^{a}	0,549
	Triplaris americana	2,13^{a}	1,42^{a}	2,25^{a}	0,787
	Bunchosia nitida	1,56^{a}	0,88^{a}	2,01^{a}	0,866
	Posoqueria sp.	2,03^{ab}	5,40^{a}	0,00^{b}	0,0581
	Guarea guidonia	0,00^{b}	1,06^{ab}	3,45^{a}	0,091
	Piptadenia sp.	0,80	-	3,87	-
Fire – sensitive regeneration	*Leguminoseae* sp4	-	2,53	2,13	-
	Triplaris melaenodendron	-	3,97	0,57	-
	Cupania sp.	-	2,65	1,69	-
	Albizia sp2.	-	2,10	1,97	-

*Duncan Test for abundance of individuals by species.

of providing proper conditions for late-successional species, and hence, these species provide the forest with a high post-fire regeneration capacity.

The second type of species is those with fire-tolerant regeneration (Chang, 1996; Cochrane, 2009; Rocca, 2009; Jaureguiberry & Díaz, 2015). This includes all species that were not negatively affected by the fire and showed similar abundances between burned and unburned sites. Pioneer and early successional species belonged to this group, such as *Eugenia procera* and *Cupania latifolia* that were dominant both in the forest gaps and burned sites (Table 4). Also, mid-late successional species such as *Astronium graveolens* were assigned to this regeneration type as it was both dominant in the burned and unburned forest sites. Other studies have also reported high density of post-fire seedlings of the *Astronium* genus (Gould et al., 2002; Kennard et al., 2002; Rodrigues et al., 2005), suggesting that this genus has successful post-fire regeneration mechanisms as it tolerates or can be stimulated by fire.

The fire-sensitive regeneration species were those species with a significant decline in their abundance in the burned forest, or those that were not recorded in the burned forest despite being well represented in the unburned one (*e.g. Posoqueria* sp., *Guarea guidonia, Triplaris melaenodendron*). The low numbers or absence in the burned forest indicates that they might be negatively affected by recurrent fires, and hence, their recruitment will depend on resprouting or seed dispersal from distant areas in order to avoid local extinction (Lloret, 2004; Fernández et al., 2010; Rutherford, Powrie, Husted, & Turner, 2011; Moreira, Arianoutsou, Corona, & De las Heras, 2012). Therefore, further studies should focus on other post-fire regeneration mechanisms such as resprouting and seed dispersal for both tolerant and sensitive species, in order to gain a better understanding of the recover capacity of TDF after wildfires.

Considerations for Forest Management

Bearing in mind that forest fires are increasing in frequency and intensity due to the changing climate, the plant species whose sexual regeneration showed a positive response to fire should be considered in restoration programs in order to increase the capacity of TDF to overcome future forest fires (Bhadouria et al. 2016; Allen et al., 2017; Bhadouria et al., 2017). For instance, these species can be used in mixed native species plantations at forest edges, in order to protect forests from the fire that comes from the agricultural matrix or other transformed areas (Kellman & Meave, 1997; Vieira & Scariot, 2006). In contrast, the maintenance of fire-sensitive species must be managed both through forest enrichment and forest corridors that promote post-fire establishment by seed dispersal.

Finally, similar to other studies that found greater diversity in unburned areas, in this study, fire affected negatively the diversity of regeneration (Salazar & Goldstein, 2014). However, other studies have reported greater diversity in burned areas (Galvão de Melo, Durigan, & Romero Gorenstein, 2007). Therefore, future studies should include other life forms such as lianas, vines and herbs that were not included in this study, and hence can under or over-estimate post-fire diversity measurements (Oliveira et al., 2014). Moreover, in this investigation the short distance of the Jaccard similarity index between burned and unburned sites suggest that, over time, the former could take a successional trajectory similar to the unburned ones due to seed dispersal processes (Keeley et al., 2003; Rodrigues et al., 2005; Cárdenas-Arévalo & Vargas-Ríos, 2008; Gandiwa, 2011; Calama et al., 2017). Consequently, longer term monitoring should be done in order to evaluate if post-fire establishment reaches pre-disturbance diversity levels or if the system develops into a different plant community (Koontz et al., 2019; Pausas et al., 2004; Rodrigues et al., 2005; Young et al., 2018).

CONCLUSION

Although research on post-fire regeneration has been done in different tropical dry forests, the results of this study are the first data on post-fire sexual regeneration of Colombian tropical dry forests. It was found that species such as *Cordia alliodora, Machaerium capote, Eugenia micrantha* and *Casearia corymbosa* have the capacity to germinate and establish after a fire. In contrast, some others, such as *Posoqueria* sp., *Guarea guidonia*, and *Piptadenia* cf. *uliginosa*, showed high sensitivity to this disturbance. Therefore, further research should focus on the effect of recurrent fire on sensitive species and how compositional and functional diversity of plant communities' changes under high fire-frequencies.

ACKNOWLEDGMENTS

This research is part of the project "Effect of fire on woody seedling establishment in a Colombian tropical dry forest" financed by the Research Center at the Universidad Distrital (CIDC, Grant 31 - 2016). The authors would like to thank Camila Pizano, Pablo Pizano and the entire staff of the RNSC Jabiru for their gracious support. Assistance in the field was provided by Alejandro Hernández, Laura Segura, Mauricio Matoma, Anilfa Ortega, Juan Carlos Henández, Cristian Bustos, Ricardo Hernández y Jordan Sosa. Many thanks to Eduardo Sastoque and the Herbario Florestal UDBC at the Universidad Distrital Francisco José Caldas for their help with specimen identification. Andrés Felipe Ortiz and Mauricio Matoma provided assistance in data analysis. Julián Diaz, Alejandro Hernández, Rocio Cortes, Anselm Rodrigo, Sandra Bravo and Robert Leal revised former versions of this manuscript. Gabriel and Kristie Suárez helped draft the English version of this text.

REFERENCES

Aguilar, J., Mendoza, H., Medina, S., Correa, D., Nieto, J., Quintana, A., & Gonzalez, R. (2016). Caracterización florística del bosque seco tropical de Honda - Méndez (Tolima). In *Expediciones Humboldt. Honda - Méndez, Tolima* (pp. 17–25). Bogotá, D. C., Colombia: Instituto de Investigaciones de Recursos Biológicos Alexander von Humboldt.

Alanís-Rodríguez, E., Aguirre-Calderón, O., Jiménez-Pérez, J., Pando-Moreno, M., Treviño-Garza, E. J., Aranda-Ramos, R., & Canizales-Velázquez, P. A. (2010). Efecto de la severidad del fuego sobre la regeneración asexual de especies leñosas de un ecosistema mixto (Pinus-Quercus) en el Parque Ecológico Chipinque, México. *Interciencia, 35*, 690–695.

Alanís-Rodríguez, E., Jiménez-Pérez, J., Valdecantos-Dema, A., González-Tagle, M. A., Aguirre-Calderón, Ó. A., & Treviño-Garza, E. J. (2012). Composición y diversidad de la regeneración natural en comunidades de Pinus- Quercus sometidas a una alta recurrencia de incendios en el noreste de México. *Revista Mexicana de Biodiversidad, 83*(4), 1208–1214. doi:10.7550/rmb.29708

Allen, K., Dupuy, J. M., Gei, M. G., Hulshof, C. M., Medvigy, D., Pizano, C., … Powers, J. S. (2017). Will seasonally dry tropical forests be sensitive or resistant to future changes in rainfall regimes? *Environmental Research Letters, 12*(2), 0–19.

Amat, J. (2016). Estadística con R. RPubs. Retrieved from https://github.com/JoaquinAmatRodrigo/Estadistica-con-R

Arenas, J., & Noy, A. (2019). *Análisis de la severidad de un incendio en un relicto de bosque seco tropical en Colombia*. Universidad Distrital Francisco José De Caldas.

Armenteras, D., Bernal, F., Gonzalez, F., Morales, M., Pabón, J., Páramo, G., & Parra-Lara, Á. del C. (2011). *Incendios de la cobertura vegetal en Colombia* (Á. del C. Parra-Lara, Ed.).

Armenteras, D., González-Alonso, F., & Aguilera, C. F. (2009). Geographic and temporal distribution of fire in Colombia using thermal anomalies data. *Caldasia, 31*(2), 303–318.

Bhadouria, R., Singh, R., Srivastava, P., & Raghubanshi, A. S. (2016). Understanding the ecology of tree-seedling growth in dry tropical environment: A management perspective. *Energy, Ecology & Environment, 1*(5), 296–309. doi:10.100740974-016-0038-3

Bhadouria, R., Srivastava, P., Singh, R., Tripathi, S., Singh, H., & Raghubanshi, A. S. (2017). Tree seedling establishment in dry tropics: An urgent need of interaction studies. *Environment Systems & Decisions, 37*(1), 88–100. doi:10.100710669-017-9625-x

Calama, R., Manso, R., Lucas-Borja, M. E., Espelta, J. M., Piqué, M., Bravo, F., ... Pardos, M. (2017). Natural regeneration in Iberian pines: A review of dynamic processes and proposals for management. *Forest Systems, 26*(2). doi:10.5424/fs/2017262-11255

Cantillo-H, E., & Rangel–CH, J. O. (2002). Caracterización estructural, ordenación y dinámica de la vegetación en la zona de captación de aguas de la microcuenca El Tigre - municipio de Norcasia, Caldas. *Colombia Forestal, 52*(1), 1–5.

Cárdenas-Arévalo, G., & Vargas-Ríos, O. (2008). Rasgos de historia de vida de especies en una comunidad vegetal alterada en un páramo húmedo (Parque Nacional Natural Chingaza). *Caldasia, 30*(2), 245–264.

Carón, M. M., Dalmasso, A. D., Ortín, A. E., & Verheyen, K. (2015). Regeneración post fuego en un bosque tropical seco del Monte en el noroeste de Argentina. *Multequina (Mendoza), 24*, 5–17.

Chang, C.-R. (1996). Ecosystem responses to fire and variations fire regimes. In Sierra Nevada Ecosystem Project: final report to Congress, vol. II, Assessments and scientific basis for management options (Vol. II, pp. 1071–1099). Davis, CA: University of California, Centers for Water and Wildland Resources.

Chao, A., Gotelli, N. J., Hsieh, T. C., Sander, E. L., Ma, K. H., Colwell, R. K., & Ellison, A. M. (2014). Rarefaction and extrapolation with Hill numbers: A framework for sampling and estimation in species diversity studies. *Ecological Monographs, 84*(1), 45–67. doi:10.1890/13-0133.1

Chao, A., & Jost, L. (2012). Coverage-based rarefaction and extrapolation: Standardizing samples by completeness rather than size. *Ecology, 93*(12), 2533–2547. doi:10.1890/11-1952.1 PMID:23431585

Cochrane, M. A. (2009). *Tropical Fire Ecology*. Climate Change, Land Use, and Ecosystem Dynamics. doi:10.1007/978-3-540-77381-8

de Oliveira, M. T., Damasceno-Junior, G. A., Pott, A., Paranhos Filho, A. C., Suarez, Y. R., & Parolin, P. (2014). Regeneration of riparian forests of the Brazilian Pantanal under flood and fire influence. *Forest Ecology and Management, 331*, 256–263. doi:10.1016/j.foreco.2014.08.011

Etter, A., Andrade, A., Amaya, P., & Arévalo, P. (2015). Estado de los ecosistemas colombianos-2014: una aplicación de la metodología de lista roja de ecosistemas - UICN.

Fernández, I., Morales, N., Olivares, L., Salvatierra, J., Gómez, M., & Montenegro, G. (2010). Restauración ecológica para ecosistemas nativos afectados por incendios forestales.

Fernández-Manso, A., Fernández-Manso, O., & Quintano, C. (2016). SENTINEL-2A red-edge spectral indices suitability for discriminating burn severity. *International Journal of Applied Earth Observation and Geoinformation, 50*, 170–175. doi:10.1016/j.jag.2016.03.005

Fredericksen, T. S., & Mostacedo, B. (2000). Regeneration of timber species following selection logging in a Bolivian tropical dry forest. *Forest Ecology and Management*, *131*(1–3), 47–55. doi:10.1016/S0378-1127(99)00199-1

Galvão de Melo, A. C., Durigan, G., & Romero Gorenstein, M. (2007). Efeito do fogo sobre o banco de sementes em faixa de borda de Floresta Estacional Semidecidual, SP, Brasil. *Acta Botanica Brasílica*, *21*(4), 927–934. doi:10.1590/S0102-33062007000400017

Gandiwa, E. (2011). Effects of repeated burning on woody vegetation structure and composition in a semiarid southern African savanna. *International Journal of Environmental Sciences*, *2*(2), 458–471.

Gómez-González, S., Paula, S., Cavieres, L. A., Pausas, J. G., Nuñez, M., Whitlock, C., & Smith-Ramírez, C. (2017). Postfire responses of the woody flora of Central Chile: Insights from a germination experiment. *PLoS One*, *12*(7). doi:10.1371/journal.pone.0180661 PMID:28704449

González-M, R., Avella, A., & Díaz-Triana, J. E. (2015). Plataformas de monitoreo para vegetación: Toma y análisis de datos. In Monitoreo a procesos de restauración ecológica aplicado a ecosistemas terrestres (pp. 87–107).

Gould, K. A., Fredericksen, T. S., Morales, F., Kennard, D., Putz, F. E., Mostacedo, B., & Toledo, M. (2002). Post-fire tree regeneration in lowland Bolivia: Implications for fire management. *Forest Ecology and Management*, *165*(1–3), 225–234. doi:10.1016/S0378-1127(01)00620-X

Griscom, H. P., & Ashton, M. S. (2011). Restoration of dry tropical forests in Central America: A review of pattern and process. *Forest Ecology and Management*, *261*(10), 1564–1579. doi:10.1016/j.foreco.2010.08.027

Hammer, O., Harper, D. A. T., & Ryan, P. D. (2005). PAST: Paleontological statistics software package for education and data analysis. *Palaeontologia Electronica* *4*(1). Retrieved from http://palaeo-electronica.org/2001_1/past/issue1_01.htm

Hardesty, J., Myers, R., & Fulks, W. (2005). Fire, ecosystems and people: A preliminary assessment of fire as a global conservation issue. *Fire Management*, *22*(4), 78–87.

Hoffmann, W. (2000). Post-Establishment Seedling Success in the Brazilian Cerrado: A Comparison of Savanna and Forest Species. *Biotropica*, *32*(1), 62–69. doi:10.1111/j.1744-7429.2000.tb00448.x

Hsieh, T. C., Ma, K. H., & Chao, A. (2016). iNEXT: An R package for rarefaction and extrapolation of species diversity (Hill numbers). *Methods in Ecology and Evolution*, *7*(12), 1451–1456. doi:10.1111/2041-210X.12613

Jaureguiberry, P., & Díaz, S. (2015). Post-burning regeneration of the Chaco seasonally dry forest: Germination response of dominant species to experimental heat shock. *Oecologia*, *177*(3), 689–699. doi:10.100700442-014-3161-x PMID:25421096

Keeley, J. E., & Fotheringham, C. J. (2000). Role of fire in regeneration from seed. Seeds: *The Ecology of Regeneration in Plant Communities*, (January), 311–330. doi:10.1079/9780851994321.0311

Keeley, J. E., Lubin, D., & Fotheringham, C. J. (2003). Fire and grazing impacts on plant diversity and alien plant invasions in the southern Sierra Nevada. *Ecological Applications*, *13*(5), 1355–1374. doi:10.1890/02-5002

Kellman, M., & Meave, J. (1997). Fire in the tropical gallery forests of Belize. *Journal of Biogeography*, *24*(1), 23–34. doi:10.1111/j.1365-2699.1997.tb00047.x

Kennard, D. K., Gould, K., Putz, F. E., Fredericksen, T. S., & Morales, F. (2002). Effect of disturbance intensity on regeneration mechanisms in a tropical dry forest. *Forest Ecology and Management*, *162*(2–3), 197–208. doi:10.1016/S0378-1127(01)00506-0

Koontz, M. J., North, M. P., Werner, C. M., Fick, S. E., & Latimer, A. M. (2019). Local variability of vegetation structure increases forest resilience to wildfire. *EcoEvoRxiv*, 1–38. doi:10.32942/OSF.IO/K72YE

Lahoreau, G., Barot, S., Gignoux, J., Hoffmann, W. A., Setterfield, S. A., & Williams, P. R. (2006). Positive effect of seed size on seedling survival in fire-prone savannas of Australia, Brazil and West Africa. *Journal of Tropical Ecology*, *22*(06), 719–722. doi:10.1017/S026646740600349X

Lezama, K. (2018). *Caracterización de la regeneración natural de bosques en tres ambientes contrastantes en el Retorno*. Guaviare, Colombia. doi:10.22201/fq.18708404e.2004.3.66178

Lloret, F. (2004). Régimen de incendios y regeneración. Ecología del bosque mediterráneo en un mundo cambiante, 101–126.

Melo, O., Fernandez-Méndez, F., & Villanueva, B. (2016). Hábitat lumínico, estructura, diversidad y dinámica de los bosques secos tropicales del alto magdalena. *Colombia Forestal*, *20*(1), 19–30. doi:10.14483/udistrital.jour.colomb.for.2017.1.a02

Melo, O., & Vargas, R. (2003). Evaluacion ecologica y silvicultural de ecosistemas boscosos.

Moreira, F., Arianoutsou, M., Corona, P., & De las Heras, J. (2012). Post-fire management and restoration of Southern European Forest.

Mostacedo, B., Fredericksen, T. S., Gould, K., & Toledo, M. (1999). *Comparacion de la respuesta de las comunidades vegetales a los incendios forestales en los Bosques Tropicales Secos y Húmedos de Bolivia. Proyecto de Manejo Forestal Sostenible BOLFOR*. Retrieved from http://pdf.usaid.gov/pdf_docs/Pnacl876.pdf

Myers, R. L. (2006). Living with Fire—Sustaining Ecosystems & Livelihoods through Integrated Fire Management. Global Fire Initiative. *The Nature Conservancy, (June), 28.* Retrieved from http://www.nature.org/initiatives/ fire/files/integrated_fire_management_myers_2006.pdf

National Natural Parks PNN. (2014). Resolución 019 del 21 de febrero de 2014 Por medio de la cual se registra la Reserva Natural de la Sociedad Civil "Jabirú."

Navarro, G., Caballero, I., Silva, G., Parra, P.-C., Vázquez, Á., & Caldeira, R. (2017). Evaluation of forest fire on Madeira Island using Sentinel-2A MSI imagery. *International Journal of Applied Earth Observation and Geoinformation*, *58*, 97–106. doi:10.1016/j.jag.2017.02.003

Norden, N., Chave, J., Belbenoit, P., Caubère, A., Châtelet, P., Forget, P. M., ... Thébaud, C. (2009). Interspecific variation in seedling responses to seed limitation and habitat conditions for 14 Neotropical woody species. *Journal of Ecology*, *97*(1), 186–197. doi:10.1111/j.1365-2745.2008.01444.x

Ocampo-Zuleta, K., & Bravo, S. (2019). Reclutamiento de especies leñosas en bosques tropicales expuestos a incendios: Una revisión. Ecosistemas. *Revista Científica y Técnica de Ecología y Medio Ambiente*, *28*(1), 106–117.

Otterstrom, S. M., Schwartz, M. W., & Velazquez-Rocha, I. (2006). Responses to Fire in Selected Tropical Dry Forest Trees. *Biotropica*, *38*(5), 592–598. doi:10.1111/j.1744-7429.2006.00188.x

Pausas, J. G., Bradstock, R., Keith, D. A., & Keeley, J. E. (2004). Plant functional traits in relation to fire in corwn-fire ecosystems. *Ecology*, *85*(4), 1085–1100. doi:10.1890/02-4094

Pausas, J. G., & Lavorel, S. (2003). A hierarchical deductive approach for functional types in disturbed ecosystems. *Journal of Vegetation Science*, *14*(3), 409–416. doi:10.1111/j.1654-1103.2003.tb02166.x

Pausas, J. G., Pratt, R. B., Keeley, J. E., Jacobsen, A. L., Ramirez, A. R., Vilagrosa, A., ... Davis, S. D. (2016). Towards understanding resprouting at the global scale. *The New Phytologist*, *209*(3), 945–954. doi:10.1111/nph.13644 PMID:26443127

Pausas, J. G., & Ribeiro, E. (2017). Fire and plant diversity at the global scale. *Global Ecology and Biogeography*, *26*(8), 889–897. doi:10.1111/geb.12596

Pizano, C., & García, H. (2014). El Bosque seco tropical en Colombia. Bogota D.C., Colombia: Instituto de Investigacion de Recursos Biologicos Alexander von Humboldt (IAvH).

Pizano, C., González-M, R., Hernández-Jaramillo, A., & García, H. (2017). Agenda de investigación y monitoreo en bosques secos de Colombia (2013-2015): Fortaleciendo redes de colaboración para su gestión integral en el territorio. *Biodiversidad En La Práctica*, *2*(1), III. doi:10.4067/S0717-97072014000100001

R Core Team. (2018). R: A language and environment for statistical computing. R Foundation for Statistical Computing, Vienna, Austria. Retrieved from https://www.R-project.org/

Rangel-Ch, J. O., & Velázquez, A. (1997). Métodos de estudio de la vegetación. In J. O. Rangel-Ch, C. Lowy, D. Petter, & M. Aguilar P. (Eds.), Colombia Diversidad Biótica II: Tipos de vegetación en Colombia (pp. 59–87). doi:10.100713398-014-0173-7.2

Rocca, M. E. (2009). Fine-Scale Patchiness in Fuel Load Can Influence Initial Post-Fire Understory Composition in a Mixed Conifer Forest, Sequoia National Park, California. *Natural Areas Journal*, *29*(2), 126–132. doi:10.3375/043.029.0204

Rodrigo, A., Retana, J., & Xavier, F. (2005). Diferencias En La Dinámica De Regeneración De Los Bosques Mediterráneos Después De Grandes Incendios: Consecuencias En El Paisaje Forestal.

Rodrigues, R. R., Martins, S. V., & Matthes, L. A. F. (2005). Post-Fire Regeneration in a Semideciduous Mesophytic Forest, South-Eastern Brazil. In A. R. Burk (Ed.), *New research on forest ecosystems* (pp. 1–19). New York, NY: Nova Science Publishers

Rodríguez-Buritica, S., Aguilar-Garavito, M., & Norden, N. (2017). Identifying a Fire Ecology Research Agenda for Colombia. *Eos (Washington, D.C.), 1–4*. doi:10.1029/2017EO065869

Rutherford, M. C., Powrie, L. W., Husted, L. B., & Turner, R. C. (2011). Early post-fire plant succession in Peninsula Sandstone Fynbos: The first three years after disturbance. *South African Journal of Botany, 77*(3), 665–674. doi:10.1016/j.sajb.2011.02.002

Safford, H. D., & Harrison, S. (2004). Fire effects on plant diversity in Serpentine vs. Sandstone Chaparral. *Ecology, 85*(2), 539–548. doi:10.1890/03-0039

Salazar, A., & Goldstein, G. (2014). Effects of fire on seedling diversity and plant reproduction (sexual vs. vegetative) in neotropical savannas differing in tree density. *Biotropica, 46*(2), 139–147. doi:10.1111/btp.12090

Shlisky, A. J., Waugh, P., Gonzalez, M., Gonzalez, M., Manta, H., Santoso, E., … Zollner, D. (2007). Fire, ecosystems &people: threats and strategies for global biodiversity conservation. GFI Technical Report 2007-2. The Nature Conservancy, 28.

Snook, L. K. (1993). *Regeneration, Growth, and Sustainability of Mahogany in Mexico's Yucatan Forests*. Big-Leaf Mahogany. doi:10.1007/0-387-21778-9_9

Vieira, D. L. M., & Scariot, A. (2006). Principles of natural regeneration of tropical dry forests for restoration. *Restoration Ecology, 14*(1), 11–20. doi:10.1111/j.1526-100X.2006.00100.x

Villanueva, B., Melo, O., & Rincón, M. (2015). Estado del conocimiento y aportes a la flora vascular del bosque seco del Tolima. *Colombia Forestal, 18*(1), 9–23. doi:10.14483/udistrital.jour.colomb.for.2015.1.a01

Young, D. J. N., Werner, C. M., Welch, K. R., Young, T. P., Safford, H. D., & Latimer, A. M. (2018). Post-fire forest regeneration shows limited climate tracking and potential for drought-induced type conversion. *Ecology, 100*(2). doi:10.1002/ecy.2571 PMID:30516290

Zuloaga-Aguilar, S. (2010). *Efecto del fuego sobre la germinación y el banco de semillas de bosques templados del occidente de México*. Instituto de Ecología A.C.

APPENDIX 1

Table 5. Floristic composition and abundance of natural regeneration of woody species in burned forest (BF), unburned forest (UBF) and forest gaps.

Family	Species	Abundance / Site			Total
		BF	UBF	FG	
Acanthaceae	*Aphelandra glabrata* Willd. ex Nees	-	2	4	6
	Acanthaceae sp1.	-	-	4	4
	Acanthaceae sp2.	6	-	-	6
Amaranthaceae	*Amaranthaceae* sp.	1	-	-	1
Anacardiaceae	*Astronium graveolens* Jacq.	44	32	12	88
Bignoniaceae	*Handroanthus chrysanthus* (Jacq.) S.O.Grose	-	-	1	1
	Handroanthus ochraceus (Cham.) Mattos	-	-	4	4
Boraginaceae	*Cordia alliodora* (Ruiz & Pav.) Oken.	56	8	25	89
	Cordia sp1.	-	9	3	12
	Cordia sp2.	-	1	1	2
Cannabaceae	*Celtis iguanaea* (Jacq.) Sarg.	-	-	4	4
	Trema micrantha (L.) Blume	1	-	-	1
Capparaceae	*Cynophalla amplissima* (Lam.) Iltis & Cornejo	-	2	-	2
Celastraceae	*Celastraceae* sp.	-	1	-	1
Euphorbiaceae	*Acalypha diversifolia* Jacq.	1	8	-	9
	Acalypha sp.	3	-	-	3
	Acalypha sp2.	1	-	-	1
	Croton schiedeanus Schltdl.	-	-	3	3
Lamiaceae	*Aegiphila* sp.	2	-	-	2
Lauraceae	*Nectandra* sp	-	1	-	1
	Lauraceae sp1	-	4	-	4
Lecythidaceae	*Gustavia* sp1.	-	6	3	9
	Gustavia speciose (Kunth) DC.	1	-	-	1

continued on following page

Table 5. Continued

Family	Species	Abundance / Site			Total
		BF	UBF	FG	
Leguminosae	*Albizia* sp.	10	26	21	57
	Albizia sp2.	-	6	7	13
	Calliandra sp.	-	3	-	3
	Chloroleucon mangense (Jacq.) Britton & Rose	1	-	-	1
	Inga sp.	-	2	-	2
	Lonchocarpus sp.	2	-	-	2
	Machaerium biovulatum Micheli	-	2	3	5
	Machaerium capote Triana ex Dugand	62	16	10	88
	Machaerium goudotii Benth.	3	-	-	3
	Machaerium microphyllum (E.Mey.) Standl.	-	-	1	1
	Piptadenia sp.	2	-	14	16
	Swartzia trianae Benth.	18	14	8	40
	Leguminoseae sp2	-	-	1	1
	Leguminoseae sp3	-	-	2	2
	Leguminoseae sp4	-	8	6	14
Malpighiaceae	*Bunchonsia nitida* (Jacq.) DC.	4	3	5	12
	Bunchosia sp.	-	11	-	11
	Hiraea sp.	-	2	-	2
	Hiraea ternifolia (Kunth) A.Juss.	1	-	-	1
	Malpighiaceae sp2.	-	-	1	1
Malvaceae	*Byttneria* sp.	2	-	-	2
	Ceiba pentandra (L.) Gaertn.	-	1	-	1
	Guazuma ulmifolia Lam.	4	-	4	8
	Pseudobombax septenatum (Jacq.) Dugand	-	-	1	1
	Malvaceae sp1.	1	-	-	1
Meliaceae	*Guarea guidonia* (L.) Sleumer.	-	3	9	12
	Guarea sp.	-	1	-	1
Moraceae	*Sorocea sprucei* (Baill.) J.F. Macbr.	1	4	4	9
Myrtaceae	*Eugenia micrantha* (Kunth) DC.	50	5	3	58
	Eugenia procera (Sw.) Poir.	26	11	3	40
Nyctaginaceae	*Neea amplifolia* Donn.Sm.	1	-	-	1
Phytolaccaceae	*Petiveria alliacea* L.	-	-	3	3
Piperaceae	*Piper angustifolium* Lam.	-	11	-	11
	Piper marginatum Jacq.	-	-	1	1

continued on following page

Table 5. Continued

Family	Species	Abundance / Site			Total
		BF	UBF	FG	
Polygonaceae	*Triplaris americana* L.	9	5	5	19
	Triplaris melaenodendron (Bertol.) Standl. & Steyerm.	-	16	1	17
Primulaceae	*Cybianthus cuatrecasasii* Pipoly	-	1	-	1
	Cybianthus sp1.	-	1	-	1
Rubiaceae	*Posoqueria* sp.	7	18	-	25
	Rubiaceae sp3	-	1	-	1
Rutaceae	*Amyris pinnata* Kunth	-	1	-	1
	Zanthoxylum rhoifolium Lam.	-	1	-	1
Salicaceae	*Casearia corymbosa* Kunth	121	54	86	261
Sapindaceae	*Cupania cinerea* Poepp. & Endl.	-	7	1	8
	Cupania latifolia Kunth	8	1	3	12
	Cupania sp.	-	8	4	12
	Sapindus saponaria L.	-	4	-	4
Solanaceae	*Capsicum* sp.	2	-	-	2
	Cestrum sp.	1	-	-	1
Ulmaceae	*Ampelocera* sp.	1	-	2	3
Verbenaceae	*Petrea rugosa*	1	-	-	1
	Verbenaceae sp.	-	-	1	1
Indeterminate Sp1		-	1	-	1
Indeterminate Sp3		-	1	-	1
Indeterminate Sp6		-	-	2	2
TOTAL		454	323	276	1053

continued on following page

Table 6. Natural regeneration index -NRI- of woody species in burned forest (BF), unburned forest (UF) and forest gaps

Family	Specie	Relative Abundance (%)			Relative Frequency (%)			Relative Size class (%)			NRI (%)		
		BF	UBF	FG	BF	UBF	FG	BF	UBF	FG	BF	UBF	FG
Acanthaceae	*Aphelandra glabrata*	-	-	1,45	-	-	0,87	-	-	1,87	-	-	1,40
	Acanthaceae sp1.	1,32	-	-	1,52	-	-	1,33	-	-	1,39	-	-
	Acanthaceae sp2.	-	0,62	1,45	-	0,68	0,87	-	0,80	1,54	-	0,70	1,29
Amaranthaceae	*Amaranthaceae* sp.	0,22	-	-	0,76	-	-	0,14	-	-	0,37	-	-
Anacardiaceae	*Astronium graveolens*	9,69	9,91	4,35	4,55	3,42	2,61	8,78	8,84	4,63	7,67	7,39	3,86
Bignoniaceae	*Handroanthus chrysanthus*	-	-	0,36	-	-	0,87	-	-	0,47	-	-	0,57
	Handroanthus ochraceus	-	-	1,45	-	-	1,74	-	-	1,87	-	-	1,69
Boraginaceae	*Cordia alliodora*	12,33	2,48	9,06	15,15	5,48	12,17	12,05	2,38	9,74	13,18	3,45	10,32
	Cordia sp1.	-	2,79	1,09	-	3,42	0,87	-	2,93	0,42	-	3,05	0,79
	Cordia sp2.	-	0,31	0,36	-	0,68	0,87	-	0,40	0,14	-	0,46	0,46
Cannabaceae	*Celtis iguanaea*	-	-	1,45	-	-	2,61	-	-	1,06	-	-	1,71
	Trema micrantha	0,22	-	-	0,76	-	-	0,04	-	-	0,34	-	-
Capparaceae	*Cynophalla amplissima*	-	0,62	-	-	1,37	-	-	0,63	-	-	0,87	-
Celastraceae	*Celastraceae* sp.	-	0,31	-	-	0,68	-	-	0,40	-	-	0,46	-
Euphorbiaceae	*Acalypha diversifolia*	0,22	2,48	-	0,76	4,11	-	0,29	0,62	-	0,42	2,40	-
	Acalypha sp.	0,66	-	-	0,76	-	-	0,72	-	-	0,71	-	-
	Acalypha sp2.	0,22	-	-	0,76	-	-	0,29	-	-	0,42	-	-
	Croton schiedeanus	-	-	1,09	-	-	2,61	-	-	1,00	-	-	1,56
Lamiaceae	*Aegiphila* sp.	0,44	-	-	1,52	-	-	0,57	-	-	0,84	-	-
Lauraceae	*Nectandra* sp	-	1,24	-	-	0,68	-	-	1,59	-	-	1,17	-
	Lauraceae sp1	-	0,31	-	-	0,68	-	-	0,40	-	-	0,46	-
Lecythidaceae	*Gustavia* sp1.	-	1,86	1,09	-	1,37	0,87	-	1,75	1,40	-	1,66	1,12
	Gustavia speciosa	0,22	-	-	0,76	-	-	0,14	-	-	0,37	-	-

continued on following page

Table 6. Continued

Family	Specie	Relative Abundance (%)			Relative Frequency (%)			Relative Size class (%)			NRI (%)		
		BF	UBF	FG	BF	UBF	FG	BF	UBF	FG	BF	UBF	FG
Leguminosae	*Albizia* sp.	2,20	8,05	7,61	3,79	6,85	2,61	2,43	9,06	7,21	2,81	7,99	5,81
	Albizia sp2.	-	1,86	2,54	-	2,05	1,74	-	2,39	1,65	-	2,10	1,97
	Calliandra sp.	-	0,93	-	-	1,37	-	-	0,87	-	-	1,06	-
	Chloroleucon mangense	0,22	-	-	0,76	-	-	0,29	-	-	0,42	-	-
	Inga sp.	-	0,62	-	-	0,68	-	-	0,80	-	-	0,70	-
	Lonchocarpus sp.	0,44	-	-	0,76	-	-	0,57	-	-	0,59	-	-
	Machaerium biovulatum	-	0,62	1,09	-	1,37	1,74	-	0,63	1,40	-	0,87	1,41
	Machaerium capote	13,66	4,95	3,62	9,85	6,16	6,96	12,88	4,91	4,27	12,13	5,34	4,95
	Machaerium goudotii	0,66	-	-	1,52	-	-	0,62	-	-	0,93	-	-
	Machaerium microphyllum	-	-	0,36	-	-	0,87	-	-	0,47	-	-	0,57
	Piptadenia sp.	0,44	-	5,07	1,52	-	2,61	0,43	-	3,93	0,80	-	3,87
	Swartzia trianae	3,96	4,33	2,90	7,58	2,74	3,48	3,19	4,12	3,42	4,91	3,73	3,26
	Leguminoseae sp2	-	-	0,36	-	-	0,87	-	-	0,47	-	-	0,57
	Leguminoseae sp3	-	-	0,72	-	-	0,87	-	-	0,53	-	-	0,71
	Leguminoseae sp4	-	2,48	2,17	-	2,74	1,74	-	2,37	2,48	-	2,53	2,13
Malpighiaceae	*Bunchonsia nitida*	0,88	0,93	1,81	3,03	0,68	2,61	0,76	1,03	1,61	1,56	0,88	2,01
	Bunchosia sp.	-	3,41	-	-	3,42	-	-	3,40	-	-	3,41	-
	Hiraea sp.	-	0,62	-	-	0,68	-	-	0,63	-	-	0,65	-
	Hiraea ternifolia	0,22	-	-	0,76	-	-	0,14	-	-	0,37	-	-
	Malpighiaceae sp2.	-	-	0,36	-	-	0,87	-	-	0,47	-	-	0,57
Malvaceae	*Byttneria* sp.	0,44	-	-	1,52	-	-	0,09	-	-	0,68	-	-
	Ceiba pentandra	-	0,31	-	-	0,68	-	-	0,40	-	-	0,46	-
	Guazuma ulmifolia	0,88	-	1,45	0,76	-	1,74	1,00	-	1,87	0,88	-	1,69
	Pseudobombax septenatum	-	-	0,36	-	-	0,87	-	-	0,47	-	-	0,57
	Malvaceae sp1.	0,22	-	-	0,76	-	-	0,04	-	-	0,34	-	-
Meliaceae	*Guarea guidonia*	-	0,93	3,26	-	1,37	4,35	-	0,87	2,74	-	1,06	3,45
	Guarea sp.	-	0,31	-	-	0,68	-	-	0,24	-	-	0,41	-
Moraceae	*Sorocea sprucei*	0,22	1,24	1,45	0,76	1,37	2,61	0,29	1,59	1,87	0,42	1,40	1,98
Myrtaceae	*Eugenia micrantha*	11,01	1,55	1,09	5,30	3,42	2,61	11,87	1,34	1,08	9,40	2,10	1,59
	Eugenia procera	5,73	3,41	1,09	4,55	6,85	2,61	6,53	3,09	0,75	5,60	4,45	1,48

continued on following page

Table 6. Continued

Family	Specie	Relative Abundance (%)			Relative Frequency (%)			Relative Size class (%)			NRI (%)		
		BF	UBF	FG	BF	UBF	FG	BF	UBF	FG	BF	UBF	FG
Nyctaginaceae	*Neea amplifolia*	0,22	-	-	0,76	-	-	0,04	-	-	0,34	-	-
Phytolaccaceae	*Petiveria alliacea*	-	-	1,09	-	-	0,87	-	-	1,40	-	-	1,12
Piperaceae	*Piper angustifolium*	-	3,41	-	-	2,74	-	-	2,78	-	-	2,97	-
	Piper marginatum	-	-	0,36	-	-	0,87	-	-	0,47	-	-	0,57
Polygonaceae	*Triplaris americana*	1,98	1,55	1,81	2,27	1,37	2,61	2,15	1,35	2,34	2,13	1,42	2,25
	Triplaris melaenodendron	-	4,95	0,36	-	2,05	0,87	-	4,91	0,47	-	3,97	0,57
Primulaceae	*Cybianthus cuatrecasasii*	-	0,31	-	-	0,68	-	-	0,40	-	-	0,46	-
	Cybianthus sp1.	-	0,31	-	-	0,68	-	-	0,24	-	-	0,41	-
Rubiaceae	*Posoqueria* sp.	1,54	5,57	-	3,03	4,11	-	1,52	6,52	-	2,03	5,40	-
	Rubiaceae sp3	-	0,31	-	-	0,68	-	-	0,40	-	-	0,46	-
Rutaceae	*Amyris pinnata*	-	0,31	-	-	0,68	-	-	0,40	-	-	0,46	-
	Zanthoxylum rhoifolium	-	0,31	-	-	0,68	-	-	0,24	-	-	0,41	-
Salicaceae	*Casearia corymbosa*	26,65	16,72	31,16	16,67	11,64	18,26	27,98	17,47	29,80	23,77	15,28	26,41
Sapindaceae	*Cupania cinerea*	-	2,17	0,36	-	2,74	0,87	-	2,63	0,47	-	2,51	0,57
	Cupania latifolia	1,76	0,31	1,09	3,03	0,68	2,61	1,91	0,40	1,00	2,23	0,46	1,56
	Cupania sp.	-	2,48	1,45	-	3,42	1,74	-	2,06	1,87	-	2,65	1,69
	Sapindus saponaria	-	1,24	-	-	0,68	-	-	1,27	-	-	1,07	-
Solanaceae	*Capsicum* sp.	0,44	-	-	1,52	-	-	0,19	-	-	0,71	-	-
	Cestrum sp.	0,22	-	-	0,76	-	-	0,14	-	-	0,37	-	-
Ulmaceae	*Ampelocera* sp.	0,22	-	0,72	0,76	-	0,87	0,29	-	0,61	0,42	-	0,73
Verbenaceae	*Petrea rugosa*	0,22	-	-	0,76	-	-	0,29	-	-	0,42	-	-
	Verbenaceae sp.	-	-	0,36	-	-	0,87	-	-	0,47	-	-	0,57
Indeterminate Sp1		-	0,31	-	-	0,68	-	-	0,24	-	-	0,41	-
Indeterminate Sp3		-	0,31	-	-	0,68	-	-	0,24	-	-	0,41	-
Indeterminate Sp6		-	-	0,72	-	-	0,87	-	-	0,28	-	-	0,63
TOTAL		100											

continued on following page

Chapter 18
Ecology of Natural Regeneration of Tropical Dry Forests of Africa and Its Implications for Their Sustainable Man

Stephen Syampungani
Copperbelt University, Zambia

Annie Namuuya Sikanwe
Copperbelt University, Zambia

Paxie W. C. Chirwa
https://orcid.org/0000-0002-7544-973X
University of Pretoria, South Africa

ABSTRACT

Recovery of African dry forests and woodlands is through either sexual or vegetative means. A sufficient number of regeneration pool (root suckers, sprouts, and seedlings) tend to occur in the herbaceous layer of the African dry forests. However, the prevailing environmental conditions such as fires and competition for nutrients and light influence the number of surviving individuals to enter the next size class. Developing restoration strategy for the African savannas requires the knowledge of regeneration mechanisms and dynamics of these woodlands which will be helpful in their management.

INTRODUCTION

Tropical dry forests (TDF) are forests occurring in tropical regions characterized by pronounced seasonality in rainfall distribution, resulting in several months of drought (Miles et al., 2006). The tropical dry forests is also referred to as Global Ecological Zone (GEZ) covering miombo and Sudanian woodlands, the African Savanna, (see FAO, 2010). These forests experience a tropical climate, with summer rains

DOI: 10.4018/978-1-7998-0014-9.ch018

and a dry period of 5 to 8 months with annual rainfall ranging from 500 to 1500 mm (Sunderland et al., 2015). The forests of GEZ share a broadly similar structure and physiognomy also in addition to occurring on mesotrophic soils. Among the many major vegetation formations in the world, the tropical dry forests are the most threatened as a result of high rates of fragmentation and fire incidences. Throughout the dry forests, population and economic growth are the major drivers of forest destruction and habitat loss. Activities such as logging, settlement, conversion to agriculture, and mining contribute to the loss and degradation of the remaining TDF fragments (Elmqvist et al., 2007). Estimates of tropical forest loss and changes in land cover are still uncertain and a 50% margin of error appears possible (Elmqvist et al., 2007). Recovery of dry forests and woodlands is through either sexual or vegetative means. Sexual regeneration is achieved through seed germination and establishment of seedlings and their recruitment into tree phase while vegetative regeneration occurs through the recruitment of sprouts or resprouts into tree phase from pre-existing trees that get cut or damaged (Timberlake et al 2010). Sufficient number of regeneration pool (root suckers, sprouts and seedlings) tend to occur in the herbaceous layer (Syampungani et al 2016). However, the prevailing environmental conditions such as, fires, competition for nutrients and light influence the number of surviving individuals to enter the next size class (Syampungani et al 2016). Seed dispersal, predation, desiccation and seedling mortality can act as strong constraints that impede dry forest and woodland recovery after disturbance (Timberlake et al 2010). The ability to sprout after severe injury from disturbances such as herbivory, fire, floods, logging or drought overcomes these barriers, as individuals bypass the seed stage and tend to have more vigorous shoots than seedlings presumably because vegetative shoots may take advantage of the extensive root system and the substantial food storage in the remaining parts of the parent plant (Timberlake et al 2010). Although studies have been undertaken to demonstrate the recovery of African Tropical Dry Forests, studies on regeneration modes are limited and scanty to some extent. There is also little knowledge about regeneration rates in terms of functional aspects of biodiversity and generation of ecosystem services that would be useful in planning for both local and regional human consumption and use (Elmqvist et al., 2007; Sunderland et al., 2015) . It is estimated that at global level, the annual regrowth area of humid tropical forest is 1 million ha compared with the annual forest loss of 5.8 million ha. Annual regeneration may correspond to roughly 20% of the total area of deforestation in humid tropics. Sustainable management of African Tropical Dry Forests requires a comprehensive understanding of the contribution of different regeneration mechanisms to forest and woodland recovery. As such, the book chapter seeks to discuss the different regeneration mechanisms of the TDF of Africa.

TROPICAL DRY FORESTS OF AFRICA

In Africa, the distribution of flora and of the major physiognomic types correspond close enough and 18 major phytogeographical areas are recognized (White, 1983).The composition of dry forests is based on White (1983) as there has been no recent work on floristic composition of the vegetation of sub-saharan Africa (Timberlake et al 2010)

Southern Africa

The Zambezian phytoregion about 8500 species of which about 4600 are endemic to this region (White, 1983). The region includes the whole of Zambia, Malawi, and Zimbabwe, large parts of Angola, Tanzania and Mozambique as well as smaller parts of the Democratic Republic of Congo, Botswana and South Africa (Geldenhuys and Golding, 2008). It is the second largest phytochorion in Africa (after the Sahara) and the largest in Southern Africa (Geldenhuys & Golding, 2008) with an expanse of 980, 000 Km^2 (Timberlake et al 2010). Within the Zambezian phytoregion, are dry forests occurring on deep, freely drained soils with enough supply of moisture in lower horizons during the dry season (Chidumayo, 1997). These dry forests most comprise dense forest or woodlands, much of which have been reduced to wooded grasslands or secondary grassland with scattered forest trees after destruction by humans and fires (Timberlake et al 2010). Within the warm dry forest and woodland area of Southern Africa are four extensive vegetation types: undifferentiated woodland, Miombo woodland, Mopane woodland and semi-arid shrubland (White 1983; Figure 1).

Western Africa

The Sudanian region stretches west-east from the Atlantic coast, across the southern fringes of the Sahara Desert to the Ethiopian Highlands and Red Sea coast (Geldenhuys and Golding, 2008). The Sudanian Savanna is an ecosystem marked by diversity and endemism. This tropical grassland is home to over 1000 endemic plants and a number of animal species (Salzmann, 2000). The area covers almost 100,000 km2 and is marked by long grasses, shrubs, and occasional trees. This ecosystem spans across a great number of Central and Southern African nations including Cameroon, Chad, Nigeria, CAR, Democratic Republic of Congo, Eritrea, Ethiopia, Kenya, Nigeria, Sudan and Uganda (Salzmann, 2000).

They can be categorized under the three broad climatic zones-sub-humid, warm mesic and warm dry- each forming a relatively narrow band with much of the area being low (under 750 m altitude with a few higher areas such as Jos Plateau in northern Nigeria(Timberlake et al 2010). It falls within the tropical summer rainfall zone, and its climate is very similar to that of the Zambezian region in terms of rainfall. Mean annual rainfall is 24^{0}C to 28^{0}C and, together with the impact of the hot and dry harmattan wind, makes the dry season more severe (Geldenhuys and Golding, 2008). Rainfall ranges from 1200-1500 mm per year in the east to 1500-2000mm per year in the west (Timberlake et al 2010). The flora contains about 2750 species with about 33% endemic, with no endemic families and a few endemic genera (Geldenhuys and Golding, 2008).

Eastern Africa

The most extensive woodland type of eastern Africa in the semi-arid zone, covering 1.6 million Km^2 (Timberlake et al 2010). They are mainly composed of deciduous microphyllous bushland and thicket dominated by species of *Vachelia, Senegalia* and *Commiphora*

Floristic Composition of Various Vegetation Formations of the African Dry Forests

The African Savannas occupy half of the African continent with approximately 13 million km^2 and lies between the equatorial forest and mid latitude deserts (Ky-Dembele et al., 2007). It is divided into the Sudanian and the Zambezian comprising mainly of woodlands, shrub savannas and grassland or steppes (Ky-Dembele et al., 2007). This tropical dry forests usually receive about 500-1000mm annual rainfall and have prolonged droughts for more than three months per year (Sunderland et al., 2015). The Su-

Figure 1.: The distribution of African Savannas (I =Guineo Congo Regional Centre of Endemism (RCE); II = Zambezian RCE; III =Sundanian RCE IV = Somalia –Masai RCE X =Guinea_Congolia/Zambezian Regional Transition Zone (RTZ); XI =Guinea-Congolia/Sudanian (RTZ); XII =Lake Victoria Regional Mosaic (RM); XIII =Zanibar-Inhambane RM; XIV =Kalahari_Highveld Transition Zone; XVI =Sahel RTZ
(Source: Geldenhuy et al 2008)

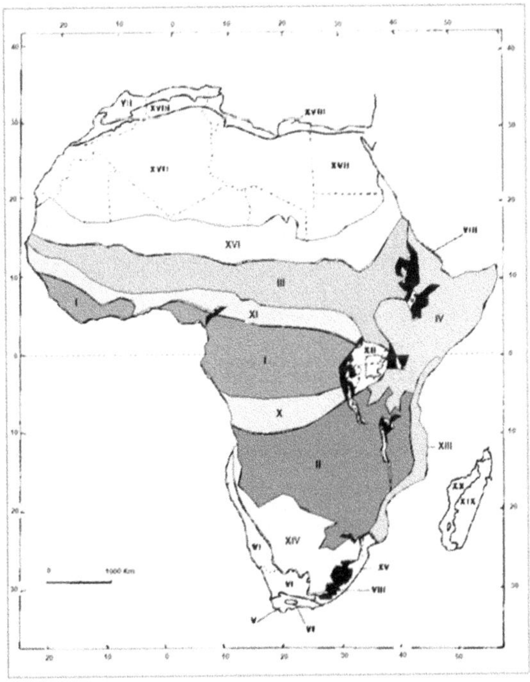

danian woodlands' altitude ranges between 260 and 360 mm. Its mean annual rainfall is about 914mm with the highest relative humidity reaching 80% with august being the wettest month. In addition to this the woodland has a mean annual temperature range from 15 to 40.8^{0}celcius with January and December being the coldest months while March and April are the hottest months in the year (Ky-Dembele et al., 2007). Soils are identified as ferruginous lixisols, which are mainly deep silty-clay (Bognounou et al., 2010). The vegetation is a tree and bush savannah with a grass layer dominated by the annual grasses *Andropogon pseudapricus*, *Loudetia togoensis* and *Pennisetum pedicellatum* and the perennial grass *Andropogon gayanus*. Members of the Mimosaceae, Fabaceae and Combretaceae families dominate the woody vegetation (Bognounou et al., 2010; Ky-Dembele et al., 2007).

The Zambezian Miombo woodlands cuts across southern central Africa and is one of the largest eco zones on the continent and home to a great variety of wildlife, including many large mammals (Syampungani et al., 2016). The region covers a large area stretching northeast from Angola, including the southeast section of the Democratic Republic of Congo, the northern half of Zambia, a large section of western Tanzania, southern Burundi, and northern and western Malawi (Sunderland et al., 2015). It occupies the largest parts of Congo, In Zambia it covers the northern half of the country above Lusaka, including the eastern and western and the Copperbelt whilst in Tanzania it covers the western inland provinces between Lake Victoria, Lake Tanganyika and Lake Malawi. The area is mostly flat plateau, and the soils are poor (Shackleton et al., 2007). Is has a tropical climate with a long dry season, up to seven months, which leaves the forest vulnerable to fires, and a rainy season from November to March. The woodland is interspersed with riverside dambo (grassy wetlands) that constitute up to thirty percent of the region. (Syampungani et al., 2016)

The woodlands contain much typical miombo flora of high trees with shrub and grassland underneath with much of the other plant life too. There are typically more evergreen trees than in most Miombo woodlands. The classic Miombo trees are of the genera *Brachystegia, Julbernardia,* and *Isoberlinia* dominate the woodlands with other tree species such as *Pterocarpus angolensis*, *Albizia* sp. and *Afzelia quanzensis*. The under story of the trees is covered by plants such as the *herbaceous Crotalaria* and *Indigofera*.

Sub-saharan Africa has a wide range of dry forests and woodland formations each with a diverse of flora (Shumba et al 2010). There is variation in genera and species among the vegetation formations (Table 1) with the Zambezian region having the highest diversity of vegetation formations

IMPORTANCE OF TROPICAL DRY FORESTS

In most African countries where the African savannas lie, the forests provide both urban and rural populations with a variety of products such as non-timber products like fruits, medicines and fodders. The freely available products provide enterprise opportunities even for the very poor and are means of poverty alleviation and economic development (Djoudi et al., 2015). For example, the African Miombo

Table 1. Diversity of vegetation types and associated genera in Africa

Phytoregion	Vegetation type	A few associated genera
Guineo-Congolian/Zambezian Regional Transitional Zone	Southern dry evergreen	Albizia, Antiaris, Canarium, Celtis, Chlorophora
Guineo-Congolian/Sudania Regional Transitional Zone	Guinea dry forests	Entadraphragma, Combretum, Salacia
Zambezian Region	Dry forest/Scrub forest	Baikaiea, Guibortia, Pterocarpus
	Zambezian wooded grassland	Parinari, Pterocarpus, Anisophyllea
	Itigi deciduous forest	Baphia, Burttia, Combretum
	Miombo woodland	Isoberlinia, Julbernadia & Brachystegia
	Mopane woodland	Colophospermum, Adansonia,
	Undifferentiated woodland	Burkea, Afzelia, Pericopsis, Dombeya
Sudanian region	Sudanian Isoberlinia woodland	Isoberlinia
	Undifferentiated woodland	Combretum, Dalbergia, Albzia
	Acacia woodland	Vachelia and Senegalia
Kalahari-Highveld Regional Transition zone	Acacia woodland	Vachelia and Senegalia
	Wooded grassland	Guibortia, Burkea, Diplorhynchus, Parinari
	Semi and shrubland	Hymenocardia, Ximenia, Vitex, Rhus
Somali-Masai Region	Acacia-Commiphora bushland	Senegalia, Vachelia and Commiphora

alone supports the livelihoods of over 100 million people in both urban and rural areas (Syampungani et al. 2009) Additionally, forests contribute towards meeting the greatest proportion of fuel requirements (Ky-Dembele et al., 2007; Shackleton et al., 2007; Shackleton et al., 2011) at both local and national levels. These products are extremely important for local food security and dietary diversity, notably in times of agricultural scarcity (Sunderland et al., 2015). Wood fuel and charcoal are the main source of energy for the majority of households in dry forested regions. Majority of the population of less developed countries cook with wood-based energy source. The production, transport, and trading of wood fuel provides employment to many people despite the price of wood being relatively low.

Dry forests also contribute towards climate change mitigation (Becknell et al., 2012). However, they store less carbon than humid forests. Dry forests are highly resistant to drought, this makes the food and livelihoods provided by dry forests play a critical role in building the adaptive capacity of communities living in their proximity in the context of threats of climate change (Djoudi et al., 2015; Sunderland et al., 2015).

Dry forests provide a wide range of ecosystem services, such as water management, livestock provisioning, pollination services, nutrient cycling and soil improvement (Becknell et al., 2012; Djoudi et al., 2015). They also harbour considerable biodiversity in terms of species richness, endemism and functional diversity of plants and animals that sometimes even exceeds that of humid forests (Miles et al., 2006). However, over exploitation of these forests continues to put these forests under pressure and often results in degradation. This calls for developing sustainable management systems for these forests. In order to develop these sustainable systems, it is important that the regeneration ecology of these forests is well understood.

REGENERATION MECHANISMS OF SPECIES OF TROPICAL DRY FORESTS

Regeneration in dry forests and woodland trees may either be asexual or asexual propagation. There are five regeneration mechanisms of trees or plants of the dry forests and woodlands namely, i) true seedling ii) Seedling sprout iii) Root sucker iv) Coppice and v) Water sprout vi) Layer (Bognounou et al., 2010; Ky-Dembele et al., 2007). Table 2 explain in detail what each of the regeneration mechanism is.

Seedling sprouts arise from root collars of plantlets of seed origins or plantlets that emerged from remaining stems that seemed dead but held some buds (Ky-Dembele et al., 2007; Traoré et al., 2008)

Table 2. Description of the different regeneration mechanisms

Regeneration mechanism	Characteristics
True seedling	An individual of seed origin that was never affected by dieback
Seedling sprout	An individual of seed origin that was affected by dieback but resprouted from the collar of the seedling
Root sucker	An individual arising vertically from superficial lateral root
Coppice	An individual arising from stumps of cut matre tree in response to logging or non-logging disturbance and with root diameter exceeds 10 cm
Water sprout	An individual developed from the base of alive mature tree
Layer	An individual developed from low hanging lateral branch layers and arising from adventitious buds

Source: Bognounou et al., 2010; Ky-Dembele et al., 2007

whilst true Seedlings are referred to as plantlets produced directly from seed germination (Traoré et al., 2008).The regeneration mechanisms may vary from species to species and also from vegetation type to vegetation type. The common mechanisms of regeneration in the dry forests and woodlands include current-year seedlings, seedling resprouts that occur after seedling shoot dieback, coppice or sprouts from stumps of mature trees and root suckers that arise from lateral roots (Figure 2; Table 3). Plantlets are woody species with a shoot height up to 100cm irrespective of the regeneration mechanisms (Ky-Dembele et al., 2007). The Savanna species have both vertical and horizontal root systems. The horizontal root systems may additionally produce root suckers once the above ground parts are removed. These root suckers are plantlets that arise vertically from superficial lateral roots and later on developed their own root systems and became autonomous. In addition to this, seedling sprouts and suckers categories are also characterized by the presence of one or more aerial stems derived from buds (Traoré et al., 2008).

The importance of each of these regeneration mechanisms depends on the floristic composition of the forest/woodland, the type of disturbance (Timberlake et al 2010) and the age of the stand since the disturbance ceased. Syampungani et al. (2015) reported variation in regeneration mechanisms among different Miombo woodland species by disturbance type (charcoal production, slash and burn agriculture and single tree selection harvesting). Syampungani (2008) also observed variation in regeneration mechanisms across disturbances; in previously timber harvested and cultivated stands of the Copperbelt Miombo, *Parinari curatellifolia* mostly developed from root suckers while *Albizia antunesiana, Brachystegia floribunda, Brachystegia longifolia, Brachystegia spiciformis, Isoberlinia angolensis, Julbernadia paniculata, Pericopsis angolensis, Pseudolachnostylis maprouneifolia* and *Pterocarpus angolensis* mostly developed from seed. Coppicing is mostly common in *Albizia antunesiana, Brachystegia floribunda, Isoberlinia angolensis, Julbernadia paniculata* and *Pseudolachnostylis maprouneifolia*. Other species, such as *Brachystegia longifolia, Brachystegia spiciformis, Parinari curatellifolia, Pericopsis angolensis* and *Pterocarpus angolensis* regenerated mostly from seed.

These different mechanisms of regeneration influence growth and survival of various species once the disturbance ceases (Kennard et al. 2002). Syampungani (2008) reported that the growth rates of *Julbernadia paniculata, Brachystegia floribunda, Isoberlinia angolensis* and *Albizia antunesiana* varied from charcoal regrowth stands to slash and burn regrowth stands. Stems of the same species were observed to grow faster in charcoal regrowth stands than in slash and burn regrowth stands. This was attributed to the fact that most of these tend to develop from the already established rootstocks as compared to regeneration from seedlings in slash and burn agriculture. Seedlings face frequent and severe

Table 3. Structure of the regeneration pool one year after clear cutting in Miombo woodland in Zambia and two years after cutting a Sudanian savanna in Burkina Faso

Regeneration type	Miombo woodland		Sudanian Savanna	
	Plants/ha	Percent	Plants/ha	Percent
True seedling	14, 575	56.1	1154	3.5
Seedling sprouts	9,870	38.1	28,269	84.5
Stump sprouts	1425	5.5	2000	6.0
Root suckers	0	0	1308	3.9
Water sprout	0	0	718	2.2

Source: Timberlake et al 2010

Figure 2.: Various regeneration mechanisms in the Southern African woodlands

fire damage that retards their recruitment into the tree layer (Timberlake et al. 2010). Between fires, seeds have to germinate and build enough root reserves to survive the next fire (Timberlake et al 2010). However, given that fires usually occur once in 2-4 years, sprouts would need to grow rapidly to escape damage (Jacobs and Schloeder, 2002). Once the African dry forest is cleared and the land abandoned, regeneration often takes place but the speed of recovery depends on the disturbance regime, the sources available for regeneration and site history (type, frequency and intensity of stress and /or disturbance) (Syampungani et al 2015; Zulu et al 2018: Timberlake et al 2010).

The initial regeneration is mainly from trunk and root sprouts, thus making the tree composition similar to that of the initial vegetation (Fanshawe, 1971; Syampungani, 2008). Consequently, the regrowth forest consists of multi-stemmed trees often with a stem density higher than in previous old growth stands it replaced (Timberlake et al 2010).

Species Composition, Seedling, and Plantlet Abundance

Miombo woodlands species are known to resprout from roots and stumps once the above ground biomass is removed or killed by harvesting or fires (Handavu et al., 2011). This makes miombo species to be productive and resilient once their stems are cut and the growth rate of this resultant coppice shoot is influences by several factors which are the size of the tree, height of cutting and root/shoot after felling. However, sprouts grow faster than the newly established seedlings due to the already established root system (Syampungani ct al., 2016). This may be attributed to the species genetic differences which tend to influence their ability to produce growth hormones, buds and food reserves for bud development. According to Handavu et al., (2011) the influence of stump size on the coppicing ability varied between species and also diameter class.

In terms of composition of regeneration mechanisms, the composition is dependent on species, the type and density of disturbance. Ky-Dembele et al (2007) observed variation in percentage of regeneration pool of the Sudanian savannas namely; seedling sprouts (83%); root sucker (4%), coppice (5%), water sprout (2%) and layering (less than 1%). True seedlings constituted a minor component (5%) of the plantlet population. The most dominant species where *Dichrostachys cinerea*, *Pteleopsis suberosa* and *Detarium microcarpum* with *Feretia apodanthera* as the only layering species encountered (Ky-Dembele et al., 2007). Ky-Dembele et al. (2007) reported that the regeneration pool in a Sudanian savanna dominated by *Detarium microcarpum, Terminalia avicenniodes* and *Acacia* species in Burkina Faso Zulu et al (2018) observed that seedlings was the most common type (48%) of regeneration mechanism followed by coppicing (40%) in areas that were previously under shifting cultivation in the Zambian *Baikiaea* forest. These sources of regeneration constituted 95% of the regeneration pool in the Zambian Miombo woodland (Chidumayo 2004).

MANAGEMENT AND RESTORATION IMPLICATIONS OF REGENERATION ECOLOGY OF TROPICAL DRY FOREST SPECIES

High diversity tropical forests are often deforested for conversion to agriculture (Aide et al., 2000). Studies show that certain types of long-disturbed forests can show a high degree of resilience, particularly if many plant species can resprout from roots (Sampaio et al., 2007). Resprouting is a key attribute of resilience and productivity of the woodland savannas (Shackleton, 2001). The ability of most of the tropical dry forest species to coppice has been observed in studies of many tropical dry forests, such as the South African Savanna (Kaschula et al 2005) and the Miombo ecoregion (Syampungani & Chirwa, 2011) and the Sudanian savanna woodland of West Africa (Dayamba et al 2011). This is mainly because most of the dry forest species generally have extensive and horizontal root systems that facilitates recuperation after cutting (Mistry, 2000).

Factors Influencing Regeneration Mechanisms

The African species may be classified into several categories based on the subject of interest. For example, Trapnell (1959) recognized four categories of species with regard to fire:

1. Fire-tolerant mostly evergreen species that occurs where there is complete protection against fires (e.g. *Parinari excelsa*)
2. Fire tender species which reduce in numbers under regular burning and under fires protection
3. Semi-tolerant species that are relatively unaffected by early dry season fires but decline under late dry season fires
4. Fire tolerant species that are able to survive late dry season fires

In terms of light, tropical dry forest species may be classified as shade tolerant and light demanding (see Syampungani 2008). Shade tolerant species are mostly understorey species that do well under shade while the light demanding species tend to do well under maximum light. Light demanding are mainly canopy species.

The regeneration mechanisms of woodland savannas are influenced by a number of factors such as the species, the disturbance type and the age of the stand. Each of the regeneration mechanisms is influenced by different factors. For example, coppicing as a regeneration mechanism is influenced by a number of factors namely; i) species ii) stand age and disturbance regime iii) Stump size and height. A number of studies have demonstrated the influence of these different factors (Shackleton 2000; Luoga, 2004; Syampungani et al 2017). In the Zambian Miombo woodland, coppice density and shoot vigour were observed to be influenced by stump size and time since the cutting of some species (Syampungani et al 2017). Similar observations were made in eastern and Southern African savannas, where the influence of species, plant size/age, stump height and the percentage of stand removed on the coppicing ability (Shackleton 2000; Luoga 2004). The influence of species on coppicing ability may be attributed to the fact that there is a trade-off between storage and growth when trees are allocating resources, which in turn may vary among species (Syampungani et al 2017). As such, resprouting is more successful for species or trees that have larger reserves to support regrowth (Knox and Clarke, 2005). Luoga et al (2004) observed that the size of the tree at the time of cutting tends to influence the resprouting ability of many African savanna species. The influence of the size or rather age of the tree at the time of cutting may be attributed to physiological changes to which the capacity for rejuvenation by vegetative means depend (Bond and Midgley, 2003).

The influence of stump height on coppice density is observed in many African woodland species such as *Terminalia sericea* (Syampungani et al 2017). Within the southern African Savannas, a number of studies have demonstrated the influence of stump height on resprouting ability of the Zimbabwean Miombo (Mushove and Makoni, 1993) and South African savanna (Shackleton, 2001) and Zambian Miombo woodland (Handavu et al, 2011). This may be attributed to the availability of more reserved food and dormant buds on longer stumps (Syampungani et al 2017).

The African savannas tend to sufficient number of regeneration pool (root suckers, sprouts, seedlings) in the herbaceous layers (Chidumayo, 1993; Ky-Dembele et al 2007). However, the prevailing environmental conditions such as fires, competition for nutrients and moisture, seedling predation, drought stress, and light influences the number of surviving individuals to enter the next size class (Syampungani et al 2017). Additionally, seedling recruitment is usually affected by pre and post dispersal seed. Repeated bush fires are an integral ecological factor in the Savanna woodland ecosystem that has a detrimental effect on seed and seedling survival and growth despite the availability of soil moisture to allow germination and successful seedling establishment (Shackleton et al., 2011; Syampungani et al., 2016). Irregular and sporadic seedling establishment exhibits bell-shaped structural profile of the woodland savannas. Irregular and sporadic seedling establishment may be attributed to a number of factors such as effects of fires, competition for nutrients and light (Geldenhuys, 1993). Repeated shoot dieback due to herbivory, frequent fire and drought are common phenomena in the Savanna woodland and accounts for the slow height growth of plantlets (Ky-Dembele et al., 2007). Deforested and abandoned sites that are usually invaded by tall grass (2.5m) usually non-native grass species such as *Saccharum spontaneum L. ssp. Spontaneum* that grows in dense and impenetrable stands halts natural regeneration (Hooper et al., 2005). In addition to this grass invasion also increases the likelihood of fire that arrests the natural regeneration in abandoned and deforested lands which impoverishes the soil, reducing seedling growth and impeding forest recovery (Hooper et al., 2005)

Regeneration Mechanisms, Restoration, and Management of Dry Forests

Ecological restoration of tropical forest is a challenging task, but given the high biodiversity of these forests and their influence on global water and carbon cycles, it is critical that we understand how to manage and restore them. Natural regeneration is an effective strategy for restoration of dry forests (Aide et al., 2000). To achieve this adequate information is required to successfully design restoration strategies for tropical forests (Bognounou et al., 2010; Hooper et al., 2005).

Previous studies have demonstrated that the African savannas are a resilient ecosystem. This is because the woodland has high regeneration pool that supports recovery. As such restoration of the African Savannas requires understanding and making use of information governing regeneration dynamics. Previous studies (Zulu et al 2017; Syampungani et al 2017; Shackleton, 2000) bring out valuable information that would contribute towards management of the African Savannas based on regeneration dynamics. For example, Syampungani et al (2017) observed the following in the Zambian miombo that would be useful for its management; i) Stump diameter and height significantly affect coppice density with marked variation among species ii) Coppice density is correlated with shoot vigour. Additionally, Zulu et al (2018) and Syampungani et al (2016) observed variation in regeneration mechanisms between disturbances. The main regeneration mechanisms of *Baikaiea* forest for sites previously under shifting cultivation are coppices and resprouts (Zulu et al 2018). However, Syampungani et al (2016) observed that coppice among *Albizia antunesiana, Brachystegia floribunda, Isoberlinia angolensis, Julbernadia paniculata* and *Pseudolachnostylis maprouneifolia* was the most common (72%) mode of regeneration while seedling was more common (61%) for *Brachystegia longifolia, Brachsytegia spiciformis, Parniari curatellifolia, Pericopsis angolensis and Pterocarpus angolensis* in stands previously under charcoal production. Therefore, restoration of such sites would focus on protecting each regeneration pool. The management of species would focus on protection of coppices and seedlings against destructive agents such as fires and rodents (in case of seeds). Interventions such as weeding and prescribed burning allow young seedlings to establish for species that main develop through seedlings. Delaying artificial seed dispersal reduces the time that seeds can be preyed upon. Most tropical dry forests also produce dormant seeds which facilitate seed storage exsitu (Sampaio et al., 2007; Vieira and Scariot, 2006) Additionally, efforts should be directed at enhancing the survival and vigour of stems through reduction of competition for water, nutrients and sunlight. This may be achieved through thinning out of stems on each stand to reduce competition.

CONCLUSION

The African savannas have a diverse of regeneration mechanisms. These mechanisms of regeneration of the African Savannas are mainly species and disturbance specific, although individual species tend to be associated with particular regeneration mechanisms with varying disturbances. The diversity of regeneration mechanisms of the African Savannas makes it a highly resilient and productive ecosystem while the survival of juvenile individuals of the African Savannas is influenced by a number of factors such as fires, size of the stump, regeneration mechanism and the species. Developing restoration strategy for the African savannas requires that the knowledge of regeneration mechanisms and dynamics of these woodlands is incorporated in the management. Management regimes should be designed to incorporate various factors that influence the regeneration mechanisms of the African Dry Forests. This would require incorporation of the knowledge of woodland recovery and factors that influence their recovery.

REFERENCES

Aide, T. M., Zimmerman, J. K., Pascarella, J. B., Rivera, L., & Marcano-Vega, H. (2000). Forest regeneration in a chronosequence of tropical abandoned pastures: Implications for restoration ecology. *Restoration Ecology*, *8*(4), 328–338. doi:10.1046/j.1526-100x.2000.80048.x

Becknell, J. M., Kucek, L. K., & Powers, J. S. (2012). Aboveground biomass in mature and secondary seasonally dry tropical forests: A literature review and global synthesis. *Forest Ecology and Management*, *276*, 88–95. doi:10.1016/j.foreco.2012.03.033

Bognounou, F., Tigabu, M., Savadogo, P., Thiombiano, A., Boussim, I. J., Oden, P. C., & Guinko, S. (2010). Regeneration of five Combretaceae species along a latitudinal gradient in Sahelo-Sudanian zone of Burkina Faso. *Annals of Forest Science*, *67*(3), 306. doi:10.1051/forest/2009119

Chidumayo, E. N. (1997). *Miombo Ecology and Management: An introduction.* Stockholm, Sweden: IT Publications.

Djoudi, H., Vergles, E., Blackie, R., Koame, C. K., & Gautier, D. (2015). Dry forests, livelihoods and poverty alleviation: Understanding current trends. *International Forestry Review*, *17*(2), 54–69. doi:10.1505/146554815815834868

<edb>Geldenhuys, C. J., & Golding, J. (2008). Resource Use Activities, Conservation and Management of Natural Resources of African Savannas. In F. G. Faleiro & A. L. F. Neto (Eds), Savannas Desafios Estrategias para o equilibrium entre sociedade, agronegocio e recursos naturais. pp 225-262, Planatltina, Embrapa CERRADOS.

Elmqvist, T., Pyykönen, M., Tengö, M., Rakotondrasoa, F., Rabakonandrianina, E., & Radimilahy, C. (2007). Patterns of loss and regeneration of tropical dry forest in Madagascar: The social institutional context. *PLoS One*, *2*(5). doi:10.1371/journal.pone.0000402 PMID:17476324

Handavu, F., Syampungani, S., & Chisanga, E. (2011). The influence of stump diameter and height on coppicing ability of selected key Miombo woodland tree species of Zambia: A guide for harvesting for charcoal production. [</edb>]. *Journal of Ecology and the Natural Environment*, *3*, 461–468.

Hooper, E., Legendre, P., & Condit, R. (2005). Barriers to forest regeneration of deforested and abandoned land in Panama. *Journal of Applied Ecology*, *42*(6), 1165–1174. doi:10.1111/j.1365-2664.2005.01106.x

Ky-Dembele, C., Tigabu, M., Bayala, J., Ouédraogo, S. J., & Odén, P. C. (2007). The relative importance of different regeneration mechanisms in a selectively cut savanna-woodland in Burkina Faso, West Africa. *Forest Ecology and Management*, *243*(1), 28–38. doi:10.1016/j.foreco.2007.01.091

Miles, L., Newton, A. C., DeFries, R. S., Ravilious, C., May, I., Blyth, S., ... Gordon, J. E. (2006). A global overview of the conservation status of tropical dry forests. *Journal of Biogeography*, *33*(3), 491–505. doi:10.1111/j.1365-2699.2005.01424.x

Salzmann, U. (2000). Are modern savannas degraded forests?-A Holocene pollen record from the Sudanian vagetation zone of NE Nigeria. *Vegetation History and Archaeobotany*, *9*(1), 1–15. doi:10.1007/BF01295010

Sampaio, A. B., Holl, K. D., & Scariot, A. (2007). Does restoration enhance regeneration of seasonal deciduous forests in pastures in central Brazil? *Restoration Ecology*, *15*(3), 462–471. doi:10.1111/j.1526-100X.2007.00242.x

Santos, J. C., Leal, I. R., Almeida-Cortez, J. S., Fernandes, G. W., & Tabarelli, M. (2011). Caatinga: The scientific negligence experienced by a dry tropical forest. *Tropical Conservation Science*, *4*(3), 276–286. doi:10.1177/194008291100400306

Shackleton, C. M., Shackleton, S. E., Buiten, E., & Bird, N. (2007). The importance of dry woodlands and forests in rural livelihoods and poverty alleviation in South Africa. *Forest Policy and Economics*, *9*(5), 558–577. doi:10.1016/j.forpol.2006.03.004

Shackleton, S., Delang, C. O., & Angelsen, A. (2011). From subsistence to safety nets and cash income: exploring the diverse values of non-timber forest products for livelihoods and poverty alleviation. In Non-timber forest products in the global context (pp. 55–81). Berlin, Germany: Springer. doi:10.1007/978-3-642-17983-9_3

Sunderland, T., Apgaua, D., Baldauf, C., Blackie, R., Colfer, C., Cunningham, A., ... Gumbo, D. (2015). Global dry forests: A prologue. *International Forestry Review*, *17*, 1–9. doi:10.1505/146554815815834813

Syampungani, S., Geldenhuys, C. J., & Chirwa, P. W. (2016). Regeneration dynamics of miombo woodland in response to different anthropogenic disturbances: Forest characterisation for sustainable management. *Agroforestry Systems*, *90*(4), 563–576. doi:10.100710457-015-9841-7

Syampungani, S., Tigabu, M., Matakala, N., Handavu, F., & Oden, P. C. (2017). Coppicing ability of dry miombo woodland species harvested for traditional charcoal production in Zambia: A win–win strategy for sustaining rural livelihoods and recovering a woodland ecosystem. *Journal of Forestry Research*, *28*(3), 549–556. doi:10.100711676-016-0307-1

Timberlake, J., Chidumayo, E., & Sawadogo, L. (2010). Distribution and Characteristics of African Dry Forests and Woodlands. In E. N. Chidumayo & D. J. Gumbo (Eds.), The Dry Forests and Woodlands of Africa: Managing for products and services. London, UK: Earthscan Publishing.

Traoré, S., Tigabu, M., Ouédraogo, S. J., Boussim, J. I., Guinko, S., & Lepage, M. G. (2008). Macrotermes mounds as sites for tree regeneration in a Sudanian woodland (Burkina Faso). *Plant Ecology*, *198*(2), 285–295. doi:10.100711258-008-9404-3

Vieira, D. L., & Scariot, A. (2006). Principles of natural regeneration of tropical dry forests for restoration. *Restoration Ecology*, *14*(1), 11–20. doi:10.1111/j.1526-100X.2006.00100.x

Wohlfart, C., Wegmann, M., & Leimgruber, P. (2014). Mapping threatened dry deciduous dipterocarp forest in South-east Asia for conservation management. *Tropical Conservation Science*, *7*(4), 597–613. doi:10.1177/194008291400700402

Zulu, F., Syampungani, S., & Fushike, P. (2018). Recovery of Baikiaea forest of South Western Zambia from shifting cultivation and its implications for sustainable management. *Journal of Forestry Research*. doi:10.100711676-018-0656-z

Chapter 19
Role of Institutional Governance in Protecting the Natural Environment:
A Case of National Capital Region, India

K. K. Yadav
Independent Researcher, India

Kumud Dhanwantri
Amity University Haryana, India

ABSTRACT

In the present age of industrialization and unregulated urbanization, the Aravali ranges in India are facing deforestation and degradation. The major reasons behind this are the needs of the poor, and greed of the rich. Therefore, part of the Aravalli Ranges falling in different sub-regions of the National Capital Region, has been taken as case study. The chapter intends to provide an insight into the scenario of forests and wildlife in the sub-regions; the challenges, responses, and immediate initiatives taken up by the constituent state governments. It also discusses ways forward to engage the governments and local communities in the protection of forests and wildlife. The conclusion strives to provide probable strategies that can be adopted to transform the transitions of Aravalli into a positive one and ways for engaging government machinery for better governance to escape the grim future we foresee.

INTRODUCTION

Being one of the most vulnerable resources and gifts of the nature, forests and wildlife play a significant role in the overall natural environment of our planet earth. However, these are scarce but renewable. The forests not only act as carbon sink, support the livelihoods by providing food, shelter, clothing, wood, timber and medicines, etc., but are also the store-house of energy as well as scarce resources such as water. They also act as a habitat for pollination and other biotechnological processes. Forests are also

DOI: 10.4018/978-1-7998-0014-9.ch019

helpful in preventing floods, droughts and deserts. Wildlife conservation, which includes protection of wild animals and plants as well as their habitat, is not only necessary for animals but also for the human beings and other living organisms. It is obvious; therefore, that there is a need to protect habitats and natural environment such as forests, biospheres, oceans and rivers (World Wide Fund, 2018).

The fact that forests and wildlife conservation is an international and interconnected issue, it needs to be seen in a wider perspective and protected at all levels i.e. local, sub-national, national and international. The conservation planning and management needs to address the terrestrial, aquatic and the atmospheric ecosystems in a wholesome and integrated manner. Mapping of interrelations within the biotic and abiotic environment shall help in overall protection efforts.

The post-industrial revolution characterized by rapid increase in population, depletion of natural resources, increased air and water pollution as well as increased consumption of fossil fuels caught the world attention towards destruction of forest and wildlife. Environmental awareness gave birth to several environmental movements and the IUCN[1] (1948), Club of Rome[2] and the British Clean Air Act, 1956. The non-governmental initiatives and efforts such as formulation of IUCN-the oldest and largest global environmental network - the Club of Rome - the Limits of Growth, the Sierra Club[3] caught public attention on environmental degradation and the demand for comprehensive environmental protection laws increased.

This awareness inspired and directed several global governments to initiate efforts to protect the environment and it was for the first time that United Nations held a conference on human environment in Stockholm, Sweden in 1972. This conference resulted into formulation of 26 principles and United Nations Environment Programme (United Nations Environment Program, 1972). This summit played a significant role not only in making the world aware but in inspiring to nurture and protect the environment especially the forest and the wildlife. Consequently, the World Conservation Strategy (1980) adopted by the United Nations, the World Commission on Environment and Development, famous for Brundtland Commission and further publication of its report on "Our Common Future" in 1987 necessitated the need of sustainable development to establish symbiotic relationship between environment and development (Brundtland Report, 1987). Its impact was so strong and wide that United Nations felt the need to hold a conference on environment and development i.e. Rio United Nations Conference on Environment and Development 1992. Reaffirming the declaration of the United Nations Conference on the Human Environment adopted at Stockholm in June, 1972, the Rio declaration sought the cooperation and equitable partnership among States in key sectors of societies and people. It emphasized the need towards international agreements respecting the interests of all and protects the integrity of the global environmental and development system for the first time with a balanced and integrated approach (United Nations Educational, Scientific and Cultural Organization, 1992). The agenda 21 was intended to be a set of guidelines on environment and development issues to be followed by States in the 21st century (United Nations Conference on Environment and Development, 1992). Thereafter, second Earth Summit, the United Nations Conference on Sustainable Development, popularly known as Rio Summit was held in 2012 (International Institute for Sustainable Development, 2012).

Indian Scenario

Enjoying varied bio-geographic regions, India is one of the front leaders in forest and wild life conservation in the world. Accounting for 2.4% of the World's surface area, sustaining the needs of 17% of human and 18% of livestock population, India covers 24.39% of land area under forest and tree cover having

10th global ranking (Ministry of Environment, Forest and Climate Change [MoEFCC], 2017). The latest 'India State of Forest Report, 2017' indicates that there is an increase of 8021 km^2. i.e. 80.20 million hectares in the total forest and tree cover area of the country in comparison to its earlier assessment in the year 2015. With 6,778 km^2 of forest cover and 1,243 km^2 tree cover, it constitutes 24.39% of the total geographical area of the country (Forest Survey of India [FSI], 2017). Among four categories of very dense forest (VDF), moderately dense forest (MDF), open forest (OF) and mangrove forest (MF), it is the VDF category which has shown much of the increase in the forest area followed by open forest during this period. It is quite significant as VDF absorbs the maximum carbon dioxide from the atmosphere.

Lakshadweep with 90.33%, Mizoram with 86.27% and Andaman & Nicobar with 81.73% of forest cover to their geographical area are the top three states/ UTs as Table 1 shows, while the states showing the highest increase in their area of forest cover from the previous report of 2015 are Andhra Pradesh (2,114 km^2) followed by Karnataka (1,801 km^2) and Kerala (1,043 km^2) (FSI, 2017).

As per Forest (Conservation) Act 1980, forest land may be diverted for non-forestry purposes like dams, industrial projects and mining in the same district in the non-forest areas (Ministry of Environment and Forest [MoEF], 2004). But the fact remains, that about 70% of compensatory afforestation across ten states was done on forest land violating the provisions of the act according to a report released by Community Forest Rights-Learning and Advocacy (CFR-LA) (Tatpati, 2015).

The state of Haryana with its dismal record in afforestation, its protection and conservation efforts, has been shown with the lowest percentage area of 3.59 under forest of its geographical area in the latest Forest Survey of India assessment report in 2017. Surprisingly, the neighbouring states of Punjab, Rajasthan and Uttar Pradesh with 3.65%, 4.54% and 6.09% of forest cover respectively, also present a similar picture, further adding to the degradation of natural environment and become a cause of concern for the National Capital Region, Delhi. Table 1 justify this cause of concern by showing, three major states of NCR fall under the lowest forest cover category, in India (FSI, 2017).

With its marginal increase in forest cover (0.02%) from the previous survey of 2015, the state of Haryana remains at the bottom among all the Indian states. Haryana, while comparing with the state of Punjab, presents higher percentage of area under very dense forest (canopy density of 70% and above) and open forest (canopy density of 10-40%) but lower status in moderately dense forest (canopy density between 40-70%). Various reports published in the daily newspapers suggest that the government has shown scarce regards towards protection of forest especially in the districts of Gurugram and Faridabad. These two districts because of their proximity to the National Capital, witness high urbanisation.

Table 1. State with highest and lowest percentage of area under forest to their geographical area, 2017

Sl. No	Highest Forest Cover		Lowest Forest Cover	
	State/UTs	Forest cover as percent of geographic area	State/UT	Forest cover as percent of geographic area
1	Lakshadweep	90.33	Haryana	3.59
2	Mizoram	86.27	Punjab	3.65
3	Andaman & Nikobar	81.73	Rajasthan	4.84
4	Arunachal Pradesh	79.96	Uttar Pradesh	6.09
5	Manipur	77.69	Gujarat	7.52

Source: Arora, The Times of India, February 15, 2018

The fact that biodiversity provides ecological, economic, aesthetic and cultural benefits, it calls for understanding spatial distribution patterns, interrelationships between different ecosystems and habitats, its losses and finally conservation approaches and strategies at local, national and global levels in a scientific manner. The uncontrolled urban sprawl in the NCR to this effect needs special attention.

The National Capital Region and Natural Environment

The need of creating The National Capital Region (NCR) was felt while the first master plan of Delhi was prepared. In the wake of increasing growth rate of population of Delhi Metropolis, it was decided to expand its area for planning purposes so as to divert increasing pressure of population and regulate the development in various towns of its region as well. This would not only decongest the National Capital but would also create opportunities for development in its demarcated region in the surroundings. It is by far one of the largest planning region in India, involving four States of National Capital Territory (NCT) of Delhi, Haryana, Uttar Pradesh and Rajasthan attempting an exercise in spatial and development planning. This region enjoys the largest and unique mosaic of ecology and biodiversity, perhaps in the world with the presence of the Aravalli hills and the river Yamuna basin.

The Aravallis in India forms parts of the sub-regions of NCT of Delhi, Haryana and Rajasthan of National Capital Region. Starting from NCT of Delhi in south-western direction, it passes through southern parts of Haryana, Rajasthan and extends up to Gujarat. Its longitudinal stretch is approximately 692 km^2 long. It was formed in the pre-Cambrian era. Broadly these ranges in the NCR make their presence in three distinct administrative areas. The first one is the north-eastern end i.e. the Delhi Ridge which is an extension of the Aravalli hills sprawling from southwestern via Alwar (Rajasthan), Faridabad and Gurugram (Haryana). The second is the Aravalli ranges in Haryana which are spread in the districts of Gurugram, Faridabad, Mewat and also located in isolation and scattered form in the districts of Mahendergarh, Rewari and Bhiwani. The third one falls in the administrative control of Alwar district of Rajasthan. Aravalli ranges in NCR are characterized by isolated rocky hills, thick and dry deciduous tropical forests, and rich biodiversity (Kohli, 2004).

Aravalli, being the oldest mountain system in India, has a rich variety of more than hundred species of medicinal plants such as Ashwagandha (*Withania Somnifera*), Sarpagandha (*Rauwolfia Serpentina*), Vach (*Acorus Calamus*), Brahmi (*Bacopa*), Chitrak (*Plumbato Zeylanica*), Safed Musali (*Chlorophytum Borivilianum*), Shatavari (*Asparagus Racemosus*), Pippali (*Exbucklandia*), Makoy (*Solanum Nigrum*), Bhumi Amlaki (*Phyllanthus Niruri*), Gwarpatha (*Aloe Vera*), Harad (*Terminalia Chebula*), Bahera (*Terminalia Bellirica*), Amla (*Phyllanthus Emblica*), Bael (*Aegle Marmelos*), Kalahari (*Dichrostachys Cinerea*), Patharchat (*Bryophyllum Pinnatum*), *Lemongrass, Licorice, Jatropha, Palmarosa* (Jha, 2015).

With rich biodiversity, the Aravalli ranges enjoy thick and dry deciduous tropical woods impacting socio-economic and cultural life of the rural and urban dwellers in its surroundings. Aravallis in Delhi, Haryana and Rajasthan has humid subtropical and hot semi-arid continental climate with very hot summers and relatively cool winters. The central Aravalli ranges in Rajasthan has an arid and dry climate. The southern Aravalli ranges in Gujarat has tropical wet and dry climate.

Aravalli hills in its north-eastern parts have been attracting people throughout history but more so after Britishers declared Delhi as the capital in the first decade of the 20th centaury. Delhi attaining its capital status and its location in the northern Indo-Gangetic plain linking north, west, east and central India attracted more economic activities. After independence it became an important hub of trade and commerce. The economic reforms in early 1990s gave it further boost. This resulted into pressure on the

Aravalli ranges in the NCT of Delhi and neighboring state of Haryana. The first pressure was in the form of mining and the farmhouses and the second was urbanization. Presently, the pressure on Aravalli ranges is so high that even its legal protector the governments have begun to dilute the legislative provisions in the favor of businessmen and the rich and powerful. The emerging nexus of 'polity-bureaucracy-business' has weakened its institutional delivery mechanism. Rampant ignoring even the orders of the Apex Court of India as mentioned in Annexure 1, with some or the other excuses by the successive governments especially the governments of Haryana are a routine, indirectly favoring the powerful and the rich.

Land, water and forest are the major natural resources, the utilization of which have wider implications on the life of people in a region. The National Capital Region (NCR) with 4.61 crores of population is facing crises due to over-exploitation of these closely related trio-resources of land, water and forest. The ever-increasing population in the NCR has brought spurt in the economic activities with trade and commerce rising with unprecedented trend. This has increased the demand of land as every socio-economic and cultural activity end up with the space i.e. land. The NCR is witnessing urban sprawl due to its natural population growth and migration. In its different forms including expansion of residential, industrial and infrastructure development, the urban sprawl is badly putting pressure on agriculture land and other sources of forest, water and biodiversity. Thousands of acres of agriculture and forest land is being converted into 'built up use' due to wrong land use planning policies. The contiguous expansion of Gurugram, Faridabad, Sonipat districts in the state of Haryana and Ghaziabad, Noida, Meerut districts in the state of Uttar Pradesh further expending along different highways in the corridor form is a clear proof of the loss of huge fertile agriculture and forest land.

Presence of Aravalli ranges in their respective sub regions of NCT Delhi, Haryana and Rajasthan of NCR play a significant role in the overall natural environment. The Aravallis with diverse landscape forms the basis and source of forest, water and other ecosystem services in this region.

Sub-Region of NCT of Delhi

The natural environment of sub region of NCT of Delhi is comprised of three segments i.e. Delhi ridge, Yamuna River and Asola & Bhati Wildlife Sanctuary . The Delhi ridge, as shown in Figure 1, merging into the Yamuna basin and the Aravalli biodiversity park in the north west of Jawaharlal Nehru University (JNU) present a unique landscape providing ideal alternative habitats for migratory and resident fauna and flora. Delhi ridge with more than 7,777 hectares of reserved forest area in the heart of a mega city with a distinctive ecosystem is a rare example of urban forest; helpful in recharging the sources of ground water bodies in Delhi. This ridge is a long rocky hills extending roughly from north-east to south-west of Delhi. The total forest cover of National Capital Territory of Delhi State spreads over an area of 179.81 km^2 which constitutes 12.1% of the Sub-region. The ridge with deciduous arid scrub forest has varied indigenous plant species such as Babul, Thorn Forest, Neem, Peepal, Banyan, Dhav, Dhak, Imli and Arjun etc. The Asola and Bhati wildlife sanctuary, one of the important open scrub jungles with a lake, is home to over 250 species of trees, 200 species of birds, 10 species each of mammals and reptiles, eight species of amphibians and about 90% species of butterflies (Sinha, 2014).

The biologically rich wetland spreading over an area of 457 acres of land near Wazirabad village is the Yamuna Biodiversity Park acting as natural conservation site apart from being a center of environmental and ecological knowledge and awareness. The Delhi Development Authority had established the Delhi Biodiversity Foundation with a purpose to develop these natural resources of ecological, aesthetical and

Figure 1. Aravalli ranges in NCR
Source: Based on Google Earth images and online GIS database, 2018

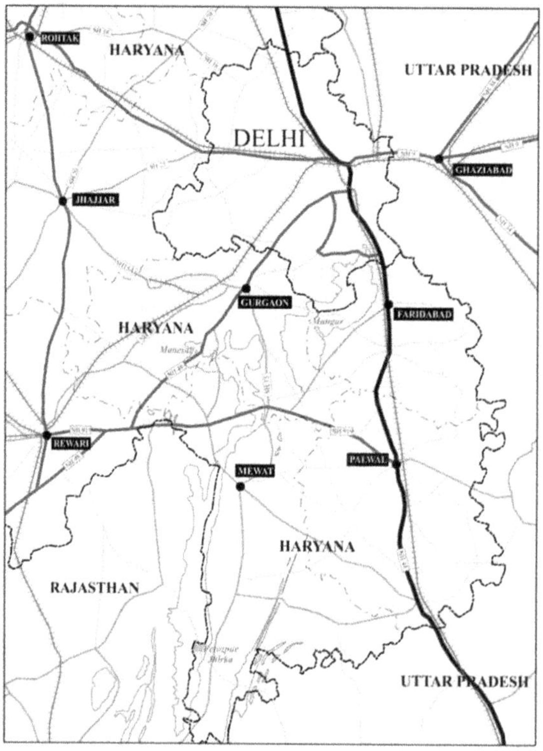

cultural quality of biodiversity sites in Delhi. Table 2 summarizes and explains about the key features of forest of Delhi ridge and Yamuna basin.

Despite of all diverse characteristics and features of the Delhi ridge, its indigenousness character is under threat. The indigenous species such as Dhav (*Anogeissus Latifolia*), Dhak (*Butea Monosperma*), Neem (*Azadirachta Indica*), Imli (*Tamarindus Indica*), and Peepal (*Ficus*) are being replaced by invasive exotic species like Vilayati Kikar (*Prosopis Juliflora*) and other garden plants like Bougainvillea hedges etc. The bifurcation of the contiguous ridge by road network and other infrastructures is playing an adverse role in conservation of its wildlife. Encroachment, unauthorized housing, illegal mining and other human interventions so far, remained the major factors behind the degradation of the ridge.

However, established in the year 2000, the 132 Eco Task Force (ETF), an infantry battalion of the Indian Army based at Delhi has considerably reduced the encroachment and other human intervention further in the Wildlife Sanctuary. Though the Wildlife Sanctuary is open to tourists through events organized by Conservation Education Centre (CEC) Delhi in a restricted manner, the Eco Task Force has proved to be a great support towards eco-restoration of the sanctuary (Sinha, 2014). While recognizing the environmental and ecological importance of the Delhi ridge, the efforts made towards its protection, preservation and conservation are meagre under multiple ineffective administrative control. Several efforts were made from time to time to handover its planning and management to one authority but all proved elusive and ignored by the decision makers.

Table 2. Key features of forest and wildlife at Delhi ridge and Yamuna basin in NCT Delhi

Sl. No.	Ridge /Basin	Notification Year /Area	Classification	Main Features
Delhi Ridge				
1	Northern Ridge	1915/ 87 hectares.	Reserved forest	• Located near Delhi University • Nearly 170 hectares. was declared as reserved forest • Presently the smallest area with 87 hectares.
2	Central Ridge	1914/ 87 hectares.	Reserved forest	• Stretches from south of Sadar Bazar to Dhaula Kuan • Some areas encroached upon
3	South Central Ridge	- / 626 hectares.	Biodiversity Park	• Located near JNU centered on Sanjay Van • Large areas have been encroached and built up • Being developed in phasing
Asola-Bhati Wildlife Sanctuary				
4	Southern Ridge	1986/ 6,200 hectares.	Wildlife Sanctuary	• Includes Asola and Bhati wildlife Sanctuary • Largest and least urbanized segment • Characterized mostly by village habitation and privately-owned farmland
Yamuna Basin				
5	Yamuna Biodiversity Park	2002/ 185 hectares.	Biodiversity Park	• Located on the flat alluvial plains of the Yamuna river • Comprises of native flora and fauna and acting as a natural conservation site for 50 endangered specific group of plants.

Source: Based on, 'An Introduction to the Delhi Ridge', 2014

Sub-Region of Haryana

The major part of Aravalli ranges in Haryana passes through districts of Gurugram, Faridabad and Mewat; some scattered and sparse ranges have their presence in the districts of Rewari, Mahendergarh and Bhiwani too. Delhi ridge, the northernmost end of the Aravalli ranges traverses to south Delhi (Asola-Bhati Wildlife Sanctuary) and meets the hills of Bandhwari - the part of Haryana Aravalli ranges constituting various isolated rocky ridges passing through southern parts of Haryana. The Aravalli ranges in Harayan sub-region are comprised of low line hard rock ridges and isolated hillocks. The forest area of this region is 461 km^2, which constitutes 3.4% of the total area of the sub-region of Haryana. Its major locations of ecological importance are Mangar, Manesar-Sohna and Firozpur Jhirka. These areas are characterized by a variety of fauna and flora.

But the species of *Prosopis Juliflora* introduced in the Aravallis during colonial times by the then government for the afforestation of the degraded areas has invaded hectares of land in the Aravallis which is threatening native biodiversity and subsequent rural sustainability. Therefore, the native characteristics of Aravalli's flora is now under threat. The indigenous species such as Dhau (*Anogeissus Pendula*), Dhak (*Butea Monosperma*), Neem (*Azadirachta Indica*), Imli (*Tamarind*), Amaltas (*Cassia Fistula*) And Peepal (*Ficus Religiosa*), Gulmohor (*Delonix Regia*), Nili Gulmohor (*Blue Jacaranda*) etc. are being replaced by invasive exotic species like *Prosopis Juliflora* (Vilayati Kikar[4]).

Furthermore, the latest survey by Wildlife Institute of India (WII) has reported that leopards occupy a 200 km^2 area in five districts of Haryana. The undulating terrain with relatively dense forests of Faridabad and Gurugram districts are the core areas of the leopards. The area of their higher population index as estimated by the survey are listed as following:

1. Ghamroj (Bhondsi)
2. Raipur (Raisina)
3. Mangar, Gothda, Badhkal
4. Kotla Kansali, Nimatpur
5. Khol and Pachnota

The sighting of varied wildlife in the Aravallis of Haryana sub region was reported by WII is listed in Table 3.

Area more than 35,000 hectares, as estimated by the forest department in the districts of Gurugram and Faridabad is the most sensitive area for wildlife. This latest survey by WII is an eye-opening report for the officials of the forest department of Haryana as they never expected and accepted such a wildlife occupancy in the region. It is because of the negligence of the state government that about ten leopards were killed in Haryana sub region in different accidents including the road accidents since 2008 (reported by various newspapers) (Kumar, 2016). The environmentalists have been emphasizing for the last one decade about the wildlife presence especially in two districts of Gurugram and Faridabad but ignored by the state government till the survey of WII.

Sub-Region of Rajasthan

The Aravalli ranges show their presence in this sub-region in the form of ridges and isolated hillocks. Sahibi, non-perennial river drains the area along with its tributaries. The sub-region with forest area of 1203 km^2 constitutes 14.4% of the total area. The forest is mostly found in small patches. The main species found in the forests are Dhok, Raunj, Khair, Asan, Bahera, Salar, Bamboo, Jhighan, Dhak, Ber, Tendu, Thor, etc.

Sariska National Park declared as Wildlife Sanctuary in 1955, was later on designated as Tiger Reserve Forests in 1978. Adapted by the Royal Bengal Tigers, it is one of the unique reserves with scrub-thorn arid forest located in Aravallis of Alwar district of Rajasthan state. A home to numerous carnivores such

Table 3. Varied wildlife reported by Wildlife Institute of India (WII), 2017

Wildlife	Numbers	Wildlife	Numbers
Leopard	7	Wild pig	14
Shipped hyena	7	Rhesus Macaque	55
Golden Jackal	9	Peafowl	57
Nilgai	5	Crested porcupine	12
Palm civet	7	-	-

Source: WII Report, 2017

as Leopard, Hyena, Tiger, Jungle Cat, Wild Dog and Jackal, the reserve also shelters species such as Chital, Nilgai, Wild Boar and Langur etc. (Choudhary, 2018).

It is also inhabited by rich and diverse wildlife such as Peacock, Sandgrouse, Tree Pie, Grey Partridge, Bush Quail and Great Indian Horned Owl. After ban on mining in the Aravallis of district Gurugram, Faridabad and Nuh district in Haryana, there was a spurt of illegal mining in the Alwar district of Rajasthan. Tijara block of Alwar district is the area witnessing large scale illegal mining despite orders of the Supreme Court. The report submitted by the Central Environmental Committee to the Supreme Court is an eye opening to the institutional set up of government responsible for protecting the natural environment. Blatant violations in the form of illegal mining and bifurcation of forest blocks in the tehsils of Tijara and Alwar are telling upon the overall health of the forests and wildlife habitat.

Therefore, it calls for an urgent decision to hand over the entire ridge to one agency for its proper management so that its further degradation and denudation is efficiently checked. With this, however, some expert groups, local civil society and NGOs should also be given powers with legal backing to make its governance more fruitful, transparent, accountable and inclusive.

Policy and Governance in NCR

The colonial regime primarily focused on maximizing its forest products and revenues through forest departments. Such a narrow and static approach continued even post-independence period resulted into a great loss of forest resources which further impacted the recharge of underground water. This also led to the loss of biodiversity and natural habitats. However, the National Forest Policy of 1988 with a paradigm shift recognized the various roles of forest including the environmental stability. The need for forest dependent communities and the people's involvement were given emphasis (MoEF, 1988).

The latest forest policy of 2018 after a period of three decades, even after a historic opportunity for the forest communities in the Forest Right Act (FRA) of 2006, could not stop profits going to corporate sector (MoEFCC, 2018). The Compensatory Afforestation Fund (CAF) at the tune of $ 7 billion is lying unused at present.

Aravalli hills have been playing an important role in the development of human settlements in terms of physical, social, economic and cultural spheres through ages. From times immemorial, the local communities have been living in a symbiotic relationship with these hills. But the modern life style with unmindful consumption of biotic and abiotic natural resources is posing threats to our rich hilly, riverine and other ecosystems in the Aravallis. The pressure and impact of urbanization and industrialization are the major concerns in the balance of environment and development. Realizing the need to control and regulate the development in the region surrounding the National Capital of Delhi, the parliament enacted the National Capital Region Planning Board act in 1985 with concurrence of the constituent states, to provide for the constitution of a planning board for the preparation of a plan for the development of the National Capital Region. This regional plan, in its para 13.1.3 specifically mentioned the hilly areas of NCT-Delhi and Aravallis at Gurugram, Faridabad and Alwar for conserving their biodiversity. Further in para 14.2 (VIII) reserved/forests other than reserved and protected forests, areas with endangered species-flora and fauna, water bodies, springs/water recharge areas and other environmental recourses areas have been proposed for their protection and conservation. Further in para 14.2 (ix) cutting of trees and construction of any clusters of dwelling units, farm houses, sheds, community centers, information centers, and any other activity connected with such constructions (including roads and part of any infrastructure relating thereto) among others have been listed to be prohibited. The NCR plan also stipulates that in the Natural Conservation Zone only activities such as agriculture and horticulture, pisciculture, social forestry/ plantations including afforestation and regional recreational activities with no construction exceeding 0.5% of the area with the permission of the competent authority are permitted (National Capital Region Planning Board [NCRPB], 2005 & 2013).

Aravalli ranges falling in three states of NCT Delhi, Haryana and Rajasthan account for major area under forest cover in the NCR. Wildlife also follow the same pattern except some additional areas such as Yamuna Biodiversity Park, Bhindawas Lake, Sultanpur National Park and other waterbodies having mainly the avian population. The first and foremost priorities, therefore, necessitates the protection and conservation of the Aravalli ranges. Despite the fact that National Capital Region is of national importance and has listed the area under forest cover, much lower than the national average, the efforts to improve it by the constituent states have been abysmal and sketchy. The dwindling area under forest and agriculture and the corresponding increase in the built up and non-forest activities becomes a cause of great concern for those urban and regional planners, policy makers and the environmentalists who have understanding and realization that the natural environment is our life-supporting-system.

The Haryana state forest policy (2006) targeted to achieve 10% of its area by 2010 and 20% by 2020 to be brought under tree cover (Govt. of Haryana, 2006). The fact is that it is far below the target. The Supreme Court in the year 1996 ordered that the state should define the area as per dictionary meaning of a forest. Even in 2014, the National Green Tribunal (NGT)[5] also ordered that the forest area be defined as forest as per dictionary meaning (NGT 2015). But the state government has been delaying the demarcation and defining the forest. The recent Gazette notification on Punjab Land Preservation Act (PLPA), 1900 by state government of Haryana to dilute its original provisions is clearly an attempt to succumbing to the pressure of builders and real estate lobby.

The apex Supreme Court of India in various cases from time to time has ordered to protect and conserve the forest, a brief description of which is shown in Annexure 1.

Sariska Wildlife Sanctuary in the district Alwar of Rajasthan is an important biodiversity hotspot of the NCR. It was declared a sanctuary in 1955 and attained the status of National Park in 1979. It is home to a variety of wildlife like Leopard, Chital, Sambar, Nilgai, Four-Horned Antelope, Wild Boar, Langur, Rhesms, Macaque, Hyena and Jungle Cat besides, Peafowl, Quail, Serpent, Eagle, Sandgrouse and Woodpeckers, Crocodile and Uger etc.

With 90% of Dhak trees, the Sariska tiger reserve is one of the important reserves in the entire Aravalli ranges (Choudhary, 2018). However, the human interventions – the presence of a temple and local rural settlements – in the reserve have been discouraging the free movements of the wildlife, especially the Tigers.

While analyzing the overall efforts on conservation of forests and wildlife in the NCR it has been noticed that the constituent state government plays an important and decisive role. The continuous pressure of the powerful and the rich on the three governments have degraded the overall natural environment, especially of the Aravalli ranges falling in the NCR.

The Supreme Court, High Courts and the National Green Tribunal (NGT) from time to time have ordered in favor of conservation and protection of the forests. One of the recent orders of NGT dated 7th August 2018 seems to be a great hope towards protection of Aravallis. The order states, *"it is (the) in-disputed legal position that the sub-region plans of the states have to be consistent with the regional plan prepared by National Capital Region Planning Board (NCRPB)."* Recently, the Supreme Court of India has shown its displeasure in case of Gazette notification on PLPA by state of Haryana may prove to be another step in restoring the degrading Aravalli ranges. Thus, the order giving primacy to the NCR plan 2021 becomes binding to the states in implementing its provisions with respect to protection and conservation of Aravallis (National Green Tribunal, 2018).

But the institutional set up and its delivery system in the states have totally failed. Rather the public institutional system favors the unscrupulous persons in the sector. The efforts of the few environmentalists and activists, including the authors have played great role in protecting the forests in the Aravalli ranges. The need of the wildlife corridor right from the NCR Delhi through south Haryana and up to Sariska in the Alwar district of Rajasthan is a dire necessity to save the wildlife in the NCR. Various highways and other roads bifurcating the forests in the Aravallis have taken a heavy toll on the wildlife. Because of deforestation, human intervention in the form of illegal mining, cutting of trees for fuelwood and the rapid rate of urbanization, most of the waterbodies, forests and wildlife have been degraded. And the fact remains, they are facing threat of wiping out (Banerjee, 2018). If serious and concerted efforts are not made at this juncture then, we will be inviting a weak and bleak immediate future leading to climate change and desertification of the NCR.

Besides orders of Supreme Court and National Green Tribunal, the Central Ministry of Environment, Forest and Climate Change (MoEFCC) and National Capital Region Planning Board (NCRPB) in their various meetings have specially dwelt and stressed upon the issue of protection and conservation of natural environment in the Aravallis falling in the states of Haryana and Rajasthan. But very surprisingly and sad is the fact that the state governments especially the Haryana and Rajasthan are not in favor of any cap on non-forest activities in the Aravallis, indirectly bestow open support to the unscrupulous real estate entities and mining mafia. NCR is, therefore, heading towards a grim future. The public institutional framework has become non-functional and destroying the age-old rich biodiversity of Aravalli ranges. All regimes (whichever political party ruling the states) have been favoring the rich and powerful for fulfilling their business interests in the Aravalli mountains.

The compensatory afforestation, a mandatory provision of the Forest (Conservation) Act, 1980 remains a futile paper and file exercise, never implemented at site. For instance, in a case of diversion of 138.45 hectares of forest land in favor of National Highway Authority of India (NHAI) for its widening to six lanes of Delhi Jaipur road from km 42.70 to 107.10 affecting the forest land falling in the districts of Gurugram and Rewari, an amount of Rs. 4,10,40,000 was received by the Haryana Forest Department in the year 2009[6]. This amount was to be spent on compensatory afforestation in Rewari division. While in Gurugram division (106.10 hectares) in Kasan, Balola and Samp ki Nangli villages, the amount received was Rs. 83,25,000 to be spend in afforestation in the same year. Thus, the total amount of Rs. 6,93,65,000 was said to be spent on afforestation in two divisions of Gurugram and Rewari. But surprisingly, no afforestation is visible at site. This is only one example of damage being done to the forests and wildlife in the Aravallis. There are other numerous examples where permissions for forest clearance have been given from time to time which speaks volume of damages of the forests and wildlife in the Aravallis.

Apart from overall degradation of the Aravallis, the indigenous species of flora is also vanishing very fast. Vilayati Kikar (*Prosopis Juliflora*) is the major invasive species responsible for the increasing wipe out of local species. However, some efforts have been made by some experts to overpower the growth of invasive species especially the *Prosopis Juliflora* in the Aravalli Biodiversity Park and Yamuna Biodiversity Park in NCT Delhi by planting the native species. This effort is also making inroads in other parts of the Delhi ridge. But the spread of *Prosopis Juliflora* in Aravalli ranges in the Haryana districts is rampant and any efforts made by local communities towards this end are discouraged by the Forest Department. Though exact estimates are not available, yet a tentative coverage of *Prosopis Juliflora* is estimated to be 75% of its forest areas of Haryana.

The serious and major issue of wildlife in the Aravallis is of habitat loss. Wildlife habitats, waterholes, dense forest and food chain should be free from human interventions. However, the fact remains that the forest areas of Delhi ridge and other Aravalli ranges in Haryana and Rajasthan upto Alwar are bifurcated by numerous road network of national highways, state highways and other village roads. Asola wildlife Park and Mangar forest areas have been bifurcated by Gurugram-Faridabad highway and other roads leading to Surajkund and crusher zone from the major road. The prominent water bodies such as Badhkal lake have completely dried up. Damdama lake is also depleting very fast. The basic need of water and food of wildlife is threatened due to development of crisscrossing of road network. Similarly, other important wildlife habitats such as Manesar-Sohna, Ferojpur Jhirka, and Sariska in Alwar district of Rajasthan have been severely affected by laying of roads without having any regard to wildlife habitats.

Considering the existing scenario of road network, at least the flyovers or underpasses, as the situation warrants, should be proposed and developed where they bifurcate the wild habitats in the Aravallis. This will considerably improve their habitat conditions.

Environmentalists have suggested a wildlife 'Corridor Plan' right from the Delhi ridge to Sariska Wildlife Sanctuary enabling the wild animals free and safe movement. However, the state of Haryana, has not initiated any action in this interconnected habitat corridor.

Urgent Government Initiatives

That the severe damage has been inflicted to the forests and wildlife in the Aravallis has been reflected in various studies and apex court orders, the state governments, especially the Haryana government has not realized the future consequences of such a damage. The definition of forest, its survey and demarcation still cry for classification. There is an urgent need for 'Environment Management Plan' which has also been emphasized in the Supreme Court orders from time to time.

The proposed environment management plan for Aravalli ranges must be prepared by the Forest Research Institute of India and Wildlife Institute of India. The team of experts from both these institutions should include the forest experts from the state of NCT Delhi, government of Haryana and Rajasthan as well as the environmentalist from the local areas. The efforts to this end had been made by NCRPB in its regional plan 2021, where in the green belts have been provided along all the road network. Besides, the NCR Plan has also envisaged the major natural features, identified as environmentally sensitive areas in the form of *Natural Conservation Zones* in all the sub regions of NCR having presence of Aravalli ranges (NCRPB, 2005). These areas also include the major lakes and other waterbodies.

This proposed environment management plan must address the following intent and issues in detail:

1. Incorporating provisions enumerated in the National Capital Region Plan-2021 in the sub regional plans of the constituent states and implemented in letter and spirit. All areas of Aravallis should be brought under the purview of Gazette Notification of 1992 issued by Union Government of India.
2. A broader land use plan of respective sub-regions indicating the permissible and non-permissible land uses, including the non-forest activities.
3. Defining the forest or forest areas, sacred forests and Natural Conservation Zones (NCZ) clearly.
4. Clearly defining and demarcating the Aravalli ranges and their immediate environs and incorporating in their respective forest and wildlife acts and be part of policies of the three states.
5. Defining the wildlife habitat and their integrated corridor along with the restricted areas, eco-sensitive zones and ecotourism.

6. Earmarking existing and proposed water bodies such as lakes, seasonal streams and channels, ponds, earthen dams, baolis and check dams etc. and their conservation plans.
7. Identification of plants and animal species in the areas; the invasive and indigenous plant species with their positive and negative role in the environment.
8. Indicating and strengthening its organisational set up for planning and implementation.
9. Performance assessment and index worked out and need to fix responsibility of the institution/ department along with their minister in-charge officers and staff involved.
10. Tracking and curbing smuggling of wildlife items, including smuggling routes and reviewing of existing laws and enforcement network.
11. Involvement of local communities in preservation and protection of forest and wildlife, including watershed management.
12. Awareness generation pertaining to the limited and optimal utilisation of ecosystem services in local communities.

In the process of degradation of the natural environment of Aravallis at the hands of states of NCT Delhi, Haryana and Rajasthan, the local communities dependent on these hills for their livelihood and delivery of ecosystem services have suffered badly as they were never made part of the government policies and programs. Having no regards to the scarce resources of the forest, water and land in the Aravallis, this process is continuously proving a serious threat. In order to avoid this serious threat, it becomes very imperative to make local communities awaked about the significance of this age-old-ecosystem as well as to engage them in the process of planning & policy formulation and corresponding implementation. With respect to this, National Capital Region Planning Board should be given powers to act as a monitoring agency with essential legal backing for stakeholders' imbibing into Policy formulation and implementation as well as Court orders' compliance.

CONCLUSION AND FUTURE RESEARCH DIRECTIONS

Conservation and restoration of tropical dry forests is very essential concern and have been comparatively less considered/taken care of than tropical rain forests in the country. But this book is exclusively focusing on tropical dry forests and their conservation. Herein, this chapter becomes imperative as it focuses primarily on an important theme which usually remained hidden or under-expressed most of the time, such as legislation, policy and governance aspects of conservation and restoration of any of the forest and corresponding effectiveness on the same. The governing issues pertaining to the protection and conservation of forests such as institutional transparency, responsibility and accountability have not been properly addressed so far in the country.

In view of this, there is a need to make legislative provisions to include the local communities in the natural resource management. More so, when the entire Aravallis are a common land in the government revenue records, the local communities in the Aravallis have been keeping balance between the consumption and conservation of natural resources and conflict free relationship between the wildlife and their habitat.

Involving and engaging local communities in the forest and wildlife conservation is in tune with their custom and tradition. Their involvement would certainly thwart the chances of government favoring the powerful and the rich. The urgent government initiatives as suggested earlier will receive a great support from the local participation of the communities in the protection and conservation of the natural environment. With the clearly defined regional mechanism in place in the form of NCR and its sub-region, the suggestions have their force in implementation. This will be in line with the present slogan of the present union government, 'Minimum Government and Maximum Governance.' And this model example of implementation will boost other states with presence of Aravallis and follow suit.

This chapter further furnishes a major opportunity to the research community to work on the very important dimension of forest conservation and restoration i.e. Institutional Governance and its Effectiveness.

ACKNOWLEDGMENT

The authors are highly obliged to Mr. Subash Yadav, General Manager, Haryana Forest Development Corporation, Gurugram, India, for his valuable comments and sharing data as well as information. We would also like to thank Mr. Ajay Sudharsan L., student, Bachelors of Planning 7th Semester, for assistance with Location map of Aravallis in NCR. This research received no specific grant from any funding agency in the public, commercial, or not-for-profit sectors.

The information related to compensatory afforestation in Gurugram and Rewari divisions was collected through RTI by K.K. Yadav. The authors have contributed equally to the conceptualization and draft writing of the paper. Both of them have reviewed and K. Dhanwantri revised drafts as per reviewer's comments. Authors have conscientiously read and approved the final manuscript.

REFERENCES

Arora, S. (2018, Feb. 15). Another dubious mark: Haryana has lowest percentage of forest cover. The Times of India. Retrieved from https://timesofindia.indiatimes.com/city/gurgaon/another-dubious-mark-state-has-lowest-percentage-of-forest-cover/articleshow/62923551.cms

Banergee, A. (2018, March 16). Aravallis: A neglected Geological Marvel. *livemint.* Retrieved from https://www.livemint.com/Leisure/f2XeUnRMnc1mIlpVZDrZ0I/Aravallis-A-geological-marvel.html

Choudhary, V. (2018). *Sariska National Park-complete detail-updated.* Retrieved from http://natureconservation.in/sariska-national-park-complete-detail-updated/

<eref>Sinha, G. N. (Ed.). (2014). An Introduction to the Delhi Ridge. Department of Forest and Wild Life, Government of National Capital Territory of Delhi. New Delhi, India. Retrieved from http://delhi.gov.in/wps/wcm/connect/1e22578043fa6dcba014e23e3c4139c7/Delhi_Ridge_Book-Sinha,+G.N.+(Ed.)+(2014.pdf?MOD=AJPERES&lmod=-804673140&CACHEID=1e22578043fa6dcba014e23e3c4139c7

<eref>United Nations Educational, Scientific and Cultural Organization. (1992). The Rio Declaration on Environment and Development. Retrieved from http://www.unesco.org/education/pdf/RIO_E.PD

Forest Survey of India. (2017). *Forest Cover.* India State of Forest Report. 21 and 30. Retrieved from http://fsi.nic.in/isfr2017/isfr-forest-cover-2017.pdf

Govt. of Haryana. (2006). *Haryana Forest Policy-2006*. Forest Department. Government of Haryana. Retrieved from http://haryanaforest.gov.in/Portals/0/forest-policy-2006.pdf

International Institute for Sustainable Development. (2012). Earth Negotiations Bulletin. Retrieved from https://enb.iisd.org/vol27/enb2751e.html

Jha, B. (2015, Nov. 18). *Uzbek team to study Aravalli plants*. Times of India. Retrieved from https://timesofindia.indiatimes.com/city/gurgaon/Uzbek-team-to-study-Aravali-plants/articleshow/49823283.cms

Kohli, M. S. (2004). *Mountains of India: Tourism, Adventure, Pilgrimage*. Indus Publishing, 29. Retrieved from https://books.google.co.in/books?id=GIs4zv17HHwC&pg=PA29&redir_esc=y#v=onepage&q&f=false

Kumar, A. (2016, May 9). *Save Aravalli to save Wildlife.* The Hindu. Retrieved from https://www.thehindu.com/news/cities/Delhi/save-aravalli-to-save-wildlife/article8574467.ece

Ministry of Environment. Forest and Climate Change. (2017). *India among top ten nations in the world in terms of Forest Area*. Retrieved from http://pib.nic.in/PressReleaseIframePage.aspx?PRID=1520284

Ministry of Environment. Forest and Climate Change. (2018). *Draft National Forest Policy*. Forest Policy Division. India. Retrieved from http://www.moef.nic.in/sites/default/files/Draft%20National%20Forest%20Policy%2C%202018.pdf

Ministry of Environment and Forest. (1988). *National Forest Policy*. India. Retrieved from http://envfor.nic.in/legis/forest/forest1.html

Ministry of Environment and Forest. (2004). *Handbook of Forest (Conservation) Act, 1980 (With Amendments made in 1988) Forest (Conservation) Rules, 2003 (With Amendments made in 12004) Guidelines & Clarifications* (Up to June, 2004). 1-2. Retrieved from http://wrd.bih.nic.in/guidelines/awadhesh02c.pdf

National Capital Region Planning Board. (2005). *Regional Land Use*. Regional Plan 2021, (NCRPB). New Delhi, India. 255-259. Retrieved from http://ncrpb.nic.in/pdf_files/Draft%20Revised%20Regional%20Plan%202021/19%20Chapter%2017%20Regional%20Land%20Use.pdf

National Capital Region Planning Board. (2005). *Salient Features of the Regional Plan-2021 for NCR*. Regional Plan-2021, notified in 2005, New Delhi, India. Retrieved from http://ncrpb.nic.in/pdf_files/Salient%20Features_RP-2021.pdf

National Capital Region Planning Board. (2013). *Draft Revised Regional Plan 2021*. Ministry of Urban Development. Govt. of India. Retrieved from http://www.indiaenvironmentportal.org.in/files/file/Draft%20Rcviscd%20Regional%20plan.pdf

National Green Tribunal. (2015). *Haryali Welfare Society Vs. Union of India & Ors.* Retrieved from http://greentribunal.gov.in/Writereaddata/Downloads/269-2013(PB-II-Judg)OA20-7-2015.pdf

National Green Tribunal. (2018). *Item No. 06*. Retrieved from http://www.greentribunal.gov.in/Writereaddata/Downloads/147-2014(PB-I)OA7-8-18.pdf

Report, B. (1987). *Report of the World Commission on Environment and Development: Our Common Future*. Retrieved from http://mom.gov.af/Content/files/Bruntland_Report.pdf

Tatpati, M. (Ed.). (2015). Citizens' Report 2015: Community Forest Rights under the Forest Rights Act. Pune, Bhubaneshwar and New Delhi: Kalpavriksh and Vasundhara in collaboration with Oxfam India on behalf of Community Forest Rights Learning and Advocacy Process. Retrieved from http://www.kalpavriksh.org/images/LawsNPolicies/CITIZENSREPORT2015.pdf</eref>

United Nations Conference on Environment and Development. (1992). *Agenda 21*. Retrieved from https://sustainabledevelopment.un.org/outcomedocuments/agenda21 and https://sustainabledevelopment.un.org/content/documents/Agenda21.pdf

United Nations Environment Program. (1972). *Declaration of the United Nations Conference on the Human Environment. Chapter 11*, 2-4. Retrieved from https://www.soas.ac.uk/cedep-demos/000_P514_IEL_K3736-Demo/treaties/media/1972%20Stockholm%201972%20-%20Declaration%20of%20the%20United%20Nations%20Conference%20on%20the%20Human%20Environment%20-%20UNEP.pdf

Wildlife Institute of India. (2017). Mapping Landuse/Landcover Patterns in Aravallis Haryana with Reference to Status of Key Wildlife Species. Retrieved from https://41ngmc3nsigz3kwwbw1kp5ic-wpengine.netdna-ssl.com/wp-content/uploads/Aravallis-Report-WII-May-2017-compressed.pdf</eref>

World Wide Fund. (2018). *WWF Report Reveals Staggering Extent of Human Impact on Planet.* Washington D.C. Retrieved from https://www.worldwildlife.org/press-releases/wwf-report-reveals-staggering-extent-of-human-impact-on-planet

ENDNOTES

1. IUCN is a global authority established in 1948 and named as International Union for Conservation of Nature.
2. Club of Rome is an organization, concerned with the future of humanity and having a mission to promote understanding of the global challenges facing by humanity and to propose solutions through scientific analysis, communication and advocacy.
3. Sierra Club is an environmental organization in the United States formed in 1892.
4. Vilayati Kikar is the local name of *Prosopis Juliflora*, commonly used in Haryana, India.
5. National Green Tribunal (NGT) is a special court on environmental protection.
6. Information received through *Rigt to Information Act* (RTI)

APPENDIX

Table 4. Related court cases; pertaining decisions and observations

Sl. No.	Court Case	Year	Major Decisions	General Observation
1	Godaverman Case	1996	In addition to notified forests areas, areas recorded as forest in government records and areas that fulfills the dictionary meaning of a forest must be considered as 'Forest', irrespective of their ownership	Defining forest
2	Samtha Case	1997	Extended the meaning of a forest by including pastures, grassland and trees	Definition of Forest extended
3	MC Mehta Case	2002	Mining and drawing groundwater in Haryana Aravallis were banned within 5 kms of Delhi border	Ban on Mining and groundwater extraction
4	MC Mehta Case	2004	• Court said Aravallis should be protected at any cost • Haryana prevented from withdrawing its affidavit on areas identified as forests (Primarily areas notified under Punjab Land Preservation Act and Plantations in Aravallis)	Protection of Aravallis emphasized
5	MC Mehta Case	2009	• Mining Banned in Gurugram and Faridabad districts • Construction in Aravallis put on hold	Mining and construction put on hold
6	Lafarge Case	2011	• Directed state to complete the process of identifying forests of Aravallis, irrespective of whether they have notified or not, • Asked MoEFCC to make criteria for identification of forests	Identification of forest areas
7	MC Mehta Case	2018	• Demolition of all constructed buildings after 1992 at Kant Enclave, village Anagpur, Faridabad district by Dec. 2018. (this housing colony was created in an eco-sensitive area covered under Punjab Land Preservation Act) • Fined the company with Rs. 5 crores	Adherence of law in an eco-sensitive zone

Source: Conceptualized by Authors based on various sources

Compilation of References

(2004). <jrn>Singh, A. N., Raghubanshi, A. S., & Singh, J. S. (2004b). Survival and growth pattern of three tropical plantations raised on coal mine spols of central India. [<jrn>]. *Indian Forester*, (April): 376–383.

<edb>Dubey, K. (2018). Restoration of Sand Stone Mining Area of Vindhyan Region through Microbial Technology. In K. Dubey (Ed.), Project Report. Prayagraj, India: FRCER.</edb>

<edb>Geldenhuys, C. J., & Golding, J. (2008). Resource Use Activities, Conservation and Management of Natural Resources of African Savannas. In F. G. Faleiro & A. L. F. Neto (Eds), Savannas Desafios Estrategias para o equilibrium entre sociedade, agronegocio e recursos naturais. pp 225-262, Planatltina, Embrapa CERRADOS.

<edb>Pachauri, R. K., Allen, M. R., Barros, V. R., Broome, J., Cramer, W., Christ, R., ... Dubash, N. K. (2014). Climate change 2014: synthesis report. Contribution of Working Groups I. II and III to the fifth assessment report of the Intergovernmental Panel on Climate Change, 151.

<eref>Food and Agriculture Organization of the United Nations. (2000). FRA 2000: Global ecological zoning for the global forest resources assessment 2000: Final report. Rome, Italy: FAO. Retrieved from http://www.fao.org/docrep/006/ad652e/ad652e00.HTM

<eref>Sinha, G. N. (Ed.). (2014). An Introduction to the Delhi Ridge. Department of Forest and Wild Life, Government of National Capital Territory of Delhi. New Delhi, India. Retrieved from http://delhi.gov.in/wps/wcm/connect/1e22578043fa6dcba014e23e3c4139c7/Delhi_Ridge_Book-Sinha,+G.N.+(Ed.)+(2014.pdf?MOD=AJPERES&lmod=-804673140&CACHEID=1e22578043fa6dcba014e23e3c4139c7

<eref>United Nations Educational, Scientific and Cultural Organization. (1992). The Rio Declaration on Environment and Development. Retrieved from http://www.unesco.org/education/pdf/RIO_E.PD

<jrn>Bernoux, M., & Chevallier, T. (2014). Carbon in dryland soils - Multiple essential functions. Montpellier, France: CSFD/Agropolis International.

<ths>Onyekwelu, J. C. (2001). Growth characteristics and management scenarios for plantation-grown Gmelina arborea and Nauclea diderrichii in south-western Nigeria (Doctoral dissertation). Available from Hieronymus Publishers, Munich, Germany.

<unknown>Thomas, V., & López, R. (2015). Global increase in climate-related disasters. Asian Development Bank Economics Working Paper Series, (466).

Aber, J. D., Neilson, R. P., McNulty, S., Lenihan, J. M., Bachelet, D., & Drapek, R. J. (2001). Forest processes and global environmental change: Predicting the effects of individual and multiple stressors. *Bioscience*, *51*(9), 735–751. doi:10.1641/0006-3568(2001)051[0735:FPAGEC]2.0.CO;2

Adams, P. W., & Sidle, R. C. (1987). Soil conditions in three recent landslides in southeast Alaska. *Forest Ecology and Management*, *18*(2), 93–102. doi:10.1016/0378-1127(87)90136-8

Adedeji, O. H., Tope-Ajayi, O. O., & Abegunde, O. L. (2015). Assessing and Predicting Changes in the Status of Gambari Forest Reserve, Nigeria Using Remote Sensing and GIS Techniques. *Journal of Geographic Information System*, *7*(03), 301–318. doi:10.4236/jgis.2015.73024

Adekunle, V. A. J., Olagoke, A. O., & Akindele, S. O. (2013). Tree species diversity and structure of a Nigerian strict nature reserve. *Tropical Ecology*, *54*(3), 275–289.

Adeyemi, A. A., & Adesoye, P. O. (2016). Tree slenderness coefficient and percent canopy cover in Oban Group Forest, Nigeria. *Journal of Natural Sciences Research*, *6*(4), 9–17.

Adger, W. N., Agrawala, S., Mirza, M. M. Q., Conde, C., Brien, K. O., Pulhin, J., ... Takahashi, K. (2007). Assessment of adaptation practices, options, constraints and capacity [Impacts, Adaptation and Vulnerability. Contribution of Working Group II to the Fourth Assessment Report of the Intergovernmental Panel on Climate Change.]. *Climatic Change*.

Adhikari, S., & Southworth, J. (2012). Simulating forest cover changes of Bannerghatta National Park based on a CA-Markov model: A remote sensing approach. *Remote Sensing*, *4*(10), 3215–3243. doi:10.3390/rs4103215

Agarwala, M., DeFries, R. S., Qureshi, Q., & Jhala, Y. V. (2016). Factors associated with long-term species composition in dry tropical forests of Central India. *Environmental Research Letters*, *11*(10). doi:10.1088/1748-9326/11/10/105008

Agarwala, M., Fries, R. S., Qureshi, Q., & Jhala, Y. V. (2016). Changes in the dry tropical forests in central India with human use. *Regional Environmental Change*, *16*(1), 5–15. doi:10.100710113-015-0903-1

Agarwala, M., Ghoshal, S., Verchot, L., Martius, C., Ahuja, R., & DeFries, R. (2017). Impact of biogas interventions on forest biomass and regeneration in southern India. *Global Ecology and Conservation*, *11*, 213–223. doi:10.1016/j.gecco.2017.06.005

Agbelade, A. D., & Akindele, S. O. (2013). Land Use Mapping and Tree species Diversity of Federal University of Technology, Akure. *American International Journal of Contemporary Research*, *3*(2), 104–113.

Agrawal, A., Chhatre, A., & Hardin, R. (2008). *Changing Governance of the World's Forests*, *320*(5882), 1460–1462.

Aguilar, J., Mendoza, H., Medina, S., Correa, D., Nieto, J., Quintana, A., & Gonzalez, R. (2016). Caracterización florística del bosque seco tropical de Honda - Méndez (Tolima). In *Expediciones Humboldt. Honda - Méndez, Tolima* (pp. 17–25). Bogotá, D. C., Colombia: Instituto de Investigaciones de Recursos Biológicos Alexander von Humboldt.

Ahern, F. J., & Leckie, D. G. (1987). Digital remote sensing for forestry: Requirements and capabilities, today and tomorrow. *Geocarto International*, *3*(3), 43–52. doi:10.1080/10106048709354107

Ahrends, A., Burgess, N. D., Milledge, S. A. H., Bulling, M. T., Fisher, B., Smart, J. C. R., & Lewis, S. L. (2010). Predictable waves of sequential forest degradation and biodiversity loss spreading from an African city. *Proceedings of the National Academy of Sciences*, *107*(33), 14556–14561. doi:10.1073/pnas.0914471107

Aide, T. M., Clark, M. L., Grau, H. R., Lopez-Carr, D., Levy, M. A., Redo, D., ... Muniz, M. (2012). Defores.tation and reforestation of Latin America and the Caribbean (2001–2010). *Biotropica*, *45*(2), 262–271. doi:10.1111/j.1744-7429.2012.00908.x

Aide, T. M., Zimmerman, J. K., Pascarella, J. B., Rivera, L., & Marcano-Vega, H. (2000). Forest regeneration in a chronosequence of tropical abandoned pastures: Implications for restoration ecology. *Restoration Ecology*, *8*(4), 328–338. doi:10.1046/j.1526-100x.2000.80048.x

Ajayi, S., & Odey, P. O. (2012). Individual tree volume equation for a plantation of *Tectona Grandis* in the Cross River University of Technology (Crutech), Obubra, Cross River State, Nigeria. *Nigerian Journal of Agriculture, Food and Environment, 8*(2), 12–17.

Akçakaya, H. R., Ferson, S., Burgman, M. A., Keith, D. A., Mace, G. M., & Todd, C. R. (2000). Making consistent IUCN classifications under uncertainty. *Conservation Biology, 14*(4), 1001–1013. doi:10.1046/j.1523-1739.2000.99125.x

Akindele, S. O. (2001). Review of urgent management requirement for Nigerian forestry. In L. Popoola, J. E. Abu, & P. I. Oni (Eds.), *Forestry and National Development: Proceedings of the 27th Annual Conference of the Forestry Association of Nigeria* (pp 75-84). Abuja, Nigeria: Forestry Association of Nigeria.

Akpodiogaga-a, P., & Odjugo, O. (2010). General overview of climate change impacts in Nigeria. *Journal of Human Ecology (Delhi, India), 29*(1), 47–55. doi:10.1080/09709274.2010.11906248

Akpor, B. O., Okoh, A. I., & Babalola, G. O. (2005). Culturable microbial population during decomposition of *Cola nitida* leaf litters in a tropical soil setting. *Akdeniz Üniversitesi Ziraat Fakültesi Dergisi, 18*, 313–319.

Alanís-Rodríguez, E., Aguirre-Calderón, O., Jiménez-Pérez, J., Pando-Moreno, M., Treviño-Garza, E. J., Aranda-Ramos, R., & Canizales-Velázquez, P. A. (2010). Efecto de la severidad del fuego sobre la regeneración asexual de especies leñosas de un ecosistema mixto (Pinus-Quercus) en el Parque Ecológico Chipinque, México. *Interciencia, 35*, 690–695.

Alanís-Rodríguez, E., Jiménez-Pérez, J., Valdecantos-Dema, A., González-Tagle, M. A., Aguirre-Calderón, Ó. A., & Treviño-Garza, E. J. (2012). Composición y diversidad de la regeneración natural en comunidades de Pinus- Quercus sometidas a una alta recurrencia de incendios en el noreste de México. *Revista Mexicana de Biodiversidad, 83*(4), 1208–1214. doi:10.7550/rmb.29708

Albert, C. H., Thuiller, W., Yoccoz, N. G., Douzet, R., Aubert, S., & Lavorel, S. (2010). A multi-trait approach reveals the structure and the relative importance of intra-vs. interspecific variability in plant traits. *Functional Ecology, 24*(6), 1192–1201. doi:10.1111/j.1365-2435.2010.01727.x

Ali, I. N., Brian, B., Richard, G. H., & Peter, C. (2014). Changing Vegetation Patterns in Yobe State Nigeria: An Analysis of the Rates of Change, Potential Causes and the Implications for Sustainable Resource Management. *International Journal of Geosciences, 5*(01), 50–62. doi:10.4236/ijg.2014.51007

Allen, K., Dupuy, J. M., Gei, M. G., Hulshof, C. M., Medvigy, D., Pizano, C., … Powers, J. S. (2017). Will seasonally dry tropical forests be sensitive or resistant to future changes in rainfall regimes? *Environmental Research Letters, 12*(2), 0–19.

Allen, C. D., Macalady, A. K., Chenchouni, H., Bachelet, D., McDowell, N., Vennetier, M., ... Cobb, N. (2010). A global overview of drought and heat induced tree mortality reveals emerging climate change risks for forests. *Forest Ecology and Management, 259*(4), 660–684. doi:10.1016/j.foreco.2009.09.001

Allen, K., Dupuy, J. M., Gei, M., Hulshof, C., Medvigy, D., Pizano, C., ... Powers, J. S. (2017). Will seasonally dry tropical forests be sensitive or resistant to future changes in rainfall regimes? *Environmental Research Letters, 12*(2). doi:10.1088/1748-9326/aa5968

Allen, W. (2000). Restoring Hawaii's Dry Forests: Research on Kona slope shows promise for native ecosystem recovery. *Bioscience, 50*(12), 1037–1041. doi:10.1641/0006-3568(2000)050[1037:RHSDF]2.0.CO;2

Alley, R. B., Marotzke, J., Nordhaus, W. D., Overpeck, J. T., Peteet, D. M., Pielke, R. A., ... Wallace, J. M. (2003). Abrupt climate change. *Science, 299*(5615), 2005–2010. doi:10.1126cience.1081056 PMID:12663908

Alvarado, A., Fernández-Moya, J., Segura, J. M., Vaides, E. E., Camacho, M., Avellán, M. J., & Ávila, C. E. (2014). Nutrient Management Improvements in Forestry Species. *Better Crops with Plant Food, 98*, 16–18.

Álvarez-Yépiz, J. C., Martínez-Yrízar, A., & Fredericksen, T. S. (2018). Resilience of tropical dry forests to extreme disturbance events.

Álvarez-Yépiz, J. C., Martínez-Yrízar, A., Búrquez, A., & Lindquist, C. (2008). Variation in vegetation structure and soil properties related to land use history of old-growth and secondary tropical dry forests in northwestern Mexico. *Forest Ecology and Management*, *256*(3), 355–366. doi:10.1016/j.foreco.2008.04.049

Amahowe, I. O., Gaoue, O. G., Natta, A. K., Piponiot, C., Zobi, I. C., & Hérault, B. (2018). Functional traits partially mediate the effects of chronic anthropogenic disturbance on the growth of a tropical tree. *AoB Plants*, *10*(3), ply036. doi:10.1093/aobpla/ply036 PMID:29991997

Amat, J. (2016). Estadística con R. RPubs. Retrieved from https://github.com/JoaquinAmatRodrigo/Estadistica-con-R

Amissah, L., Mohren, G. M., Kyereh, B., Agyeman, V. K., & Poorter, L. (2018). Rainfall seasonality and drought performance shape the distribution of tropical tree species in Ghana. *Ecology and Evolution*, *8*(16), 8582–8597. doi:10.1002/ece3.4384 PMID:30250725

Anderegg, W. R. L., Schwalm, C., Biondi, F., Camarero, J., Koch, G., Litvak, M., ... Pacala, S. (2015). Pervasive drought legacies in forest ecosystems and their implications for carbon cycle models. *Science*, *349*(6247), 528–532. doi:10.1126cience.aab1833 PMID:26228147

Anderegg, W. R., Hicke, J. A., Fisher, R. A., Allen, C. D., Aukema, J., Bentz, B., ... Pan, Y. (2015). Tree mortality from drought, insects, and their interactions in a changing climate. *The New Phytologist*, *208*(3), 674–683. doi:10.1111/nph.13477 PMID:26058406

Anderson, D. R., Burnham, K. P., Lubow, B. C., Thomas, L., Corn, P. S., Medica, P. A., & Marlow, R. W. (2001). Field trials of Line Transect Methods Applied to Estimation of Desert Tortoise Abundance. *The Journal of Wildlife Management*, *65*(3), 583–597. doi:10.2307/3803111

Andrade-Lima, D. (1981). The caatingas dominium. *Revista Brasileira de Botanica. Brazilian Journal of Botany*, *4*, 149–163.

Angeler, D., & Allen, C. (2016). Quantifying resilience. *Journal of Applied Ecology*, *53*(3), 617–624. doi:10.1111/1365-2664.12649

Angert, A. L., Huxman, T. E., Chesson, P., & Venable, D. L. (2009). Functional tradeoffs determine species coexistence via the storage effect. *Proceedings of the National Academy of Sciences of the United States of America*, *106*(28), 11641–11645. doi:10.1073/pnas.0904512106 PMID:19571002

Annual Administrative Report. (2017-2018). Department of Forest, West Bengal.

Anonyme. (2012a). Qualitative study on the causes of deforestation and forest degradation in the Democratic Republic of Congo Final Report REDD+ Task Force. 149p

Anonyme. (2015d). Project idea note "Conservation of the Ouro-doukoudje massive and reforestation of the western banks of Lake Lagdo" Sept. 7, 2015. 24 p

Anonymous. (2007). RPCB Report 2007. Retrieved from www.rpcb.nic.in

Anonymous. (2011). Current Human Impacts. Tropical Dry Forests of Mexico. Retrieved from https://tropicaldryforestsofmexico.blogspot.com/

Anonymous. (2011a). *Manda National Park Development Plan 2011- 2021 Provisional version November 2010*. National Parks, Wildlife Reserves and Hunting Department.

Anyamba, A., & Tucker, C. J. (2005). Analysis of Sahelian Vegetation Dynamics Using Noaa- AVHRR NDVI Data from 1981-2003. *Journal of Arid Environments, 63*(3), 596–614. doi:10.1016/j.jaridenv.2005.03.007

Aponte, C., Garcia, L. V., & Maranon, T. (2012). *Tree Species Effect on Litter Decomposition and Nutrient Release in Mediterranean Oak Forests Changes Over Time.* Ecosystem. doi:10.100710021-012-9577-4

Aponte, C., Garcıa, L. V., Pe'rez-Ramos, I. M., Gutierrez, E., & Maranon, T. (2011). Oak trees and soil interactions in Mediterranean forests: A positive feedback model. *Journal of Vegetable Science, 22*(5), 856–867. doi:10.1111/j.1654-1103.2011.01298.x

Appropedia. (2012). Amazon deforestation, climate change, and possible mitigation. Retrieved from http://www.appropedia.org/Amazon_deforestation,_climate_change,_and_possible_mitigation

Araújo, M. B., & Rahbek, C. (2006). How does climate change affect biodiversity? *Science, 313*(5792), 1396–1397. doi:10.1126cience.1131758 PMID:16959994

Arenas, J., & Noy, A. (2019). *Análisis de la severidad de un incendio en un relicto de bosque seco tropical en Colombia.* Universidad Distrital Francisco José De Caldas.

Armenteras, D., Bernal, F., Gonzalez, F., Morales, M., Pabón, J., Páramo, G., & Parra-Lara, Á. del C. (2011). *Incendios de la cobertura vegetal en Colombia* (Á. del C. Parra-Lara, Ed.).

Armenteras, D., González-Alonso, F., & Aguilera, C. F. (2009). Geographic and temporal distribution of fire in Colombia using thermal anomalies data. *Caldasia, 31*(2), 303–318.

Arora, S. (2018, Feb. 15). Another dubious mark: Haryana has lowest percentage of forest cover. The Times of India. Retrieved from https://timesofindia.indiatimes.com/city/gurgaon/another-dubious-mark-state-has-lowest-percentage-of-forest-cover/articleshow/62923551.cms

Arouna, O., & Djogbenou, C. P. (2006). Evaluation du plan d'aménagement participatif des forets classées de goun-goun, de la Sota et de la rôneraie de Goroubi au Benin : critères et indicateurs pertinents de réussite. In the *Colloque International sur la Gestion Concertée des Ressources Naturelles et de l'Environnement*, Cotonou, Benin.

Asase, A., Wade, S. A., Ofori-Frimpong, K., Hadley, P., & Norris, K. (2008). Carbon storage and the health of cocoa agroforestry ecosystems in southeastern Ghana. *FAO Journal World Soil Resources Report 104 "Africa and the Carbon Cycle"*. pp. 131-144.

Ashton, E. C., Hogarth, P. J., & Ormond, R. (1999). Breakdown of mangrove leaf litter in a managed mangrove forest in Peninsular Malaysia. In *Diversity and function in mangrove ecosystems* (pp. 77–88). Dordrecht, The Netherlands: Springer. doi:10.1007/978-94-011-4078-2_8

Asner, G. P., Rudel, T. K., Aide, T. M., Defries, R., & Emerson, R. (2018). *A Contemporary Assessment of Change in Humid Tropical Forests. 23*(6), 1386–1395.

Aubin, I., Munson, A. D., Cardou, F., Burton, P. J., Isabel, N., Pedlar, J. H., ... Messier, C. (2016). Traits to stay, traits to move: A review of functional traits to assess sensitivity and adaptive capacity of temperate and boreal trees to climate change. *Environmental Reviews, 24*(2), 164–186. doi:10.1139/er-2015-0072

Augeri, D. M. (2005). On the Biogeographic Ecology of the Malaysian Sun Bear. (Doctoral thesis), School of Biological Sciences, University of Cambridge, UK.

Augeri, D. M. (n.d.). In Prep. Malayan Sun Bear (*Helarctos malayanus*) Occupancy and Distribution Across the Landscape Matrix in Indonesia. *Ursus. International Association of Bear Research and Management,* Missoula, MT.

Augusto, L., Ranger, J., Binkley, D., & Rothe, A. (2002). Impact of several common tree species of European temperate forests on soil fertility. *Annals of Forest Science*, *59*(3), 233–253. doi:10.1051/forest:2002020

Austin, A. T., & Vivanco, L. (2006). Plant litter decomposition in a semi-arid ecosystem controlled by photo degradation. *Nature*, *442*(7102), 555–558. doi:10.1038/nature05038 PMID:16885982

Axel, A. C. (2018). Burned Area Mapping of an Escaped Fire into Tropical Dry Forest in Western Madagascar Using Multi-Season Landsat OLI Data. *Remote Sensing*, *10*(3), 371. doi:10.3390/rs10030371

Ayres, E., Steltzer, H., Simmons, B. L., Simpson, R. T., Steinweg, J. M., Wallenstein, M. D., ... Wall, D. H. (2009). Home-field advantage accelerates leaf litter decomposition in forests. *Soil Biology & Biochemistry*, *41*(3), 606–610. doi:10.1016/j.soilbio.2008.12.022

Bailey, R. (2018). Land Biomes: The World's Major Habitats. Retrieved from https://www.thoughtco.com/land-biomes-373501

Baithalu, M. S., Anbarashan, M., & Parthasarathy, N. (2013). Two-decadal changes in forest structure and tree diversity in a tropical dry evergreen forest on the Coromandel Coast of India. *Tropical Ecology*, *54*(3), 397–403.

Bajracharya, S. B., Mool, P. K., & Shrestha, B. R. (2007). *Impact of Climate Change on Himalaya Glaciers and Glacial Lakes: Case Studies on GLOF and Associated Hazards in Nepal and Bhutan*. Kathmandu, Nepal: ICIMOD.

Baker, P. J., Bunyavejchewin, S., Oliver, C. D., & Ashton, P. S. (2005). Disturbance history and historical stand dynamics of a seasonal tropical forest in western Thailand. *Ecological Monographs*, *75*(3), 317–343. doi:10.1890/04-0488

Baker, T. R., Burslem, D. F., & Swaine, M. D. (2003). Associations between tree growth, soil fertility and water availability at local and regional scales in Ghanaian tropical rain forest. *Journal of Tropical Ecology*, *19*(2), 109–125. doi:10.1017/S0266467403003146

Bala, G., Caldeira, K., Wickett, M., Phillips, T. J., Lobell, D. B., Delire, C., & Mirin, A. (2007). Combined climate and carbon-cycle effects of large-scale deforestation. *Proceedings of the National Academy of Sciences of the United States of America*, *104*(16), 6550–6555. doi:10.1073/pnas.0608998104 PMID:17420463

Balaguru, B., Britto, S. J., Nagamurugan, N., Natarajan, D., & Soosairaj, S. (2006). Identifying conservation priority zones for effective management of tropical forests in Eastern Ghats of India. *Biodiversity and Conservation*, *15*(4), 1529–1543. doi:10.100710531-004-6678-1

Baland, J. M., Bardhan, P., Das, S., & Mookherjee, D. (2010). Forests to the people: Decentralization and forest degradation in the Indian Himalayas. *World Development*, *38*(11), 1642–1656. doi:10.1016/j.worlddev.2010.03.007

Balvanera, P., Quijas, S., & Pérez-Jiménez, A. (2011). Distribution patterns of tropical dry forest trees along a mesoscale water availability gradient. *Biotropica*, *43*(4), 414–422. doi:10.1111/j.1744-7429.2010.00712.x

Banergee, A. (2018, March 16). Aravallis: A neglected Geological Marvel. *livemint*. Retrieved from https://www.livemint.com/Leisure/f2XeUnRMnc1mIlpVZDrZ0I/Aravallis-A-geological-marvel.html

Bannari, A., Morin, D., Bonn, F., & Huete, A. R. (1995). A review of vegetation indices'. *Remote Sensing Reviews*, *13*(1), 95–120. doi:10.1080/02757259509532298

Banoho, K. (2013). *Influences of human activities on the wood resource in the periphery of Waza National Park*. (Master's thesis in Biodiversity and Forest Productions), University of Yaoundé I. 89p

Baobab Tree Facts. (2015). 9 Fascinating baobab tree facts, Africa Geographic Editorial, Research, Travel & Lifestyle, Wildlife. *The Decoding Science*. Retrieved from https://africageographic.com/blog/9-fascinating-baobab-tree-facts/

Baral, M. (2009). Water Induced Disasters, Flood Hazard Mapping & Koshi Flood Disaster of Nepal. In *Report prepared for East*. Manila, Philippines: Southeast Asia Regional Seminar on Flood Hazard Mapping.

Baraloto, C., Timothy Paine, C. E., Patiño, S., Bonal, D., Hérault, B., & Chave, J. (2010). Functional trait variation and sampling strategies in species-rich plant communities. *Functional Ecology*, *24*(1), 208–216. doi:10.1111/j.1365-2435.2009.01600.x

Baret, F. (1986): Contribution au suivi radiometrique de cultures de cereals. These de Doctorat, Universite Paris-Sud Orsay, France, 182.

Bargali, S. S., Shukla, K., Singh, L., Ghosh, L., & Lakhera, M. L. (2015). Leaf litter decomposition and nutrient dynamics in four tree species of dry deciduous forest. *Tropical Ecology*, *56*, 191–200.

Barlow, J., & Peres, C. A. (2004). Ecological responses to El Niño–induced surface fires in central Brazilian Amazonia: Management implications for flammable tropical forests. *Philosophical Transactions of the Royal Society of London. Series B, Biological Sciences*, *359*(1443), 367–380. doi:10.1098/rstb.2003.1423 PMID:15212091

Barnes, B. V., Zak, D. R., Denton, S. R., & Spurr, S. H. (1998). *Forest Ecology* (4th ed.). New. York, NY: John Wiley and Sons.

Barradas, V., & Fanjul, L. (1985). Equilibrio hídrico y evapotranspiración en una selva baja caducifolia de la costa de Jalisco, México. *Biótica*, *10*, 199–218.

Barrett, C. B., Travis, A. J., & Dasgupta, P. (2011). On biodiversity conservation and poverty traps. *Proceedings of the National Academy of Sciences of the United States of America*, *108*(34), 13907–13912. doi:10.1073/pnas.1011521108 PMID:21873176

Bash, B. (2002). Tree of life: The world of the African Baobab. Layton, UT: Gibbs Smith.

Basu, A. K. (1986). Geology of parts of the Bundelkhand Granite massif, Central India. *Records of the Geological Survey of India*, *117*(II), 61–124.

Bates, B., Kundzewicz, Z. W., Wu, S., & Palutikof, J. (2008). *Climate change and Water: Technical paper VI. Intergovernmental Panel on Climate Change*. IPCC.

Bawa, K. S., & Dayanandan, S. (1997). Socioeconomic factors and tropical deforestation. *Nature*, *386*(6625), 562–563. doi:10.1038/386562a0

Becknell, J. M., Kucek, L. K., & Powers, J. S. (2012). Aboveground biomass in mature and secondary seasonally dry tropical forests: A literature review and global synthesis. *Forest Ecology and Management*, *276*, 88–95. doi:10.1016/j.foreco.2012.03.033

Bellhouse, D. R. (1988). Systematic Sampling. In P. R. Krishnaiah & C. R. Rao (Eds.), Handbook of Statistics. (6:125-145). Amsterdam, The Netherlands: Elsevier Science Publishers B. V. doi:10.1016/S0169-7161(88)06008-0

Benhammou, F., Baillon, J., Senotier, J.-L., Chalbert, J.-P., Lécrivain, E., Meuret, M., & Reynes, A. (2005). Pastoralisme et biodiversité. *La Voie Du Loup*, *22*, 10–23.

Berg, B., & McClaugherty, C. (2008). Plant Litter: Decomposition, Humus Formation, Carbon Sequestration. Berlin, Germany: Springer Science & Business Media. doi:10.1007/978-3-540-74923-3

Berg, B., Davey, M., De Marco, A., Emmett, B., Faituri, M., Hobbie, S., ... De Virzo Santo, A. (2010). Factors influencing limit values for pine needle litter decomposition: A synthesis for boreal and temperate pine forest systems. *Biogeochemistry*, *100*(1-3), 57–73. doi:10.100710533-009-9404-y

Bertrand, G., & Lagnaba, K. (2011). *Potentialities and constraints of rural development in the central, eastern and southern regions of Chad. Internal review on the rural sector in Chad.* French Development Agency-World Bank.

BfN. (2012). Payment for Ecosystem Services Towards an Implementation Strategy. Florian Carius (Ed).

Bhadouria, R., Singh, R., Srivastava, P., & Raghubanshi, A. S. (2016). Understanding the ecology of tree-seeding growth in dry tropical environment: A management perspective. *Energy. Ecology & Environment, 1*, 296–309.

Bhadouria, R., Singh, R., Srivastava, P., & Raghubanshi, A. S. (2016). Understanding the ecology of tree-seedling growth in dry tropical environment: A management perspective. *Energy. Ecology & Environment, 1*(5), 296–309.

Bhadouria, R., Singh, R., Srivastava, P., Tripathi, S., & Raghubanshi, A. S. (2017b). Interactive effect of water and nutrient on survival and growth of tree seedlings of four dry tropical tree species under grass competition. *Tropical Ecology, 58*(3), 611–621.

Bhadouria, R., Srivastava, P., Singh, R., Tripathi, S., Singh, H., & Raghubanshi, A. S. (2017a). Tree seedling establishment in dry tropics: An urgent need of interaction studies. *Environment Systems & Decisions, 37*(1), 88–100. doi:10.100710669-017-9625-x

Bhadouria, R., Srivastava, P., Singh, S., Singh, R., Raghubanshi, A. S., & Singh, J. S. (2018). Effects of light, nutrient and grass competition on growth of seedlings of four tropical tree species. *Indian Forester, 144*(1), 54–65.

Bhadra, B. K., Gupta, A. K., Sharma, J. R., & Choudhary, B. R. (2007). Mining activity and its impact on the environment: Study from Makrana marble and Jodhpur sandstone mining areas of Rajasthan. *Journal of the Geological Society of India, 70*, 557–570.

Bhagwat, S. A., Kushalappa, C. G., Williams, P. H., & Brown, N. D. (2005). A landscape approach to biodiversity conservation of sacred groves in the Western Ghats of India. *Conservation Biology, 19*(6), 1853–1862. doi:10.1111/j.1523-1739.2005.00248.x

Bhardwaj, S. D., & Pankaj, P. (2000). Impact of mining and rehabilitation of mined areas. *Advances in Forestry Research in India, 22*, 151–172.

Bierregaard, R. O., Jr., Laurance, W. F., Gascon, C., Benitez-Malvido, J., Fearnside, P. M., Fonseca, C. R., Ganade, G., … Williamson, B. (2001a). Principles of forest fragmentation and conservation in the Amazon. In: R. O. Bierregaard, Jr., C. Gascon, T. E. Lovejoy, & R. Mesquita (Eds.), Lessons from the Amazon: The Ecology and Conservation of a Fragmented Forest. New Haven, CT: Yale University Press.

Bierregaard, R. O. Jr, Gascon, C., Lovejoy, T. E., & Mesquita, R. (2001). *Lessons from the Amazon: The Ecology and Conservation of a Fragmented Forest.* New Haven, CT: Yale University Press.

Bijalwan, A., Swamy, S. L., Shanna, C. M., Sharma, N. K., & Tiwari, A. K. (2010). Land-use, biomass and carbon estimation in dry tropical forest of Chhattisgarh region in India using satellite remote sensing and GIS. [</jrn>]. *Journal of Forestry Research, 21*(2), 161–170.

Binkley, D. 1995. The influence of tree species on forest soils: processes and patterns. In D. J. Mead & I. S. Cornforth (Orgs.), *Proceedings of the trees and soil workshop*, Lincoln University, Agronomy society of New Zealand special publications number 10. Lincoln, New Zealand: Lincoln University Press. pp. 1-33.

Binkley, D., & Giardina, C. (1998). Why do tree species affect soils? The warp and woof of tree-soil interactions. *Biogeochemistry, 42*(1/2), 89–196. doi:10.1023/A:1005948126251

Birch, P., Bentham, H., & Harris, J. A. (1991). Soil microbial ecosystems: importance for the effective restoration of mined lands. In W. Oaks & J. Bosden (Orgs.), *Proc. 1991 Natl. Meet. ASSMR,* May 14-17, Durango, CO. pp. 631-640.

Bisht, V. K.; Nautiyal, B. P.; Kuniyal, C. P.; Prasad, P., & Sundriyal, R. C. (2014). Litter Production, Decomposition, and Nutrient Release in Subalpine Forest Communities of the Northwest Himalaya. *Journal of Ecosystems*. doi:. doi:10.1155/2014/294867

Biswas, S., Biswas, A., Sinha, S., & Khan, D. K. (2012). Major nutrient cycling of two different tropical dry deciduous forest of West Bengal, India. *Journal of Ecology and Environmental Sciences*, *3*, 77–81.

Blackie, R., Baldauf, C., Gautier, D., Gumbo, D., Kassa, H., Parthasarathy, N., . . . Sunderland, T. (2014). Tropical dry forests: The state of global knowledge and recommendations for future research. (Discussion paper). Bogor, Indonesia: CIFOR.</unknown>

Blackmore, M., & Vitousek, P. M. (2000). Cattle grazing, forest loss, and fuel loading in a dry forest ecosystem at Pu'u Wa'aWa'a Ranch, Hawai'i 1. *Biotropica*, *32*(44a), 625–632. doi:10.1646/0006-3606(2000)032[0625:CGFLAF]2.0.CO;2

Blake, S., & Hedges, S. (2004). Sinking the Flagship: The Case for Forest Elephants in Asia and Africa. *Conservation Biology*, *18*(5), 1191–1202. doi:10.1111/j.1523-1739.2004.01860.x

Blanfort, V. (2011). Impacts et services environnementaux de l'élevage en régions chaudes. In Elevage en régions chaudes. Inra prod.

Blasco, F., Whitmore, T. C., & Gers, C. (2000). A framework for the worldwide comparison of tropical woody vegetation types. *Biological Conservation*, *95*(2), 175–189. doi:10.1016/S0006-3207(00)00032-X

Boboye, O., & Jimoh, S. O. (2016). Pattern of plant species diversity in a dry forest ecosystem of Nigeria. *Journal of Forestry Research and Management*, *13*, 31–47.

Bognounou, F., Thiombiano, A., Oden, P. C., & Guinko, S. (2010). Seed provenance and latitudinal gradient effects on seed germination capacity and seedling establishment of five indigenous species in Burkina Faso. *Tropical Ecology*, *51*(2), 207.

Bognounou, F., Tigabu, M., Savadogo, P., Thiombiano, A., Boussim, I. J., Oden, P. C., & Guinko, S. (2010). Regeneration of five Combretaceae species along a latitudinal gradient in Sahelo-Sudanian zone of Burkina Faso. *Annals of Forest Science*, *67*(3), 306. doi:10.1051/forest/2009119

Bohlman, S. A. (2010). Landscape patterns and environmental controls of deciduousness in forests of central Panama. *Global Ecology and Biogeography*, *19*(3), 376–385. doi:10.1111/j.1466-8238.2009.00518.x

Boisvenue, C., & Running, S. W. (2006). Impacts of climate change on natural forest productivity—Evidence since the middle of the 20th century. *Global Change Biology*, *12*(5), 862–882. doi:10.1111/j.1365-2486.2006.01134.x

Bond-Lamberty, B., & Thomson, A. M. (2010). Temperature-associated increases in the global soil respiration record. *Nature*, *464*(7288), 579–582. doi:10.1038/nature08930 PMID:20336143

Bond, W. J., Woodward, F. I., & Midgley, G. F. (2005). The global distribution of ecosystems in a world without fire. *The New Phytologist*, *165*(2), 525–538. doi:10.1111/j.1469-8137.2004.01252.x PMID:15720663

Bonfiglioli, A. M. (1990). Pastoralisme, agro-pastoralisme et retour: Itinéraires sahéliens. *Cahier Des Sciences Humaines*, *26*(1–2), 255–266.

Bonjour, S., Adair-Rohani, H., Wolf, J., Bruce, N. G., Mehta, S., Prüss-Ustün, A., ... Smith, K. R. (2013). Solid Fuel Use for Household Cooking: Country and Regional Estimates for 1980–2010. *Environmental Health Perspectives*, *121*(7), 784–790. doi:10.1289/ehp.1205987 PMID:23674502

Borchert, R. (1994). Soil and stem water storage determine phenology and distribution of tropical dry forest trees. *Ecology, 75*(5), 1437–1449. doi:10.2307/1937467

Borchert, R. (1999). Climatic periodicity, phenology, and cambium activity in tropical dry forest trees. *IAWA Journal, 20*(3), 239–247. doi:10.1163/22941932-90000687

Both, S., Riutta, T., Paine, C. T., Elias, D. M., Cruz, R. S., Jain, A., ... Mielke, N. (2019). Logging and soil nutrients independently explain plant trait expression in tropical forests. *The New Phytologist, 221*(4), 1853–1865. doi:10.1111/nph.15444 PMID:30238458

Bouchet, P. H., Jaffré, T., & Veillon, J. M. (1995). Plant extinction in New Caledonia: Protection of sclerophyll forests urgently needed. *Biodiversity and Conservation, 4*(4), 415–428. doi:10.1007/BF00058425

Boulmane, M., Santa Regina, I., Khia, A., & Oubrahim, H. (2012). Estimation of the organic carbon stock in iliasis of the Moroccan Middle Atlas. 24pp.

Bradford, M. A., Berg, B., Maynard, D. S., Wieder, W. R., & Wood, S. A. (2016). Understanding the dominant controls on litter decomposition. *Journal of Ecology, 104*(1), 229–238. doi:10.1111/1365-2745.12507

Bradshaw, C. J. A., Sodhi, N. S., & Brook, B. W. (2009). Tropical turmoil: A biodiversity tragedy in progress. *Frontiers in Ecology and the Environment, 7*(2), 79–87. doi:10.1890/070193

Brady, N. C., & Weil, R. R. (2002). *The nature and properties of soils* (13th ed.). Upper Saddle River, NJ: Prentice Hall.

Brechet, L. M., Lopez-Sangil, L., George, C., Birkett, A. J., Baxendale, C., Trujillo, B. C., & Sayer, E. J. (2018). Distinct responses of soil respiration to experimental litter manipulation in temperate woodland and tropical forest. *Ecology and Evolution, 8*(7), 3787–3796. doi:10.1002/ece3.3945 PMID:29686858

Brenes-Arguedas, T., Coley, P. D., & Kursar, T. A. (2009). Pests vs. drought as determinants of plant distribution along a tropical rainfall gradient. *Ecology, 90*(7), 1751–1761. doi:10.1890/08-1271.1 PMID:19694125

Brodie, J., Post, E., & Doak, D. (Eds.). (2012). *(a). Climate change and wildlife conservation.* Chicago, IL: University Chicago Press. doi:10.7208/chicago/9780226074641.001.0001

Brodie, J., Post, E., & Laurance, W. F. (2012). (b). Climate change and tropical biodiversity: A new focus. *Trends in Ecology & Evolution, 27*(3), 145–150. doi:10.1016/j.tree.2011.09.008 PMID:21975172

Brodribb, T. J., & Holbrook, N. M. (2003a). Changes in leaf hydraulic conductance during leaf shedding in seasonally dry tropical forest. *The New Phytologist, 158*(2), 295–303. doi:10.1046/j.1469-8137.2003.00736.x

Brodribb, T. J., & Holbrook, N. M. (2003b). Stomatal closure during leaf dehydration, correlation with other leaf physiological traits. *Plant Physiology, 132*(4), 2166–2173. doi:10.1104/pp.103.023879 PMID:12913171

Brodribb, T. J., Holbrook, N. M., Edwards, E. J., & Gutierrez, M. V. (2003). Relations between stomatal closure, leaf turgor and xylem vulnerability in eight tropical dry forest trees. *Plant, Cell & Environment, 26*(3), 443–450. doi:10.1046/j.1365-3040.2003.00975.x

Brooks, N., Bhojvaid, V., Jeuland, M. A., Lewis, J. J., Patange, O., & Pattanayak, S. K. (2016). How much do alternative cookstoves reduce biomass fuel use? Evidence from North India. *Resource and Energy Economics, 43*, 153–171. doi:10.1016/j.reseneeco.2015.12.001

Brooks, T. M., Pimm, S. L., & Collar, N. J. (1997). Deforestation predicts the number of threatened birds in insular Southeast Asia. *Conservation Biology, 11*(2), 382–394. doi:10.1046/j.1523-1739.1997.95493.x

Brown, T. (1999). Tom Brown's Science and Art of Tracking. New York, NY: Berkley Trade.

Brown, T. H., Simangunsong, B. C. H., Sukadri, D., Brown, D. W., Subarudi, S., Dermawan, A., & Rufi'ie, A. (2005). Restructuring and Revitalization of Indonesia's Wood-based Industry: Synthesis of Three Major Studies. Ministry of Forestry, CIFOR, and DFID-MFP. Jakarta.

Brown, S., & Lugo, A. E. (1982). The storage and production of organic matter in tropical forests and their role in the global carbon cycle. *Biotropica*, *14*(3), 161–187. doi:10.2307/2388024

Bruegmann, M. M. (1996). Hawaii's dry forests. *Endangered Species Bulletin*, *11*, 26–27.

Bruelheide, H., Dengler, J., Purschke, O., Lenoir, J., Jiménez-Alfaro, B., Hennekens, S. M., ... & Kattge, J. (2018). Global trait–environment relationships of plant communities. *Nature Ecology & Evolution*, *2*(12), 1906.

Buckland, S. T., Anderson, D. R., Burnham, K. P., Laake, J. L., Borchers, D. L., & Thomas, L. (2001). Introduction to Distance Sampling: Estimating Abundance of Biological Populations. Oxford, UK: Oxford University Press.

Buckland, S. T., Anderson, D. R., Burnham, K. P., & Laake, J. L. (1993). *Distance Sampling: Estimating Abundance of Biological Populations*. London, UK: Chapman and Hall.

Buckley, M. (2013). A molecular phylogeny of Plesiorycteropus reassigns the extinct mammalian order 'Bibymalagasia'. *PLoS One*, *8*(3). doi:10.1371/journal.pone.0059614 PMID:23555726

Bullock, S. H., Mooney, H. A., & Medina, E. (Eds.). (1995). *Seasonally dry tropical forests*. Cambridge, UK: Cambridge University Press. doi:10.1017/CBO9780511753398

Bunyavejchewin, S. (1983). Analysis of the tropical dry deciduous forest of Thailand. I. Characteristics of the dominance-types. *Natural History Bulletin of the Siam Society*, *31*, 109–122.

Burapapol, K., & Nagasawa, R. (2016). Mapping wildfire fuel load distribution using Landsat 8 Operational Land Imager (OLI) data in Sri Lanna National Park, northern Thailand. *Journal of the Japanese Agricultural Systems Society*, *32*(4), 133–145.

Burney, D. A., DeCandido, R. V., Burney, L. P., Kostel-Hughes, F. N., Stafford, T. W. Jr, & James, H. F. (1995). A Holocene record of climate change, fire ecology and human activity from montane Flat Top Bog, Maui. *Journal of Paleolimnology*, *13*(3), 209–217. doi:10.1007/BF00682765

Burney, D. A., Vasey, N., & Godfrey, L. R., Jungers, W. L., Ramarolahy, M., & Raharivony, L. (2008). New findings at Andrahomana Cave, southeastern Madagascar. *Journal of Caves and Karst Studies*, *70*(1), 13–24.

Burney, L. P., & Burney, D. A. (2003). Charcoal stratigraphies for Kaua'i and the timing of human arrival. *Pacific Science*, *57*(2), 211–226. doi:10.1353/psc.2003.0011

Burnham, K. P., & Anderson, D. R. (2002). *Model Selection and Multimodel Inference: A Practical Information-Theoretic Approach* (2nd ed.). New York, NY: Springer-Verlag.

Buytaert, W., Cuesta-Camacho, F., & Tobón, C. (2011). Potential impacts of climate change on the environmental services of humid tropical alpine regions. *Global Ecology and Biogeography*, *20*(1), 19–33. doi:10.1111/j.1466-8238.2010.00585.x

Buzzard, V., Hulshof, C. M., Birt, T., Violle, C., & Enquist, B. J. (2016). Re-growing a tropical dry forest: Functional plant trait composition and community assembly during succession. *Functional Ecology*, *30*(6), 1006–1013. doi:10.1111/1365-2435.12579

Cabin, R. J., Weller, S. G., Lorence, D. H., Flynn, T. W., Sakai, A. K., Sandquist, D., & Hadway, L. J. (2000). Effects of long-term ungulate exclusion and recent alien species control on the preservation and restoration of a Hawaiian tropical dry forest. *Conservation Biology*, *14*(2), 439–453. doi:10.1046/j.1523-1739.2000.99006.x

Cabral, P., & Zamyatin, A. (2009). Markov processes in modeling land use and land cover changes in Sintra-Cascais, Portugal. *Dyna (Bilbao)*, *158*, 191–198.

Cadotte, M., Albert, C. H., & Walker, S. C. (2013). The ecology of differences: Assessing community assembly with trait and evolutionary distances. *Ecology Letters*, *16*(10), 1234–1244. doi:10.1111/ele.12161 PMID:23910526

Cain, M., Bowman, W., & Hacker, S. (2014). The biosphere. Ecology (3rd ed.). Sunderland, MA: Sinauer, 51.

Calama, R., Manso, R., Lucas-Borja, M. E., Espelta, J. M., Piqué, M., Bravo, F., ... Pardos, M. (2017). Natural regeneration in Iberian pines: A review of dynamic processes and proposals for management. *Forest Systems*, *26*(2). doi:10.5424/fs/2017262-11255

Campbell, J. B. (1987). *Introduction to Remote Sensing* (p. 551). New York, NY: The Guilford Press.

Campbell, N. A. (1996). *Biology* (4th ed.). Menlo Park, CA: The Benjamin/Cummings Publishing Company

Campo, J., & Merino, A. (2016). Variations in soil carbon sequestration and their determinants along a precipitation gradient in seasonally dry tropical forests. *Global Change Biology*, *22*(5), 1942–1956. doi:10.1111/gcb.13244 PMID:26913708

Cantillo-H, E., & Rangel–CH, J. O. (2002). Caracterización estructural, ordenación y dinámica de la vegetación en la zona de captación de aguas de la microcuenca El Tigre - municipio de Norcasia, Caldas. *Colombia Forestal*, *52*(1), 1–5.

Cárdenas-Arévalo, G., & Vargas-Ríos, O. (2008). Rasgos de historia de vida de especies en una comunidad vegetal alterada en un páramo húmedo (Parque Nacional Natural Chingaza). *Caldasia*, *30*(2), 245–264.

Cardinale, B. J., Duffy, J. E., Gonzalez, A., Hooper, D. U., Perrings, C., Venail, P., ... Kinzig, A. P. (2012). Biodiversity loss and its impact on humanity. *Nature*, *486*(7401), 59–67. doi:10.1038/nature11148 PMID:22678280

Carón, M. M., Dalmasso, A. D., Ortín, A. E., & Verheyen, K. (2015). Regeneración post fuego en un bosque tropical seco del Monte en el noroeste de Argentina. *Multequina (Mendoza)*, *24*, 5–17.

Carrera, A. L., Bertilier, M. B., & Larreguy, C. (2008). Leaf litter fall, fine root production, and decomposition in shrub lands with different canopy structure induced by grazing in the Pantagonian Monte, Argentina. *Plant and Soil*, *311*(1-2), 39–50. doi:10.100711104-008-9655-8

Carriere, M., & Toutain, B. (1995). Utilisation des terres de parcours par l'élevage et interactions avec l'environnement. Montpellier, France: CIRAD.

Carriere, M. (1996). *Impact des systemes d'elevage pastoraux sur l'environnement en afrique et en asie tropicale et sub-tropical aride et sub aride*. CIRAD.

Carrión, H. X., & Southworth, J. (2018). Understanding Land Cover Change in a Fragmented Forest Landscape in a Biodiversity Hotspot of Coastal Ecuador. *Remote Sensing*.

Casanova, J. L., Calle, A., Romo, A., & Sanz, J. A. (2015): Forest fire detection and monitoring by means of an integrated MODIS-MSG system. 409–414.

Casanoves, F., Pla, L., Di Rienzo, J. A., & Dıaz, S. (2011). F Diversity, a software package for the integrated analysis of functional diversity. *Methods in Ecology and Evolution*, *2*(3), 233–237. doi:10.1111/j.2041-210X.2010.00082.x

Castillo, A., Magana, A., Pujadas, A., Martinez, L., & Godı, C. (2005). *Understanding the Interaction of Rural People with Ecosystems: A Case Study in a Tropical Dry Forest of Mexico. 8*, 630–643. doi:10.100710021-005-0127-1

CBD. (2009). Connecting Biodiversity and Climate Change Mitigation and Adaptation: Report of the Second Ad Hoc Technical Expert Group on Biodiversity and Climate Change. Technical Series No. 41, Secretariat of the Convention on Biological Diversity, Montreal, Canada, 126.

CBS. (2004). *Handbook of Environment Statistics 2003*. Kathmandu, Nepal: Central Bureau of Statistics.

Centritto, M., Loreto, F., Massacci, A., Pietrini, F., Villani, M. C., & Zacchini, M. (2000). Improved growth and water use efficiency of cherry saplings under reduced light intensity. *Ecological Research*, *15*(4), 385–392. doi:10.1046/j.1440-1703.2000.00359.x

Cernusak, L. A., Hutley, L. B., Beringer, J., Holtum, J. A., & Turner, B. L. (2011). Photosynthetic physiology of eucalypts along a sub-continental rainfall gradient in northern Australia. *Agricultural and Forest Meteorology*, *151*(11), 1462–1470. doi:10.1016/j.agrformet.2011.01.006

Chabot, B. F., & Hicks, D. J. (1982). The ecology of leaf life spans. *Annual Review of Ecology and Systematics*, *13*(1), 229–259. doi:10.1146/annurev.es.13.110182.001305

Chadwick, R., Good, P., Martin, G., & Rowell, D. P. (2015). Large rainfall changes consistently projected over substantial areas of tropical land. *Nature Climate Change*, *6*(2), 177–181. doi:10.1038/nclimate2805

Champion, H. G. & Seth, S. K. (1968). *Revised survey of the forest types of India*. New Delhi, India: Manager of Publications.

Champion, S. H., & Seth, S. K. (1968). A revised survey of the forest types of India. *A revised survey of the forest types of India.*

Champion, H. G., & Seth, S. K. (1968). *A revised survey of forest types of India*. New Delhi, IN: Govt. of India Press.

Chang, C.-R. (1996). Ecosystem responses to fire and variations fire regimes. In Sierra Nevada Ecosystem Project: final report to Congress, vol. II, Assessments and scientific basis for management options (Vol. II, pp. 1071–1099). Davis, CA: University of California, Centers for Water and Wildland Resources.

Change, G. C. (2007). Impacts and Adaptation. *Nature Conservancy*, *434*, 951–0569.

Chan, K. M. A., Pringle, R. M., Ranganathan, J. A. I., Boggs, C. L., Chan, Y. L., Ehrlich, P. R., ... Macmynowski, D. P. (2007). When Agendas Collide: Human Welfare and Biological Conservation. *Conservation Biology*, *21*(1), 59–68. doi:10.1111/j.1523-1739.2006.00570.x PMID:17298511

Chao, A., Gotelli, N. J., Hsieh, T. C., Sander, E. L., Ma, K. H., Colwell, R. K., & Ellison, A. M. (2014). Rarefaction and extrapolation with Hill numbers: A framework for sampling and estimation in species diversity studies. *Ecological Monographs*, *84*(1), 45–67. doi:10.1890/13-0133.1

Chao, A., & Jost, L. (2012). Coverage-based rarefaction and extrapolation: Standardizing samples by completeness rather than size. *Ecology*, *93*(12), 2533–2547. doi:10.1890/11-1952.1 PMID:23431585

Chapin, F. S. III, Rupp, T. S., Starfield, A. M., DeWilde, L. O., Zavaleta, E. S., Fresco, N., ... McGuire, A. D. (2003). Planning for resilience: Modeling change in human–fire interactions in the Alaskan boreal forest. *Frontiers in Ecology and the Environment*, *1*(5), 255–261. doi:10.1890/1540-9295(2003)001[0255:PFRMCI]2.0.CO;2

Chapin, F. S. III, Zaveleta, E. S., Eviner, V. T., Naylor, R. L., Vitousek, P. M., Lavorel, S., ... Diaz, S. (2000). Consequences of changing biotic diversity. *Nature*, *405*(6783), 234–242. doi:10.1038/35012241 PMID:10821284

Chatterjee, M., & Bose, B. (2003). Marred by mines: Afforestation in nutrient deficient inhospitable lands. *Wastelands News*, *18*(3), 44–49.

Chaturvedi, R. K., & Singh, J. S. (2017). Restoration of mine spoil in a dry tropical region: a review. In Proceedings of Indian National Science Academy, 83(4), 789-844. doi:10.16943/ptinsa/2017/49123

Chaturvedi, R. K., Gopalakrishnan, R., Jayaraman, M., Bala, G., Joshi, N. V., Sukumar, R., & Ravindranath, N. H. (2011). Impact of climate change on Indian forests: A dynamic vegetation modelling approach. *Mitigation and Adaptation Strategies for Global Change*, *16*(2), 119–142. doi:10.100711027-010-9257-7

Chaturvedi, R. K., Raghubanshi, A. S., & Singh, J. S. (2011). Plant functional traits with particular reference to tropical deciduous forests: A review. *Journal of Biosciences*, *36*(5), 963–981. doi:10.100712038-011-9159-1 PMID:22116295

Chaturvedi, R. K., Raghubanshi, A. S., & Singh, J. S. (2011b). Leaf attributes and treegrowth in a tropical dry forest. *Journal of Vegetation Science*, *22*(5), 917–931. doi:10.1111/j.1654-1103.2011.01299.x

Chave, J., Brown, S., Cairns, M. A., Chambers, J. Q., Eamus, D., Folster, H., Fromard, F., … Yamakura, T. (2005). Tree allometry and improved estimates of carbon stock and balance in tropical forest. *Oecologia, 145*, 87-99.

Chave, J., Coomes, D., Jansen, S., Lewis, S. L., Swenson, N. G., & Zanne, A. E. (2009). Towards a worldwide wood economics spectrum. *Ecology Letters*, *12*(4), 351–366. doi:10.1111/j.1461-0248.2009.01285.x PMID:19243406

Chazdon, R. (2014). Second Growth: The Promise of Tropical Forest Regeneration in an Age of Deforestation. Chicago, IL: University of Chicago Press. doi:10.7208/chicago/9780226118109.001.0001

Chazdon, R. L., Harvey, C. A., Komar, O., Griffith, D. M., Bruce, G., Martínez-Ramos, M., … Philpo, S. M. (2009). Beyond Reserves : A Research Agenda for Conserving Biodiversity in Human-Modified Tropical Landscapes. Association for Tropical Biology and Conservation Stable Retrieved from http://www.jstor.org/stable/20492544 *Biotropica, 41*(2), 142–153.

Chazdon, R. L. (2003). Tropical forest recovery: Legacies of human impact and natural disturbances. *Perspectives in Plant Ecology, Evolution and Systematics*, *6*(1–2), 51–71. doi:10.1078/1433-8319-00042

Chazdon, R. L. (2008). Beyond Deforestation : Restoring Forest and Ecosystem Services on Degraded Lands. *Science*, *320*(June), 1458–1460. doi:10.1126cience.1155365 PMID:18556551

Chen, J. (2006) The combined use of chemical and organic fertilizer and or bio-fertilizer for crop growth and soil fertility. International Workshop on "*Sustained management of the Soil-rhizosphere system for efficient crop production and fertilizer use*". 37-43.

Chen, J. M., & Black, T. A. (1992). Defining leaf area index for non-flat leaves. *Plant, Cell & Environment*, *15*(4), 421–429. doi:10.1111/j.1365-3040.1992.tb00992.x

Cheong, S.-M., Brown, D. G., Kok, K., & Lopez-Carr, D. (2012). Mixed Methods in Land Change Research: Towards Integration. *Transactions of the Institute of British Geographers*, *37*(1), 8–12. doi:10.1111/j.1475-5661.2011.00482.x

Cheruto, M. C., Kauti, M. K., Kisangau, P. D., & Kariuki, P. (2016). Assessment of Land Use and Land Cover Change Using GIS and Remote Sensing Techniques: A Case Study of Makueni County, Kenya. *J. Remote Sensing and GIS*, *5*, 175. doi:10.4175/2469-4134.1000175

Chidumayo, E. N. (1997). *Miombo Ecology and Management: An introduction.* Stockholm, Sweden: IT Publications.

Choat, B., Jansen, S., Brodribb, T. J., Cochard, H., Delzon, S., Bhaskar, R., ... Jacobsen, A. L. (2012). Global convergence in the vulnerability of forests to drought. *Nature*, *491*(7426), 752–755. doi:10.1038/nature11688 PMID:23172141

Choudhary, V. (2018). *Sariska National Park-complete detail-updated.* Retrieved from http://natureconservation.in/sariska-national-park-complete-detail-updated/

Choudhury, A., Lahiri Choudhury, D. K., Desai, A., Duckworth, J. W., Easa, P. S., Johnsingh, A. J. T., Fernando, P., … Wikramanayake, E. (IUCN SSC Asian Elephant Specialist Group). (2008). *Elephas maximus*. The IUCN Red List of Threatened Species.

Chukwu, O. (2017). *Stand variables prediction from stump diameter for Tectona grandis Linn. f. Stands in Omo Forest Reserve, Nigeria.* (Unpublished master's thesis), University of Ibadan, Nigeria.

Chukwu, O., & Chenge, I. B. (2018). Stability assessment and composition of tree species in Nigerian University. *World Scientific News: An International Scientific Journal*, *95*, 235–245.

Chun-jiang, L., Ilvesniemi, H., Berg, B., Kutsch, W., Yu-sheng, Y., Xiang-qing, M., & Westman, C. J. (2003). Aboveground litterfall in Eurasian forests. *Journal of Forestry Research*, *14*(1), 27–34. doi:10.1007/BF02856758

Chuvieco, E., Cocero, D., Riano, D., Martin, P., Martinez, J., De la Riva, J., & Perez, F. (2004). Combining NDVI and surface temperature for the estimation of live fuel moisture content in forest fire danger rating. *Remote Sensing of Environment*, *92*(3), 322–331. doi:10.1016/j.rse.2004.01.019

Cincotta, R. P., Wisnewski, J., & Engelman, R. (2000). Human population in the biodiversity hotspots. *Nature*, *404*(April), 1305–1307. doi:10.1038/35010105 PMID:10801126

CITES. (2019). *Conventional International Trade of Endangered Species of Wild Fauna and Flora*. Geneva, Switzerland.

Clark, J. A., & Covey, K. R. (2012). Forest Ecology and Management Tree species richness and the logging of natural forests : A meta-analysis. *Forest Ecology and Management*, *276*, 146–153. doi:10.1016/j.foreco.2012.04.001

Clinebell, R. R. II, Phillips, O. L., Gentry, A. H., Stark, N., & Zuuring, H. (1995). Prediction of neotropical tree and liana species richness from soil and climatic data. *Biodiversity and Conservation*, *4*(1), 56–90. doi:10.1007/BF00115314

Clutter, J. L., Fortson, J. C., Pienaar, L. V., Brister, G. H., & Bailey, R. L. (1993). *Timber Management: A Quantitative Approach*. New York, NY: John Wiley & Sons.

Cochrane, M. A. (2009). *Tropical Fire Ecology*. Climate Change, Land Use, and Ecosystem Dynamics. doi:10.1007/978-3-540-77381-8

Collins, D. N., Davenport, J. L., & Stoffel, M. (2013). Climatic variation and runoff from partially-glacierised himalayan tributary basins of the Ganges. *The Science of the Total Environment*, 468–469. PMID:24296050

Colwell, R. K., Brehm, G., Cardelús, C. L., Gilman, A. C., & Longino, J. T. (2008). Global warming, elevational range shifts, and lowland biotic attrition in the wet tropics. *Science*, *322*(5899), 258–261. doi:10.1126cience.1162547 PMID:18845754

Connell, J. H., & Slatyer, R. O. (1977). Mechanisms of succession in natural communities and their role in community stability and organization. *American Naturalist*, *111*(982), 1119–1144. doi:10.1086/283241

Control Invasive Plants. (2019). Retrieved from https://www.coastalisc.com/how-to-control-invasive-plants

Control methods. (2019). Retrieved from https://dnr.wi.gov/topic/Invasives/control.html

Corbet, G. B., & Hill, J. E. (1992). *The mammals of the Indomalayan region: a systematic review*. Oxford, UK: Oxford University Press.

Cornelissen, J. H. C., Lavorel, S., Garnier, E., Diaz, S., Buchmann, N., Gurvich, D. E., ... Pausas, J. G. (2003). A handbook of protocols for standardised and easy measurement of plant functional traits worldwide. *Australian Journal of Botany*, *51*(4), 335–380. doi:10.1071/BT02124

Cornwell, W. K., Cornelissen, J. H., Amatangelo, K., Dorrepaal, E., Eviner, V. T., Godoy, O., & Quested, H. M. (2008). Plant species traits are the predominant control on litter decomposition rates within biomes worldwide. *Ecology Letters*, *11*(10), 1065–1071. doi:10.1111/j.1461-0248.2008.01219.x PMID:18627410

Cortois, R., Schröder-Georgi, T., Weigelt, A., van der Putten, W. H., & De Deyn, G. B. (2016). Plant–soil feedbacks: Role of plant functional group and plant traits. *Journal of Ecology*, *104*(6), 1608–1617. doi:10.1111/1365-2745.12643

Côté, L., Brown, S., Paré, D., Fyles, J., & Bauhus, J. (2000). Dynamics of carbon and nitrogen mineralization in relation to stand type, stand age and soil texture in the boreal mixed wood. *Soil Biology & Biochemistry*, *32*(8-9), 1079–1090. doi:10.1016/S0038-0717(00)00017-1

Cowen, D. V. (1952). *Flowering Trees and Shrubs in India*. Bombay, India: Thacker and Co., Ltd.

Cowling, R. M., & Hilton-Taylor, C. (1994). Plant diversity and endemism in southern Africa: an overview. In B. J. Huntley (Ed.), Botanical Diversity in Southern Africa (pp. 31-52). National Botanical Institute, Kirst-enbosch.

Craine, J. M., & Dybzinski, R. (2013). Mechanisms of plant competition for nutrients, water and light. *Functional Ecology*, *27*(4), 833–840. doi:10.1111/1365-2435.12081

Cramer, W., Yohe, G. W., Auffhammer, M., Huggel, C., Molau, U., Dias, M. A. F. S., . . . Tibig, L. (2014). Detection and attribution of observed impacts. In C. B. Field, V. R. Barros, D. J. Dokken, K. J. Mach, M. D. Mastrandrea, T. E. Bilir, . . . L. L. White. (Eds.), Climate Change (2014): Impacts, Adaptation, and Vulnerability. Part A: Global and Sectoral Aspects. Contribution of Working Group II to the Fifth Assessment Report of the Intergovernmental Panel on Climate Change (pp. 979–1037). Cambridge, UK: Cambridge University Press.

Crowley, T. J. (2000). Causes of climate change over the past 1000 years. *Science*, *289*(5477), 270–277. doi:10.1126cience.289.5477.270 PMID:10894770

Crozier, M., & Friedberg, E. (1977). L'acteur et le système (Editions du seuil). Paris, France.

Cuddihy, L. W. (1989). Vegetation zones of the Hawaiian Islands. In C. P. Stone & D. B. Stone (Eds.), Conservation biology in Hawaii (pp. 27-37). Cooperative National Park Resources Studies Unit, University of Hawaii, Honolulu.

Cuddihy, L. W., & Stone, C. P. (1990). *Alteration of native Hawaiian vegetation. Cooperative National Park Resources Studies Unit*. Honolulu, HI: University of Hawaii.

Cuevas, E., Brown, S., & Lugo, A. E. (1991). Above and below ground organic matter storage and production in a tropical pine plantation and a paired broadleaf secondary forest. *Plant and Soil*, *135*(2), 257–268. doi:10.1007/BF00010914

Cunningham, A., German, L., Paumgarten, F., Chikakula, M., Barr, C., Obidzinski, K., . . . Puntodewo, A. (2008). *Sustainable trade and management of forest products and services in the COMESA region: an issue paper*. Centre for International Forestry Research, Borgor, Indonesia. Retrieved from http://www.cifor.org/publications/pdf_files/Books/BCunningham0801.pdf

Cunningham, W. P., Cunningham, M. A., & Saigo, B. C. (2005). *Environmental Science, A Global Concern* (8th ed., pp. 124–125, 240). New York, NY: McGraw Hill Publishers.

Curtis, E. M., Gollan, J., Murray, B. R., & Leigh, A. (2016). Native microhabitats better predict tolerance to warming than latitudinal macro-climatic variables in arid-zone plants. *Journal of Biogeography*, *43*(6), 1156–1165. doi:10.1111/jbi.12713

D'avalos, L. M., Sanchez, K. M., & Armenteras, D. (2016). Deforestation and coca cultivation rooted in twentieth-century development projects. *Bioscience*, *66*(11), 974–982. doi:10.1093/biosci/biw118

Da Silva, J. M. C., & Oren, D. C. (1992). Notes on *Knipolegus franciscanus* Snethlage 1928 (Aves: Tyrannidae), an endemic of central Brazilian dry forests. *Goeldiana Zoologia, 16*, 1–9.

Da Silva, J. M. C., & Oren, D. C. (1997). Geographic variation and conservation of the Moustached Woodcreeper *Xiphocolaptes falcirostris*, an endemic and threatened species of north-eastern Brazil. *Bird Conservation International, 7*(3), 263–274. doi:10.1017/S0959270900001568

Dalmolin, Â. C., Lobo, F. A., Vourlitis, G., Silva, P. R., Dalmagro, P. R., Antunes, M. Z. Jr, & Ortíz, C. E. R. (2015). Is the dry season an important driver of phenology and growth for two Brazilian savannah tree species with contrasting leaf habits? *Plant Ecology, 216*(3), 407–417. doi:10.100711258-014-0445-5

Dangi, R. (2012). REDD+: Issues and challenges from a Nepalese perspective. *Climate Change and UNFCCC Negotiation Process,* 61.

Daniel, T., Jim, O., Jeff, W., & Ollivier, G. (2014). Tropical dry forest under threat & under-researched. CIFOR and CGIAR. Retrieved from http://copa.acguanacaste.ac.cr:8080/bitstream/handle/11606/590/Tropical%20Dry%20Forest%20Under%20Threat%20&%20Under-Researched.pdf?sequence=1

Dash, P. K., Mohapatra, P. P., & Giri Rao, Y. (2009). Diversity and distribution pattern of tree species Niyamgiri hill Ranges, Orissa, India. *Indian Forester, 135*(7), 922–942.

Dau, J. H., & Chenge, B. I. (2016). Growth space requirements models for *Prosopis Africana* (Guill & Perr) Taub tree species in Makurdi, Nigeria. *European Journal of Biological Research, 6*(3), 209–217.

Dau, J. H., Mati, A., & Dawaki, S. A. (2015). Role of forest inventory in sustainable forest management: A review. *International Journal of Forestry and Horticulture, 1*(2), 33–40.

Dau, J. H., Vange, T., & Amonum, J. I. (2016). Growth space requirements models for *Daniellia Oliverii* (Rolfe) Hutch and Daviz Tree in Makurdi, Nigeria. *International Journal of Forestry and Horticulture, 2*(3), 31–39.

Davidar, P., Sahoo, S., Mammen, P. C., Acharya, P., Puyravaud, J. P., Arjunan, M., ... Roessingh, K. (2010). Assessing the extent and causes of forest degradation in India: Where do we stand? *Biological Conservation, 143*(12), 2937–2944. doi:10.1016/j.biocon.2010.04.032

Daws, M. I., Mullins, C. E., Burslem, D. F., Paton, S. R., & Dalling, J. W. (2002). Topographic position affects the water regime in a semideciduous tropical forest in Panama. *Plant and Soil, 238*(1), 79–89. doi:10.1023/A:1014289930621

Dawson, T. E., & Pate, J. S. (1996). Seasonal water uptake and movement in root systems of Australian phraeatophytic plants of dimorphic root morphology: A stable isotope investigation. *Oecologia, 107*(1), 13–20. doi:10.1007/BF00582230 PMID:28307187

Day, M., Gumbo, D., Moombe, K., Wijaya, A., & Sunderland, T. (2014). Zambia Country Profile: Monitoring, reporting and verification for REDD+.CIFOR Occasional Paper # 113.

De Angelis, A., Bajocco, S., & Ricotta, C. (2012). Modelling the phenological niche of large fires with remotely sensed NDVI profiles. *Ecological Modelling, 228*, 106–111. doi:10.1016/j.ecolmodel.2012.01.003

de Bello, F., Lavorel, S., Díaz, S., Harrington, R., Cornelissen, J. H., Bardgett, R. D., ... da Silva, P. M. (2010). Towards an assessment of multiple ecosystem processes and services via functional traits. *Biodiversity and Conservation, 19*(10), 2873–2893. doi:10.100710531-010-9850-9

De Deyn, G. B., Cornelissen, J. H., & Bardgett, R. D. (2008). Plant functional traits and soil carbon sequestration in contrasting biomes. *Ecology Letters, 11*(5), 516–531. doi:10.1111/j.1461-0248.2008.01164.x PMID:18279352 de Hann, S. (1977). Humus, its formation, its relation with the mineral part of the soil and its significance for soil productivity. In *Organic matter studies* (Vol. 1, pp. 21–30). Vienna, Austria: International Atomic Energy Agency.

de Oliveira, M. T., Damasceno-Junior, G. A., Pott, A., Paranhos Filho, A. C., Suarez, Y. R., & Parolin, P. (2014). Regeneration of riparian forests of the Brazilian Pantanal under flood and fire influence. *Forest Ecology and Management*, *331*, 256–263. doi:10.1016/j.foreco.2014.08.011

De Urioste, J., & Linares, M. (2004). Canary Islands dry woodlands and forests. In N. Burgess, J. D'Amico Hales, E. Underwood, E. Dinerstein, D. Olson, I. Itoua, … K. Newman (Eds.), Terrestrial ecoregions of Africa and Madagascar: a conservation assessment (pp. xxiii + 501). World Wildlife Fund, United States. Washington, DC: Island Press.

Debinski, D. M., Jakubauskas, M. E., & Kindscher, K. (1999). A remote sensing and GIS-based model of habitats and biodiversity in the Greater Yellowstone Ecosystem. *International Journal of Remote Sensing*, *20*(17), 3281–3292. doi:10.1080/014311699211336

Deb, J. C., Phinn, S., Butt, N., & Alpine, C. A. (2018). Climate change impacts on tropical forests: Identifying risks for tropical Asia. *Journal of Tropical Forest Science*, *30*(2), 182–194. doi:10.26525/jtfs2018.30.2.182194

Decree N° 243/PR/CSPS/PNR of 23 October 1967 modifying the boundaries of the Manda National Park

Defries, R. S., Hansen, M. C., Townshend, J. R. G., Janetos, A. C., & Loveland, T. R. (2000). A new global 1-km dataset of percentage tree cover derived from remote sensing. *Global Change Biology*, *6*(2), 247–254. doi:10.1046/j.1365-2486.2000.00296.x

DeFries, R., Achard, F., Brown, S., Herold, M., Murdiyarso, D., Schlamadinger, B., & de Souza, C. Jr. (2007). Earth observations for estimating greenhouse gas emissions from deforestation in developing countries. *Environmental Science & Policy*, *10*(4), 385–394. doi:10.1016/j.envsci.2007.01.010

DeFries, R., Hansen, A., Newton, A. C., & Hansen, M. C. (2005). Increasing isolation of protected areas in tropical forests over the past twenty years. *Ecological Applications*, *15*(1), 19–26. doi:10.1890/03-5258

DeFries, R., & Pandey, D. (2010). Urbanization, the energy ladder and forest transitions in India's emerging economy. *Land Use Policy*, *27*(2), 130–138. doi:10.1016/j.landusepol.2009.07.003

DeFries, R., & Townshend, J. (1994). NDVI-derived land cover classifications at a global scale. *International Journal of Remote Sensing*, *15*(17), 3567–3586. doi:10.1080/01431169408954345

Delgado-Baquerizo, M., Maestre, F. T., Gallardo, A., Bowker, M. A., Wallenstein, M. D., Quero, J. L., ... Zaady, E. (2013). Decoupling of soil nutrient cycles as a function of aridity in global drylands. *Nature*, *502*(7473), 672–676. doi:10.1038/nature12670 PMID:24172979

Department of Forest Research and Survey. (2015). *State of Nepal's Forests*. Government of Nepal.

Deshingkar, P. (1997). *Adapting to Climate Change in a forest based land use system- A case study of Himachal Pradesh, India. Atmospheric Environmental issues in Developing Countries*. Stockholm, Sweden: Stockholm Environment Institute.

Despland, E., Rosenberg, J., & Simpson, S. J. (2004). Landscape structure and locust swarming: A satellite's eye view. *Ecography*, *27*(3), 381–391. doi:10.1111/j.0906-7590.2004.03779.x

Desreumaux, A. (2014, May). Présentation générale des théories des organisations. iae LILLE Ecole Universitaire de Management. Retrieved from bricks.univ-lille1.fr

Dexter, K. G., Smart, B., Baldauf, C., Baker, T. R., Bessike Balinga, M. P., & Brienen, R. J. W., & Pennington, R. T. (2015). Floristics and biogeography of vegetation in season-ally dry tropical regions. International Forestry Review – Special Issue. *Global Dry Forests*, *17*(S2), 10–32.

Dhyani, S., & Dhyani, D. (2016). Significance of provisioning ecosystem services from moist temperate forest ecosystems: Lessons from upper Kedarnath Valley, Garhwal, India. *Energy, Ecology & Environment*, *1*(2), 109–121.

Dia, A., & Duponnois, R. (Eds.). (2013). Le pastoralisme en Afrique subsaharienne. In La Grande Muraille Verte : Capitalisation des recherches et valorisation des savoirs locaux (pp. 12–31). Retrieved from http://books.openedition.org/irdeditions/3336

Diallo, Y., Guangdao, H., & Xingping, W. (2009). Applications of Remote Sensing in LULC Change Detection in Puer and Simao Counties. *Marsland Press Journal of American Science*, *5*(4), 157–166.

Diamond, J. M. (1975). Assembly of Species Communities. In M. L. Cody & J. M. Diamond (Eds.), *Ecology and Evolution of Communities* (pp. 342–444). Cambridge, MA: Harvard University Press.

Díaz, S., & Cabido, M. (1997). Plant functional types and ecosystem function in relation to global change. *Journal of Vegetation Science*, *8*(4), 463–474. doi:10.2307/3237198

Diaz, S., Hodgson, J. G., Thompson, K., Cabido, M., Cornelissen, J. H., Jalili, A., ... Band, S. R. (2004). The plant traits that drive ecosystems: Evidence from three continents. *Journal of Vegetation Science*, *15*(3), 295–304. doi:10.1111/j.1654-1103.2004.tb02266.x

Díaz, S., Lavorel, S., de Bello, F., Quétier, F., Grigulis, K., & Robson, T. M. (2007). Incorporating plant functional diversity effects in ecosystem service assessments. *Proceedings of the National Academy of Sciences of the United States of America*, *104*(52), 20684–20689. doi:10.1073/pnas.0704716104 PMID:18093933

Dicus, C., & Dean, T. J. (1998). Stand density effects on biomass allocation patterns and subsequent soil nitrogen demand. In *Proceedings of the Ninth Biennial Southern Silivicultural Research Conference*.

Didion, M., Repo, A., Liski, J., Forsius, M., Bierbaumer, M., & Djukic, I. (2016). Towards harmonizing leaf litter decomposition studies using standard tea bags- a field study and model application. *Forests*, *7*(12), 167. doi:10.3390/f7080167

Dinerstein, E., Olson, D. J., Graham, D. M., Webster, A. L., Primm, S. A., Bookbinder, M. P., & Ledec, G. (1995). *A conservation assessment of the terrestrial ecoregions of Latin America and the Caribbean*. Washington, DC: The World Bank. doi:10.1596/0-8213-3295-3

Ding, Y., Ren, G., Zhao, Z., Xu, Y., Luo, Y., Li, Q., & Zhang, J. (2007). Detection, causes and projection of climate change over China: An overview of recent progress. *Advances in Atmospheric Sciences*, *24*(6), 954–971. doi:10.100700376-007-0954-4

Diouf, A., & Lambin, E. (2001). Monitoring land-cover changes in semi-arid regions: Remote sensing data and field observations in the Ferlo, Senegal. *Journal of Arid Environments*, *48*(2), 129–148. doi:10.1006/jare.2000.0744

Discover. (2019). The Baobab Tree: Africa's Iconic "Tree of Life". Retrieved from https://aduna.com/blogs/learn/the-baobab-tree

Dixit, A. (2009). Koshi embankment breach in Nepal: Need for a paradigm shift in responding to floods. *Economic and Political Weekly*, 70–78.

Djogbenou, C. P., Glèlè Kakaï, R., Arouna, O., & Sinsin, B. (2011). Analyse des perceptions locales des aménagements forestiers participatifs au Bénin. VertigO - La Revue Électronique En Sciences de l'environnement, 11(1).

Djomo, A. N., Ibrahima, A., Saborowski, J., & Gravenhorst, G. (2010). Allometric equations for biomass estimates in Cameroon and tropical moist equations including biomass data from Africa. *Forest Ecology and Management*, *260*, 1873–1885. doi:10.1016/j.foreco.2010.08.034

Djoudi, H., Vergles, E., Blackie, R., Koame, C. K., & Gautier, D. (2015). Dry forests, livelihoods and poverty alleviation: Understanding current trends. *International Forestry Review*, *17*(2), 54–69. doi:10.1505/146554815815834868

Djukic, I. K., Kepfer-Rojas, S., Schmidt, I. K., Larsen, K. S., Beier, C., Berg, B., Verheyend, K., & & Tóth, Z. (2018). Early stage litter decomposition across biomes. *The Science of the Total Environment*, *626*. doi:10.1016/j.scitotenv.2018.01.012 PMID:30045558

Dobson, A. P., Bradshaw, A. D., & Baker, A. J. M. (1997). Hopes for the Future: Restoration Ecology and Conservation Biology. *Science*, *277*(5325), 515–522. doi:10.1126cience.277.5325.515

DOHS. (2005). *Annual Report 2003/2004*. Kathmandu, Nepal: Department of Health Services.

Don, A., Schumacher, J., & Freibauer, A. (2011). Impact of tropical land-use change on soil organic carbon stocks – a meta-analysis. *Global Change Biology*, *17*(4), 1658–1670. doi:10.1111/j.1365-2486.2010.02336.x

Donovan, L. A., Maherali, H., Caruso, C. M., Huber, H., & de Kroon, H. (2011). The evolution of the worldwide leaf economics spectrum. *Trends in Ecology & Evolution*, *26*(2), 88–95. doi:10.1016/j.tree.2010.11.011 PMID:21196061

Down, C. G. (1974). The relationship between colliery waste particle sizes and plant growth. *Environmental Conservation*, *1*(4), 29–40. doi:10.1017/S0376892900004902

Dozier, J. (1981). A method for satellite identification of surface temperature fields of subpixel resolution. *Remote Sensing of Environment*, *11*, 221–229. doi:10.1016/0034-4257(81)90021-3

Drenovsky, R. E., Grewell, B. J., D'antonio, C. M., Funk, J. L., James, J. J., Molinari, N., & Richards, C. L. (2012). A functional trait perspective on plant invasion. *Annals of Botany*, *110*(1), 141–153. doi:10.1093/aob/mcs100 PMID:22589328

Dry Forest Ecology. (2019). Retrieved from http://staging.ceiba.org/wp-content/uploads/Dry-Forest-Ecology.pdf

Dubey, K, Dubey, K. P., Pandey, A., & Tripathi, P. (2017b). Microbial Biofertilizer Interventions in Augmenting Agroforestry. In Probiotics. (pp. 421-442). Berlin, Germany: Springer.

Dubey, K. (2017). Socio Economic Impact Study of Mining and Mining Polices on the Livelihoods of Local Population in the Vindhyan Region of Uttar Pradesh. Report. Delhi, India: NITI Aayog.

Dubey, K. (2019). Restoration of Sand Stone Mining Area through Microbial Technology. Germany: Lambert Academic Publishing.

Dubey, K., Pandey, A., Tripathi, P., & Dubey, K. P. (2017a). Sustainable Management of Waterlogged Areas through Biodrainage & Microbial Agro-ecosystem. In Probiotics. (pp. 387-406). Berlin, Germany: Springer.

Dubuis, A. (2013). Predicting spatial Patterns of Plant Biodiversity: from Species to Communities. (Doctoral thesis), University of Lausanne, p. 295.

Duffy, J. E., Cardinale, B. J., France, K. E., McIntyre, P. B., Thébault, E., & Loreau, M. (2007). The functional role of biodiversity in ecosystems: Incorporating trophic complexity. *Ecology Letters*, *10*(6), 522–538. doi:10.1111/j.1461-0248.2007.01037.x PMID:17498151

Dumortier, P., Rabier, F., Backers, Y., Vanlierde, A., Jérôme, E., & Mathor, M. (2013). Elevage et gaz à effet de serre : le bilan des émissions de l ' animal à la filière. CRA-W et GxABT-Carrefour Productions Animales, pp. 41–52.

Dupuy, D., & Mille, D. (1993). FAO Forestry Paper: Vol. 98. *Timber plantations in the humid tropics of Africa*. Rome, Italy.

Durán, E. (2004). Estructura, diversidad y mortalidad de los árboles del dosel del bosque tropical seco. (Dissertation). Universidad Nacional Autónoma de México, México, D.F.

Duro, D. C., Coops, N. C., Wulder, M. A., & Han, T. (2007). Development of a large area biodiversity monitoring system driven by remote sensing. *Progress in Physical Geography*, *31*(3), 235–260. doi:10.1177/0309133307079054

Dutta, R. (2006). Assessment of tea bush health and yield using geospatial techniques. (Published Master's Thesis), ITC, Enschede, The Netherlands. GIS Development: *The Geospatial Resource Portal.*

Edenhofer, O., Pichs-Madruga, R., Sokona, Y., Farahani, E., Kadner, S., Seyboth, K., Adler, A., ... von Stechow, C. (2014). *Climate Change 2014: Mitigation of Climate Change. Contribution of Working Group III to the Fifth Assessment Report of the Intergovernmental Panel on Climate Change.* (Zwickel, T., & Minx, J. C., Eds.). Cambridge, UK: Cambridge University Press.

Edward, R. (2003). Dendroclimatic signals in long tree-ring chronologies from the Himalayas of Nepal. *International Journal of Climatology*, *23*(7), 707–732. doi:10.1002/joc.911

Edwards, D., Larsen, T. H., Docherty, T. D. S., Hamer, K. C., Ansell, F. A., Hsu, W. W., ... Wilcove, D. S. (2011). *Degraded lands worth protecting : the biological importance of Southeast Asia's repeatedly logged forests*, (278), 82–90. doi:10.1098/rspb.2010.1062

Eggleston, S., Buendia, L., Miwa, K., Ngara, T., & Tanabe, K. (Eds.). (2006). *2006 IPCC guidelines for national greenhouse gas inventories* (Vol. 5). Hayama, Japan: Institute for Global Environmental Strategies.

Elbrock, M. (2003). Mammal Tracks and Sign: A Guide to North American Species. Mechanicsburg, PA: Stackpole Books.

Elliott, S. D., Blakesley, D., & Hardwick, K. (2013). *Restoring tropical forests: a practical guide* (p. 344). Kew, UK: Royal Botanic Gardens.

Elmqvist, T., Pyykönen, M., Tengö, M., Rakotondrasoa, F., Rabakonandrianina, E., & Radimilahy, C. (2007). Patterns of loss and regeneration of tropical dry forest in Madagascar: The social institutional context. *PLoS One*, *2*(5). doi:10.1371/journal.pone.0000402 PMID:17476324

Elvidge, C. D., & Chen, Z. (1995). Comparison of broad-band and narrow-band red and near-infrared vegetation indices. *Remote Sensing of Environment*, *54*(1), 38–48. doi:10.1016/0034-4257(95)00132-K

Elvidge, C. D., & Lyon, R. J. (1985). Influence of rock-soil spectral variation on the assessment of green biomass. *Remote Sensing of Environment*, *17*(3), 265–279. doi:10.1016/0034-4257(85)90099-9

Enquist, B. J., & Enquist, C. A. F. (2011). Long-term change within a Neotropical forest: Assessing differential functional and floristic responses to disturbance and drought. *Global Change Biology*, *17*(3), 1408–1424. doi:10.1111/j.1365-2486.2010.02326.x

Enquist, C. A. F. (2002). Predicted regional impacts of climate change on the geographical distribution and diversity of tropical forests in Costa Rica. *Journal of Biogeography*, *29*(4), 519–534. doi:10.1046/j.1365-2699.2002.00695.x

Environment Information System (2015). *Pudukkottai district*. Environmental Information System Centre, State Environment and related issues, Department of Environment, Government of Tamil Nadu.

Esperschutz, J., Zimmermann, C., Dumig, A., Welzl, G., Buegger, F., Elmer, M., ... Schloter, M. (2013). Dynamics of microbial communities during decomposition of litter from pioneering plants in initial soil ecosystems. *Biogeosciences*, *10*(7), 5115–5124. doi:10.5194/bg-10-5115-2013

Esquivel-Muelbert, A., Baker, T. R., Dexter, K. G., Lewis, S. L., Brienen, R. J., Feldpausch, T. R., ... Higuchi, N. (2019). Compositional response of Amazon forests to climate change. *Global Change Biology*, *25*(1), 39–56. doi:10.1111/gcb.14413 PMID:30406962

ESRI. (2005). *ARC GIS, version 9.1*. Redlands, CA: ESRI.

Estrada, A., & Coates-Estrada, R. (2002). Dung Beetles in Continuous Forest, Forest Fragments and in an Agricultural Mosaic Habitat Island at Los Tuxtlas, Mexico. *Biodiversity and Conservation*, *11*(11), 1903–1918. doi:10.1023/A:1020896928578

Etter, A., Andrade, A., Amaya, P., & Arévalo, P. (2015). Estado de los ecosistemas colombianos-2014: una aplicación de la metodología de lista roja de ecosistemas - UICN.

Ewel, J. J. (1999). Natural systems as models for the design of sustainable systems of land use. *Agroforestry Systems*, *45*(1-3), 1–21. doi:10.1023/A:1006219721151

Faboya, I. O., Adebola, S. I., & Awotoye, O. O. (2015). Assessment of decomposition rate and soil nutrient status under different woody species combination in a tree plantation. *Agriculture, Forestry and Fisheries*, *4*, 46–54. doi:10.11648/j.aff.20150402.14

Fajardo, L., Gonzalez, V., Nassar, J. M., Lacabana, P., Portillo, Q. C. A., Carrasquel, F., & Rodriguez, J. P. (2005). Tropical Dry Forests of Venezuela. *Characterization and Current Conservation Status*, *37*(4), 531–546.

Fajardo, L., Gonzalez, V., Nassar, J., Lacabana, P., Portillo, C., Carrasquel, F., & Rodríguez, J. P. (2005). Tropical dry forests of Venezuela: Characterization and current conservation status. *Biotropica*, *37*(4), 531–546. doi:10.1111/j.1744-7429.2005.00071.x

Fajardo, L., Rodríguez, J. P., González, V., & Briceño-Linares, J. P. (2013). Restoration of a degraded tropical dry forest in Macanao, Venezuela. *Journal of Arid Environments*, *88*, 236–243. doi:10.1016/j.jaridenv.2012.08.009

Fallas-Cedeño, L., Michele Holbrook, N., Rocha, O. J., Vásquez, N., & Gutiérrez-Soto, M. V. (2010). Phenology, lignotubers, and water relations of Cochlospermum vitifolium, a pioneer tropical dry forest tree in Costa Rica. *Biotropica*, *42*(1), 104–111. doi:10.1111/j.1744-7429.2009.00539.x

Fang, X., Zhao, L., Zhou, G., Huang, W., & Liu, J. (2015). Increased litter input increases litter decomposition and soil respiration but has minor effects on soil organic carbon in subtropical forests. *Plant and Soil*, *392*(1-2), 139–153. doi:10.100711104-015-2450-4

FAO. (2001). FAO Forestry Paper: Vol. 140. *Global forest resources assessment. Main report*. Rome, Italy: Food and Agriculture Organization of the United Nations.

FAO. (2002). *Proceedings: second expert meeting on harmonizing forest-related definitions for use by various stakeholders. Rome.*

FAO. (2005). Global forestry resource assessment. Food and Agriculture Organization (FAO) Forestry Paper 147, Rome, Italy.

FAO. (2008). *FAOSTAT Online Statistical Service, Food and Agriculture Organization of the United Nations*. Rome, Italy: FAO.

FAO. (2008). REDD-Workshop: monitoring, assessment and verification. Rome, Italy: FAO. (FAO. Working paper, 2).

FAO. (2009). La situation mondiale de l'alimentation et de l'agriculture, le point sur l'élevage.

FAO. (2009). *State of the world's forests 2009*. Rome, Italy: Food and Agriculture Organization of the United Nations.

FAO. (2010). Global forest resources assessment 2010: main report. Rome: FAO, 2010. 340 p. *FAO Forestry Paper*, 163.

FAO. (2016). *State of the world's forest*. Rome, Italy: FAO.

FAO. (n.d.). Pastoralisme et gestion des parcours.

Fatubarin, A. R., & Olojugba, M. R. (2014). The influences of forest fire on the vegetation and some soil properties of a savanna ecosystem in Nigeria. *Journal of Soil Science and Environmental Management*, *5*(2), 28–34. doi:10.5897/JSSEM13.0413

Feng, X., Porporato, A., & Rodriguez-Iturbe, I. (2013). Changes in rainfall seasonality in the tropics. *Nature Climate Change*, *3*(9), 811–815. doi:10.1038/nclimate1907

Fernandes, E. C. M., Biot, Y., Castilla, C., Canto, A., & Matos, J. C. S. (1997). The impact of selective logging and forest conversion for subsistence agriculture and pastures on terrestrial nutrient dynamics in the Amazon. *CieciaCultura*, *49*, 34–47.

Fernández, I., Morales, N., Olivares, L., Salvatierra, J., Gómez, M., & Montenegro, G. (2010). Restauración ecológica para ecosistemas nativos afectados por incendios forestales.

Fernández-Manso, A., Fernández-Manso, O., & Quintano, C. (2016). SENTINEL-2A red-edge spectral indices suitability for discriminating burn severity. *International Journal of Applied Earth Observation and Geoinformation*, *50*, 170–175. doi:10.1016/j.jag.2016.03.005

Fernando, P., Wikramanayake, E. D., Janaka, H. K., Jayasinghe, L. K. A., Gunawardena, M., Kotagama, S. W., ... Pastorini, J. (2008). Ranging Behavior in Sri Lankan Elephants. *Mammalian Biology*, *73*(1), 2–13. doi:10.1016/j.mambio.2007.07.007

Fernando, P., Wikramanayake, E., Weerakoon, D., Jayasinghe, L. K. A., Gunawardena, M., & Janaka, H. K. (2005). Perceptions and Patterns of Human Elephant Conflict in Old and New Settlements in Sri Lanka: Insights for Mitigation and Management. *Biodiversity and Conservation*, *14*(10), 2465–2481. doi:10.100710531-004-0216-z

Ferreira, C. A., Bellote, A. F. J., & da Silva, H. D. (1993). Concentracao de nutrients minerais no lenho de Eucalyptus salignaesuarelacao com aaplicacao de fertilizantes. In Congress FlorestalBrasilicro 7 curitiba. Anais. Swo Paulo: SBS/SBEF 1: 227-230.

Ferreira, L. G., & Huete, A. R. (2004). Assessing the seasonal dynamics of the Brazilian Cerrado vegetation through the use of spectral vegetation indices. *International Journal of Remote Sensing*, *25*(10), 1837–1860. doi:10.1080/0143116031000101530

Filella, I., & Penuelas, J. (1994). The red edge position and shape as indicators of plant chlorophyll content, biomass and hydric status. *International Journal of Remote Sensing*, *15*(7), 1459–1470. doi:10.1080/01431169408954177

Fischer, G., Shah, M., & Velthuezin, V. (2002). *Climate change and Agricultural Vulnerability. IISA report for the World Summit on Sustainable Development, Johannesburg*. Vienna, Austria: IISA Publications Department.

Fisher, R. F., & Binkley, D. D. (2000). *Ecology and management of forest soils*. New York, NY: Wiley.

Fleischer, R. C., Perry, E. A., Muralidharan, K., Stevens, E. E., & Wemmer, C. M. (2001). Phylogeography of the Asian elephant (*Elephas maximus*) Based on Mitochondrial DNA. *Evolution; International Journal of Organic Evolution*, *55*(9), 1882–1892. doi:10.1111/j.0014-3820.2001.tb00837.x PMID:11681743

Flow, E. S. F., & Change, G. (2004). Reference Module in Earth Systems and Environmental Sciences.

Food and Agriculture Organization of the United Nations. (2010). Global Forest Resources Assessment 2010: Main Report. Rome, Italy: FAO. Retrieved from http://www.fao.org/docrep/013/i1757e/i1757e.pdf</eref>

Food and Agriculture Organization of the United Nations. (2015). Forests, Trees and Disasters, Unasylva, No. 243/244. Vol. 66 2015/1–2.

Food and Agriculture Organization of the United Nations. (2015). *Global Forest Resources Assessment 2015: How are the world's forests changing?* Second Edition. Rome, Italy. Available at http:// www.fao.org/3/a-i4793e.pdf)

Food and Agriculture Organization of the United Nations. (2018). *Potential implications of corporate zero-net deforestation commitments for the forest industry.* Discussion paper prepared for the 58th session of the FAO Advisory Committee on Sustainable Forest-based Industries. Retrieved from http://www.fao.org/forestry/46928-0203e234 d855d4d-c97a7e7aabfbd2f282.pdf

Forest restoration. (2019). Retrieved from https://en.wikipedia.org/wiki/Forest_restoration#cite_note-10

Forest Survey of India. (2011). *State of Forest Report.* Dehra Dun, India.

Forest Survey of India. (2017). *Forest Cover.* India State of Forest Report. 21 and 30. Retrieved from http://fsi.nic.in/isfr2017/isfr-forest-cover-2017.pdf

Forest Survey of India. (2017). The State of Forest Report of India, India.</edb>

Fortunel, C., Garnier, E., Joffre, R., Kazakou, E., Quested, H., Grigulis, K., ... Zarovali, M. (2009). Leaf traits capture the effects of land use changes and climate on litter decomposability of grasslands across Europe. *Ecology*, *90*(3), 598–611. doi:10.1890/08-0418.1 PMID:19341132

Frankel, O. H., & Soulé, M. E. (1992). *Conservation and Evolution.* Cambridge, UK: Cambridge University Press.

Franklin, S. E. (2010). *Remote Sensing for Biodiversity and Wildlife Management: Synthesi and Applications.* New York, NY: McGraw-Hill Companies

Frayer, W. E., & Furnival, G. M. (1999). Forest survey sampling designs: A history. *Journal of Forestry*, *97*(12), 4–10.

Fredericksen, T. S., & Mostacedo, B. (2000). Regeneration of timber species following selection logging in a Bolivian tropical dry forest. *Forest Ecology and Management*, *131*(1–3), 47–55. doi:10.1016/S0378-1127(99)00199-1

Freitas, A. D. S., Sampaio, E. V. S. B., Santos, C. E. R. S., & Fernandes, A. R. (2010). Biological nitrogen fixation in tree legumes of the Brazilian semi-arid caatinga. *Journal of Arid Environments*, *74*(3), 344–349. doi:10.1016/j.jaridenv.2009.09.018

Freschet, G. T., Aerts, R., & Cornelissen, J. H. (2012). A plant economics spectrum of litter decomposability. *Functional Ecology*, *26*(1), 56–65. doi:10.1111/j.1365-2435.2011.01913.x

Freschet, G. T., Valverde-Barrantes, O. J., Tucker, C. M., Craine, J. M., McCormack, M. L., Violle, C., ... Bonis, A. (2017). Climate, soil and plant functional types as drivers of global fine-root trait variation. *Journal of Ecology*, *105*(5), 1182–1196. doi:10.1111/1365-2745.12769

Friedlingstein, P., Cox, P., Betts, R., Bopp, L., Von Bloh, W., Brovkin, V., & Bala, G. (2006). Climate- carbon cycle feedback analysis: Results from the C4MIP model intercomparison. *Journal of Climate*, *19*(14), 3337–3353. doi:10.1175/JCLI3800.1

Frieler, K., Lange, S., Piontek, F., Reyer, C., Schewe, J., Warszawski, L., ... Yamagata, Y. (2017). Assessing the impacts of 1.5°C global warming - simulation protocol of the Inter-Sectoral Impact Model Inter comparison Project (ISIMIP2b). *Geoscientific Model Development*, *10*(12), 4321–4345. doi:10.5194/gmd-10-4321-2017

FSI. (2011). India State of Forest Report 2011. Retrieved from http://fsi.nic.in/cover_2011/uttarapradesh.pdf

Fukami, T., Bezemer, T. M., Mortimer, S. R., & van der Putten, W. H. (2005). Species divergence and trait convergence in experimental plant community assembly. *Ecology Letters*, *8*(12), 1283–1290. doi:10.1111/j.1461-0248.2005.00829.x

Funk, J. L. (2008). Differences in plasticity between invasive and native plants from a low resource environment. *Journal of Ecology*, *96*(6), 1162–1173. doi:10.1111/j.1365-2745.2008.01435.x

Funk, J. L., Larson, J. E., Ames, G. M., Butterfield, B. J., Cavender-Bares, J., Firn, J., ... Wright, J. (2017). Revisiting the H oly G rail: Using plant functional traits to understand ecological processes. *Biological Reviews of the Cambridge Philosophical Society*, *92*(2), 1156–1173. doi:10.1111/brv.12275 PMID:27103505

Furley, P. (1992). Edaphic changes at the forest-savanna boundary with particular reference to the Neotropics. In P. Furley, J. Ratter, & J. Proctor (Eds.), *Nature and Dynamics of Forest-Savanna Boundaries* (pp. 91–117). London, UK: Chapman and Hall.

Gadgil, M., & Meher-Homji, V. M. (1986). Localities of great significance to conservation of India's biological diversity. *Proceedings of the Indiana Academy of Sciences*, (Suppl.), 165–180.

Gairola, S., Sharma, C. M., Ghildiyal, S. K., & Suyal, S. (2012). Chemical properties of soils in relation to forest composition in moist temperate valley slopes of Garhwal Himalaya, India. *The Environmentalist*, *32*(4), 512–523. doi:10.100710669-012-9420-7

Galicia, L., Zarco-Arista, A. E., Mendoza-Robles, K. I., Palacio-Prieto, J. L., & García-Romero, A. (2008). Land use/cover, landforms and fragmentation patterns in a tropical dry forest in the southern Pacific region of Mexico. *Singapore Journal of Tropical Geography*, *29*(2), 137–154. doi:10.1111/j.1467-9493.2008.00326.x

Galvão de Melo, A. C., Durigan, G., & Romero Gorenstein, M. (2007). Efeito do fogo sobre o banco de sementes em faixa de borda de Floresta Estacional Semidecidual, SP, Brasil. *Acta Botanica Brasílica*, *21*(4), 927–934. doi:10.1590/S0102-33062007000400017

Gamble, J. S. & Fischer, C.E.C. (1915–1936). *Flora of the Presidency of Madras*. London, UK: Newman and Adlard.

Gamon, J. A., & Surfus, J. S. (1999). Assessing leaf pigment content and activity with a reflectometer. *The New Phytologist*, *143*(1), 105–117. doi:10.1046/j.1469-8137.1999.00424.x

Gandhi, D. S., & Sundarapandian, S. (2014). Inventory of trees in tropical dry deciduous forests of Tiruvannamalai district, Tamil Nadu, India. *Biodiversitas (Surakarta)*, *15*(2), 169–179. doi:10.13057/biodiv/d150208

Gandiwa, E. (2011). Effects of repeated burning on woody vegetation structure and composition in a semiarid southern African savanna. *International Journal of Environmental Sciences*, *2*(2), 458–471.

Ganesan, B. (2000). *Extraction of Non-Timber Forest Products, including Fodder and Fuelwood, in Mudumalai, India*, *47*(3), 268–274.

García Palacios, P., Shaw, E. A., Wall, D. H., & Hättenschwiler, S. (2016). Temporal dynamics of biotic and abiotic drivers of litter decomposition. *Ecology Letters*, *19*(5), 554–563. doi:10.1111/ele.12590 PMID:26947573

García-Gutiérrez, T., Jiménez-Alfaro, B., Fernández-Pascual, E., & Müller, J. V. (2018). Functional diversity and ecological requirements of alpine vegetation types in a bio geographical transition zone. *Phytocology*, *48*, 77–89.

Gardner, T. A., Barlow, J., Chazdon, R., Ewers, R. M., Harvey, C. A., Peres, C. A., & Sodhi, N. S. (2009). Prospects for tropical forest biodiversity in a human-modified world. *Ecology Letters*, *12*(6), 561–582. doi:10.1111/j.1461-0248.2009.01294.x PMID:19504750

Garg, S., Shukla, P. R., & Kapshe, M. (2007). From Climate Change Impacts to Adaptation: A Development Perspective for India. *Natural Resources Forum*, *31*(2), 132–141. doi:10.1111/j.1477-8947.2007.00142.x

Garnier, E., & Navas, M. L. (2011). 1 5 Assessing the Functional Role of Plant Diversity in Grasslands: a Trait-based Approach. Grassland Productivity and Ecosystem Services, 138.

Garnier, E., Cortez, J., Billès, G., Navas, M. L., Roumet, C., Debussche, M., & Neill, C. (2004). Plant functional markers capture ecosystem properties during secondary succession. *Ecology*, *85*(9), 2630–2637. doi:10.1890/03-0799

Garnier, E., & Laurent, G. (1994). Leaf anatomy, specific mass and water content in congeneric annual and perennial grass species. *The New Phytologist*, *128*(4), 725–736. doi:10.1111/j.1469-8137.1994.tb04036.x

Garnier, E., Shipley, B., Roumet, C., & Laurent, G. (2001). A standardized protocol for the determination of specific leaf area and leaf dry matter content. *Functional Ecology*, *15*(5), 688–695. doi:10.1046/j.0269-8463.2001.00563.x

Gashaw, T., Mebrat, W., Hagos, D., & Nigussie, A. (2014). Climate change adaptation and mitigation measures in Ethiopia. *J Biol Agric Healthc*, 2225-093.

Gautam, D. K., & Dulal, K. (2013). Determination of Threshold Runoff for Flood Warning in Nepalese Rivers. *Journal of Integrated Disaster Risk Management*, *3*(1), 125–136. doi:10.5595/idrim.2013.0061

Gei, M. G., & Powers, J. S. (2015). The influence of seasonality and species effects on surface fine roots and nodulation in tropical legume tree plantations. *Plant and Soil*, *388*(1-2), 187–196. doi:10.100711104-014-2324-1

Genc, H., Genc, L., Turhan, H., Smith, S. E., & Nation, J. L. (2008). Vegetation indices as indicators of damage by the sunn pest (Hemiptera: Scutelleridae) to field grown wheat. *African Journal of Biotechnology*, *7*, 2.

Gentry, A. H. (1982). Patterns of Neotropical plant species diversity. In M. K. Hecht, B. Wallace, & G. T. Prance (Eds.), *Evolutionary Biology* (pp. 1–84). Boston, MA: Springer. doi:10.1007/978-1-4615-6968-8_1

Gentry, A. H. (1988). Changes in plant community diversity and floristic composition on environmental and geographical gradients. *Annals of the Missouri Botanical Garden*, *75*(1), 1–34. doi:10.2307/2399464

Gentry, A. H. (1995). Diversity and floristic composition of Neotropical dry forests. In S. H. Bullock, H. A. Mooney, & E. Medina (Eds.), *Seasonally Dry Tropical Forests* (pp. 146–194). Cambridge, UK: Cambridge University Press. doi:10.1017/CBO9780511753398.007

Gerhardt, K., & Hytteborn, H. (1992). Natural dynamics and regeneration methods in tropical dry forests-an introduction. *Journal of Vegetation Science*, *3*(3), 361–364. doi:10.2307/3235761

Getzin, S., Dean, Ch., He, F., Trofymow, J. A., Wiegand, K., & Wiegand, T. (2006). Spatial patterns and competition of tree species in a douglas-fir Vancouver Island. *Ecography*, *29*, 671–682. doi:10.1111/j.2006.0906-7590.04675.x

Ghazoul, J., Burivalova, Z., Garcia-Ulloa, J., & King, L. A. (2015). Conceptualizing Forest Degradation. *Trends in Ecology & Evolution*, *30*(10), 622–632. doi:10.1016/j.tree.2015.08.001 PMID:26411619

Gholz, H. L., Wedin, D. A., Smitherman, S. M., Harmon, M. E., & Parton, W. J. (2000). Long-term dynamics of pine and hardwood litter in contrasting environments: Toward a global model of decomposition. *Global Change Biology*, *6*(7), 751–765. doi:10.1046/j.1365-2486.2000.00349.x

Ghosh, P., Mukhopadhyay, A., Chanda, A., Mondal, P., Akhand, A., Mukherjee, S., ... Hazra, S. (2017). Application of Cellular automata and Markov-chain model in geospatial environmental modeling-A review. *Remote Sensing Applications: Society and Environment*, *5*, 64–77. doi:10.1016/j.rsase.2017.01.005

Gibbs, H. K., Brown, S., Niles, J. O., & Foley, J. A. (2007). Monitoring and estimating tropical forest carbon stocks: Making REDD a reality. *Environmental Research Letters*, *2*(4). doi:10.1088/1748-9326/2/4/045023

Gillespie, T. W., Foody, G. M., Rocchini, D., Giorgi, A. P., & Saatchi, S. (2008). Measuring and modelling biodiversity from space. *Progress in Physical Geography*, *32*(2), 203–221. doi:10.1177/0309133308093606

Gillespie, T. W., Grijalva, A., & Farris, C. N. (2000). Diversity, composition, and structure of tropical dry forests in Central America. *Plant Ecology*, *147*(1), 37–47. doi:10.1023/A:1009848525399

Gillespie, T. W., & Jaffré, T. (2003). Tropical dry forests in New Caledonia. *Biodiversity and Conservation*, *12*(8), 1687–1697. doi:10.1023/A:1023649831355

Gillespie, T., Lipkin, B., Sullivan, L., Benowitz, D., Pau, S., & Keppel, G. (2012). The rarest and least protected forests in biodiversity hotspots. *Biodiversity and Conservation*, *21*(14), 3597–3611. doi:10.100710531-012-0384-1

Giraldo, J. P., & Holbrook, N. M. (2011). Physiological mechanisms underlying the seasonality of leaf senescence and renewal in seasonally dry tropical forest trees. In *Seasonally Dry Tropical Forests* (pp. 129-140). Washington, DC: Island Press.

Gleason, S. M., Butler, D. W., Ziemińska, K., Waryszak, P., & Westoby, M. (2012). Stem xylem conductivity is key to plant water balance across Australian angiosperm species. *Functional Ecology*, *26*(2), 343–352. doi:10.1111/j.1365-2435.2012.01962.x

Glenday, J. (2008). Carbon storage and emissions offset potential in an African dry forest, the Arabuko-Sokoke Forest, Kenya. *Environmental Monitoring and Assessment*, *142*(1-3), 85–95. doi:10.100710661-007-9910-0 PMID:17882526

GOFC-GOLD. (2017). *A sourcebook of methods and procedures for monitoring essential biodiversity variables in tropical forests with remote sensing*. The Netherlands: GOFC-GOLD Land Cover Project Office, Wageningen University.

Gómez-González, S., Paula, S., Cavieres, L. A., Pausas, J. G., Nuñez, M., Whitlock, C., & Smith-Ramírez, C. (2017). Postfire responses of the woody flora of Central Chile: Insights from a germination experiment. *PLoS One*, *12*(7). doi:10.1371/journal.pone.0180661 PMID:28704449

Gon, S., & Olson, D. (2019). Hawaii tropical dry forests. Tropical and subtropical dry broadleaf forests. World Wildlife Fund (WWF). Washington. Retrieved from https://www.worldwildlife.org/ecoregions/oc0202

GoN/MoFSC. (2014). *Nepal Biodiversity Strategy and Action Plan 2014-2020*. Kathmandu, Nepal: Government of Nepal, Ministry of Forests and Soil Conservation.

Gong, P., Pu, R., & Yu, B. (2001). Conifer species recognition: Effects of data transformation. *International Journal of Remote Sensing*, *17*, 2471–2481.

González-M, R., Avella, A., & Díaz-Triana, J. E. (2015). Plataformas de monitoreo para vegetación: Toma y análisis de datos. In Monitoreo a procesos de restauración ecológica aplicado a ecosistemas terrestres (pp. 87–107).

Gonzalez-ruis, T., Jaramillo, V. J., Cabriales, J., & Flores, A. (2008). Nodulation Dynamics and Nodule Activity in Leguminous Tree Species of a Mexican Tropical Dry Forest. *Journal of Tropical Ecology*, *24*(1), 107–110. doi:10.1017/S0266467407004634

Goodman, S. M., & Benstead, J. P. (2005). Updated estimates of biotic diversity and endemism for Madagascar. *Oryx*, *39*(1), 73–77. doi:10.1017/S0030605305000128

Goodman, S. M., Vasey, N., & Burney, D. A. (2007). Description of a new species of subfossil shrew tenrec (Afrosoricida: Tenrecidae: Microgale) from cave deposits in southeastern Madagascar. *Proceedings of the Biological Society of Washington*, *120*(4), 367–377. doi:10.2988/0006-324X(2007)120[367:DOANSO]2.0.CO;2

Gopala, A., & Hadian, O., *Sunarto, S. A., Williams, A., Leimgruber, P., Chambliss, S. E., & Gunaryadi, D. (2011). Elephas maximus ssp. sumatranus.* The IUCN Red List of Threatened Species.

Gotsch, S. G., Geiger, E. L., Franco, A. C., Goldstein, G., Meinzer, F. C., & Hoffmann, W. A. (2010). Allocation to leaf area and sapwood area affects water relations of co-occurring savanna and forest trees. *Oecologia*, *163*(2), 291–301. doi:10.100700442-009-1543-2 PMID:20058025

Gott, R. (1993). Land without evil: utopian journeys across the South American watershed. Brooklyn, NY: Verso.

Gould, K. A., Fredericksen, T. S., Morales, F., Kennard, D., Putz, F. E., Mostacedo, B., & Toledo, M. (2002). Post-fire tree regeneration in lowland Bolivia: Implications for fire management. *Forest Ecology and Management*, *165*(1–3), 225–234. doi:10.1016/S0378-1127(01)00620-X

Government of India. (2011). Census of India, 2001: Tables on houses, household amenities, and assets.

Govt. of Haryana. (2006). *Haryana Forest Policy-2006*. Forest Department. Government of Haryana. Retrieved from http://haryanaforest.gov.in/Portals/0/forest-policy-2006.pdf

Govt., K. (2015). Chapter 4: Land Degradation-Mining and Quarrying. Retrieved from https://www.karnataka.gov.in/empri/Documents/CHAPTER-4.%20Land%20degradation-%20Mining%20and%20Quarrying-DS.pdf

Grant, C. D. (2006). State-and-Transition Successional Model for Bauxite Mining Rehabilitation in the Jarrah Forest of Western Australia. *Restoration Ecology*, *14*(1), 28–37. doi:10.1111/j.1526-100X.2006.00102.x

Grassi, G., & Magnani, F. (2005). Stomatal, mesophyll conductance and biochemical limitations to photosynthesis as affected by drought and leaf ontogeny in ash and oak trees. *Plant, Cell & Environment*, *28*(7), 834–849. doi:10.1111/j.1365-3040.2005.01333.x

Greve, P., Orlowsky, B., Mueller, B., Sheffield, J., Reichstein, M., & Seneviratne, S. I. (2014). Global assessment of trends in wetting and drying over land. *Nature Geoscience*, *7*(10), 716–721. doi:10.1038/ngeo2247

Grime, J. P. (1979). *Plant strategies and vegetation processes.* Plant strategies and vegetation processes.

Grime, J. P. (1977). Evidence for the existence of three primary strategies in plants and its relevance to ecological and evolutionary theory. *American Naturalist*, *111*(982), 1169–1194. doi:10.1086/283244

Grime, J. P. (1998). Benefits of plant diversity to ecosystems: Immediate, filter and founder effects. *Journal of Ecology*, *86*(6), 902–910. doi:10.1046/j.1365-2745.1998.00306.x

Griscom, H. P., & Ashton, M. S. (2011). Restoration of dry tropical forests in Central America: A review of pattern and process. *Forest Ecology and Management*, *261*(10), 1564–1579. doi:10.1016/j.forcco.2010.08.027

Grove, R. H. (1994). A historical review of early institutional and conservationist responses to fears of artificially induced global climate change: The deforestation-desiccation discourse 1500–1860. *Chemosphere*, *29*(5), 1001–1013. doi:10.1016/0045-6535(94)90165-1

Grubb, P. J. (1977). Control of forest growth and distribution on wet tropical mountains: With special reference to mineral nutrition. *Annual Review of Ecology and Systematics*, *8*(1), 83–107. doi:10.1146/annurev.es.08.110177.000503

Guendehou, G. H. S., Liski, J., Tuomi, M., Moudachirou, M., Sinsin, B., & Mäkipää, R. (2014). Decomposition and changes in chemical composition of leaf litter of five dominant tree species in a West African tropical forest. *Tropical Ecology, 55*, 207–220.

Guerschman, J. P., Hill, M. J., Renzullo, L. J., Barrett, D. J., Marks, A. S., & Botha, E. J. (2009). Estimating fractional cover of photosynthetic vegetation, non-photosynthetic vegetation and bare soil in the Australian tropical savanna region upscaling the EO-1 Hyperion and MODIS sensors. *Remote Sensing of Environment, 113*(5), 928–945. doi:10.1016/j.rse.2009.01.006

Gumbo, D. J., & Chidumayo, E. N. (Eds.). (2010). The Dry forests and woodlands of Africa: managing for products and services. London, UK: Earthscan Ltd., Dunstan House.

Gupta, J. P., Sharma, M. P., & Gupta, G. D. (2001). Characterization of Kandi Belt Soils of Jammu Region as Affected by Different Land Use Patterns. *Journal of the Indian Society of Soil Science, 49*, 770–773.

Hagen-Thorn, A., Varnagiryte, I., Nihlgård, B., & Armolaitis, K. (2006). Autumn nutrient resorption and losses in four deciduous forest tree species. *Forest Ecology and Management, 228*(1-3), 33–39. doi:10.1016/j.foreco.2006.02.021

Halfpenny, J. C. (1988). *A Field Guide to Mammal Tracking in North America* (1st ed.). Boulder, CO: Johnson Books.

Halley, E. (1694). An Account of the Watery Circulation of the Sea, and of the Cause of Springs. *Philosophical Transactions of the Royal Society, 142*(7), 468–472.

Hallik, L., Niinemets, Ü., & Wright, I. J. (2009). Are species shade and drought tolerance reflected in leaf-level structural and functional differentiation in Northern Hemisphere temperate woody flora? *The New Phytologist, 184*(1), 257–274. doi:10.1111/j.1469-8137.2009.02918.x PMID:19674334

Hamid, A. F., & Darjo, L. (1993). Laporan kegiatan Monitoring Proses Pelapukan Kotoran Gajah di Stasiun Penelitian Sekundur Ecosystem Leuser. Program Pengembangan Leuser, EU, and the Indonesian government, Division of Research, Monitoring, and Information, Medan.

Hammer, O., Harper, D. A. T., & Ryan, P. D. (2005). PAST: Paleontological statistics software package for education and data analysis. *Palaeontologia Electronica 4*(1). Retrieved from http://palaeo-electronica.org/2001_1/past/issue1_01.htm

Han, T.; Huang, W.; Liu, J.; Zhou, G., & Xiao, Y. (2015). Different soil respiration responses to litter manipulation in three subtropical successional forests. *Scientific Reports-Nature* 5: 18166.

Handavu, F., Syampungani, S., & Chisanga, E. (2011). The influence of stump diameter and height on coppicing ability of selected key Miombo woodland tree species of Zambia: A guide for harvesting for charcoal production. [</edb>]. *Journal of Ecology and the Natural Environment, 3*, 461–468.

Hansen, K., Vesterdal, L., Schmidt, I. K., Gundersen, P., Sevel, L., Bastrup-Birk, A., ... Bille-Hansen, J. (2009). Litterfall and nutrient return in five tree species in a common garden experiment. *Forest Ecology and Management, 257*(10), 2133–2144. doi:10.1016/j.foreco.2009.02.021

Hansen, M. C., DeFries, R. S., Townshend, J. R. G., & Sohlberg, R. A. (2000). Global land cover classification at 1 km spatial resolution using a classification tree approach. *International Journal of Remote Sensing, 21*(6-7), 1331–1364. doi:10.1080/014311600210209

Hanski, I. A., & Gilpin, M. E. (1997). *Metapopulation Biology: Ecology, Genetics, and Evolution*. Academic Press.

Hardesty, J., Myers, R., & Fulks, W. (2005). Fire, ecosystems and people: A preliminary assessment of fire as a global conservation issue. *Fire Management, 22*(4), 78–87.

Hardin, G. (1968). The Tragedy of the Commons. *Science*, *162*(June), 1243–1248. doi:10.1126cience.162.3859.1243 PMID:5699198

Harris, A. (2019). How Does Climate Change Affect the Temperature in a Deciduous Forest? Hearst newspaper, Hearst Seattle Media, LLC. Retrieved from https://education.seattlepi.com/climate-change-affect-temperature-deciduous-forest-6755.html

Harris, L. D. (1984). The Fragmented Forest: Island Biogeography Theory and the Preservation of Biotic Diversity. Chicago, IL: University of Chicago Press. doi:10.7208/chicago/9780226219950.001.0001

Hasnat, G. N. T., Kabir, M. A., & Hossain, M. A. (2018). Major environmental issues and problems of South Asia, Particularly Bangladesh. Handbook of Environmental Materials Management, 109-148.

Hättenschwiler, S., & Gasser, P. (2005). Soil animals alter plant litter diversity effects on decomposition. *Proceedings of the National Academy of Sciences of the United States of America*, *102*(5), 1519–1524. doi:10.1073/pnas.0404977102 PMID:15671172

Hayden, B., Greene, D. F., & Quesada, M. (2010). A field experiment to determine the effect of dry-season precipitation on annual ring formation and leaf phenology in a seasonally dry tropical forest. *Journal of Tropical Ecology*, *26*(2), 237–242. doi:10.1017/S0266467409990563

Haywood, A. M., Dowsett, H., & Dolan, A. M. (2016). Integrating geological archives and climate models for the mid-Pliocene warm period. *Nature Communications*, *7*(1). doi:10.1038/ncomms10646 PMID:26879640

Hedges, S. (Ed.). (2005). Monitoring the Illegal Killing of Elephants; Dung Count Population Survey Standards for the MIKE Programme. Final draft (V1d) July 18, 2005. CITES MIKE Programme, Nairobi, Kenya.

Hedges, S., & Lawson, D. (2006). Dung Survey Standards for the MIKE Programme. CITES MIKE Programme, Central Coordinating Unit, PO Box 68200, Nairobi, Kenya.

Hedges, S. (2006). Conservation. In M. E. Fowler & S. K. Mikota (Eds.), *Biology, Medicine and Surgery of Elephants* (pp. 475–490). Oxford, UK: Blackwell Publishing. doi:10.1002/9780470344484.ch36

Hedges, S., & Tyson, M. (2002). *Some Thoughts on Counting Elephants in SE Asian forests, with Particular Reference to the CITES Monitoring of the Illegal Killing of Elephants Program.* New York, NY: Wildlife Conservation Society.

Hedges, S., Tyson, M. J., Sitompul, A. F., Kinnaird, M. F., Gunaryadi, A., & Aslan. (2005). Distribution, Status, and Conservation Needs of Asian elephants (*Elephas maximus*) in Lampung Province, Sumatra, Indonesia. *Biological Conservation*, *124*(1), 35–48. doi:10.1016/j.biocon.2005.01.004

Heim, A., & Frey, B. (2004). Early stage litter decomposition rates for Swiss forests. *Biogeochemistry*, *70*(3), 299–313. doi:10.100710533-003-0844-5

Hein, G., Justice, O., & Andrisano, A. (2010): From monospecific to mixed forest after fire? An early forecast for the montane belt of Majella, Italy. *Forest Ecology and Management.* 259, 433-439. doi:10.1016/j.foreco.2009.10.040

Helms, J. A. (1998). *The dictionary of forestry*. Bethesda, MD: Society of American Foresters.

Heltberg, R., Arndt, T. C., & Sekhar, N. U. (2000). Fuelwood Consumption and Forest Degradation: A Household Model for Domestic Energy Substitution in Rural India. *Land Economics*, *76*(2), 213. doi:10.2307/3147225

Henricksen, B., & Durkin, J. (1986). Growing period and drought early warning in Africa using satellite data. *International Journal of Remote Sensing*, *7*(11), 1583–1608. doi:10.1080/01431168608948955

Henry, J. K. (2008). Sacred groves – the need for their conservation in Eastern Ghats region. *ENVIS newsletter*, *14*(3), 3–5.

Hérault, B., Bachelot, B., Poorter, L., Rossi, V., Bongers, F., Chave, J., ... Baraloto, C. (2011). Functional traits shape ontogenetic growth trajectories of rain forest tree species. *Journal of Ecology*, *99*(6), 1431–1440. doi:10.1111/j.1365-2745.2011.01883.x

Herold, M., Román-Cuesta, R. M., Mollicone, D., Hirata, Y., Van Laake, P., Asner, G. P., & MacDicken, K. (2011). Options for monitoring and estimating historical carbon emissions from forest degradation in the context of REDD+. *Carbon Balance and Management*, *6*(1), 13. doi:10.1186/1750-0680-6-13 PMID:22115360

Herrera, R., Jordan, C. F., Klinge, H., & Medina, E. (1978). Amazon ecosystems. Their structure and functioning with particular emphasis on nutrients. *Interciencia*, *3*, 223–231.

Herrmann, S. M., Anyamba, A., & Tucker, C. J. (2005). Recent trends in vegetation dynamics in the African Sahel and their relationship to climate. *Global Environmental Change*, *15*(4), 394–404. doi:10.1016/j.gloenvcha.2005.08.004

Heywood, V. H., & Iriondo, J. M. (2003). Plant conservation: Old problems, new perspectives. *Biological Conservation*, *113*(3), 321–335. doi:10.1016/S0006-3207(03)00121-6

Hiby, L., & Lovell, P. (1991). DUNGSURV– A Program for Estimating Elephant Density from Dung Density Without Assuming 'Steady State'. Censusing Elephants in Forests. In *Proceedings of an International Workshop (Eds.* U. Ramakrishnan, J. A. Santosh,, & R. Sukumar*)*, pp. 73–79. Asian Elephant Conservation Centre, Bangalore, India.

Hines, J. (2004). Contrast, v.2. Patuxent Wildlife Research Center, United States Geological Survey. US Department of Interior, Laurel, MD.

Hines, J. (2006). PRESENCE v.8.3 (updated 2015). Software to Estimate Patch Occupancy and Related Parameters. USGS Patuxant Wildlife Research Center, Laurel, MD.

Hiya Maidawa, M., Andres, L., Yamba, B., & Lebailly, P. (n.d.). Mobilité pastorale au Sahel et en Afrique de l'Ouest: essai de synthèse.

Hobbs, R. J. (2016). Degraded or just different? Perceptions and value judgements in restoration decisions. *Restoration Ecology*, *24*(2), 153–158. doi:10.1111/rec.12336

Hobbs, R. J., Higgs, E., & Harris, J. A. (2009). Novel ecosystems: Implications for conservation and restoration. *Trends in Ecology & Evolution*, *24*(11), 599–605. doi:10.1016/j.tree.2009.05.012 PMID:19683830

Hoekstra, J. M., Boucher, T. M., Ricketts, T. H., & Roberts, C. (2005). Confronting a biome crisis: Global disparities of habitat loss and protection. *Ecology Letters*, *8*(1), 23–29. doi:10.1111/j.1461-0248.2004.00686.x

Hoffmann, W. (2000). Post-Establishment Seedling Success in the Brazilian Cerrado: A Comparison of Savanna and Forest Species. *Biotropica*, *32*(1), 62–69. doi:10.1111/j.1744-7429.2000.tb00448.x

Hogberg, N. M. (2004). *Carbon and nitrogen relations among soils, microbes and plants in boreal forests.* (Unpublished doctoral thesis), Swedish University of Agricultural Sciences.

Holdaway, R. J., Richardson, S. J., Dickie, I. A., Peltzer, D. A., & Coomes, D. A. (2011). Species-and community-level patterns in fine root traits along a 120,000-year soil chronosequence in temperate rain forest. *Journal of Ecology*, *99*(4), 954–963. doi:10.1111/j.1365-2745.2011.01821.x

Holdridge, L. R. (1967). Life zone ecology. Life zone ecology, (rev. ed.).

Holdridge, L. R. (1967). Life zone ecology. Life zone ecology., (rev. ed.).

Holdridge, L. R. (1947). Determination of world plant formations from simple climatic data. *Science*, *342*(2727), 367–368. doi:10.1126cience.105.2727.367 PMID:17800882

Holdridge, L. R. (1967). *Life Zone Ecology*. San Jose, Costa Rica: Tropical Science Center.

Holl, K. D. (2002). Long-term vegetation recovery on reclaimed coal surface mines in the eastern USA. *Journal of Applied Ecology*, *39*(6), 960–970. doi:10.1046/j.1365-2664.2002.00767.x

Holmes, D. A. (2002). The Predicted Extinction of Lowland Forests in Indonesia. In E. Wickramanayake, E. Dinerstein, C. J. Loucks, D. M. Olson, J. Morrison, J. Lamoreux, & ... P. Hedao (Eds.), *Terrestrial Ecoregions of the Indo-Pacific: A Conservation Assessment* (pp. 7–13). Washington, D.C.: Island Press.

Holm, J. A., Bloem, S. J. V., Larocque, G. R., & Shugart, H. H. (2017). Shifts in biomass and productivity for a subtropical dry forest in response to simulated elevated hurricane disturbances. *Environmental Research Letters*, *12*(2). doi:10.1088/1748-9326/aa583c

Holzman, B. A. (2008). Tropical forest & woodland biomes. In S. L. Woodward (General Ed.), Greenwood guide to biomes of the world. Westport, CT: Greenwood Press.

Hooper, D. U., Chapin, F. S., Ewel, J. J., Hector, A., Inchausti, P., Lavorel, S., & Schmid, B. (2005). Effects of biodiversity on ecosystem functioning: A consensus of current knowledge. *Ecological Monographs*, *75*(1), 3–35. doi:10.1890/04-0922

Hooper, E., Legendre, P., & Condit, R. (2005). Barriers to forest regeneration of deforested and abandoned land in Panama. *Journal of Applied Ecology*, *42*(6), 1165–1174. doi:10.1111/j.1365-2664.2005.01106.x

Horion, S., Fensholt, R., Tagesson, T., & Ehammer, A. (2014). Using earth observation-based dry season NDVI trends for assessment of changes in tree cover in the Sahel. *International Journal of Remote Sensing*, *35*(7), 2493–2515. doi: 10.1080/01431161.2014.883104

Horneck, D. A., & Miller, R. O. (1997). Determination of total nitrogen in plant tissue. In Handbook of reference methods for plant analysis (pp. 85–93). Boca Raton, FL: CRC Press.

Hosonuma, N., Herold, M., De Sy, V., De Fries, R. S., Brockhaus, M., Verchot, L., ... Romijn, E. (2012). An assessment of deforestation and forest degradation drivers in developing countries. *Environmental Research Letters*, *7*(4). doi:10.1088/1748-9326/7/4/044009

Houghton, R. A. (2007). Balancing the global carbon budget. *Annual Review of Earth and Planetary Sciences*, *35*(1), 313–347. doi:10.1146/annurev.earth.35.031306.140057

Hsieh, T. C., Ma, K. H., & Chao, A. (2016). iNEXT: An R package for rarefaction and extrapolation of species diversity (Hill numbers). *Methods in Ecology and Evolution*, *7*(12), 1451–1456. doi:10.1111/2041-210X.12613

Hubbell, S. P. (1983). Diversity of canopy trees in a neotropical forest and implications for conservation. *Tropical rain forest: ecology and management*, 25-41.

Hubbell, S. P. (1979). Tree dispersion, abundance, and diversity in a tropical dry forest. *Science*, *203*(4387), 1299–1309. doi:10.1126cience.203.4387.1299 PMID:17780463

Huete, A. R. (1988). A soil-adjusted vegetation index (SAVI). *Remote Sensing of Environment*, *25*(3), 295–309. doi:10.1016/0034-4257(88)90106-X

Huete, A. R., Jackson, R. D., & Post, D. F. (1985). Spectral response of a plant canopy with different soil backgrounds. *Remote Sensing of Environment*, *17*(1), 37–53. doi:10.1016/0034-4257(85)90111-7

Huete, A., Didan, K., Miura, T., Rodriguez, E. P., Gao, X., & Ferreira, L. G. (2002). Overview of the radiometric and biophysical performance of the MODIS vegetation indices. *Remote Sensing of Environment*, *83*(1-2), 195–213. doi:10.1016/S0034-4257(02)00096-2

Hulme, M., & Viner, D. (1995). *A climatic change scenario for assessing the impact of climate change on tropical rain forests* (p. 34). US: Climatic Research Unit for WWF.

Hulshof, C. M., Mart'ınez-Yr'ızar, A., Burquez, A., Boyle, B., & Enquist, B. J. (2013) Functional trait variation in tropical dry forests: a review and synthesis. Tropical Dry Forests in the Americas: Ecology, Conservation, and Management. Boca Raton, FL: CRC Press. pp 129–40.

Hulshof, C. M., & Swenson, N. G. (2010). Variation in leaf functional trait values within and across individuals and species: An example from a Costa Rican dry forest. *Functional Ecology*, *24*(1), 217–223. doi:10.1111/j.1365-2435.2009.01614.x

Husch, B., Beers, T. W., & Kershaw, J. A. Jr. (2003). *Forest mensuration* (4th ed.). Hoboken, NJ: John Wiley & Sons

Hussner, A., Stiers, I., Verhofstad, M. J. J. M., Bakker, E. S., Grutters, B. M. C., Haury, J., ... Anderson, L. W. J. (2017). Management and control methods of invasive alien freshwater aquatic plants: A review. *Aquatic Botany*, *136*, 112–137. doi:10.1016/j.aquabot.2016.08.002

Hutchinson, C. (1991). Uses of satellite data for famine early warning in sub-Saharan Africa. *International Journal of Remote Sensing*, *12*(6), 1405–1421. doi:10.1080/01431169108929733

Hu, Z., Fu, Y., Xiao, W., Zhao, Y., & Wei, T. (2015). Ecological restoration plan for abandoned underground coal mine site in Eastern China. *International Journal of Mining, Reclamation and Environment*, *29*(4), 316–330. doi:10.1080/17480930.2014.1000645

Ibarra, J. M. N., & de las Heras, M. M. (2005). Open-Cast Mining Reclamation. In *Forest Restoration in Landscapes*. Berlin, Germany: Springer. doi:10.1007/0-387-29112-1_53

Ibisch, P. L., Columba, K., & Reichle, S. (2002). *Plan de Conservación y Desarrollo Sostenible para el Bosque seco Chiquitabno, cedrrado y Pantanal Boliviano. Fundación para la Conservacióndel Bosque Chiquitano, Editoral F.A.N.* Santa Cruz , Bolivia.

Idrissou, L. (2012). Dynamics of Conflict in Participatory Forest Management in Benin. A Framing Perspective [Doctoral thesis]. Wageningen University, The Netherlands.

India, N. B. S. A. P. (1999). *National Policy and Macro Level Action on Biodiversity*. New Delhi, India: Government of India, Ministry of Environment and Forests.

Indorante, A. J., Jansen, I. J., & Boast, C. W. (1981). Surface mining and reclamation: Initial changes in soil character. *Journal of Soil and Water Conservation*, *36*(6), 347–350.

Indriyanto. (2009). Produksiseresahpadakomunitashutan yang dikelolapetanidalam Register 19 Provinsi Lampung. In *ProsidingPenelitian-penelitianAgroforestri di Indonesia*. INAFE Punlisher, Tahun Lampung. pp 75-83.

Intergovernmental Panel on Climate Change (IPCC) (2007). The fourth assessment report (AR4). Synthesis report for policymakers.

Intergovernmental Panel on Climate Change. (2001). *Climate change 2001: The Scientific Basis: Contribution of working group I to the third assessment report of the intergovernmental panel on climate change (IPCC)*. Cambridge, UK: Cambridge University Press.

International Energy Agency. (2016). *World Energy Outlook*. Paris, France.

International Institute for Sustainable Development. (2012). Earth Negotiations Bulletin. Retrieved from https://enb.iisd.org/vol27/enb2751e.html

International Union for Conservation of Nature. (2017). Forests and Climate Change. Retrieved from https://www.iucn.org/sites/dev/files/forests_and_climate_change_issues_brief.pdf

IPCC. (2006). Guide for national greenhouse gas inventory agriculture, forestry and other land use. *Institute for Global Environmental Strategies, Japan*, *4*, 46–52.

IPCC. (2007). Summary for policy makers. In S. Solomon et al. (Eds.), Climate Change 2007: the Physical Science Basis (pp. 1-18). Contribution of working group I to the fourth assessment report of the Intergovernmental Panel on Climate Change. Cambridge, UK: Cambridge University Press.

ITTO. (2002). ITTO guidelines for the restoration, management and rehabilitation of degraded and secondary tropical forests. In ITTO guidelines for the restoration, management and rehabilitation of degraded and secondary tropical forests.

IUCN. (2001). IUCN Red List categories and criteria: Version 3.1. IUCN Species Survival Commission. Gland, Switzerland: IUCN.

IUCN. (2003). Guidelines for application of IUCN Red List criteria at regional levels: Version 3.0. IUCN Species Survival Commission. Gland, Switzerland: IUCN.

IUCN. (2019). The IUCN Red List of Threatened Species 2011: e.T199856A9129626. doi:. doi:10.2305/IUCN.UK.2011-2.RLTS.T199856A9129626.en

IUCN/UNEP/WWF. (1987). *Madagascar, an environmental profile*. Gland, Switzerland: IUCN.

Jackel, H., Rudner, M., & Deil, U. (2013). Density, special pattern and relief features of sacred sites in northern Morocco. *Landscape Online*, *32*, 1–16. doi:10.3097/LO.201332

Jaffré, T., Morat, P., Veillon, J. M., Rigault, F., & Dagostini, G. (2001). Composition et caractérisation de la flore indigène de Nouvelle-Calédonie= Composition and characterisation of the native flora of New Caledonia, Centre IRD, 121. (Documents Scientifiques et Techniques - IRD: II; 4). Retrieved from http://horizon.documentation.ird.fr/exl-doc/pleins_textes/pleins_textes_7/divers3/010027139.pdf

Jaffre, T., Bouchet, P., & Veillon, J. M. (1998). Threatened plants of New Caledonia: Is the system of protected areas adequate? *Biodiversity and Conservation*, *7*(1), 109–135. doi:10.1023/A:1008815930865

Janzen, D. (1988). *Tropical dry forests. The most endangered.*

Janzen, D. H. (1988): Tropical Dry Forests The Most Endangered Major Tropical Ecosystem. In E. O. Wilson & F. M. Peter (Eds.) *Biodiversity.* Washington DC: National Academies Press. pp. 1-6. Available at https://www.ncbi.nlm.nih.gov/books/NBK219281/?report=printable

Janzen, D. H. (1967). Synchronization of sexual reproduction of trees within the dry season in Central America. *Evolution; International Journal of Organic Evolution*, *21*(3), 620–637. doi:10.1111/j.1558-5646.1967.tb03416.x PMID:28563685

Janzen, D. H. (1982). (a). Cenizero tree (Leguminosae: *Pithecellobium saman*) delayed fruit development in Costa Rican deciduous forests. *American Journal of Botany*, *69*(8), 1269–1276. doi:10.1002/j.1537-2197.1982.tb13372.x

Janzen, D. H. (1982). (b). Variation in average seed size and fruit seediness in a fruit crop of a Guanacaste tree (Leguminosae: *Enterolobium cyclocarpum*). *American Journal of Botany*, *69*(7), 1169–1178. doi:10.1002/j.1537-2197.1982.tb13361.x

Janzen, D. H. (1983). *Costa Rican Natural History*. Chicago, IL: University of Chicago Press. doi:10.7208/chicago/9780226161204.001.0001

Janzen, D. H. (1986). *Guanacaste National Park: tropical ecological and cultural restoration (No. AS 50907). Fundación Tinkes, New York (EUA) FPN.* San José, CA: UNED.

Janzen, D. H. (1988). Management of Habitat Fragments in a Tropical Dry Forest: Growth. *Annals of the Missouri Botanical Garden, 75*(1), 105–116. doi:10.2307/2399468

Janzen, D. H. (1988). Tropical dry forest. The most endangered major tropical ecosystems. In E. O. Wilson (Ed.), *Biodiversity* (pp. 130–137). Washington, DC: National Academy Press.

Janzen, D. H. (1988). *Tropical dry forest: the most endangered major tropical ecosystem. In E. O. Wilson (Ed.), Biodiversity.*

Janzen, D. H. (1988). Tropical dry forest: the most endangered major tropical ecosystems. In E. O. Wilson (Ed.), *Biodiversity* (pp. 130–137). Washington, DC: National Academy Press.

Janzen, D. H. (1988). Tropical dry forests. *Biodiversity (Nepean)*, 130–137.

Janzen, D. H. (1998). Tropical dry forests: the most endangered major tropical ecosystem. In E. O. Wilson (Ed.), *Biodiversity National.* Washington, DC: Academy Press.

Japan Association of Remote Sensing. (2010). *Sensors.* Available at http://www.jars1974.net/pdf/03_Chapter02.pdf

Jaramillo, V. J., MartÍnez-YrÍzar, A., & Sanford, R. L. (2011). Primary productivity and biogeochemistry of seasonally dry tropical forests. In Seasonally dry tropical forests (pp. 109-128). Washington, DC: Island Press.

Jaureguiberry, P., & Díaz, S. (2015). Post-burning regeneration of the Chaco seasonally dry forest: Germination response of dominant species to experimental heat shock. *Oecologia, 177*(3), 689–699. doi:10.100700442-014-3161-x PMID:25421096

Jensen, J. (2007). Remote Sensing of the Environment. Upper Saddle River, NJ: Pearson Prentice Hall.

Jeuland, M. A., & Pattanayak, S. K. (2012). Benefits and Costs of Improved Cookstoves: Assessing the Implications of Variability in Health. *Forest and Climate Impacts., 7*(2). doi:10.1371/journal.pone.0030338 PMID:22348005

Jha, B. (2015, Nov. 18). *Uzbek team to study Aravalli plants.* Times of India. Retrieved from https://timesofindia.indiatimes.com/city/gurgaon/Uzbek-team-to-study-Aravali-plants/articleshow/49823283.cms

Jha, M. N., & Dimri, B. M. (1991). Soil nutrients and forest productivity in four natural stands. *Journal of the Indian Society of Soil Science, 39*, 735–738.

Jiagho, E., Zapfack L., Banoho, L., Tsayem-Demaze, M., Corbonnois, J. & Tchawa, P. (2016). Diversity of woody flora on the outskirts of Waza National Park (Cameroon). *VertigO - the electronic journal in environmental sciences [Online], 16*(1). Retrieved from http://vertigo.revues.org/17249

Jiang, Z., Huete, A. R., Didan, K., & Miura, T. (2008). Development of a two-band enhanced vegetation index without a blue band. *Remote Sensing of Environment, 112*(10), 3833–3845. doi:10.1016/j.rse.2008.06.006

Jitendra, S. K. (2019). *Aravallis broken beyond repair.* Retrieved from Error! Hyperlink reference not valid.

Johnson, D. W., & Ball, J. T. (1996). Interactions between CO_2 and nitrogen in forests: can we extrapolate from the seedling to the stand level? In G. W. Koch & H. A. Mooney (Eds.), *Carbon dioxide and terrestrial ecosystems.* San Diego, CA: Academic Press. doi:10.1016/B978-012505295-5/50017-3

Johnson, K. P., Adler, F. R., & Cheery, J. L. (2000). Genetic and Phylogenetic Consequences of Island Biogeography. *Evolution; International Journal of Organic Evolution, 54*(2), 387–396. doi:10.1111/j.0014-3820.2000.tb00041.x PMID:10937215

Johnston, C. A. (1998). *Geographic information systems in ecology*. Oxford, UK: Blackwell Science.

Julian, A. (2018). The Adaptations of the Baobab Tree. Retrieved from https://sciencing.com/adaptations-baobab-tree-8344865.html

Juvik, S. P., & Juvik, J. O. (1998). *Atlas of Hawaii* (3rd ed.). Honolulu, HI: University of Hawaii Press.

Kabeer, K. A. A., & Nair, V. J. (2009). *Flora of Tamil Nadu–Grasses* (pp. 1–524). Kolkata, India: Botanical Survey of India.

Kacholi, D. S. (2014). Analysis of Structure and Diversity of the Kilengwe Forest in the Morogoro Region, Tanzania. *International Journal of Biodiversity*, *2014*, 1–8. doi:10.1155/2014/516840

Kanemasu, E. T. (1974). Seasonal canopy reflectance patterns of wheat, sorghum, and soybean. *Remote Sensing of Environment*, *3*(1), 43–47. doi:10.1016/0034-4257(74)90037-6

Kangas, A., Gove, J. H., & Scott, C. T. (2006). Introduction. In A. Kangas & M. Maltamo (Eds.), Managing Forest Ecosystems: Forest Inventory Methodology and Applications. (10: 3-11). Switzerland: Springer Nature.

Karanth, K. U., Gopalaswamy, A. M., Kumar, N. S., Vaidyanathan, S., Nichols, J. D., & MacKenzie, D. I. (2011). Monitoring Carnivore Populations at the Landscape Scale: Occupancy Modelling of Tigers from Sign Surveys. *Journal of Applied Ecology*, *48*(4), 1048–1056. doi:10.1111/j.1365-2664.2011.02002.x

Kareiva, P., Watts, S., Mcdonald, R., & Boucher, T. (2007). Domesticated Nature : Shaping for Human Welfare. *Science*, *316*(June), 2005–2008. doi:10.1126cience.1140170 PMID:17600209

Karna, Y. K., Hussin, Y. A., Gilani, H., Bronsveld, M. C., Murthy, M. S. R., Qamer, F. M., ... Baniya, C. B. (2015). Integration of WorldView-2 and airborne LiDAR data for tree species level carbon stock mapping in Kayar Khola watershed, Nepal. *International Journal of Applied Earth Observation and Geoinformation*, *38*, 280–291. doi:10.1016/j.jag.2015.01.011

Kavvadias, V. A., Alifragis, D., Tsiontsis, A., Brosfas, G., & Stamatelos, G. (2001). Litterfall, litter accumulation and litter decomposition rates in four forest ecosystems in northern Greece. *Forest Ecology and Management*, *144*(1-3), 113–127. doi:10.1016/S0378-1127(00)00365-0

Kawahara, T., & Tsutsuni, T. (1972). Studies on the circulation of Carbon and Nitrogen in forest ecosystem. *Bulletin of Kyoto University of Forestry*, *44*, 141–158.

Kazakou, E., Violle, C., Roumet, C., Navas, M. L., Vile, D., Kattge, J., & Garnier, E. (2014). Are trait-based species rankings consistent across data sets and spatial scales? *Journal of Vegetation Science*, *25*(1), 235–247. doi:10.1111/jvs.12066

Keeley, J. E., & Fotheringham, C. J. (2000). Role of fire in regeneration from seed. Seeds: *The Ecology of Regeneration in Plant Communities*, (January), 311–330. doi:10.1079/9780851994321.0311

Keeley, J. E., Lubin, D., & Fotheringham, C. J. (2003). Fire and grazing impacts on plant diversity and alien plant invasions in the southern Sierra Nevada. *Ecological Applications*, *13*(5), 1355–1374. doi:10.1890/02-5002

Keeling, C. D., Chin, J. F. S., & Whorf, T. P. (1996). Increased activity of northern vegetation inferred from atmospheric CO_2 measurements. *Nature*, *382*(6587), 146–149. doi:10.1038/382146a0

Keith, H., Mackey, B. G., & Lindenmayer, D. B. (2009). Re-evaluation of forest biomass carbon stocks and lessons from the world's most carbon-dense forests. *Proceedings of the National Academy of Sciences of the United States of America*, *106*(28), 11635–11640. doi:10.1073/pnas.0901970106 PMID:19553199

Kellman, M., & Meave, J. (1997). Fire in the tropical gallery forests of Belize. *Journal of Biogeography*, *24*(1), 23–34. doi:10.1111/j.1365-2699.1997.tb00047.x

Kemf, E., & Santiapillai, C. (2000). *Asian Elephants in the Wild. A WWF Species Status Report.* Gland, Switzerland: World Wide Fund for Nature.

Kennard, D. K., Gould, K., Putz, F. E., Fredericksen, T. S., & Morales, F. (2002). Effect of disturbance intensity on regeneration mechanisms in a tropical dry forest. *Forest Ecology and Management, 162*(2–3), 197–208. doi:10.1016/S0378-1127(01)00506-0

Keramitsoglou, I., Kontoes, C., Sykioti, O., Sifakis, N., & Xofis, P. (2008). Reliable, accurate and timely forest mapping for wildfire management using ASTER and Hyperion satellite imagery. *Forest Ecology and Management, 255*(10), 3556–3562. doi:10.1016/j.foreco.2008.01.077

Khanal, S. (2015). Wildfire trends in Nepal based on MODIS burnt-area data. *Banko Janakari, 25*(1), 76–79. doi:10.3126/banko.v25i1.13477

Khan, M. L., Khumbongmayum, A. D., & Tripathi, R. S. (2008). Sacred groves and their significance in conserving biodiversity: An overview. *International Journal of Ecology and Environmental Sciences, 34*, 277–291.

Khayani, A. H., Gould, W. A., Harmsen, E., Terando, A., Quinones, M., & Collazo, J. A. (2016). Climate change implications for Tropical Islands: Interpolating and interpreting statistically downscaled GCM projections for management and planning. *Journal of Applied Meteorology, 55*(2), 265–282. doi:10.1175/JAMC-D-15-0182.1

Khoshoo, T. N. (1991). Environmental concern and strategies (3rd ed.). Punjabi Bagh, Delhi, India: Ashish Publishing House.

Khurana, E., Sagar, R., & Singh, J. S. (2006). Seed size: A key trait determining species distribution and diversity of dry tropical forest in northern India. *Acta Oecologica, 29*(2), 196–204. doi:10.1016/j.actao.2005.10.003

Khurana, E., & Singh, J. S. (2001). Ecology of seed and seedling growth for conservation and restoration of tropical dry forest: A review. *Environmental Conservation, 28*(1), 39–52. doi:10.1017/S0376892901000042

Khurana, E., & Singh, J. S. (2004). Germination and seedling growth of five tree species from tropical dry forest in relation to water stress: Impact of seed size. *Journal of Tropical Ecology, 20*(4), 385–396. doi:10.1017/S026646740400135X

Killeen, T. J., Chavez, E., Peña-Claros, M., Toledo, M., Arroyo, L., Caballero, J., ... Soria, L. (2006). The Chiquitano dry forest, the transition between humid and dry forest in eastern lowland Bolivia. In R. T. Pennington & J. A. Ratter (Eds.), *Neotropical savannas and seasonally dry forests: Plant Diversity, Biogeography, and Conservation* (pp. 225–245). Boca Raton, FL: CRC Press. doi:10.1201/9781420004496-9

King, E. D. I. O., Viji, C., & Narasimhan, D. (1997). Sacred groves: Traditional ecological heritage. *International Journal of Ecology and Environmental Sciences, 23*, 463–470.

Kinnaird, M. F., Sanderson, E. W., O'Brien, T. G., Wibisono, H. T., & Woolmer, G. (2003). Deforestation Trends in a Tropical Landscape and Implications for Endangered Large Mammals. *Conservation Biology, 17*(1), 245–257. doi:10.1046/j.1523-1739.2003.02040.x

Kinver, M. (2016). Dry tropical forests 'overlooked and under threat'. Science and Environment, BBC News. Retrieved from https://www.bbc.com/news/science-environment-37440485

Kirch, P. V. (1982). The impact of prehistoric Polynesian on the Hawaiian ecosystem. *Pacific Science, 36*(1), 1–14.

Kirilenko, A. P., & Sedjo, R. A. (2007). Climate change impacts on forestry. *Proceedings of the National Academy of Sciences of the United States of America, 104*(50), 19697–19702. doi:10.1073/pnas.0701424104 PMID:18077403

Kitajima, K., Mulkey, S. S., & Wright, S. J. (1997). Decline of photosynthetic capacity with leaf age in relation to leaf longevities for five tropical canopy tree species. *American Journal of Botany*, *84*(5), 702–708. doi:10.2307/2445906 PMID:21708622

Kleyer, M., Dray, S., Bello, F., Lepš, J., Pakeman, R. J., Strauss, B., ... Lavorel, S. (2012). Assessing species and community functional responses to environmental gradients: Which multivariate methods? *Journal of Vegetation Science*, *23*(5), 805–821. doi:10.1111/j.1654-1103.2012.01402.x

Kohli, M. S. (2004). *Mountains of India: Tourism, Adventure, Pilgrimage*. Indus Publishing, 29. Retrieved from https://books.google.co.in/books?id=GIs4zv17HHwC&pg=PA29&redir_esc=y#v=onepage&q&f=false

Kokaly, R. F., & Clark, R. N. (1999). Spectroscopic determination of leaf biochemistry using band-depth analysis of absorption features and stepwise linear regression. *Remote Sensing of Environment*, *67*(3), 267–287. doi:10.1016/S0034-4257(98)00084-4

Kolb, T. E., Wagner, M. R., & Covington, W. W. (1994). Concepts of forest health- Utilitarian and ecosystem perspectives. *Journal of Forestry*, *92*(7), 10–15.

Kolmagne, M. N. (2000). *Study of the settlement of fauna and flora in Manda Park Report of the Ministry of Environment and Water*.

Kong, D., Ma, C., Zhang, Q., Li, L., Chen, X., Zeng, H., & Guo, D. (2014). Leading dimensions in absorptive root trait variation across 96 subtropical species. *The New Phytologist*, *203*(3), 863–672. doi:10.1111/nph.12842 PMID:24824672

Koontz, M. J., North, M. P., Werner, C. M., Fick, S. E., & Latimer, A. M. (2019). Local variability of vegetation structure increases forest resilience to wildfire. *EcoEvoRxiv*, 1–38. doi:10.32942/OSF.IO/K72YE

Koul, D. N. (2004). *Carbon sequestration estimates of various land uses in Terai Zone of West Bengal*. (Unpublished Master of Science Thesis), Uttar Banga Krishi Viswavidyalaya, Pundibari, West Bengal.

Kraft, N. J., Valencia, R., & Ackerly, D. D. (2008). Functional traits and niche-based tree community assembly in an Amazonian forest. *Science*, *322*(5901), 580–582. doi:10.1126cience.1160662 PMID:18948539

Kramer-Walter, K. R., Bellingham, P. J., Millar, T. R., Smissen, R. D., Richardson, S. J., & Laughlin, D. C. (2016). Root traits are multidimensional: Specific root length is independent from root tissue density and the plant economic spectrum. *Journal of Ecology*, *104*(5), 1299–1310. doi:10.1111/1365-2745.12562

Krause, G. H., Winter, K., Krause, B., Jahns, P., García, M., Aranda, J., & Virgo, A. (2010). High-temperature tolerance of a tropical tree, *Ficus insipida*: Methodological reassessment and climate change considerations. *Functional Plant Biology*, *37*(9), 890–900. doi:10.1071/FP10034

Krishna, M. P., & Mohan, H. (2017). Litter decomposition in forest ecosystems: A review. *Energy. Ecology & Environment*, *2*, 236–249.

Krishnaswamy, J., Bonell, M., Venkatesh, B., Purandara, B. K., Rakesh, K. N., Lele, S., ... Badiger, S. (2013). The groundwater recharge response and hydrologic services of tropical humid forest ecosystems to use and reforestation: Support for the 'infiltration–evapotranspiration trade-off hypothesis. *Journal of Hydrology (Amsterdam)*, *498*, 191–209. doi:10.1016/j.jhydrol.2013.06.034

Kumar, A. (2016, May 9). *Save Aravalli to save Wildlife*. The Hindu. Retrieved from https://www.thehindu.com/news/cities/Delhi/save-aravalli-to-save-wildlife/article8574467.ece

Kumud, D. (2010). Bioreclamation of Silica Mining Area of Shankargarh, Allahabad. (Unpublished doctoral thesis). Forest Research Institute, Dehradun, India.

Kuster, E. C., Bierman, S. M., Klotz, S., & Kuhn, I. (2010). Modelling the impact of climate and land use change on the geographical distribution of leaf anatomy in a temperate flora. *Ecography*, *12*, 2001–2012.

Ky-Dembele, C., Tigabu, M., Bayala, J., Ouédraogo, S. J., & Odén, P. C. (2007). The relative importance of different regeneration mechanisms in a selectively cut savanna-woodland in Burkina Faso, West Africa. *Forest Ecology and Management*, *243*(1), 28–38. doi:10.1016/j.foreco.2007.01.091

Lahoreau, G., Barot, S., Gignoux, J., Hoffmann, W. A., Setterfield, S. A., & Williams, P. R. (2006). Positive effect of seed size on seedling survival in fire-prone savannas of Australia, Brazil and West Africa. *Journal of Tropical Ecology*, *22*(06), 719–722. doi:10.1017/S026646740600349X

Laidlaw, R. K. (2000). Effects of Habitat Disturbance and Protected Areas on Mammals of Peninsular Malaysia. *Conservation Biology*, *14*(6), 1639–1648. doi:10.1111/j.1523-1739.2000.99073.x

Laing, S. E., Buckland, S. T., Burn, R. W., Lambie, D., & Amphlett, A. (2003). Dung and Nest Surveys: Estimating Decay Rates. *Journal of Applied Ecology*, *40*(6), 1102–1111. doi:10.1111/j.1365-2664.2003.00861.x

Lal, J. B. (1992). *Forest Ecology*. Dehradun, India: Natraj Publishers.

Lal, R. (2004). Soil carbon sequestration impacts on global climate change and food security. *Science*, *304*(5677), 1623–1627. doi:10.1126cience.1097396 PMID:15192216

Lambers, H. A. N. S., & Poorter, H. (1992). Inherent variation in growth rate between higher plants: A search for physiological causes and ecological consequences. [Academic Press]. *Advances in Ecological Research*, *23*, 187–261. doi:10.1016/S0065-2504(08)60148-8

Lambin, E. F. (1999). Monitoring Forest Degradation in Tropical Regions by Remote Sensing: Some Methodological Issues. *Global Ecology and Biogeography*, *8*(3), 191–198. doi:10.1046/j.1365-2699.1999.00123.x

Landais, É. (1994). Système d'Elevage D'une intuition holiste à une méthode de recherche, le cheminement d'un concept. In *Blanc-Pamard et Boutais (1994): A la croisée des parcours: Pasteurs, éleveurs, cultivateurs (OSTROM). In Dynamique des Systèmes agraires (OSTROM)*. Paris, France: ORSTROM.

Landauer, G. (1989): Detection, classification and mapping of forest decline by means of multispectral remote sensing. In Proceedings, Global Natural Resource Monitoring and Assessment: Preparing for the 21st Century, Vol. 1.

Lansina, T. (2013). *Influence of climate and protection on woody vegetation in western Burkina Faso*. (Doctoral thesis), Applied Biological Sciences, specialty: Botany and Phytoecology, University of Ouagadougou.

Laughlin, D. C. (2014). The intrinsic dimensionality of plant traits and its relevance to community assembly. *Journal of Ecology*, *102*(1), 186–193. doi:10.1111/1365-2745.12187

Laughlin, D. C., Leppert, J. J., Moore, M. M., & Sieg, C. H. (2010). A multi-trait test of the leaf-height-seed plant strategy scheme with 133 species from a pine forest flora. *Functional Ecology*, *24*(3), 493–501. doi:10.1111/j.1365-2435.2009.01672.x

Laurance, W. F. (2003). *Slow burn : the insidious effects of surface fires on tropical forests*. *18*(5), 209–212. doi:10.1016/S0169-5347(03)00073-9

Laurance, W. F., & Bierregaard, R. O. (Eds.). (1997). Tropical Rainforest Remnants: Ecology, Management, and Conservation of Fragmented Communities. Chicago, IL: University of Chicago Press.

Laurance, W. F. (2008). Can carbon trading save vanishing forests? *Bioscience*, *58*(4), 286–287. doi:10.1641/B580402

Laurance, W. F., Sayer, J., & Cassman, K. G. (2014). Agricultural expansion and its impacts on tropical nature. *Trends in Ecology & Evolution*, *29*(2), 107–116. doi:10.1016/j.tree.2013.12.001 PMID:24388286

Laurance, W. F., Useche, D. C., Shoo, L. P., Herzog, S. K., Kessler, M., Escobar, F., ... Hietz, P. (2011). Global warming, elevational ranges and the vulnerability of tropical biota. *Biological Conservation*, *144*(1), 548–557. doi:10.1016/j.biocon.2010.10.010

Lavorel, S., Díaz, S., Cornelissen, J. H. C., Garnier, E., Harrison, S. P., McIntyre, S., & Urcelay, C. (2007). Plant functional types: are we getting any closer to the Holy Grail? In *Terrestrial ecosystems in a changing world* (pp. 149–164). Berlin, Germany: Springer. doi:10.1007/978-3-540-32730-1_13

Lavorel, S., & Garnier, É. (2002). Predicting changes in community composition and ecosystem functioning from plant traits: Revisiting the Holy Grail. *Functional Ecology*, *16*(5), 545–556. doi:10.1046/j.1365-2435.2002.00664.x

Lavorel, S., McIntyre, S., Landsberg, J., & Forbes, T. D. A. (1997). Plant functional classifications: From general groups to specific groups based on response to disturbance. *Trends in Ecology & Evolution*, *12*(12), 474–478. doi:10.1016/S0169-5347(97)01219-6 PMID:21238163

Layton, P. A., Guynn, S. T., & Guynn, D. C. (2004). Biodiversity metrics in Sustainable Certification Programs. *Journal of Forestry*, *102*(3), 46–52.

LEAD/FAO. (2006). L'ombre Portée De l' élevage, impacts environnementaux et options pour leur atténuation.

Leal, C. G., & de Gusmão Câmara, I. (Eds.). (2003). *The Atlantic Forest of South America: biodiversity status, threats, and outlook* (Vol. 1). Washington, DC: Island Press.

Lebrija-Trejos, E., Pérez-García, E. A., Meave, J. A., Poorter, L., & Bongers, F. (2011). Environmental changes during secondary succession in a tropical dry forest in Mexico. *Journal of Tropical Ecology*, *27*(5), 477–489. doi:10.1017/S0266467411000253

Lee, H. S., & Jarvis, P. G. (1996). Effects of tree maturity on some responses to elevated CO_2 in Sitka Spruce (*Picea sitchensis* Bong. Carr.). In G. W. Koch & H. A. Mooney (Eds.), *Carbon dioxide and terrestrial ecosystems* (pp. 53–70). San Diego, CA: Academic Press. doi:10.1016/B978-012505295-5/50005-7

Leemans, R. (1992). Global holdridge life zone classifications. Digital raster data on a 0.5 degree cartesian orthonormal geodetic (lat/long) 360x720 grid. In Global ecosystems database version 2.0. Boulder, CO: NOAA National Geophysical Data Center. Two independent single-attribute spatial layers. 537,430 bytes in 8 files. [first published in 1989]

Leemans, R. (1999). *Possible changes in natural vegetation patterns due to global warming*. Laxenburg, Austria: International Institute of Applied Systems Analysis.

Lee, Z., Casey, B., Arnone, R., Weidemann, A., Parsons, R., Montes, M. J., ... Dye, J. (2007). Water and bottom properties of a coastal environment derived from Hyperion data measured from the EO-1 spacecraft platform. *Journal of Applied Remote Sensing*, *1*(1), 011502. doi:10.1117/1.2822610

Leimgruber, P., Gagnon, J. B., Wemmer, C. M., Kelly, D. S., Songer, M. A., & Selig, E. R. (2003). Fragmentation of Asia's remaining wildlands: Implications for Asian elephant conservation. *Animal Conservation*, *6*(4), 347–359. doi:10.1017/S1367943003003421

Leon, J. R. R., van Leeuwen, W. J., & Casady, G. M. (2012). Using MODIS-NDVI for the modeling of post-wildfire vegetation response as a function of environmental conditions and pre-fire restoration treatments. *Remote Sensing*, *4*(3), 598–621. doi:10.3390/rs4030598

Leskiw, L. A. (1998). *Land capability classification for forest ecosystem in the oil stands region*. Edmonton, Canada: Alberta Environmental Protection.

Lezama, K. (2018). *Caracterización de la regeneración natural de bosques en tres ambientes contrastantes en el Retorno*. Guaviare, Colombia. doi:10.22201/fq.18708404e.2004.3.66178

Lieth, H. (1975). Primary production of the major vegetation units of the world. In Primary productivity of the biosphere (pp. 203-215). Berlin, Germany: Springer.

LIF (Leuser International Foundation). (n.d.). Unpublished Data.

Liu, B., Li, H., Zhu, B., Eissenstat, D. M., & Guo, D. (2015). Complementarity of absorptive fine roots and arbuscular mycorrhizal fungi in nutrient foraging strategies across 14 coexisting subtropical tree species. *The New Phytologist*, *208*(1), 125–136. doi:10.1111/nph.13434 PMID:25925733

Liu, H., Park Williams, A., Allen, C. D., Guo, D., Wu, X., Anenkhonov, O. A., ... Badmaeva, N. K. (2013). Rapid warming accelerates tree growth decline in semi-arid forests of Inner Asia. *Global Change Biology*, *19*(8), 2500–2510. doi:10.1111/gcb.12217 PMID:23564688

Liu, M., Hu, Y., Chang, Y., He, X., & Zhang, W. (2009). Land use and land cover change analysis and prediction in the upper reaches of the Minjiang River, China. *Environmental Management*, *43*(5), 899–907. doi:10.100700267-008-9263-7 PMID:19290569

Liu, Z. H., Tang, Z. Z., Huang, J. P., Huang, Z. Y., Huang, H., Li, Y. S., & ... . (2009). Survey on environmental lead pollution in rural areas around lead/zinc mining area in Guangxi, 2008. *Journal of Environmental Health*, *26*(8), 708–710.

Li, Y., & Ye, Y. (2014). Dynamics of decomposition and nutrient release of leaf litter in *Kandelia obovata* mangrove forests with different ages in Jiulongjiang Estuary, China. *Ecological Engineering*, *73*, 454–460. doi:10.1016/j.ecoleng.2014.09.102

Lloret, F. (2004). Régimen de incendios y regeneración. Ecología del bosque mediterráneo en un mundo cambiante, 101–126.

Lloyd, J., & Farquhar, G. D. (2008). Effects of rising temperatures and [CO_2] on the physiology of tropical forest trees. *Philosophical Transactions of the Royal Society of London. Series B, Biological Sciences*, *363*(1498), 1811–1817. doi:10.1098/rstb.2007.0032 PMID:18267901

Locatelli, B., & Pramova, E. (2011). Beyond mitigation: Forest-based adaptation to climate change. FAO and UNEP Meeting on Forests and Climate Change Adaptation in Asia, Oct. 26, 2011, Bangkok, Thailand. Retrieved from http://agritrop.cirad.fr/564596/1/document_564596.pdf

Loetsch, F., & Haller, K. E. (1973). *Forest inventory* (Vol. I). Mu¨nchen, Germany: BLV Verlagsgesellschaft.

Lohbeck, M., Poorter, L., Lebrija-Trejos, E., Martínez-Ramos, M., Meave, J. A., Paz, H., ... Bongers, F. (2013). Successional changes in functional composition contrast for dry and wet tropical forest. *Ecology*, *94*(6), 1211–1216. doi:10.1890/12-1850.1 PMID:23923479

Loranger, J., & Shipley, B. (2010). Interspecific covariation between stomatal density and other functional leaf traits in a local flora. *Botany*, *88*(1), 30–38. doi:10.1139/B09-103

Lovejoy, T. E., Bierregaard, B. O. Jr, Rylands, A. B., Malcolm, J. R., Quintela, C. E., Harper, L. H., ... Hays, M. B. (1986). Edge and Other Effects of Isolation on Amazon Forest Fragments. In M. E. Soulé (Ed.), *Conservation Biology; the Science of Scarcity and Diversity* (pp. 257–285). Sunderland, MA: Sinauer Associates

Lovett, G. M., Christenson, L. M., Groffman, P. M., Jones, C. G., Hart, J. E., & Mitchell, M. J. (2002). Insect Defoliation and Nitrogen Cycling in Forests: Laboratory, plot, and watershed studies indicate that most of the nitrogen released from forest foliage as a result of defoliation by insects is redistributed within the ecosystem, whereas only a small fraction of nitrogen is lost by leaching. *Bioscience*, *52*(4), 335–341. doi:10.1641/0006-3568(2002)052[0335:IDANCI]2.0.CO;2

Lugo, A. E., & Helmer, E. (2004). Emerging forests on abandoned land: Puerto Rico's new forests. *Forest Ecology and Management*, *190*(2-3), 145–161. doi:10.1016/j.foreco.2003.09.012

Lugo, A. E., Medina, E., Trejo-Torres, J. C., & Helmer, E. (2006). Botanical and ecological basis for the resilience of Antillean dry forests. In R. T. Pennington, G. P. Lewis, & J. A. Ratter (Eds.), *Neotropical savannas and seasonally dry forests: Plant diversity, biogeography and conservation* (pp. 359–381). Boca Raton, FL: CRC Press.

Lund, H. G. (2012). National definitions of forest/forestland listed by country. Forest information service. Retrieved from http://home.comcast.net/gyde/lundpub.htm

Lunetta, R. S., Knight, J. F., Ediriwickrema, J., Lyon, J. G., & Worthy, L. D. (2006). Land-cover change detection using multi-temporal MODIS NDVI data. *Remote Sensing of Environment*, *105*(2), 142–154. doi:10.1016/j.rse.2006.06.018

Luther, J. E., Fournier, R. A., Piercey, D. E., Guindon, L., & Hall, R. J. (2006). Biomass mapping using forest type and structure derived from Landsat TM imagery. *International Journal of Applied Earth Observation and Geoinformation*, *8*(3), 173–187. doi:10.1016/j.jag.2005.09.002

Luttge, U. (1997). Physiological ecology of tropical plants (2nd ed.). Berlin, Germany: Springer-Verlag. doi:10.1007/978-3-662-03340-1

Maass, J. M., Balvanera, P., Castillo, A., Daily, G. C., Mooney, H. A., Ehrlich, P., ... Martínez-Yrizar, A. (2005). Ecosystem services of tropical dry forests: insights from long term ecological and social research on the Pacific Coast of Mexico. *Ecology and society: a journal of integrative science for resilience and sustainability, 10*(1), 1-23.

Maass, J. M., Balvanera, P., Castillo, A., Daily, G. C., Mooney, H. A., Ehrlich, P., Quesada, M., … Martínez-Yrizar, A. (2005). Ecosystem services of tropical dry forests: insights from longterm ecological and social research on the Pacific Coast of Mexico. *Ecology and society: a journal of integrative science for resilience and sustainability, 10*(1), 1-23.

Maass, J. M., Balvanera, P., Castillo, A., Daily, G. C., Mooney, H. A., Ehrlich, P., ... Sarukhan, J. (2005). Ecosystem services of tropical dry forests: Insights from long-term ecological and social research on the pacific coast of Mexico. *Ecology and Society*, *10*(1), 17. doi:10.5751/ES-01219-100117

Maass, M., & Burgos, A. (2011). Water dynamics at the ecosystem level in seasonally dry tropical forests. In *Seasonally dry tropical forests* (pp. 141–156). Washington, DC: Island Press. doi:10.5822/978-1-61091-021-7_9

MacArthur, R. H. (1972). *Geographical Ecology. Patterns in the Distribution of Species.* Princeton, NJ: Princeton University Press.

MacArthur, R. H., & Pianka, E. R. (1966). On Optimal usc of a Patchy Environment. *American Naturalist*, *100*(916), 604–609. doi:10.1086/282454

MacArthur, R. H., & Wilson, E. O. (1963). An Equilibrium Theory of Insular Zoogeography. *Evolution; International Journal of Organic Evolution*, *17*(4), 373–387. doi:10.1111/j.1558-5646.1963.tb03295.x

MacArthur, R. H., & Wilson, E. O. (1967). *The Theory of Island Biogeography.* Princeton, NJ: Princeton University Press.

Mace, G. M., & Lande, R. (1991). Assessing extinction threats: Toward a reevaluation of IUCN threatened species categories. *Conservation Biology*, *5*(2), 148–157. doi:10.1111/j.1523-1739.1991.tb00119.x

MacKenzie, D. I., Nichols, J., & Royle, J. A., Pollock, K. H., Bailey, L., & Hines, J. (2006). Occupancy Estimation and Modeling. Inferring Patterns and Dynamics of Species Occurrence. Academic Press.

Maiti, S. K, & Ghose, M. K. (2005). Ecological restoration of acidic coalmine overburden dumps-An Indian case study. *Land Contamination & Reclamation, 13*(4), 1-10.

Maiti, S. K. (2006). Ecorestoration of coalmine overburden dumps in Korba, Gevra and Kusmunda area of SECL, India. *International Symposium on Environmental issues of Mineral industries,* VNIT, India.

Maiti, S. K. (2003). Eco restoration problems of coalmine overburden dumps. *International Conference on Ecorestoration*, Dehradun, India.

Maiti, S. K., & Das, D. (2004). Ecorestoration of coalmine overburden dumps. *Environment and Ecology, 22*(2), 329–333.

Maleki, K., Kiviste, A., & Korjus, H. (2015). Analysis of Individual Tree Competition Effect on Diameter Growth of Silver Birch in Estonia; *Forest Systems, 24*(2).

Malhi, Y., Roberts, J. T., Betts, R. A., Killeen, T. J., Li, W., & Nobre, C. A. (2008). Climate change, deforestation, and the fate of the Amazon. *Science, 319*(5860), 169–172. doi:10.1126cience.1146961 PMID:18048654

Malik, A., Shazia, L., Lone, F. A., Malik, M. A., Dar, Z. M. et al. (2018). Liquid Bio-Fertilizers and Sustainable Forestry: A Mini Review. *Open Access J. ARTOAJ 15*(3). doi:. 2018. 15. 555958 doi:10.19080/ARTOAJ

Marenco, R. A., Antezana-Vera, S. A., & Nascimento, H. C. S. (2009). Relationship between specific leaf area, leaf thickness, leaf water content and SPAD-502 readings in six Amazonian tree species. *Photosynthetica, 47*(2), 184–190. doi:10.100711099-009-0031-6

Marengo, J., Nobre, C. A., Betts, R. A., Cox, P. M., Sampaio, G., & Salazar, L. (2009). Global warming and climate change in Amazonia: Climate-vegetation feedback and impacts on water resources. *Amazonia and Global Change, 186*, 273–292. doi:10.1029/2008GM000743

Margono, B.A., Turubanova, S., Zhuravleva, I., Potapov, P., Tyukavina, A., Baccini, A., Goetz, S., & Hansen, M. C. (2012). Mapping and Monitoring Deforestation and Forest Degradation in Sumatra (Indonesia) Using Landsat Time Series Data Sets from 1990 to 2010. Environ. Res. Lett. doi:. doi:10.1088/1748-9326/7/3/034010

Margono, B. A., Potapov, P. V., Turubanova, S., Stolle, F., & Hansen, M. C. (2014). Primary forest cover loss in Indonesia over 2000–2012. *Nature Climate Change, 29*(June), 1–6. doi:10.1038/NCLIMATE2277

Marino, G., Aqil, M., & Shipley, B. (2010). The leaf economics spectrum and the prediction of photosynthetic light–response curves. *Functional Ecology, 24*(2), 263–272. doi:10.1111/j.1365-2435.2009.01630.x

Markesteijn, L., Iraipi, J., Bongers, F., & Poorter, L. (2010). Seasonal variation in soil and plant water potentials in a Bolivian tropical moist and dry forest. *Journal of Tropical Ecology, 26*(5), 497–508. doi:10.1017/S0266467410000271

Markesteijn, L., & Poorter, L. (2009). Seedling root morphology and biomass allocation of 62 tropical tree species in relation to drought-and shade-tolerance. *Journal of Ecology, 97*(2), 311–325. doi:10.1111/j.1365-2745.2008.01466.x

Markesteijn, L., Poorter, L., & Bongers, F. (2007). Light-dependent leaf trait variation in 43 tropical dry forest tree species. *American Journal of Botany, 94*(4), 515–525. doi:10.3732/ajb.94.4.515 PMID:21636421

Markesteijn, L., Poorter, L., Bongers, F., Paz, H., & Sack, L. (2011). Hydraulics and life history of tropical dry forest tree species: Coordination of species' drought and shade tolerance. *The New Phytologist, 191*(2), 480–495. doi:10.1111/j.1469-8137.2011.03708.x PMID:21477008

Martínez-Garza, C., Bongers, F., & Poorter, L. (2013). Are functional traits good predictors of species performance in restoration plantings in tropical abandoned pastures? *Forest Ecology and Management*, *303*, 35–45. doi:10.1016/j.foreco.2013.03.046

Martinez-Garza, C., Tobon, W., Campo, J., & Howe, H. (2013). Drought mortality of tree seedlings in an eroded tropical pasture L. *Degraded Development*, *24*(3), 287–295. doi:10.1002/ldr.1127

Martinez-Yrizar, A., Burquez, A., Maass, M., Robichaux, R., & Yetman, D. (2000). Structure and functioning of tropical deciduous forest in western Mexico. *The Tropical Deciduous Forest of Southern Sonora Mexico: Ecology and Conservation of a Threatened Ecosystem.*

Martínez-Yrízar, A., Núñez, S., Miranda, H., & Búrquez, A. (1999). Temporal and spatial variation of litter production in Sonoran Desert communities. *Plant Ecology*, *145*(1), 37–48. doi:10.1023/A:1009896201047

Martin, M. E., Newman, S. D., Aber, J. D., & Congalton, R. G. (1998). Determining forest species composition using high spectral resolution remote sensing data. *Remote Sensing of Environment*, *65*(3), 249–254. doi:10.1016/S0034-4257(98)00035-2

Martinuzzi, S., Gould, W. A., Gonzalez, O. M. R., Robles, A. M., Maldonado, P. C., Pérez-Buitrago, N., & Caban, J. J. F. (2008): Mapping tropical dry forest habitats integrating Landsat NDVI, Ikonos imagery, and topographic information in the Caribbean Island of Mona. *Rev. Biol. Trop., 56*(2), 625-639.

Martinuzzi, S., Gould, W. A., Vierling, L. A., Hudak, A. T., Nelson, R. F., & Evans, J. S. (2013). Quantifying tropical dry forest type and succession: Substantial improvement with LiDAR. *Biotropica*, *45*(2), 135–146. doi:10.1111/j.1744-7429.2012.00904.x

Marty, A., Bonnet, B., Guibert, B., & Swift, J. (2006, July). La mobilité pastorale et sa viabilité, entre atouts et défis (No. 3). IRAM.

Mascha, J. K. V.; Andrea Polle, F. M., & Thomas. (2010). Leaf litter decomposition in temperate deciduous forest stands with a decreasing fraction of beech (*Fagus sylvatica*). *Oecologia* 164: 1083-1094.

Mason-Romo, E. D., Ceballos, G., Lima, M., Martínez-Yrízar, A., Jaramillo, V. J., & Maass, M. (2018). Long-term population dynamics of small mammals in tropical dry forests, effects of unusual climate events, and implications for management and conservation. *Forest Ecology and Management*, *426*, 123–133. doi:10.1016/j.foreco.2018.05.058

Masoodi, T. H., & Soni, P. (1999). Physical properties of lime stone mine soil restored for forest land use. *Indian Journal of Forestry*, *22*(1-2), 30–36.

Matin, M. A., Chitale, V. S., Murthy, M. S., Uddin, K., Bajracharya, B., & Pradhan, S. (2017). Understanding forest fire patterns and risk in Nepal using remote sensing, geographic information system and historical fire data. *International Journal of Wildland Fire*, *26*(4), 276–286. doi:10.1071/WF16056

Matthew, K. M. (1981). *Materials for the flora of Tamil Nadu Carnatic. The Rapinat Herbarium.* Tiruchirappalli, India: St. Joseph's College.

Mauro, C. A., Dantas, M., & Roso, F. A. (1982). Geomorfologia. In S. D. Folha (Ed.), *Brasil, MME/SG/ Projeto RAD-AMBRASIL, Brasília: geologia, geomorfologia, pedologia, vegetação e usopotencial da terra* (pp. 205–296). Rio de Janeiro, Brazil: Ministério das Minas e Energia.

McCormack, M. L., Dickie, I. A., Eissenstat, D. M., Fahey, T. J., Fernandez, C. W., Guo, D., ... Leppälammi-Kujansuu, J. (2015). Redefining fine roots improves understanding of below-ground contributions to terrestrial biosphere processes. *The New Phytologist*, *207*(3), 505–518. doi:10.1111/nph.13363 PMID:25756288

McGill, B. J., Enquist, B. J., Weiher, E., & Westoby, M. (2006). Rebuilding community ecology from functional traits. *Trends in Ecology & Evolution*, *21*(4), 178–185. doi:10.1016/j.tree.2006.02.002 PMID:16701083

McGuire, K. L., & Treseder, K. K. (2010). Microbial communities and their relevance for ecosystem models: Decomposition as a case study. *Soil Biology & Biochemistry*, *42*(4), 529–535. doi:10.1016/j.soilbio.2009.11.016

McKown, A. D., Guy, R. D., Azam, M. S., Drewes, E. C., & Quamme, L. K. (2013). Seasonality and phenology alter functional leaf traits. *Oecologia*, *172*(3), 653–665. doi:10.100700442-012-2531-5 PMID:23212534

McLaren, J. R., & Turkington, R. (2010). Ecosystem properties determined by plant functional group identity. *Journal of Ecology*, *98*(2), 459–469. doi:10.1111/j.1365-2745.2009.01630.x

McLaren, K. P., & McDonald, M. A. (2003). The effects of moisture and shade on seed germination and seedling survival in a tropical dry forest in Jamaica. *Forest Ecology and Management*, *183*(1-3), 61–75. doi:10.1016/S0378-1127(03)00100-2

McRoberts, R. E. (2010a). Probability and model-based approaches to inference for proportion forest using satellite imagery as ancillary data. *Remote Sensing of Environment*, *114*(5), 1017–1025. doi:10.1016/j.rse.2009.12.013

McRoberts, R. E. (2010b). The effects of rectification and Global Positioning System errors on satellite image-based estimates of forest area. *Remote Sensing of Environment*, *114*(8), 1710–1717. doi:10.1016/j.rse.2010.03.001

McRoberts, R. E., & Tomppo, E. O. (2007). Remote sensing support for national forest inventories. *Remote Sensing of Environment*, *110*(4), 412–419. doi:10.1016/j.rse.2006.09.034

McRoberts, R. E., Wendt, D. G., Nelson, M. D., & Hansen, M. H. (2002). Using a land cover classification based on satellite imagery to improve the precision of forest inventory area estimates. *Remote Sensing of Environment*, *81*(1), 36–44. doi:10.1016/S0034-4257(01)00330-3

Meave, J., Flores-Rodriguez, C., Perez-Garcia, E., & Romero-Romero, M. (2012). Edaphic and seasonal heterogeneity of seed banks in agricultural fields of a tropical dry forest region in southern Mexico. *Botanical Sciences*, *90*(3), 313–329. doi:10.17129/botsci.393

Medina, E. (1983). Adaptations of tropical trees to moisture stress. In F. B. Golley (Ed.), *Tropical rain forest ecosystems, A structure and function* (pp. 225–237). Amsterdam, The Netherlands: Elsevier Scientific Company Publishing.

Medina, E. (1983). Adaptations of tropical trees to moisture stress. In F. B. Golley (Ed.), *Tropical rain forest ecosystems/ A structure and function* (pp. 225–237). Amsterdam, The Netherlands: Elsevier Scientific Company Publishing.

Medina, E. (1995). Diversity of life forms plants in neotropical dry forest. In H. A. Mooney, S. H. Bullock, & E. Medina (Eds.), *Seasonally dry tropical forests*. Cambridge, UK: University of Cambridge Press. doi:10.1017/CBO9780511753398.009

Meentemeyer, V. (1984). The geography of organic decomposition rates. *Annals of the Association of American Geographers*, *74*(4), 551–560. doi:10.1111/j.1467-8306.1984.tb01473.x

Mehta, V. K., Sullivan, P. J., Walter, M. T., Krishnaswamy, J., & De Gloria, S. D. (2008). Ecosystem impacts of disturbance in a dry tropical forests in southern India. *Ecohydrology*, *1*(2), 149–160. doi:10.1002/eco.14

MEHU. (2012, October). Cadre fonctionnel du Programme de Gestion des Forêts et Terroirs Riverains (PGFTR).

Meijaard, E., Sheil, D., Nasi, R., Augeri, D., Rosenbaum, B., Iskandar, D., ... O'Brien, T. (2005). *Life After Logging: Reconciling wildlife conservation and production forestry in Indonesian Borneo*. Bogor, Indonesia: Center for International Forestry Research.

Meir, P., & Pennington, R. T. (2011). Climate change and seasonal dry tropical forests. In R. Dirzo, H. Young, H. Mooney, & G. Ceballos (Eds.), *Seasonally dry tropical forests: ecology and conservation* (pp. 294–315). London, UK: Island Press. doi:10.5822/978-1-61091-021-7_16

Meli, F., & Meli, V. (2015). Modelling Drivers of Forest Cover Change in the Santchou Wildlife Reserve, West Cameroon using Remote Sensing and Land Use Dynamic Degree Indexes. *Canadian Journal of Tropical Geography*, *2*(2), 29–42.

Melkania, N. P. (1987). Allelopathy and its significance on the production of agroforestry plant associations. In P. K. Khosla & D. K. Khurana (Eds.), *Agroforestry for Rural Needs* (pp. 221–224). Solan, India: ISTS.

Melkania, N. P. (1991). Woody fodder species in Indian mountain ecosystems. *Range Management and Agroforestry*, *12*(2), 179–194.

Melo, O., & Vargas, R. (2003). Evaluacion ecologica y silvicultural de ecosistemas boscosos.

Melo, O., Fernandez-Méndez, F., & Villanueva, B. (2016). Hábitat lumínico, estructura, diversidad y dinámica de los bosques secos tropicales del alto magdalena. *Colombia Forestal*, *20*(1), 19–30. doi:10.14483/udistrital.jour.colomb.for.2017.1.a02

Méndez-Alonzo, R., Paz, H., Zuluaga, R. C., Rosell, J. A., & Olson, M. E. (2012). Coordinated evolution of leaf and stem economics in tropical dry forest trees. *Ecology*, *93*(11), 2397–2406. doi:10.1890/11-1213.1 PMID:23236911

MEPN. (2010). Plan d'Aménagement Participatif de la Forêt classée des trois rivières 2010-2019.

Mermet, L. (1991). *Dans quel sens pouvons-nous gérer l'environnement?* (pp. 68–81). Annales Des Mines/Gérer & Comprendre.

Mermet, L. (1992). *Stratégies pour la gestion de l'environnement: la nature comme jeu de société?* Paris, France: L'Harmattan.

Mermet, L., Billé, R., Leroy, M., Narcy, J.-B., & Poux, X. (2005). L'analyse stratégique de la gestion environnementale : Un cadre théorique pour penser l'efficacité en matière d'environnement. *Nature Sciences Sociétés*, *13*(2), 127–137. doi:10.1051/nss:2005018

Meziane, D., & Shipley, B. (1999). Interacting determinants of specific leaf area in 22 herbaceous species: Effects of irradiance and nutrient availability. *Plant, Cell & Environment*, *22*(5), 447–459. doi:10.1046/j.1365-3040.1999.00423.x

Miles, L., Newton, A. C., DeFries, R. S., Ravilious, C., May, I., Blyth, S., ... Gordon, J. E. (2006). A global overview of the conservation status of tropical dry forests. *Journal of Biogeography*, *33*(3), 491–505. doi:10.1111/j.1365-2699.2005.01424.x

Miles, L., Newton, A., Defries, R. S., Ravilious, C., May, I., Blyth, S., ... Gordon, J. (2006). A Global Overview of the Conservation Status of Tropical Dry Forests. *Journal of Biogeography*, *16*(3), 201–202.

Millar, C. S. (1974). Decomposition of coniferous leaf litter. In C. H. Dickinson & G. J. F. Pugh (Eds.), *Biology of plant litter decomposition* (pp. 105–128). London, UK: Academic Press. doi:10.1016/B978-0-12-215001-2.50010-6

MINENVIS. (2006). *Newsletter of the ENVIS Centre on Environmental Problems of Mining Areas. Number 49. June 2006*. Published by Professor Gurdeep Singh Head and Co-ordinator. ENVIS Nodal Centre of Mining Environment Indian School of Mines Dhanbad.

Ministry of Environment and Forest. (1988). *National Forest Policy*. India. Retrieved from http://envfor.nic.in/legis/forest/forest1.html

Ministry of Environment and Forest. (2004). *Handbook of Forest (Conservation) Act, 1980 (With Amendments made in 1988) Forest (Conservation) Rules, 2003 (With Amendments made in 12004) Guidelines & Clarifications* (Up to June, 2004). 1-2. Retrieved from http://wrd.bih.nic.in/guidelines/awadhesh02c.pdf

Ministry of Environment and Forests. (1999) National Policy and Macrolevel Action Strategy on Biodiversity. New Delhi, India: Government of India, Ministry of Environment and Forests.

Ministry of Environment. Forest and Climate Change. (2017). *India among top ten nations in the world in terms of Forest Area*. Retrieved from http://pib.nic.in/PressReleaseIframePage.aspx?PRID=1520284

Ministry of Environment. Forest and Climate Change. (2018). *Draft National Forest Policy*. Forest Policy Division. India. Retrieved from http://www.moef.nic.in/sites/default/files/Draft%20National%20Forest%20Policy%2C%202018.pdf

Mishra, A. K., & Singh, V. P. (2010). A review of drought concepts. *Journal of Hydrology (Amsterdam)*, *391*(1–2), 202–216. doi:10.1016/j.jhydrol.2010.07.012

Mitchell, T. D., & Jones, P. D. (2005). An improved method of constructing a database of monthly climate observations and associated high-resolution grids. *International Journal of Climatology*, *25*(6), 693–712. doi:10.1002/joc.1181

Mittermeier, R. A., Turner, W. R., Larsen, F. W., Brooks, T. M., & Gascon, C. (2011). Global Biodiversity Conservation: The Critical Role of Hotspots. In F. E. Zachos & J. C. Habel (Eds.), *Biodiversity Hotspots* (pp. 3–22). Berlin, Germany: Springer. doi:10.1007/978-3-642-20992-5_1

Moat, J., & Smith, P. P. (2007). *Atlas of the vegetation of Madagascar*. Kew, UK: Royal Botanic Gardens.

Mohammad, R., Romana, A., Mirajul, I., Kazi, M. M., & Mohammad, S. H. C. (2015). Estimating growing stock volume in a Bangladesh forest site using Landsat TM and field-measured data. *International Journal of Geomatics and Geosciences*, *6*(2), 1607–1619.

Mokany, K., Ash, J., & Roxburgh, S. (2008). Functional identity is more important than diversity in influencing ecosystem processes in a temperate native grassland. *Journal of Ecology*, *96*(5), 884–893. doi:10.1111/j.1365-2745.2008.01395.x

Moline, A. (1999). *Tropical dry forest restoration in the Guanacaste Conservation Area*. Costa Rica.

Mondal, P., & Southworth, J. (2010). Evaluation of conservation interventions using a cellular automata-Markov model. *Forest Ecology and Management*, *260*(10), 1716–1725. doi:10.1016/j.foreco.2010.08.017

Monserud, A. R. (2003). Evaluating Forest Models in a Sustainable Forest Management Context; *Forest Biometry. Modeling and Information Sciences*, *1*, 35–47.

Mooney, H. A., Bullock, S. H., & Medina, E. (1995). Introduction. In S. H. Bullock, H. A. Mooney, & E. Medina (Eds.), *Seasonally dry tropical forests* (pp. 1–8). Cambridge, UK: Cambridge University Press. doi:10.1017/CBO9780511753398.001

Mooney, H. A., Lubchenco, J., Dirzo, R., & Sala, O. E. (1995). Biodiversity and ecosystem functioning: basic principles. In V. H. Heywood (Ed.), *Global biodiversity assessment. United Nations Environment Programme* (pp. 275–325). Cambridge, UK: Cambridge University Press.

Moores, L. J., Pittman, B., & Kitchen, G. (1996). Forest ecological classification and mapping: Their application for ecosystem management in Newfoundland. *Environmental Monitoring and Assessment*, *39*(1–3), 571–577. doi:10.1007/BF00396169 PMID:24198030

Moreira, F., Arianoutsou, M., Corona, P., & De las Heras, J. (2012). Post-fire management and restoration of Southern European Forest.

Mori, A. S., Lertzman, K. P., & Gustafsson, L. (2017). Biodiversity and ecosystem services in forest ecosystems: A research agenda for applied forest ecology. *Journal of Applied Ecology*, *54*(1), 12–27. doi:10.1111/1365-2664.12669 PMID:29749183

Mostacedo, B., Fredericksen, T. S., Gould, K., & Toledo, M. (1999). *Comparacion de la respuesta de las comunidades vegetales a los incendios forestales en los Bosques Tropicales Secos y Húmedos de Bolivia. Proyecto de Manejo Forestal Sostenible BOLFOR*. Retrieved from http://pdf.usaid.gov/pdf_docs/Pnacl876.pdf

Mueller-Dombois, D. (1981) Fire in tropical ecosystems. In *Proceedings of the Conference: Fire Regimes and Ecosystem Properties*. Washington, DC: U.S. Department of Agriculture, General Technical Report WO-26, 137–176.

Mueller-Dombois, D., & Ellenberg, H. (1947). *Aims and methods of vegetation ecology*. New York, NY: John Wiley and Sons.

Mugnier, A., Cassagne, B., Bayos, N., & Lafon, C. (2009). Estimation of carbon stocks in Congo Basin forests for REDD: Comparative study conducted on 22 forest types, 4 countries and a management plan 4.8 million ha. Retrieved from https//www.observatoire-comifac.net/µdocs/edf2008/EN/state-of-forests

Mukhopadhyay, A., Mondal, P., Barik, J., Chowdhury, S. M., Ghosh, T., & Hazra, S. (2015). Changes in mangrove species assemblages and future prediction of the Bangladesh Sundarbans using Markov chain model and cellular automata. *Environmental Science. Processes & Impacts*, *17*(6), 1111–1117. doi:10.1039/C4EM00611A PMID:25719448

Muller, E. (2002). Reintegrate secondary forests into the landscape. ITTO.

Muoghalu, J. I., & Odiwe, A. I. (2011). Litter production and decomposition in cacao (*Theobroma cacao*) and kolanut (*Cola nitida*) plantations. *Ecotropica (Bonn)*, *17*, 79–90.

Murdiyarso, D., & Kauffman, J. B. (2011). *Addressing climate change adaptation and mitigation in tropical wetland ecosystems of Indonesia* (Vol. 41). Center for International Forestry Research (CIFOR). Retrieved from http://citeseerx.ist.psu.edu/viewdoc/download?doi=10.1.1.372.448&rep=rep1&type=pdf

Murdiyarso, D., Hergoualc'h, K., & Verchot, L. V. (2010). Opportunities for reducing greenhouse gas emissions in tropical peatlands. *Proceedings of the National Academy of Sciences of the United States of America*, *107*(46), 19655–19660. doi:10.1073/pnas.0911966107 PMID:21081702

Murie, O. J., Elbrock, M., & Peterson, R. T. (2005). Peterson Field Guide to Animal Tracks (3rd ed.). Boston, MA: Houghton Mifflin Harcourt.

Murphy, P. G., & Lugo, A. E. (1986). Ecology of tropical dry forest. *Annual Review of Ecology and Systematics*, *17*(1), 67–88. doi:10.1146/annurev.es.17.110186.000435

Murthy, I. K., Tiwari, R., & Ravindranath, N. H. (2011). Climate change and forests in India: Adaptation opportunities and challenges. *Mitigation and Adaptation Strategies for Global Change*, *16*(2), 161–175. doi:10.100711027-010-9261-y

Myers, R. L. (2006). Living with Fire—Sustaining Ecosystems & Livelihoods through Integrated Fire Management. Global Fire Initiative. *The Nature Conservancy, (June), 28.* Retrieved from http://www.nature.org/initiatives/ fire/files/integrated_fire_management_myers_2006.pdf

Myers, J. H., & Bazely, D. (2003). *Ecology and control of introduced plants* (Vol. 313). Cambridge, UK: Cambridge University Press. doi:10.1017/CBO9780511606564

Myers, N., Mittermeier, R. A., Mittermeier, C. G., Da Fonseca, G. A., & Kent, J. (2000). Biodiversity hotspots for conservation priorities. *Nature*, *403*(6772), 853–858. doi:10.1038/35002501 PMID:10706275

Naeem, S., Duffy, J. E., & Zavaleta, E. (2012). The functions of biological diversity in an age of extinction. *Science*, *336*(6087), 1401–1406. doi:10.1126cience.1215855 PMID:22700920

Naeem, S., & Wright, J. P. (2003). Disentangling biodiversity effects on ecosystem functioning: Deriving solutions to a seemingly insurmountable problem. *Ecology Letters*, *6*(6), 567–579. doi:10.1046/j.1461-0248.2003.00471.x

Nagendra, H. (2012). Assessing relatedness and redundancy of forest monitoring and change indicators. *Journal of Environmental Management*, *95*(1), 108–113. doi:10.1016/j.jenvman.2011.10.002 PMID:22115515

Nagendra, H., Rocchini, D., & Ghate, R. (2010). Beyond parks as monoliths: Spatially differentiating park-people relationships in the Tadoba Andhari Tiger Reserve in India. *Biological Conservation*, *143*(12), 2900–2908. doi:10.1016/j.biocon.2010.04.050

Naidu, M. T., & Kumar, O. A. (2016). Tree diversity, stand structure, and community composition of tropical forests in Eastern Ghats of Andhra Pradesh, India. *Journal of Asia-Pacific Biodiversity*, *9*(3), 328–334. doi:10.1016/j.japb.2016.03.019

Natarajan, D., Britto, S. J., Balaguru, B., Nagamurugan, N., Soosairaj, S., & Arockiasamy, D. I. (2004). Identification of conservation priority sites using remote sensing and GIS – A case study from Chitteri Hills, Eastern Ghats, Tamil Nadu. *Current Science*, *86*(9), 1316–1323.

National Capital Region Planning Board. (2005). *Regional Land Use*. Regional Plan 2021, (NCRPB). New Delhi, India. 255-259. Retrieved from http://ncrpb.nic.in/pdf_files/Draft%20Revised%20Regional%20Plan%202021/19%20Chapter%2017%20Regional%20Land%20Use.pdf

National Capital Region Planning Board. (2005). *Salient Features of the Regional Plan-2021 for NCR*. Regional Plan-2021, notified in 2005, New Delhi, India. Retrieved from http://ncrpb.nic.in/pdf_files/Salient%20Features_RP-2021.pdf

National Capital Region Planning Board. (2013). *Draft Revised Regional Plan 2021*. Ministry of Urban Development. Govt. of India. Retrieved from http://www.indiaenvironmentportal.org.in/files/file/Draft%20Revised%20Regional%20plan.pdf

National Green Tribunal. (2015). *Haryali Welfare Society Vs. Union of India & Ors.* Retrieved from http://greentribunal.gov.in/Writereaddata/Downloads/269-2013(PB-II-Judg)OA20-7-2015.pdf

National Green Tribunal. (2018). *Item No. 06.* Retrieved from http://www.greentribunal.gov.in/Writereaddata/Downloads/147-2014(PB-I)OA7-8-18.pdf

National Natural Parks PNN. (2014). Resolución 019 del 21 de febrero de 2014 Por medio de la cual se registra la Reserva Natural de la Sociedad Civil "Jabirú."

Natural Resources Canada. (2008): Fundamentals of Sensing. Retrieved from ww.nrcan.gc.ca

Navarro, G., Caballero, I., Silva, G., Parra, P.-C., Vázquez, Á., & Caldeira, R. (2017). Evaluation of forest fire on Madeira Island using Sentinel-2A MSI imagery. *International Journal of Applied Earth Observation and Geoinformation*, *58*, 97–106. doi:10.1016/j.jag.2017.02.003

NCVST. (2017). *Vulnerability through the eyes of the vulnerable: climate change induced uncertainties and Nepal's development predicaments, Institute for Social and Environmental Transition – Nepal (ISET-N)*. Kathmandu & Boulder, CO: Institute for Social and Environmental Transition.

NDRI. (2010). *United Nations World Food Programme (WFP); Food Security Monitoring Task Force of National Planning Commission (NPC). The food Security Atlas of Nepal; Nepal Development Research Institute*. Patan, Nepal: NDRI.

Nelson, H. P., Devenish-Nelson, E. S., Rusk, B. L., Geary, M., & Lawrence, A. J. (2018). A call to action for climate change research on Caribbean dry forests. *Regional Environmental Change*, *18*(5), 1337–1342. doi:10.100710113-018-1334-6

Nepal, M., Nepal, A., & Grimsrud, K. (2010). *Unbelievable but improved cookstoves are not helpful in reducing firewood demand in Nepal*, *16*, 1–23. doi:10.1017/S1355770X10000409

Nepstad, D. C., de Carvalho, C. R., Davidson, E. A., Jipp, P. H., Lefebvre, P. A., Negreiros, G. H., ... Vieira, S. (1994). The role of deep roots in the hydrological and carbon cycles of Amazonian forests and pastures. *Nature*, *372*(6507), 666–669. doi:10.1038/372666a0

Nepstad, D. C., Stickler, C. M., Filho, B. S., & Merry, F. (2008). Interactions among Amazon land use, forests and climate: Prospects for a near-term forest tipping point. *Philosophical Transactions of the Royal Society of London. Series B, Biological Sciences*, *363*(1498), 1737–1746. doi:10.1098/rstb.2007.0036 PMID:18267897

Nepstad, D., Lefebvre, P., Lopes da Silva, U., Tomasella, J., Schlesinger, P., Solórzano, L., ... Guerreira Benito, J. (2004). Amazon drought and its implications for forest flammability and tree growth: A basin-wide analysis. *Global Change Biology*, *10*(5), 704–717. doi:10.1111/j.1529-8817.2003.00772.x

Ngomand, A., Lebamba, J., Engone-Obiang, N., Lepengue, N., & M'Batchi, B. (1997). 2013. Characterization of Dry Biomass of forest-savannah mosaics in Gabon. *Journal of Applied Bioscience*, *68*, 5417–5425.

Ngueguim, J. R., Gatchui, H. C., Ayobami, S. T., & Orimoogunje, O. O. (2009). Evaluation of logging impacts on tropical rainforest in Eastern Cameroon using Remote Sensing and GIS techniques. *International Journal of Biological and Chemical Sciences*, *3*(4), 771–785. doi:10.4314/ijbcs.v3i4.47163

Nichols, E., Larsen, T., Spector, S., Davis, A. L., Escobar, F., Favila, M., & Vulinec, K. (2007). Global Dung Beetle Response to Tropical Forest Modification and Fragmentation: A Quantitative Literature Review and Meta-Analysis. *Biological Conservation*, *137*(1), 1–19. doi:10.1016/j.biocon.2007.01.023

Nicotra, A. B., Leigh, A., Boyce, C. K., Jones, C. S., Niklas, K. J., Royer, D. L., & Tsukaya, H. (2011). The evolution and functional significance of leaf shape in the angiosperms. *Functional Plant Biology*, *38*(7), 535–552. doi:10.1071/FP11057

Niinemets, Ü. (1999). Research review. Components of leaf dry mass per area–thickness and density–alter leaf photosynthetic capacity in reverse directions in woody plants. *The New Phytologist*, *144*(1), 35–47. doi:10.1046/j.1469-8137.1999.00466.x

Niinemets, U. (2010). A review of light interception in plant stands from leaf to canopy in different plant functional types and in species with varying shade tolerance. *Ecological Research*, *25*(4), 693–714. doi:10.100711284-010-0712-4

Niinemets, Ü., Díaz-Espejo, A., Flexas, J., Galmés, J., & Warren, C. R. (2009). Role of mesophyll diffusion conductance in constraining potential photosynthetic productivity in the field. *Journal of Experimental Botany*, *60*(8), 2249–2270. doi:10.1093/jxb/erp036 PMID:19395391

Niinemets, U., Keenan, T. F., & Hallik, L. (2015). A worldwide analysis of within-canopy variations in leaf structural, chemical and physiological traits across plant functional types. *The New Phytologist*, *205*(3), 973–993. doi:10.1111/nph.13096 PMID:25318596

Niinemets, Ü., & Tenhunen, J. D. (1997). A model separating leaf structural and physiological effects on carbon gain along light gradients for the shade-tolerant species Acer saccharum. *Plant, Cell & Environment*, *20*(7), 845–866. doi:10.1046/j.1365-3040.1997.d01-133.x

Nirmal Kumar, J. I., Kumar, R. N., Hiren Kumar, B. S., & Amit, N. G. (2000). Preliminary investigations of plant diversity of Khatana and Waghai forests of North western Ghats, South Gujarat. *International Journal of Ecology, Environmental Conservation*, *6*, 87–92.

Nirmal Kumar, J. I., Kumar, R. N., & Joseph, S. (2001). Tree species diversity of Khatana forest of south Gujarat. *Journal of Nature Conservation*, *13*, 149–166.

Nirmal Kumar, J. I., Kumar, R. N., & Joseph, S. (2002). Tree species diversity of Waghai forest of the Northern part of Western Ghats. *International Journal of Ecology, Environmental Conservation, 8*, 235–248.

Nkongmeneck, B. A. (1999). *The Boumba-Bek and Nki forest reserves: botany and ethnobotany WWF-Cameroon report.*

Norden, N., Angarita, H. A., Bongers, F., Martínez-Ramos, M., Granzow-de la Cerda, I., Van Breugel, M., & Finegan, B. (2015). Successional dynamics in Neotropical forests are as uncertain as they are predictable. *Proceedings of the National Academy of Sciences of the United States of America, 112*(26), 8013–8018. doi:10.1073/pnas.1500403112 PMID:26080411

Norden, N., Chave, J., Belbenoit, P., Caubère, A., Châtelet, P., Forget, P. M., ... Thébaud, C. (2009). Interspecific variation in seedling responses to seed limitation and habitat conditions for 14 Neotropical woody species. *Journal of Ecology, 97*(1), 186–197. doi:10.1111/j.1365-2745.2008.01444.x

Ntonmen, Y. (2013). Biodiversity and carbon stock of plant formations from Akomnyada and Nkologok (Mbalmayo). (Master thesis), University of Yaounde I.

Ocampo-Zuleta, K., & Bravo, S. (2019). Reclutamiento de especies leñosas en bosques tropicales expuestos a incendios: Una revisi6n. Ecosistemas. *Revista Científica y Técnica de Ecología y Medio Ambiente, 28*(1), 106–117.

Odihi, J. (2003). Deforestation in Afforestation Priority Zone in Sudano-Sahelian Nigeria. *Applied Geography (Sevenoaks, England), 23*(4), 227–259. doi:10.1016/j.apgeog.2003.08.004

Odiogor, H. (2010). *Special Report on Desertification in Nigeria: The Sun Eats Our Land.* Vanguard Newspaper. Retrieved from http://odili.net/news/source/2010/may/4/300.html

Oguntala, A. B. (1982): *Wildland fires particularly in tropical regions.* World Meteorological Organization. CAgM Report No.10.

Oindo, B. O., de By, R. A., & Skidmore, A. K. (2000). Interannual variability of NDVI and bird species diversity in Kenya. *International Journal of Applied Earth Observation and Geoinformation, 2*(3), 172–180. doi:10.1016/S0303-2434(00)85011-4

Olivares, E., & Medina, E. (1992). Water and nutrient relations of woody perennials from tropical dry forests. *Journal of Vegetation Science, 3*(3), 383–392. doi:10.2307/3235764

Oliveira-Filho, A. T., Jarenkow, J. A., & Rodal, M. J. N. (2006). Floristic relationships of seasonally dry forests of eastern South America based on tree species distribution patterns. In R. T. Pennington, G. P. Lewis, & J. A. Ratter (Eds.), *Neotropical savannas and seasonally dry forests: Plant diversity, biogeography and conservation* (pp. 159–192). Boca Raton, FL: CRC Press. doi:10.1201/9781420004496-7

Oliveira, R. S., Dawson, T. E., Burgess, S. S., & Nepstad, D. C. (2005). Hydraulic redistribution in three Amazonian trees. *Oecologia, 145*(3), 354–363. doi:10.100700442-005-0108-2 PMID:16091971

Olivier, R. C. D. (1978). *On the ecology of the Asian elephant, Elephas maximus Linn. With particular reference to Malaya and Sri Lanka.* [Doctoral thesis], University of Cambridge, Cambridge.

Olson, D. M., & Dinerstein, E. (1998). The Global 200: A representation approach to conserving the Earth's most biologically valuable ecoregions. *Conservation Biology, 12*(3), 502–515. doi:10.1046/j.1523-1739.1998.012003502.x

Olson, D. M., & Dinerstein, E. (2002). The Global 200: Priority ecoregions for global conservation. *Annals of the Missouri Botanical Garden, 89*(2), 199–224. doi:10.2307/3298564

Olson, D. M., Dinerstein, E., Abell, R., Allnutt, T., Carpenter, C., McClenachan, L., ... Thieme, M. (2000). *The global 200: a representation approach to conserving the Earth's distinctive ecoregions*. Washington, DC: Conservation Science Program, World Wildlife Fund-US.

Olson, D. M., Dinerstein, E., Wikramanayake, E. D., Burgess, N. D., Powell, G. V., Underwood, E. C., ... Loucks, C. J. (2001). Terrestrial ecoregions of the world: A new map of life on earth. *Bioscience*, *51*(11), 933–938. doi:10.1641/0006-3568(2001)051[0933:TEOTWA]2.0.CO;2

Onojeghuo, A. O., & Onojeghuo, A. R. (2015). Mapping forest transition trends in Okomu reserve using Landsat and UK-DMC-2 satellite data. *South African Journal of Geomatics*, *4*(4), 486–501. doi:10.4314ajg.v4i4.9

Onyekwelu, J. C. (2003). Choosing appropriate index age for estimating site index of Gmelina arborea timber plantations in Oluwa forest reserve, Nigeria. [</ths>]. *Journal of Food Agriculture and Environment*, *1*(3 & 4), 286–290.

Onyekwelu, J. C. (2005). Site index curves for site quality assessment of *Nauclea diderrichii* monoculture plantations in Omo Forest Reserve, Nigeria. *Journal of Tropical Forest Science*, *17*(4), 532–542.

Oostra, S.; Majdi, H., & Olsson, M. (2006). Impact of tree species on soil carbon stocks and soil acidity in southern Sweden. *Scandinavian Journal of Forest Research,* 21, 364-371. doi: 27580600950172. doi:10.1080/028

Oseia, A., Edmund C. M., & Yaw, A. T. (2005): Use of Remote Sensing Data to Detect Environmental Degradation in the Coastal Region of Southern Nigeria. 1-6.

Osher, L. J., Matson, P. A., & Amundson, R. (2003). Effect of land use change on soil carbon in Hawaii. *Biogeochemistry*, *65*(2), 213–232. doi:10.1023/A:1026048612540

Ostrom, E. (2009). *A General Framework for Analyzing Sustainability of Social-Ecological Systems, 325*(5939), 419–422.

Ostrom, E. (2009). A General Framework for Analyzing Sustainability of Social-Ecological Systems. *Science*, *325*(5939), 419–422. doi:10.1126cience.1172133 PMID:19628857

Otterstrom, S. M., Schwartz, M. W., & Velazquez-Rocha, I. (2006). Responses to Fire in Selected Tropical Dry Forest Trees. *Biotropica*, *38*(5), 592–598. doi:10.1111/j.1744-7429.2006.00188.x

Ouaba. (2006). Phytoecological and socio-economic study of the classified forests of Burkina Faso: Case of the Niangoloko classified forest. (Doctoral thesis), University of Ouagadougou.

Ouya Bondoro. (2010). *Conservation and Sustainable Use of Biodiversity Around and in Protected Areas in South-East Chad: The Case of Manda Park and the Djoli-Kera Forest Uses and Risks, What Strategies for Upgrading and Local Development?* (Doctoral thesis in geography and spatial planning), University Paul-Valery of Montpelier III.

Pacheco, P., & Mertens, B. (2004). Land use change and agricultural development in Santa Cruz, Bolivia. *CEP*, *66095*, 780.

Pain, A. (2009). What is driving change in the Nilgiri Biosphere Reserve and what effects might such change have on the role of NTFP in the livelihoods of indigenous people? In R. Dutt, J. Seeley, & P. Roy (Orgs.), *Proceedings of the Biodiversity and Livelihoods Conference, Coonoor, India.*

Paine, C. T., Amissah, L., Auge, H., Baraloto, C., Baruffol, M., Bourland, N., ... Doust, S. (2015). Globally, functional traits are weak predictors of juvenile tree growth, and we do not know why. *Journal of Ecology*, *103*(4), 978–989. doi:10.1111/1365-2745.12401

Paine, C. T., Baraloto, C., Chave, J., & Hérault, B. (2011). Functional traits of individual trees reveal ecological constraints on community assembly in tropical rain forests. *Oikos*, *120*(5), 720–727. doi:10.1111/j.1600-0706.2010.19110.x

Pan, C., Zhao, H., Zhao, X., Han, H., Wang, Y., & Li, J. (2013). Biophysical properties as determinants for soil organic carbon and total nitrogen in grassland salinization. *PLoS One*, *8*(1), e54827. doi:10.1371/journal.pone.0054827 PMID:23372776

Panda, T., Panda, B., & Mishra, N. (2007). Comparative study of *Penicillia* from soil, leaf, litter and air in coastal sandy belt of Orissa. *Journal of Phytological Research*, *20*, 335–336.

Pande, P. K. (2001). Litter nutrient dynamics of *Shorea robusta* Gaertn. plantation at Doon Valley (Uttaranchal), India. *Indian Forester*, *127*, 980–994.

Pande, P. K., Meshram, P. B., & Banarjee, S. K. (2002). Litter production and nutrient return in tropical dry deciduous teak forests of Satpura plateau in central India. *Tropical Ecology*, *43*, 337–344.

Pandey, D. N., Chaubey, A. C., Gupta, A. K., & Vardhan, H. (2005). Mine Spoil Restoration: A strategy combining rainwater harvesting and adaptation to random recurrence of droughts in Rajasthan. *International Network on Ethnoforestry*, p. 25.

Pandey, P. C., Anand, A., & Srivastava, P. K. (2019). Spatial distribution of mangrove forest species and biomass assessment using field inventory and earth observation hyperspectral data. *Biodiversity and Conservation*, 1–20.

Parajuli, A., Chand, D. B., Rayamajhi, B., Khanal, R., Baral, S., Malla, Y., & Poudel, S. (2015). Spatial and temporal distribution of forest fires in Nepal. In *XIV World Forestry Congress*, Durban, South Africa (pp. 7-11).

Parker, T. A., III, & Carr, J. L. (Eds.). (1992). Status of forest remnants in the Cordillera de la Costa and adjacent areas of southwestern Ecuador. RAP Working Papers 2 (p. 172). Conservation International, Washington, DC.

Parker, T. A., III, Gentry, A. H., Foster, R. B., Emmons, L. H., & Remsen, J. V., Jr. (1993). The lowland dry forests of Santa Cruz, Bolivia: A global conservation priority. RAP Working Papers 4 (p. 104). Conservation International, Washington, DC.

Parmesan, C. (2006). Ecological and evolutionary responses to recent climate change. *Annual Review of Ecology Evolution and Systematics*, *37*(1), 637–669. doi:10.1146/annurev.ecolsys.37.091305.110100

Parry, J., Hammill, A., & Drexhage, J. (2005). *Climate Change and Adaptation. International Institute for Sustainable Development*. Winnipeg, Canada: IISD.

Pascoe, J. (2018). Tropical dry forests 'canary in the coal mine' on climate change. Retrieved from https://phys.org/news/2018-03-tropical-forests-canary-coal-climate.html

Pastor, J., Gardner, R. H., Dale, V. H., & Post, W. M. (1987). Successional changes in nitrogen availability as a potential factor contributing to spruce declines in boreal North America. *Canadian Journal of Forest Research*, *17*(11), 1394–1400. doi:10.1139/x87-216

Patricio, M. S., Nunes, L. F., & Pereira, E. L. (2012). Litterfall and litter decomposition in chestnut high forest stands in northern Portugal. *Forest Systems*, *21*(2), 259–271. doi:10.5424/fs/2012212-02711

Pausas, J. G., Bradstock, R., Keith, D. A., & Keeley, J. E. (2004). Plant functional traits in relation to fire in corwn-fire ecosystems. *Ecology*, *85*(4), 1085–1100. doi:10.1890/02-4094

Pausas, J. G., & Lavorel, S. (2003). A hierarchical deductive approach for functional types in disturbed ecosystems. *Journal of Vegetation Science*, *14*(3), 409–416. doi:10.1111/j.1654-1103.2003.tb02166.x

Pausas, J. G., Pratt, R. B., Keeley, J. E., Jacobsen, A. L., Ramirez, A. R., Vilagrosa, A., ... Davis, S. D. (2016). Towards understanding resprouting at the global scale. *The New Phytologist*, *209*(3), 945–954. doi:10.1111/nph.13644 PMID:26443127

Pausas, J. G., & Ribeiro, E. (2017). Fire and plant diversity at the global scale. *Global Ecology and Biogeography, 26*(8), 889–897. doi:10.1111/geb.12596

Pearson, R. L., & Miller, L. D. (1972). Remote mapping of standing crop biomass for estimation of the productivity of the shortgrass prairie. ed. *Remote Sensing of Environment, VIII*, 1355.

Pekkarinen, A., Reithmaier, L., & Strobl, P. (2009). Pan-European forest/non-forest mapping with Landsat ETM+ and CORINE Land Cover 2000 data. *ISPRS Journal of Photogrammetry and Remote Sensing, 64*(2), 171–183. doi:10.1016/j.isprsjprs.2008.09.004

Penman, J., Gytarsky, M., Hiraishi, T., Krug, T., Kruger, D., & Pipatti, R. … Wagner, F. (2003). Good practice guidance for land use, land-use change and forestry. Kanagawa Prefecture, Japan: Institute for Global Environmental Strategies.

Penman, J., Gytarsky, M., Hiraishi, T., Krug, T., Kruger, D., & Pipatti, R. … Wagner, F. (Eds.) (2003). Good practice guidance for land use, land-use change and forestry. Intergovernmental Panel on Climate Change: IPCC National Greenhouse Gas Inventories Programme. Kanagawa Japan: Institute for Global Environmental Strategies (IGES) for the IPCC. Retrieved from http://www.ipcc-nggip.iges.or.jp/public/gpglulucf/gpglulucf_contents.htm

Pennington, R. T., Lavin, M., & Oliveira-Filho, A. (2009). Woody plant diversity, evolution, and ecology in the tropics: Perspectives from seasonally dry tropical forests. *Annual Review of Ecology Evolution and Systematics, 40*(1), 437–457. doi:10.1146/annurev.ecolsys.110308.120327

Pennington, R. T., Lavin, M., Prado, D. E., Pendry, C. A., Pell, S. K., & Butterworth, C. A. (2004). Historical climate change and speciation: Neotropical seasonally dry forest plants show patterns of both tertiary and quaternary diversification. *Philosophical Transactions of the Royal Society of London. Series B, Biological Sciences, 359*(1443), 515–538. doi:10.1098/rstb.2003.1435 PMID:15212100

Pennington, R. T., Lewis, G. P., & Ratter, J. A. (2006). An overview of the plant diversity, biogeography and conservation of neotropical savannas and seasonally dry forests. In R. T. Pennington, G. P. Lewis, & J. A. Ratter (Eds.), *Neotropical savannas and seasonally dry forests: Plant diversity, biogeography and conservation* (pp. 1–29). Boca Raton, FL: CRC Press. doi:10.1201/9781420004496-1

Pennington, T., Lewis, G., & Ratter, J. (2006). *Neotropical Savannas and Seasonally Dry Forests: Plant Diversity, Biogeography and Conservation.* Boca Raton, FL: CRC Press. doi:10.1201/9781420004496

Pennington, T., Prado, D. E., & Pendry, C. A. (2000). Neotropical seasonally dry forests and quaternary vegetation changes. *Journal of Biogeography, 27*(2), 261–273. doi:10.1046/j.1365-2699.2000.00397.x

Pereira Marinho, F., Guilherme Gerhardt, G., Pellegrini Manhaes, A., Weisser, W. W., & Ganade, G. (2016). *Effects of past and present land use on vegetation cover and regeneration in a tropical dryland forest, 132*, 26–33. doi:10.1016/j.jaridenv.2016.04.006

Peres, C. A., Baider, C., Zuidema, P. A., Wadt, H. O., Kainer, K. A., Gomes-Silva, D. A. P., … Freckleton, R. P. (2004). *Demographic Threats to the Sustainability of Brazil Nut Exploitation, 302*(December 2003).

Peres, C. A., Barlow, J., & Laurance, W. F. (2016). *Detecting anthropogenic disturbance in tropical forests. 21*(March), 3–6. doi:10.1016/j.tree.2006.03.007

Perez-Harguindeguy, N., Diaz, S., Garnier, E., Lavorel, S., Poorter, H., Jaureguiberry, P., ... Urcelay, C. (2016). Corrigendum to: New handbook for standardised measurement of plant functional traits worldwide. *Australian Journal of Botany, 64*(8), 715–716. doi:10.1071/BT12225_CO

Petchey, O. L., & Gaston, K. J. (2002). Functional diversity (FD), species richness and community composition. *Ecology Letters*, *5*(3), 402–411. doi:10.1046/j.1461-0248.2002.00339.x

Petchey, O. L., & Gaston, K. J. (2006). Functional diversity: Back to basics and looking forward. *Ecology Letters*, *9*(6), 741–758. doi:10.1111/j.1461-0248.2006.00924.x PMID:16706917

Peters, E. B., Wythers, K. R., Zhang, S., Bradford, J. B., & Reich, P. B. (2013). Potential climate change impacts on temperate forest ecosystem processes. *Canadian Journal of Forest Research*, *43*(10), 939–950. doi:10.1139/cjfr-2013-0013

Peterson, A. T., & Kluza, D. A. (2005). Ecological niche modeling as a new paradigm for large-scale investigations of diversity and distribution of birds. In J. C. Ralph & T. D. Rich (Eds.), Proceedings of the Third International Partners in Flight Conference, 2002, Asilomar (pp. 1201-1204). Albany: U.S. Department of Agriculture Forest Service.

Peterson, A. T., Sánchez-Cordero, V., Soberón, J., Bartley, J., Buddemeier, R. W., & Navarro-Sigüenza, A. G. (2001). Effects of global climate change on geographic distributions of Mexican Cracidae. *Ecological Modelling*, *144*(1), 21–30. doi:10.1016/S0304-3800(01)00345-3

Pettorelli, N., Safi, K., & Turner, W. (2014). Satellite remote sensing, biodiversity research and conservation of the future. *Philosophical Transactions of the Royal Society of London. Series B, Biological Sciences*, *369*(1643). doi:10.1098/rstb.2013.0190 PMID:24733945

Phillips, O. L., Van Der Heijden, G., Lewis, S. L., López-González, G., Aragão, L. E., Lloyd, J., ... Amaral, I. (2010). Drought–mortality relationships for tropical forests. *The New Phytologist*, *187*(3), 631–646. doi:10.1111/j.1469-8137.2010.03359.x PMID:20659252

Phillipson, P. B., Schatz, G. E., Lowry, P. P., II, & Labat, J. N. (2006). A catalogue of the vascular plants of Madagascar. In S. A. Ghazanfar & H. J. Beentje (Eds.), Taxonomy and ecology of African plants: their conservation and sustainable use (pp. 613–627). *Proceedings xVIIth AETFAT Congress*. Kew, UK: Royal Botanic Gardens.

Piana, R. P., & Marsden, S. J. (2014). *Impacts of cattle grazing on forest structure and raptor distribution within a neotropical protected area*, 559–572. doi:10.100710531-013-0616-z

Piao, S., Fang, J., Ciais, P., Peylin, P., Huang, Y., Sitch, S., & Wang, T. (2009). The carbon balance of terrestrial ecosystems in China. *Nature*, *458*(7241), 1009–1013. doi:10.1038/nature07944 PMID:19396142

Picard, N. (2006). Forest inventory method. Participatory Rural Development Project in the Middle Atlas Central (Khenifra Project).

Piha, M., Mpepereki, S. M. (Ed.), & Makonese, F. T. (1998). Use of legumes in the revegetation of mine wastes. Harnessing biological nitrogen fixation in African agriculture: challenges and opportunities. *Sixth International Conference of the African Association for Biological Nitrogen Fixation, Harare, Zimbabwe*, Sept. 12-17, 1994. Selected papers. 251-254.

Pimentel, D., Harvey, C., Resosudarmo, P., Sinclair, K., Kurz, D., McNair, M., ... Blair, R. (1995). Environmental and economic costs of soil erosion and conservation benefits. *Science*, *267*(5201), 1117–1123. doi:10.1126cience.267.5201.1117 PMID:17789193

Pimm, S. L., Alibhai, S., Bergl, R., Dehgan, A., Giri, C., Jewell, Z., ... Loarie, S. (2015). Emerging Technologies to Conserve Biodiversity. *Trends in Ecology & Evolution*, *30*(11), 685–696. doi:10.1016/j.tree.2015.08.008 PMID:26437636

Pimm, S. L., & Raven, P. (2000). Biodiversity: Extinction by numbers. *Nature*, *403*(6772), 843–845. doi:10.1038/35002708 PMID:10706267

Pinto, O. B. Jr, Vourlitis, G. L., & Carneiro, E. M. (2018). Interactions between vegetation, hydrology, and litter inputs on decomposition and soil CO_2 efflux of tropical forests in the Brazilian Pantanal. *Forests*, *9*(5), 281–297. doi:10.3390/f9050281

Piper, F. I., & Fajardo, A. (2014). Foliar habit, tolerance to defoliation and their link to carbon and nitrogen storage. *Journal of Ecology*, *102*(5), 1101–1111. doi:10.1111/1365-2745.12284

Piperno, D. R., & Pearsall, D. M. (2000). *Origins of Agriculture in the Neotropics*. Academic Press.

Pizano, C., & García, H. (2014). El Bosque seco tropical en Colombia. Bogota D.C., Colombia: Instituto de Investigacion de Recursos Biologicos Alexander von Humboldt (IAvH).

Pizano, C., González-M, R., Hernández-Jaramillo, A., & García, H. (2017). Agenda de investigación y monitoreo en bosques secos de Colombia (2013-2015): Fortaleciendo redes de colaboración para su gestión integral en el territorio. *Biodiversidad En La Práctica*, *2*(1), III. doi:10.4067/S0717-97072014000100001

Polansky, C. (2003). Participatory Forest Management in Africa: Lessons not learned. *International Journal of Sustainable Development and World Ecology*, *10*(2), 109–118. doi:10.1080/13504500309469790

Polyakova, O., & Billor, N. (2007). Impact of deciduous tree species on litterfall quality, decomposition rates and nutrient circulation in pine stands. *Forest Ecology and Management*, *253*(1-3), 11–18. doi:10.1016/j.foreco.2007.06.049

Pontius, R. G. J., & Malanson, J. (2005). Comparison of the structure and accuracy of two land change models. *International Journal of Geographical Information Science*, *19*(2), 243–265. doi:10.1080/13658810410001713434

Poorter, H., Niinemets, Ü., Poorter, L., Wright, I. J., & Villar, R. (2009). Causes and consequences of variation in leaf mass per area (LMA): A meta-analysis. *The New Phytologist*, *182*(3), 565–588. doi:10.1111/j.1469-8137.2009.02830.x PMID:19434804

Poorter, L. (2009). Leaf traits show different relationships with shade tolerance in moist versus dry tropical forests. *The New Phytologist*, *181*(4), 890–900. doi:10.1111/j.1469-8137.2008.02715.x PMID:19140935

Poorter, L., & Bongers, F. (2006). Leaf traits are good predictors of plant performance across 53 rain forest species. *Ecology*, *87*(7), 1733–1743. doi:10.1890/0012-9658(2006)87[1733:LTAGPO]2.0.CO;2 PMID:16922323

Poorter, L., Wright, S. J., Paz, H., Ackerly, D. D., Condit, R., Ibarra-Manríquez, G., ... Muller-Landau, H. C. (2008). Are functional traits good predictors of demographic rates? Evidence from five neotropical forests. *Ecology*, *89*(7), 1908–1920. doi:10.1890/07-0207.1 PMID:18705377

Portillo-Quintero, C., & Smith, V. (2018). Emerging trends of tropical dry forests loss in North & Central America during 2001–2013: The role of contextual and underlying drivers. *Applied Geography, 94*(October 2017), 58–70. doi:10.1016/j.apgeog.2018.03.011

Portillo-Quintero, C. A., & Sánchez-Azofeifa, G. A. (2010). Extent and conservation of tropical dry forests in the Americas. *Biological Conservation*, *143*(1), 144–155. doi:10.1016/j.biocon.2009.09.020

Portillo-Quintero, C. P., Sanchez-Azofeifa, G., Calvo-Alvarado, J. C., Quesada, M., & do Espirito Santo, M. M. (2015). The role of tropical dry forests for biodiversity, carbon and water conservation in the neotropics: Lessons learned for its sustainable management. *Regional Environmental Change*, *15*(6), 1039–1049. doi:10.100710113-014-0689-6

Posada, J. M., Lechowicz, M. J., & Kitajima, K. (2009). Optimal photosynthetic use of light by tropical tree crowns achieved by adjustment of individual leaf angles and nitrogen content. *Annals of Botany*, *103*(5), 795–805. doi:10.1093/aob/mcn265 PMID:19151040

Post, E., Forchhammer, M. C., Bret-Harte, M. S., Callaghan, T. V., Christensen, T. R., Elberling, B., ... Ims, R. A. (2009). Ecological dynamics across the Arctic associated with recent climate change. *Science*, *325*(5946), 1355–1358. doi:10.1126cience.1173113 PMID:19745143

Pouyat, R. V., Setälä, H., Szlavecz, K., Yesilonis, I. D., Cilliers, S., Hornung, E., ... Whitlow, T. H. (2017). Introducing GLUSEEN: A new open access and experimental network in urban soil ecology. *Journal of Urban Economics*, *3*(1). doi:10.1093/jue/jux002

Powers, R. F. (2001). Assessing Potential Sustainable Wood Yield. In J. Evans (Ed.). The Forestry Handbook: Applying Forest Science for Sustainable Management (2: 105-128). Oxford, UK: Blackwell Science, Ltd. doi:10.1002/9780470757079.ch5

Powers, J. S., Becknell, J. M., Irving, J., & Perez-Aviles, D. (2009). Diversity and structure of regenerating tropical dry forests in Costa Rica: Geographic patterns and environmental drivers. *Forest Ecology and Management*, *258*(6), 959–970. doi:10.1016/j.foreco.2008.10.036

Powers, J. S., Montgomery, R. A., Adair, E. C., Brearley, F. Q., DeWalt, S. J., Castanho, C. T., & González-Iturbe, J. A. (2009). Decomposition in tropical forests: A pan-tropical study of the effects of litter type, litter placement and mesofaunal exclusion across a precipitation gradient. *Journal of Ecology*, *97*(4), 801–811. doi:10.1111/j.1365-2745.2009.01515.x

Powers, J. S., & Tiffin, P. (2010). Plant functional type classifications in tropical dry forests in Costa Rica: Leaf habit versus taxonomic approaches. *Functional Ecology*, *24*(4), 927–936. doi:10.1111/j.1365-2435.2010.01701.x

Prado, D. E. (2000). Seasonally dry forests of tropical South America: From forgotten ecosystems to a new phytogeographic unit. *Edinburgh Journal of Botany*, *57*(3), 437–461. doi:10.1017/S096042860000041X

Prado, D. E., & Gibbs, P. E. (1993). Patterns of species distributions in the dry seasonal forests of South America. *Annals of the Missouri Botanical Garden*, *80*(4), 902–927. doi:10.2307/2399937

Pragasan, L. A., & Parthasarathy, N. (2005). Litter production in tropical dry evergreen forests of south India in relation to season, plant life-forms and physiognomic groups. *Current Science*, *88*, 1255–1263.

Prescott, C. E. (2010). Litter decomposition: What controls it and how can we alter it to sequester more carbon in forest soils? *Biogeochemistry*, *101*(1-3), 133–149. doi:10.100710533-010-9439-0

Prescott, C. E., Reid, A., Wu, S. Y., & Nilsson, M. C. (2017). Decomposition rates of surface and buried forest-floor material. *Canadian Journal of Forest Research*, *47*(8), 1140–1144. doi:10.1139/cjfr-2016-0533

Preston-Mafham, K. (1991). *Madagascar: A natural history*. Oxford, UK: Facts on File.

Prieto-Torres, D. A., Navarro-Sigüenza, A. G., Santiago-Alarcon, D., & Rojas-Soto, O. R. (2016). Response of the endangered tropical dry forests to climate change and the role of Mexican Protected Areas for their conservation. *Global Change Biology*, *22*(1), 364–379. doi:10.1111/gcb.13090 PMID:26367278

Pulla, S., Ramaswami, G., Mondal, N., Chitra-Tarak, R., Suresh, H. S., Dattaraja, H. S., ... Sukumar, R. (2015). Assessing the resilience of global seasonally dry tropical forests. *International Forestry Review*, *17*(2), 91–113. doi:10.1505/146554815815834796

Purahong, W., Kapturska, D., Pecyna, M. J., Schulz, E., Schloter, M., Buscot, F., ... Krüger, D. (2014). Influence of different forest system management practices on leaf litter decomposition rates, nutrient dynamics and the activity of ligninolytic enzymes: A case study from Central European forests. *PLoS One*, *9*(4). doi:10.1371/journal.pone.0093700 PMID:24699676

Purkis, S. J., & Klemas, V. V. (2011). *Remote sensing and global environmental change*. Hoboken, NJ: John Wiley & Sons. doi:10.1002/9781118687659

Putz, F. E., & Redford, K. H. (2010). The Importance of Defining 'Forest': Tropical Forest Degradation, Deforestation, Long-Term Phase Shifts, and Further Transitions. Association for Tropical Biology and Conservation Stable. Retrieved from http://www.jstor.org/stable/27742857 *The Imp. 42*(1), 10–20.

Putz, F. E., Zuidema, P. A., Synnott, T., Pe, M., Pinard, M. A., Sheil, D., … Zagt, R. (2012). *Sustaining conservation values in selectively logged tropical forests: the attained and the attainable, 5*, 296–303. doi:10.1111/j.1755-263X.2012.00242.x

Putz, F. E., Zuidema, P. A., Pinard, M. A., Boot, R. G. A., Sayer, J. A., Sheil, D., ... Vanclay, J. K. (2008). *Improved Tropical Forest Management for Carbon Retention, 6*(7), 1368–1369. doi:10.1371/journal.pbio.0060166

Puyravaud, J. P., Davidar, P., & Laurance, W. F. (2010). Cryptic destruction of India's native forests. *Conservation Letters, 3*(6), 390–394. doi:10.1111/j.1755-263X.2010.00141.x

Quarmby, N. A., Milnes, M., Hindle, T. L., & Silleos, G. N. (1993). The use of multi-temporal NDVI measurements from AVHRR data for crop yield estimation and prediction. *International Journal of Remote Sensing, 14*(2), 199–210. doi:10.1080/01431169308904332

Quattrochi, D. A., & Luvall, J. C. (2009). Thermal remote sensing in Earth science research. In T. A. Warner, M. D. Nellis, & G. M. Foody (Eds.), *The SAGE handbook of remote sensing* (pp. 64–78). London, UK: SAGE Publications Ltd. doi:10.4135/9780857021052.n5

Qu, J. F., Hou, Y. L., Meng, Y. G., Wang, K., Liu, S., Zhang, S.-L., ... Chen, F. (2017). Carbon Dynamics of Reclaimed Coal Mine Soil under Agricultural Use: A Chronosequence Study in the Dongtan Mining Area, Shandong Province, China. *Sustainability, 9*(4), 629. doi:10.3390u9040629

R Core Team. (2018). R: A language and environment for statistical computing. R Foundation for Statistical Computing, Vienna, Austria. Retrieved from https://www.R-project.org/

Raghubanshi, A. S., & Tripathi, A. (2009). Effect of disturbance, habitat fragmentation and alien invasive plants on floral diversity in dry tropical forests of Vindhyan highland: A review. *Tropical Ecology, 50*, 57–69.

Raghuvanshi, D. S., Verma, N., & Gupta, A. (2019). I2/K2S2O8: An Unprecedented Deoxygenating System for N-Oxides and Sulfoxides. *ChemistrySelect, 4*(7), 2075–2078. doi:10.1002lct.201801838

Raharimahefa, T. (2012). Geoconservation and geodiversity for sustainable development in Madagascar. *Madagascar Conservation and Development, 7*(3), 126–134.

Rai, A., Singha, A. K., Ghosal, N., & Singh, N. (2016). Understanding the effectiveness of litter from tropical dry forests for the restoration of degraded lands. *Ecological Engineering, 93*, 76–81. doi:10.1016/j.ecoleng.2016.05.014

Raij, B. (1981). *Avaliacao da fertilidade do solo* (p. 142). Piracicaba, Brazil: Instituto da Potassa and Fosfato.

Rajagopal, K., Buvaneswaran, C., Subramanian, V., & George, M. (2005). Nutrient cycling in young teak plantation I-Restitution of nutrients through litter and rain wash. *Indian Forester, 131*, 221–228.

Ramachandra, T. V., & Kumar, U. (2004): Geographic Resources Decision Support System for Land Use, Land Cover Dynamics Analysis. *Proceedings of the FOSS/GRASS Users Conference*, Bangkok, Thailand. 12-14.

Ramanujam, M. P., & Kadamban, D. (2001). Plant biodiversity of two tropical dry evergreen forests in the Pondicherry region of South India and the role of belief systems in their conservation. *Biodiversity and Conservation, 10*(7), 1203–1217. doi:10.1023/A:1016637627623

Randriamalala, H., & Liu, Z. (2010). Rosewood of Madagascar: Between democracy and conservation. *Madagascar Conservation and Development, 5*(1), 11–22. doi:10.4314/mcd.v5i1.57336

Rangel-Ch, J. O., & Velázquez, A. (1997). Métodos de estudio de la vegetación. In J. O. Rangel-Ch, C. Lowy, D. Petter, & M. Aguilar P. (Eds.), Colombia Diversidad Biótica II: Tipos de vegetación en Colombia (pp. 59–87). doi:10.100713398-014-0173-7.2

Raposo, M. A. (1997). A new species of Arremon (Passeriformes: Emberizidae) from Brazil. *Ararajuba*, *5*(1), 3–9.

Ratter, J. A., Askew, G. P., Montgomery, R. F., & Gifford, D. R. (1978). Observations on forests of some mesotrophic soils in central Brazil. *Revista Brasileira de Botânical*, *1*(1), 47-58.

Ratter, J. A., Richards, P. W., Argent, G., & Gifford, D. R. (1973). Observations on the vegetation of northeastern Mato Grosso: I. The woody vegetation types of the Xavantina-Cachimbo Expedition area. *Philosophical Transactions of the Royal Society of London. Series B, Biological Sciences*, *266*(880), 449–492. doi:10.1098/rstb.1973.0053

Raven, J. A. (1993). The evolution of vascular plants in relation to quantitative functioning of dead water-conducting cells and stomata. *Biological Reviews of the Cambridge Philosophical Society*, *68*(3), 337–363. doi:10.1111/j.1469-185X.1993.tb00735.x

Ravindranath, N. H., Joshi, N. V., Sukumar, R., & Saxena, A. (2006). Impact of climate change on forest in India. *Current Science*, *90*(3), 354–361.

Ravindranath, N. H., Murthy, I. K., Priya, J., Upgupta, S., Mehra, S., & Nalin, S. (2014). Forest area estimation and reporting: Implications for conservation, management and REDD. *Current Science*, *106*(9), 1201–1206.

Ravindranath, N. H., & Sukumar, R. (1998). Climate Change and Tropical Forests of India. *Climatic Change*, *39*(2/3), 563–581. doi:10.1023/A:1005394505216

Ravindranath, N. H., Sukumar, R., & Deshingkar, P. (1997). *Climate change and forests: Impacts and adaptations. Regional assessment for the western ghats India. Atmospheric Environmental Issues in Developing Countries*. Stockholm, Sweden: Stockholm Environment Institute.

Rawat, N., Nautiyal, B. P., & Nautiyal, M. C. (2010). Annual nutrients budget for the grazed and ungrazed sites of an alpine expanse in North-West Himalaya, India. *The Environmentalist*, *30*(1), 54–66. doi:10.100710669-009-9243-3

Rawat, N., Nautiyal, B. P., & Nautiyal, M. C. (2010). Litter decomposition rate and nutrient release from different litter forms in a Himalayan alpine ecosystem. *The Environmentalist*, *30*(3), 279–288. doi:10.100710669-010-9275-8

Reddy, C. S., Dutta, K., & Jha, C. S. (2013). Analysing the gross and net deforestation rates in India. *Current Science*, *105*(11), 1492–1500.

Reddy, C. S., Jha, C. S., Diwakar, P. G., & Dadhwal, V. K. (2015). Nationwide classification of forest types of India using remote sensing and GIS. *Environmental Monitoring and Assessment*, *187*(12), 777. doi:10.100710661-015-4990-8 PMID:26615560

Reeder, J. D., & Sabey, B. (1987). Nitrogen. pp. 155-184. In R. D. Williams & G. E. Schuman (Eds.), Reclaiming mine soils and overburden in the western United States: Analytical Parameters and procedures. Soil and Water Conservation Society, Ankeney, IA.

Regmi, B., & Paudyal, A. (2009). *Climate Change and Agro biodiversity in Nepal: Opportunities to include agrobiodiversity maintenance to support Nepal's National Adaptation Programme of Action*. NAPA.

Reich, P. B. (2014). The world-wide 'fast–slow' plant economics spectrum: A traits manifesto. *Journal of Ecology*, *102*(2), 275–301. doi:10.1111/1365-2745.12211

Reich, P. B., Luo, Y., Bradford, J. B., Poorter, H., Perry, C. H., & Oleksyn, J. (2014). Temperature drives global patterns in forest biomass distribution in leaves, stems, and roots. *Proceedings of the National Academy of Sciences of the United States of America*, *111*(38), 13721–13726. doi:10.1073/pnas.1216053111 PMID:25225412

Reich, P. B., Walters, M. B., & Ellsworth, D. S. (1992). Leaf life-span in relation to leaf, plant, and stand characteristics among diverse ecosystems. *Ecological Monographs*, *62*(3), 365–392. doi:10.2307/2937116

Reich, P. B., Walters, M. B., & Ellsworth, D. S. (1997). From tropics to tundra: Global convergence in plant functioning. *Proceedings of the National Academy of Sciences of the United States of America*, *94*(25), 13730–13734. doi:10.1073/pnas.94.25.13730 PMID:9391094

Reich, P. B., Walters, M. B., Kloeppel, B. D., & Ellsworth, D. S. (1995). Different photosynthesis-nitrogen relations in deciduous hardwood and evergreen coniferous tree species. *Oecologia*, *104*(1), 24–30. doi:10.1007/BF00365558 PMID:28306909

Reifsnyder, W. E. (1978). *Systems for evaluating and predicting the effects of weather and climate on wildland fires.* World Meteorological Organization Special Report No. 11.

Rentería, L. Y., & Jaramillo, V. J. (2011). Rainfall drives leaf traits and leaf nutrient resorption in a tropical dry forest in Mexico. *Oecologia*, *165*(1), 201–211. doi:10.100700442-010-1704-3 PMID:20607297

Report, B. (1987). *Report of the World Commission on Environment and Development: Our Common Future.* Retrieved from http://mom.gov.af/Content/files/Bruntland_Report.pdf

Reynolds, J. D., & Mace, G. M. (1999). Risk assessments of threatened species. *Trends in Ecology & Evolution*, *14*(6), 215–217. doi:10.1016/S0169-5347(99)01629-8

Rezendes, P. (1999). Tracking and the Art of Seeing: How to Read Animal Tracks and Sign. New York, NY: Collins Reference.

Richard, P. (1980). PIELOU, E. C. (1979): *Biogeography*. John Wiley & Sons, ix + 351 p., ill., index, 26, $90. *Géographie physique et Quaternaire*, *34*(2), 263–263. doi:10.7202/1000411ar

Richardson, J. A. (1975). Physical problems of growing plants o colliery wastes. In M. J. Chadwick & G. T. Goodman (Eds.), *Ecology of resource degradation and renewal* (pp. 275–285). Oxford, UK: Blackwell Scientific Publication.

Richards, P. W. (1952). *The Tropical Rain Forest: an Ecological Study*. Cambridge, UK: Cambridge University Press.

Ricketts, T. H., Dinerstein, E., Olson, D. M., Eichbaum, W., Loucks, C. J., Kavanaugh, K., ... Carney, K. (1999). *Terrestrial ecoregions of North America: a conservation assessment* (Vol. 1). Washington, DC: Island Press.

Rimal, R. K., Maharjan, R., Khanal, K., Koirala, S., Karki, B., Nepal, S. M., & Shrestha, H. L. (2017). Detection, assessment, and updating the maps of encroached forest areas: A case study from Bara district, Nepal. *Banko Janakari*, *27*(1), 65–71. doi:10.3126/banko.v27i1.18554

Rist, L., Shanley, P., Sunderland, T., Sheil, D., Ndoye, O., Liswanti, N., & Tieguhong, J. (2012). The impacts of selective logging on non-timber forest products of livelihood importance. *Forest Ecology and Management*, *268*, 57–69. doi:10.1016/j.foreco.2011.04.037

Ritika, B., & Uptal, D. (2014). Bio-fertilizer a way towards organic agriculture: A Review. *Academic Journals.*, *8*(24), 2332–2342.

Ritter, E. (2005). Litter decomposition and nitrogen mineralization in newly formed gaps in a Danish beech (*Fagus sylvatica*) forest. *Soil Biology & Biochemistry*, *37*(7), 1237–1247. doi:10.1016/j.soilbio.2004.11.020

Rocca, M. E. (2009). Fine-Scale Patchiness in Fuel Load Can Influence Initial Post-Fire Understory Composition in a Mixed Conifer Forest, Sequoia National Park, California. *Natural Areas Journal*, *29*(2), 126–132. doi:10.3375/043.029.0204

Rodin, L. E., & Basilevic, N. I. (1968). World distribution of plant biomass. Functioning of terrestrial ecosystems at the primary production level. UNESCO, Paris, 45-52.

Rodrigo, A., Retana, J., & Xavier, F. (2005). Diferencias En La Dinámica De Regeneración De Los Bosques Mediterráneos Después De Grandes Incendios: Consecuencias En El Paisaje Forestal.

Rodrigues, P. M. S., Silva, J. O., Eisenlohr, P. V., & Schaefer, C. E. G. R. (2015). Climate change effects on the geographic distribution of specialist tree species of the Brazilian tropical dry forests. *Brazilian Journal of Biology*, *75*(3), 679–684. doi:10.1590/1519-6984.20913 PMID:26465729

Rodrigues, R. R., Martins, S. V., & Matthes, L. A. F. (2005). Post-Fire Regeneration in a Semideciduous Mesophytic Forest, South-Eastern Brazil. In A. R. Burk (Ed.), *New research on forest ecosystems* (pp. 1–19). New York, NY: Nova Science Publishers

Rodríguez-Buritica, S., Aguilar-Garavito, M., & Norden, N. (2017). Identifying a Fire Ecology Research Agenda for Colombia. *Eos (Washington, D.C.)*, *1–4*. doi:10.1029/2017EO065869

Rohr, T., Manzoni, S., Feng, X., Menezes, R. S. C., & Porporato, A. (2013). Effect of rainfall seasonality on carbon storage in tropical dry ecosystems. *Journal of Geophysical Research. Biogeosciences*, *118*(3), 1156–1167. doi:10.1002/jgrg.20091

Rondeaux, G., Steven, M., & Baret, F. (1996). Optimization of soil-adjusted vegetation indices. *Remote Sensing of Environment*, *55*(2), 95–107. doi:10.1016/0034-4257(95)00186-7

Roper, J. E. (2018). What animals live in the tropical dry forest? USA Today. Retrieved from https://traveltips.usatoday.com/animals-live-tropical-dry-forest-100072.html

Rossini, M., Panigada, C., Meroni, M., & Colombo, R. (2006). Assessment of oak forest condition based on leaf biochemical variables and chlorophyll fluroscence. *Tree Physiology*, *26*(11), 1487–1496. doi:10.1093/treephys/26.11.1487 PMID:16877333

Rottmann, N., Dyckmans, J., & Joergensen, R. G. (2010). Microbial use and decomposition of maize leaf straw incubated in packed soil columns at different depths. *European Journal of Soil Biology*, *46*(1), 27–33. doi:10.1016/j.ejsobi.2009.09.004

Roumet, C., Birouste, M., Picon-Cochard, C., Ghestem, M., Osman, N., Vrignon-Brenas, S., ... Stokes, A. (2016). Root structure–function relationships in 74 species: Evidence of a root economics spectrum related to carbon economy. *The New Phytologist*, *210*(3), 815–826. doi:10.1111/nph.13828 PMID:26765311

Rouse, J. W. Jr, Haas, R. H., Schell, J. A., & Deering, D. W. (1974). Monitoring vegetation systems in the Great Plains with ERTS. *NASA Special Publication*, *351*, 309.

Royle, J. A., & Nichols, J. D. (2003). Estimating Abundance from Repeated Presence-Absence Data or Point Counts. *Ecology*, *84*(3), 777–790. doi:10.1890/0012-9658(2003)084[0777:EAFRPA]2.0.CO;2

Rozendaal, D. M. A., Hurtado, V. H., & Poorter, L. (2006). Plasticity in leaf traits of 38 tropical tree species in response to light; relationships with light demand and adult stature. *Functional Ecology*, *20*(2), 207–216. doi:10.1111/j.1365-2435.2006.01105.x

Rutherford, M. C., Powrie, L. W., Husted, L. B., & Turner, R. C. (2011). Early post-fire plant succession in Peninsula Sandstone Fynbos: The first three years after disturbance. *South African Journal of Botany*, *77*(3), 665–674. doi:10.1016/j.sajb.2011.02.002

Safford, H. D., & Harrison, S. (2004). Fire effects on plant diversity in Serpentine vs. Sandstone Chaparral. *Ecology*, *85*(2), 539–548. doi:10.1890/03-0039

Sagar, R., Archita, S., & Singh, J. S. (2008). Differential effect of woody plant canopies on species composition and diversity of ground vegetation: A case study. *Tropical Ecology*, *49*, 189–197.

Sagar, R., Pandey, A., & Singh, J. S. (2012). Composition, species diversity, and biomass of the herbaceous community in dry tropical forest of northern India inrelation to soil moisture and light intensity. *The Environmentalist*, *32*(4), 485–493. doi:10.100710669-012-9414-5

Saigal, S. (2008). *Non-Timber Forest Products and Forest Governance Forest Governance Learning Group – India.*

Salazar, A., & Goldstein, G. (2014). Effects of fire on seedling diversity and plant reproduction (sexual vs. vegetative) in neotropical savannas differing in tree density. *Biotropica*, *46*(2), 139–147. doi:10.1111/btp.12090

Salazar, L. F., Nobre, C. A., & Oyama, M. D. (2007). Climate change consequences on the biome distribution in tropical South America. *Geophysical Research Letters*, *34*(9). doi:10.1029/2007GL029695

Salgado-Negret, B., Canessa, R., Valladares, F., Armesto, J. J., & Pérez, F. (2015). Functional traits variation explains the distribution of Aextoxicon punctatum (Aextoxicaceae) in pronounced moisture gradients within fog-dependent forest fragments. *Frontiers in Plant Science*, *6*, 511. doi:10.3389/fpls.2015.00511 PMID:26257746

Salzmann, U. (2000). Are modern savannas degraded forests?-A Holocene pollen record from the Sudanian vagetation zone of NE Nigeria. *Vegetation History and Archaeobotany*, *9*(1), 1–15. doi:10.1007/BF01295010

Sampaio, A. B., Holl, K. D., & Scariot, A. (2007). Does restoration enhance regeneration of seasonal deciduous forests in pastures in central Brazil? *Restoration Ecology*, *15*(3), 462–471. doi:10.1111/j.1526-100X.2007.00242.x

Sanches, L., Valentini, C. M. A., Júnior, O. B. P., de Souza Nogueira, J., Vourlitis, G. L., Biudes, M. S., & de Almeida Lobo, F. (2008). Seasonal and interannual litter dynamics of a tropical semideciduous forest of the southern Amazon Basin, Brazil. *Journal of Geophysical Research. Biogeosciences*, *113*(G4).

Sanchez-Azofeifa, A., Portillo-Quintero, C., Wilson-Fernandes, G., Stoner, K., & Shimizu, T. (2013). *The policy process for land use/cover change and forest degradation in the semi-arid Latin American/Caribbean region: perspectives and opportunities*. A literature review prepared for the Inter-American Development Bank.

Sanchez-Azofeifa, A., Quesada, M., Rodriguez, J. P., Nassar, J. M., Stoner, K. E., Castillo, A., ... John, A. (2005). Research priorities for neotropical dry forests. *Biotropica, 37*(4), 477–8-5.

Sanchez-Azofeifa, A., Kalacska, M. E., Quesada, M., Stoner, K. E., Lobo, J. A., & Arroyo-Mora, P. (2003). Tropical dry climates. In M. D. Schwartz (Ed.), *Phenology: an integrative environmental science* (pp. 121–137). Dordrecht, The Netherlands: Springer. doi:10.1007/978-94-007-0632-3_9

Sánchez-Azofeifa, G. A., Quesada, M., Rodríguez, J. P., Nassar, J. M., Stoner, K. E., Castillo, A., ... Cuevas-Reyes, P. (2005). Research priorities for Neotropical dry forests. *Biotropica: The Journal of Biology and Conservation*, *37*(4), 477–485.

Sanchez-Azofeifa, G. A., Quesada, M., Rodriguez, J. P., Nassar, J. M., Stoner, K. E., Castillo, A., ... Kalacska, M. E. R. (2005). Research Priorities for Neotropical Dry Forests. *Biotropica*, *37*, 477–485.

Sangha, K. K., Jalota, R. K., & Midmore, D. J. (2006). Litter production, decomposition and nutrient release in cleared and uncleared pasture systems of central Queensland, Australia. *Journal of Tropical Ecology*, *22*(2), 177–189. doi:10.1017/S0266467405003020

Sani, A. R. (2009). Biophysical Characterization of Wood Resources in a Reverdi Site and a Degraded Site in the Department of MIRRIAH. (End of Cycle Master's Degree in Water and Forest Engineering Diploma), Abdou Moumounide Niamey University.

Santiago, L. S., Goldstein, G., Meinzer, F. C., Fisher, J. B., Machado, K., Woodruff, D., & Jones, T. (2004). Leaf photosynthetic traits scale with hydraulic conductivity and wood density in Panamanian forest canopy trees. *Oecologia*, *140*(4), 543–550. doi:10.100700442-004-1624-1 PMID:15232729

Santos, J. C., Leal, I. R., Almeida-Cortez, J. S., Fernandes, G. W., & Tabarelli, M. (2011). Caatinga: The scientific negligence experienced by a dry tropical forest. *Tropical Conservation Science*, *4*(3), 276–286. doi:10.1177/194008291100400306

Santos, M. G., Oliveira, M. T., Figueiredo, K. V., Falcão, H. M., Arruda, E. C. P., Almeida-Cortez, J., ... Antonino, A. C. D. (2014). Caatinga, the Brazilian dry tropical forest: Can it tolerate climate changes? *Theoretical and Experimental Plant Physiology*, *26*(1), 83–99. doi:10.100740626-014-0008-0

Saradoum, G., Ngom, Fm., Diallo, A., & Guissé, A. (2012). Characterization of the herbaceous vegetation of the National Park of Manda in Chad. *International Journal of Science and Advanced Technology.*

Sariyildiz, T., & Anderson, J. M. (2005). Variation in the chemical composition of green leaves and leaf litters from three deciduous tree species growing on different soil types. *Forest Ecology and Management,* 210, 303-319. doi:eco.2005.02.043. doi:10.1016/j.for

Sariyildiz, T.; Anderson, J. M., & Kucuk, M. (2005). Effects of tree species and topography on soil chemistry, litter quality and decomposition in Northeast Turkey. *Soil Biology and Biochemistry,* 37, 1695-1706. doi:.02.004. doi:10.1016/j.soilbio.2005

Sariyildiz, T., & Anderson, J. M. (2003). Interactions between litter quality, decomposition and soil fertility: A laboratory study. *Soil Biology & Biochemistry*, *35*(3), 391–399. doi:10.1016/S0038-0717(02)00290-0

Sariyildiz, T., & Küçük, M. (2008). Litter mass loss rates in deciduous and coniferous trees in Artvin, northeast Turkey: Relationships with litter quality, microclimate, and soil characteristics. *Turkish Journal of Agriculture and Forestry*, *32*, 547–559.

Sariyildiz, T., Savaci, G., & Kravkaz, I. S. (2015). Effects of tree species, stand age and land-use change on soil carbon and nitrogen stock rates in northwestern Turkey. *IForest (Viterbo)*, *9*(1), 165–170. doi:10.3832/ifor1567-008

Sarjubala Devi, A., & Yadava, P. S. (2007). Wood and leaf litter decomposition of *Dipterocarpus tuberculatus* Roxb. in a tropical deciduous forest of Manipur, Northeast India. *Current Science*, *93*, 243–246.

Sarma, K., & Yadav, P. K. (2013). Relentless Mining in Meghalaya, India. *Conservation Science*, *1*(1), 5–12. doi:10.3126/cs.v1i1.8578

SAS. (2011). *SAS version 9.2*. Cary, NC: SAS Institute

Sasaki, N., & Putz, F. E. (2009). Critical need for new definitions of "forest" and "forest degradation" in global climate change agreements. *Conservation Letters*, *2*(5), 226–232. doi:10.1111/j.1755-263X.2009.00067.x

Sauer, J. R., & Williams, B. K. (1989). Generalized Procedures for Testing Hypotheses About Survival or Recovery Rates. *The Journal of Wildlife Management*, *53*(1), 137–142. doi:10.2307/3801320

Saxena, S. K., & Chatterji, P. C. (1988a). Ecological imbalances caused by mining in Rajasthan and their restoration. Mining and Environment in India. Joshi and Bhattacharya (Eds.). HRG Publication Series, Nainital, India. pp. 223-241.

Saxena, S. K., & Chatterji, P. C. (1988b). In K. A. Shankarnarayan (Ed.), *Mining activity and creation of wasteland in western Rajasthan. Wasteland development and their utilization.* (pp. 41–61). Jodhpur, India: Scientific Publisher.

Schafer, W. M., Nielsenand, G. A., & Nettleton, W. D. (1980). Mine spoil genesis and morphology I a spoil chronosequence in Montana. *Soil Science Society of America Journal, 44*(4), 802–807. doi:10.2136ssaj1980.03615995004400040029x

Scheffer, M., Carpenter, S., Foley, J. A., Folke, C., & Walker, B. (2001). Catastrophic shifts in ecosystems. *Nature, 413*(6856), 591–596. doi:10.1038/35098000 PMID:11595939

Schimper, A. F. W. (1898). Pflanzen-geographie auf physiologischerGrundlage. Germany: Gustav Fischer.

Schmerbeck, J. (2011). Linking dynamics and locally important ecosystem services of South Indian dry forests: An approach. *Resources, Energy, and Development, 8*(2), 149–172.

Schmitt, C. B., Belokurov, A., Besancon, C., Boisrobert, L., Burgess, N. D., Campbell, A. … Winkel, G. (2008). Global Ecological Forest Classification and Forest Protected Area Gap Analysis. Analyses and recommendations in view of the 10% target for forest protection under the Convention on Biological Diversity (CBD). In I. Perfecto & J. Vandermeer (Eds.), Biodiversity conservation in tropical agroecosystems (1134, 173–200). Annals of the New York Academy of Sciences, University of Freiburg, Germany.

Schmitz, O. J., Post, E., Burns, C. E., & Johnston, K. M. (2003). Ecosystem responses to global climate change: Moving beyond color mapping. *Bioscience, 53*(12), 1199–1205. doi:10.1641/0006-3568(2003)053[1199:ERTGCC]2.0.CO;2

Scholander, P. F., Bradstreet, E. D., Hammel, H. T., & Hemmingsen, E. A. (1966). Sap concentrations in halophytes and some other plants. *Plant Physiology, 41*(3), 529–532. doi:10.1104/pp.41.3.529 PMID:5906381

Schönbeck, L., Lohbeck, M., Bongers, F., Ramos, M., & Sterck, F. (2015). How do light and water acquisition strategies affect species selection during secondary succession in moist tropical forests? *Forests, 6*(6), 2047–2065. doi:10.3390/f6062047

Schreckenberg, K., Luttrell, C., & Moss, C. (2006). Participatory forest management: An overview. Forest Policy and Environment Programme: Grey Literature.

Schulz, K., Voigt, K., Beusch, C., Almeida-Cortez, J. S., Kowarik, I., Walz, A., & Cierjacks, A. (2016). Grazing deteriorates the soil carbon stocks of Caatinga forest ecosystems in Brazil. *Forest Ecology and Management, 367*, 62–70. doi:10.1016/j.foreco.2016.02.011

Schwalm, C., & Ek, A. R. (2001). Climate change and site: Relevant mechanisms and modeling techniques. *Forest Ecology and Management, 150*(3), 241–257. doi:10.1016/S0378-1127(00)00568-5

Scott, N. A., & Binkley, D. (1997). Foliage litter quality and annual net N mineralization: Comparison across North America forest site. *Oecologia, 111*(2), 151–159. doi:10.1007004420050219 PMID:28307988

Second Working Plan, Puruliya District. (2014-2015) to (2023-2024), Volume 1.

Second Working Plan, Puruliya District. (n.d.). Volume 2.

Secretariat of the Convention on Biological Diversity. (2002). *Review of the status and trends of, and major threats to, the forest biological diversity.*

Seppälä, R., Buck, A., & Katila, P. (Eds.). (2009). *Adaptation of forests and people to climate change: A Global Assessment Report* (Vol. 22). Helsinki, Finland: IUFRO World Series.

Septer, J. D. (2019). Tropical dry forest plants. Retrieved from https://www.hunker.com/13427136/tropical-dry-forest-plants

Serraj, R., Siclair, T. R., & Purcell, L. C. (1999). Symbiotic N_2 fixation response to drought. *Journal of Experimental Botany*, *50*, 143–155.

Setalaphruk, C., & Prince, L. (2007). Children's Traditional Ecological Knowledge of Wild Food Resources: A Case Study in a Rural Village, Northeast of Thailand. *Journal of Ethnobiology and Ethnomedicine*, *3*(1), 33. doi:10.1186/1746-4269-3-33 PMID:17937791

Settele, J., Scholes, R., Betts, R., Bunn, S., Leadley, P., & Nepstad, D. (2014). Terrestrial and inland water systems In: C. B. Field, V. R. Barros, D. J. Dokken, K. J. Mach, M. D. Mastrandrea, T. E. Bilir, et al. (Eds.), Climate Change 2014: Impacts, Adaptation, and Vulnerability Part A: Global and Sectoral Aspects. Contribution of Working Group II to the Fifth Assessment Report of the Intergovernmental Panel on Climate Change. Cambridge, UK: Cambridge University Press. 271-359.

Sfair, J. C., de Bello, F., de França, T. Q., Baldauf, C., & Tabarelli, M. (2018). Chronic human disturbance affects plant trait distribution in a seasonally dry tropical forest. *Environmental Research Letters*, *13*(2). doi:10.1088/1748-9326/aa9f5e

Shackleton, S., Delang, C. O., & Angelsen, A. (2011). From subsistence to safety nets and cash income: exploring the diverse values of non-timber forest products for livelihoods and poverty alleviation. In Non-timber forest products in the global context (pp. 55–81). Berlin, Germany: Springer. doi:10.1007/978-3-642-17983-9_3

Shackleton, C. M., Shackleton, S. E., Buiten, E., & Bird, N. (2007). The importance of dry woodlands and forests in rural livelihoods and poverty alleviation in South Africa. *Forest Policy and Economics*, *9*(5), 558–577. doi:10.1016/j.forpol.2006.03.004

Shamaki, S. B., & Akindele, S. O. (2013). Volume estimation models from stump diameter for teak (*Tectona grandis* Linn f.) plantation in Nimbia forest reserve, Nigeria. *Journal of Environmental Science and Water Resources*, *2*(3), 89–94.

Shanahan, M., Shubert, W., Scherer, C., & Corcoran, T. (2013). Climate change in Africa: a guidebook for journalists. In F. Banda (Ed.), *UNESCO Series on Journalism Education.*

Sharkey, T. D., & Schrader, S. M. (2006). High temperature stress. In *K. V. M. Rao, A. S. Raghavendra, & K. J. Reddy (Link), Physiology and molecular biology of stress tolerance in plants* (pp. 101–129). Dordrecht, The Netherlands: Springer. doi:10.1007/1-4020-4225-6_4

Sharma, K. D., Kumar, S., & Gough, L. P. (2000). Rehabilitation of lands mined for limestone in the Indian desert. *Land Degradation & Development*, *11*(6), 563–574. doi:10.1002/1099-145X(200011/12)11:6<563::AID-LDR414>3.0.CO;2-I

Sharma, M., Oerwald, R. G., & Amateis, R. L. (2000). A consistent system of equations for tree and stand volume. *Northern Journal of Applied Forestry*, *54*(2), 110–125.

Sharma, V., & Chaudhry, S. (2018). Vegetation composition and plant diversity in mining disturbed tropical thorn forest of Asola-Bhatti Wildlife Sanctuary, Northern India. *Taiwania*, *63*(3), 267–280.

Shipley, B., & Vu, T. T. (2002). Dry matter content as a measure of dry matter concentration in plants and their parts. *The New Phytologist*, *153*(2), 359–364. doi:10.1046/j.0028-646X.2001.00320.x

Shippert, P. (2004). Why use hyperspectral imagery? *Photogrammetric Engineering and Remote Sensing*, *70*(4), 377–396.

Shlisky, A. J., Waugh, P., Gonzalez, M., Gonzalez, M., Manta, H., Santoso, E., … Zollner, D. (2007). Fire, ecosystems &people: threats and strategies for global biodiversity conservation. GFI Technical Report 2007-2. The Nature Conservancy, 28.

Shono, K., Cadaweng, E. A., & Durst, P. B. (2007). Application of assisted natural regeneration to restore degraded tropical forestlands. *Restoration Ecology*, *15*(4), 620–626. doi:10.1111/j.1526-100X.2007.00274.x

Short, T. A., Menzies, N. W., Mulligan, D. R., & Loch, R. (Ed.), & Jasper, D. (2000). Mining disturbance alters phosphorus fractions in northern Australian soils. Special Issue: Soil research for mine reclamation. Australian Journal of Soil Research, 38(2): 411-421.

Shoshani, J., & Tassy, P. (2005). Advances in Proboscidean Taxonomy and Classification, Anatomy and Physiology, and Ecology and Behavior. *Quaternary International, 126-128*, 5–20. doi:10.1016/j.quaint.2004.04.011

Shrestha, H. L., & Dhonju, H. K. (2013). Satellite Based Forest Fire Burned Area Assessment in Nepal from 2000 to 2013. *Journal of the Institute of Forestry,* Nepal, 98.

Shrestha, H. L., Uddin, K., Gilani, H., Pradhan, S., Shrestha, B. R., & Murthy, M. S. R. (2015). Forest Carbon Flux Assessment in Nepal Using the Gain-Loss Method. In M. S. R. Murthy, S. Wesselman, & H. Gilani (Eds.), Multi-scale forest biomass assessment and monitoring in the Hindu Kush Himalayan region: A geospatial perspective. Kathmandu, Nepal: ICIMOD

Shuaibu, R. B., & Dagba, B. I. (2013). Challenges of forest inventory and its effects on forest management and planning in Nigeria. In L. Popoola, O. Y. Ogunsanwo, V. A. J. Adekunle, I. O. Azeez, & N. O. Adewole (Eds.), *The green economy: Balancing: environmental sustainability and livelihoods in an emerging economy. Proceedings of the 36thAnnual Conference of Forestry Association of Nigeria.* Uyo, Nigeria: Forestry Association of Nigeria.

Shukla, G. (2010). *Biomass production and vegetation analysis of Chilapatta reserve forest ecosystem of West Bengal.* [Published doctoral thesis], Uttar Banga Krishi Viswavidyalaya, Pundibari.

Shukla, A. K., Srivastava, P. K., Singh, B., Behera, S. K., & Thomas, T. (2017). Litterfall patterns and soil nutrient chemistry in varied tropical deciduous forests. *International Journal of Chemical Studies, 5*, 1203–1210.

Siddiqui, S., Bharati, L., Pant, M., Gurung, P., & Rakhal, B. (2012). Nepal: building climate resilience of watersheds in mountain eco-regions climate change and vulnerability mapping in watersheds in middle and high mountains of Nepal. ADB Technical Assistance Consultant's Report for Department of Soil Conservation and Watershed Management (DSCWM), Government of Nepal. Kathmandu, Nepal: Asian Development Bank (ADB). 96p.

Silori, C. S., & Mishra, B. K. (2001). Assessment of livestock grazing pressure in and around the elephant corridors in Mudumalai Wildlife Sanctuary, south India. *Biodiversity & Conservation, 10*(12), 2181–2195.

Silver, W. L. (1998). The potential effects of elevated CO_2 and climate change on tropical forest soils and biogeochemical cycling. *Climatic Change, 39*(2/3), 337–361. doi:10.1023/A:1005396714941

Singh, A. K., Parsad, A., & Singh, B. (1986). Availability of phosphorus and potassium and its relationship with physicochemical properties of some forest soils of Pali-range (Shahdol, MP). *Indian Forester, 112*, 1094–1104.

Singh, A. N., Raghubanshi, A. S., & Singh, J. S. (2002). Plantations as a tool for mine spoil restoration. *Current Science, 82*(12), 1436–1441.

Singh, A. N., Raghubanshi, A. S., & Singh, J. S. (2004a). Impact of native tree plantations on mine spoil in Adry Tropical Environment. *Forest Ecology and Management, 187*(1), 49–60. doi:10.1016/S0378-1127(03)00309-8

Singh, J. S., & Gupta, S. R. (1977). Plant decomposition and soil respiration in terrestrial ecosystems. *Botanical Review, 43*(4), 449–528. doi:10.1007/BF02860844

Singh, J. S., Raghubanshi, A. S., Singh, R. S., & Srivastava, S. C. (1989). Microbial biomass acts as a source of plant nutrients in dry tropical forests and savanna. *Nature, 238*, 449–500.

Singh, J. S., & Singh, V. K. (1992). Phenology of seasonally dry tropical forest. *Current Science, 63*, 684–689.

Singh, K. P., & Kushwaha, C. P. (2005). Emerging paradigms of tree phenology in dry tropics. *Current Science*, *89*(6), 964–975.

Singh, L., & Singh, J. S. (1991). Species structure, dry matter dynamics and carbon flux of a dry tropical forest in India. *Annals of Botany*, *68*(3), 263–273. doi:10.1093/oxfordjournals.aob.a088252

Singh, R., Sagar, R., Srivastava, P., Singh, P., & Singh, J. S. (2017). Herbaceous species diversity and soil attributes along a forest-savanna-grassland continuum in a dry tropical region. *Ecological Engineering*, *103*, 226–235. doi:10.1016/j.ecoleng.2017.04.020

Singh, R., Singh, H., Singh, S., Afreen, T., Upadhyay, S., Singh, A. K., ... Raghubanshi, A. S. (2017). Riparian land uses affect the dry season soil CO_2 efflux under dry tropical ecosystems. *Ecological Engineering*, *100*, 291–300. doi:10.1016/j.ecoleng.2017.01.002

Sist, P., Sheil, D., Kartawinata, K., & Priyadi, H. (2002). Reduced-impact logging in Indonesian Borneo: Some results confirming the need for new silvicultural prescriptions. In Forest, Science and Sustainability: The Bulungan Model Forest, Technical Report Phase I 1997-2001. ITTO Project PD 12/97 REV.1 (F).

Sivakumar, M. V., & Stefanski, R. (2007): Climate and land degradation—an overview. Climate and Land Degradation. Springer, 105-135.

Skutsch, M. M., & Ba, L. (2010). Crediting carbon in dry forests: The potential for community forest management in West Africa. *Forest Policy and Economics*, *12*(4), 264–270. doi:10.1016/j.forpol.2009.12.003

Snook, L. K. (1993). *Regeneration, Growth, and Sustainability of Mahogany in Mexico's Yucatan Forests*. Big-Leaf Mahogany. doi:10.1007/0-387-21778-9_9

Sobrado, M. A. (1991). Cost-benefit relationships in deciduous and evergreen leaves of tropical dry forest species. *Functional Ecology*, *5*(5), 608–616. doi:10.2307/2389479

Sohmer, S. H., & Gustafson, R. (1987). Plants and flowers of Hawaii. Honolulu, HI: University of Hawaii Press.

Soil Survey Staff. (1999). *Soil taxonomy: A basic system of soil classification for making and interpreting soil surveys* (2nd ed.). Washington, DC: USDA Natural Resources Conservation Service.

Soni, P. (2003). Climate change and restoration of tropical forests. *Indian Forester*, *129*, 865–873.

Soosairaj, S., Britto, S. J., Balaguru, B., Nagamurugan, N., & Natarajan, D. (2004). Mapping forest types of Pacchaimalai hills, Eastern Ghats, India – using remote sensing and GIS. *Eco. Env. & Cons.*, *10*(2), 131–135.

Sotta, E. D., Meir, P., Malhi, Y., Nobre, A. D., & Grace, J. (2004). Soil CO_2 efflux in a tropical forest in the central Amazon. *Global Change Biology*, *10*(5), 601–617. doi:10.1111/j.1529-8817.2003.00761.x

Soulé, M. E. (1980). Thresholds for Survival: Maintaining Fitness and Evolutionary Potential. In M. E. Soulé & B. A. Wilcox (Eds.), *Conservation Biology: An Evolutionary-Ecological Perspective* (pp. 153–169). Sunderland, MA: Sinaur Associates.

Soulé, M. E., & Wilcox, B. A. (Eds.). (1980). *Conservation Biology: An Evolutionary-Ecological Perspective*. Sunderland, MA: Sinaur Associates

Souza, M. C., Franco, A. C., Haridasan, M., Rossatto, D. R., Araújo, J. F. D., Morellato, L. P. C., & Habermann, G. (2015). The length of the dry season may be associated with leaf scleromorphism in cerrado plants. *Anais da Academia Brasileira de Ciências*, *87*(3), 1691–1699. doi:10.1590/0001-376520150381 PMID:26221991

Specht, M. J., Pinto, S. R. R., Albuqueque, U. P., Tabarelli, M., & Melo, F. P. L. (2015). Burning biodiversity: Fuelwood harvesting causes forest degradation in human-dominated tropical landscapes. *Global Ecology and Conservation*, *3*, 200–209. doi:10.1016/j.gecco.2014.12.002

Springer, C. J., & Ward, J. K. (2007). Flowering time and elevated CO_2. *The New Phytologist*, *176*(2), 243–255. doi:10.1111/j.1469-8137.2007.02196.x PMID:17822407

Srivastava, P. K., George, P. P., Manika, G., Sudhir, K. S., Tanvir, I., & Dimitra, L. (2018). Deriving forest fire probability maps from the fusion of visible/infrared satellite data and geospatial data mining. *Modeling Earth Systems and Environment*, 1–17. doi:10.100740808-018-0555-5

Srivastava, S. C. (1999). Effect of coal mining on microbial biomass and nutrient availability in dry tropical forest of Vindhyan Hill region. *Journal of Tropical Forestry*, *15*(1), 15–23.

StatPlus. (2010). StatPlus version. Walnut, CA: AnalystSoft

Stavins, R., & Richards, K. (2005). *The cost of U.S. forest-based carbon sequestration*. USA: Prepared for the Pew Center on Global Climate Change.

Stearns, S. C. (1992). *The Evolution of Life Histories*. Oxford, UK: Oxford University Press.

Steinheim, G., Wegge, P., Fjellstad, J. I., Jnawali, S. R., & Weladji, R. B. (2005). Dry Season Diets and Habitat use of Sympatric Asian elephants (*Elephas maximus*) and Greater One-horned Rhinoceros (*Rhinocerus unicornis*) in Nepal. *Journal of Zoology (London, England)*, *265*(4), 377–385. doi:10.1017/S0952836905006448

Steininger, M. K., Tucker, C. J., Ersts, P., Killeen, T. J., Villegas, Z., & Hecht, S. B. (2001a). Clearance and fragmentation of tropical deciduous forest in the Tierras Bajas, Santa Cruz, Bolivia. *Conservation Biology*, *15*(4), 856–866. doi:10.1046/j.1523-1739.2001.015004856.x

Steininger, M. K., Tucker, C. J., Townshend, J. R., Killeen, T. J., Desch, A., Bell, V., & Ersts, P. (2001b). Tropical deforestation in the Bolivian Amazon. *Environmental Conservation*, *28*(2), 127–134. doi:10.1017/S0376892901000133

Steinke, T. D., Holland, A. J., & Singh, Y. (1993). Leaching losses during decomposition of mangrove leaf litter. *South African Journal of Botany*, *59*(1), 21–25. doi:10.1016/S0254-6299(16)30770-0

Sterck, F., Markesteijn, L., Schieving, F., & Poorter, L. (2011). Functional traits determine trade-offs and niches in a tropical forest community. *Proceedings of the National Academy of Sciences of the United States of America*, *108*(51), 20627–20632. doi:10.1073/pnas.1106950108 PMID:22106283

Sternberg, T., Tsolmon, R., Middleton, N., & Thomas, D. S. G. (2011). Tracking desertification on the Mongolian steppe through NDVI and field-survey data. *International Journal of Digital Earth*, *4*(1), 50–64. doi:10.1080/17538940903506006

Stewart-Cox, B. (1995). *Wild Thailand*. Cambridge, MA: MIT Press.

Stibig, H. J., Achard, F., Carboni, S., Raši, R., & Miettinen, J. (2014). Change in tropical forest cover of Southeast Asia from 1990 to 2010. *Biogeosciences*, *11*(2), 247–258. doi:10.5194/bg-11-247-2014

Stoms, D. M., & Estes, J. E. (1993). A remote sensing research agenda for mapping and monitoring biodiversity. *International Journal of Remote Sensing*, *14*(10), 1839–1860. doi:10.1080/01431169308954007

Stone, C. P., Smith, C. W., & Tunison, J. T. (Eds.). (1992). *Alien plant invasions in native ecosystems of Hawaii: management and research. Cooperative National Park Resources Studies Unit*. Honolulu, HI: University of Hawaii.

Stow, D. A., Hope, A., McGuire, D., Verbyla, D., Gamon, J., Huemmrich, F., ... Myneni, R. (2004). Remote sensing of vegetation and land-cover change in Arctic Tundra Ecosystems. *Remote Sensing of Environment, 89*(3), 281–308. doi:10.1016/j.rse.2003.10.018

Subedi, B. P., Gauli, K., Joshi, N. R., Pandey, A., Charmakar, S., Poudel, A., ... Khanal, S. C. (2015). *Forest Carbon Assessment in Chitwan-Annapurna Landscape. Study Report, WWF Nepal Hariyo Ban Program.* Kathmandu, Nepal: Baluwatar.

Subedi, B. P., Ghimire, P. L., Koontz, A., Khanal, S. C., Katwal, P., Sthapit, K. R., & Mishra, S. K. (2014). *Private Sector Involvement and Investment in Nepal's Forestry: Status, Prospects and Ways Forward. Study Report, Multi Stakeholder Forestry Programme - Services Support Unit.* Kathmandu, Nepal: Babarmahal.

Suding, K. N., Gross, K. L., & Houseman, G. R. (2004). Alternative states and positive feedbacks in restoration ecology. *Trends in Ecology & Evolution, 19*(1), 46–53. doi:10.1016/j.tree.2003.10.005 PMID:16701225

Sukumar, R. (2000). Asian Elephant. In R. P. Reading & B. Miller (Eds.), Endangered Animals (pp. 34–39). Westport, CT: Greenwood Press.

Sukumar, R. (1989). Ecology of the Asian Elephant in Southern India. I. Movement and Habitat Utilization Patterns. *Journal of Tropical Ecology, 5*(1), 1–18. doi:10.1017/S0266467400003175

Sukumar, R. (1990). Ecology of the Asian Elephant in Southern India. II. Feeding habits and Crop Raiding Patterns. *Journal of Tropical Ecology, 6*(1), 33–53. doi:10.1017/S0266467400004004

Sukumar, R. (1992). *The Asian Elephant: Ecology and Management* (2nd ed.). Cambridge, UK: Cambridge University Press.

Sukumar, R., & Santiapillai, C. (1996). *Elephas maximus*: Status and Distribution. In J. Shoshani & P. Tassy (Eds.), *The Proboscidea: Evolution and Paleoecology of Elephants and Their Relatives* (pp. 327–331). Oxford, UK: Oxford University Press.

Sunderland, T., Apgaua, D., Baldauf, C., Blackie, R., Colfer, C., Cunningham, A. B., ... Ickowitz, A. (2015). Global dry forests: A prologue. *International Forestry Review, 17*(2), 1–9. doi:10.1505/146554815815834813

Suresh, H., Dattaraja, H., & Sukumar, R. (2010). Relationship between annual rainfall and tree mortality in a tropical dry forest: Results of a 19-year study at Mudumalai southern India. *Forest Ecology and Management, 259*(4), 762–776. doi:10.1016/j.foreco.2009.09.025

Suttle, K. B., Thomsen, M. A., & Power, M. E. (2007). Species interactions reverse grassland responses to changing climate. *Science, 315*(5812), 640–642. doi:10.1126cience.1136401 PMID:17272720

Swart, E. R. (1963). Age of the Baobab tree. *Nature, 198*(4881), 708–709. doi:10.1038/198708b0

Swathi, V. (2010). *The use and benefits of bio-fertilizer and biochar on agricultural soils.* (Unpublished thesis). Department of Chemical and Biological Engineering, Chalmers University of Technology, Sweden. 20-24.

Swenson, N. G. (2013). The assembly of tropical tree communities–the advances and shortcomings of phylogenetic and functional trait analyses. *Ecography, 36*(3), 264–276. doi:10.1111/j.1600-0587.2012.00121.x

Swenson, N. G., & Enquist, B. J. (2007). Ecological and evolutionary determinants of a key plant functional trait: Wood density and its community-wide variation across latitude and elevation. *American Journal of Botany, 94*(3), 451–459. doi:10.3732/ajb.94.3.451 PMID:21636415

Swenson, N. G., Enquist, B. J., Pither, J., Kerkhoff, A. J., Boyle, B., Weiser, M. D., ... Kraft, N. J. (2012). The biogeography and filtering of woody plant functional diversity in North and South America. *Global Ecology and Biogeography*, *21*(8), 798–808. doi:10.1111/j.1466-8238.2011.00727.x

Swift, M. J., Heal, O. W., & Anderson, J. M. (1979). *Decomposition in terrestrial ecosystems*. Oxford, UK: Blackwell Scientific Publications.

Syampungani, S., Geldenhuys, C. J., & Chirwa, P. W. (2016). Regeneration dynamics of miombo woodland in response to different anthropogenic disturbances: Forest characterisation for sustainable management. *Agroforestry Systems*, *90*(4), 563–576. doi:10.100710457-015-9841-7

Syampungani, S., Tigabu, M., Matakala, N., Handavu, F., & Oden, P. C. (2017). Coppicing ability of dry miombo woodland species harvested for traditional charcoal production in Zambia: A win–win strategy for sustaining rural livelihoods and recovering a woodland ecosystem. *Journal of Forestry Research*, *28*(3), 549–556. doi:10.100711676-016-0307-1

Taiz, L., & Zeiger, E. (2006). Plant Physiology. Sunderland, MA: Sinauer Associates.

Tatpati, M. (Ed.). (2015). Citizens' Report 2015: Community Forest Rights under the Forest Rights Act. Pune, Bhubaneshwar and New Delhi: Kalpavriksh and Vasundhara in collaboration with Oxfam India on behalf of Community Forest Rights Learning and Advocacy Process. Retrieved from http://www.kalpavriksh.org/images/LawsNPolicies/CITIZENSREPORT2015.pdf</eref>

Tattoni, C., Ciolli, M., & Ferretti, F. (2011). The fate of priority areas for conservation in protected areas: A fine-scale Markov chain approach. *Environmental Management*, *47*(2), 263–278. doi:10.100700267-010-9601-4 PMID:21190021

Taylor, B., & Parkinson, D. (1988). A new microcosm approach to litter decomposition studies. *Canadian Journal of Botany*, *66*(10), 1933–1939. doi:10.1139/b88-265

TDF. (2019). Tropical Dry Forest. Slater Museum of Natural History, University of Puget Sound, Tacoma. Retrieved from https://www.pugetsound.edu/academics/academic-resources/slater-museum/biodiversity-resources/world-biomes/characteristics-of-bioclimatic/tropical-dry-forest/

Telwala, Y., Brook, B. W., Manish, K., & Pandit, M. K. (2013). Climate-Induced Elevational Range Shifts and Increase in Plant Species Richness in a Himalayan Biodiversity Epicentre. *PLoS One*, *8*(2). doi:10.1371/journal.pone.0057103 PMID:23437322

Temminghoff, E. E., & Houba, V. J. (Eds.). (2004). *Plant analysis procedures* (Vol. 179). Dordrecht, The Netherlands: Kluwer Academic Publishers. doi:10.1007/978-1-4020-2976-9

Tewksbury, J. J., Huey, R. B., & Deutsch, C. A. (2008). Putting the heat on tropical animals. *Science*, *320*(5881), 1296–1297. doi:10.1126cience.1159328 PMID:18535231

Thomas, V., & López, R. (2015). *Global increase in climate-related disasters*. ADB Economics Working Paper Series, No. 466. Asian Development Bank, Metro Manila, Philippines.

Thomas, W. W., & Britton, E. G. (2008). Atlantic coastal forest of Northeastern Brazil. Bronx, NY: The New York Botanical Garden Press.

Thomas, L., Laake, J. L., Strindberg, S., Marques, F. F. C., Buckland, S. T., Borchers, D. L., ... Marques, T. A. (2006). *Distance 5.0. Release 2. Research Unit for Wildlife Population Assessment*. UK: University of St. Andrews.

Thompson, I. D., Guariguata, M. R., Okabe, K., Bahamondez, C., Nasi, R., Heymell, V., & Sabogal, C. (2013). An Operational Framework for Defining and Monitoring Forest Degradation. *Ecology and Society*, *18*(2), 307–329. doi:10.5751/ES-05443-180220

Thuiller, W. (2007). Climate change and the ecologist. *Nature*, *448*(7153), 550–552. doi:10.1038/448550a PMID:17671497

Ticktin, T., Ganesan, R., Parmesha, M., & Shetty, S. (2012). Disentangling the effects of multiple anthropogenic drivers on the decline of two tropical dry forest trees. *Journal of Applied Ecology*, *49*(4), 774–784. doi:10.1111/j.1365-2664.2012.02156.x

Tilman, D. (1987). Secondary succession and the patterns of plant dominance along experimental nitrogen gradients. *Ecological Monographs*, *57*(3), 189–214. doi:10.2307/2937080

Timberlake, J., Chidumayo, E., & Sawadogo, L. (2010). Distribution and Characteristics of African Dry Forests and Woodlands. In E. N. Chidumayo & D. J. Gumbo (Eds.), The Dry Forests and Woodlands of Africa: Managing for products and services. London, UK: Earthscan Publishing.

Todd, M. C. L., Adams, M. A., Grierson, P. F., Loch, R. (Ed.), & Jasper, D. (2000). Mineralisation of nitrogen in a chronosequence of rehabilitated bauxite mines. Special Issue: Soil research for mine reclamation. Australian Journal of Soil Research, 38(2), 435-451.

Toko Mouhamadou, I., & Ozer, A. (2007). Évolution de l'occupation du sol dans les zones périphériques de la forêt classée des Trois Rivières (Bénin) entre 1949 et 1986. In Quelles aires protégées pour l'Afrique de l'Ouest? Conservation de la biodiversité et développement (IRD).

Tomppo, E., Haakana, M., Katilia, M., & Peräsaari, J. (2008). *Multi-source national forest inventory (373)*. Dordrecht, The Netherlands: Springer.

Tonin, A. M., Gonçalves Jr.; J. F., Bambi, P., Couceiro, S. R. M., Feitoza, L. A. M., Fontana, L. E., Hamada, N., … Boyero, L. (2017). Plant litter dynamics in the forest stream interface: Precipitation is a major control across tropical biomes. *Scientific Reports*, *7*. doi:10.103841598-017-10576-8 PMID:28883445

Toy, J. W., & Shay, D. (1987). Comparision of some soil properties on natural and reclaimed hill slopes. *Soil Science*, *143*(4), 264–277. doi:10.1097/00010694-198704000-00004

Traoré, S., Tigabu, M., Ouédraogo, S. J., Boussim, J. I., Guinko, S., & Lepage, M. G. (2008). Macrotermes mounds as sites for tree regeneration in a Sudanian woodland (Burkina Faso). *Plant Ecology*, *198*(2), 285–295. doi:10.100711258-008-9404-3

Trejo, I., & Dirzo, R. (2000). Deforestation of seasonally dry tropical forest: A national and local analysis in Mexico. *Biological Conservation*, *94*(2), 133–142. doi:10.1016/S0006-3207(99)00188-3

Trejo, I., & Dirzo, R. (2002). Floristic diversity of Mexican seasonally dry tropical forests. *Biodiversity and Conservation*, *11*(11), 2063–2084. doi:10.1023/A:1020876316013

Tripathi, S., Bhadauria, R., Srivastava, P., Singh, R., & Raghubanshi, A. S. (2016). Abiotic determinants of tree seedling growth in tropical dry forests. *Advances in plant physiology (Bethesda, MD)*, *17*, 119–131.

Tropical Dry Forest. (2019). Retrieved from https://www.pugetsound.edu/academics/academic-resources/slater-museum/biodiversity-resources/world-biomes/characteristics-of-bioclimatic/tropical-dry-forest/

Tropical Dry Forests. (2019). Retrieved from https://ceiba.org/conservation/dry-forests/

Tudun-Wada, M. I., Tukur, Y. M., & Hussaini, Y. (2014). Analysis of forest cover changes in Nimbia Forest Reserve, Kaduna State, Nigeria using Geographic Information System and Remote Sensing techniques. *International Journal of Environmental Monitoring and Analysis*, *2*(2), 91–99. doi:10.11648/j.ijema.20140202.15

Turner, B. L., Brenes-Arguedas, T., & Condit, R. (2018). Pervasive phosphorus limitation of tree species but not communities in tropical forests. *Nature*, *555*(7696), 367–370. doi:10.1038/nature25789 PMID:29513656

Turner, N. C. (1988). Measurement of plant water status by the pressure chamber technique. *Irrigation Science*, *9*(4), 289–308. doi:10.1007/BF00296704

Turner, W., Spector, S., Gardiner, N., Fladeland, M., Sterling, E., & Steininger, M. (2003). Remote sensing for biodiversity science and conservation. *Trends in Ecology & Evolution*, *18*(6), 306–314. doi:10.1016/S0169-5347(03)00070-3

Tyson, M. J., Hedges, S., & Sitompul, A. F. (2002). Elephant Defecation Rate Study, Way Kambas National Park 2000/2001: Final Report to WWF, WCS, and PHKA, Jan. 5, 2002. Wildlife Conservation Society — Indonesia Program, Bogor, Indonesia.

Udayalakshmi, V., Murthy, M. S. R., & Dutt, C. B. S. (1998). Efficient forest resources management through GIS and remote sensing. *Current Science*, *75*, 272–282.

Uddin, K., Shrestha, H. L., Murthy, M. S. R., Bajracharya, B., Shrestha, B., Gilani, H., ... Dangol, B. (2015). Development of 2010 national land cover database for the Nepal. *Journal of Environmental Management*, *148*, 82–90. doi:10.1016/j.jenvman.2014.07.047 PMID:25181944

Udvardy, M. D. F. (1975). A classification of the biogeographical provinces of the world. IUCN Occasional Paper 18. International Union of Conservation of Nature and Natural Resources, Morges, Switzerland.

UNCCD. (1994): *Elaboration of an International Convention to Combat Desertification in Countries Experiencing Serious Drought and/or Desertification, Particularly in Africa.* Paris, France: United Nations Convention to Combat Desertification.

UNEP. (2012). *Sahel Atlas of Changing Landscapes: Tracing trends and variations in vegetation cover and soil condition.* Nairobi, Kenya: United Nations Environment Programme.

UNEP-WCMC Forest Programme. (2011). *Global statistics*. Retrieved from www.unep-wcmc.org

UNFCC. (2002). *Report of the Conference of the Parties on its Seventh Session, held at Marrakesh from Oct. 29 to Nov. 10, 2001.*

UNFCCC. (1992). United Nations framework convention on climate change. Climate change secretariat, Bonn, Germany. Retrieved from http://unfccc.int/

United Nations Conference on Environment and Development. (1992). *Agenda 21*. Retrieved from https://sustainabledevelopment.un.org/outcomedocuments/agenda21 and https://sustainabledevelopment.un.org/content/documents/Agenda21.pdf

United Nations Environment Program. (1972). *Declaration of the United Nations Conference on the Human Environment. Chapter 11*, 2-4. Retrieved from https://www.soas.ac.uk/cedep-demos/000_P514_IEL_K3736-Demo/treaties/media/1972%20Stockholm%201972%20-%20Declaration%20of%20the%20United%20Nations%20Conference%20on%20the%20Human%20Environment%20-%20UNEP.pdf

Upadhyay, N., & Verma, SPratap Singh, ADevi, SVishwakarma, KKumar, NPandey, A… Sharma, S. (2016). Soil ecophysiological and microbiological indices of soil health: A study of coal mining site in Sonbhadra, Uttar Pradesh. *Journal of Soil Science and Plant Nutrition*, *16*(3), 778–800.

Uriarte, M., Turner, B. L., Thompson, J., & Zimmerman, J. K. (2015). Linking spatial patterns of leaf litterfall and soil nutrients in a tropical forest: A neighborhood approach. *Ecological Applications*, *25*(7), 2022–2034. doi:10.1890/15-0112.1 PMID:26591466

Uryu, Y., Mott, C., Foead, N., Yulianto, K., Budiman, A. S., Takakai, F., Nursamsu, S., … Stüwe, M. (2007). Deforestation, forest degradation, biodiversity loss and CO2 emissions in Riau, Sumatra, Indonesia. WWF Indonesia, Jakarta.

Uryu, Y., Purastuti, E., Laumonier, Y., Sunarto, S., Budiman, A., Yulianto, K., Sudibyo, A., … Stüwe, M. (2010). Sumatra's Forests, Their Wildlife and the Climate. Windows in Time: 1985, 1990, 2000 and 2009. A quantitative assessment of some of Sumatra's natural resources submitted as technical report by invitation to the National Forestry Council (DKN) of Indonesia. WWF Indonesia, Jakarta.

Vaieretti, M. V., Díaz, S., Vile, D., & Garnier, E. (2007). Two measurement methods of leaf dry matter content produce similar results in a broad range of species. *Annals of Botany*, *99*(5), 955–958. doi:10.1093/aob/mcm022 PMID:17353207

van Schaik, C., & Supriatha, J. (1996). *Leuser: a Sumatran Sanctuary*. Depok, Indonesia: Yayasan Leuser Internasional.

Vargas, R., Allen, M. F., & Allen, E. B. (2008). Biomass and carbon accumulation in a fire chronosequence of a seasonally dry tropical forest. *Global Change Biology*, *14*(1), 109–124.

Varghese, A., Ticktin, T., Mandle, L., & Nath, S. (2015). *Assessing the Effects of Multiple Stressors on the Recruitment of Fruit Harvested Trees in a Tropical Dry Forest, Western Ghats, India*, *10*(3), 1–17. doi:10.1371/journal.pone.0119634

Veldman, J. W., & Putz, F. E. (2010). *Long-distance Dispersal of Invasive Grasses by Logging Vehicles in a Tropical Dry Forest*, *42*(6), 697–703.

Velho, N., Defries, R. S., Tolonen, A., Srinivasan, U., & Patil, A. (2018). Aligning conservation efforts with resource use around protected areas. *Ambio*. doi:10.100713280-018-1064-5 PMID:29949080

Vendrami, J. L., Jurinitz, C. F., Castanho, C. T., Lorenzo, L., & Oliveira, A. A. (2012). Litterfall and leaf decomposition in forest fragments under different successional phases on the Atlantic Plateau of the state of Sao Paulo, Brazil. *Biota Neotropica, 12.* Retrieved from http://www.biotaneotropica.org.br/v12n3/en/abstract?article+bn03312032012.

Venkateswaran, R., & Parthasarathy, N. (2005). Tree population changes in a tropical dry evergreen forest of south India over a decade (1992–2002). *Biodiversity and Conservation*, *14*(6), 1335–1344. doi:10.100710531-004-9649-7

Verma, S., Singh, A. P., Devi, S., Mewaram, R. R., Sharma, S., & Dubey, K. (2014). Assessment of Microbial Community and Soil Enzyme Activity of Coal Mine Dumps of Sonbhadra Uttar Pradesh, India. In *Proc. of the Intl. Conf. on Advances In Bio-Informatics, Bio-Technology and Environmental Engineering,* pp. 12-15.

Vico, G. (2017). Climate changes in the tropics may be an advantage to drought-deciduous plants, Tropical dry forests. Sveriges lantbruks universitet (SLU) news. Retrieved from https://www.slu.se/en/ew-news/2017/7/tropical-dry-forests/

Vieira, D. L., & Scariot, A. (2006). Principles of natural regeneration of tropical dry forests for restoration. *Restoration Ecology*, *14*(1), 11–20. doi:10.1111/j.1526-100X.2006.00100.x

Vile, D., Garnier, E., Shipley, B., Laurent, G., Navas, M. L., Roumet, C., ... Midgley, G. F. (2005). Specific leaf area and dry matter content estimate thickness in laminar leaves. *Annals of Botany*, *96*(6), 1129–1136. doi:10.1093/aob/mci264 PMID:16159941

Villa, F., Rossi, O., & Sartore, F. (1992). Understanding the Role of Chronic Environmental Disturbance in the Context of Island Biogeographic Theory. *Environmental Management*, *16*(5), 653–666. doi:10.1007/BF02589019

Villanueva, B., Melo, O., & Rincón, M. (2015). Estado del conocimiento y aportes a la flora vascular del bosque seco del Tolima. *Colombia Forestal*, *18*(1), 9–23. doi:10.14483/udistrital.jour.colomb.for.2015.1.a01

Villar, R., Ruiz-Robleto, J., Ubera, J. L., & Poorter, H. (2013). Exploring variation in leaf mass per area (LMA) from leaf to cell: An anatomical analysis of 26 woody species. *American Journal of Botany, 100*(10), 1969–1980. doi:10.3732/ajb.1200562 PMID:24107583

Villeger, S., Mason, N. W. H., & Mouillot, D. (2008). New multidimensional functional diversity indices for a multifaceted framework in functional ecology. *Ecology, 89*(8), 2290–2301. doi:10.1890/07-1206.1 PMID:18724739

Viña, A., Gitelson, A. A., Nguy-Robertson, A. L., & Peng, Y. (2011). Comparison of different vegetation indices for the remote assessment of green leaf area index of crops. *Remote Sensing of Environment, 115*(12), 3468–3478. doi:10.1016/j.rse.2011.08.010

Vincini, M., Frazzi, E., & D'alessio, P. (2007). Comparison of narrow-band and broad-band vegetation indexes for canopy chlorophyll density estimation in sugar beet. In *Precision Agriculture '07: Proceedings of the 6th European Conference on Precision Agriculture* (189-196).

Vincini, M., Frazzi, E. R. M. E. S., & D'Alessio, P. A. O. L. O. (2008). A broad-band leaf chlorophyll vegetation index at the canopy scale. *Precision Agriculture, 9*(5), 303–319. doi:10.100711119-008-9075-z

Violle, C., Navas, M. L., Vile, D., Kazakou, E., Fortunel, C., Hummel, I., & Garnier, E. (2007). Let the concept of trait be functional! *Oikos, 116*(5), 882–892. doi:10.1111/j.0030-1299.2007.15559.x

Vishal, K. D., & Abhishek, C. (2014). Isolation and characterization of Rhizobium leguminosarum from root nodules of *Pisums sativum L. Journal of Academic and Industrial Research, 2*(8), 464–467.

Vitousek, P. M., Aber, J. D., Howarth, R. W., Likens, G. E., Matson, P. A., Schindler, D. W., ... Tilman, D. G. (1997). Human alteration of the global nitrogen cycle: Sources and consequences. *Ecological Applications, 7*(3), 737–750.

Vivanco, L., & Austin, A. T. (2008). Tree species identity alters forest litter decomposition through long-term plant and soil interactions in Patagonia, Argentina. *Journal of Ecology, 96*(4), 727–736. doi:10.1111/j.1365-2745.2008.01393.x

Vogelman, T. C., Nishio, J. N., & Smith, W. K. (1996). Leaves and light capture: Light propagation and gradients of carbon fixation within leaves. *Trends in Plant Science, 1*(2), 65–70. doi:10.1016/S1360-1385(96)80031-8

Vogt, K. A., Grier, C. C., & Vogt, D. J. (1986). Production, turnover and nutrient dynamics of above and belowground detritus of world forests. *Advances in Ecological Research, 15*, 303–377. doi:10.1016/S0065-2504(08)60122-1

Vose, J. M., & Maass, J. M. (1999). A Comparative analysis of hydrologic responses of tropical deciduou s and temperate deciduous watershed ecosystems to climatic change. In C. Aguirre-Bravo, & C. Rodríguez-Franco (Eds.), *Proceedings of the North American Science Symposium: Toward a Unified Framework for Inventorying and Monitoring Forest Ecosystem Resources. RMRSP-12*:292–298. Guadalajara, México: USDA Forest Service.

Vos, J., & Biemond, H. (1992). Effects of nitrogen on the development and growth of the potato plant. 1. Leaf appearance, expansion growth, life spans of leaves and stem branching. *Annals of Botany, 70*(1), 27–35. doi:10.1093/oxfordjournals.aob.a088435

VSNI. (2012). *Genstat, version 15.2.0.8801. VSN International.* Hemel Hempstead, UK: LTD.

Wachman, M. (2019). Tropical Dry Forest Biome Plants. Retrieved from https://www.gardenguides.com/12451445-tropical-dry-forest-biome-plants.html

Waeber, P., Ramesh, B., Parthasarathy, N., Pulla, S., & Garcia, C. (2012). Seasonally dry tropical forests in South Asia: A research agenda. In A research agenda to contribute to the discussions on "Key Issues for the Global Dry Forests" workshop organized by CIFOR, Zurich, Switzerland.

Wagner, N. (2018). Types of Plants & Animals in a Tropical Dry Forest. USA Today. Retrieved from https://traveltips.usatoday.com/types-plants-animals-tropical-dry-forest-61382.html

Walker, L. R., & del Moral, R. (2003). *Primary Succession and Ecosystem Rehabilitation*. New York, NY: Cambridge University Press. doi:10.1017/CBO9780511615078

Walker, R. F. (2002a). Organic amendment, fertilizer, and lime effects on bare root Jeffrey pine out planted on a Sierra Nevada surface mine. *Journal of Sustainable Forestry*, *15*(3), 29–55. doi:10.1300/J091v15n03_02

Walker, R. F. (2002b). Reestablishment of Jeffrey pine on an acidic Sierra Nevada surface mine: Influence of fertilizer and lime amendments on early growth and nutrition. *Journal of Sustainable Forestry*, *15*(3), 1–27. doi:10.1300/J091v15n03_01

Walker, R. F. (2002c). Responses of Jeffrey pine on a surface mine site to fertilizer and lime. *Restoration Ecology*, *10*(2), 204–212. doi:10.1046/j.1526-100X.2002.00070.x

Walker, S., & Desanker, P. (2004). The impact of land use on soil carbon in Miombo Woodlands of Malawi. *Forest Ecology and Management*, *203*(1-3), 345–360. doi:10.1016/j.foreco.2004.08.004

Wanger, Y., Bonynge, G., Nugranad, J., Traber, M., Ngusaru, A., Lynne, H., ... Makota, V. (2003). Remote sensing of mangrove change along the Tanzania coast. *Marine Geodesy*, *26*, 1–14.

Wang, K., Franklin, S. E., Guo, X., & Cattet, M. (2010). Remote sensing of ecology, biodiversity and conservation: A review from the perspective of remote sensing specialists. *Sensors (Basel)*, *10*(11), 9647–9667. doi:10.3390101109647 PMID:22163432

Wang, Y., Titus, S. J., & Lemay, V. M. (1998). Relationship between tree slenderness coefficients and tree or stand characteristics for major species in Boreal mixed Forest. *Canadian Journal of Forest Research*, *28*(8), 1171–1183. doi:10.1139/x98-092

Wani, R. A., Baba, U. A., Afeefa, Q. S., & Imran, K. (2016). *Conserving Biological Diversity: A Multiscaled Approach. Published by* (pp. 133–158). New Delhi, India: Research India Publications.

Wardle, D. A., Bardgett, R. D., Klironomos, J. N., Setälä, H., Van Der Putten, W. H., & Wall, D. H. (2004). Ecological linkages between aboveground and belowground biota. *Science*, *304*(5677), 1629–1633. doi:10.1126cience.1094875 PMID:15192218

Waring, R. H., & Running, S. W. (2010). Forest ecosystems: analysis at multiple scales. Amsterdam, The Netherlands: Elsevier.

Warszawski, L., Frieler, K., Huber, V., Piontek, F., Serdeczny, O., & Schewe, J. (2014). The Inter-Sectoral Impact Model Intercomparison Project (ISI-MIP): Project framework. *Proceedings of the National Academy of Sciences of the United States of America*, *111*(9), 3228–3232. doi:10.1073/pnas.1312330110 PMID:24344316

Warszawski, L., Friend, A., Ostberg, S., Frieler, K., Lucht, W., Schaphoff, S., & Schellnhuber, H. J. (2013). A multi-model analysis of risk of ecosystem shifts under climate change. *Environmental Research Letters*, *8*(4). doi:10.1088/1748-9326/8/4/044018

Watson, R. T., Zinyowera, M. C., & Moss, R. H. (1995). *Climate Change 1995: Impacts, Adaptations and Mitigation of Climate Change: Scientific-Technical Analyses; Contribution of Working Group II to the Second Assessment of the Intergovernmental Panel on Climate Change (878)*. Cambridge, UK: Cambridge University Press.

WCS. (2010). Nigeria: World Conservation Society (WCS). Retrieved from www.wsc.org>wherewework>Africa.

Webb, C. O., Ackerly, D. D., McPeek, M. A., & Donoghue, M. J. (2002). Phylogenies and community ecology. *Annual Review of Ecology and Systematics*, *33*(1), 475–505. doi:10.1146/annurev.ecolsys.33.010802.150448

Weiher, E. (2011). A primer of trait and functional diversity. *Biological diversity: frontiers in measurement and assessment*, 175-193.

Weiher, E., Werf, A., Thompson, K., Roderick, M., Garnier, E., & Eriksson, O. (1999). Challenging Theophrastus: A common core list of plant traits for functional ecology. *Journal of Vegetation Science*, *10*(5), 609–620. doi:10.2307/3237076

Werneck, F. P., Costa, G. C., Colli, G. R., Prado, D. E., & Sites, J. W. Jr. (2011). Revisiting the historical distribution of Seasonally Dry Tropical Forests: New insights based on palaeodistribution modelling and palynological evidencegeb. *Global Ecology and Biogeography*, *20*(2), 272–288. doi:10.1111/j.1466-8238.2010.00596.x

West, G. B., Brown, J. H., & Enquist, B. J. (1999). A general model for the structure and allometry of plant vascular systems. *Nature*, *400*(6745), 664–667. doi:10.1038/23251

Westhoff, V., & van der Maarel, E. (1978). The Braun Blanquet approach. In R. H. Whittaker (Ed.), *Classification of plant communities*. The Hague, The Netherlands: Junk. doi:10.1007/978-94-009-9183-5_9

Westoby, M. (1998). A leaf-height-seed (LHS) plant ecology strategy scheme. *Plant and Soil*, *199*(2), 213–227. doi:10.1023/A:1004327224729

Westoby, M., Falster, D. S., Moles, A. T., Vesk, P. A., & Wright, I. J. (2002). Plant ecological strategies: Some leading dimensions of variation between species. *Annual Review of Ecology and Systematics*, *33*(1), 125–159. doi:10.1146/annurev.ecolsys.33.010802.150452

Westoby, M., & Wright, I. J. (2006). Land-plant ecology on the basis of functional traits. *Trends in Ecology & Evolution*, *21*(5), 261–268. doi:10.1016/j.tree.2006.02.004 PMID:16697912

WhatA. S. T. D. F. P. (2019). Retrieved from https://www.reference.com/science/tropical-dry-forest-plants-9b9362d1a1fd-6f2a

Whisenant, S. (2005). First steps in erosion control. In S. Mansourian, D. Vallauri, & N. Dudley (Eds.), *Forest restoration in landscapes: beyond planting trees*. New York, NY: Springer. doi:10.1007/0-387-29112-1_50

White, J. W., & Montes-r, C. (2005). Variation in parameters related to leaf thickness in common bean (Phaseolus vulgaris L.). *Field Crops Research*, *91*(1), 7–21. doi:10.1016/j.fcr.2004.05.001

Whittaker, R. H. (1975). *Communities and ecosystems* (2nd ed.). New York, NY: Macmillan Publishing Co.

Whitten, T., Damanik, S. J., Anwar, J., & Hisyam, N. (Eds.). (2000). *The Ecology of Sumatra. Periplus Editions*. Singapore: Berkeley Books Pte. Ltd.

Wiens, J. A. (1997). Metapopulation Dynamics and Landscape Ecology. In I. A. Hanski & M. E. Gilpin (Eds.), *Metapopulation Biology: Ecology, Genetics, and Evolution* (pp. 43–62). San Diego, CA: Academic Press. doi:10.1016/B978-012323445-2/50005-5

Wikramanayake, E., Dinerstein, E., Loucks, C. J., Olson, D. M., Morrison, J., Lamoreux, J., ... Hedao, P. (2002). *Terrestrial ecoregions of the Indo-Pacific. A conservation assessment. Conservation Science Program, WWF-US, Washington, DC*: Island Press.

Wilcove, D. S., Giam, X., Edwards, D. P., Fisher, B., & Koh, L. P. (2013). Navjot's nightmare revisited: Logging, agriculture, and biodiversity in Southeast Asia. *Trends in Ecology & Evolution*, *28*(9), 531–540. doi:10.1016/j.tree.2013.04.005 PMID:23764258

Wildlife Institute of India. (2017). Mapping Landuse/Landcover Patterns in Aravallis Haryana with Reference to Status of Key Wildlife Species. Retrieved from https://41ngmc3nsigz3kwwbw1kp5ic-wpengine.netdna-ssl.com/wp-content/uploads/Aravallis-Report-WII-May-2017-compressed.pdf</eref>

Wilkie, D. S., & Mark, C. T. (1993). Biogeophysical Setting and Global Climate Change. In Central Africa: Global Climate Change and Development. Technical Repon. Washington, DC: Biodiversity Support Program (a consortium of World Wildlife Fund, The Nature Conservancy, and World Resources Institute).

Williams, B. K., Nichols, J. D., & Conroy, M. J. (2002). *Analysis and Management of Animal Populations* (1st ed.). San Diego, CA: Academic Press.

Williams, G. L., & Aquino, C. A. (2010). Tropical Dry Forest Landscape Restoration in Central Veracruz, Mexico. *Ecological Restoration*, *28*(3), 259–261. doi:10.3368/er.28.3.259

Williams, L. E. (1987). Growth of 'Thompson Seedless' grapevines: I. Leaf area development and dry weight distribution. *Journal of the American Society for Horticultural Science*, *112*, 325–330.

Willkomm, M., Vierneisel, B., & Dannenberg, P. (2016). Land use change dynamics in the Mt. Kenya region – a remotely sensed analysis using RapidEye satellite images. *Zbl. Geol. Paläont. Teil 1*, *1*, 23–40.

Wilson, E. O. (1999). Preface. In R. H. MacArthur and E. O. Wilson (Eds.), The Theory of Island Biogeography. Princeton, NJ: Princeton University Press.

Wilson, G., Desai, A. A., Sim, D. A., Gruber, M. A. M., & Lester, P. J. (2014). *The association between invasive Lantana camara and seedlings/saplings of a plant community in Mudumalai Tiger Reserve, India*, 551–563. doi:10.1017/S026646741400039X

Wilson, B., & Wang, S. (1999). Sustainable forestry--the policy prescription in British Columbia. In A. Yoshimoto & K. Yukutake (Eds.), *Global Concerns for Forest Resource Utilization-- Sustainable Use and Management* (pp. 35–45). London, UK: Kluwer Academic Publishers. doi:10.1007/978-94-017-6397-4_4

Wilson, E. O. (1988). The current state of biological diversity. *Biodiversity (Nepean)*, *521*(1), 3–18.

Wilson, E. O. (2016). *Half-Earth: Our Planet's Fight for Life.* New York, N.Y.: Liveright Publishing Corporation.

Wilson, K. B., Baldocchi, D. D., & Hanson, P. J. (2001). Leaf age affects the seasonal pattern of photosynthetic capacity and net ecosystem exchange of carbon in a deciduous forest. *Plant, Cell & Environment*, *24*(6), 571–583. doi:10.1046/j.0016-8025.2001.00706.x

Witkowski, E. T. F., Lamont, B. B., Walton, C. S., & Radford, S. (1992). Leaf demography, sclerophylly and ecophysiology of two Banksias with contrasting leaf life spans. *Australian Journal of Botany*, *40*(6), 849–862. doi:10.1071/BT9920849

Wohlfart, C., Wegmann, M., & Limgruber, P. (2014). Research Article Mapping threatened dry deciduous Dipterocarp forest in South-east Asia for conservation management. *Tropical Conservation Science*, *7*(4), 597–613. doi:10.1177/194008291400700402

Wolkovich, E. M., & Cleland, E. E. (2014). Phenological niches and the future of invaded ecosystems with climate change. *AoB Plants*, 6. PMID:24876295

World Wide Fund. (2018). *WWF Report Reveals Staggering Extent of Human Impact on Planet.* Washington D.C. Retrieved from https://www.worldwildlife.org/press-releases/wwf-report-reveals-staggering-extent-of-human-impact-on-planet

Worthy, S. J., & Swenson, N. G. (2019). Functional perspectives on tropical tree demography and forest dynamics. *Ecological Processes*, *8*(1), 1. doi:10.118613717-018-0154-4

WRI (World Resources Institute). (2019). Global Forest Watch. Retrieved from http://globalforestwatch.org/Dashboards/country/IDN?treeLossTsc=eyJoaWdobGlnaHRlZCI6ZmFsc2V9

Wright, J. H., Hill, N. A. O., Roe, D., Rowcliffe, J. M., K, F., Day, M., & Booker, F. (2015). Reframing the concept of alternative livelihoods. *Conservation biology, 30*(1), 7–13. doi:10.1111/cobi.12607

Wright, I. J., Ackerly, D. D., Bongers, F., Harms, K. E., Ibarra-Manriquez, G., Martinez-Ramos, M., ... Wright, S. J. (2007). Relationships among ecologically important dimensions of plant trait variation in seven neotropical forests. *Annals of Botany*, *99*(5), 1003–1015. doi:10.1093/aob/mcl066 PMID:16595553

Wright, I. J., Reich, P. B., Westoby, M., Ackerly, D. D., Baruch, Z., Bongers, F., ... Flexas, J. (2004). The worldwide leaf economics spectrum. *Nature*, *428*(6985), 821–827. doi:10.1038/nature02403 PMID:15103368

Wright, I. J., Westoby, M., & Reich, P. B. (2002). Convergence towards higher leaf mass per area in dry and nutrient-poor habitats has different consequences for leaf life span. *Journal of Ecology*, *90*(3), 534–543. doi:10.1046/j.1365-2745.2002.00689.x

Wright, J. W., Davies, K. F., Lau, J. A., McCall, A. C., & McKay, J. K. (2006). Experimental verification of ecological niche modeling in a heterogeneous environment. *Ecology*, *87*(10), 2433–2439. doi:10.1890/0012-9658(2006)87[2433:EVOENM]2.0.CO;2 PMID:17089652

Wright, S. J., Kitajima, K., Kraft, N. J. B., Reich, P. B., Wright, I. J., Bunker, D. E., ... Zanne, A. E. (2010). Functional traits and the growth–mortality trade-off in tropical trees. *Ecology*, *91*(12), 3664–367. doi:10.1890/09-2335.1 PMID:21302837

Wright, S. J., Muller-Landau, H. C., & Schipper, J. A. N. (2009). The future of tropical species on a warmer planet. *Conservation Biology*, *23*(6), 1418–1426. doi:10.1111/j.1523-1739.2009.01337.x PMID:20078642

Wu, J.-S., Jiang, P.-K., Chang, S. X., Xu, Q.-F., & Lin, Y. (2009). Dissolved soil organic carbon and nitrogen were affected by conversion of native forests to plantations in subtropical China. *Canadian Journal of Soil Science*, *90*(1), 27–36. doi:10.4141/CJSS09030

WWF (World Wildlife Fund). (2019). Knowledge Hub: Priority Places: Sumatra: Species at Risk. Retrieved from http://wwf.panda.org/knowledge_hub/where_we_work/sumatra/

WWF, & IUCN. (1994). Centres of plant diversity: A guide and strategy for their conservation. 3 Volumes. Cambridge, UK: IUCN Publications Unit.

WWF. (2009). *Climate Change: Confronting a Global Challenge*. USA: World Wildlife Fund.

WWF. (2016a). Major habitat types. Retrieved from https://web.archive.org/web/20161017075416/http://wwf.panda.org/about_our_earth/ecoregions/about/habitat_types/

WWF. (2016b). Tropical and subtropical dry broadleaf forest ecoregions. Retrieved from https://web.archive.org/web/20161023161710/http://wwf.panda.org:80/about_our_earth/ecoregions/about/habitat_types/selecting_terrestrial_ecoregions/habitat02.cfm

WWF. (2018). *Building a future in which humans live in harmony with nature*. USA: World Wildlife Fund.

WWF. (2019). Tropical and subtropical dry broadleaf forests, World Wildlife Fund, Washington, DC. Retrieved from https://www.worldwildlife.org/biomes/tropical-and-subtropical-dry-broadleaf-forests

Xie, Y., Sha, Z., & Yu, M. (2008). Remote sensing imagery in vegetation [94] mapping: A review. *Journal of Plant Ecology*, *1*(1), 9–23. doi:10.1093/jpe/rtm005

XLSTAT. (2012). XLSTAT version 2012.6.02. Addinsoft, New York, NY.

Xu, X., & Hirata, E. (2005). Decomposition patterns of leaf litter of seven common canopy species in a subtropical forest: N and P dynamics. *Plant and Soil, 273*(1-2), 279–289. doi:10.100711104-004-8069-5

Yadav, R. K., & Yadav, A. S. (2008). Phenology of selected woody species in a tropical dry deciduous forest in Rajasthan, India. *Tropical Ecology, 49*(1), 25.

Yadav, S., Yadav, J. P., Arya, V., & Panghal, M. (2010). Sacred groves in conservation of plant biodiversity in Mahendergarh district of Haryana. *Indian Journal of Traditional Knowledge, 9*(4), 693–700.

Yan, H., Wang, S., Wang, C., Zhang, G., & Patel, N. (2005). Losses of soil organic carbon under wind erosion in China. *Global Change Biology, 11*(5), 828–840. doi:10.1111/j.1365-2486.2005.00950.x

Yoshio, I. (2003). Synergy of Remote Sensing and Modeling for Estimating Ecophysiological Processes in Plant Production. *Plant Production Science, 6*, 1, 3–16. doi:10.1626/ pps.6.3

Young, D. J. N., Werner, C. M., Welch, K. R., Young, T. P., Safford, H. D., & Latimer, A. M. (2018). Post-fire forest regeneration shows limited climate tracking and potential for drought-induced type conversion. *Ecology, 100*(2). doi:10.1002/ecy.2571 PMID:30516290

Yuan, Z. Y., & Chen, H. Y. (2010). Fine root biomass, production, turnover rates, and nutrient contents in boreal forest ecosystems in relation to species, climate, fertility, and stand age: Literature review and meta-analyses. *Critical Reviews in Plant Sciences, 29*(4), 204–221. doi:10.1080/07352689.2010.483579

Zapfack, L. (2005). The impact of shifting cultivation on burning on plant biodiversity and carbon sequestration. (Doctoral Thesis), University of Yaoundé I, Yaoundé.

Zarco-Tejada, P. J., Miller, J. R., Noland, T. L., Mohammed, G. H., & Sampson, P. H. (2001). Scaling-up and model inversion methods with narrowband optical indices for chlorophyll content estimation in closed forest canopies with hyperspectral data. *IEEE Transactions on Geoscience and Remote Sensing, 39*(7), 1491–1507. doi:10.1109/36.934080

Zawila-Niedzqiecki, T. (1989). Satellite images for forest decline assessment. In *Proceedings, Global Natural Resource Monitoring and Assessment: Preparing for the 21st Century* (Vol. 1, pp. 473–478). Bethesda, MD: American Society for Photogrammetry and Remote Sensing.

Zedler. J. B. (2005). Ecological restoration: guidance from theory. *San Francisco Estuary and Watershed Science 3*(2) [September 2005], Art. 4.

Zerihun, A., & Yemir, T. (2013). Forest Inventory and Management in the Context of SFM & REDD. Training Manual (1-67), Hawassa University, Ethiopia: Wondo Genet College of Forestry and Natural Resources.

Zhao, X., Li, F., Zhang, W., & Ai, Z. (2016). Contribution of aboveground litter to soil respiration in *Populus davdiana* Dode plantations at different stand ages. *Journal of Mountain Science, 13*(6), 1000–1012. doi:10.100711629-015-3777-2

Zhou, H. C., Tam, N. F. Y., Lin, Y. M., Ding, Z. H., Chai, W. M., & Wei, S. D. (2014). Relationships between degree of polymerization and antioxidant activities: A study on proanthocyanidins from the leaves of a medicinal mangrove plant Ceriopstagal. *PLoS One, 9*(10). doi:10.1371/journal.pone.0107606 PMID:25313980

Zhu, L., Suomalainen, J., Liu, J., Hyyppä, J., Kaartinen, H., & Haggren, H. (2018). A Review: Remote Sensing Sensors. In R. B. Rustamov, S. Hasanova, & M. H. Zeynalova (Eds.), *Multi-purposeful Application of Geospatial Data* (pp. 19–42). Haggren, IntechOpen; doi:10.5772/intechopen.71049

Zhu, S. D., & Cao, K. F. (2010). Contrasting cost–benefit strategy between lianas and trees in a tropical seasonal rain forest in southwestern China. *Oecologia, 163*(3), 591–599. doi:10.100700442-010-1579-3 PMID:20191291

Zuloaga-Aguilar, S. (2010). *Efecto del fuego sobre la germinación y el banco de semillas de bosques templados del occidente de México*. Instituto de Ecología A.C.

Zulu, F., Syampungani, S., & Fushike, P. (2018). Recovery of Baikiaea forest of South Western Zambia from shifting cultivation and its implications for sustainable management. *Journal of Forestry Research*. doi:10.100711676-018-0656-z

About the Contributors

Rahul Bhadouria is working as Post-Doctoral Researcher at University of Delhi, New Delhi, India. He obtained his doctorate degree from Department of Botany, Banaras Hindu University, Varanasi, in 2015. The area of his doctoral research was performance evaluation of tree seedling growth under dry tropical environment. He has published more than 15 papers, five book chapters, and a book in the internationally-reputed journals of ecology/reputed publishers. Current research areas are 'Management of Soil C Dynamics to Mitigate Climate Change', 'A perspective on tree seedling survival and growth attributes in tropical dry forests under the realm of Climate Change' and 'Plant community assembly, functional diversity and soil attributes along forest-savanna-grassland continuum in India'.

Sachchidanand Tripathi did his post graduation and Ph.D. from Department of Botany, Banaras Hindu University (BHU), Varanasi, India. His areas of interest are forest ecology, restoration ecology, aquatic ecology, and eco-physiology. He has published more than 12 research papers in peer reviewed reputed international journals, five books with international publishers, 7 book chapters in books published by Elsevier, Springer etc. He has also edited a conference proceedings. He is interested in collaborating with scholars from the same fields as well as from other disciplines of interest.

Pratap Srivastava has recently completed his doctoral degree in Botany from Department of Botany, Banaras Hindu University. His research topic incorporates studies of soil microbial ecology in relation to SOC dynamics for the improvement of soil quality and for mitigation of global climate change.

Pardeep Singh is presently working as Assistant professor (Department of Environmental Science PGDAV College University of Delhi New Delhi India. He has obtained his master degree from Department of Environmental Science Banaras Hindu University, Varanasi India in 2011. He obtained his doctorate degree from Indian Institute of Technology (Banaras Hindu University) Varanasi in the year of 2017. The area of his doctoral research is the degradation of organic pollutants through various indigenous isolated microbes and by using various types of photocatalytic. He has published more than 40 papers in the international journals in the fields of waste management.

* * *

Dave Augeri is Founder, Director, and Principal Scientist for Biodiversity Unlimited Research and Consulting Group and is an Adjunct Professor in the College of Environment and Life Sciences at the University of Rhode Island. He holds a Ph.D. in Wildlife Ecology, Biodiversity, and Conservation Science from the University of Cambridge. Dr. Augeri has conducted more than 30 years of terrestrial and marine research and conservation work on six continents in most of the major ecosystems on Earth on a diverse array of species, taxa, and conservation science issues. He is a member of and Senior Expert/ Advisor to Species Survival Commission Specialist and Expert Groups and is/has been a Senior Expert, Advisor, Member, and/or consultant to the IUCN, the World Commission on Protected Areas, World Heritage Site Advisory Panel, the United Nations, USAID, the World Bank, the Indonesian Institute of Sciences (LIPI), a variety of government agencies, and other entities, such as the Royal Society of London, Royal Geographic Society, Smithsonian Institute, UNESCO, WWF, WCS, and others.

Mohamed Nasser Baco is a sociologist with 18 years of experience in rural and environmental sociology. His previous work focused on the analysis of farming systems, seed systems, socio-economic evaluation of agricultural innovations (soil fertility management technologies, …) value chain of yam, and maize. In recent years, his work is in the socio-anthropological field and multidisciplinary confrontations of environmental issues involving rural areas, natural resource management, biodiversity, and nature conservation. Dr. Baco is interested in environmental policies and interactions between society and nature, the management of agrobiodiversity, and the diversity of associated knowledge.

Idrissou Bako is Ph. D. student at research laboratory on Innovation for agricultural development, University of Parakou, Parakou, Benin. His specialization is environmental science.

Balaguru Balakrishnan is an Assistant Professor in the Department of Botany spl: Conservation Biology.

Masuma Begum is a Deputy Ranger/Forester, Department of Forest, Government of West Bengal. She is a working professional in the field of Geo-informatics. Her research interests lie in forestry, ecology, and environment.

Sumit Chakravarty is a Professor in Department of Forestry, Uttar Banga Krishi Viswavidyalaya, Pundibari, West Bengal, India. His field of interest is forest ecology, agroforestry, medicinal plants, and carbon sequestration.

Paxie W. C. Chirwa is a Professor of Forest Science at the University of Pretoria. He specializes in Forest Management and rural livelihood of Southern and Eastern Africa.

Pooja Choksi is a Ph.D. student in the Department of Ecology, Evolution, and Environmental Biology at Columbia University, USA. She has previously worked in the tropical dry forests in the Kanha-Pench region on a community-based wildlife conservation model. She also has experience working in the Carribean region of Colombia, South America, and Belize where she studied the co-management framework of protected areas to explore the replicability of the framework to her home country, India. Pooja's research examines the patterns of forest degradation and their connection to land uses to create coexistence landscapes in central India.

Onyekachi Chukwu is a Lecturer in the Department of Forestry and Wildlife, Nnamdi Azikiwe University, Awka, Nigeria. He is also a PhD candidate at the Department of Social and Environmental

Forestry, University of Ibadan, Nigeria. His PhD examines different statistical methods and applications of remote sensing (RS)/geographic information system (GIS) technologies in modelling forest stand structure and geospatial attributes of a tropical rainforest in Nigeria. He has published his work extensively and reviewed in local, regional, and international peer-reviewed academic journals. Onyekachi is a member of the Forestry Association of Nigeria (FAN), African Forest Forum (AFF), Forests and Forest Products Society (FFPS), Wildlife Society of Nigeria (WISON), and Nigerian Society of Environmental Conservation (NSEC). His particular areas of interest lie in the application of mathematical, statistical and RS/GIS principles to proffer solutions to forest, wildlife, and environmental problems.

Japheth H. Dau is a forest Biometrician and Environmentalist; a PhD candidate in Forest Biometrics and Inventory (The Federal University of Technology, Akure, Nigeria). His main research area of interest are: Forest Establishment and Management, Forest Assessment & inventory, Analysis and Modeling, Environmental Management, Environmental Impact Assessment, Forest and Climate change, assessment of biomass and carbon sequential and ecological assessment, among others. He served with College of Agriculture Kano (Nigeria). He has received polytechnic Award (2005), University Award (2012), and Community Service Award (2013) in Nigeria. He authored and coauthored more than 20 research articles published in reputable Journals (International and National).

Kumud Dhanwantri is an Assistant Professor at Amity School of Architecture and Planning, Amity University, Gurugram, Haryana, India, since 2013. She is pursuing her Ph.D. in the sphere of efficacy of environmental laws and policies in a hilly ecosystem of Aravalis in National Capital Region, India. Her major fields of study have been geography, regional planning, and applications of GIS. She has published and presented her work in various national and international journals of repute. Her research interests are in the areas of rural-urban continuum, regional and rural development, gender issues in planning, as well as environmental laws and policies.

K. P. Dubey is an officer of Indian Forest Service of 1987 batch from Uttar Pradesh Cadre. Presently, he is PCCF in UP Forest Department and posted as APCCF, Project. He has prepared working plans of four forest divisions and has expertise in forestry research, administration, and management. He published several research papers in reputed national and international journals. He has been awarded by the Chief Minister, Uttar Pradesh for his outstanding contribution in organization of Kumbh-2013.

Kumud Dubey is Scientist in Indian council of Forestry Research and Education, Dehradun, an Autonomous Council of Ministry of Environment, Forest, and Climate Change, Government of India, New Delhi, India. She is presently a Senior Scientist in Forest Research Centre for Ecorehabilitation, Prayagraj, India. She has expertise in Forest Ecology and Reclamation, Restoration, Bioremediation through Probiotics, Social Forestry, and Agroforestry. She published several research papers in reputed national and international journals. She has been awarded by different organizations for her outstanding contribution in Environment conservation.

Mbayngone Elisée is a Searcher and Responsible for the Laboratory of Systematic Botany and Plant Ecology, University of N'Djamena, N'Djamena, Chad.

Jacinta U. Ezenwenyi is a lecturer in the Department of Forestry and Wildlife, Nnamdi Azikiwe University, Awka, Nigeria. She is also a PhD candidate at University of Ibadan, Nigeria. Her Ph.D. focuses on modelling yield and volume ratio for Mixed Tree Species in the Moist Tropical Rainforest, Nigeria. She had a Master of Science degree in Forest Biometrics and Remote Sensing at the Department of Forest Resources Management, University of Ibadan, Nigeria and Bachelor of Forestry and Wildlife Management (Forestry Option) at the Department of Forestry and Wildlife Management, Federal University of Agriculture Abeokuta, Nigeria. She teaches; researches and supervises research students in Forest and Wildlife Biometrics, inventory, mensuration, remote sensing and geographic information system. Ezenwenyi has reviewed for and published her work in reputable academic journals. She is a life member of Forestry Association of Nigeria (FAN), full member of the African Forest Forum (AFF) and Nigerian Society of Environmental Conservation (NSEC). Her research interests relate to forest modelling, application of Geographic Information System/Remote Sensing (GIS/RS) technologies in Renewable Natural Resources, forest health and environmental conservation and restoration.

Anghy Gutiérrez Rincón obtained a BSc. in forestry engineering at the Universidad Distrital Francisco José de Caldas. She is finishing her Master studies on Forest Conservation and Management at the Universidad Distrital Francisco Jose de Caldas. Her thesis focused on post-fire regeneration and establishment of tropical dry forests. She also carries out research and consulting on landscape ecology and land-use planning, including fire risk assessments.

G. N. Tanjina Hasnat is a Lecturer at the Institute of Forestry and Environmental Sciences, University of Chittagong, and former Lecturer of Department of Land Administration, Faculty of Land Management and Administration, Patuakhali Science and Technology University, Dumki, Patuakhal, Bangladesh. She completed her B.Sc. (Hons.) and MS (Thesis) in Forestry from Institute of Forestry and Environmental Sciences, University of Chittagong, Bangladesh. She has nine published research articles from six different international and three national journals, and one book chapter in book published by Springer. Four research articles are in review by four different international and national journals.

Mohammed Kamal Hossain is Professor of Institute of Forestry and Environmental Sciences, Chittagong University, Chittagong, Bangladesh. He has contributed more than 151 scientific articles to national and international refereed journals and more than 42 publications in proceedings and in book chapters in the fields of Silviculture, biodiversity, and plantation forestry. He is the lead/co-author of 12 books in the field of Silviculture and Environment. In the Institute, he supervised more than 102 B.Sc (Hons), 66 M.Sc/M.S., and 4 PhD students.

Zapfack Louis is a professor and Senior expert in Evaluation of Carbon Storage, University Professor. Searcher, Chief of Botany/Ecology Laboratory and responsible for Research Unity on Systematic/ Ecology and Carbon Stock evaluation of the Department of Plant Biology, University of Yaounde I.

Sayani Dutta Majumder is a research scholar working at Jadavpur University.

Ismaïl Moumouni is working as an associate professor in rural sociology at research Laboratory on Innovation for Agricultural Development, University of Parakou, Parakou, Benin.

Anirban Mukhopadhyay is working on hazards and vulnerability for the last 10 years. He is an expert in geospatial analysis and modeling.

Djekota Christophe Ngarmari is a researcher at the University of N'Djamena, Faculty of Exact and Applied Sciences, Laboratory of Systematic Botany and Plant Ecology. His research interest is plant ecology.

Adewole Olagoke holds a PhD in forest ecology and modelling. His research centers on gaining insights into factors (both biotic and abiotic) and mechanisms underlying species distribution, structural dynamics and ecological restoration in tropical forest systems. His research approach combines field experiments, remote sensing data, and statistical- and Individual-based modelling techniques.

Ramoudane Orou Sannou is an early career researcher from Benin, with an academic background in agricultural economics and environmental management. He has research experience in natural resource and climate change governance with a keen interest in the Science-Policy interface. Ramoudane works as Researcher (Data Analyst) for the African Union Commission – Human Resources, Science & Technology Department, on a research program providing evidence that supports the harness of Demographic Dividends across the African continent.

Raja P is working as an assistant professor at J.J. College of Arts and Science (Autonomous), Pudukkottai, India. His specialization is taxonomy.

Nazir Pala is an Assistant Professor in the Department of Forestry, Uttar Banga Krishi Viswavidyalaya, Pundibari, West Bengal, India. His fields of interest are forest ecology, agroforestry, medicinal plants, and carbon sequestration.

Angela Parrado-Rosselli holds a Ph.D. on Natural Sciences from the University of Amsterdam. She worked for more than 10 years with the Dutch cooperation for research and conservation of tropical Amazonian forests. Currently, she is an associate professor in ecology and conservation of tropical ecosystems at the Universidad Distrital Francisco Jose de Caldas. Her research interests focus on the regeneration processes of tropical forests and the effect of disturbances on forest dynamics. She also works on collaborative monitoring and forest conservation by local communities, ecotourism, and protected areas. She has coauthored various scientific articles on these topics. Nowadays, she is investigating fire ecology of Colombian tropical dry forests and its capacity to recover after this disturbance.

Niloy Pramanick is a researcher working on Geospatial modeling. His research interests are groundwater, subsidence, dry forest, etc.

Akhilesh Raghubanshi is working as Professor of Environmental Studies and Director at Institute of Environment & Sustainable Development, Banaras Hindu University, Varanasi. His research interests are in ecological understanding of structure, functioning, and sustainable management of tropical terrestrial ecosystems, in particular landscape ecology using remote sensing and GIS tools, understanding plant diversity and distributions, ecosystem functioning including nutrient dynamics, greenhouse gas emissions, and restoration of degraded ecosystems. He is currently working on (i) urban disaster risk-reduction project funded by Ministry of Foreign Affairs, Government of Japan; (ii) Crop discrimination using hyperspectral remote sensing funded by Indian Space Research Organization (ISRO). He has published more than 130 research papers including in leading peer-reviewed science journals such as Nature, AMBIO, Journal of Applied Ecology, Soil Biology and Biochemistry, Biology and Fertility of Soils, Applied Soil Ecology, New Forests, Forest Ecology and Management, Journal of Environmental Management, Environmental Monitoring and Assessment, Ecological Indicators, etc.

Prakash Rai just completed his Ph.D. Forestry from Uttar Banga Krishi Viswavidyalaya, Pundibari, West Bengal, India. He is working as Assistant Professor in Forestry at Zee Himgiri University, Dehra Dun, India.

Kabelong Banoho Louis-Paul Roger is a researcher at the University of Yaoundé I, Faculty of Sciences. His research interest is forestry and biodiversity.

Soosairaj S. is working as an assistant professor at the Department of Botany, St.Joseph's College (Autonomous), Tiruchirappalli, India. research interest is the taxonomy.

Him Lal Shrestha completed his PhD degree in 2015 from Kathmandu University and is working as Associate Professor at Kathmandu Forestry College. He also has an academic degree of MSc in Geoinformatics. He has wide experience on application of Remote Sensing and GIS for the forest management and its related aspects. He has published a number of peer-reviewed journal papers and national/regional journal papers. He also attended an international workshop, seminar, and conventions to present his paper and posters. He is also accustomed for the tailor-made training on GIS, RS, and GPS application in mapping, forestry, and allied subjects.

Gopal Shukla is an Assistant Professor in Department of Forestry, Uttar Banga Krishi Viswavidyalaya, Pundibari, West Bengal, India. His fields of interest are forest ecology, agroforestry, medicinal plants, and carbon sequestration.

Annie Sikanwe is a Natural Resources expert specializing in Wildlife monitoring, GIS, and vegetation interaction.

Shipra Singh is a Ph.D. scholar in the School of Environmental Sciences, JNU, under the supervision of Prof. K. G. Saxena. Ms. Singh works on the role of plant functional traits in Western Himalayan forest ecosystem at an altitudinal gradient of 3000m. She has completed her M.Phil. on how vegetation and soil properties vary along the slope aspect with increasing elevation.

Pooja Gokhale Sinha is working as Assistant Professor, Department of Botany, Sri Venkateswara College, University of Delhi. She has obtained her Ph. D. from the University of Delhi and her area of specialization is climate change, the impact of climate change of crop plants, carbon footprint, environmental monitoring and assessment of urban ecosystems. She has more than 15 years of experience in research and undergraduate teaching. She has been a member of the organizing a committee of several national and international conferences and seminars. She has conducted an Add-on Course on Climate Change: Issues concerns and challenges for undergraduate and post graduate students.

Stephen Syampungani specializes in Woodland Ecology and management. His research links plant species functional traits and management.

Ngaba Waye Taroum Caleb is a senior expert in Evaluation of Carbon Storage, University Professor. He is researcher, Chief of Botany/Ecology Laboratory, and responsible for Research Unity on Systematic/ Ecology and Carbon Stock evaluation of the Department of Plant Biology, University of Yaounde I.

Abhishek Verma has submitted his PhD recently in School of Environmental Sciences, Jawaharlal Nehru University, New Delhi. His research is focused on structure and functions of Himalayan ecosystem.

Pramit Verma is working in the Institute of Environment & Sustainable Development, Banaras Hindu University, India. His research interests lie in urban ecology, social-ecological systems, urban metabolism, remote sensing and GIS. He is studying the complex relationship between human drivers and ecological responses in urban ecosystems.

Vineeta is an Assistant Professor in the Department of Forestry, Uttar Banga Krishi Viswavidyalaya, Pundibari, West Bengal, India. Her research interest is forestry and environmental science.

Abhinav Yadav is currently working in Institute of Environment & Sustainable Development, Banaras Hindu University, Varanasi. His major research area is the impact of climate change and human disturbance on plant functional traits and soil nutrient balance in tropical dry forests."

Krishan Kumar Yadav is a former Chief Town Planner. His major fields of studies are Urban and Regional Planning and Protection and Conservation of Natural Environment. He has extensively published his work in different national and international journals as well as presented in reputed conferences in Poland, Japan, and Dubai. "Tourism and Eco Development in the Aravalli Ranges near Delhi"; "Planning for a Smart City with a Human Face in Developing India". His research interests are mainly focused on the Urbanization and developmental challenges faced by Indian cities, governance issues, and the conservation of natural environment, safety, and safe movement of wildlife in Aravalis, NCR, India. Prof. Yadav in the capacity of Chairman for the Haryana Regional Chapter, Institute of Town Planners, India. He is the Fellow member of the Institute of Town Planners of India, New Delhi.

Index

S

T

V

W

www.ingramcontent.com/pod-product-compliance
Lightning Source LLC
LaVergne TN
LVHW062106300926
849161LV00010B/2385

* 9 7 8 1 7 9 9 8 0 0 1 4 9 *